Finite Inelastic Deformations – Theory and Applications

International Union of Theoretical and Applied Mechanics

D. Besdo · E. Stein (Eds.)

Finite Inelastic Deformations – Theory and Applications

IUTAM Symposium Hannover, Germany 1991

Springer-Verlag
Berlin Heidelberg New York
London Paris Tokyo
Hong Kong Barcelona Budapest

Prof. Dr.-Ing. Dieter Besdo

Institut für Mechanik
Appelstraße 11
D-3000 Hannover 1

Prof. Dr.-Ing. Erwin Stein

Institut für Baumechanik
und Numerische Mechanik
Appelstraße 9A
D-3000 Hannover 1

ISBN-13:978-3-642-84835-3 e-ISBN-13:978-3-642-84833-9
DOI: 10.1007/978-3-642-84833-9

Softcover reprint of the hardcover 1st edition 1992

Typesetting: Camera ready by authors

61/3020-5 4 3 2 1 0 – Printed on acid-free paper.

Scientific Commitee

S. N. Atluri, Atlanta, USA
D. Besdo, Hannover, Germany (co-chairman)
J. F. Besseling, Delft, Netherlands
H. Günther, Chemnitz, Germany
P. Ladeveze, Cachan, France
J. Lemaitre, Cachan, France
G. Maier, Milano, Italy
R. M. McMeeking, Santa Barbara, USA
G. K. Mikhailov, Moscow, USSR
M. Z. Mróz, Warsaw, Poland
A. Samuelsson, Göteborg, Sweden
E. Stein, Hannover, Germany (co-chairman)
M.V. Tvergaard, Lyngby, Denmark
F. Ziegler, Vienna, Austria

List of Participants

Altenbach, J.	Magdeburg, Germany
Andrä, H.	Karlsruhe, Germany
Aravas, N.	Philadelphia, USA
Bammann, D. J.	Livermoore, USA
Bellet, M.	Valbonne, France
Besdo, D.	Hannover, Germany
Bertram, A.	Berlin, Germany
Bonn, R.	Kassel, Germany
de Borst, R.	Delft, Netherlands
Boucher, M.	Cachan, France
Bruhns, O. T.	Bochum, Germany
Celigoj, C.	Graz, Austria

Chabrand, P.	Marseille, France
Chatzikonstantinou, T.	Stuttgart, Germany
Cordebois, J. P.	Cachan, France
Crisfield, M. A.	London, U. K.
Cristescu, N.	Bucharest, Romania
Dafalias, Y. F.	Davis, USA
Davini, C.	Udine, Italy
Doege, E.	Hannover, Germany
Doltsinis, I. St.	Stuttgart, Germany
Duszek-Perzyna, M. K.	Warsaw, Poland
Estrin, Y.	Hamburg, Germany
Fornengo, E.	Rivarolo, Italy
Fotiu, P.	Vienna, Austria
Gambin, W.	Warsaw, Poland
Gelin, J. C.	Besancon, France
Glaser, H.	Stuttgart, Germany
Häggblad, B.	Stockholm, Sweden
Haupt, P.	Kassel, Germany
Huétink, J.	Twente, Netherlands
Hunsche, U. E.	Hannover, Germany
Irschik, H.	Vienna, Austria
Johnson, C.	Göteborg, Sweden
Kihara, J.	Tokyo, Japan
Kim, K. S.	Providence, USA
Kleiber, M.	Tokyo, Japan
Kollmann, F. G.	Darmstadt, Germany
Kröplin, B. H.	Stuttgart, Germany
Kuhn, G.	Erlangen, Germany
Kukudzanov, V. N.	Moscow, USSR
Lehmann, Th.	Hannover, Germany
Mathiak, F.	Berlin, Germany
Mahrenholtz, O.	Hamburg, Germany
Miehe, C.	Hannover, Germany
Mróz, Z.	Warsaw, Poland
Müller, I.	Berlin, Germany
Müller, M.	Hannover, Germany
Müller-Hoeppe, N.	Hannover, Germany
Onat, E. T.	New Heaven, USA
Ortiz, M.	Providence, USA
Pawelski, O.	Düsseldorf, Germany
Pecherski, R. B.,	Warsaw, Poland
Perić, D.	Swansea, U. K.
Perzyna, P.	Warsaw, Poland

Petryk, H.	Warsaw, Poland
Roelandt, J. M.	Compiègne, France
Rougee, P.	Cachan, France
Samuelsson, A.	Göteborg, Sweden
Simo, J. C.	Stanford, USA
Skrzypek, J.	Krakow, Poland
Steck, E. A.	Braunschweig, Germany
Stein, E.	Hannover, Germany
Steinmann, P.	Karlsruhe, Germany
Storåkers, B.	Stockholm, Sweden
Szuwalski, K.	Krakow, Poland
Tamme, A.	Chemnitz, Germany
Tomita, Y.	Kobe, Japan
Tvergaard, V.	Lyngby, Denmark
Watanabe, O.	Ibaraki, Japan
Wriggers, P.	Darmstadt, Germany
Wunderlich, W.	München, Germany
Xia, Z.	Edmonton, Canada
Zbib, H. M.	Houghton, USA
Ziegler, F.	Vienna, Austria

Preface

The IUTAM-Symposium on "Finite Inelastic Deformations – Theory and Applications" took place from August 19 to 23, 1991, at the University of Hannover, Germany, with 75 participants from 14 countries.

Scope of the symposium was a fundamental treatment of new developments in plasticity and visco-plasticity at finite strains. This covered the phenomenological material theory based on continuum mechanics as well as the treatment of microstructural phenomena detected by precise experimental datas. In a restricted number, lectures on new experimental facilities for measuring finite strains were also implemented into the symposium. Another important topic of the symposium was the treatment of reliable and effective computational methods for solving engineering problems with finite inelastic strains. Within this context it was an essential feature that theory, numerical and computational analysis were be seen in an integrated way.

In total 9 sessions with 37 lectures, many of them given by well known keynote-lecturers, and a poster session with 10 contributions met fully our expectations of a high ranking up-to-date forum for the interaction of four topics, namely the *physical* and *mathematical modelling of finite strain inelastic deformations* including localizations and damage as well as the achievements in the *numerical analysis and implementation* and the *solution of complicated engineering systems.* Special and important features were reliable material datas from macroscopic and microscopic tests as well as test results of complex engineering problems, like deep drawing and extrusion.

The organization should like to thank the IUTAM for delegating this task to the two instituts of mechanics in the faculties of mechanical and civil engineering of the University of Hannover. We are glad that so many outstanding scientists followed our invitation for giving lecture- and poster-contributions, and we acknowledge especially the excellent keynote-lecturers.

We are indebt to the Deutsche Forschungsgemeinschaft (DFG), the land of Lower Saxony, the IUTAM and the following companies for financial support: "AUDI AG", Ingolstadt, "Bayrische Motoren Werke (BMW)", München, "Innovationsgesellschaft für fortgeschrittene Produktionssysteme in der Fahrzeugindustrie m.b.H. (INPRO)", Berlin, "Wilhelm Karmann GmbH", Osnabrück, "Maschinenfabrik Müller-Weingarten AG", Weingarten, "L. Schuler Pressen GmbH", Göppingen.

The organizers express their warm thanks to the University of Hannover and especially to its President, Professor Dr. h.c. mult. Hinrich Seidel, for providung us the opportunity to use the nice lecture hall and other facilities in the University of Hannover as well as for giving the President's reception on Monday, August 19, 1991.

Furthermore we are thankful to the Lord Mayor of Hannover, Mr. Herbert Schmalstieg, granting a reception in the New Town Hall on Tuesday, August 20, 1991. We appreciated especially the presence and the welcome addresses of the Assistent Director Dr. Christian Hodler, representing the Minister of Science and Culture of Lower Saxony, Mrs. Helga Schuchardt, of the Lord Mayor of Hannover, Mr. Herbert Schmalstieg, and the President of the University of Hannover, Professor Dr. h.c. mult. Hinrich Seidel.

Last but not least it is our pleasant duty to thank the secretary of the Symposium, Dr.-Ing. C. Miehe, furthermore the secretaries of both institutes, Mrs. Christa Ernst and Mrs. Vera Halfar, and all contributing staff of our institutes for their drive and motivation in preparing and performing the symposium.

Hannover, January 1992

Dieter Besdo
Institut für Mechanik

Erwin Stein
Institut für Baumechanik
und Numerische Mechanik

Authors

Contents

Numerical Analysis and Computational Aspects

Experiments and Applications

Physical Phenomena and Mathematical Modelling

Elastic-Viscoplastic Analysis of Ductile Fracture

VIGGO TVERGAARD[+] and ALAN NEEDLEMAN[++]

[+] Department of Solid Mechanics
The Technical University of Denmark, DK–2800 Lyngby, Denmark

[++] Division of Engineering
Brown University, Providence, RI 02912, U.S.A.

Abstract
Analyses of ductile failure are discussed where the material's constitutive description allows for the possibility of a complete loss of stress carrying capacity, with the associated creation of new free surface. No additional failure criterion is employed so that fracture arises as a natural outcome of the deformation history. Attention is directed toward circumstances where the inherent rate sensitivity of plastic flow plays a role. Numerical computations of ductile failure in notched bars and of dynamic crack growth are reviewed.

1. Introduction

A review is given of some computational studies of ductile fracture phenomena based on constitutive relations that incorporate models of the physical micro–rupture process. The possibility of a complete loss of stress carrying capacity, with the associated creation of new free surface, is allowed for and fracture arises as a natural outcome of the deformation history. This contrasts with the usual approach to failure analysis, where the constitutive characterization of the material and a fracture criterion are specified separately.

Attention in this paper is on analyses of failure processes in rate dependent plastic solids. A constitutive framework due to Gurson [1,2] for a progressively cavitating solid is described, generalized to allow for both strain rate hardening [3] and for thermal softening due to adiabatic heating [4]. Within this constitutive framework, ductile failure by the nucleation, growth and coalescence of voids is represented in terms of the evolution of a single internal variable representing the void volume concentration. Failure by cleavage is modelled approximately in terms of a critical maximum principal stress.

Relatively simple infinite band, material stability analyses, see e.g. [5], can be used to analyze the emergence of localization from homogeneous deformation states. Such

D. Besdo, E. Stein (Eds.)
Finite Inelastic Deformations –
Theory and Applications
IUTAM Symposium Hannover/Germany 1991

analyses reveal the role played by various constitutive features. However, the initiation and propagation of localization bands from strain concentrators, e.g. notches, requires a full numerical solution. Various analyses of ductile failure in notched bars, including analyses of the effect of strain rate and temperature (through the temperature dependence of the flow strength) on the failure mode transition in the Charpy V–notch specimen, are reviewed. In addition, calculations are presented that illustrate the convergence, with respect to mesh refinement, of dynamic ductile failure predictions.

Recent numerical analyses of dynamic crack growth are discussed where the speed and mode of crack growth is directly determined by modelling the micromechanisms of failure. The crack growth mechanism analyzed involves two populations of void nucleating particles; discretely modelled large particles that nucleate voids at relatively small strains and uniformly distributed smaller particles that nucleate voids at much larger strains. The effect of the distribution and spacing of the large particles on crack speed and on crack path are illustrated.

2. Constitutive Law

The rate–sensitive version of the Gurson model to be discussed here was introduced by Pan, Saje and Needleman [3]. Subsequently, cleavage fracture was included in the model by Tvergaard and Needleman [6], and thermal softening effects were included by Needleman and Tvergaard [4]. The material model is presented in the context of a Lagrangian formulation of the field equations, in which a material point is identified by the coordinates x^i in the reference configuration. The metric tensors in the current configuration and the reference configuration are denoted by G_{ij} and g_{ij}, respectively, with determinants G and g, and η_{ij} denotes the Lagrangian strain tensor. The contravariant components of the Cauchy stress tensor σ^{ij} and the Kirchhoff stress tensor τ^{ij} on the embedded deformed coordinates are related by the expression $\tau^{ij} = \sqrt{G/g}\,\sigma^{ij}$. Indices range from 1 to 3, and the summation convention is adopted for repeated indices.

The rate–independent constitutive relations proposed by Gurson [1,2] make use of an approximate yield condition of the form $\Phi(\sigma^{ij}, \sigma_M, f) = 0$, where σ^{ij} is the average macroscopic Cauchy stress tensor, σ_M is the matrix flow strength, representing the average microscopic stress state in the matrix material, and f is the current void volume fraction. In the viscoplastic model there is no yield condition; but the same function Φ is used as a plastic potential. Thus, the plastic part of the macroscopic strain–rate is specified by an expression of the form

$$\dot{\eta}^{\rm P}_{\rm ij} = \Lambda \frac{\partial \Phi}{\partial \sigma^{\rm ij}} \, . \tag{1}$$

The function to be used here as a plastic potential is

$$\Phi = \frac{\sigma_{\rm e}^2}{\sigma_{\rm M}^2} + 2{\rm q}_1{\rm f}^*\cosh\left[\frac{{\rm q}_2\,\sigma^{\rm k}_{\rm k}}{2\sigma_{\rm M}}\right] - 1 - ({\rm q}_1{\rm f}^*)^2 = 0 \, , \tag{2}$$

where $\sigma_{\rm e} = (3{\rm s}_{\rm ij}{\rm s}^{\rm ij}/2)^{\frac{1}{2}}$ is the macroscopic Mises stress, ${\rm s}^{\rm ij} = \sigma^{\rm ij} - {\rm G}^{\rm ij}\sigma^{\rm k}_{\rm k}/3$ is the stress deviator, and $\sigma^{\rm k}_{\rm k}/3$ is the macroscopic mean stress. The parameters q_1 and q_2 were introduced by Tvergaard [7,8] to bring predictions of the model at low void volume fractions in closer agreement with full numerical analyses for periodic arrays of voids. The function $f^*(f)$ was proposed by Tvergaard and Needleman [9] to model the loss of stress–carrying capacity associated with void coalescence. Modifying Φ by $f^*(f)$ is only relevant for void volume fractions larger than a certain critical value f_C, and this function is taken as

$$f^* = \begin{cases} f & \text{for } f \le f_C \\ f_C + (f_U^* - f_C)(f - f_C)/(f_F - f_C) & \text{for } f \ge f_C \end{cases} \tag{3}$$

where $f_U^* = 1/q_1$ and $f^*(f_F) = f_U^*$. It is noted that f_F is the void volume fraction, at which the stress–carrying capacity is lost completely. Based on experimental and numerical studies the two failure parameters in (3) have been chosen as $f_C = 0.15$, or somewhat smaller, and $f_F = 0.25$.

For the thermally softening viscoplastic solid the matrix plastic strain rate is taken to be given by the power law relation

$$\dot{\epsilon}^{\rm P}_{\rm M} = \dot{\epsilon}_0 \left[\frac{\sigma_{\rm M}}{{\rm g}(\epsilon^{\rm P}_{\rm M},{\rm T})}\right]^{1/{\rm m}} \tag{4}$$

where T is the temperature, m is the strain–rate hardening exponent, and $\dot{\epsilon}_0$ is a reference strain–rate. If strain hardening follows a power law with exponent N and the flow stress dependence on temperature is approximated as linear in the range of interest, the function $g(\epsilon^P_M, T)$ may be expressed as

$$g(\epsilon^{\rm P}_{\rm M},{\rm T}) = \sigma_0\left[1 - \beta({\rm T}-{\rm T}_0)\right]\left[1 + \epsilon^{\rm P}_{\rm M}/\epsilon_0\right]^{\rm N} \tag{5}$$

By setting the macroscopic plastic work rate equal to the matrix dissipation

$$\sigma^{\rm ij}\dot{\eta}^{\rm P}_{\rm ij} = (1-{\rm f})\sigma_{\rm M}\dot{\epsilon}^{\rm P}_{\rm M} \tag{6}$$

the parameter Λ in (1) is found to be

$$\Lambda = (1-f)\sigma_M \dot{\epsilon}_M^P \left[\sigma^{ij} \frac{\partial \Phi}{\partial \sigma^{ij}}\right]^{-1} . \tag{7}$$

The total strain–rate is taken to be the sum of an elastic part, a plastic part and a part due to thermal expansion, $\dot{\eta}_{ij} = \dot{\eta}_{ij}^E + \dot{\eta}_{ij}^P + \dot{\eta}_{ij}^T$. Assuming small thermal and elastic strains, the elastic, incremental stress–strain relationship and the thermal expansion are given by

$$\overset{\nabla}{\sigma}{}^{ij} = \mathcal{R}^{ijkl}\dot{\eta}_{kl}^E , \qquad \dot{\eta}_{ij}^T = \alpha G_{ij}\dot{T} \tag{8}$$

Then, the constitutive relations for the viscoplastic material can be written in the rate form

$$\overset{\nabla}{\sigma}{}^{ij} = \mathcal{R}^{ijkl}\left(\dot{\eta}_{kl} - \dot{\eta}_{kl}^P - \dot{\eta}_{kl}^T\right) \tag{9}$$

where $\overset{\nabla}{\sigma}{}^{ij}$ is the Jaumann (co–rotational) rate of the Cauchy stress tensor. The rate of increase of the void volume fraction is given by

$$\dot{f} = (1-f)G^{ij}\dot{\eta}_{ij}^P + \mathcal{B}\left(\dot{\sigma}_M + (\sigma_k^k)^{\cdot}/3\right) + \mathcal{D}\dot{\epsilon}_M^P \tag{10}$$

where the first term represents the growth of existing voids and the last two terms represent stress controlled nucleation and plastic strain controlled nucleation, respectively (Needleman and Rice [10]).

If heating due to plastic dissipation is assumed to occur under adiabatic conditions the rate of temperature increase is obtained from balance of energy

$$\rho\, c_p \frac{\partial T}{\partial t} = \chi\, \tau^{ij}\dot{\eta}_{ij}^P \tag{11}$$

where ρ is the mass density in the reference configuration, c_p is the heat capacity, and χ is the fraction of plastic work converted to hear (often $\chi \simeq 0.9$ for metals, Taylor and Quinney [11]).

Cleavage fracture in b.c.c. metals has been investigated by many authors (e.g., Cottrell [12], Hahn [13]) and, based on these investigations, a constant critical value σ_c of the maximum principal tensile stress is regarded as a sufficiently good approximate criterion for slip induced cleavage failure.

3. Localization Predictions

For rate independent solids, the onset of localization can be formulated as a bifurcation problem within the framework of a material instability analysis, as discussed by Rice [5], where references to the literature can be found. For a three dimensional solid the mode of localization corresponds to a shear band, whereas in the context of two dimensional plane stress theory the bifurcation mode is one of localized sheet necking. In contrast, for viscoplastic solids a localization bifurcation is essentially precluded, since the viscoplastic part of the rate of deformation tensor depends on the stress state and on the current values of internal variables, but not on any rate quantity, see [4]. As a consequence, the relevant moduli for bifurcation are the elastic moduli. Initial imperfections can trigger localization in viscoplastic solids, where localization is identified with the ratio of strain rate in the shear band (or localized neck) to that outside becoming arbitrarily large. Under isothermal conditions strain rate hardening retards the development of localization significantly, even for small amounts of strain rate hardening, Hutchinson and Neale [14], since the material stiffens as the strain rate concentrates in the incipient band. Nevertheless, the phenomenology of localization in viscoplastic solids can be the same as in rate independent plastic solids.

Within this material instability framework, an element of a solid is considered subject to all around displacement boundary conditions. An initial imperfection is specified in the form of a band of material having different properties from the surrounding material. The deformation history can be calculated by a relatively simple numerical procedure. Such analyses have been used to explore the effects of various constitutive features on localization in porous solids under isothermal conditions. For example, because plane strain states are more susceptible to localization than axisymmetric states, localization occurs rather early in plane strain tension and can occur with a relatively small increase in porosity. Accordingly, the prediction of localization in plane strain tension can be rather sensitive to the adopted void nucleation criterion, while this sensitivity is much reduced in axisymmetric tension, Xu and Needleman [15]. Also, as found by Pan *et al.* [3], for a given material rate sensitivity, the retardation effect on flow localization is greater for plane strain tension than for axisymmetric tension. The delay in localization due to material rate sensitivity can result in the void volume fraction in the band reaching the value at which the material's stress carrying capacity is lost before localization occurs. This is illustrated by Needleman and Tvergaard [16] in the context of plane stress sheet necking.

In the more general case localization and failure occur in nonhomogeneous deformation fields, e.g. around stress concentrations such as notches, and determining their development requires a full boundary value problem solution. Notched plane strain and axisymmetric bars, consisting of an elastic–viscoplastic material and subject to dynamic loading, have been analyzed by Tvergaard and Needleman [17], accounting for adiabatic heating due to plastic dissipation and the associated thermal softening. For the axisymmetric bars, failure initiates at the center and propagates along the midsection, whereas in plane strain failure takes place in a shear band. The greater resistance of the axisymmetric configuration to shear band localization is as expected from the material stability analyses. Even in axisymmetric configurations, shear localization plays a role on a local scale, through zig–zag crack growth, and near free surfaces where the axisymmetric constraints are relaxed, as shown in the cup–cone fracture analysis of Tvergaard and Needleman [9]. However, the meshes used in [17] were too coarse to resolve such behavior.

The separate effects of progressive microrupture and thermal softening were considered in [17]. In plane strain, when either thermal softening or void nucleation and growth, or both, are included in the analysis the deformations localize into a shear band. By way of contrast, for the fully dense solid under isothermal conditions no such localization takes place. With thermal softening alone, the initiation of localization is only slightly delayed and the initial stress drop is nearly as rapid as when both progressive microrupture and thermal softening are accounted for, but separation does not occur since failure is not included in the analysis. For a porous solid under isothermal conditions separation does occur, but with the times for the initiation of localization and for separation delayed by approximately 30 per cent from those when both effects are accounted for.

An important issue in numerical, grid based localization and failure calculations is the sensitivity of the predictions to the choice of discretization. This was addressed in Tvergaard and Needleman [17] by repeating one calculation with a finer mesh. However, the fine mesh results shown in Fig. 5 of [17] are not the correct results for the stated problem, so that the comparison in [17] is not an appropriate one. Figure 1 shows a comparison of the average stress (force/unit original area at the impact end) versus time for one plane strain calculation using three discretizations, with material and loading parameters as specified in [17]. Failure takes place in a shear band. The results for the 544 quadrilateral mesh coincide with the corresponding results in [17], Fig. 5. The overall stress–strain response is essentially the same for all three meshes up to and including the initiation of localization. Mesh refinement leads to a narrower shear band and to a more abrupt drop in the specimen stress carrying capacity. The stress carrying capacity vanishes at $t = 35.0\ \mu sec$, $33.6\ \mu sec$ and $32.6\ \mu sec$ for the 544, 1696 and 6784 quadrilateral meshes (each quadrilateral consists of four "crossed" triangular linear

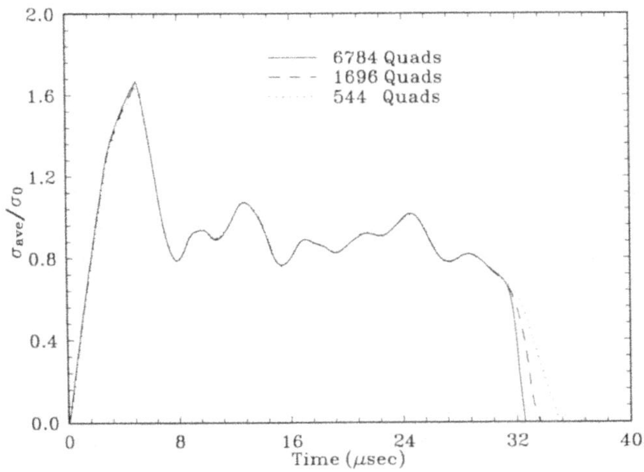

Fig. 1. Effect of mesh refinement on average stresses at the loaded end of a plane strain notched bar, for $V_1 = 80$ m/sec and $t_{rise} = 5\mu$sec .

displacement elements), respectively. Figure 1 shows convergence for the value of the time at which the stress carrying capacity vanishes. In this regard, note that because time increases monotonically there is a limit to how abruptly the specimen stress carrying capacity can drop.

4. Brittle–Ductile Transition

The energy absorbed to fracture in the Charpy V–notch test is widely used to evaluate the fracture resistance of materials. For b.c.c. metals fracture occurs by cleavage when the temperature is sufficiently low, so that the yield stress is high. At somewhat higher temperatures the yield stress is so much reduced that the critical stress for cleavage is not reached during plastic yielding, and then fracture occurs by ductile hole growth, leading to a much larger absorbed energy. When the specimen is impact loaded, the stress level is increased due to material strain–rate sensitivity, and thus the likelihood of cleavage failure is increased. The material model presented in section 2 is ideally suited for analysing the brittle–ductile transition, since the two competing failure mechanisms as well as the effect of strain–rate sensitivity are incorporated.

Numerical analyses for the Charpy V–notch test have been carried out by Tvergaard and Needleman [6,18]. First the effect of different strain rates at a fixed temperature was analysed [6], taking into account the temperature dependence of the flow strength,

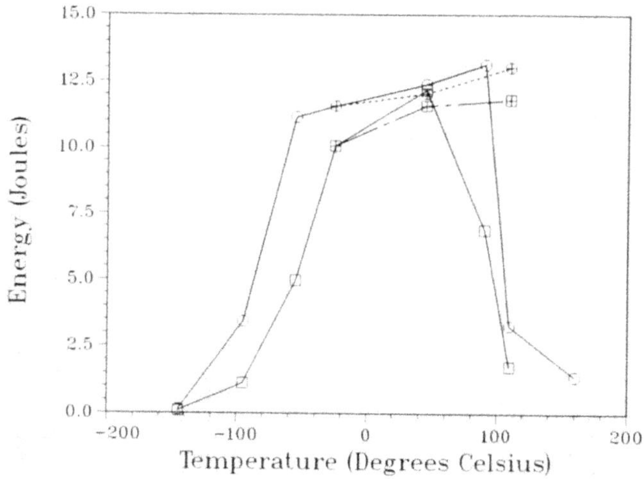

Fig. 2. Computed curves of Charpy V–notch energies vs. temperature; o and □ marks imposed velocities of 5×10^{-4} m/sec and 5 m/sec. Points marked with + are for a comparison material. (from [18]).

but not deformation induced thermal softening. For a temperature in the transition regime, at which slow bending gave predominantly ductile failure, a 10^6 times higher loading rate gave early cleavage and a significantly lower absorbed energy. Thus, the strong influence of the strain rate on the brittle–ductile transition is clearly illustrated by the material model.

In a subsequent investigation [18] the temperature dependence of the absorbed energy was analysed for the high nitrogen mild steel investigated by Ritchie, Knott and Rice [19]. The results in Fig. 2 for slow loading and impact loading, respectively, show the transition from very low absorbed energies at -145^oC to much higher absorbed energies at -25^oC. For temperatures higher than 45^oC the behaviour of this particular steel is rather untypical due to the occurrence of blue brittleness. In terms of the material model this additional brittle regime for $T > 45^oC$ is caused by a rapid increase in work hardening rate with temperature, which results in stress levels high enough to reach the cleavage stress prior to ductile failure. The distributions of void volume fraction and maximum principal stress at the onset of cleavage are shown in Fig. 3 for the specimen subject to slow bending at -25^oC, to illustrate the two competing failure mechanisms. It is seen that significant void growth has occurred at the notch tip where the strains grow large, whereas the peak stress occurs somewhat below the notch tip where stress triaxiality is high.

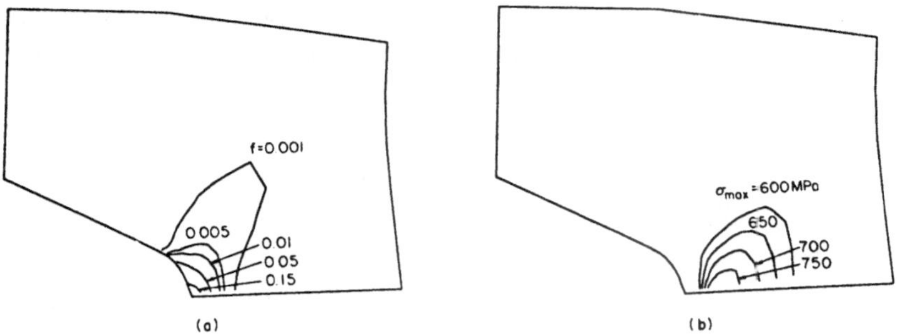

Fig. 3. Contours of field quantities for Charpy V–notch specimen. (a) Void volume fraction. (b) Maximum principal stress. (from [18]).

5. Dynamic Fracture

The elastic–viscoplastic material model with thermal softening has been applied by Needleman and Tvergaard [4,20,21] to analyse dynamic, ductile crack growth. Here the ductile porous material model has been used to represent two populations of void nucleating particles; large particles (inclusions) that nucleate voids at relatively small strains and smaller particles that nucleate voids at much larger strains. The large particles are modelled as "islands" of the amplitude of the stress controlled void nucleation, while the small particles are taken to be uniformly distributed and to nucleate by a plastic strain controlled mechanism. Furthermore, adiabatic heating due to plastic dissipation and the corresponding thermal softening is accounted for in the numerical analyses.

Previous analyses of dynamic crack growth have been based on specifying a fracture criterion in terms of a selected crack tip characterizing parameter, such as a critical value of the crack–tip–opening displacement, the crack–tip–opening angle, the dynamic stress intensity factor, or a conservation integral. In contrast to this type of approach, the crack growth speeds predicted in [4,20,21] are directly determined by the micromechanisms of failure incorporated in the material model. Voids nucleate and grow near the crack tip, where high levels of stress and strain occur, and the crack advances when failure by void coalescence is predicted. These analyses have focussed on the range where failure occurs solely by the void coalescence mechanism. Of course, for many materials undergoing rapid crack growth cleavage is an important failure mechanism.

The numerical analyses in [4] were carried out for a plane strain double edge cracked specimen subject to axial loading. Symmetric loading at the two ends was assumed, e.g.

resulting from two simultaneously triggered explosive charges, and a row of uniformly spaced larger inclusions were taken to lie along the initial crack line so that the crack grows straight ahead. Initially when the stress wave has hit the crack plane some crack tip blunting develops, associated with rather large strains and some void nucleation in a region around the initial crack tip. Subsequently the straining leading to crack growth is rather localized, and an essentially constant crack tip opening angle is predicted in each case. Fig. 4 shows the crack growth vs. time for four different sets of material or loading parameters. It is seen that during the initial crack tip blunting the speed of crack growth increases gradually, and subsequently the predicted crack growth speed remains nearly constant in each case. With material parameters representative of a structural steel the crack growth speeds determined in Fig. 4 range from 0.056 to 0.21 times the shear wave speed. For a non–uniform distribution of the larger inclusions in front of the crack it has been found [20] that the crack speed oscillates according to local inclusion spacings, and the average crack speed decays with increasing deviation from a uniform inclusion spacing.

The possibility of dynamic crack growth along a zig–zag path or a curved path has been analysed in [21] for an edge cracked specimen, which models the configuration in experiments of Prakash and Clifton [22]. Here the loading is unsymmetric and the full specimen is analysed numerically for different distributions of the larger inclusions. The contours of constant void volume fraction in Fig. 5a show that in a square array of inclusions the crack prefers to grow straight ahead. In Fig. 5b the array of inclusions is

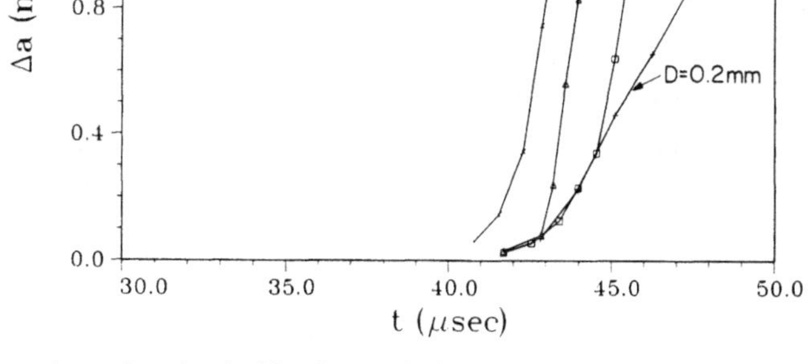

Fig. 4. Crack growth vs. time for double edge cracked specimen under dynamic tensile loading. (from [4]).

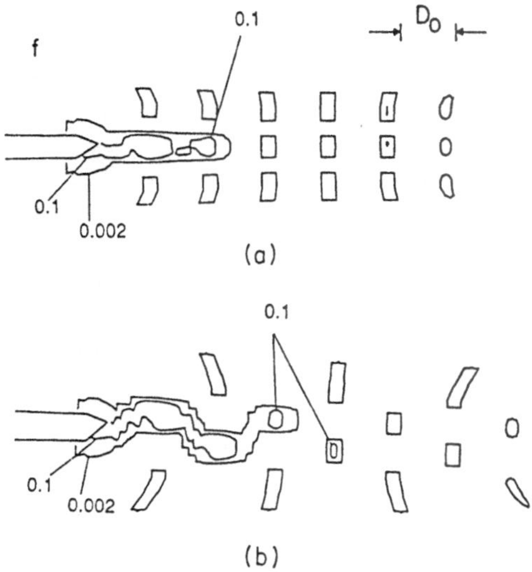

Fig. 5. Contours of constant void volume fraction during dynamic crack growth in edge cracked specimen. (a) Square array of inclusions. (b) A staggered array of inclusions. (from [21]).

chosen such that different zig–zag paths appear to be possible, but the solution shows that the crack grows along the path between the inclusions lying closest to the initial crack line. The average crack growth speed predicted for the zig–zag growth in Fig. 5b is nearly identical to that for straight ahead crack growth in Fig. 5a. It is worth emphasizing that in these calculations, the crack path and the crack speed are entirely determined by the interaction of the assumed failure mechanisms, including the distribution of the larger inclusions, in the near tip stress and strain fields induced by propagating stress waves.

Acknowledgements — A.N. is grateful for support provided by the Office of Naval Research through grant N00014–89–J–3054. Computations reported on here were carried out on the Cray YMP at the Pittsburgh Supercomputer Center.

References

1. Gurson, A.L.: Plastic flow and fracture behavior of ductile materials incorporating void nucleation, growth and interaction. Ph.D. Thesis, Brown University (1975).
2. Gurson, A.L.: Continuum theory of ductile rupture by void nucleation and growth I. Yield criteria and flow rules for porous ductile media. J. Engng. Materials Technol. 99 (1977) 2–15.
3. Pan, J.; Saje, M.; Needleman, A.: Localization of deformation in rate sensitive porous plastic solids. Int. J. Fracture 21 (1983) 261–278.

4. Needleman, A.; Tvergaard, V.: An analysis of dynamic, ductile crack growth in a double edge cracked specimen. Int. J. Fracture 49 (1991) 41–67.

5. Rice, J.R.: The localization of plastic deformation. in Theoretical and Applied Mechanics (ed. W.T. Koiter), 207–220, North–Holland (1977).

6. Tvergaard, V.; Needleman, A.: Effect of material rate sensitivity on failure modes in the Charpy V–notch test. J. Mech. Phys. Solids 34 (1986) 213–241.

7. Tvergaard, V.: Influence of voids on shear band instabilities under plane strain conditions. Int. J. Fracture 17 (1981) 389–407.

8. Tvergaard, V.: On localization in ductile materials containing spherical voids. Int. J. Fracture 18 (1982) 237–252.

9. Tvergaard, V.; Needleman, A.: Analysis of the cup–cone fracture in a round tensile bar. Acta Metallurgica 32 (1984) 157–169.

10. Needleman, A.; Rice, J.R.: Limits to ductility set by plastic flow localization. in Mechanics of Sheet Metal Forming (eds. D.P. Koistinen *et al.*), 237–267, Plenum Publishing Corporation (1978).

11. Taylor, G.I.; Quinney, H.: The latent energy remaining in a metal after cold working. Proc. Royal Soc. London A143 (1934) 307–326.

12. Cottrell, A.H.: Theory of brittle fracture in steel and similar metals. Transactions AIME 212 (1958) 192–203.

13. Hahn, G.T.: The influence of microstructure on brittle fracture toughness. Metallurgical Transactions 15A (1984) 947–959.

14. Hutchinson, J.W.; Neale, K.W.: Influence of strain rate sensitivity on necking under uniaxial tension. Acta Metall. 25 (1977) 839–846.

15. Xu, X.–P.; Needleman, A.: The influence of nucleation criterion on shear localization in rate sensitive porous plastic solids. Int. J. Plast. (in press).

16. Needleman, A.; Tvergaard, V.: Limits to formability in rate–sensitive metal sheets. in Mechanical Behavior of Materials – IV (eds. J. Carlson and N.G. Ohlson), 51–65, Pergamon Press (1984).

17. Tvergaard, V.; Needleman, A.: Ductile failure modes in dynamically loaded notched bars. in Damage Mechanics in Engineering Materials (eds. J.W. Ju, D. Krajcinovic, H.L. Schreyer), AMD–Vol. 109 (and MD–Vol. 24), 117–128, American Society of Mechanical Engineers, New York, NY (1990).

18. Tvergaard, V.; Needleman, A.: Temperature and rate dependence of Charpy V–notch energies for a high nitrogen steel. Int. J. Fracture 37 (1988) 197–215.

19. Ritchie, R.O.; Knott, J.F.; Rice, J.R.: On the relationship between critical tensile stress and fracture toughness in mild steel. J. Mech. Phys. Solids 21 (1973) 395–410.

20. Needleman, A.; Tvergaard, V.: A numerical study of void distribution effects on dynamic ductile crack growth. Engng. Fracture Mechanics 38 (1991) 157–173.

21. Tvergaard, V.; Needleman, A.: Effect of crack meandering on dynamic, ductile fracture. Danish Center for Appl. Math. and Mech., Report No. 419 (1991).

22. Prakash, V.; Clifton, R.J.: Experimental and analytical investigation of dynamic fracture under conditions of plane strain. Proc. 22nd Nat. Symp. Fract. Mech. (in press).

On the Stability of Finite Plastic Deformations

H.M. Zbib

Washington State University
Pullman, WA 99164-2920, USA, and

E.C. Aifantis

Michigan Technological University
Hougton, MI 49931, USA and
Aristotle University of Thessaloniki
Thessaloniki 54006, GREECE

Summary
The stability of inelastic deformation and the formation of shear bands are examined by considering a recently developed gradient-dependent theory of elasto-viscoplasticity. The main issues addressed in the present work are the role of viscosity, inertia and higher order strain gradients in providing internal length scales, as well as the effect of back stress (deformation induced anisotropy) and plastic spin (texture development) on the structure of the shear bands in the post localization regime.

Introduction

The problems of constitutive modelling of finite plastic deformations and the associated phenomena of deformation pattern-formation and shear banding have been the subjects of interest among many investigators during the past two decades; recent reviews can be found in the works of Rice [1], Aifantis [2a,b] and Zbib and Aifantis [3a,b]. When examining these phenomena, the crucial questions that arise are the identification of the basic scaling mechanisms, the associated microstructural aspects (see for example, Korbel et al. [4]), the appropriate internal variables that a constitutive theory should include and the way the macroscopic constitutive response is related to the physical microprocesses (see for example, Aifantis [2] and Drucker [5]).

The stability of plastic flow and the shear banding phenomenon, in particular, have gained the attention of many researchers in both the metallurgical and mechanics communities [1-11]. While the main mathematical and physical aspects pertaining to the understanding of the mechanisms leading to the onset of the phenomenon have been examined, work on the characterization and evolution of shear bands including the determination of band widths and spacings has only recently begun. The main difficulty in this problem lies in the proper modelling of material behavior in the post-localization regime once the shear bands are formed. This can be attributed to the fact that in this regime the deformation field becomes highly inhomogeneous and, therefore, constitutive equations originally proposed for homogeneous or "near homogeneous" fields become insufficient. To overcome such difficulties, it was suggested in the work of Aifantis and co-workers [2,3, 12-18] to include the higher order strain gradients

D. Besdo, E. Stein (Eds.)
Finite Inelastic Deformations –
Theory and Applications
IUTAM Symposium Hannover/Germany 1991

into the constitutive theory of plasticity to account for the heterogeneity of plastic flow and capture internal deformation wavelengths.

In the present work we first review the "scale-invariance" approach outlined in [2] and used in [3] to obtain physically based constitutive equations for the plastic stretching, back stress (modelling anisotropy) and plastic spin (modelling texture development). Such a theory may be viewed as a compromise between "multislip" crystal plasticity averaging models and purely continuum plasticity models. The theory is then modified to account for the heterogeneity of plastic flow occurring during the shear banding process. This, in fact, can be directly related to microstructural inhomogeneities resulting, for example, from tangled cell structures of dislocations, texture development, etc. This, in turn, leads to the development of strain heterogeneities causing the initiation of localized deformation bands. The macroscopic manifestation of such microscopic events is assumed here to be the development of higher order strain gradients which would enter into the form of macroscopic phenomenological theory. In fact, we retain the structure of the classical theory of plasticity, including the yield surface (for rate-independent materials), the normality condition and the flow rule, but introduce higher order strain-gradients into the equation for the flow stress.

In order to examine the role of higher order strain gradients on the material behavior, we consider a simplified one-dimensional problem and investigate the effect of inertia, viscosity and gradients on the structure of the shear band. It is shown that for the quasi-static loading case, higher order strain gradients are the only source to provide an internal length scale to the problem for both rate-independent and rate-dependent materials. For the dynamic loading case, however, it i shown that viscosity and inertia, together, are sufficient to provide a natural length scale to the problem without resorting to higher order strain gradients, but when viscosity is dropped, higher order gradients become again necessary.

In order to examine the effect of plastic spin and anisotropy on the development of shear bands we consider a two-dimensional problem. The problem is treated numerically using the finite element method. It is shown that for the case of shear banding from a ground-state of pure extension (no shearing), the plastic spin has no influence on the onset of instability, but it does affect the development of severe localization. In fact, an increase in the degree of plastic spin decreases "ductility," in the sense that the value of average strain to severe localization and "load collapse" decreases.

A Gradient-Dependent Theory of Elasto-Plasticity

We consider a continuous medium under dynamic loading conditions whose motion is governed by the linear momentum equation,

$$\operatorname{div} \boldsymbol{\sigma} = \rho\, \dot{\mathbf{v}}, \tag{1}$$

where σ is the Cauchy stress tensor, ρ is the mass density, and $\mathbf{v}$ is the velocity vector field. Assuming that the solid undergoes large deformations, we can decompose the deformation gradient $\mathbf{F}$ into elastic $\mathbf{F}^e$, plastic $\mathbf{F}^p$ and rotation $\mathbf{R}^m$ parts such that $\mathbf{F} = \mathbf{R}^m \mathbf{F}^e \mathbf{F}^p$ as proposed by Zbib and Aifantis [3a]. Following their method, one can then obtain the decomposition

$$\mathbf{D} = \mathbf{D}^e + \mathbf{D}^p\,,\ \omega = \mathbf{W} - \mathbf{W}^p, \tag{2}$$

where the stretching $\mathbf{D}$ is the symmetric part of the velocity gradient, $\mathbf{D}^e$ and $\mathbf{D}^p$ are the elastic and plastic stretching respectively, ω is the spin of the material (substructure), $\mathbf{W}$ is the antisymmetric part of the velocity gradient (spin of the continuum) and $\mathbf{W}^p$ is the plastic spin. The elastic strain rate $\mathbf{D}^e$ is given by Hooke's law, modified for large deformations, as

$$\mathbf{D}^e = [\mathbf{C}^e]^{-1}\overset{\circ}{\mathbf{S}}\ ;\ \overset{\circ}{\mathbf{S}} = \dot{\mathbf{S}} - \omega\mathbf{S} + \mathbf{S}\omega, \tag{3}$$

where $[\mathbf{C}^e]$ is the elasticity tensor and $\overset{\circ}{\mathbf{S}}$ is the corotational stress rate.

Using a scale-invariance argument and a maximization procedure, Aifantis and co-workers [e.g. 2,3] have been able to derive explicit expressions for the plastic stretching $\mathbf{D}^p$ and plastic spin $\mathbf{W}^p$ based on the process of crystal slip and dislocation glide. For kinematic hardening plasticity models it turns out that the plastic anisotropy can be represented by the back stress $\boldsymbol{\alpha}$ whose evolution is, in fact, determined by the plastic spin $\mathbf{W}^p$. As a special case of this formulation one obtains the following constitutive equations

$$\mathbf{D}^P = \frac{\dot{\bar{\gamma}}}{2\bar{\tau}}(\mathbf{S}'-\boldsymbol{\alpha}'),\ \overset{\circ}{\boldsymbol{\alpha}} = h\mathbf{D}^P - c\dot{\bar{\gamma}}\,\boldsymbol{\alpha};\ \overset{\circ}{\boldsymbol{\alpha}} = \dot{\boldsymbol{\alpha}} - \omega\boldsymbol{\alpha} + \boldsymbol{\alpha}\omega; \tag{4}$$
$$\omega = \mathbf{W} - \mathbf{W}^P,\ \mathbf{W}^P = \zeta(\boldsymbol{\alpha}\mathbf{D}^P - \mathbf{D}^P\boldsymbol{\alpha}),$$

where prime indicates deviatoric part while, h, c and ζ are material parameters. [As usual, $\dot{\bar{\gamma}} = (2\,\mathbf{D}^p\,.\,\mathbf{D}^p)^{1/2}$ denotes the effective strain and $\bar{\tau} = (\mathbf{S}'\,.\,\mathbf{S}'/2)^{1/2}$ the effective stress].

Equation $(4)_5$ suggests that the plastic spin is a measure of noncoaxiality between the back stress and the stretching. Its implication to large deformation has been examined in detail by the authors [3]. In particular, it was shown that $\mathbf{W}^p$ has a considerable effect on the development of axial stress (or strain) in a fixed (or free) end torsion of a cylindrical bar and that the value of the parameter ζ can be obtained from such torsion tests. [The results of Zbib and Aifantis [3] suggested that ζ has very small values; in fact, by defining $\zeta = a/\sigma_y$, where σ_y is the initial yield stress, they obtained, for example, a = 4.5 for AL-1100]. The implication of the plastic spin to localization is examined in the last section of this paper in connection with the numerical analysis.

Rate-Dependent Materials

The flow stress $\bar{\tau}$ appearing in equation $(4)_1$ and its dependence on strain and strain rate determines whether the material is rate-independent or rate-dependent. In addition to this classical interpretation of the flow stress, it was proposed in [2,3] that higher order gradients of micro (dislocation densities) or macro (strain) variables be included in the respective constitutive equation in order to capture the details of deformation patterning and shear banding phenomena. For the case of macro shear bands, for example, the following gradient-dependent expression for the flow stress was proposed

$$\bar{\tau} = \bar{\tau}_H - \bar{\tau}_I \,; \; \bar{\tau}_H = \kappa(\bar{\gamma}, \dot{\bar{\gamma}}), \; \bar{\tau}_I = C_1 \nabla^2 \bar{\gamma} + C_2 \, |\nabla \bar{\gamma}|^2, \tag{5}$$

where $\kappa\,(\bar{\gamma}, \dot{\bar{\gamma}})$ is the usual homogeneous stress and $C_1\,(\bar{\gamma})$ and $C_2\,(\bar{\gamma})$ are the so-called gradient coefficients. This simple expression provides an internal length to the theory of plasticity and, hence, makes possible the investigation of pattern-forming instabilities in the deformation field.

Upon combining equations (3) and (4) with (2) we obtain the following gradient-dependent constitutive equation for rate-dependent elastic-viscoplastic materials with isotropic/kinematic hardening

$$\overset{\circ}{\mathbf{S}} = [\mathbf{C}^e]\mathbf{D} + \frac{1}{2\mu}[\mathbf{C}^e](\mathbf{S}' - \boldsymbol{\alpha}'); \qquad \mu \equiv \mu_H - \mu_I \,; \; \mu_H = \frac{\bar{\tau}_H}{\dot{\bar{\gamma}}}, \;\; \mu_I = \frac{\bar{\tau}_I}{\dot{\bar{\gamma}}}, \tag{6}$$

where μ can be defined as the nonlinear total viscosity, with a homogeneous part μ_H and an inhomogeneous part μ_I.

When elastic effects are neglected ($\mathbf{D}^e \to 0$) for large plastic deformations, then we obtain the following constitutive relation for a gradient-dependent viscoplastic material with isotropic/kinematic hardening

$$\mathbf{S} = 2\mu\mathbf{D} + \boldsymbol{\alpha} - p\mathbf{1}; \tag{7}$$

where $p \equiv -\,\mathrm{tr}\mathbf{S}/3$ is the hydrostatic pressure. This equation is reminiscent to that describing a Newtonian viscous fluid (without α and a gradient-dependent viscosity).

Rate-Independent Materials

The dependence of the flow stress κ on $\dot{\gamma}$ indicates viscoplastic behavior. Plastic behavior is obtained by dropping in (5) the dependence on $\dot{\gamma}$. In this case $\dot{\gamma}$ and the condition of loading or unloading are obtained from the yield function F and the consistency condition $\dot{F} = 0$. The loading-unloading conditions are given by

$$F=\sqrt{J_2'}-\bar{\tau}, \quad \begin{cases} F=0 \text{ and } \dot{\bar{\gamma}}>0 \text{ or } \dot{\bar{\gamma}}=0 \rightarrow \\ \qquad \text{loading or neutral loading} \\ F=0 \text{ and } \dot{\bar{\gamma}}<0 \rightarrow \text{ unloading} \end{cases} \tag{8}$$

where $J_2' = (\mathbf{S}'-\boldsymbol{\alpha}')\cdot(\mathbf{S}'-\boldsymbol{\alpha}')/2$. The consistency condition $\dot{F}=0$ then yields

$$C_1\nabla^2\dot{\bar{\gamma}} + 2C_2\nabla\bar{\gamma}\cdot\nabla\dot{\bar{\gamma}} - \bar{h}\dot{\bar{\gamma}} = -\frac{(\mathbf{S}'-\boldsymbol{\alpha}')\cdot\dot{\mathbf{S}}'}{2\bar{\tau}}; \tag{9}$$

where the "effective" hardening modulus $\bar{h}$ is given by

$$\bar{h}=\frac{d\bar{\tau}_H}{d\bar{\gamma}}+\frac{h}{2}-\frac{c}{2\bar{\tau}}\boldsymbol{\alpha}'\cdot(\mathbf{S}'-\boldsymbol{\alpha}')-\frac{dC_1}{d\bar{\gamma}}\nabla^2\bar{\gamma}-\frac{dC_2}{d\bar{\gamma}}\nabla^2\bar{\gamma}-\frac{dC_2}{d\bar{\gamma}}|\nabla\bar{\gamma}|^2. \tag{10}$$

Relation (9) is a differential equation for the loading index $\dot{\bar{\gamma}}$. This is a major difference between the present theory and the classical theory of plasticity. The loading index now should be treated as an independent variable governed by equation (9). [This could be treated numerically and incorporated in usual finite element programs where equations (9) can be solved independently at each integration increment].

<u>The One-Dimensional Problem: Length and Time Scales</u>

In order to illustrate the implications of the theory outlined in the previous section to the shear banding problem, we consider the simple shearing of an infinite block in the x direction. We assume that the state of stress is pure shear such that $S_{xy} = S_{yx} = \tau(t)$ are the only nonzero components of $\mathbf{S}$. Similarly, $\dot{\varepsilon}_{xy} = \dot{\varepsilon}_{yx} = \dot{\gamma}/2$ are the only nonzero components of $\mathbf{D}$. Moreover, we assume isotropic hardening with no back stress. Then it can be shown that equations (1)-(3) reduce to

$$\frac{\partial\tau}{\partial y}=\rho\dot{v}\;,\;\dot{\tau}=G(\dot{\gamma}-\dot{\gamma}^p), \tag{11}$$

where G is the elastic shear modulus, $\dot{\gamma}^p$ is the plastic strain rate, and v is the velocity field in the x direction. Various classes of material behavior are considered below.

Elastic-Plastic Materials

a) No gradients: $C_1 = C_2 = 0$.

For this case we drop the dependence on $\dot{\bar{\gamma}}$ from equation (5) and obtain $\dot{\gamma}^p = \dot{\tau}/H$, where $H = \partial\kappa/\partial\gamma^p$. Then equations (11) (with $\dot{\gamma} = v_y$) yield a rather familiar partial differential equation

$$\ddot{v}=c^2 v_{yy}+\frac{1}{\rho}\left(\frac{GH}{H+G}\right)_y v_y\;;\quad c^2=\frac{GH}{\rho(H+G)}, \tag{12}$$

where the subscript y indicates partial derivative. The type of this differential equation is governed by the sign of H. If H > 0 the equation is hyperbolic and travelling waves with a speed equal to c exist. However, when H < 0 the equation becomes elliptic and wave trapping occurs [19]. In this respect the problem is "ill-posed" due to the change of type occurring in the governing differential equation as the material enter the softening regime (H < 0).

b) With gradients: C_1 = constant, C_2 = constant.

The problem of "ill-posedness" is eliminated by introducing higher order gradients. This can be seen by considering equation (5) (with no dependence on $\dot{\bar{\gamma}}$) which upon neglecting elasticity from $(11)_2$ $(\dot{\gamma}=\dot{\gamma}^p)$ and combining with $(11)_1$ gives

$$\rho\ddot{u}=Hu_{yy}-C_1u_{yyyy}-2C_2u_{yy}u_{yyy}, \tag{13}$$

with u denoting the displacement. Now the character of this differential equation is independent of H and is determined by C_1 which is a constant. Therefore, the system does not change type. Moreover, equation (13) can be nondimensionalized by introducing the following "natural" time (η) and length (l) scales

$$\eta=\sqrt{\frac{\rho C_1}{H^2}}\ ,\ l=\sqrt{\frac{C_1}{|H|}}\ . \tag{14}$$

Then with the nondimensional variables

$$\bar{y}=\frac{y}{l},\quad \bar{u}=\frac{u}{l},\quad \bar{t}=\frac{t}{\eta}, \tag{15}$$

equation (13) becomes

$$\bar{u}_{\bar{t}\bar{t}}=\bar{u}_{\bar{y}\bar{y}}-\bar{u}_{\bar{y}\bar{y}\bar{y}\bar{y}}-\left(\frac{2C_2}{C_1}\right)\bar{u}_{\bar{y}\bar{y}}\bar{u}_{\bar{y}\bar{y}\bar{y}} \tag{16}$$

This suggests that the width of the shear band (w) is proportional to l, i.e.,

$$w\propto\left(\frac{C_1}{|H|}\right)^{1/2}. \tag{17}$$

When H > 0, it is shown in [3] that the homogeneous state is stable and, therefore, shear bands do not initiate. They do, however, when $H \le 0$; i.e. according to (17) at H = 0, $w \to \infty$, and as $|H| \to \infty$ (in the softening regime) $w \to 0$, corresponding to severe localization. This is consistent with the quantitative result given in [12,13] where equation (16) is solved numerically with the inertia term on the left hand side set equal to zero. The result for 70/30 brass is shown in Figure 1 and compared with the experimental ones obtained recently by Joshi et al. [20]. Note that such nondimensionalization cannot be performed on equation (12) for which a "natural" length scale cannot be defined.

Elastic-viscoplastic Materials

For this case we drop the gradient terms from equation (5) and assume that $(5)_2$ can be inverted to formally read

$$\dot{\gamma}^p=f(\tau,\gamma^p). \tag{18}$$

Then equation (18) can be substituted into $(11)_2$ which combined with $(11)_1$ yields

$$\ddot{v}+G\frac{\partial f}{\partial \tau}\dot{v}=c^2 v_{yy} - c^2\frac{\partial f}{\partial \gamma^p}\gamma^p_y\,,\quad c^2=\frac{G}{\rho}. \tag{19}$$

This partial differential equation is always hyperbolic since the speed of the propagating wave c is real and equal to the speed of the elastic wave (regardless of the sign of the hardening modulus $\partial f/\partial\tau$). This result for elastic-viscoplastic materials is different than that for the elastic-plastic materials given by (12) where the wave speed and the character of the governing differential equation is determined by the sign of H. Moreover, in this case one can also define "natural" length and time scales as

$$\eta=\left(G\frac{\partial f}{\partial \tau}\right)^{-1}\quad,\quad l=c\eta\;, \tag{20}$$

which with the aid of the nondimensional variables (15) can reduce equation (19) into the form

$$\bar{v}_{\bar{t}\bar{t}} + \bar{v}_{\bar{t}} = \bar{v}_{\bar{y}\bar{y}} -\left(\eta\frac{\partial f}{\partial \gamma^p}\right)\gamma^p_{\bar{y}}\,, \tag{21}$$

where $\bar{v}\equiv v/c$. This nondimensionalization implies that the band width is proportional to l or in view of (20) and $(19)_2$,

$$w \propto \frac{1}{(\partial f/\partial\tau)\sqrt{G\rho}}. \tag{22}$$

In the static case, however, equation $(11)_1$ implies that τ is homogeneous in y and, therefore, one has only equation (18) which has no length scale. Hence, higher order gradients should be considered and the original equation (5) yields

$$\tau(t)=\kappa(\gamma,\dot{\gamma})-C_1\gamma_{yy}-C_2\gamma_y^2, \tag{23}$$

where $\tau(t)$ is a function of time only since $\partial\tau/\partial y = 0$ for the static case. Equation (23) has been solved in [13] numerically where it is shown that the band width is independent of the mesh size and depends only on the values of C_1 and C_2 as can be seen from Figure 2.

Effect of Plastic Spin: Numerical Analysis

For the numerical analysis considered here, we neglect elasticity and assume that the material is viscoplastic exhibiting kinematic/isotropic strain hardening and softening according to the constitutive equations (4), (5) and (7). This, in turn, results into a convenient viscous flow FEM formulation [21] with a nonlinear secant stiffness matrix, and eliminates the severe numerical limitations associated with usual elasto-plastic formulations. In fact, equation (7) and the momentum equation (1), with the inertia term neglected, along with the usual finite element formulation yield [21]

$$[\mathbf{K}(\mathbf{X},\mathbf{V},\bar{\tau}_{,l})]\mathbf{V}=\mathbf{F}_s+\mathbf{F}_\alpha, \tag{24}$$

where $\mathbf{K}$ is the secant stiffness matrix which is a nonlinear function of the nodal velocity vector $\mathbf{V}$, nodal position vector $\mathbf{X}$ and strain gradients $\bar{\tau}_{,l}$, $\mathbf{F}_s$ is the boundary force vector and $\mathbf{F}_\alpha$ is the body force vector arising from the back stress $\boldsymbol{\alpha}$. The nonlinear dependence on $\mathbf{V}$ arises from the fact that the viscosity μ is nonlinearly related to $\dot{\bar{\gamma}}$.

Equation (24) is solved for V using the secant method. An explicit integration scheme with a constant time step using the Newton's forward method is used to integrate the velocity and the evolution equation for the back stress $(4)_2$. This integration algorithm along with the secant method to handle nonlinearities in V and by continuously updating the geometry at each time increment to handle the geometric nonlinearities, proved to be very efficient and produced stable results.

For the numerical analysis we use the following power law expression to model the isotropic hardening and softening

$$\kappa(\bar{\gamma},\dot{\bar{\gamma}})=\tau_0\left(\frac{\dot{\bar{\gamma}}}{\dot{\gamma}_0}\right)^m\left(1+\frac{\bar{\gamma}}{\gamma_0}\right)^n\left(1+\frac{\bar{\gamma}}{\gamma_1}\right)^\nu , \tag{25}$$

where τ_0 is the initial yield stress, $m > 0$ and $n > 0$ are the strain rate sensitivity and strain hardening exponents respectively, $\nu < 0$ is the "strain softening" exponent, and $\dot{\gamma}_0, \gamma_0$ and γ_1 are material constants [21]. For a typical structural steel we use the values $\tau_0=400/\sqrt{3}$ MPa, $h = 400$ MPA, $C = 4.0$, $n = 0.1$, $m = 0.01$, $\nu = -2.0$, $\gamma_0=2\times10^{-3}$, $\dot{\gamma}_0=1.0\ \text{sec}^{-1}$, $\gamma_1=2.0$. It is noted that for the present case the gradient effects are dropped by setting $C_1 = C_2 = 0$.

The numerical method is used to analyze the development of shear bands in a plane-strain tensile test. Due to symmetry, only one-quarter of the tensile specimen is examined. The one-quarter is meshed into 350 rectangular elements each consisting of four cross-triangular elements. The specimen is deformed in tension with a constant cross-head speed of V = 0.01 cm/sec. A length to width aspect ratio of 3 is used. A brief summary of the results are shown in Figures 3-7 which are sufficient to illustrate the effects of viscosity and plastic spin on the shear bands development. Figure 3 shows the mesh of the specimen with two "cross" shear band clearly formed. A closer look at the bands is given in Figure 4 where only one-quarter of the specimen is shown. It can be seen from the figure that the width of the band spans over at least two elements. The corresponding strain contours are shown in Figure 5 indicating the severity of the strain in the band and the fact that the strain field varies only across the band with constant contours along it. The effect of the plastic spin is examined by varying the value of $\zeta=a/\sigma_0$. [When $\zeta=0$ the corotational rate given by $(4)_3$ reduces to the usual Jaumman rate]. Figure 6 shows P/P_y versus U/L for various values of a, where P is load, P_y is yield load, U is elongation and L is specimen length. The sharp drop in the load-displacement curve for U/L > 0.2 corresponds to the development of the shear band. The band seems to initiate around U/L ~ 0.2. It seems that the parameter a has no influence on the curve until the band becomes severe around U/L ~ 0.25. This is expected since the plastic spin influences the shearing mode which (for the present problem of initially stretching model) becomes significant only when the band forms. The result, thus, suggests that the plastic spin does not influence the initiation of shear bands from a stretching mode (see also Tvergaard and Giessen [22]), it may however from a shearing mode. An increase in the degree of the plastic spin seems to decrease "ductility" in the sense that the value of the average strain to final load "collapse" decreases as *a* increases.

Finally, Figure 7 shows the variation of the kinetic energy E/E_0, where $E_0 \equiv mV^2/2$, m is the total mass, and V is the imposed boundary velocity. It can be seen from the figure that the kinetic energy remains almost constant in the early stages of deformation then increases steadily until the shear band initiates around $U/L \cong 0.2$, resulting to a sharp increase in the kinetic energy. Thereafter, the kinetic energy reaches another "steady" state once the band is completely formed. Note that E/E_0 is slightly less than 1.0. This is expected since the material outside the band moves rigidly with velocity V.

Acknowledgement

The support of the US Army Research Office Under Contract DAAL03-90-G-0151 (ECA and HMZ), and that of the US National Science Foundation under grants MSS-8907748 (HMZ), CES-8800459 (ECA), and MSS-8920700 (ECA) is gratefully acknowledged.

References

1. Rice, J: The localization of plastic deformation in: Theoretical and Applied Mechanics, ed. Koiter, W.T., North-Holland, pp. 207-220, 1976.

2. a) Aifantis, E.C.: On the microstructural origin of certain inelastic models. ASME, J. Eng. Mat. Tech. 106. (1984) 326-330.
b) Aifantis, E.C.: The physics of plastic deformation. Int J. Plasticity 3 (1987) 211-247.

3. a) Zbib, H.M.; Aifantis, E.C.: On the concept of relative and plastic spins and its implications to large deformation theories, I & II. Acta Mechanica 75 (1988) 15-33, 35-56.
b) Zbib, H.M.; Aifantis, E.C.: On the localization and postlocalization behavior of plastic deformation, I, II & III. Res Mechanica 23 (1988) 261-277, 279-292 & 293-305.

4. Korbel, A.; Embury, J.D.; Hatherly, M.; Martin, P.L.; Erbsloh, H.W.: Microstructural aspects of strain localization in Al-Mg alloys. Acta Metall. 34 (1986) 1999-2009.

5. Drucker, D.C.: Material response and continuum relations; or from microscales to macroscales. ASME, J. Eng. Mat. Tech. 106 (1984) 286-289.

6. Hill, R.; Hutchinson, J.H.: Bifurcation phenomena in the plane tension test. J. Mech. Phys. Solids 23 (1975) 239-264.

7. Tvergaard, V.: Influence of voids on shear band instabilities under plane strain conditions. Int. J. Fracture 17 (1981), 389-407.

8. Hutchinson, J.H. (ed.), Viewpoint on Shear Bands. Scripta Metall. 18 (1984) 421-458.

9. Anand, L.; Kim, K.H.; and Shawki, T.G.: Onset of shear localization in viscoplastic solids. J. Mech. Phys. Solids 35 (1987) 407-429.

10. Batra, R.C.: The initiation and growth of, and the interaction among, adiabatic shear bands in simple and dipolar materials. Int. J. Plasticity 3 (1987) 75-89.

11. Déve, H.E.; Asaro, R.J.: The development of plastic failure modes in crystalline materials: Shear bands in fcc polycrystals. Metall. Trans. 20A (1989) 579-597.

12. Zbib, H.M.; Aifantis, E.C.: On the structure and width of shear bands. Scripta Metall. 22 (1988) 703-708.

13. Zbib, H.M.; Aifantis, E.C.: On the gradient-dependent theory of plasticity and shear banding. Acta Mechanica (in press).

14. Zbib, H.M.; Aifantis, E.C.: A gradient-dependent flow theory of plasticity: Application to metal and soil instabilities. Appl. Mech. Review 42 (1989) 295-304.

15. Vardoulakis, I.; Aifantis, E.C.: Gradient-dependent dilatancy and its implications to shear banding and liquefaction. Ing. Arch. 59 (1989) 197-208.

16. Mühlhaus, H.B.; Aifantis, E.C.: The influence of microstructure-induced gradients on the localization of deformation in viscoplastic materials. Acta Mech. (in press).

17. Charalambakis, N.C.; Aifantis, E.C.: On stress controlled thermoviscoplastic shearing and higher order strain gradients. Acta Mech. 81 (1990) 109-114.

18. Oka, F.; Yashima A.; Adachi T.; Aifantis, E.C.: A gradient-dependent viscoplastic model for clay and its application to FEM consolidation analysis in: Constitutive Laws for Engineering Materials - Theory & Application, ed. Desai, C.S., ASME Press, pp. 313-316, 1991.

19. Wu, F.H.; Freund, L.B.: Deformation trapping due to thermoplastic instability in one-dimensional wave propagation. J. Mech. Phys. Solids 32 (1984) 119-132.

20. Joshi, R.B.; Bayoumi, A.E.; Zbib, H.M.: Evaluation of macroscopic shear banding using a digital image processing technique. Scripta Metall. et Mater. 24 (1990) 1747-1752.

21. Zbib, H.M.: Shear banding in viscoplastic materials exhibiting isotropic hardening/softening. MME Report No. 9/-01, WSU, Pullman, (1991).

22. Tvergaard, V.; van der Giessen, E.: Effect of plastic spin on localization predictions for a porous ductile material. Report No. 402, The Tech. Univ. of Denmark (1990).

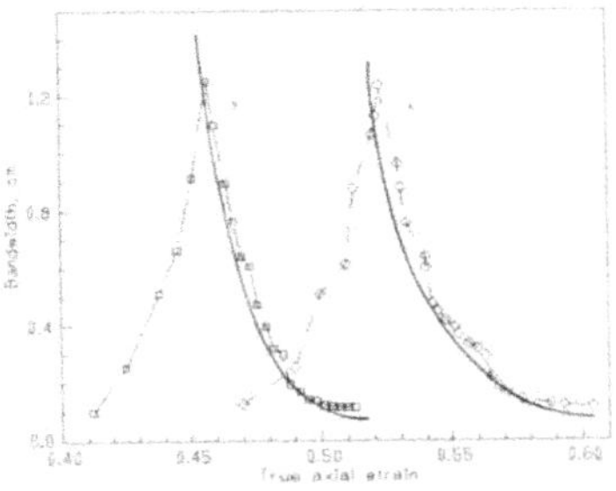

Figure 1. Shear bandwidth versus Strain in 70/30 Brass,
A) rolling angle of 45°,
B) rolling angle up 0°.

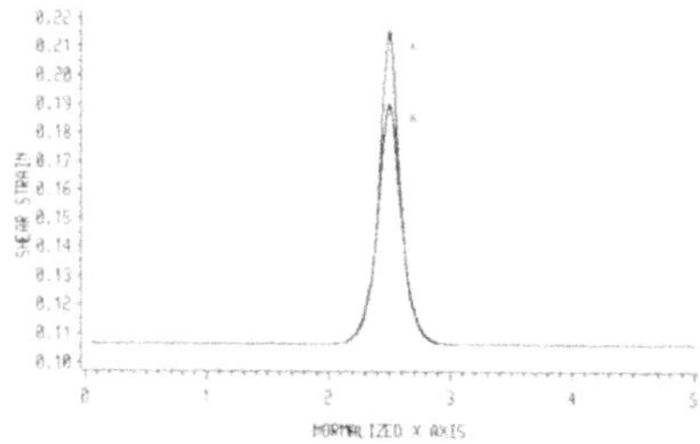

Figure 2. Effect of mesh size
A) $\Delta\bar{y} = 0.1$, B) $\Delta\bar{y} = 0.5$.

Figure 3. Deformed Mesh.

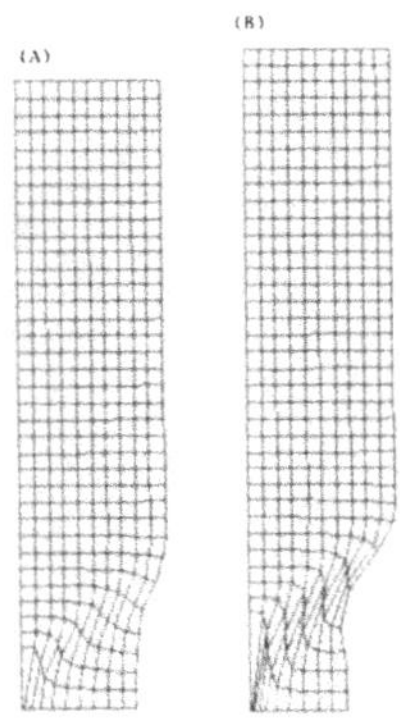

Figure 4. Deformed meshes.

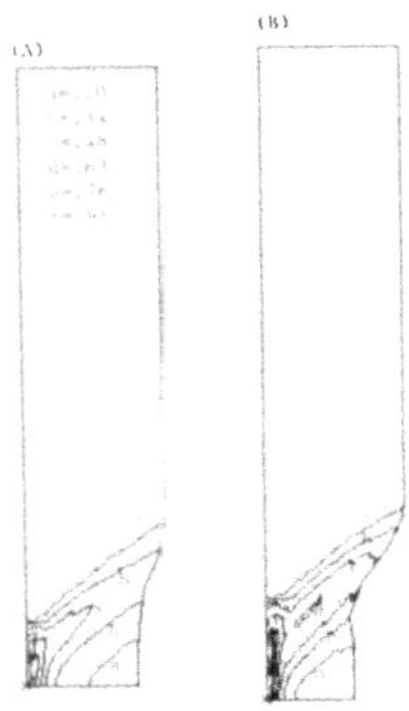

Figure 5. Strain Contours.

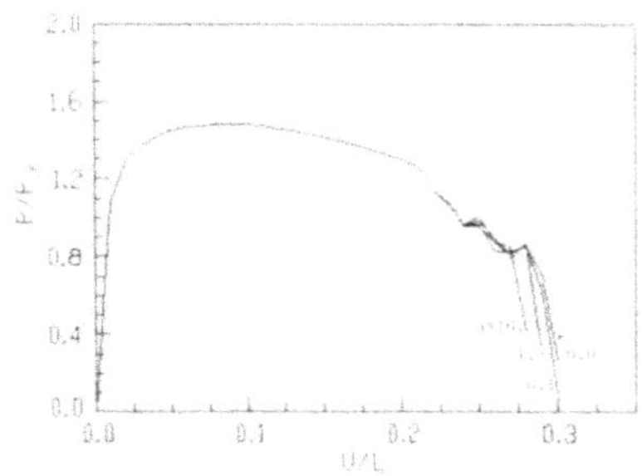

Figure 6. Effect of plastic spin on Load-extension curve.

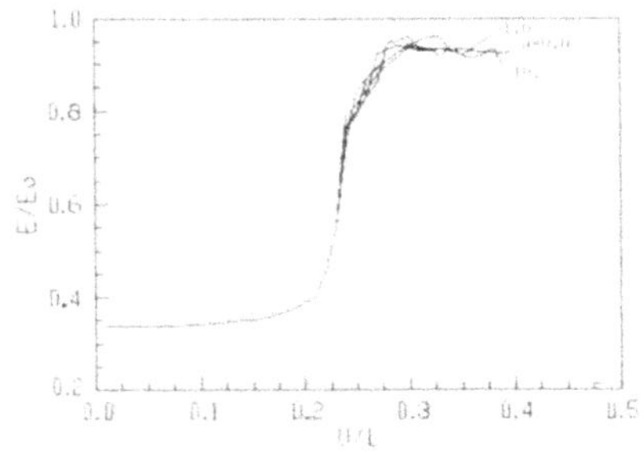

Figure 7. Effect of plastic spin on Kinetic energy.

On Constitutive Spins at Large Inelastic Deformations

Y.F. DAFALIAS

Department of Civil Engineering, University of California, Davis, CA, USA, and
Division of Mechanics, National Technical University of Athens, HELLAS

Summary

A general constitutive framework for large inelastic deformations is presented in a Eulerian setting, employing multiple constitutive spins for the corotational rate equations of evolution of the internal variables. Related issues for single or multiple spin formulations are critically addressed.

Introduction

The constitutive spins used in the corotational rate equations of evolution of the tensorial internal variables and for the stress rate in its relation to the rate of deformation, is an essential aspect of macroscopic large inelastic deformation theories in a Eulerian framework. Their purpose is not only to satisfy invariance requirements under superposed rigid body rotations, but also to attribute a physical meaning to the orientational aspects of a constitutive law. It is mainly the aspect of the constitutive spins which differentiates theories according to the material in question, and which has been the focus of numerous recent works. The main objective of this article is to examine systematically for the first time the role of multiple constitutive and plastic spins within a general constitutive framework for models which employ multiple tensorial internal variables, and address critically some issues related to both single and multiple constitutive spins formulations. Along the lines of presentation some general guidelines will be set in relation to metallic and elastomeric materials. The notation will be standard (bold face for tensors), and the summation convention over repeated indices is adopted, unless otherwise stated in specific cases.

General Constitutive Framework

Considering the multiplicative decomposition of the deformation gradient $\mathbf{F} = \mathbf{VP}$ into a symmetric elastic part $\mathbf{V} = \mathbf{V}^T$ and a plastic part $\mathbf{P}$, Lee [1], an observer's frame spinning by $\boldsymbol{\omega}$ according to Mandel's concept of director vectors [2], and making in addition the assumption of small elastic strains $\boldsymbol{\varepsilon}^e = \varepsilon\mathbf{V}'$ based on $\mathbf{V} = \mathbf{I} + \varepsilon\mathbf{V}'$ with $\mathbf{V}'$ of order 1 and $\varepsilon \ll 1$, the symmetric (subscript s) and antisymmetric (subscript a) parts of the velocity gradient are given by [3,4]

D. Besdo, E. Stein (Eds.)
Finite Inelastic Deformations –
Theory and Applications
IUTAM Symposium Hannover/Germany 1991

$$(\dot{\mathbf{F}}\mathbf{F}^{-1})_s = \mathbf{D} = \varepsilon \overset{o}{\mathbf{V}}' + (\overset{o}{\mathbf{P}}\mathbf{P}^{-1})_s = \mathbf{D}^e + \mathbf{D}^p = \mathbf{D}^e + <\lambda> \mathbf{N}^p \tag{1a}$$

$$(\dot{\mathbf{F}}\mathbf{F}^{-1})_a = \mathbf{W} = \omega + (\overset{o}{\mathbf{P}}\mathbf{P}^{-1})_a = \omega + \mathbf{W}^p = \omega + <\lambda> \boldsymbol{\Omega}^p \tag{1b}$$

with $\mathbf{D}$, $\mathbf{D}^e$ and $\mathbf{D}^p$ the total, elastic and plastic rate of deformation tensors, correspondingly, $\mathbf{W}$ and $\mathbf{W}^p$ the material and plastic spin tensors, respectively, and a superposed o denoting the corotational rate with respect to ω as defined in [2,3,4]. The scalar-valued loading index λ ($< >$ are the Macauley brackets) and tensor valued functions $\mathbf{N}^p$ and $\boldsymbol{\Omega}^p$ provide constitutive relations for $\mathbf{D}^p$ and $\mathbf{W}^p$, respectively. Due to invariance requirements under superposed rigid body rotation, $\mathbf{N}^p$ and $\boldsymbol{\Omega}^p$ are isotropic functions of the state variables which are the Cauchy stress σ and a set of internal variables $\mathbf{a}_i$ [3], considered for simplicity to be second order symmetric tensors and scalars. The $\mathbf{a}_i$ collectively represent the macroscopic manifestation of the material substructure. The constitutive rate equation of evolution for each $\mathbf{a}_i$ is expressed by

$$\overset{c}{\mathbf{a}}_i = \dot{\mathbf{a}}_i - \omega_i \mathbf{a}_i + \mathbf{a}_i \omega_i = <\lambda> \bar{\mathbf{a}}_i \qquad \text{(no sum over i)} \tag{2}$$

in terms of a corotational rate denoted by a superposed c in relation to a constitutive spin ω_i, which under superposed rigid-body rotation $\mathbf{Q}(t)$ is required to satisfy the transformation $\omega_i^* = \mathbf{Q}\omega_i\mathbf{Q}^T + \dot{\mathbf{Q}}\mathbf{Q}^T$. It is assumed that the $\mathbf{a}_i$ evolve only when plastic deformation occurs, hence, the common loading index λ with Eq. (1). Each constitutive function $\bar{\mathbf{a}}_i$ is isotropic functions of σ and all $\mathbf{a}_i$, due to invariance requirements and the foregoing property of ω_i. For a purely orientational variable, e.g. $\mathbf{a}_i = \mathbf{n}_i \otimes \mathbf{n}_i$ (no sum over i) with $\mathbf{n}_i$ a unit direction, the corresponding $\bar{\mathbf{a}}_i \equiv \mathbf{0}$. The determination of the different ω_i is the key issue of Eq. (2), addressing the "orientational" part of the constitutive law, while $\bar{\mathbf{a}}_i$ addresses the "extensional" part. A word of caution is pertinent here; the spin of the eigenvectors of $\mathbf{a}_i$ is not in general identical to ω_i, but of course it depends on ω_i and $\bar{\mathbf{a}}_i$ according to Eq. (2) (in case of $\bar{\mathbf{a}}_i \equiv \mathbf{0}$ the two spins are identical). One can define the relative plastic spin $\mathbf{W}_i^p$ associated with $\mathbf{a}_i$ by the equation

$$\mathbf{W} = \omega_i + \mathbf{W}_i^p = \omega_i + <\lambda> \boldsymbol{\Omega}_i^p \tag{3}$$

The transformation of $\mathbf{W}$ and ω_i under superposed rigid body rotation, implies from Eq. (3) that $\boldsymbol{\Omega}_i^p$ is an isotropic function of σ and all $\mathbf{a}_i$, not excluding the possibility to use σ and only the $\mathbf{a}_i$ relevant to Eqs. (2) and (3). An important observation is pertinent here. If the $\boldsymbol{\Omega}_i^p$ is exclusively a function of σ and $\mathbf{a}_i$ and the latter have zero deviatoric values (isotropy) at a certain stage of deformation, it follows that $\boldsymbol{\Omega}_i^p = \mathbf{0}$ and $\omega_i = \mathbf{W}$. But if in the definition of $\mathbf{W}_i^p$ (or ω_i) one includes also spins of other entities, e.g. the Eulerian spin $\boldsymbol{\Omega}^E$, the foregoing conclusion is not necessarily true.

The next step is the determination of the loading index λ. Considering the equation of the yield surface $f(\sigma,\mathbf{a}_i) = 0$, isotropic functions of σ and $\mathbf{a}_i$ due to invariance requirements, the consistency condition $\dot{f} = 0$ yields according to Eq. (A2) of the Appendix in [3] with respect to any spin

$$\dot{f} = (\partial f/\partial \sigma):\overset{o}{\sigma} + (\partial f/\partial \mathbf{a}_i):\overset{o}{\mathbf{a}}_i = 0 \tag{4}$$

where : denotes the trace of the product of the tensors (recall $\mathbf{a}_i = \mathbf{a}_i^T$). Choosing now the spin entering Eq. (4) to be ω, it is a matter of straightforward algebraic manipulations to obtain using Eqs. (1b), (2), (3), (4) and the immediate consequence of the arbitrariness of the spin in Eq. (A2) of the Appendix in [3], which reads $\sigma(\partial f/\partial\sigma) - (\partial f/\partial\sigma)\sigma + \mathbf{a}_i(\partial f/\partial \mathbf{a}_i) - (\partial f/\partial \mathbf{a}_i)\mathbf{a}_i \equiv \mathbf{0}$, the following equivalent expressions for λ

$$\lambda = (\mathbf{N}^n : \overset{o}{\sigma}) / H_o = (\mathbf{N}^n : \overset{\nabla}{\sigma}) / H_\nabla \tag{5}$$

$$H_o = H + (\sigma\mathbf{N}^n - \mathbf{N}^n\sigma) : \Omega^p + [\mathbf{a}_i(\partial f/\partial \mathbf{a}_i) - (\partial f/\partial \mathbf{a}_i)\mathbf{a}_i] : \Omega_i^p \tag{6a}$$

$$H_\nabla = H + [\mathbf{a}_i(\partial f/\partial \mathbf{a}_i) - (\partial f/\partial \mathbf{a}_i)\mathbf{a}_i] : \Omega_i^p \tag{6b}$$

$$\mathbf{N}^n = \partial f/\partial\sigma \quad ; \qquad H = -(\partial f/\partial \mathbf{a}_i) : \bar{\mathbf{a}}_i \tag{7}$$

with H the hardening modulus, the sum over i implied even when i appears three times, and a superposed ∇ denoting the Jaumann rate with respect to $\mathbf{W}$. Eq. (6b) can be obtained directly from Eq. (4) using the spin $\mathbf{W}$ for the corotational rates, or equivalently from Eq. (6a) by setting $\Omega^p = \mathbf{0}$. Observe that both $\bar{\mathbf{a}}_i$ ("extensional" part) and Ω_i^p ("orientational" part) of the constitutive laws for $\mathbf{a}_i$ enter Eqs. (6). It is remarkable that if all ω_i are equal, one can choose $\omega = \omega_i$ which entails $\Omega^p = \Omega_i^p$ and based on the aforementioned consequence of Eq. (A2) of the Appendix in [3] one obtains from Eq. (6a) $H_o = H$, and from Eq. (6b) $H_\nabla = H + \mathbf{N}^n : (\sigma\Omega^p - \Omega^p\sigma)$, results already obtained in Dafalias [4]. This is in fact the classical case of Mandel's theory [2], where the spins used in the corotational rate equations of evolution for all $\mathbf{a}_i$ are identical to the spin of the director vectors' triad. With g now an elastic potential, which due to invariance is an isotropic function of $\varepsilon^e = \varepsilon\mathbf{V}'$ and $\mathbf{a}_i$, one has $\sigma = \partial g/\partial\varepsilon^e$. Based on Eq. (1a) for the definition of $\mathbf{D}^e$ and Eq. (A1) of the Appendix in [3], it follows that

$$\overset{o}{\sigma} = L : \mathbf{D}^e + \mathbf{C}_i : \overset{o}{\mathbf{a}}_i \tag{8}$$

where $L = \partial^2 g/\partial\varepsilon^e \otimes \partial\varepsilon^e$ are the elastic moduli of $O(\sigma/\varepsilon)$, and $\mathbf{C}_i = \partial^2 g/\partial\varepsilon^e \otimes \partial\mathbf{a}_i$ are the elastoplastic coupling moduli of $O(\sigma/\mathbf{a}_i)$. Based on the definition of λ in the first part of Eq. (5) and using Eqs. (1), (2), (3), (6a) and (8), one can derive by standard procedures the inverse relation between $\overset{o}{\sigma}$ and $\mathbf{D}$ as

$$\overset{\circ}{\sigma} = [\, L - [(L : \mathbf{N}) \otimes (\mathbf{N}^n : L)] / H_D] : \mathbf{D} \tag{9}$$

$$\mathbf{N} = \mathbf{N}^p - L^{-1} : \mathbf{C}_i : \mathbf{R}_i \;; \qquad \mathbf{R}_i = \bar{\mathbf{a}}_i + \mathbf{a}_i(\Omega_i^p - \Omega^p) - (\Omega_i^p - \Omega^p)\mathbf{a}_i \tag{10}$$

$$H_D = H_o + \mathbf{N}^n : L : \mathbf{N}^p - \mathbf{N}^n : \mathbf{C}_i : \mathbf{R}_i \tag{11}$$

where no sum over i takes place in the definition of $\mathbf{R}_i$, and the loading index $\lambda = \mathbf{N}^n : L : \mathbf{D} / H_D$. Observe that if the $\mathbf{a}_i$ entering g are purely orientational, $\bar{\mathbf{a}}_i \equiv \mathbf{0}$ but still $\mathbf{R}_i \neq \mathbf{0}$. If in addition for these orientational $\mathbf{a}_i$ the Ω_i^p is common (i.e., ω_i is common), one can choose $\Omega^p = \Omega_i^p$ (i.e., $\omega = \omega_i$), hence, $\mathbf{R}_i = \mathbf{0}$ and $\mathbf{N} = \mathbf{N}^p$ (if also $\mathbf{N}^p = \mathbf{N}^n$ one has symmetric elastoplastic moduli). If the Ω_i^p is common for all $\mathbf{a}_i$ (not only the ones which enter g), and in addition the $\bar{\mathbf{a}}_i \equiv \mathbf{0}$ for those $\mathbf{a}_i$ which enter g, then by choosing $\Omega^p = \Omega_i^p$ it follows again that $\mathbf{R}_i = \mathbf{0}$ and $\mathbf{N} = \mathbf{N}^p$, and from Eq. (6a), (11) and the aforementioned consequence of Eq. (A2) of the Appendix in [3] that $H_D = H + \mathbf{N}^n : L : \mathbf{N}^p$, i.e. the classical result. This result can also be obtained without the foregoing assumptions about Ω_i^p and $\bar{\mathbf{a}}_i$ if terms of order ε are neglected compared to 1, recalling the order of L and $\mathbf{C}_i$ and rendering by proper normalization f of $O(\sigma)$ (hence $\mathbf{N}^n$ of $O(1)$), and $\mathbf{N}^p$, Ω^p and Ω_i^p of $O(1)$. It may be observed that if one had chosen $\omega = \mathbf{W}$ in writing Eq. (8), hence $\Omega^p = \mathbf{0}$, $\overset{\circ}{\sigma} = \overset{\nabla}{\sigma}$ and from Eq. (9) no dependence on ω appears. However, this would have implied that $\mathbf{D}^e = \overset{\nabla}{\varepsilon}{}^e = \varepsilon\overset{\nabla}{\mathbf{V}}{}'$ instead of $\mathbf{D}^e = \varepsilon\overset{\circ}{\mathbf{V}}{}'$, which is correct only within order ε compared to the plastic transformation rate since $\overset{\nabla}{\varepsilon}{}^e - \overset{\circ}{\varepsilon}{}^e = \varepsilon\lambda(\mathbf{V}'\Omega^p - \Omega^p\mathbf{V}')$. In fact, this question of the proper definition of the elastic rate of deformation has occupied two of the last works of Mandel [5,6], who eventually concluded that the choice of ω cannot be made arbitrarily but becomes an important constitutive assumption. We will return to this point later. The $\overset{\nabla}{\sigma}$ can correctly substitute for $\overset{\circ}{\sigma}$ in Eq. (9), by simultaneously adding the term $-L^{-1} : (\sigma\Omega^p - \Omega^p\sigma)$ in the right-hand side of Eq. $(10)_1$, as shown in [4]. It is also important to observe that the choice of ω has no effect on the definition of $\mathbf{D}^p = \langle\lambda\rangle\, \mathbf{N}^p$, since $(\overset{\circ}{\mathbf{P}}\mathbf{P}^{-1})_s = (\dot{\mathbf{P}}\mathbf{P}^{-1})_s$ for any ω, based on $\overset{\circ}{\mathbf{P}} = \dot{\mathbf{P}} - \omega\mathbf{P}$ [2,3]. A corresponding theory of viscoplasticity can be developed by substituting in the foregoing equations for the loading index λ proper overstress functions Φ for $\mathbf{D}^p$ and $\mathbf{W}^p$, and Φ_i for $\overset{c}{\mathbf{a}}_i$ and $\mathbf{W}_i^p$ as shown in [7], without the additional complication of having to determine λ. The choice of ω is still a constitutive issue in Eqs. (8) and (9).

Single and Multiple Spins

When a common ω_i is assumed for all $\mathbf{a}_i$ in Eqs. (2) and (3), one sets $\omega = \omega_i$, hence, $\Omega^p = \Omega_i^p$. Dafalias [8,9] and Loret [10] first proposed specific expressions for Ω^p using the representation theorems for antisymmetric tensor-valued isotropic functions. In the case of a single symmetric internal variable $\mathbf{a}$ the simplest possible such expression takes the form

$$\Omega^p = \eta(a\sigma - \sigma a) \tag{12}$$

with η the plastic spin coefficient and where the deviatoric parts α and $\mathbf{s}$ of $\mathbf{a}$ and σ, respectively, can be used instead. Equation (12) has been used extensively for kinematic hardening induced anisotropy with α being the back-stress tensor, or rotational symmetries around a unit direction $\mathbf{n}$, in which case $\mathbf{a} = \mathbf{n} \otimes \mathbf{n}$ (similarly for orthotropic symmetries). In the former case, a rearrangement of Eq. (12) in conjunction with a Mises type kinematic hardening model and Eq. (1b) yields $\mathbf{W}^p = (1/2)\rho(\alpha\mathbf{D}^p - \mathbf{D}^p\alpha)$ [3,4,9], where ρ is a normalized form of η. When applied to the case of simple shear γ along x_1 in the x_1 - x_2 plane, the foregoing yields

$$\omega_{12} = (1/2)\dot{\gamma}(1 - \rho\alpha_{11}) \tag{13}$$

where $W^p_{12} = (1/2)\dot{\gamma}\rho\alpha_{11}$. For rotational symmetries around $\mathbf{n}$ (similarly for orthotropic) Eqs. (2), (3) and (12) yield for simple shear [3,11].

$$\dot{\theta} = -\omega_{12} = (1/2)\dot{\gamma}(\eta\cos2\theta - 1) \tag{14}$$

with $W^p_{12} = (1/2)\dot{\gamma}\eta\cos2\theta$, and θ the angle of $\mathbf{n}$ with the x_1 axis. The simplicity of Eqs. (12), (13) and (14) cannot be overestimated. If properly treated, they yield important insight in the material response and interpretation of ρ and η. Since this has been already reported [3,4,7,12], it suffices to state that for homogeneous deformations (particularly in simple shear), the foregoing equations provide specific analytical expressions in closed form for limiting values of the tensorial and substructural orientations (in what can be considered to be the counterpart of ideal orientations in polycrystal analysis), and the stress components. In a recent work [13] the effect of Eq. (14) was also studied for inhomogeneous deformations related to fold formations.

While most of the foregoing analyses were performed for a constant plastic spin coefficient η or ρ, a number of works appeared with variable ρ. A word of caution is pertinent at this point, which has been mentioned in [3] but not taken into account by some authors. Without accounting for the important observation made after Eq. (3), and within the present theoretical framework, Eq. (12) must yield continuously $\Omega^p = \mathbf{0}$ as $\alpha s - s\alpha$ tends to $\mathbf{0}$, which occurs when α and $\mathbf{s}$ are co-axial (this includes the special case $\alpha = \mathbf{0}$). This imposes restrictions on the specifications of η or ρ. In [14] for example, the suggestion $\rho = (2/\mathrm{tr}\alpha^2)^{1/2} = (\alpha_{11}^2 + \alpha_{12}^2)^{-1/2}$ (the last expression for simple shear) was made. This violates the foregoing continuity requirement as $\alpha \to \mathbf{0}$ with constant eigenvectors' orientation, as it can be seen easier from Eq. (13) where $W^p_{12} = (1/2)\dot{\gamma}\alpha_{11}/(\alpha_{11}^2+\alpha_{12}^2)^{1/2}$. Also in [15] the proposition $\eta = y(\varepsilon_{eq})/[\mathrm{tr}(\alpha s - s\alpha)^2]^{1/2}$ (y a function of equivalent strain ε_{eq}) was made, which besides an apparent error in sign (it appears that the absolute value of the trace must be used), violates the continuity requirement as $\alpha \to \mathbf{0}$ or α tends to become co-axial with $\mathbf{s}$, because the $\eta(\alpha \mathbf{s} - \mathbf{s}\alpha)$ is a non-zero function of $y(\varepsilon_{eq})$. In a different setting, Dafalias and Aifantis [16] proposed $\rho = \rho_o \exp(c_r \varepsilon^p_{eq})$ motivated by micromechanical considerations, which yields $\rho \to \infty$ as the equivalent plastic strain $\varepsilon^p_{eq} \to \infty$. It was shown rigorously in [7] that for simple shear $\alpha_{11} \to 0$ and $\alpha_{12} \to$ the same finite value as $\gamma \to +\infty$ for any value of ρ_o (since $\rho \to \infty$), corroborating the numerical plots in [16], a result which may not be always desirable. In contrast, for the simpler case of ρ = constant > 0, the limit, as $\gamma \to \infty$, of α_{11} decreases and of α_{12} increases with increasing ρ, towards specific values [3], a very desirable feature from the point of view of data simulation and calibration of model parameters. Finally, it should be pointed out that the key Eq. (12) is based on the symmetry of $\mathbf{a}$. The use of asymmetric internal variables has been reported in [16,17], but this aspect will be addressed elsewhere.

In a different setting for large viscoelastic deformations of elastomers (no multiplicative decomposition of **F**), but pertinent to the present discussion, the author used the concept of an internal stress α as the macroscopic manifestation of the molecular chain network resistance to deform affinely instantaneously [18]. With the polar decomposition **F** = **VR**, a rate equation of evolution was given for α in relation to a constitutive spin ω_i defined by

$$\omega_i = \Omega^E - \Omega^I \;; \qquad \Omega^I = \eta_r (\alpha \mathbf{V} - \mathbf{V}\alpha) \tag{15}$$

The Ω^E is the Eulerian spin (spin of eigenvectors of **V**) and Ω^I the novel concept of an internal spin, whose definition reflects the tendency of α to become co-axial with **V** since the former is associated with the transient state of the molecular chain network which tends to deform affinely with the continuum at equilibrium. A closer examination, however, showed a possible weakness of the definition of ω_i in terms of Ω^E. If a pre-existing small tension or compression along x_2, which establishes the eigenvectors of **V** along x_1 and x_2, is followed by simple shear along x_1, a violent rate of rotation of these eigenvectors will occur in the neighborhood of $\gamma = 0$, i.e. a very large Ω^E occurs, which may cause unrealistic oscillations of the α components at $\gamma \sim 0$. It is only when $\alpha = \mathbf{0}$ at $\gamma = 0$ that no such problem arises. Thus, it is proposed to substitute **W** for Ω^E in Eq. $(15)_1$, which will not affect the desirable properties stemming from Ω^I, the rest in [18] remaining the same. This should also provide a word of caution for the foregoing consequences to other researchers who use Ω^E as part of the definition of a constitutive spin.

While the general development with multiple spins and related issues appears to be first presented in the present work, some models with multiple spins were considered in recent works. Within the setting of rigid-plastic response, Zbib and Aifantis [19] and Dafalias and Cho [20] presented two such cases. In the former work [19] a two-back stress model was used, postulating for each back stress a different Ω^p of the form of Eq. (12) with different η. The results under simple shear show a qualitative resemblance to experimental data of fixed-end torsion [21] for the normal stress component, but it appears that the model cannot describe the shear stress response. It is believed this is due to the fact that single or multiple back-stresses can properly account only up to a point for texture development, particularly if the latter is very intense. In contrast, the work in [20] introduced a combination of kinematic and orthotropic hardening, the latter restricted to cubical symmetries. The kinematic hardening is associated with a constitutive spin ω_i following from Eq. (3) and a corresponding plastic spin according to Eq. (12) by substituting $\partial J/\partial\sigma$ for σ, with J the viscoplastic potential function. In fact for simple shear it takes the form of Eq. (13). On the other hand, the orthotropic directions and intensity develop along the eigenvectors of α, which rotate by a different spin from the ω_i, as observed earlier. The application of the model achieved for the first time a macroscopic quantitatively accurate simulation of experimental data in simple shear for both normal and shear stresses at different rates, as reported in [21], and in fact provided semi-analytical expressions for the stress components. The successful simulation can be attributed to the fact that kinematic hardening and the development of orthotropic symmetries, considered simultaneously, account for both residual stress and texture development in a macroscopic way.

It is instructive to comment now on some aspects of the work in [20], which have not been explicitly mentioned. First, Fig. 1, taken from [20], shows a comparison between the model simulation and the experimental data for fixed-end torsion of a solid cylindrical iron α-Fe specimen [21]. The experimental stress values refer to an element on the outer surface of the solid specimen, and were obtained from the measured average stresses by proper calculations proposed in [21]. The simulation was performed under the assumption of simple shear γ along x_1 on the x_1 - x_2 plane, and the corresponding equations and model constants are reported in [20]. Both shear and normal stress components are seen to match satisfactorily the experimental data. Of particular interest is the successful simulation of the change of sign of σ_{22} from negative (compressive) to positive (tensile), while the σ_{12} shows a monotonic increase without oscillations. In order to understand this kind of response, a succession of configurations of the dynamic yield surface corresponding to the event of Fig. 1 (the model is cast in the viscoplastic framework) is shown in Fig. 2 for the values of the shear $\gamma = 0, 1, 5, 20$, according to the analytical expressions developed in [20]. In fact, what is shown in Fig. 2 is the projection on the σ_{11} - σ_{12} plane of the intersection of the successive positions of the yield surface with the plane $\sigma_{11} + \sigma_{22} = 0$, which eloquently demonstrate the effect of kinematic hardening and orthotropic texture development on the evolution of the surface. Three kinds of hardening mechanisms determine such evolution. First, the kinematic hardening shifts the "center" of the surface; second, the development of orthotropic intensity causes an "ovalization" of the surface; and third, the rotation of the orthotropic axes, as described earlier, causes a corresponding rotation of the ovalized surface, which otherwise would have kept its major axis parallel to the

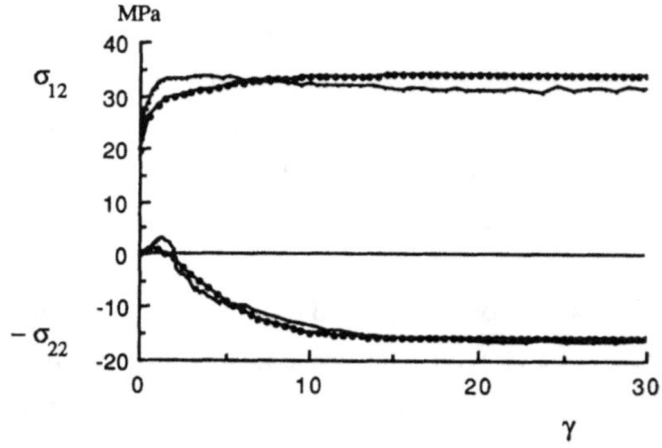

Fig. 1. Experimental data after [21] shown with thin-dotted line, and superposed model simulation after [20] shown with thick-dotted line for large simple shear.

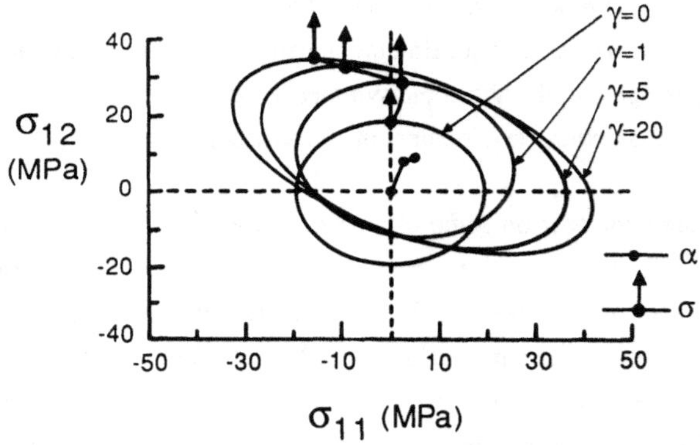

Fig. 2. The evolution of the dynamic yield surface and associated stress path for the data shown in Fig. 1 due to kinematic hardening, orthotropic texture development and rotation of orthotropic axes.

σ_{11} axis. This third mechanism can be called orientational hardening, and is intimately related to the definition of the constitutive and plastic spins used in the model. The next step is to observe the orbit of σ. Since rigid-plastic response and associative flow rule are assumed, for simple shear the σ must remain on the point of the evolving yield surface at which the outward normal (shown by arrows in Fig. 2) is parallel to the σ_{12} axis. Following these positions of σ in Fig. 2, it is easy to understand the stress response shown in Fig. 1. While the σ_{12} keeps increasing, the σ_{11} becomes first tensile as the initially predominant effect of kinematic hardening shifts the surface to the right (recall that the σ_{22} shown in Fig. 1 equals $-\sigma_{11}$). Then, σ_{11} turns left and going through zero becomes increasingly compressive as the ovalization and rotation of the surface dominates over the effect of kinematic hardening at later stages of deformation. Finally, the evolution of the yield surface reaches a saturation stage from all three points of view (kinematic, orthotropic intensity and orientation). As a final note to this subject, it must be understood that this is an initial attempt to describe texture development and its effect on stress-strain response at large strains by macroscopic means; as such it is subject to future modifications as, for example, the consideration of orthotropic symmetries more general than cubical, and/or a definition of the spin of the orthotropic axes, as orthotropic intensity develops, according to Eq. (14) with a variable η. What is important is the underlying concept and its demonstrated capability to provide reasonably good simulation of experimental data in a much simpler way than micromechanical polycrystalline models. Yet, the latter can shed much light in future developments of similar macroscopic models.

It remains finally the subject of specification of ω or equivalently Ω^p for a case with multiple spins and multiple internal variables, a subject not considered in [19,20] due to the rigid-plastic assumption. One suggestion could be to consider ω (or Ω^p) a properly weighted average of the different ω_i (or Ω_i^p). A simpler macroscopic approach would be to consider the physical significance of each internal variable in a specific model, and then decide accordingly. For example, in a combined kinematic hardening/orthotropic development model, as in [20], it is clear that the orthotropic development is closely associated with the texture development, which will effect also the elastic properties. Hence, it is proposed to identify in such a case the ω with the ω_i of orthotropy (or the Ω^p with the Ω_i^p), while leaving the kinematic hardening to have its own constitutive spin. Then the general development presented earlier can be immediately applied using the complete set of equations. Nevertheless recall that if terms of order ε are neglected compared to 1, the Ω^p (or ω) does not appear in the stress rate-rate of deformation relations. Further research is in progress along these lines and will be reported elsewhere.

Acknowledgement

The support by NSF Grants No. MSM-86192298 and MSS-8918531 is acknowledged.

References

1. Lee, E.H.: Elastic-plastic deformations at finite strains. ASME J. Appl. Mech. 36 (1969) 1-6.

2. Mandel, J.: Plasticité classique et viscoplasticité. Courses and Lectures No. 97, ICMS Udine: Springer 1971.

3. Dafalias, Y.F.: The plastic spin. ASME J. Appl. Mech. 52 (1985) 865-871.

4. Dafalias, Y.F.: A missing link in the macroscopic constitutive formulation of large plastic deformations. Plasticity Today, Sawczuk, A. and Bianchi, G. (eds.). Essex: Elsevier 1985 135-151.

5. Mandel, J.: Sur la définition de la vitesse de déformation élastique et sa relation avec la vitesse de contrainte. Int. J. Solids Structures 17 (1981) 873-878.

6. Mandel, J.: Sur la définition de la vitesse de déformation élastique en grande transformation elastoplastique. Int. J. Solids Structures 19 (1983) 573-578.

7. Dafalias, Y.F.: The plastic spin in viscoplasticity. Int. J. Solids Structures 26 (1990) 149-163.

8. Dafalias, Y.F.: On the evolution of structure variables in anisotropic yield criteria at large plastic transformations. Critères de Rupture des Matériaux a Structure Interne Orientée, Boehler, J.P. (ed.). Paris: C.N.R.S. No. 351 June 1983, in press.

9. Dafalias, Y.F.: Corotational rates for kinematic hardening at large plastic deformations. ASME J. Appl. Mech. 50 (1983) 561-565.

10. Loret, B.: On the effects of plastic rotation in finite deformation of anisotropic elastoplastic materials. Mech. of Materials 2 (1983) 287-304.

11. Dafalias, Y.F.: The plastic spin concept and a simple illustration of its role in finite plastic transformations. Mech. of Materials 3 (1984) 223-233.

12. Dafalias, Y.F.; Rashid, M.M.: The effect of plastic spin on anisotropic material behavior. Int. J. Plasticity 5 (1989) 227-246.

13. Loret, B.; Dafalias, Y.F.: The effect of anisotropy and plastic spin on fold formations. J. Mechanics Physics Solids, in press.

14. Lee, E.H.; Mallet, R.L.; Wertheimer, T.B.: Stress analysis for anisotropic hardening in finite-deformation plasticity. ASME J. Appl. Mech. 50 (1983) 554-560.

15. Paulun, J.E.; Pecherski, R.B.: On the application of the plastic spin concept for the description of anisotropic hardening in finite deformation plasticity. Int. J. Plasticity 3 (1987) 303-314.

16. Dafalias, Y.F.; Aifantis, E.C.: On the macroscopic origin of the plastic spin. Acta Mechanica 82 (1990) 31-48.

17. Van der Giessen, E.: Continuum models of large deformation plasticity. Part II: a kinematic hardening model and the concept of a plastically induced orientational structure. Eur. J. Mech., A/Solids 8 (1989) 89-108.

18. Dafalias, Y.F.: Constitutive model for large viscoelastic deformations of elastomeric materials. Mech. Res. Comm. 18 (1991) 61-66.

19. Zbib, H.M.; Aifantis, E.C.: On the concept of relative and plastic spins and its implications to large deformation theories. Part II: anisotropic hardening plasticity. Acta Mechanica 75 (1988) 35-56.

20. Dafalias, Y.F.; Cho, H.W.: Verification of the plastic spin concept in viscoplasticity. Advances in Plasticity 1989, Khan, A.S. and Tokuda, M. (eds.). New York: Pergamon 1989 287-290.

21. Montheillet, F.; Cohen, M.; Jonas, J.J.: Axial stresses and texture development during the torsion testing of Al, Cu and α-Fe. Acta metall. 32 (1984) 2077-2089.

Comparison of Two Constitutive Models with One- and Multiaxial Experiments

F. KUBLIK, E. STECK

Institut fuer Allgemeine Mechanik
und Festigkeitslehre
Technische Universitaet Braunschweig, Germany

Summary

The stochastic model by Steck [1,2] will be presented as well as Miller's original model MATMOD [3,4]. Both theories are based on creep processes. The numerical techniques for the solution of the constitutive equations and the parameter identification are discussed. Finally both theories will be applied to uniaxial data of High Purity Aluminum for the determination of the material parameters.
Some experimental results are presented for multiaxial loading cases in the high temperature area. Using a photographical technique the strain field on flat surfaces can be recorded. A plate with a central hole is chosen as specimen.

Introduction

In the area of high temperature applications creep is very important for any prediction of lifetime in engineering structures. To account for this in the last years a strong effort was put on the development of constitutive relations to describe the material behaviour. Several new theories have been developed where the mayority of them lead to a set of differential equations. Within this scope the importance for the comparison of different theories has increased. Main topics for investigation in this field should be:

- Numerical integration effort for the constitutive equations.
- Parameter identification for different models.
- Multiaxial loading cases in experiment and simulation.

Some work in this field has already been done within a NASA project by Chan, Lindholm and Bodner [5]. They did some investigation for the application of the models of Bodner-Partom and Walker in the structural analysis of hot section components of gas turbine engines. Some additional work was done by Hartmann [6] in the analysis of the models of Hart, Miller/Schmidt and Walker. His aim was the comparison based on published parameter sets.

D. Besdo, E. Stein (Eds.)
Finite Inelastic Deformations –
Theory and Applications
IUTAM Symposium Hannover/Germany 1991

Stochastic Mean Value Model

Deformation processes can be described by a movement of dislocations. Starting at this physical level the stochastic model uses flow-units that can be seen as dislocations or dislocation packages. These flow-units can overcome internal barriers formed by solute atoms, grain boundaries, dispersoid particles or entangled dislocations. For this process the flow-units need some activation entered by temperature or applied stress. According to Ilschner [7] the Arrhenius function gives a good representation of the kinetic equation in the high temperature area as it is mainly controlled by diffusion processes. In figure 1 a representation is given of the three components of activation by self-diffusion U_0, the energy of the internal barrier u_i, and the part of the external applied stress $\mu(\sigma)$.

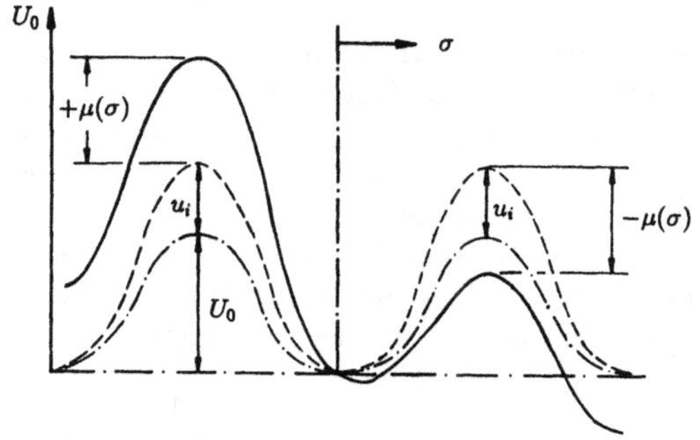

Fig. 1: Energy balance for a flow-unit

The kinetic equation results from the sum of all single deformation steps in positive and negative direction of the applied stress. Assuming that the number of steps in positive direction are equal to the number of steps in negative direction this results in (R represents the gas constant in this paper):

$$\dot{\varepsilon}^{pl} = \lambda^* \cdot c_1 \cdot Z \cdot \exp\left[-\frac{U_0}{RT}\right] \cdot 2 \cdot \sinh\left[\frac{\mu(\sigma)}{RT}\right] \sum_{i=1}^{\infty} n_i(t) \exp\left[-\frac{u_i}{RT}\right] \tag{1}$$

with Z : total number of flow-units

λ^*, c_1 : material parameters

Over time the flow-units situated each in a class of internal barrier energy u_i can move into another class. If they step into a class of higher energy this results in hardening, if they step into a class of lower energy it gives recovery. It also exists the possibility that they stay in the original class. Each of these processes is described by a transition probability, where hardening is activated by stress and temperature as follows:

$$V_{i,i+1} = c_1 \cdot \Delta t \cdot Z \cdot 2 \cdot \sinh\left[\frac{\mu(\sigma)}{RT}\right] \cdot \exp\left[-\frac{U_0 + u_i}{RT}\right] \tag{2}$$

Recovery is only a function of temperature given as:

$$E_{i,i-1} = c_2 \cdot \Delta t \cdot \exp\left[-\frac{\alpha U_0 - \beta u_i}{RT}\right] \tag{3}$$

In figure 2 the possibilities for the development of each flow-unit is illustrated.

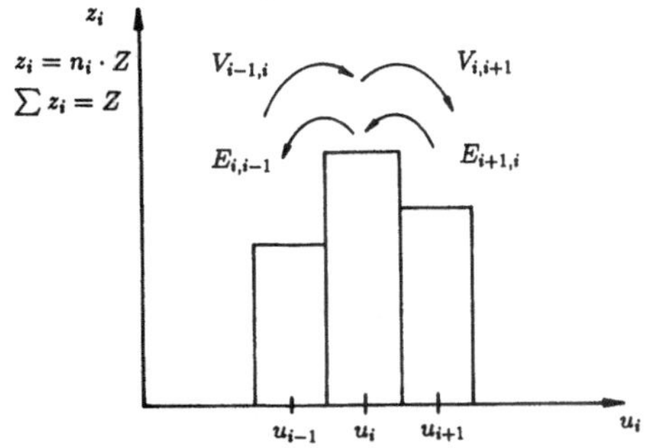

Fig. 2: Possibilities for the flow-unit development

In the original model by Steck [1,2] the flow-units are located in a discrete number of energy classes u_i. The process over time is then described as a Markov-chain. Looking at the development of the internal energy in the classes, it is obvious that the mean value plays an important role. A reduction of all the classes of internal barrier-energy to one, describing the mean value, seems plausible. Such a formulation of the model leads to a set of two differential equations, describing the uniaxial behaviour of the model used in this paper (Schettler-Koehler [9]; Steck/Kublik, [10]). The resulting model is the stochastic mean value model. There is one internal variable F in the model describing isotropic hardening, which is correlated to the mean internal barrier-energy.
Finally the set equations is given as:

$$\dot{\varepsilon}^{pl} = \lambda' \cdot c_1' \cdot \exp\left[-\left(1 - \frac{\alpha - 1}{\kappa}\right)\frac{U_0}{RT}\right] \cdot \left[2 \cdot \sinh\left[\frac{\Delta V \sigma}{RT}\right]\right]^{1+1/\kappa} \cdot \exp\left[-\frac{F}{RT}\right] \tag{4}$$

$$\dot{F} = \frac{1}{\lambda'} \cdot \dot{\varepsilon}^{pl} - c_2' \cdot \exp\left[-\frac{\alpha U_0 - \beta F}{RT}\right] \tag{5}$$

with	$\lambda', c_1', c_2', \alpha, \beta, \kappa, \Delta V$	- material parameters
	F_0	- initial value for the internal variable F
	U_0	- activation energy for self-diffusion
	F	- internal variable - isotropic hardening

Material Model by Miller (MATMOD)

Miller started to develop his model early in 1976. The resulting model was at that time one of the first so called "unified models", which means that there is no distinction between plasticity, creep and relaxation. The basis for the development of Miller's equations were the experimental investigations of Garofalo, who determined in several experiments that steady-state creep for low and high steady-state stresses can be described in a wide temperature range using a hyperbolic sine function. His measurements gave a relation between the steady-state strain rate $\dot{\varepsilon}^{pl}_{ss}$ and the steady-state stress σ_{ss} as follows:

$$\dot{\varepsilon}^{pl}_{ss} = B' \left[\sinh(A \cdot \sigma_{ss}) \right]^n \tag{6}$$

This equation was the starting point for Miller. His aim was the formulation of steady-state behaviour as described by equation (6). Additional he wanted to give a good representation of the primary creep region, assuming that the governing kinetic equation will still be valid. He introduced two internal state variables for the description of isotropic and kinematic hardening. The isotropic internal variable D is independent of the loading direction and mainly dependent on the internal formation of subgrains, which is changing in the primary creep region. He introduced this internal varibale in the denominator to ensure that an increase of isotropic hardening reduces the plastic strain rate. The second internal variable, the kinematic hardening variable G, accounts for hardening which is dependent on the loading direction. It has its origin in the pileup and the bowing of dislocations before obstacles and leads to a backstress. This process reduces the external applied stress, so that Miller finally got the following interaction between the internal and external variables:

$$\dot{\varepsilon}^{pl} = f\left(\frac{\sigma - G}{D}\right) \tag{7}$$

The evolution equations for the internal variables consist of a hardening term governed by the plastic strain rate and a recovery term dependent on the internal variable and temperature. For the description of cyclic processes Miller introduced an additional term in the evolution equation for the isotropic hardening variable to distinguish between isotropic and kinematic hardening, keeping the steady-state conditions unaltered.

Miller developed his equations from experimental findings in creep processes in the high temperature field resulting in a phenomenological theory. The equations are applicable as well in the low temperature area using the experimental observations that the activation energy decreases nearly linear for temperatures below 0.5 T_m (T_m: melting temperature in Kelvin). But for simulations in this area the correlation between model prediction and experimental results was not satisfactory. In the following, Miller's model is only used for high temperatures.

In the kinetic equation the difference of the applied stress and the backstress results in an effective stress σ_{eff} which is responsible for the activation of the dislocations, resulting in the general set of equations given by:

$$\dot{\varepsilon}^{pl} = B' \left[\sinh \left[\left(\frac{|\sigma - G|}{D} \right)^{1.5} \right] \right]^n \text{sgn}(\sigma - G) \tag{8}$$

$$\dot{G} = H_1 \, \dot{\varepsilon}_{pl} - H_1 B' \left[\sinh(A_1 \,|\, G \,|) \right]^n \text{sgn}(G) \tag{9}$$

$$\dot{D} = H_2 \, \dot{\varepsilon}_{pl} \, (C_2 + |\, G \,| - \frac{A_2}{A_1} D^3) - H_2 C_2 B' \left[\sinh(A_2 D^3) \right]^n \tag{10}$$

with $B' = B \cdot \exp[-Q_0/RT]$

$A_1, A_2, B, C_2, H_1, H_2, n$ - material parameters

Q_0 - activation energy

$D_0, G_0 = 0$ - initial values for the internal variables

D, G - internal variables

- isotropic / kinematic hardening

With Miller's model it is possible to simulate inelastic behaviour of metals that show considerable hardening like aluminum and stainless steel. Isotropic and kinematic hardening and cylic phenomena like the Bauschinger effect and the cyclic saturation state can also be represented using this model. In comparison to the stochastic model the background is more phenomenological while the resulting kinetic equation is in both cases the hyperbolic sine function leading to the same steady-state behaviour. The model includes two internal variables (D and G) and eight material parameters (inclusive the activation energy Q_0) plus one initial value for the internal variable D.

Integration of the Constitutive Equations

The constitutive theories explained before result in a set of differential equations. The number of differential equations depends on the phenomena described. Commonly one internal variable is used to account for isotropic hardening and one for kinematic hardening. In one-dimensional simulations the equations have to be solved numerically as analytical solutions are not available. In the case of Miller's model there are two internal variables D and G that have to be considered. For creep experiments Miller [3] showed that the constitutive equations can be numerically "stiff" in some regions. This occurs when both internal variables develop in opposite directions with different gradients. Small errors in the numerical integration can then result in a large error for the material response. To reduce the error in the numerical integration Miller used a multistep Gear method. In the case of the stochastic mean value model there is only one internal variable so that the model doesn't show "stiffness".

For the integration of Miller's constitutive model several techniques were employed and compared to each other. The idea was to find a "simple" one-step-integration method that can be employed as well in a finite element code. For that purpose a fully implicit Euler technique with a timestep control was used and showed good results with respect to CPU-time, number of calls for the constitutive equations and accuracy (Kublik [8]). In comparison to a Gear method the number of function calls and the CPU-time was about five times smaller using the Euler method.

The timestep control (Kublik [8]) needs two different results for the integration step. In a first calculation a fully implicit step is made. In a second step the implicit and the explicit solution for the gradients are combined for the integration like in the Predictor-Corrector method. The difference between these two calculations is used to estimate the relative integration error. If this error is below a given tolerance then in the next timestep the stepwidth is enlarged. If it is above the given value then the whole timestep is repeated with a reduced stepwidth. This method showed good results in the test case where a creep experiment with stress changes was used. In figure 3 a representation is given of the number of function calls over time in the test case. It can be seen that the number of function calls is very high in the situations of strong changes in the internal and external variables, i.e. in the regions of stress changes, and low in the steady state condition.

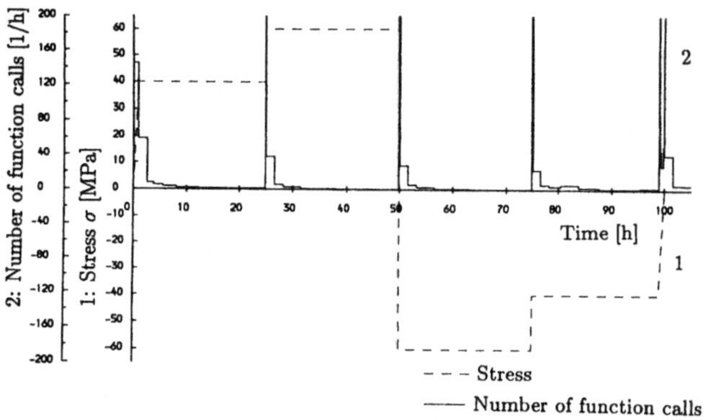

Fig. 3: Function calls over time using a timestep control

Parameter Identification

Contrary to several publications in the literature in this paper numerical techniques were employed for the parameter identification. In the majority of the published constitutive theories parameter findings were just done by "trial and error". To compensate for the problems occuring in the input for these parameter determinations the same numerical techniques were used for the comparison of the mentioned constitutive models.

One of the methods employed for the parameter estimation is the evolution strategy (Kublik [11]). This is a stochastic method which has the advantage that it is easy to handle, general valid, a local minimum can be overcome and any boundary condition can be included without restricting the search for a new parameter set. In comparison to the Monte-Carlo method several parameter sets can be stored in a "memory" and be combined in any way. The disadvantage is the big amount of CPU time as several parameter sets lead to no result.

The quality of each parameter set is determined by comparison with uniaxial experiments using weight functions for the quadratic deviation of the plastic strain and strain rate. In the case of Miller's model additional weight functions for the internal variables were introduced to ensure that these variables will come to a steady state.

Additional some deterministic techniques as the Simplex method and the gradient strategy were looked at and applied to the results out of the evolution strategy. In these cases it was not possible to achieve results with a better quadratic deviation.

Results for High Purity Aluminum Al 99.999

Several parameter studies have been made for High Purity Aluminum as good experimental data by Mecking [12] was available. As both models are based on high temperature creep, the parameters were found for creep tests at several stress levels and at temperatures between 550 K and 700 K. First the response of the models at one temperature was determined, then the response at different temperatures was studied before an overall identification of the material parameters was done. In the figures 4 and 5 some of the results are shown. Using this parameter set a simulation is given for a creep test with a positive change in the stress level. In figure 6 it can be seen that the result of the simulations correspond well to the experimental findings.

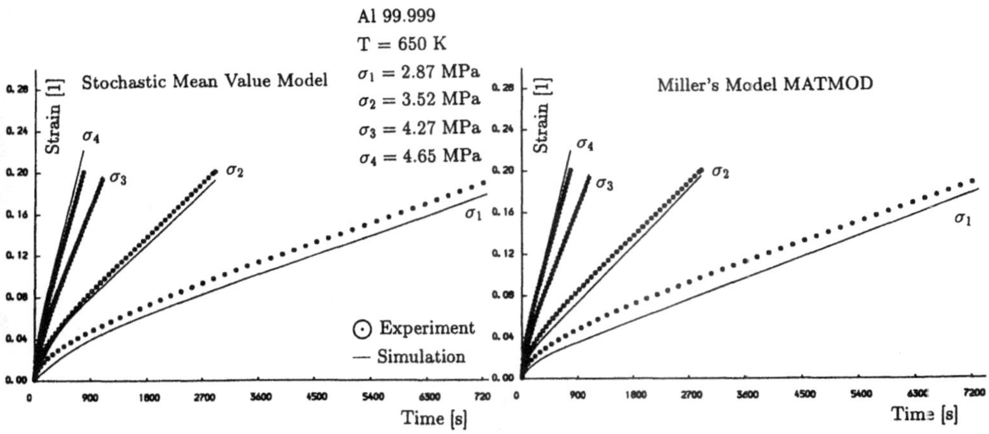

Fig. 4: Representation of experimental creep data of Aluminum by the models

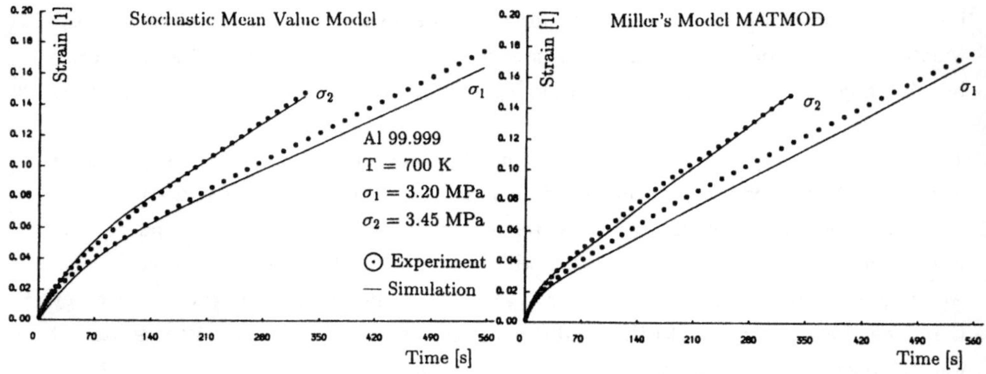

Fig. 5: Representation of experimental creep data of Aluminum by the models

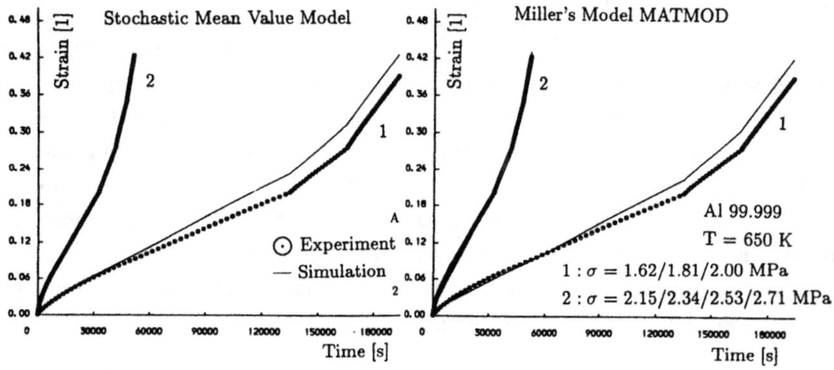

Fig. 6: Changes in the stress levels of creep data simulated by the models

Multiaxial Loading

In the area of constitutive equations the presented models have mainly been connected to uniaxial experimental findings. The area of multiaxial loading cases was only done in simulations using finite elements as numerical technique. The measurement of multiaxial strain fields is difficult as strain gauges have a limited temperature range and inductive techniques are inconvenient for heating equipment.

A photographical technique could be developed within a special research project at the Technical University Braunschweig. The experiments, carried out mainly in the room temperature area (Andresen, Ritter, Steck [13]), were enlarged to the high temperature level. Using a plate with a central hole, a heat resistent grid of 5–10 lines/mm out of TiO_2 was applied to the specimen. During the creep experiment at a constant load several photographs were taken that could be processed by a computer controlled analysis resulting first in the displacements and then in the strain field around the hole. The comparison of the experimental results with the numerical simulations is shown in figure 7 for the cross-section at the hole. The performance of these experiments is still at the beginning but the results are promising

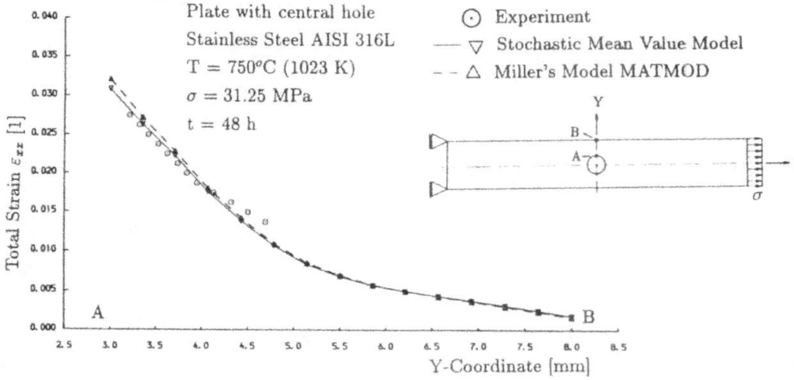

Fig. 7: Comparison of the multiaxial experimental results and the predictions given by the different models

Conclusions

Both models presented are based on high temperature creep processes. The kinetic equation is given as a hyperbolic sine function of the applied stress, temperature and the internal variables. In the stochastic model a mean value formulation is employed which is able to represent material behaviour including isotropic hardening. Miller's material model (MATMOD) has two internal variables accounting for isotropic and kinematic hardening. Contrary to Miller's model the stochastic model describes physical mechanisms in the material represented by flow-units and internal barriers.

Numerical techniques were applied for the solution of the differential equations as well as for the parameter identification. Experimental data of High Purity Aluminum creep curves at different stress and temperature levels are represented well by both models.

For a plate with a central hole the multiaxial loading case was studied from the experimental point of view, applying a photographical technique, and then compared to predictions by both models using the finite element method.

Both models show good prediction capabilities so that the kinetic equation together with the physical explanation in the stochastic model seems well founded. For further development of the models the stochastic model with the physical background is most promising.

Acknowledgement

The presented work was supported by the Deutsche Forschungsgemeinschaft within the "Sonderforschungsbereich 319 (SFB 319): Stoffgesetze fuer das inelastische Verhalten metallischer Werkstoffe - Entwicklung und technische Anwendung".

References

1. Steck, E.A.: A Stochastic Model for the High Temperature Plasticity of Metals. Int. Journal of Plasticity, Vol.1, pp. 243-258, 1985.

2. Steck, E.A.: A Stochastic Model for the Interaction of Plasticity and Creep in Metals. Nuclear Engineering and Design, North-Holland, Amsterdam, Vol. 114, pp. 285-294, 1989.

3. Miller, A.: An Inelastic Constitutive Model for Monotonic, Cyclic and Creep Deformation: Part I - Equations Development and Analytical Procedures. Transactions of the ASME, Journal of Engineering Materials and Technology, Vol.98, pp.97-105, 1976.

4. Miller, A.: An Inelastic Constitutive Model for Monotonic, Cyclic and Creep Deformation: Part II - Application to Type 304 Stainless Steel. Transactions of the ASME, Journal of Engineering Materials and Technology, Vol.98, pp.106-113, 1976.

5. Chan, K.S.; Lindholm, U.S.; Bodner, S.R.: Constitutive Modelling For Isotropic Materials (HOST). Final Report, NASA CR-182132, 1988.

6. Hartmann, G.: Vergleich des einachsigen Verhaltens dreier inelastischer Werkstoffmodelle mit internen Zustandsvariablen durch numerische Experimente. Dr.-Ing. Thesis, Technical University Darmstadt 1988.

7. Ilschner, B.: Hochtemperaturplastizitaet, Springer Berlin 1973.

8. Kublik, F.: Vergleich zweier Werkstoffmodelle bei ein- und mehrachsigen Versuchsfuehrungen im Hochtemperaturbereich. Dr.-Ing. Thesis, Technical University Braunschweig (will be published in 1991).

9. Schettler-Koehler, R.-W.: Entwicklung eines makroskopischen Kriechgesetzes fuer Metalle aus einem stochastischen Kriechmodell. Dr.-Ing. Thesis, Technical University Braunschweig 1985.

10. Steck, E.A.; Kublik, F.: Berechnungsverfahren fuer metallische Bauteile bei Beanspruchungen im Hochtemperaturbereich. Report B1, SFB 319, Technical University Braunschweig 1990.

11. Kublik, F.: Numerische Simulation von einachsigen Kriechversuchen. Proceedings of: 'Numerische Methoden der Plastomechanik', Institute of Mechanics Hannover, pp.142-156, 1989.

12. Mecking, H.: Bestimmung der Werkstoffparameter fuer das Kriechen von Aluminium-Legierungen. Final Report for the DFG research project Me 428/7, Technical University Hamburg-Harburg, January 1989.

13. Andresen, K.; Ritter, R.; Steck, E.: Theoretical and Experimental Investigations of Crack Extension by FEM- and Grating Method. Proceedings on the European Symposium on Elastic-Plastic Fracture Mechanics: Element of Defect Assessment, 1989.

A Continuum Damage Model for the Description of High Strain Rate Deformations

O.T. BRUHNS

Lehrstuhl für Mechanik I
Ruhr-Universität Bochum, D-4630 Bochum

Summary

The behaviour of metals under dynamic loading is described by introducing a continuum damage theory, which is also valid for nonisothermal processes. Microstructural changes are modelled by internal variables, e.g. damage due to shear bands is described with a second order tensor. Dislocation induced viscoplastic deformations are modelled using a flow rule of the overstress-type.

1. Introduction

In many methods for the analysis of impact loaded structures S, a near field S_n in the vicinity of the point of impact and a far field S_f are distinguished, which are separated by the boundary ∂S_{nf}. In general, S_n is deformed inelastically, the state of stress is three dimensional and propagation phenomena are of importance. On the contrary, elastic behaviour prevails in S_f; concerning the state of stress the approximations of beam, plate or shell theory are valid and the characteristic times of the structural behaviour are large compared with L/c, where L is a typical length of the structure and c is the velocity of propagation of acceleration waves. Hence the solution in S_f can be expanded in eigenfunctions corresponding to the lower eigenfrequencies, provided plausible assumptions are made about the stresses on ∂S_{nf}. Structural failure under dynamic loading now frequently depends on the behaviour in S_n. According to what has been said about the state of stress in S_n, a complete continuum mechanics description is required.

In a vertical impact e.g. of a blunt steel projectile against a steel target, the plate fails by "plugging" due to shear band formation and by "spalling" due to cracks, which are nucleated by tensile waves reflected on the rear side. Further investigations have shown that two types of mechanisms contribute to the inelastic behaviour , i.e. dislocation processes and (with a larger characteristic length) damage

D. Besdo, E. Stein (Eds.)
Finite Inelastic Deformations -
Theory and Applications
IUTAM Symposium Hannover/Germany 1991

processes. The latter comprise nucleation, growth and coalescence of defects such as microvoids, microcracks or micro shear bands. In vertical impact problems, shear band formation has been found the most important of these damage mechanisms.

A defect here will be called a microdefect if its extension is small compared with the linear dimension of a representative volume element ΔV of S. During a typical loading process, the nucleation and the growth of these microdefects may finally result in the nucleation of a macrodefect at some instant time $t = t^*$. The time interval $[0, t^*]$ will be called phase 1. Only this phase can be described within the frame of continuum mechanics. For phase 2., i.e. for $t > t^*$, we have to deal with the growth of macrodefects (either macrocracks or macro shear bands) in a damaged solid and thus elements of fracture mechanics have to be included in order to complete the constitutive relations. However, even for phase 2 an accurate description of phase 1 is necessary since otherwise the failure criteria (fracture, localization etc.) would be based on physically unrealistic values of state variables Cauchy stress $\boldsymbol{\sigma}$, temperature Θ etc.

In this paper we therefore will concentrate on the description of phase 1 processes. To this end a continuum damage theory for moderate and high strain rate deformations is developed, which takes into consideration damage due to shear bands [1, 2]. This theory will be valid for strain rates between 10^{-4} sec^{-1} and 10^5 sec^{-1} and for arbitrarily large deformations and rotations. The model has been implemented into 1-D and 3-D axisymmetric [3] finite element codes. Some numerical results for vertical impact problems are given in [4].

2. General Considerations

Consider a representative volume element ΔV as a polycrystal containing a finite number of defects like stress concentrators etc. The motion of this element may be defined by averaging values over ΔV, e.g.

$$\mathbf{D} := \text{averaging strain rates} \tag{1}$$

the rate of deformation tensor is the average effect of the different mechanisms that contribute to the motion in the disproportionately stressed grains of ΔV.

The continuum damage theory is now characterized by the additive decomposition of this rate of deformation tensor $\mathbf{D}$ according to

$$\mathbf{D} = \mathbf{D}_r + \mathbf{D}_i \qquad \text{with} \qquad \mathbf{D}_i = \mathbf{D}_{da} + \mathbf{D}_{dt} + \mathbf{D}_{sb} \tag{2}$$

where $\mathbf{D}_r$ and $\mathbf{D}_i$ are reversible and irreversible (inelastic) parts, respectively. $\mathbf{D}_i$ contains contributions $\mathbf{D}_{da}$ of the athermal dislocation motion, $\mathbf{D}_{dt}$ of thermally activated dislocation motions and $\mathbf{D}_{sb}$ of the micro shear band formation. Additional variables are introduced in order to describe microstructural changes and to formulate necessary yield and loading conditions. The identification of these variables is based on a homogenization method which is explained in [1]. We here simply refer to the results of this method. Let

$$\mathcal{Q} = \{\mathbf{X}, \varkappa, \boldsymbol{\tau}, \vartheta, \mathbf{Z}, \boldsymbol{\alpha}_q\} \;, \qquad \bar{\mathcal{Q}} = \{u, \mathbf{A}_{dl}\} \tag{3}$$

be the sets of internal and process variables, respectively, where $\mathbf{X}$ is a kinematic hardening tensor and $\varkappa$ is an isotropic hardening parameter related to the total dislocation density N. The microstress concentration tensor $\boldsymbol{\tau}$ represents an average of local stress concentrations in the vicinity of microdefects and ϑ is an average of local temperature concentrations. Further $\mathbf{Z}$ is a symmetric positive definite damage tensor and $u = N_{mob}/N$ is the quotient of mobile and total dislocation densities. $\mathbf{A}_{dl}$ in (3) is an average of local dislocation induced deformations in the vicinity of microdefects and $\boldsymbol{\alpha}_q$ is a vector-valued thermal variable which is related to the heat flux vector $\mathbf{q}$.

For the description of nonisothermal processes the specific free enthalpy g is introduced as a function

$$g = g(\boldsymbol{\sigma}, \Theta; \mathcal{Q}) \;, \tag{4}$$

while the process variables $\bar{\mathcal{Q}}$ do not alter the energy of the solid.

3. The Constitutive Law

The constitutive law comprehends

i) the state function for the specific free enthalpy, from which the reversible strains and the specific entropy are deduced,

ii) evolution laws for the internal as well as the process variables $\mathcal{Q}$ and $\bar{\mathcal{Q}}$, respectively, and finally

iii) flow rules for $\mathbf{D}_{da}$, $\mathbf{D}_{dt}$, $\mathbf{D}_{sb}$ and the inelastic spin $\mathbf{W}_i$.

According to the nonlinear inelastic behaviour of the material some of the flow rules and evolution laws are subjected to yield and/or loading conditions.

The decomposition (2) is motivated by the introduction of a so-called intermediate configuration $\hat{\chi}_t$ (cf. e.g. Lehmann [5]) with the following properties

$$\left.\begin{aligned} &\mathbf{F} = \mathbf{F}_r \mathbf{F}_i \\ &\mathbf{F}_i : \chi_o \longrightarrow \hat{\chi}_t, \quad \mathbf{F}_r : \hat{\chi}_t \longrightarrow \chi_t \end{aligned}\right\} \tag{5}$$

of the reversible and the irreversible parts, respectively, of the deformation gradient $\mathbf{F}$, where χ_o denotes the initial configuration and χ_t the actual configuration of the considered body in motion.

Contrary to Lehmann [6], however, we here adopt the concept of isoclinic intermediate configurations, which has the advantage that a linear relation between $\mathbf{D}_r$ and an objective rate of the reversible Green's strain tensor $\mathbf{E}_r$ can be formulated irrespective of any isotropy conditions, and where the objective time derivative (Jaumann -type rate), e.g.

$$\overset{o}{\mathbf{A}} := \dot{\mathbf{A}} - (\mathbf{W} - \mathbf{W}_i)\mathbf{A} + \mathbf{A}(\mathbf{W} - \mathbf{W}_i) \tag{6}$$

which has been introduced in the intermediate configuration has the chain rule property [4]. The consequence of this assumption, however, is that an additional constitutive law is needed for the inelastic spin $\mathbf{W}_i$ (cf. Dafalias [6, 7]).

The evolution laws and flow rules are first formulated for variables with respect to this intermediate configuration and thereafter transformed to the actual configuration. This conversion includes multiplications with $\mathbf{F}_r$ and is simplified in the present case since in metals only small reversible changes of the shape occur

$$\mathbf{F}_r \approx \Delta \mathbf{1}\,, \quad \Delta = \det(\mathbf{F}_r) \;:\; \hat{\chi}_t \longrightarrow \chi_t\,. \tag{7}$$

The formulation of constitutive laws with respect to the actual configuration is particularly helpful for an updated Lagrangian formulation of the finite element code.

i) Reversible part of deformations

Contrary to what is usually adopted in the formulation of elastic-plastic constitutive relations we here no longer maintain the assumption that the inelastic deformations will not influence the reversible material behaviour. This is especially due to the fact that the damage will influence the material properties in the elastic domain. The reversible strains thus follow from the enthalpy function $g(\cdot)$

$$\mathbf{E}_r = \frac{f(\zeta)}{2\mu\Delta}\left\{\sigma - \frac{\nu}{1+\nu}\,\mathrm{tr}(\sigma)\,\mathbf{1}\right\} + \alpha_t(\Theta - \Theta_o)\,\mathbf{1} \tag{8}$$

and the rate of deformation tensor becomes

$$\mathbf{D}_r = \frac{f(\zeta)}{2\mu\Delta^3}\left\{\overset{\circ}{\sigma} - \frac{\nu}{1+\nu}\,\mathrm{tr}(\overset{\circ}{\sigma})\,\mathbf{K}(\cdot)\right\} + \frac{1}{\Delta^2}\,\alpha_t\dot{\Theta}\,\mathbf{B}(\cdot) \tag{9}$$

for an instantaneously isotropic material. Herein μ (shear modulus), ν (Poisson's ratio) and α_t (thermal expansion coefficient) are the material parameters of the undamaged elastic material. The influence of the damage tensor $\mathbf{Z}$ on the elastic response is introduced by functions

$$\Delta = \Delta(\sigma, \Theta, \mathbf{Z}), \qquad f = f(\zeta) \tag{10}$$

wherein $\zeta := \|\mathbf{Z}\|$ and material tensors $\mathbf{K}(\sigma, \Theta, \mathbf{Z})$ and $\mathbf{B}(\sigma, \Theta, \mathbf{Z})$ of eqn. (9). Neglecting in these material tensors now some higher order terms [2] we finally can arrive at $\mathbf{K} = \mathbf{1}$ and $\mathbf{B} = \mathbf{1}$. Thus $\mathbf{K}(\cdot)$ and $\mathbf{B}(\cdot)$ describe higher order anisotropy effects due to damage of the material, which will be disregarded in the following.

ii) Flow rules - part 1: Dislocation induced inelastic deformations

Neglecting for moderate up to high strain rates the contribution of the athermal dislocation motion the dislocation induced part of the inelastic rate of deformation tensor becomes

$$\left.\begin{aligned} &\mathbf{D}_d = \mathbf{D}_{dt} = \gamma_t(u,\ldots)\left[\Phi_t\right]_t \mathbf{N}_t \qquad \text{if} \\ &F_t = \frac{1}{\Delta^2}(\sigma' - \mathbf{X}')\cdot(\sigma' - \mathbf{X}') - g_t(x, \Theta, u) \geq 0\,, \end{aligned}\right\} \tag{11}$$

where $\mathbf{N}_t$ is the normal to the surface $F_t = \text{const.}$, $F_t \geq 0$ is the yield condition and the square brackets here and in the following mean

$$[x]_y = \begin{cases} x & \text{if } F_y \geq 0, \\ 0 & \text{otherwise} \end{cases} \tag{12}$$

$F_y \geq 0$ is a yield condition for changes of a variable $y \in (\mathcal{Q} \cup \bar{\mathcal{Q}})$. The prime $(\ )'$ in

(11) as usual denotes the deviatoric part of a tensor. Here and in what follows functions γ_y, Φ_y and g_y have been introduced as in usual viscoplasticity except an additional dependency of some of these functions on the process variable u, which has been taken to allow for necessary transitions of $\mathbf{D}_i$ according to the rate dependency of the different contributions. Moreover different functions have been introduced for the description of the different mechanisms that contribute to the motion.

iii) Flow rules - part 2: Shear band induced processes

For the shear band induced motion we find in agreement with the second law of thermodynamics

$$\mathbf{D}_{sb} = \left[\Phi_s\right]_s \langle\langle\langle LC_s \rangle\rangle\rangle \frac{\sigma'}{\|\sigma'\|} \quad \text{if} \tag{13}$$

$$\left.\begin{aligned} F_s &= \frac{1}{\Delta^2}\,\tau'\cdot\tau' - \frac{\nu_s}{\Delta}\,\mathrm{tr}(\tau) - g_s(\zeta) \geq 0 \quad \text{and} \\ LC_s &= \frac{1}{\Delta}\left(\frac{2}{\Delta}\,\tau' - \nu_s \mathbf{1}\right)\cdot\left(\overset{\#}{\tau} + m\tau\right) > 0. \end{aligned}\right\} \tag{14}$$

where $\langle\langle\langle\,\cdot\,\rangle\rangle\rangle$ denotes a Maccauly bracket with respect to the loading condition $LC_s > 0$. We note that herein and in what follows the necessary yield and loading conditions for damage processes are formulated in (τ, ϑ)-space rather than in (σ, Θ)-space as it is the case for the dislocation processes of eqn. (11), since the damage here is connected with local stress and temperature concentrations. Moreover, damage processes depend on the local hydrostatic pressure through $\mathrm{tr}(\tau)$. Furthermore

$$\overset{\#}{\tau} := \overset{\circ}{\tau} - \tau\,\mathrm{tr}(\mathbf{D}) \tag{15}$$

is a second objective time derivative which has to be introduced in addition to (6) in connection with the transformation to the actual configuration.

iv) Flow rules - part 3: Inelastic spin

Following some ideas of Mandel [8], Dafalias [6, 7] has introduced the concept of plastic spin where with

$$\mathbf{W} := \mathbf{\Omega} + \mathbf{W}_i \tag{16}$$

$\mathbf{W}$ is the total spin, $\mathbf{W}_i$ is the inelastic spin defined as antisymmetric part of the inelastic velocity gradient and $\mathbf{\Omega}$ is explained as rigid body spin of the substructure, which is different from the material spin $\mathbf{W}$. We note that herein the reversible part of the total spin has been neglected for the assumed small reversible distortions. According to this concept the inelastic spin or $\mathbf{\Omega}$, respectively, have to be given by a proper constitutive law. We note further that to our mind there is no necessity for explaining $\mathbf{\Omega}$ as mentioned above, since $\mathbf{\Omega}$ is defined through (16) once a constitutive relation for the inelastic spin has been formulated.

Therefore similar as in (2) $\mathbf{W}_i$ is decomposed into

$$\mathbf{W}_i = \mathbf{W}_{dt} + \mathbf{W}_{sb} \tag{17}$$

the different parts that contribute to the motion, where here the simplification of (11) has already been taken into consideration. Since the contribution of the damage to the motion can be assumed small compared with the thermally activated dislocation motion $\mathbf{W}_{sb}$ will be disregarded and thus

$$\mathbf{W}_i = \gamma_w(u)\left[\Phi_t\right]_t \frac{(\mathbf{X}\boldsymbol{\sigma}' - \boldsymbol{\sigma}'\mathbf{X})}{\|\boldsymbol{\sigma}'\|\|\mathbf{X}\|} \quad ; \qquad \text{if } F_t \geq 0 \tag{18}$$

$\mathbf{W}_i$ is introduced in a form of the same type as (11).

v) Evolution laws - part 1: Dislocation variables

The evolution of the hardening of the material is here described in an analogous way as in usual viscoplasticity (cf. Perzyna and Wojno [9], Chaboche [10])

$$\overset{\#}{\mathbf{X}} = \Delta c_t(\varkappa)\, \mathbf{D}_d \ , \qquad \dot{\varkappa} = k(\varkappa)\, \boldsymbol{\sigma} \cdot \mathbf{D}_d \tag{19}$$

where $c_t(\cdot)$ and $k(\cdot)$ are functions of the average dislocation density of ΔV.

vi) Evolution laws - part 2: Damage variables

It is known from the theory of (elastic) composites that the local stress concentrations $\boldsymbol{\tau}$ can be related to the overall stresses $\boldsymbol{\sigma}$ e.g. by

$$\overset{\#}{\boldsymbol{\tau}} = \mathbb{D}\, \overset{\#}{\boldsymbol{\sigma}} \tag{20}$$

where $\mathbb{D}$ is a positive definite material tensor [11]. This relation here is somewhat generalized to the form

$$\overset{\#}{\boldsymbol{\tau}} = \mathbb{D}\, \overset{\#}{\boldsymbol{\sigma}} + d_4 \dot{\varkappa}\, \boldsymbol{\tau} - m \boldsymbol{\tau} \tag{21}$$

where

$$\mathbf{D}\overset{\#}{\sigma} = d_1\overset{\#}{\sigma} + \frac{d_2}{\Delta}(\tau\overset{\#}{\sigma} + \overset{\#}{\sigma}\tau) + \frac{d_3}{\Delta}(\mathbf{Z}\overset{\#}{\sigma} + \overset{\#}{\sigma}\mathbf{Z}) \tag{22}$$

and $m(\cdot)$ and $d_1(\cdot)$ to $d_4(\cdot)$ are yet undetermined functions of $\|\tau\|$ and the process variables u and $\|\mathbf{A}_{dl}\|$, respectively.

For the evolution of the local temperature concentrations we adopt a relation

$$\dot{\vartheta} = \left[\Phi_\Theta\right]_v \quad \text{if } F_v = \frac{1}{\Delta^2}\tau'\cdot\tau' - g_v \geq 0 \tag{23}$$

i.e we here neglect possible contributions of damage induced deformations to the evolution of the temperature concentration. The evolution of the damage tensor $\mathbf{Z}$ is characterized by nucleation and growth of the damage. We thus state

$$\overset{\#}{\mathbf{Z}} = \left[\Phi_z\right]_z \langle\langle LC_z \rangle\rangle (\mathbf{N}_n + \mathbf{N}_g) \quad \text{if} \tag{24}$$

$$\left.\begin{aligned} F_z &= \frac{1}{\Delta^2}\tau'\cdot\tau' - \frac{\nu_z}{\Delta}\,\mathrm{tr}\,\tau - g_z(\vartheta) \geq 0 \quad \text{and} \\ LC_z &= \frac{1}{\Delta}\left(\frac{2}{\Delta}\tau' - \nu_z\mathbf{1}\right)\cdot\left(\overset{\#}{\tau} + m\tau\right) > 0. \end{aligned}\right\} \tag{25}$$

where now $\langle\langle\cdot\rangle\rangle$ is a second Maccauly bracket with respect to the loading condition $LC_z > 0$. Herein $\mathbf{N}_n$ and $\mathbf{N}_g$ describe the changes of $\mathbf{Z}$ due to shear band nucleation and shear band growth, respectively. $\mathbf{Z}$ itself contains informations about average shear band size and average shear band orientation and can be used in formulating failure criteria.

vii) Evolution laws - part 3: Thermal variables
We introduce

$$\overset{\circ}{\boldsymbol{\alpha}}_q = \Delta\,\mathrm{grad}\,\Theta - \frac{1}{\tau_q}\boldsymbol{\alpha}_q\,, \tag{26}$$

where the objective rate $(\overset{\circ}{\cdot})$ is according to (6) defined as

$$\overset{\circ}{\boldsymbol{\alpha}}_q = \dot{\boldsymbol{\alpha}}_q - (\mathbf{W} - \mathbf{W}_i)\,\boldsymbol{\alpha}_q \tag{27}$$

for a vector-valued variable. With an appropriate choice of the specific free enthalpy function g, we can derive $\mathbf{q} = -(\boldsymbol{k}/\tau_q)\boldsymbol{\alpha}_q$ from the entropy inequality (cf. Kosinski [12]) and thus find Maxwell-Cattaneo's law of heat conduction

$$\tau_q\overset{\circ}{\mathbf{q}} + \mathbf{q} = -\Delta\,\boldsymbol{k}\,\mathrm{grad}\,\Theta, \tag{28}$$

where $\boldsymbol{k}$ is the heat conductivity.

viii) Evolution laws - part 4: Process variables $\bar{\mathcal{Q}}$

The evolution law for the process variables u has to be adopted in the form of a second order differential equation

$$\ddot{u} + c\dot{u} + \frac{c^2}{4}(u - u_{eq}) = 0 \; ; \qquad \text{with } c = c_0(\alpha + v_p) \tag{29}$$

to fulfill the above mentioned transition conditions. Here $v_p := \max\{0, \|\mathbf{D}\| + \chi\dot{\Theta}\}$ is a sufficient measure to describe the process velocity and $u_{eq}(\cdot)$ is the equilibrium value of u which for different values of v_p can be determined from experiments.

Finally the influence of the local dislocation motion on the deformations is described by a most simple rate dependent relation of the overstress-type

$$\overset{\ast}{\mathbf{A}}_{dl} = \Delta^2 \gamma_{dl} \left[\Phi_{dl}\right]_v \frac{\tau'}{\|\tau'\|} + tr(\mathbf{D})\mathbf{A}_{dl} \; ; \qquad \text{if } F_v \geq 0 \, . \tag{30}$$

4. Identification of Material Parameters and Conclusions

A strategy for the determination of the material parameters is described by Fornefeld [3]. Since experiments with high strain rate deformations are characterized by non-homogeneous states of stress and strain, the interpretation of experiments is difficult and fitting the parameters to experimental curves requires the numerical solution of an initial-boundary value problem. Therefore the identification process represents an iterative procedure where an initial set $\mathcal{C}$ of parameters is chosen such that the simulation of fictitious homogeneous deformations gives plausible results. $\mathcal{C}$ is modified until the numerical simulation coincides with experimental results.

A useful mathematical tool for this iteration is the evolutionary strategy, a statistical optimization method for engineering systems, which imitates the principles of biological evolution such as mutation and selection and which was introduced by Rechenberg [13].

The investigation of the material data for an armour steel is currently under progress. So far we have defined a so-called reference material, whose behaviour is representative for a wide class of metals and alloys. This reference material is defined by prescribing σ-ε-curves at different constant strain rates, creep curves, relaxation curves etc.

The main features of the model have been examined and were found in agreement with experimental results [4]. Further predictions of the model in the case of cyclic tests, strain rate jump tests etc. can be found in [2] together with predictions for combined tension-torsion tests.

References

1. Bruhns, O.T. and Diehl, H.; An Internal Variable Theory of Inelastic Behaviour at High Rates of Strain", Arch. Mech. **41** (1989), 427-460

2. Diehl, H.; Ein Materialmodell zur Berechnung von Hochgeschwindigkeitsdeformationen metallischer Werkstoffe unter besonderer Berücksichtigung der Schädigung durch Scherbänder, Report No. 66, Institute of Mechanics, RU Bochum 1989

3. Fornefeld, W.; Zur Parameteridentifikation und Berechnung von Hochgeschwindigkeitsdeformationen metallischer Werkstoffe anhand eines Kontinuums-Damage -Modells, Report No. 73, Institute of Mechanics, RU Bochum 1990

4. Bruhns, O.T.; Shear Band Formation in Impact Problems, in Wriggers, P. and Wagner, W. (eds.), Nonlinear Computational Mechanics, Springer 1991

5. Lehmann, T., The Constitutive Law in Thermoplasticity, CISM Courses and Lectures No. 281, Springer-Verlag 1984

6. Dafalias, Y.N.; The Plastic Spin Concept and a Simple Illustration of its Role in Finite Plastic Transformations, Mech. Materials **3** (1984), 223-233

7. Dafalias, Y.N.; The Plastic Spin, J. Appl. Mech. **52** (1985), 865-871

8. Mandel, J.; Plasticitè classique et viscoplasticitè, CISM Courses and Lectures No. 97, Springer-Verlag 1971

9. Perzyna, P. and Wojno, W.; Unified Constitutive Equations for Elastic-Viscoplastic Material, Bull. Acad. Polon. Sci., Sèr. Sci. Techn. **24** (1976), 85-94

10. Chaboche, J.L.; Constitutive Equations for Cyclic Plasticity and Cyclic Viscoplasticity, Int. J. Plasticity **5** (1989), 247-302

11. Dvorak, G.J.; Mechanical Properties of Composites, in Boehler, J.P. (ed.), Mechanical Behaviour of Anisotropic Solids, M. Nijhoff Publ. 1982

12. Kosinski, W.; Thermal Waves in Inelastic Bodies, Arch. Mech. **27** (1975), 733-748

13. Rechenberg, I.; Evolutionsstrategie, Stuttgart: F. Frommann Verlag 1973

Asymptotic Analysis and Numerical Simulation of Deformation Processing of Porous Metals

R.M. Govindarajan and N. Aravas
Department of Mechanical Engineering and Applied Mechanics
University of Pennsylvania, Philadelphia PA 19104 USA

Abstract

Plane strain extrusion of porous metals is analyzed using asymptotic techniques. The asymptotic expansions are based on a small parameter ϵ defined to be the ratio of the height reduction to the length of the die in the reduction region. Gurson's plasticity model is used to describe the constitutive behavior of the material. It is shown that the effects of porosity enter as an $O(\epsilon)$ correction. The effects of strain hardening of the matrix and friction at the die-metal interface are accounted for. The asymptotic solutions developed are verified using detailed finite element calculations.

1 Introduction

Extrusion is a deformation process in which a billet of material is converted into a continuous product of uniform cross-section by forcing it through a suitably shaped die. The most commonly used methods of analysis of such forming processes are slip line solutions (Hill [1]), upper bound techniques (Avitzur [2]), and finite element methods (Lee *et al.* [3]; Kobayashi *et al.* [4]).

The use of asymptotic methods has been suggested recently as an alternative technique for the solution of metal forming problems such as extrusion, drawing, and rolling. Johnson [5] considered a rigid plastic material that obeys the von Mises yield criterion with associated flow rule and used asymptotic methods to analyze axisymmetric extrusion. Aravas and McMeeking [6] used a combination of asymptotic and finite element techniques to analyze three-dimensional extrusion. Durban and Mear [7] using a "radial" velocity field presented an asymptotic solution for axisymmetric extrusion of porous metals. Govindarajan and Aravas [8] carried out an asymptotic analysis for plane strain extrusion of perfectly plastic fully dense and porous metals.

In this paper an asymptotic analysis of plane strain extrusion of power law hardening porous metals is presented. The asymptotic expansions are based on a small parameter ϵ defined to be the ratio of the reduction to the length of the die, which implies that the slope of the die is small. Gurson's [9] yield criterion with associated flow rule is employed to describe the plastic behavior of the porous metal and Coulomb's law is used to model the frictional forces along the metal-die interface. It is shown that the effects of porosity enter as an $O(\epsilon)$ correction and results of the asymptotic analysis are compared with those of detailed finite element calculations.

D. Besdo, E. Stein (Eds.)
Finite Inelastic Deformations -
Theory and Applications
IUTAM Symposium Hannover/Germany 1991

2 Extrusion of porous metals

In this section we briefly discuss Gurson's pressure dependent plasticity model which is used to describe the constitutive behavior of the porous medium. Based on a rigid-plastic upper bound solution for spherically symmetric deformations of a single spherical void, Gurson [9] proposed yield condition of the following form

$$\Phi = \hat{\tau}_e^2 + \frac{\hat{\sigma}_m^2}{3}\left[2\hat{f}\cosh\left(\frac{3}{2}\frac{\hat{p}}{\hat{\sigma}_m}\right) - \left(1+\hat{f}^2\right)\right] = 0, \tag{1}$$

where $\hat{\sigma}_{ij}$ is the Cauchy (true) stress tensor, $\hat{\tau}_e = \sqrt{\frac{1}{2}\hat{S}_{ij}\hat{S}_{ij}}$, $\hat{S}_{ij}$ is the deviatoric part of $\hat{\sigma}_{ij}$, $\hat{p} = \frac{1}{3}\hat{\sigma}_{kk}$ is the hydrostatic stress, $\hat{\sigma}_m$ is the microscopic equivalent stress in the matrix material, and $\hat{f}$ is the void volume fraction. The yield stress of the matrix $\hat{\sigma}_m$ is assumed to vary as

$$\hat{\sigma}_m = \hat{\sigma}_0\left(1 + \frac{\hat{\epsilon}_m^p}{\hat{\epsilon}_0}\right)^{1/n}, \tag{2}$$

where $\hat{\sigma}_0$ is the initial yield stress of the material, $\hat{\epsilon}_0$ is the yield strain, $\hat{\epsilon}_m^p$ is the microscopic equivalent plastic strain, and n is the hardening exponent. As $n \to \infty$, we have a perfectly plastic matrix material. For a fully dense medium ($\hat{f} = 0$), the above yield condition reduces to that of von Mises and is independent of pressure.

The aforementioned yield condition is also used as a plastic potential and the plastic part of the deformation-rate tensor is written as

$$\hat{D}_{ij}^p = \hat{\lambda}\frac{\partial\Phi}{\partial\hat{\sigma}_{ij}}, \tag{3}$$

where $\hat{\lambda}$ is a non-negative flow parameter. In Gurson's model the microscopic equivalent plastic strain $\hat{\epsilon}_m^p$ in the matrix material is assumed to vary according to the equivalent plastic work expression

$$(1-\hat{f})\hat{\sigma}_m\dot{\hat{\epsilon}}_m^p = \hat{\sigma}_{ij}\hat{D}_{ij}^p, \tag{4}$$

where the superposed dot indicates the material time derivative. The rate of change of porosity is related to the deformation rate as

$$\dot{\hat{f}} = (1-\hat{f})\hat{D}_{kk}^p. \tag{5}$$

2.1 Formulation of the problem

We consider plane strain extrusion of porous materials through a gradually tapering die. The shape of the die in the reduction region is given by the function $\hat{y} = \hat{h}(\hat{x})$, where $\hat{x}$ is the direction of extrusion. The assumption that the dies taper at a maximum angle which is of order ϵ (a small positive number), forms the heart of this formulation, *i.e.*,

$$\frac{d\hat{h}}{d\hat{x}} = O(\epsilon). \tag{6}$$

In this paper we present the formulation only for the plastic part of the deformation of the material. A detailed discussion on the effects of elasticity on the elasto-plastic solution can be found in Govindarajan and Aravas [8], where it is also shown that such effects enter the solution only to order ϵ^2 or higher.

The formulation of the problem is given in the following. The only non-zero stress and velocity components are $\hat{\sigma}_{xx}$, $\hat{\sigma}_{yy}$, $\hat{\sigma}_{zz}$, $\hat{\sigma}_{xy}$, $\hat{v}_x$ and $\hat{v}_y$, where $\hat{z}$ is the coordinate axis normal to the plane of deformation.

The equations of body-force-free quasi-static equilibrium are

$$\frac{\partial \hat{\sigma}_{xx}}{\partial \hat{x}} + \frac{\partial \hat{\sigma}_{xy}}{\partial \hat{y}} = 0, \qquad \text{and} \qquad \frac{\partial \hat{\sigma}_{xy}}{\partial \hat{x}} + \frac{\partial \hat{\sigma}_{yy}}{\partial \hat{y}} = 0. \tag{7}$$

The condition of plane strain requires that

$$\hat{S}_{xx} + \hat{S}_{yy} = -\hat{S}_{zz} = \frac{1}{3}\hat{\sigma}_m \hat{f} \sinh\left(\frac{\hat{\sigma}_{kk}}{2\hat{\sigma}_m}\right). \tag{8}$$

It is assumed that the porous material yields *everywhere* inside the reduction region of the die and the flow rule in that region, derived from Gurson's yield condition, is written as

$$\frac{\partial \hat{v}_x}{\partial \hat{x}} = \hat{\lambda}(2\hat{S}_{xx} + \hat{S}_{yy}), \tag{9}$$

$$\frac{\partial \hat{v}_x}{\partial \hat{x}} + \frac{\partial \hat{v}_y}{\partial \hat{y}} = 3\hat{\lambda}(\hat{S}_{xx} + \hat{S}_{yy}), \tag{10}$$

$$\frac{\partial \hat{v}_x}{\partial \hat{y}} + \frac{\partial \hat{v}_y}{\partial \hat{x}} = 2\hat{\lambda}\hat{\sigma}_{xy}. \tag{11}$$

Note that, in the flow rule given above, the hyperbolic terms are replaced with the in-plane deviatoric stress components, in accordance with (8). We rewrite the yield condition in the following form:

$$\hat{S}_{xx}^2 + \hat{S}_{yy}^2 + \hat{S}_{xx}\hat{S}_{yy} + \hat{\sigma}_{xy}^2 + \frac{\hat{\sigma}_m^2}{3}\left[2\hat{f}\cosh\left(\frac{\hat{\sigma}_{kk}}{2\hat{\sigma}_m}\right) - \left(1+\hat{f}^2\right)\right] = 0. \tag{12}$$

The equivalent plastic work expression gives the following differential equation for the microscopic equivalent stress:

$$(1-\hat{f})\hat{\sigma}_m\left(\frac{\partial \hat{\sigma}_m}{\partial \hat{x}}\hat{v}_x + \frac{\partial \hat{\sigma}_m}{\partial \hat{y}}\hat{v}_y\right) = \hat{H}\left[\hat{\sigma}_{xx}\frac{\partial \hat{v}_x}{\partial \hat{x}} + \hat{\sigma}_{yy}\frac{\partial \hat{v}_y}{\partial y} + \hat{\sigma}_{xy}\left(\frac{\partial \hat{v}_x}{\partial \hat{y}} + \frac{\partial \hat{v}_y}{\partial \hat{x}}\right)\right], \tag{13}$$

where

$$\hat{H} = \frac{d\hat{\sigma}_m}{d\hat{\epsilon}_m^p} = \frac{\hat{\sigma}_0}{n\hat{\epsilon}_0}\left(\frac{\hat{\sigma}_m}{\hat{\sigma}_0}\right)^{1-n} \tag{14}$$

is the tangent modulus of the uniaxial stress-strain curve of the matrix material in the strain hardening region.

The boundary conditions of zero normal velocity and Coulomb's law of friction lead to the following expressions along the die-metal interface $\hat{y} = \hat{h}(\hat{x})$:

$$\hat{v}_y = \frac{d\hat{h}}{d\hat{x}}\hat{v}_x, \tag{15}$$

and

$$\hat{\sigma}_{xy} = \hat{\mu}\left[\hat{\sigma}_{yy} - 2\frac{d\hat{h}}{d\hat{x}}\hat{\sigma}_{xy} + \left(\frac{d\hat{h}}{d\hat{x}}\right)^2\hat{\sigma}_{xx}\right] + \frac{d\hat{h}}{d\hat{x}}(\hat{S}_{xx} - \hat{S}_{yy}) + \left(\frac{d\hat{h}}{d\hat{x}}\right)^2\hat{\sigma}_{xy}, \tag{16}$$

where $\hat{\mu}$ is the coefficient of friction. The symmetry of the problem about the $\hat{x}$ axis, implies that the shear stress and the transverse velocity vanish on $\hat{y} = 0$. A constant flux of material flowing through a given cross-section $\hat{x}$ leads to the following condition:

$$\int_0^{\hat{h}(\hat{x})}(1 - \hat{f})\,\hat{v}_x(\hat{x}, \hat{y})\,d\hat{y} = (1 - \hat{f}_0)\,\hat{h}_0\hat{u}_0, \tag{17}$$

where $\hat{h}_0 = \hat{h}(0)$ and $\hat{u}_0$ and $\hat{f}_0$ are the velocity and porosity of the entering billet. The boundary condition that the net force at exit is zero requires that

$$\int_0^{\hat{h}_1}\hat{\sigma}_{xx}(\hat{L}, \hat{y})\,d\hat{y} = 0, \tag{18}$$

where $\hat{h}_1 = \hat{h}(\hat{L})$, $\hat{L}$ being the length of the die in the reduction region. Also the initial yielding and the entrance porosity conditions are written as

$$\hat{\sigma}_m(0, \hat{y}) = \hat{\sigma}_0, \qquad \text{and} \qquad \hat{f}(0, \hat{y}) = \hat{f}_0. \tag{19}$$

For gradually tapering dies the problem is simplified by stretching one coordinate direction with respect to the other (Van Dyke [10]); $\hat{y}$ is normalized by the height reduction $\Delta\hat{h}$, and distance down the extrusion axis is made dimensionless by $\hat{L}$, *i.e.*,

$$x = \frac{\hat{x}}{\hat{L}} \qquad \text{and} \qquad y = \frac{\hat{y}}{\Delta\hat{h}} = \frac{\hat{y}}{\epsilon\hat{L}}, \tag{20}$$

where $\epsilon = \Delta\hat{h}/\hat{L}$. The following normalizations are also introduced:

$$\sigma_{xx} = \frac{\hat{\sigma}_{xx}}{\hat{\sigma}_0}, \qquad \sigma_{yy} = \frac{\hat{\sigma}_{yy}}{\hat{\sigma}_0}, \qquad \sigma_{xy} = \frac{\hat{\sigma}_{xy}}{\epsilon\hat{\sigma}_0}, \qquad \sigma_m = \frac{\hat{\sigma}_m}{\hat{\sigma}_0}, \tag{21}$$

$$v_x = \frac{\hat{v}_x}{\hat{u}_0}, \qquad v_y = \frac{\hat{v}_y}{\epsilon\hat{u}_0} \qquad \text{and} \qquad \lambda = \frac{\hat{L}\hat{\sigma}_0}{\hat{u}_0}\hat{\lambda}. \tag{22}$$

We assume that the porosity in the metal and the coefficient of friction at the die-metal interface are $O(\epsilon)$ and define

$$f = \frac{\hat{f}}{\epsilon}, \qquad f_0 = \frac{\hat{f}_0}{\epsilon} \qquad \text{and} \qquad \mu = \frac{\hat{\mu}}{\epsilon}, \tag{23}$$

where f, f_0 and μ are now $O(1)$.

The governing differential equations and boundary conditions are now written in terms of the non-dimensional variables. The equilibrium equations are:

$$\frac{\partial\sigma_{xx}}{\partial x} + \frac{\partial\sigma_{xy}}{\partial y} = 0, \qquad \text{and} \qquad \epsilon^2\frac{\partial\sigma_{xy}}{\partial x} + \frac{\partial\sigma_{yy}}{\partial y} = 0. \tag{24}$$

The condition of plane strain translates to

$$S_{xx} + S_{yy} = -S_{zz} = \frac{1}{3}\epsilon\sigma_m f\sinh\left(\frac{\sigma_{kk}}{2\sigma_m}\right). \tag{25}$$

Using the above equation we can write the flow rule and yield condition as follows

$$\begin{aligned} \frac{\partial v_x}{\partial x} &= \lambda(2S_{xx}+S_{yy}), & (26) \\ \frac{\partial v_x}{\partial x}+\frac{\partial v_y}{\partial y} &= 3\lambda(S_{xx}+S_{yy}), & (27) \\ \frac{\partial v_x}{\partial y}+\epsilon^2\frac{\partial v_y}{\partial x} &= 2\epsilon^2\lambda\sigma_{xy}, & (28) \end{aligned}$$

and

$$3(S_{xx}^2+S_{yy}^2+S_{xx}S_{yy}+\epsilon^2\sigma_{xy}^2)+\sigma_m^2\left[2\epsilon f\cosh\left(\frac{\sigma_{kk}}{2\sigma_m}\right)-(1+\epsilon^2 f^2)\right]=0. \quad (29)$$

A note of caution is relevant at this point. When $\sigma_{kk}/(2\sigma_m)\approx 3$ or larger, the first term in the square bracket of equation (29) is no longer $O(\epsilon)$ and enters the problem to leading order. A discussion of the effects of such exponential terms is given in Section 3.1 where comparisons with finite element solutions are also made.

The equivalent plastic work expression and the evolution of porosity can be written as

$$(1-\epsilon f)\sigma_m^n\left(\frac{\partial\sigma_m}{\partial x}v_x+\frac{\partial\sigma_m}{\partial y}v_y\right)=\frac{1}{n\hat{\epsilon}_0}\left[\sigma_{xx}\frac{\partial v_x}{\partial x}+\sigma_{yy}\frac{\partial v_y}{\partial y}+\sigma_{xy}\left(\frac{\partial v_x}{\partial y}+\epsilon^2\frac{\partial v_y}{\partial x}\right)\right], \quad (30)$$

$$\epsilon\frac{\partial f}{\partial x}v_x+\epsilon\frac{\partial f}{\partial y}v_y=(1-\epsilon f)\left(\frac{\partial v_x}{\partial x}+\frac{\partial v_y}{\partial y}\right). \quad (31)$$

The boundary conditions on the die-metal interface $y=h(x)$ are given as

$$v_y=h'v_x, \quad (32)$$

$$\sigma_{xy}=\mu\sigma_{yy}+h'(S_{xx}-S_{yy})+\mu\epsilon^2 h'(h'\sigma_{xx}-2\sigma_{xy})+\epsilon^2 h'^2\sigma_{xy}. \quad (33)$$

The constant flux and zero-exit-force boundary conditions are written as

$$\int_0^{h(x)}[1-\epsilon f(x,y)]\,v_x(x,y)\,dy=h_0(1-\epsilon f_0), \qquad \text{and} \qquad \int_0^{h_1}\sigma_{xx}(1,y)\,dy=0. \quad (34)$$

Also, the conditions of initial yielding and entrance porosity are written as

$$\sigma_m(0,y)=1, \qquad \text{and} \qquad f(0,y)=f_0. \quad (35)$$

Equations (24)–(35) show that the solution is of the form $\mathcal{F}=\mathcal{F}(x,y,\mu,f_0,\epsilon)$, where $\mathcal{F}$ is a solution variable of the normalized problem, namely, $\boldsymbol{v}$, $\boldsymbol{\sigma}$, σ_m, f or λ.

2.2 Solution to the problem

We attempt a perturbation expansion of the solution in the form

$$\mathcal{F}=\mathcal{F}^{(0)}+\epsilon\mathcal{F}^{(1)}+O(\epsilon^2), \quad (36)$$

where $\mathcal{F}$ is $\boldsymbol{v}$, $\boldsymbol{\sigma}$, σ_m, f or λ.

Substituting the above expansions into the equations (24)–(35), we obtain a hierarchy of problems. The complete problem and the solution methodology upto order ϵ can be found in Govindarajan [11]. In this paper, we give only the solution to the leading order and $O(\epsilon)$ problems.

The leading order velocity field is given as

$$v_x^{(0)}(x) = \frac{h_0}{h(x)}, \tag{37}$$

$$v_y^{(0)}(x,y) = v_x^{(0)}(x)\frac{h'(x)}{h(x)}y. \tag{38}$$

The first of the above equations shows that the leading order solution involves "slab flow" as assumed by von Kármán [12] in the analysis of plane strain rolling. Also, for the case of a straight die, equations (37) and (38) show that the leading order velocity field is "radial" (Durban and Mear [7]) and provide a theoretical justification for the use of such velocity fields in the analysis of extrusion through dies with moderate slopes.

The complete leading order solution is as follows

$$\begin{aligned}
S_{xx}^{(0)} &= -S_{yy}^{(0)} = \frac{\sigma_m^{(0)}}{\sqrt{3}}, && (39)\\
\sigma_m^{(0)}(x) &= \left[1 + \frac{2}{\sqrt{3}n\hat{\epsilon}_0}\ln\left(\frac{h_0}{h(x)}\right)\right]^{1/n}, && (40)\\
\sigma_{xx}^{(0)}(x) &= 2\left[\int_x^1 S_{xx}^{(0)}(\xi)\left(\frac{h'(\xi)-\mu}{h(\xi)}\right)\exp\left(-\mu\int_\xi^1 \frac{d\eta}{h(\eta)}\right)d\xi\right]\exp\left[\mu\int_x^1 \frac{d\eta}{h(\eta)}\right], && (41)\\
\sigma_{yy}^{(0)}(x) &= \sigma_{xx}^{(0)}(x) - 2S_{xx}^{(0)}(x), && (42)\\
\sigma_{xy}^{(0)}(x,y) &= \left[\mu\sigma_{xx}^{(0)}(x) + 2S_{xx}^{(0)}\left(h'(x)-\mu\right)\right]\frac{y}{h(x)}, && (43)
\end{aligned}$$

It is interesting to note that the normal stress components are, to leading order, uniform over any transverse section and the shear stress $\hat{\sigma}_{xy}$, which is order $\epsilon\hat{\sigma}_0$, is linear in y. It is also seen that the presence of friction does not affect, to leading order, the velocity field or the deviatoric stress components, but it does affect the hydrostatic stress in an exponential way.

Here we note that Johnson [5] obtains expressions similar to (40–43) while analyzing the axisymmetric extrusion of fully dense metals. He computes the extent of hardening by integrating along the leading order streamlines to determine the equivalent plastic strain and hence $\hat{\sigma}_m$. It should be noted, however, that the use of plastic work expression of the form (30) gives the extent of hardening rather directly.

It should be noted that the leading order problem and solution corresponds to that of a fully dense metal and that the effects of porosity enter the solution as a first order correction. The porosity itself, however, is a function of the leading order solution and is found to be

$$f^{(0)}(x) = f_0 \exp\left[-\sqrt{3}\int_0^x \frac{h'(\eta)}{h(\eta)}\sinh\left(\frac{\sigma_{kk}^{(0)}(\eta)}{2\sigma_m^{(0)}(\eta)}\right)d\eta\right]. \tag{44}$$

The velocity and stress fields upto $O(\epsilon)$ are given as

$$v_x^{(1)}(x) = v_x^{(0)}(x)\left[f^{(0)}(x) - f_0\right], \tag{45}$$

$$v_y^{(1)}(x,y) = v_x^{(1)}(x)\frac{h'(x)}{h(x)}y, \tag{46}$$

and

$$\sigma_m^{(1)}(x) = \left[\int_0^x g_2(\xi)\exp\left(\int_0^\xi g_1(\eta)\,d\eta\right)d\xi\right]\exp\left(-\int_0^x g_1(\eta)\,d\eta\right), \tag{47}$$

$$\sigma_{xx}^{(1)}(x) = \left[\int_x^1 \left(\frac{h'(\xi)-\mu}{h(\xi)}\right)a(\xi)\exp\left(-\mu\int_\xi^1 \frac{d\eta}{h(\eta)}\right)d\xi\right]\exp\left(\mu\int_x^1 \frac{d\eta}{h(\eta)}\right), \tag{48}$$

$$\sigma_{yy}^{(1)}(x) = \sigma_{xx}^{(1)}(x) - a(x), \tag{49}$$

$$\sigma_{xy}^{(1)}(x,y) = \left[\mu\sigma_{xx}^{(1)}(x) + a(x)\,(h'(x)-\mu)\right]\frac{y}{h(x)}, \tag{50}$$

where

$$g_1(x) = \frac{2(1-n)}{\sqrt{3}n\hat{\epsilon}_0}\frac{h'(x)}{h(x)}\frac{1}{\sigma_m^{(0)^n}(x)}, \tag{51}$$

$$g_2(x) = -\frac{1}{n\hat{\epsilon}_0}\frac{h'(x)}{h(x)}\frac{f^{(0)}(x)}{\sigma_m^{(0)^n}(x)} \times\left[\sqrt{3}\sigma_{xx}^{(0)}(x)\sinh\left(\frac{\sigma_{kk}^{(0)}(x)}{2\sigma_m^{(0)}(x)}\right) + 2S_{xx}^{(0)}(x)\left\{1-\cosh\left(\frac{\sigma_{kk}^{(0)}(x)}{2\sigma_m^{(0)}(x)}\right)\right\}\right], \tag{52}$$

$$a(x) = \frac{2}{\sqrt{3}}\left[\sigma_m^{(1)}(x) - \sigma_m^{(0)}(x)f^{(0)}(x)\cosh\left(\frac{\sigma_{kk}^{(0)}(x)}{2\sigma_m^{(0)}(x)}\right)\right]. \tag{53}$$

It should be noted that for a fully dense material ($f_0 = 0$) the above expressions reduce to $\boldsymbol{\sigma}^{(1)} = \mathbf{0}$ and $\boldsymbol{v}^{(1)} = \mathbf{0}$. The next term in the porosity expansion is found to be ([11])

$$f^{(1)}(x) = \left[\int_0^x F_2(\xi)\exp\left(\int_0^\xi F_1(\eta)\,d\eta\right)d\xi\right]\exp\left(-\int_0^x F_1(\eta)\,d\eta\right), \tag{54}$$

where

$$F_1(x) = \sqrt{3}\frac{h'(x)}{h(x)}\sinh\left(\frac{\sigma_{kk}^{(0)}(x)}{2\sigma_m^{(0)}(x)}\right), \tag{55}$$

$$F_2(x) = -\frac{\sqrt{3}}{\sigma_m^{(0)^2}(x)}\frac{h'(x)}{h(x)}\left[3b(x)\left(\frac{\sqrt{3}}{2}\{3b(x)-a(x)\} + \sigma_m^{(1)}(x) - \sigma_m^{(0)}(x)f^{(0)}(x)\right) - \frac{1}{2}\left(\sigma_{kk}^{(1)}(x) - \sigma_{kk}^{(0)}(x)\frac{\sigma_m^{(1)}(x)}{\sigma_m^{(0)}(x)}\right)\left(\frac{\sqrt{3}}{2}a(x) - \sigma_m^{(1)}(x)\right)\right], \tag{56}$$

$$b(x) = \frac{1}{3}\sigma_m^{(0)}(x)f^{(0)}(x)\sinh\left(\frac{\sigma_{kk}^{(0)}(x)}{2\sigma_m^{(0)}(x)}\right). \tag{57}$$

Also, we note that in most cases the contribution of the second term $f^{(1)}$ in the porosity expansion (36) is not insignificant and should be taken into account. This becomes clear in Section 3.1, where comparisons with detailed finite element calculations are made.

3 Finite element solutions

The asymptotic solutions developed in the previous section are now compared with the results of detailed finite element calculations, which are performed using the ABAQUS general purpose finite element program (Hibbitt [13]). The analysis is carried out incrementally using an updated Lagrangian formulation and Newton's method is used to solve the discretized equilibrium equations. Four-node isoparametric elements with 2×2 Gauss integration points are used for the constitutive calculations at material points. The integration of the elastoplastic equations for Gurson's model and the computation of the corresponding "linearization moduli" are implemented in ABAQUS using the method presented by Aravas [14].

The constitutive parameters used in the calculations are $\hat{E}/\hat{\sigma}_0 = 300$, $\nu = 0.49$, $n = 5$, and $\hat{f}_0 = 0.10$. For the die, it is assumed that the area reduction is $r = \Delta\hat{h}/\hat{h}_0 = 0.25$, and the length of the reduction region is $\hat{L} = 2.5\hat{h}_0$, so that $\epsilon = \Delta\hat{h}/\hat{L} = 0.1$. The reduction area of the die is shaped in the form of a fifth order polynomial with zero slope and curvature at both ends. The coefficient of friction $\hat{\mu}$ along the die-metal interface is assumed to be 0.05.

3.1 Results and comparisons

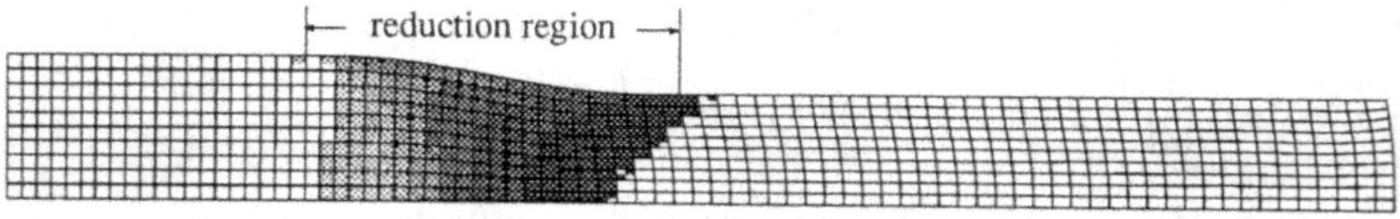

Figure 1: Finite element mesh of the extruded billet

Figure 1 shows the finite element mesh at the end of the calculation. The mesh shows the plastic zone and the extent of hardening by means of differing shades of gray, with darker shade pointing to a higher hardening. A visual effect of friction on the deformation process can also be seen in the form of the large amount of shear induced in the deformed mesh.

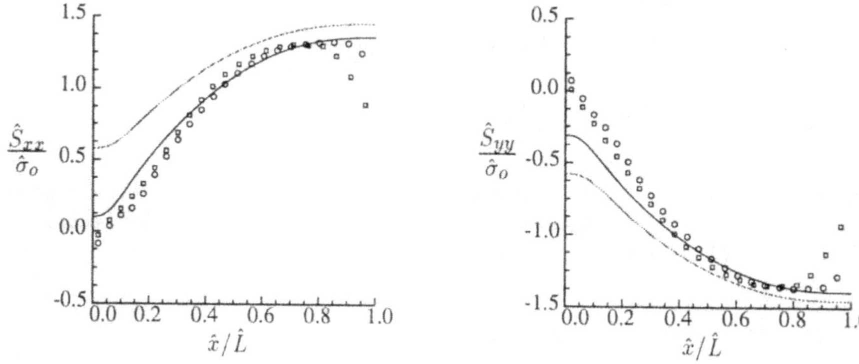

Figure 2: Variation of deviatoric stresses in the reduction region

Figure 2 shows the variation of the deviatoric stresses $\hat{S}_{xx}/\hat{\sigma}_0$ and $\hat{S}_{yy}/\hat{\sigma}_0$ along the direction of extrusion inside the reduction region. In this figure, the finite element solutions are denoted by a square on $y = 0$ and a circle on $y = h(x)/2$. Broken lines are the predictions of the leading term in the asymptotic expansion of the solution, whereas solid lines are the sum of the first and the second terms in the expansion. This convention applies to Fig. 3 also. As the billet enters the die, it deforms elastically until the stresses are high enough to cause yielding; such elastic effects are not taken into account in the asymptotic solution and this is the reason for the deviation of the two-term asymptotic deviatoric stresses from the finite element solution near the entrance.

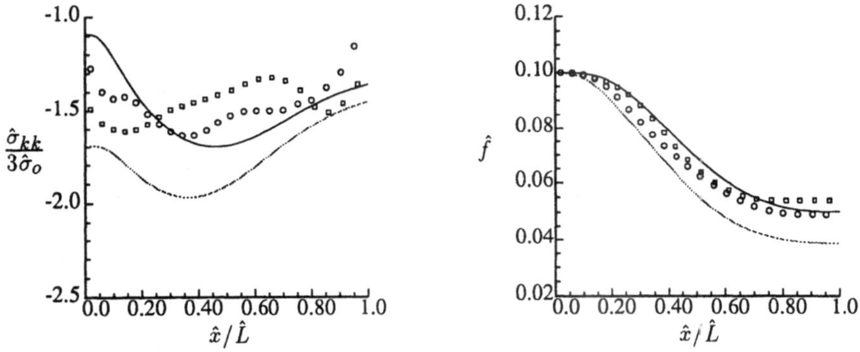

Figure 3: Variation of pressure and porosity in the reduction region

Figure 3 shows the variation of the hydrostatic stresses $\hat{\sigma}_{kk}/(3\hat{\sigma}_0)$ and the porosity $\hat{f}$ in the reduction region. The pressure variations indicate that the correction added to the leading order term of the asymptotic solution is not significant; this can be easily traced back to the hyperbolic terms in the solution, whose arguments are these pressure terms. Figure 3 suggests that higher order terms must be calculated in order to obtain an accurate estimate for the pressure distribution in the reduction region.

Figures 2 and 3 show that when the effects of the second term in the expansion of the solution are taken into account, the predictions of the asymptotic solutions are improved and agree well with the results of the finite element analysis.

In conclusion, we mention that this fairly simple and elegant analysis gives an overview of a rather complex deformation process and provides a basis for the testing of the applicability of extensive numerical procedures in solving such problems.

References

[1] Hill, R.: *The Mathematical Theory of Plasticity,* Oxford: Oxford University Press 1950.

[2] Avitzur, B.: *Metal Forming — The Application of Limit Analysis,* New York: Marcel Dekker, Inc. 1980.

[3] Lee, E.H.; Mallet, R.L.; Yang, W.H.: Stress and deformation analysis of the metal extrusion process, *Comp. Meth. App. Mech. Engng.* **10** (1977) 339-353.

[4] Kobayashi, S.; Oh, S-I.; Altan, T.: *Metal Forming and the Finite-Element Method,* Oxford: Oxford University Press 1989.

[5] Johnson, R.E.: Conical extrusion of a work-hardening material: An asymptotic analysis, *J. Engng. Math.* **21** (1987) 295-329.

[6] Aravas, N.; McMeeking, R.M.: An asymptotic analysis of three-dimensional extrusion, *ASME J. App. Mech.* **56** (1989) 519–526.

[7] Durban, D.; Mear, M.E.: Asymptotic analysis for extrusion of sintered powder metals, *University of Texas at Austin Report* 1989.

[8] Govindarajan, R.M.; Aravas, N.: Asymptotic analysis of extrusion of porous metals, *Int. J. Mech. Sci.* (1991), to appear.

[9] Gurson, A.L.: Continuum theory of ductile rupture by void nucleation and growth: Part I—Yield criteria and flow rules for porous ductile materials, *J. Eng. Matl. Tech.* **99** (1977) 2–15.

[10] Van Dyke, M.: *Perturbation Methods in Fluid Mechanics*, Stanford: The Parabolic Press 1975.

[11] Govindarajan, R.M.: Ph. D. Thesis (in preparation), Department of Mechanical Engineering and Applied Mechanics, University of Pennsylvania, Philadelphia, PA, USA 1991.

[12] von Kármán, T.: Beitrag zur Theorie des Walzvorganges, *Z. angew. Math. Mech.* **5** (1925) 139–141.

[13] Hibbitt, H.D.: ABAQUS/EPGEN - A general purpose finite element code with emphasis on nonlinear applications, *Nucl. Eng. Des.* **77** (1984) 271–297.

[14] Aravas, N.: On the numerical integration of a class of pressure-dependent plasticity models, *Int. J. Num. Meth. Engng.* **24** (1987) 1395–1416.

Large Dynamic Deflections of Elastic-Plastic Structures Including Material Damage due to Void Growth

P.A. FOTIU, Department of Applied Mechanics and Engineering Sciences, University of California, San Diego, CA 92093
H. IRSCHIK, University of Linz, Austria, A-4040
F. ZIEGLER, Department of Civil Engineering, Technical University of Vienna, Austria, A-1040

Summary

Inelastic deformations are in general a result of dissipative processes that take place in the materials microstructure. Therefore, the microscopic behavior is included in an averaged sense in the formulation of the macroscopic constitutive relations. A method of dynamic, rate-dependent plasticity based on such a concept that describes in a consistent manner the microscopic as well as the macroscopic behavior of ductile structures made of polycrystalline solids with micro-cavities, and that includes also geometric nonlinearities of large deflections, is developed in this paper. Dissipative mechanisms and the geometric nonlinearities are considered as defects in a background material of time-invariant linear elastic constitutive properties. Defects are mechanically described by fields of distorsions (or eigenstrains in the notion of Mura [1]). Response due to this kind of agency is determined by means of the linear Green's function approach.

Introduction

Severely excited slender structures like elastic-visco-plastic beams vibrate with amplitudes of an order of magnitude that requires a geometrically nonlinear formulation. The kind of this nonlinearity is determined by the axial boundary conditions of the problem. In the case of free axial motion of the supports, higher order terms of the expression of the curvature are to be included in the analysis, but the beam's central fiber may be assumed to be inextensible. Otherwise, having both ends pinned, the influence of the membrane stress due to the stretching of the central fiber has to be accounted for.

Elastic nonlinear natural vibrations of pinned-pinned beams have been investigated by Woinowsky-Krieger [2], and more recently by Elishakoff *et al* [3] among others. Forced vibrations and their stability are discussed, eg by Mettler [4] and Eisley [5]. An account of thermal stresses is given by Ziegler and Rammerstorfer [6]. These problems are approximately solved by the Ritz-Galerkin method together with proper assumptions on the deflection modes. Commonly the continuous parameter problem is projected onto a set of (coupled) Duffing-type nonlinear differential equations. Solutions of the forced Duffing oscillator, their stability and the bifurcation characteristics are discussed in detail in various textbooks and papers, see eg Hayashi [7], Nayfeh and Mook [8], Guckenheimer and Holmes [9], Holmes and Rand [10], and with respect to a unifying result on plates, see Heuer *et al* [11]. However, all of these

D. Besdo, E. Stein (Eds.)
Finite Inelastic Deformations –
Theory and Applications
IUTAM Symposium Hannover/Germany 1991

results are based on elastic vibrations with dissipative effects restricted to linear viscous damping.

Plastic deformations may have a grave effect on the nonlinear dynamic behavior of beams, that has been shown by Symonds and Yu [12] in a paradox of a beam having a permanent plastic deflection opposite to the direction of the lateral loading. That behavior is caused by a snap-through effect due to the plastic elongation of the central fiber. Some attempts have been made to investigate the chaotic motion of simple rigid-plastic systems, see, eg Poddar *et al* [13]. Moreover, with the inclusion of material damage even small deflections may in some aspects exhibit chaos-like behavior, see Fotiu *et al* [14].

In this paper we consider the dynamic behavior of elastic-plastic hardening beams including a gradual degradation of strength due to continuum damage and we account for the development of plastic zones. Effects of inelastic material behavior on the response are regarded as additional fields of eigenstrains acting on the associated linear elastic system. This has been effectively done in several papers on small elastic-plastic vibrations of beams (Irschik and Ziegler [15] ; Irschik [16]; Fotiu and Irschik [17]). Within such a theory, however, it seems obvious to represent geometric nonlinearities by some type of additional loading, too. Consequently, the total solution of the problem is found as the superposition of the associated linear elastic solution and the solution due to all additional effects accounting for physical and geometric nonlinearities. A corresponding integral representation based on a Lagrangean approach has been derived by Irschik and Ziegler [18]. Mukherjee and Chandra [19] established a boundary integral description of large strain elastic-viscoplastic problems, where they also treated geometrically nonlinear contributions as fictitious loadings. In the present analysis we consider an elastic-plastic material incorporating ductile damage due to void growth.

Analysis is based on an integral equation formulation with a modal representation of the dynamic solution. The unknown increments of the additional loadings are evaluated in an iterative procedure especially designed for the rather involved constitutive model. It is emphasized that only such a procedure allows for a modal formulation, that relies on the use of linear differential operators in the description of the problem.

Equivalent Fields of Eigenstrain

It is well known that the effects of irregularities or defects in a body on the structural behavior can be represented by eigenstrains (Mura, [1]). A dynamic field of eigenstrains $\bar{\boldsymbol{\varepsilon}}(\mathbf{x},t)$ in a continuum of volume V produces displacements according to the relation, see, eg Fotiu [20])

$$u_i(\mathbf{x},t) = \int_V \left[\tilde{\sigma}_{\alpha\beta i}(\boldsymbol{\xi},\mathbf{x}) * \bar{\varepsilon}_{\alpha\beta}(\boldsymbol{\xi})\right](t)\, dV + \rho \int_V \left(\tilde{u}_{\alpha i}(\boldsymbol{\xi},\mathbf{x},t)\, \dot{u}_\alpha(\boldsymbol{\xi},0) + \dot{\tilde{u}}_{\alpha i}(\boldsymbol{\xi},\mathbf{x},t)\, u_\alpha(\boldsymbol{\xi},0)\right) dV \ , \tag{1}$$

where ρ is the mass density and the dynamic Green´s functions $\tilde{\sigma}_{\alpha\beta i}(\boldsymbol{\xi},\mathbf{x},t)$, $\tilde{u}_{\alpha i}(\boldsymbol{\xi},\mathbf{x},t)$ are defined as the stress $\sigma_{\alpha\beta}(\boldsymbol{\xi},t)$ and displacement $u_\alpha(\boldsymbol{\xi},t)$ due to a singular unit force at the point $\mathbf{x}$ applied at time t=0, pointing in the direction of $\mathbf{e}_i$. Here, $\mathbf{e}_1,\mathbf{e}_2,\mathbf{e}_3$ denote the unit base vectors of a cartesian coordinate system. The symbol $*$ designates the time-convolution.

Stresses due to eigenstrains are given by

$$\sigma_{ij} = E_{ijkl}\,\varepsilon^e_{kl} = E_{ijkl}\,(\varepsilon_{kl} - \bar{\varepsilon}_{kl}) \,, \tag{2}$$

where $\mathbf{E}$ is the elasticity tensor of the background material.

Exemplarily, we consider a beam of length l with its axis in the x-direction and deforming in the x-z plane. The beam´s cross-section with area A and moment of inertia J is assumed to be spanwise constant. The longitudinal total strain $\varepsilon(x,z)$ is derived from the axial displacement $u^{(z)}(x,z)$ and the transverse deflection $w(x)$

$$\varepsilon(x,z) = u^{(z)}_{,x}(x,z) + \frac{1}{2}(w_{,x})^2 \quad . \tag{3}$$

Introducing the Bernoulli-Euler hypothesis

$$u^{(z)}(x,z) = u(x) - z\, w_{,x} \,, \tag{4}$$

where $u(x)$ denotes the longitudinal displacement of the central axis, we obtain

$$\varepsilon(x,z) = u_{,x}(x) + \frac{1}{2}(w_{,x})^2 - z\, w_{,xx} \quad . \tag{5}$$

In the following we shall treat the case of a beam on immovable supports. If the beam is loaded by eigenstrains $\bar{\varepsilon}$ as well as transverse and axial forces q and p, respectively, we have the following expressions for the strain energy U, the kinetic energy T and the loading potential W

$$U = \frac{1}{2}\int_V E(\varepsilon^e)^2\, dV = \frac{1}{2}\int_V E\left(u_{,x} + \frac{1}{2}(w_{,x})^2 - z\, w_{,xx} - \bar{\varepsilon}\right)^2 dV \,, \tag{6}$$

$$T = \frac{1}{2}\int_l \mu\, \dot{w}^2\, dx \,, \tag{7a}$$

$$W = \int_l (q\, w + p\, u)\, dx \,, \tag{7b}$$

where μ = const. is the beam´s mass per unit length and a dot designates differentiation with respect to time. Application of Hamilton´s principle $\delta\int_{t_1}^{t_2}(T-U-W)\,dt = 0$ yields, after some algebra, the equations of motion

$$N_{,x} = -p \quad , \tag{8a}$$

$$EJ\, w_{,xxxx} - (N\, w_{,x})_{,x} + \mu \dot{w} = q - EJ\, \bar{\kappa}_{,xx} \quad , \tag{8b}$$

where the axial force N is given by

$$N = EA \left[u_{,x} + \frac{1}{2} (w_{,x})^2 - \bar{e} \right] \; , \tag{9}$$

and

$$\bar{e} = \frac{1}{A} \int_A \bar{\varepsilon}\, dA \;\; , \quad \bar{\kappa} = \frac{1}{J} \int_A \bar{\varepsilon} z\, dA \; . \tag{10}$$

From now on we put p=0. That renders N spanwise constant due to Eq. (8a). Since $u(0) = u(l) = 0$, we obtain from Eq. (9) the following expression for the axial force

$$N = \frac{EA}{l} \int_l \left[\frac{1}{2} (w_{,x})^2 - \bar{e} \right] dx \; . \tag{11}$$

According to the remarks stated in the introduction we proceed, formulating the coupled nonlinear problem of inelastic deformation and large deflection as a solution of an associated linear elastic background problem with additional equivalent loadings. Such eigenstrains, which are equivalent to physical nonlinearities shall be denoted by $\bar{\varepsilon}$, while a hat $(\hat{\;})$ identifies contributions emanating from geometric nonlinearities. Consequently, the total solution w is the sum of two parts

$$w = w^0 + w^*, \tag{12}$$

where, see Fotiu [20]

$$w^0(x,t) = \int_l [\tilde{w}(\xi,x) * q(\xi)](t)\, d\xi + \mu \int_l \left[\tilde{w}(\xi,x,t)\, \dot{w}^0(\xi,0) + \dot{\tilde{w}}(\xi,x,t)\, w^0(\xi,0) \right] d\xi \tag{13}$$

is the portion due to the given external forces, while

$$w^*(x,t) = \int_l \left[\tilde{M}(\xi,x) * \left(\bar{\kappa}(\xi) + \hat{\kappa}(\xi) \right) \right](t)\, d\xi + \mu \int_l \left[\tilde{w}(\xi,x,t)\, \dot{w}^*(\xi,0) + \dot{\tilde{w}}(\xi,x,t)\, w^*(\xi,0) \right] d\xi \tag{14}$$

is the deflection due to additional (internal) loadings with

$$\hat{\kappa}(\xi,t) = -\frac{N(t)}{EJ}\, w(\xi,t) \; . \tag{15}$$

In analogy to the statement made subsequently to Eq. (1), the dynamic Green´s functions $\tilde{w}(\xi,x,t)$, $\tilde{M}(\xi,x,t)$ are deflection and moment at ξ and time t, respectively, produced by a transverse unit force at x applied at t=0.

Finally, curvatures $\kappa(x,t)$, moments $M(x,t)$ and strains are found as follows

$$\kappa^0(x,t) = -\, w^0_{,xx}(x,t)\ ,\quad \kappa^*(x,t) = -\, w^*_{,xx}(x,t)\ , \tag{16a}$$

$$M^0(x,t) = EJ\ \kappa^0(x,t)\ ,\quad M^*(x,t) = EJ\left(\kappa^*(x,t) - \bar{\kappa}(x,t)\right), \tag{16b}$$

$$\varepsilon^0(x,z,t) = z\ \kappa^0(x,t)\ ,\quad \varepsilon^*(x,z,t) = z\ \kappa^*(x,t) + e^*(x,t)\ ,\ e^*(x,t) = \frac{N(t)}{EA} + \bar{e}(x,t)\ . \tag{16c}$$

Incremental Formulation

According to the nonlinearity of the problem, an incremental solution procedure has to be adopted. Assuming the state of the system to be known at time t_a, the increments of deflections w^0 and w^* in the interval $\Delta t = t_{a+1} - t_a$ are, according to Eqs. (13) and (14)

$$\Delta w^0(x,t) = \int_l \left[\tilde{w}(\xi,x) \ast \Delta q(\xi)\right](\Delta t)\, d\xi + \mu \int_l \left[\tilde{w}(\xi,x,\Delta t)\ \dot{w}^0(\xi,t_a) + \tilde{\dot{w}}(\xi,x,\Delta t)\ w^0(\xi,t_a)\right] d\xi\ , \tag{17}$$

$$\Delta w^*(x,t) = \int_l \left[\tilde{M}(\xi,x) \ast \left(\Delta\bar{\kappa}(\xi)+\Delta\hat{\kappa}(\xi)\right)\right](\Delta t)\, d\xi + \mu \int_l \left[\tilde{w}(\xi,x,\Delta t)\ \dot{w}^*(\xi,t_a) + \tilde{\dot{w}}(\xi,x,\Delta t)\ w^*(\xi,t_a)\right] d\xi\ , \tag{18}$$

where

$$\Delta\hat{\kappa}(\xi) = -\frac{1}{EJ}\left[(N_a+\Delta N)\Delta w(\xi) + \Delta N\ w_a(\xi)\right]\ , \tag{19}$$

$$\Delta N = \frac{EA}{l} \int_l \left[(w_{,x})_a\ \Delta w_{,x} + \frac{1}{2}(\Delta w_{,x})^2 - \Delta\bar{e}\right] dx\ , \tag{20}$$

and the subscript $(\ )_a$ denotes values at time t_a. The convolution within the time interval takes the form

$$[a \ast b](\Delta t) = \int_0^{\Delta t} a(\Delta t-\bar{\tau})\, b(\bar{\tau})\, d\bar{\tau}\ ,\quad \bar{\tau} = \tau - t_a\ . \tag{21}$$

Within Δt we approximately assume a linear variation of all the excitation functions $\Delta q(\xi,\bar{\tau})$, $\Delta\bar{\kappa}(\xi,\bar{\tau})$ and $\Delta\hat{\kappa}(\xi,\bar{\tau})$, ie

$$\Delta q(\xi,\bar{\tau}) = g(\bar{\tau},\Delta t)\ \Delta q(\xi),\quad \Delta\bar{\kappa}(\xi,\bar{\tau}) = g(\bar{\tau},\Delta t)\ \Delta\bar{\kappa}(\xi),\quad \Delta\hat{\kappa}(\xi,\bar{\tau}) = g(\bar{\tau},\Delta t)\ \Delta\hat{\kappa}(\xi)\ , \tag{22}$$

where

$$g(\bar{\tau},\Delta t) = \begin{cases} 1, & \bar{\tau} \geq \Delta t \\ \bar{\tau}/\Delta t, & 0 \leq \bar{\tau} \leq \Delta t \\ 0, & \bar{\tau} \leq 0 \end{cases}\ . \tag{23}$$

In general, dynamic Green´s functions are given by modal series, eg

$$\tilde{w}(\xi,x,t-\tau) = \sum_{n=1}^{\infty} \frac{\Phi^{(n)}(x)\,\Phi^{(n)}(\xi)}{\mu\,\omega_{nD}} \exp\left(\zeta_n\omega_n(t-\tau)\right) \sin\omega_{nD}(t-\tau)\,,\quad \omega_{nD}=\sqrt{1-\zeta_n^2}\,\omega_n\,, \tag{24}$$

where $\Phi^{(n)}$, ω_n are the n-th normalized mode and the n-th undamped linear eigenfrequency, respectively, and ζ_n is the given linear damping coefficient of the n-th vibrational mode. However, in the case of beams, the quasistatic part $\tilde{w}^S(\xi,x)$ of the function $\tilde{w}(\xi,x,t-\tau)$ exists in a closed form as a solution of

$$EJ\,\tilde{w}^S_{,\xi\xi\xi\xi}(\xi,x) = \delta(\xi-x)\,, \tag{25}$$

where $\delta(\xi-x)$ is the Dirac-delta function. Therefore, we express each solution as a sum of the quasistatic part and the dynamic part,

$$\Delta w^{0,*}(x) = \Delta w_S^{0,*}(x) + \Delta w_D^{0,*}(x)\,, \tag{26}$$

where the quasistatic increments are given exactly by

$$\Delta w_S^0(x) = \int_l \tilde{w}^S(\xi,x)\,\Delta q(\xi)\,d\xi\,,\quad \Delta w_S^*(x) = \int_l \tilde{M}^S(\xi,x)\left(\Delta\bar{\kappa}(\xi)+\Delta\hat{\kappa}(\xi)\right)d\xi \tag{27}$$

with $\tilde{M}^S(\xi,x) = -EJ\,\tilde{w}^S_{,\xi\xi}(\xi,x)$. Performing a twofold integration by parts of Eqs. (13) and (14), we may extract from them the quasistatic contributions, thereby obtaining an equation for the dynamic deflection (Fotiu [20]), eg

$$w_D^0(x,t) = \int_l \left[\tilde{w}^D(\xi,x) * \ddot{q}(\xi)\right](t)\,d\xi + \mu\int_l \left[\tilde{w}(\xi,x,t)\,\dot{w}_D^0(\xi,0) + \dot{\tilde{w}}(\xi,x,t)\,w_D^0(\xi,0)\right]d\xi\,. \tag{28}$$

With the shape function $g(\bar{\tau},\Delta t)$ according to Eq. (23) the time integration in Eq. (28) can be carried out within Δt, yielding for the increment

$$\Delta w_D^0(x) = \int_l \tilde{w}^D(\xi,x,\Delta t)\,\Delta q(\xi)\,d\xi + \mu\int_l \left[\tilde{w}(\xi,x,\Delta t)\,\dot{w}_{Da}^0(\xi) + \dot{\tilde{w}}(\xi,x,\Delta t)\,w_{Da}^0(\xi)\right]d\xi \tag{29}$$

with

$$\tilde{w}^D(\xi,x,\Delta t) = -\sum_{n=1}^{\infty} \frac{\Phi^{(n)}(x)\,\Phi^{(n)}(\xi)}{\mu\,\omega_n^2}\,\mathcal{T}(\Delta t)\,, \tag{30}$$

$$\mathcal{T}(\Delta t) = \exp(-\zeta_n\omega_n\Delta t)\,\frac{\sin\omega_n\Delta t}{\omega_n\Delta t}\,. \tag{31}$$

Light linear modal damping, $\zeta_n \ll 1$, is assumed in the approximate time factor in Eq. (31).

The derivation of the increments $\Delta w_D^*(x)$ follows exactly the same lines with the kernel function

$$\widetilde{M}^D(\xi,x,\Delta t) = -\, EJ\, \widetilde{w}^D_{,\xi\xi}(\xi,x,\Delta t). \tag{32}$$

The use of modal analysis guarantees an accurate solution in the time domain, since only the forcing functions are approximated by the shape function $g(\bar{\tau},\Delta t)$ within the time interval and not the solution likewise to the common case, eg in the direct integration method. It is worth mentioning that the inclusion of linear viscous damping may be easily performed by assigning proper values to the parameters ζ_n. This appears to be more general than the proportional (Rayleigh-) damping, which yields for continuous structures: $\zeta_n=r_1/(2\mu\omega_n)$, mass-proportional damping; $\zeta_n=r_2\,\omega_n/2$, stiffness-proportional damping, with r_1 and r_2 assigned material constants.

Constitutive Equations

From the constitutive equations we ought to determine the necessary amount of equivalent eigenstrain $\Delta\bar{\varepsilon}$ as a function of $\Delta\varepsilon$. We consider a viscoplastic porous material, that is described by a dynamic yield surface (Fotiu [20])

$$F = \widetilde{\sigma}_e^2 + \frac{9}{4} D\, \widetilde{\sigma}_m^2 - k^2 \ , \tag{33}$$

with

$$\widetilde{\sigma}_e = \frac{3}{2}\, \widetilde{s}_{ij}\widetilde{s}_{ij} \ , \quad \widetilde{s}_{ij} = \widetilde{s}_{ij} - \widetilde{\sigma}_m\delta_{ij} \ , \quad \widetilde{\sigma}_m = \widetilde{\sigma}_{ii}/3 \ , \tag{34}$$

where δ_{ij} is the Kronecker symbol, a tilde denotes the net-stresses, eg $\widetilde{\boldsymbol{\sigma}} = \boldsymbol{\sigma}/(1-D)$ and D is the damage parameter. The plastic strain rate is derived from F by

$$\dot{\varepsilon}^p_{ij} = \frac{\Gamma}{k} \left\langle \Phi(F/k^2) \right\rangle \frac{\partial F}{\partial \sigma_{ij}} = 3\lambda \left(\widetilde{s}_{ij} + \frac{D}{2}\, \widetilde{\sigma}_m\, \delta_{ij} \right), \ \lambda = \frac{\Gamma \left\langle \Phi(F/k^2) \right\rangle}{k\,(1-D)} \ , \tag{35}$$

where Γ is a viscosity constant and Φ is an increasing function of its argument. The brackets $\langle\cdot\rangle$ have the meaning $\langle F \rangle = F\, H(F)$, where H denotes the Heaviside function. Assuming linear isotropic hardening, the radius of the yield surface varies according to

$$\dot{k} = \mathscr{K}\, \dot{\varepsilon}^p_e \ , \tag{36}$$

where $\dot{\varepsilon}^p_e$ is the equivalent plastic strain rate and $\mathscr{K}$ is a constant hardening modulus.

Up to a critical value of damage D_{cr} the porosity f can be assumed to be proportional to the amount of damage, ie $D = \vartheta f$, with values of ϑ between 1.5 and 2.5. This is in basic agreement with experimental results (Spitzig *et al.* [21]). Moreover, a factor of $\vartheta \approx 2$ can be verified theoretically by a self-consistent analysis of a continuum containing spherical voids

(Fotiu *et al* [22]). We describe ductile void growth by the following evolution equation for the damage parameter

$$\dot{D} = D\,(1-f)\left[\alpha\,\tilde{\sigma}_m^2\right]^n \langle\dot{\varepsilon}_e^p\rangle = \mathscr{G}\langle\dot{\varepsilon}_e^p\rangle, \quad D<D_{cr}\,, \tag{37}$$

where α and n are material constants. The factor $\mathscr{G}$ in Eq. (37) comes from a micromechanical analysis conducted by Davison *et al* [23] with an evolution equation for the void volume V_ν

$$\dot{V}_\nu = \left[\alpha\,\tilde{\sigma}_m^2\right]^n V_\nu \langle\dot{\varepsilon}_e^p\rangle\,. \tag{38}$$

A dependence of the void growth rate on the hydrostatic stress σ_m has also been included in constitutive laws by Barbee *et al.* [24] and Seaman *et al.* [25] considering spall damage in impacted metal specimens. However, with the appearence of the equivalent plastic strain rate Eqs. (37) and (38) become rate independent and, consequently, they can be used for quasistatic loading conditions, too. Such a proportionality between $\dot{V}_\nu / V_\nu$ and $\dot{\varepsilon}_e^p$ has been found by Rice and Tracey [26] in a theoretical analysis of a spherical void in an ideal elastic plastic matrix. According to Eq. (38) ductile void growth is produced only by tensile plastic strains, while loading in compression has no influence on the amount of damage. For values $D>D_{cr}$ plastic localizations develop between the voids, leading to an overproportional decrease of strength compared to the increase of the void volume fraction. In this case Eq. (37) will be no longer valid. In the present paper, however, we will only consider the subcritical range $D<D_{cr}$.

In the uniaxial case with $\sigma_{xx}=\sigma$, $\varepsilon_{xx}=\varepsilon$, $\dot{\varepsilon}_e^p=\dot{\varepsilon}^p$ Eqs. (35) and (37) take on the simple incremental form within the finite time interval $\Delta t = t_{a+1}-t_a$

$$\Delta\varepsilon_{xx}^p = \Delta\varepsilon^p = (2+D_\theta/2)\,\lambda_\theta\,\tilde{\sigma}_\theta\,\Delta t\,, \tag{39}$$

$$\Delta D = \mathscr{G}_a\langle\Delta\varepsilon^p\rangle\,. \tag{40}$$

In Eqs. (39) and (40), subscripts a and θ denote values at time t_a and $t_\theta=t_a+\theta\Delta t$, respectively. Values at t_θ are expanded in Taylor-series at t_a, regarding only terms up to the first order of θ, eg $\tilde{\sigma}_\theta = \tilde{\sigma}_a + \theta\Delta\tilde{\sigma}$. This renders the right hand side of Eq. (35) linear in the increments $\Delta\tilde{\sigma}$, ΔD and $\Delta k = \mathscr{K}|\Delta\varepsilon^p|$. Introducing Hooke´s law $\Delta\tilde{\sigma} = E\,(\Delta\varepsilon-\Delta\varepsilon^p)$, we finally obtain a relation of the form $\Delta\varepsilon^p = \Delta\varepsilon^p(\Delta\varepsilon)$ (see Fotiu [20] for details) and the equivalent eigenstrain $\Delta\bar{\varepsilon}$ is given by

$$\Delta\bar{\varepsilon} = (1-D_a-\Delta D)\,\Delta\varepsilon^p + (D_a+\Delta D)\,\Delta\varepsilon + (\varepsilon_a - \varepsilon_a^p)\,\Delta D\,. \tag{41}$$

Thereby, Eq. (41) is obtained from the equivalence $\Delta\left[(1-D)\,(\varepsilon-\varepsilon^p)\right] = \Delta\varepsilon - \Delta\bar{\varepsilon}$.

The use of a strain controlled algorithm $\Delta\varepsilon^p = \Delta\varepsilon^p(\Delta\varepsilon)$ is essential for numerical stability in elastoplastic or elasto-viscoplastic problems, particularly when damage effects are included.

Optimal stability conditions are achieved for $\theta=1$. That value will be used subsequently for all numerical calculations.

Examples

We consider the problem of an impulsively loaded clamped-clamped beam made of mild steel, that has been treated by Symonds and Jones [27]. They have assumed a rate dependent rigid ideal-plastic material behavior and then calculated the permanent deflections on the basis of a plastic hinge theory. They used a material law, that has been proposed by Bodner and Symonds [28],

$$\dot{\varepsilon}^p = \beta\left(\bar{\sigma}-1\right)^s , \quad \bar{\sigma}=\sigma/k_0 , \tag{42}$$

where k_0 is the constant yield stress and the material constants are assigned the following values: $\beta=40s^{-1}$, $s=5$. Assuming $\Phi(\cdot)=(\cdot)^m$ and with $D=\mathscr{K}=0$, Eq. (35) becomes in the uniaxial case

$$\dot{\varepsilon}^p = 2\Gamma\left(\bar{\sigma}^2-1\right)^m \bar{\sigma} . \tag{43}$$

For a sufficiently wide range of $\dot{\varepsilon}$, Eq. (42) can be closely approximated by Eq. (43) taking the parameters $\Gamma=0.2$, $m=3.3$ (see Fig. 1).

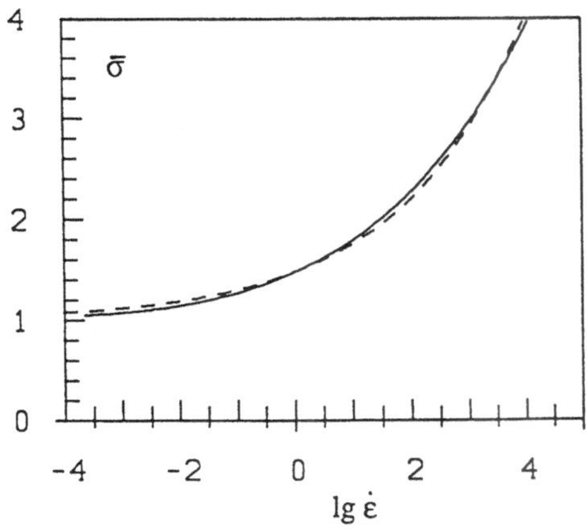

Fig.1. Strain rate dependence of yield stress. (——), Eq. (43), $\Gamma=0.2$, $m=3.3$; (- - - -) Eq. (42), $\beta=40$, $s=5$. $\bar{\sigma}=\sigma/k_0$.

The beam with a rectangular crossection A is loaded by an uniformly distributed impulse $I=\mu V_0$, where V_0 is the equivalent initial velocity. Dimensions of the common test beam and material parameters of mild steel are taken from [27]: $h=0.11$ in, $A=0.0413$ in^2, $l=5$ in, $\mu=3.024\cdot10^{-8}$ ksi$\cdot$s^2, $E=3\cdot10^4$ ksi, $k_0=30$ ksi. Symonds and Jones [27] deduced the following relation for the permanent deflection at midspan $\overline{w}=w(x=l/2, t=\infty)$

$$\frac{\overline{w}}{h}=\frac{1}{2}\left\{\sqrt{\frac{\Lambda}{1+\left(\frac{c}{\beta}\Lambda^{3/2}\right)^{1/s}}}-1\right\},\ c=\frac{h^2}{4l^3}\sqrt{\frac{k_o A}{\mu}},\ \Lambda=\frac{\mu}{k_o A}\frac{l^2}{h^2}V_0^2. \tag{44}$$

This curve is shown in Fig. 2 together with experimental results due to loading by a sheet explosive and numerical calculations with the present viscoplastic model for three different initial velocities V_0=1340 in/s, 2080 in/s, 2308 in/s. It can be seen in Fig. 2 that Eq. (44)

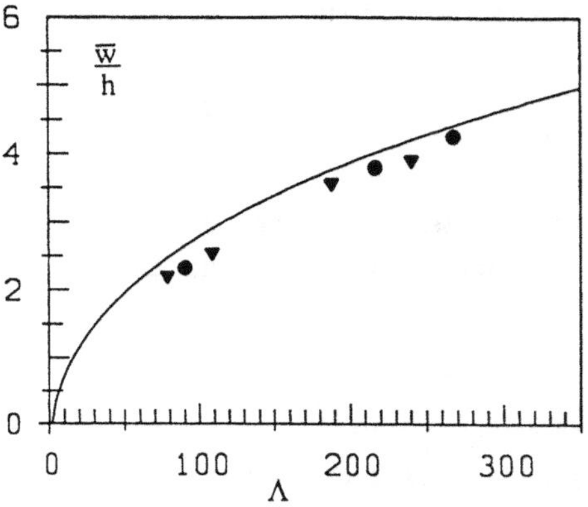

Fig.2: Permanent midspan deflection of a clamped-clamped beam.(——), Eq. (44); ▼ Experiments (Symonds and Jones [27]); ● Numerical results.

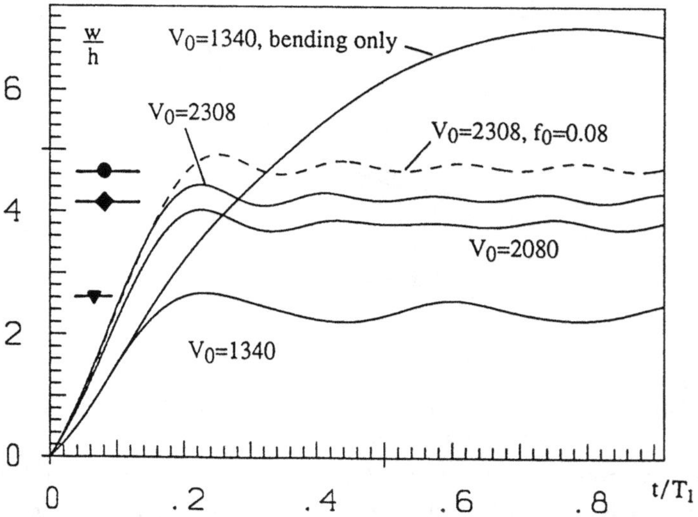

Fig.3: Temporal variation of midspan deflection for different initial velocities V_0=1340 in/s, 2080 in/s, 2308 in/s. Solid curves: f_0=0; Dashed curve: f_0=0.08. Permanent deflections computed by Symonds and Jones [27]: ▼ V_0=1340 in/s; ● V_0=2080 in/s; ◆ V_0=2308 in/s. Fundamental period T_1 = 1092 μs.

slightly overestimates the midspan deflection, whereas the numerical results agree well with experiments. Symonds and Jones [27] performed also numerical calculations by a finite difference method using these three initial velocities. Their results for the permanent deflections given in Fig. 3 are also slightly higher when compared with results of the present calculations.

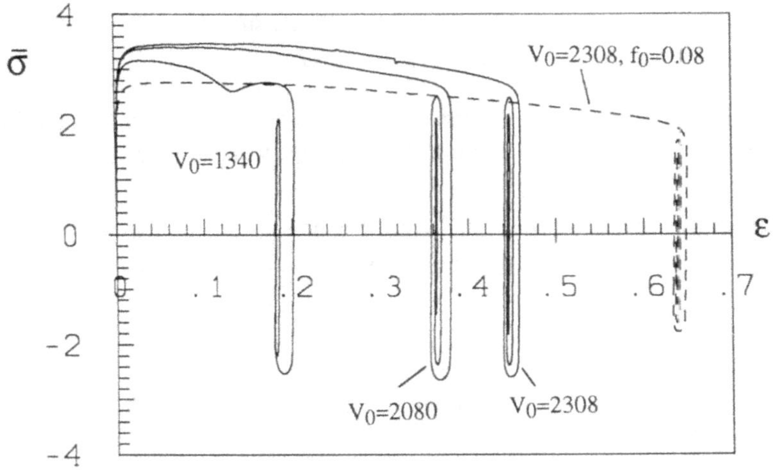

Fig.4. Temporal variation of maximum tensile stresses at the clamped edge for different initial velocities V_0=1340 in/s, 2080 in/s, 2308 in/s. Solid curves: f_0=0; Dashed curve: f_0=0.08. $\bar{\sigma}=\sigma/k_0$.

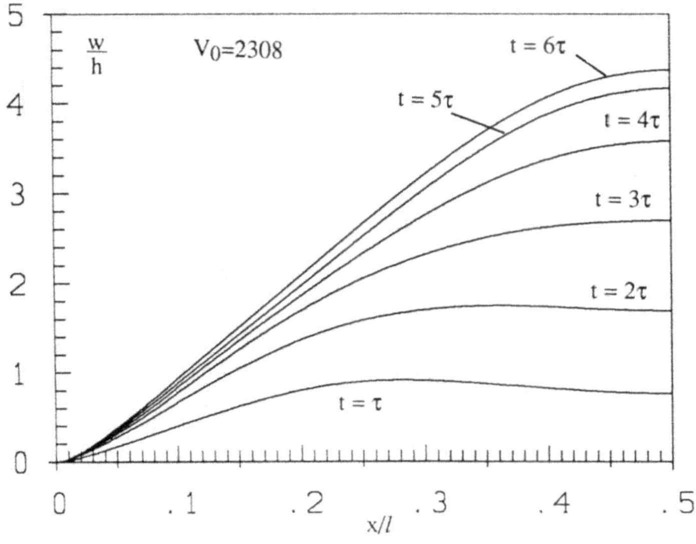

Fig.5. Development of deflection in early stages of deformation. Initial velocity V_0=2308 in/s. τ = 40 μs.

It can also be seen from Fig. 3 that neglecting the membrane stresses yields unrealistically large deformations about 200% above the measured values. Maximum tensile stresses at the clamped edge are depicted in Fig. 4. In Fig. 5 the development of the beam's deflection during the first 240 microseconds is presented. In the early phase of motion negative curvatures arise in the central section of the beam and the location of maximum positive curvature moves from the clamped edge towards the centre. This supports the ad hoc assumptions of a travelling plastic hinge in the rigid plastic analysis of Symonds and Jones [27].

Calculations have also been performed for V_0=2308 in/s and an initial porosity f_0=0.08 with material constants αk_0=0.2846, n=3 and ϑ=2. In comparison to the response of the fully dense material we observe a decrease in the yield limit and a considerable increase of permanent plastic strain at the clamped edge (Fig. 4). The effect of porosity on the dynamic deflection can be seen in Fig. 3. Finally, Fig. 6 shows the evolution of porosity at the uppermost fiber at the

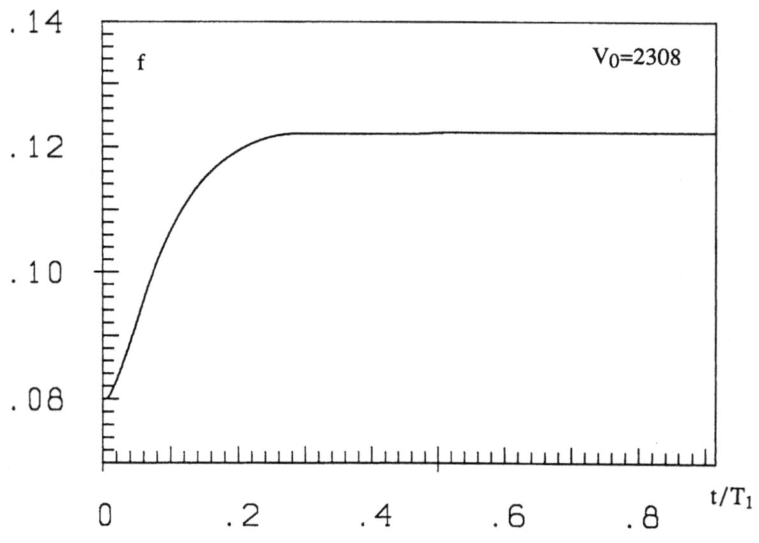

Fig.6. Temporal variation of porosity in outer tensile fiber at the clamped edge. Initial velocity V_0=2308 in/s. Fundamental period T_1 = 1092 μs.

clamped edge where the maximum tensile stresses appear.

Conclusions

It has been demonstrated that the plastic source method of distorsions can be extended to include nonlinear geometric effects as well. In the case of a laterally loaded beam additional internal loadings must be considered. In a time-stepping manner the intensities and distributions of the distorsions due to viscoplasticity, void-growth and large deflections are to be determined and updated. The powerful methods of linear elastodynamics account for all types of loadings most efficiently and computer-oriented.

Acknowledgement: P.A. Fotiu acknowledges a post-doctoral Schrödinger-Fellowship 1991/92 through the Austrian National Science Foundation FWF.

References

1. Mura, T. *Micromechanics of Defects in Solids*. 2nd ed. Dordrecht: Martinus Nijhoff 1987.

2. Woinowsky-Krieger, S.: The effect of axial force on the vibration of hinged bars. *J. Appl. Mech.* **17** (1950) 35–36.

3. Elishakoff, I.; Birman, V.; Singer, J.: Influence of initial imperfections on nonlinear free vibration of elastic bars. *Acta Mech.* **55** (1985) 191–202.

4. Mettler, E. : Zum Problem der Stabilität erzwungener Schwingungen elastischer Körper. *Z. Angew. Math. Mech.* **31** (1951) 263–264.

5. Eisley, J.G.: Nonlinear vibration of beams and rectangular plates. *Z. Angew. Math. Phys.* **15** (1964) 167–174.

6. Ziegler, F.; Rammerstorfer F.G.: Thermoelastic Stability. *Thermal Stresses III*, Ch. 2 (Ed. R.B. Hetnarski). Amsterdam: North-Holland 1989, 108-189.

7. Hayashi, C.: *Nonlinear Oscillations in Physical Systems*. New York McGraw-Hill 1964.

8. Nayfeh, A.H.; Mook, D.T.: *Nonlinear Oscillations*. New York: Wiley-Interscience 1979.

9. Guckenheimer, J.; Holmes, P.J.: *Nonlinear Oscillations, Dynamical Systems and Bifurcations of Vector Fields*. New York: Springer-Verlag 1983 .

10. Holmes, P.J.; Rand, D.A.: The bifurcations of Duffing's equation : an application of catastrophe theory. *J. Sound Vib.* **44** (1976) 237–253.

11. Heuer, R.; Irschik, H.; Ziegler, F.: Large Amplitude Random Vibrations of Polygonal Plates. Proc. IUTAM Symposium on *Nonlinear Stochastic Mechanics* (Eds. N. Bellomo and F. Casciati), Berlin: Springer-Verlag, (in press).

12. Symonds, P.S.; Yu, T.X.: Counterintuitive behavior in a problem of elastic-plastic beam dynamics. *J. Appl. Mech.* **52** (1985) 517–522.

13. Poddar, B.; Moon, F.C.; Mukherjee, S.: Chaotic motion of an elastic-plastic beam. *J. Appl. Mech.* **55** (1988) 185–189.

14. Fotiu, P.; Irschik, H; Ziegler, F.: Dynamic plasticity: Structural drift and modal projections. In Proc. IUTAM-Symposium on *Nonlinear Dynamics in Engineering Systems* (Ed. W. Schiehlen), (1990) 75–82, Berlin: Springer-Verlag.

15. Irschik, H.; Ziegler, F.: Thermal shock loading of elastoplastic beams. *J. Thermal Stresses* **8** (1985) 53–69.

16. Irschik, H.: Biaxial dynamic bending of elastoplastic beams. *Acta Mech.* **62** (1986) 155–167.

17. Fotiu, P.; Irschik, H.: Modal analysis of vibrating viscoplastic composite beams on multiple supports. *Earthquake Eng. Struct. Dynamics* **18** (1989) 1053-1064.

18. Irschik, H.; Ziegler, F. (1988): Dynamics of linear elastic structures with selfstress: A unified treatment for linear and nonlinear problems. *Z. angew. Math. Mech. (ZAMM)* **68** (1988) 199–205.

19. Mukherjee, S.; Chandra, A.: Boundary element formulations for large strain–large deformation problems of plasticity and viscoplasticity. In *Developments in Boundary Element Methods-3* (Eds. P.K. Banerjee and S. Mukherjee). London: Elsevier Appl. Science Publ. (1984) 27–58.

20. Fotiu, P.A.: Die dynamische Wirkung dissipativer Prozesse in Materialien mit Mikrodefekten mit Anwendungen auf Schwingungsprobleme inelastischer Plattentragwerke. Doctoral thesis, Technical University of Vienna 1990.

21. Spitzig, W.A.; Smelser, R.E; Richmond, O.: The evolution of damage and fracture in iron compacts with various initial porosities. *Acta Metall.* **36** (1988) 1201–1211.

22. Fotiu, P.; Irschik, H.; Ziegler, F.: Micromechanical foundations of dynamic plasticity with applications to damaging structures. In: Advances in Continuum Mechanics (Eds. O. Brueller, V. Manni, J. Najer), 338-349, Springer-Verlag, Berlin, 1991.

23. Davison, L.; Stevens, A.L; Kipp, M.E.: Theory of spall damage accumulation in ductile metals. *J. Mech. Phys. Solids* **25** (1977) 11–28.

24. Barbee, T.W.,Jr.; Seaman, L.; Crewdson, R.; Curran, D.R.: Dynamic fracture criteria for ductile and brittle materials. *J. Materials* **7** (1972) 393–401.

25. Seaman, L.; Curran, D.R.; Shockey, D.A.: Computational models for ductile and brittle fracture. *J. Appl. Phys.* **47** (1976) 4814–4826.

26. Rice, J.R.; Tracey, D.M.: On the ductile enlargement of voids in triaxial stress fields. *J. Mech. Phys. Solids* **17** (1969) 201 – 217.

27. Symonds, P.S.; Jones, N.: Impulsive loading of fully clamped beams with finite plastic deflections and strain-rate sensitivity. *Int. J. Mech. Sci.* **14** (1972) 49–69.

28. Bodner, S.R.; Symonds, P.S.: Experimental and theoretical investigation of the plastic deformation of cantilever beams subjected to impulsive loading. *J. Appl. Mech.* **29** (1962) 719–728.

A Methodology for Formulating Large Strain Viscoplastic Constitutive Equations with Application to Simple Shear

H.-P. HACKENBERG and F.G. KOLLMANN

Technische Hochschule Darmstadt
Fachgebiet Maschinenelemente und Maschinenakustik

SUMMARY

In this paper, a methodology for the formulation of large strain constitutive equations for rate dependent solids is developed. An attempt is made to obtain a consistent frame which allows the use of different unified constitutive models and which can also be used to extend models formulated for small strains to finite deformations. For a particular inelastic constitutive model, different elastic constitutive equations are compared for the case of simple shear. In addition, the simplification of the inelastic constitutive equations based on the assumption of small elastic strains is investigated.

INTRODUCTION

The formulation of constitutive equations for large viscoplastic strains is still discussed controversially in the research literature. For small strains viscoplastic models based on the concept of internal state variables [2,6] are widely accepted and have reached a stage of significant maturity. In this paper, therefore, the concept of internal state variables shall be adopted for large strain viscoplastic deformations.

There exist relatively few viscoplastic constitutive models which have been formulated *a priori* for large strains (e.g. [1]). In this paper we base our work on the multiplicative decomposition of the deformation gradient $\mathcal{F}$ [10]. It is well known, that there exists an intensive debate on the uniqueness of this decomposition. Many authors (e.g. Haupt [7]) argue that the $\hat{\mathcal{F}}_e\mathcal{F}_p$-decompostion is not unique, since an arbitrary orthogonal tensor $\hat{\mathbf{Q}}$ can be placed between the elastic part $\hat{\mathcal{F}}_e$ and the inelastic one $\mathcal{F}_p$. Based on physical arguments given by Mandel [11], Cleja–Ţigoiu and Sóos [4] introduce the *constitutive assumption* that the multiplicative decomposition of the deformation gradient is unique. This approach requires that not only an evolution equation for the plastic part of the deformation rate tensor but also the evolution of the plastic spin tensor is prescribed in the frame of the constitutive model. We follow the lines of Mandel and Cleja–Ţigoiu and Sóos despite the fact, that a fully satisfactory physical motivation for the concept of plastic spin for polycristalline materials is still lacking

D. Besdo, E. Stein (Eds.)
Finite Inelastic Deformations –
Theory and Applications
IUTAM Symposium Hannover/Germany 1991

KINEMATICS OF INELASTICALLY DEFORMED SOLIDS

In this paper we adopt tensor calculus on manifolds as introduced into continuum mechanics by Marsden and Hughes [12]. We restrict all considerations to simple bodies at homogeneous and constant temperature. Let the reference configuration $\mathcal{B} \subset \mathbb{R}^3$ be a representation of the material body. We assume that $\mathcal{B}$ is a differentiable n-manifold and that it is mapped by a family of at least C^1 diffeomorphisms on a family $\mathcal{S}_t \subset \mathbb{R}^3$ of differentiable threedimensional manifolds. Here the time $t \in \mathbb{R}$ is a parameter and $\mathcal{S}_t$ is the family of current configurations. Therefore, we have the family of maps $\phi(t) : \mathcal{B} \longrightarrow \mathcal{S}_t$ as motion of the body. Since no ambiguity can arise, we will not indicate the dependence on time t in the sequel.

As mentioned in the introduction we assume the multiplicative decomposition of the total deformation gradient $\mathcal{F}$

$$\mathcal{F} = \hat{\mathcal{F}}_e \mathcal{F}_p \,. \tag{1}$$

This decomposition can be motivated by the introduction of an intermediate configuration $\hat{\mathcal{B}}$ which is defined by relaxing each material element from the current configuration $\mathcal{S}$ to a stress free state. In general, the intermediate configuration is incompatible and therefore can not be reached physically.

The kinematics of inelastically deformed bodies has been studied in detail by Kollmann and Hackenberg [8]. In the following we use the notions of push-forward- (ϕ_*) and pull-back-operators (ϕ^*) as in [12]. If these operators are formed with the elastic or plastic part of the deformation gradient we use notations as e.g. ϕ_*^e and ϕ_e^*. We denote Riemannian metrics on $\mathcal{B}$, $\hat{\mathcal{B}}$ and $\mathcal{S}$ by $\mathbf{G}$, $\hat{\mathbf{G}}$ and $\mathbf{g}$, respectively. Their inverses are indicated by $\sharp$ as e.g. $\mathbf{g}^\sharp$.

Pushing forward the metric $\hat{\mathbf{G}}$ of the intermediate to the current configuration leads to the following deformation tensor

$$\tilde{\mathbf{c}}^\flat := \phi_*^e(\hat{\mathbf{G}}) \,. \tag{2}$$

Its inverse is a Finger type deformation tensor

$$\tilde{\mathbf{b}}^\sharp := (\tilde{\mathbf{c}}^\flat)^{-1} = \phi_*^e(\hat{\mathbf{G}}^\sharp) = \hat{\mathcal{F}}_e \hat{\mathbf{G}}^\sharp \hat{\mathcal{F}}_e^T \,. \tag{3}$$

We introduce the spatial velocity gradient

$$\mathbf{l} := \operatorname{grad} \boldsymbol{v} = \dot{\mathcal{F}} \mathcal{F}^{-1} \,. \tag{4}$$

Next, we define the covariant spatial deformation rate tensor

$$\mathbf{d}^\flat := {}^s(\mathbf{g}\mathbf{l}) \tag{5}$$

where ${}^s(\cdot)$ denotes the symmetric part of the term in brackets. Correspondingly ${}^a(\cdot)$ denotes the skewsymmetric part. Inserting (4) into (5) leads under consideration of (1) to the additive decomposition

$$\mathbf{d}^\flat = \mathbf{d}_e^\flat + \mathbf{d}_p^\flat \tag{6}$$

$$\mathbf{d}_e^\flat := {}^s(\mathbf{g}\, \dot{\hat{\mathcal{F}}}_e\, \hat{\mathcal{F}}_e^{-1}) \tag{7}$$

$$\mathbf{d}_p^\flat := {}^s(\mathbf{g}\, \hat{\mathcal{F}}_e\, \dot{\mathcal{F}}_p\, \mathcal{F}_p^{-1} \hat{\mathcal{F}}_e^{-1}) \,. \tag{8}$$

However, the decomposition (6) - (8) is not unique. A second decomposition can be derived from the fact, that the strain rate tensor $\mathbf{d}^\flat$ can also be obtained as Lie derivative [12] of Almansi's strain tensor

$$\mathbf{e}^\flat = \frac{1}{2}(\mathbf{g} - \phi_*(\mathbf{G})) = \frac{1}{2}(\mathbf{g} - \mathcal{F}^{-T}\,\mathbf{G}\,\mathcal{F}^{-1}) \tag{9}$$

$$\mathbf{d}^\flat = \mathcal{L}_{\boldsymbol{v}}(\mathbf{e}^\flat)\,, \tag{10}$$

where the operator $\mathcal{L}_{\boldsymbol{v}}(\cdot)$ indicates the Lie derivative formed with respect to the reference configuration. For Almansi's strain tensor the following decomposition holds [8]

$$\mathbf{e}^\flat = \mathbf{e}_e^\flat + \mathbf{e}_p^\flat \tag{11}$$

$$\mathbf{e}_e^\flat := \frac{1}{2}(\mathbf{g} - \tilde{\mathbf{c}}^\flat) \tag{12}$$

$$\mathbf{e}_p^\flat := \frac{1}{2}(\tilde{\mathbf{c}}^\flat - \phi_*(\mathbf{G}))\,. \tag{13}$$

From (11) and (10) we obtain a second decomposition of the strain rate tensor $\mathbf{d}^\flat$

$$\mathbf{d}^\flat = \mathbf{d}_e^{*\flat} + \mathbf{d}_p^{*\flat} \tag{14}$$

$$\mathbf{d}_e^{*\flat} := \mathcal{L}_{\boldsymbol{v}}(\mathbf{e}_e^\flat) \tag{15}$$

$$\mathbf{d}_p^{*\flat} := \mathcal{L}_{\boldsymbol{v}}(\mathbf{e}_p^\flat)\,. \tag{16}$$

Next, we establish a relation between the two plastic parts $\mathbf{d}_p^\flat$ and $\mathbf{d}_p^{*\flat}$, respectively, of the rate of deformation tensor $\mathbf{d}^\flat$. Eq.(8) motivates the definiton of a plastic tensor

$$\hat{\mathbf{L}}_p := \dot{\mathcal{F}}_p\;\mathcal{F}_p^{-1}\,. \tag{17}$$

The tensor $\hat{\mathbf{L}}_p^\flat = \hat{\mathbf{G}}\,\hat{\mathbf{L}}_p$ can be decomposed into a symmetric and a skewsymmetric part

$$\hat{\mathbf{L}}_p^\flat = \hat{\mathbf{D}}_p^\flat + \hat{\mathbf{W}}_p^\flat \tag{18}$$

$$\hat{\mathbf{D}}_p^\flat := {}^s(\hat{\mathbf{G}}\;\dot{\mathcal{F}}_p\;\mathcal{F}_p^{-1}) \tag{19}$$

$$\hat{\mathbf{W}}_p^\flat := {}^a(\hat{\mathbf{G}}\;\dot{\mathcal{F}}_p\;\mathcal{F}_p^{-1})\,. \tag{20}$$

It can be shown that [8]

$$\mathbf{d}_p^{*\flat} = \phi_*^e(\hat{\mathbf{D}}_p^\flat)\,. \tag{21}$$

Therefore, we define

$$\mathbf{w}_p^{*\flat} := \phi_*^e(\hat{\mathbf{W}}_p^\flat)\,. \tag{22}$$

Then the required relation follows as [5]

$$\mathbf{d}_p^\flat = {}^s(\mathbf{g}\,\tilde{\mathbf{b}}^\sharp\,(\mathbf{d}_p^{*\flat} + \mathbf{w}_p^{*\flat}))\,. \tag{23}$$

We denote the tensors $\hat{\mathbf{W}}_p^\flat$ and $\mathbf{w}_p^{*\flat}$, as plastic spin tensors on the intermediate and current configuration, respectively.

CONSTITUTIVE EQUATIONS

In this section we formulate constitutive equations for the elastic and the plastic part of the deformation.

Elastic constitutive models

We start with a *hyperelastic* model as originally proposed by Simo, Taylor and Pister [15]. We split the total deformation gradient into a volumetric and an isochoric part

$$\mathcal{F} = J^{\frac{1}{3}} \widehat{\mathcal{F}} , \tag{24}$$

where J is the determinant of $\mathcal{F}$. It can be shown, that the tensor $\tilde{\mathbf{b}}^{\sharp}$ can be split as

$$\tilde{\mathbf{b}}^{\sharp} = J^{\frac{2}{3}} \ \widehat{\tilde{\mathbf{b}}}^{\sharp} . \tag{25}$$

The hyperelastic model takes the form

$$\boldsymbol{\tau}^{\sharp} = K \ln J \, \mathbf{g}^{\sharp} + \mu \operatorname{dev} \widehat{\tilde{\mathbf{b}}}^{\sharp} , \tag{26}$$

where $\boldsymbol{\tau}^{\sharp}$ is Kirchhoff's stress tensor, K the bulk modulus and μ the shear modulus. The deviator of an arbitrary contravariant tensor $\mathbf{t}^{\sharp}$ is defined as

$$\operatorname{dev} \mathbf{t}^{\sharp} := \mathbf{t}^{\sharp} - \frac{1}{3} (\mathbf{t}^{\sharp} : \mathbf{g}) \, \mathbf{g}^{\sharp} . \tag{27}$$

A special case of a hyperelastic constitutive model, which is valid for *small elastic strains* is the St. Venant-Kirchoff model. The notion of small elastic strains will be made precise later. We formulate it on the intermediate configuration. Pulling back the Kirchhoff stress tensor to the intermediate configuration we define

$$\hat{\mathbf{T}}^{\sharp} := \phi_e^*(\boldsymbol{\tau}^{\sharp}) . \tag{28}$$

Then the St. Venant-Kirchhoff model is given as

$$\hat{\mathbf{T}}^{\sharp} = 2\mu \, \hat{\mathbf{G}}^{\sharp} \, \hat{\mathbf{E}}_e^{\flat} \, \hat{\mathbf{G}}^{\sharp} + \lambda (\hat{\mathbf{E}}_e^{\flat} : \hat{\mathbf{G}}^{\sharp}) \, \hat{\mathbf{G}}^{\sharp} , \tag{29}$$

where $\hat{\mathbf{E}}_e^{\flat}$ is the elastic part of Green's strain tensor on the intermediate configuration [8]. Eq. (29) can be transformed to the current configuration

$$\boldsymbol{\tau}^{\sharp} = 2\mu \, \tilde{\mathbf{b}}^{\sharp} \, \mathbf{e}_e^{\flat} \, \tilde{\mathbf{b}}^{\sharp} + \lambda (\mathbf{e}_e^{\flat} : \tilde{\mathbf{b}}^{\sharp}) \, \tilde{\mathbf{b}}^{\sharp} . \tag{30}$$

Inelastic constitutive models

Our basic approach is to formmulate the inelastic constitutive equations on the intermediate configuration and then to transform it to the current configuration. As additional constitutive assumption we presuppose inelastic incompressibility

$$\det \mathcal{F}_p = 1 . \tag{31}$$

From (31) it can be concluded [8] that

$$\hat{\mathbf{D}}_p^{\flat} : \hat{\mathbf{G}}^{\sharp} = 0 \, . \tag{32}$$

Pushing forward (32) yields on the current configuration

$$\mathbf{d}_p^{*\flat} : \tilde{\mathbf{b}}^{\sharp} = 0 \, , \tag{33}$$

where (3) has been observed. Eq. (33) indicates, that the tensor $\mathbf{d}_p^{*\flat}$ is not a deviatior.

Next, we formulate constitutive equations for the plastic parts of the deformation rate and the spin tensor, respectively. Consider as an example the plastic part of the deformation rate tensor. In the isothermal case the tensor $\hat{\mathbf{D}}_p^{\flat}$ depends on the list $\{\hat{\mathbf{T}}^{\sharp}, \hat{\mathbf{G}}, \hat{\mathbf{Z}}_i\}$, where the quantities $\hat{\mathbf{Z}}_i$, $i = 1, 2, ... n$ are a set of suitably defined internal variables. We stipulate the following constitutive equation

$$\hat{\mathbf{D}}_p^{\flat} = \hat{\mathbf{D}}_p^{\flat}(\operatorname{dev} \hat{\mathbf{T}}^{\sharp}, \hat{\mathbf{G}}, \hat{\mathbf{Z}}_i) \, . \tag{34}$$

Onat [13] has shown that the set of internal variables $\hat{\mathbf{Z}}_i$ can be restricted to scalars and tensors of even rank. Since the tensor $\hat{\mathbf{D}}_p^{\flat}$ is deviatoric, it is reasonable to assume that it depends on the deviator

$$\operatorname{dev} \hat{\mathbf{T}}^{\sharp} = \hat{\mathbf{T}}^{\sharp} - \frac{1}{3}(\hat{\mathbf{T}}^{\sharp} : \hat{\mathbf{G}})\, \hat{\mathbf{G}}^{\sharp} \tag{35}$$

of the stress $\hat{\mathbf{T}}^{\sharp}$. Considering (2) and (21) we obtain from (34)

$$\mathbf{d}_p^{*\flat} = \mathbf{d}_p^{*\flat}(\operatorname{dev}^* \boldsymbol{\tau}^{\sharp}, \tilde{\mathbf{c}}^{\flat}, \mathbf{z}_i) \, . \tag{36}$$

with

$$\operatorname{dev}^* \boldsymbol{\tau}^{\sharp} := \phi_*^e(\operatorname{dev} \hat{\mathbf{T}}^{\sharp}) = \boldsymbol{\tau}^{\sharp} - \frac{1}{3}(\boldsymbol{\tau}^{\sharp} : \tilde{\mathbf{c}}^{\flat})\, \tilde{\mathbf{b}}^{\sharp} \tag{37}$$

$$\mathbf{z}_i := \phi_*^e(\hat{\mathbf{Z}}_i) \, . \tag{38}$$

A completely corresponding constitutive equation can be obtained for the plastic spin tensor $\mathbf{w}_p^{*\flat}$ which is omitted for the sake of brevity.

It has to be emphasized that the plastic constitutive equation (36) in combination with the elastic constitutive equations (26) or (30) is not sufficient but an evolution equation for the tensor $\tilde{\mathbf{b}}^{\sharp}$ is required. It can be shown [5] that

$$\mathcal{L}_v(\tilde{\mathbf{b}}^{\sharp}) = -2\, \tilde{\mathbf{b}}^{\sharp}\, \mathbf{d}_p^{*\flat}\, \tilde{\mathbf{b}}^{\sharp} \, , \, . \tag{39}$$

Considering $\tilde{\mathbf{b}}^{\sharp} = (\tilde{\mathbf{c}}^{\flat})^{-1}$ we obtain from (36)

$$\mathcal{L}_v(\tilde{\mathbf{b}}^{\sharp}) = \mathbf{f}\,(\operatorname{dev}^* \boldsymbol{\tau}^{\sharp}, \tilde{\mathbf{b}}^{\sharp}, \mathbf{z}_i) \, , \tag{40}$$

where $\mathbf{f}$ is a tensorvalued function of its arguments.

Since we will use the plastic constitutive equation (40) we have to rewrite the condition of plastic incompressibility [5]

$$\mathcal{L}_v(\tilde{\mathbf{b}}^{\sharp}) : \tilde{\mathbf{c}}^{\flat} = 0 \, . \tag{41}$$

Finally, we have to formulate evolution equations for the internal variables $\mathbf{z}_i$. In an exemplaric manner we confine our considerations to the case, that the internal variables are a scalar $\hat{S}$ and a second order tensor $\hat{\mathbf{Z}}^\sharp$ on the intermediate configuration. The scalar $\hat{S}$ models isotropic hardening. The second order tensor $\hat{\mathbf{Z}}^\sharp$ can be interpreted as a back stress to describe kinematic hardening. The evolution equation for the scalar variable is

$$\dot{\hat{S}} = \dot{\hat{S}}\,(\operatorname{dev}\hat{\mathbf{T}}^\sharp, \hat{\mathbf{G}}, \hat{S}, \hat{\mathbf{Z}}^\sharp)\,. \tag{42}$$

Pushing forward (42) to the current configuration with $s = \phi_*^e(\hat{S}) = \hat{S}$ and $\mathbf{z}^\sharp = \phi_*^e(\hat{\mathbf{Z}}^\sharp)$ yields

$$\dot{s} = \dot{s}\,(\operatorname{dev}^*\boldsymbol{\tau}^\sharp, \tilde{\mathbf{c}}^\flat, s, \mathbf{z}^\sharp)\,. \tag{43}$$

The evolution equation for tensorial variable is

$$\dot{\hat{\mathbf{Z}}}^\sharp = \dot{\hat{\mathbf{Z}}}^\sharp\,(\operatorname{dev}\hat{\mathbf{T}}^\sharp, \hat{\mathbf{G}}, s, \hat{\mathbf{Z}}^\sharp)\,. \tag{44}$$

By push-forward with the elastic part of the deformation gradient we obtain

$$\mathcal{L}_{\mathbf{v}}^e(\mathbf{z}^\sharp) \;=\; \mathbf{h}\,(\operatorname{dev}^*\boldsymbol{\tau}^\sharp, \tilde{\mathbf{c}}^\flat, s, \mathbf{z}^\sharp)\,, \tag{45}$$

where the elastic Lie derivative $\mathcal{L}_{\mathbf{v}}^e(\cdot)$ has been introduced in [8]. In (45) $\mathbf{h}$ again is a tensor-valued function of its arguments.

Next we specialize our inelastic constitutive model for a simple isotropic model as originally derived by Anand [3] on the current configuration. With persisting isotropy due to an objectivity argument [9]

$$\mathbf{w}_p^{*\flat} \;=\; \mathbf{0} \tag{46}$$

has to be prescribed as constitutive equation for the plastic spin tensor. The flow rule on the intermediate configuration can only depend on the scalar $\hat{S}$ and it is assumed as

$$\hat{\mathbf{D}}_p^\flat \;= \dot{\varphi}\,(\hat{I\!I}, \hat{S})\,\hat{\mathbf{N}}^\flat\,, \tag{47}$$

where $\dot{\varphi}$ is a scalar function of its arguments, which has been specified by Anand. The quantity $\hat{I\!I}$ is the root of the second invariant of the deviator of $\hat{\mathbf{T}}^\sharp$ on $\hat{\mathcal{B}}$

$$\hat{I\!I} \;=\; \sqrt{\frac{3}{2}(\operatorname{dev}\hat{\mathbf{T}}^\sharp) : (\hat{\mathbf{G}}\operatorname{dev}\hat{\mathbf{T}}^\sharp\,\hat{\mathbf{G}})} \tag{48}$$

and

$$\hat{\mathbf{N}}^\flat \;=\; \frac{3}{2}\hat{\mathbf{G}}\,\frac{\operatorname{dev}\hat{\mathbf{T}}^\sharp}{\hat{I\!I}}\,\hat{\mathbf{G}}\,. \tag{49}$$

Pushing forward (47)-(49) we obtain

$$\mathbf{d}_p^{*\flat} \;=\; \dot{\varphi}\,(\tilde{I\!I}, s)\,\tilde{\mathbf{n}}^\flat \tag{50}$$

$$\tilde{I\!I} \;=\; \sqrt{\frac{3}{2}(\operatorname{dev}^*\boldsymbol{\tau}^\sharp) : (\tilde{\mathbf{c}}^\flat\operatorname{dev}^*\boldsymbol{\tau}^\sharp\,\tilde{\mathbf{c}}^\flat)} \tag{51}$$

$$\tilde{\mathbf{n}}^\flat \;=\; \frac{3}{2}\tilde{\mathbf{c}}^\flat\,\frac{\operatorname{dev}^*\boldsymbol{\tau}^\sharp}{\tilde{I\!I}}\,\tilde{\mathbf{c}}^\flat\,. \tag{52}$$

However, since in the elastic constitutive equations (26) and (30), respectively, the tensor $\tilde{\mathbf{b}}^{\sharp}$ is involved as driving kinematic quantity for the stress $\boldsymbol{\tau}^{\sharp}$, eq. (50) is not adequate in the present context. Using (39) we can express the tensor $\mathbf{d}_p^{*\flat}$ by the tensor $\mathcal{L}_{\boldsymbol{v}}(\tilde{\mathbf{b}}^{\sharp})$ and obtain

$$\mathcal{L}_{\boldsymbol{v}}(\tilde{\mathbf{b}}^{\sharp}) = -2\,\dot{\varphi}\,(\tilde{I\!I},s)\,\mathbf{n}^{*\sharp} \tag{53}$$

$$\mathbf{n}^{*\sharp} = \frac{3}{2}\frac{\operatorname{dev}^{*}\boldsymbol{\tau}^{\sharp}}{\tilde{I\!I}}\,. \tag{54}$$

For the internal variable s we take from [3] the evolution equation

$$\dot{s} = h(\tilde{I\!I},s)\;\dot{\varphi}\,(\tilde{I\!I},s)\,. \tag{55}$$

The functions $\dot{\varphi}\,(\cdot)$ and $h(\cdot)$ are given by Anand.

Small elastic strains

For metals the concept of small elastic strains is well established from experiments. To make the notion of small elastic strain precise we start from the polar decomposition

$$\hat{\mathcal{F}}_e = \mathbf{v}_e\,\hat{\mathcal{R}}_e\,, \tag{56}$$

where the elastic left stretch tensor $\mathbf{v}_e$ lives on the current configuration and $\hat{\mathcal{R}}_e$ is an orthogonal two-point tensor spanned by the intermediate and current configurations, respectively. We decompose the tensor $\mathbf{v}_e$ additively into

$$\mathbf{v}_e = \mathbf{i} + \boldsymbol{\epsilon}\,, \tag{57}$$

with $\mathbf{i}$ being the identity tensor on the current configuration and $\boldsymbol{\epsilon}$ a symmetric tensor. The elastic strains are small if in a suitably defined norm

$$\sup_{\forall x\in\mathcal{S}} \|\boldsymbol{\epsilon}\| = \epsilon \ll 1\,. \tag{58}$$

We use symbols $O(\epsilon)$ and $\mathbf{O}(\epsilon)$, respectively, to indicate that terms vanish as ϵ if $\epsilon \longrightarrow 0$. The following order of magnitude relations can be established [5,8]

$$\tilde{\mathbf{b}}^{\sharp} = \mathbf{g}^{\sharp} + \mathbf{O}(\epsilon) \tag{59}$$

$$\tilde{\mathbf{c}}^{\flat} = \mathbf{g} + \mathbf{O}(\epsilon) \tag{60}$$

$$\operatorname{dev}^{*}\boldsymbol{\tau}^{\sharp} = \operatorname{dev}\boldsymbol{\tau}^{\sharp} + \mathbf{O}(\epsilon)\,. \tag{61}$$

We will confine the following considerations to the plastic incompressibility condition (41). Using (59) and (60)

$$\mathcal{L}_{\boldsymbol{v}}(\tilde{\mathbf{b}}^{\sharp}) : \mathbf{g} + O(\epsilon) = 0 \tag{62}$$

can be concluded. Some authors (e.g. [14]) assume (62), where the term $O(\epsilon)$ is neglected. Whether this approximation can be used depends on the specific form of the elastic constitutive model [5]. Therefore, we propose an approximation which yields reasonable results for all elastic models investigated. We decompose the Lie derivative into three parts

$$\mathcal{L}_v(\tilde{\mathbf{b}}^\sharp) = [\mathcal{L}_v(\tilde{\mathbf{b}}^\sharp)]_{\text{dev}} + \mathbf{O}_{\text{dev}}(\epsilon) + \mathbf{O}_{\text{vol}}(\epsilon) \tag{63}$$

Here $[\mathcal{L}_v(\tilde{\mathbf{b}}^\sharp)]_{\text{dev}}$ is the purely deviatoric approximation of the Lie derivative $\mathcal{L}_v(\tilde{\mathbf{b}}^\sharp)$. The quantities $\mathbf{O}_{\text{dev}}(\epsilon)$ and $\mathbf{O}_{\text{vol}}(\epsilon)$ represent additional deviatoric and volumetric contributions. Inserting (63) and (60) into (41) yields after some manipulations

$$\mathcal{L}_v(\tilde{\mathbf{b}}^\sharp) : \mathbf{g} = [\mathcal{L}_v(\tilde{\mathbf{b}}^\sharp)]_{\text{dev}} : (\mathbf{g}\,\tilde{\mathbf{b}}^\sharp\,\mathbf{g})\,, \tag{64}$$

where terms $O(\epsilon^2)$ have been neglected. This is a better approximation than eq. (62).

Without prove we note the following approximations [5]

$$\tilde{I\!I} = I\!I + O(\epsilon)\,; \qquad I\!I = \sqrt{\frac{3}{2}(\text{dev}\,\boldsymbol{\tau}^\sharp) : (\mathbf{g}\,\text{dev}\,\boldsymbol{\tau}^\sharp\,\mathbf{g})} \tag{65}$$

$$\mathbf{n}^{*\sharp} = \mathbf{n}^\sharp + \mathbf{O}(\epsilon)\,; \qquad \mathbf{n}^\sharp = \frac{3}{2}\frac{\text{dev}\,\boldsymbol{\tau}^\sharp}{I\!I}\,. \tag{66}$$

NUMERICAL EXPERIMENTS FOR SIMPLE SHEAR

In this section we give three different constitutive models. In the first two models the elastic constitutive equation (26) is used. Both models differ in the inelastic part. In model 1 the plastic flow rule (53) in combination with (54) and (51) is used. In model 2 we use the approximation

$$\mathcal{L}_v(\tilde{\mathbf{b}}^\sharp) = -2\,\dot{\varphi}\,(I\!I, s)\,\mathbf{n}^\sharp\,, \tag{67}$$

where $I\!I$ is defined in (65) and $\mathbf{n}^\sharp$ in (66). Model 3 applies the same plastic flow rule as model 1. However, the St. Venant-Kirchhoff model (30) is used as elastic constitutive equation.

All three constitutive models have been applied on simple shear. The material parameters for a steel Fe-2% Si at 800 °C have been taken from Anand [3]. For time integration Euler's forward formula has been used with very small time steps to maintain numerical stability and accuracy. All simulations have been performed until a maximum shear of $\kappa = 2$ with $\dot{\kappa} = 1\text{s}^{-1}$.

Fig. 1 presents the dependence of the normal stresses and of the shear stress on the shear κ for all three constitutive models. First we note, that for all three models the normal stresses are very small in comparison to the shear stress. While models 1 and 2 within the limits of numerical accuracy give the same stresses, model 3 leads to remarkably different normal stresses. The normal stress σ_{11} is at maximum shear κ by the factor 2.63 larger than for models 1 and 2. Furthermore, model 3 results in positive normal stresses σ_{22} and σ_{33}, while for models 1 and 2 these stresses are negative. It has been observed that these second order effects are purely elastic. This can be seen, if the formulations are repeated with inelastic deformations neglected.

Fig. 2 gives the evolution of the components of the deformation tensor $\tilde{\mathbf{b}}^\sharp$ with the shear κ. In the undeformed reference configuration this tensor is an unit tensor. Due to (59) this tensor remains close to the identity tensor for small elastic strains. It can be seen that $\tilde{b}_{12}$ approaches a value of 0.0015 for all three models. For models 1 and 3 the normal components of $\tilde{\mathbf{b}}^\sharp$ remain approximatly 1. However, with model 2 these components drift to 1.002. This is due to a deviation from plastic incompressibility.

Therefore, we have performed additional computations with model 1 and 2. In model 1a and 2a the elastic constitutive equation (26) has been replaced in model 1 and 2, respectively, by

$$\boldsymbol{\tau}^\sharp = \lambda \ln J \, \mathbf{g}^\sharp + \mu(\tilde{\mathbf{b}}^\sharp - \mathbf{g}^\sharp) . \tag{68}$$

Model 2b is identical with model 2, however, the approximation (64) for the plastic incompressibility constraint is included. Model 2c is identical with model 2a, where again the plastic incompressibility condition (64) is included.

Since (with the exception of model 2a) the graphs of stress and strain components are indistinguishable, we present the results for $\kappa = 2$ in tables 1 and 2. It can be seen from table 2 that models 2 and 2a lead to too large values of the normal strain components. Furthermore, it has to be observed, that the normal stresses in formulation 2a are completely unrealistic. Notice, that this dramatic change of the normal stresses is caused by a 'marginal' change of the elastic constitutive model. This is due to a violation of the condition of plastic incompressibility. Models 2b and 2c with the improved approximation of the plastic incompressibility condition yield satisfactory normal stresses. Note, that formulation 2c corresponds with formulation 1a, while model 2b gives the same results as model 1.

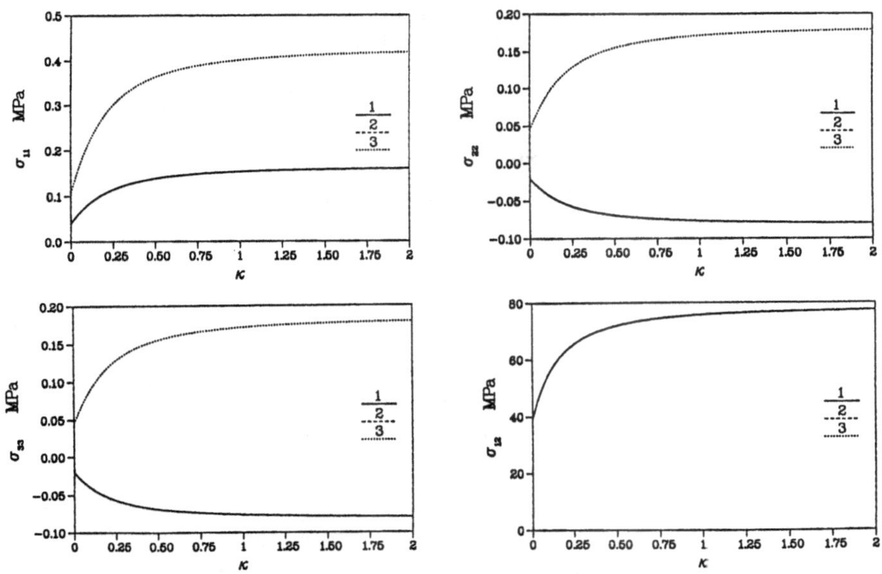

Fig. 1: Stresses in simple shear

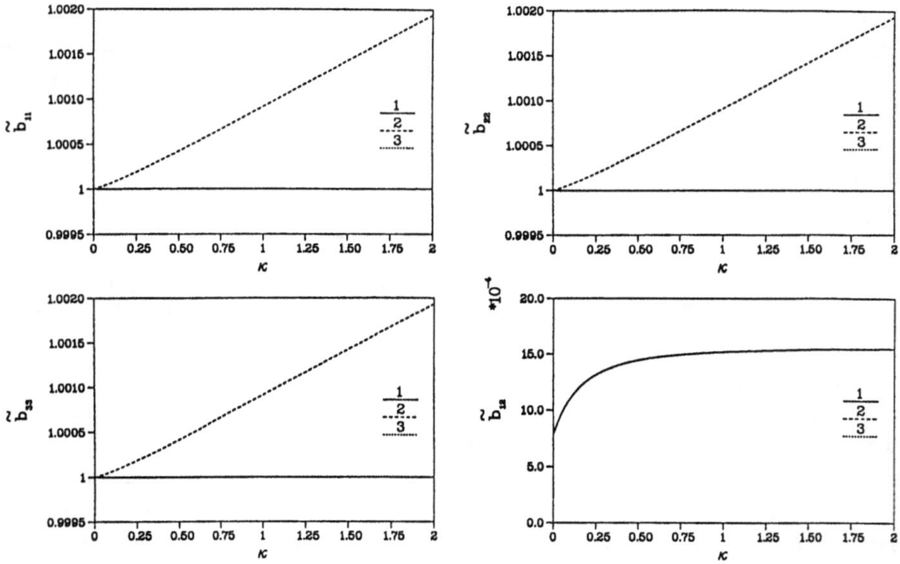

Fig. 2: Components of $\tilde{b}^{I}$ in simple shear

model	σ_{11}	σ_{22}	σ_{33}	σ_{12}
1	0.1597597	-.0798798	-.0798798	77.40197
1a	0.1995815	-.0400584	-.0400584	77.40203
2	0.1595404	-.0797702	-.0797702	77.42363
2a	97.02141	96.78209	96.78209	77.42363
2b	0.1597597	-.0798798	-.0798798	77.40197
2c	0.1993715	-.0402680	-.0402680	77.40197

Table 1: Stresses at $\kappa = 200\%$ in N/mm^2

model	$\tilde{b}_{11}$	$\tilde{b}_{22}$	$\tilde{b}_{33}$	$\tilde{b}_{12}$
1	1.000004	0.9999992	0.9999992	0.001548039
1a	1.000004	0.9999992	0.9999992	0.001548041
2	1.001940	1.001936	1.001936	0.001548473
2a	1.001940	1.001936	1.001936	0.001548473
2b	1.000004	0.9999992	0.9999992	0.001548039
2c	1.000004	0.9999992	0.9999992	0.001548039

Table 2: Components of $\tilde{b}^{I}$ at $\kappa = 200\%$

REFERENCES

[1] Bammann, D. J.; Aifantis, E. C.: A model for finite-deformation plasticity. *Acta Mechanica* **69**, 97–117 (1987).

[2] Bodner, S. R.; Partom, Y.: Constitutive Equations for elastic-viscoplastic strain-hardening materials. *Trans. ASME, J. Appl. Mech.* **42**, 385–389 (1975).

[3] Brown, S. B.; Kim, K. H.; Anand, L.: An internal variable constitutive model for hot working of metals. *Int. J. Plasticity* **5**, 95–130 (1989).

[4] Cleja-Ţigoiu, S.; Soós, E.: Elastoviscoplastic models with relaxed configurations and internal state variables. *ASME, Appl. Mech. Rev.* **43**(7), 131–151 (1990).

[5] Hackenberg, H. - P.: *Ein Vergleich verschiedener inelastischer Stoffgesetzformulierungen für finite Deformationen am Beispiel des ebenen Scherversuchs.* Unpublished report MuM-Report 91/2, Technische Hochschule Darmstadt, Darmstadt, Federal Republic of Germany 1991.

[6] Hart, E. W.: Constitutive relations for the non-elastic deformation of metals. *Trans. ASME, J. Eng. Mat. Tech.* **98**, 193–202 (1976).

[7] Haupt, P.: On the concept of an intermediate configuration and its application to a representation of viscoelastic-plastic material behavior. *Int. J. Plasticity* **1**, 303–316 (1988).

[8] Kollmann, F. G.; Hackenberg, H. - P.: *Kinematics and rate-type elastic constitutive equations for finite rate dependent inelastic deformation of materials with small elastic strains.* Unpublished report MuM-Report 90/2, Technische Hochschule Darmstadt, Darmstadt, Federal Republic of Germany 1990.

[9] Kratochvil, J.: On a finite strain theory of elastic-plastic materials. *Acta Mechanica* **16**, 127–142 (1973).

[10] Lee, E. H.: Elastic-plastic deformation at finite strains. *Trans. ASME, J. Appl. Mech.* **36**, 1–6 (1969).

[11] Mandel, J.: Equations constitutives et directeurs dans les milieux plastiques et viscoplastiques. *Int. J. Solids Structures* **9**, 725–740 (1973).

[12] Marsden, E.; Hughes, T. J. R.: *Mathematical foundations of elasticity.* Englewood Cliffs: Prentice Hall 1983.

[13] Onat, E. T.: Representation of inelastic behavior in the presence of anisotropy and finite deformations. In: *Plasticity of metals: Theory, computations and experiment* (E. H. Lee; R. L. Mallet, eds.), pp. 519–544. Stanford, USA and Troy, USA: Stanford University and Rensslaer Polytechnical Institute 1982.

[14] Simo, J. C.: A framework for finite strain elastoplasticity based on maximum plastic dissipation and the multiplicative decomposition: Part I. Continuum formulation. *Comp. Meths. Appl. Mech. Engrg.* **66**, 199–219 (1988).

[15] Simo, J. C.; Taylor, R. L.; Pister, K. S.: Variational and projection methods for the volume constraint in finite deformation elastoplasticity. *Comp. Meth. Appl. Mech. Engrg.* **51**, 177–208 (1985).

An Unified Approach to Modelling and Numerical Solution of Coupled Field Problems in Nonlinear Solid Mechanics

A.Tamme, Institut für Mechanik, Pf 408, D-O 9026 Chemnitz, Germany

Abstract

Recently, two features have been observed in the field of the unified description of inelastic material behaviour. The first approach is based on the development of a general material law describing the different inelastic properties ([1]). Another approach, which is used by Bergander ([2]) consists in a unified representation of current constitutive laws. All these considerations are characterized by a pure constitutive analysis, without any reference to the field equations.

The aim of this paper is a generalization of Bergander's *standard formulation* ([2]) under the aspect of a uniform treatment of both, the field equations and the constitutive relations. Furthermore, this approach should allow us to develop a uniform numerical solution strategy consisting of finite element discretization, implicit time integration and a consistent linearization as a generalization of the well- known return-algorithm.

1 Principles of Unified Modelling

1.1 Field Equations

Extending some considerations of G.A. Maugin ([6]) we consider a *generalized placement* definition as a supervector

$$\mathbf{U} = (\mathbf{U}^{(0)}, \cdots, \mathbf{U}^{(N_u)})^T$$

consisting of the coefficients of p_k-order tensors $\mathbf{U}^{(k)} = \mathbf{U}^{(k)}(X,t)$ $(k = 1, \cdots, N_u)$ (e.g. the classical displacement vector, the temperature etc.).

The other variables of continuum physics can be classified (see also [3]) in the following manner

- The *generalized independent variables* (with regard to the placement $\mathbf{U}^{(k)}$) are in general nonlinear tensor functions of the placement gradients

  $$\mathbf{E}^{(i,k)} = \mathbf{E}^{(i,k)}(\mathbf{U}^{(k)}) = \mathcal{E}^{(i,k)}(\mathbf{U}^{(k)}, \nabla\mathbf{U}^{(k)}, \cdots \nabla^{(m_i)}\mathbf{U}^{(k)}) \quad (i = 0, \cdots, M_k)$$

  forming (with it's coefficients) the vector

  $$\mathbf{E} = (\mathbf{E}^{(0)}, \cdots, \mathbf{E}^{(N_u)})^T \text{ with } \mathbf{E}^{(k)} = (\mathbf{E}^{(0,k)}, \cdots, \mathbf{E}^{(M_k,k)}) \qquad (k = 0, \cdots, N_u)\,.$$

- The duals $\mathbf{S}^{(i,k)}$ of the independent variables $\mathbf{E}^{(i,k)}$ are denoted as *dependent variables* and form the vector

  $$\mathbf{S} = (\mathbf{S}^{(0)}, \cdots, \mathbf{S}^{(N_u)})^T \text{ with } \mathbf{S}^{(k)} = (\mathbf{S}^{(0,k)}, \cdots, \mathbf{S}^{(M_k,k)})\,.$$

- The (scalar or tensorial) internal variables $\mathbf{Q}^{(k)} = (\mathbf{Q}^{(0,k)}, \cdots, \mathbf{Q}^{(M_k,k)})$ form the vector $\mathbf{Q} = (\mathbf{Q}^{(0)}, \cdots, \mathbf{Q}^{(N_u)})^T$.

D. Besdo, E. Stein (Eds.)
Finite Inelastic Deformations – Theory and Applications
IUTAM Symposium Hannover/Germany 1991

Examples for independent variables in the above sense are the temperature (as zero order gradient), the temperature gradient, the Almansi deformation tensor, the electromotive intensity or the magnetic induction. Their duals (i.e. the dependent variables) are the internal energy, the energy flux, the 2nd Piola-Kirchhoff stress tensor, the electric current and the magnetization.
Let the initial body manifold $\mathcal{B}_0$ be the reference placement and let $\mathcal{W}^{(k)}(\mathcal{B}_0)$ $(k = 0, \cdots, N_u)$ denote the Sobolev type spaces

$$\mathcal{W}^{(k)}(\mathcal{B}_0) = \left\{ \mathbf{U}^{(k)} : \ \mathcal{E}^{(i,\,k)}(\mathbf{U}^{(k)}, \cdots \nabla^{(m_i)}\mathbf{U}^{(k)}) \in [L^2(\mathcal{B}_0)]^{3^{m_i}} \ \ i = 0, \cdots, M_k \right\} .$$

The prescribed boundary placements define the spaces of admissible placements

$$\mathcal{V}^{(k)} = \left\{ \mathbf{U}^{(k)} \in \mathcal{W}^{(k)}(\mathcal{B}_0) : \quad \mathbf{U}^{(k)} = \overline{\mathbf{U}}^{(k)} \quad on \ \ \Gamma_k \subset \partial\mathcal{B}_0 \right\}$$

and the spaces of test functions

$$\mathcal{V}_0^{(k)} = \left\{ \mathbf{U}^{(k)} \in \mathcal{W}^{(k)}(\mathcal{B}_0) : \quad \mathbf{U}^{(k)} = \mathbf{O} \quad on \ \ \Gamma_k \subset \partial\mathcal{B}_0 \right\} .$$

In the absence of acceleration forces ([6]) the following hypothesis on the field equations is acceptable

(HF) There are given energy functionals $\Pi^{(k)} = \Pi^{(k)}(\dot{\mathbf{U}}, \mathbf{U}, \dot{\mathbf{S}}, \mathbf{S}) \in [\mathcal{V}_0^{(k)}]^*$, and load functionals $\mathcal{L}^{(k)} \in [\mathcal{V}_0^{(k)}]^*$ $(k = 0, \cdots, N_u)$ such that the field equations can be formulated by the identities

$$\left\langle \Pi^{(k)}(\dot{\mathbf{U}}, \mathbf{U}, \dot{\mathbf{S}}, \mathbf{S}) \, , \, \mathbf{V}^{(k)} \right\rangle = \left\langle \mathcal{L}^{(k)} \, , \, \mathbf{V}^{(k)} \right\rangle \qquad \forall \mathbf{V}^{(k)} \in \mathcal{V}_0^{(k)} .$$

1.2 Constitutive relations

To describe the thermodynamical state of a medium at a point and at fixed time t it is often assumed that

- the independent variables $\mathbf{E}^{(i,\,k)}$ are the sum of a reversible part $\mathbf{E}_R^{(i,\,k)}$ and a residual (irreversible) contribution $\mathbf{E}_I^{(i,\,k)}$;
- there exists a free energy Ψ per unit volume having the follwing dependence on the state variables

$$\Psi = \Psi(\mathbf{E}_R, \mathbf{Q}) \quad ; \qquad \mathbf{E}_R = \mathbf{E} - \mathbf{E}_I \tag{1}$$

such that

$$\mathbf{S} = \frac{\partial \Psi(\mathbf{E}_R, \mathbf{Q})}{\partial \mathbf{E}_R} \ ; \tag{2}$$

- the irreversible parts $\mathbf{E}_I^{(i,k)}$ are controlled by a *state parameter* $\lambda^{(i,k)}$ and a *state function* $F^{(i,k)} = F^{(i,k)}(\mathbf{B},\mathbf{S},\mathbf{Q})$ with the help of a generalized loading criterion[1]

$$\Phi_\lambda^{(i,k)}(\mathbf{B},\lambda^{i,k},\mathbf{S},\mathbf{Q}) \equiv \widehat{\Phi}_\lambda^{(i,k)}(\lambda^{i,k}, F^{(i,k)}(\mathbf{B},\mathbf{S},\mathbf{Q})) = 0 \tag{3}$$

  where $\mathbf{B}$ is the inverse Cauchy-Green Tensor ;

- the irreversible parts have to fulfill a generalized *hardening rule*[2]

$$\dot{\mathbf{E}}_I^{(i,k)} = \lambda^{(i,k)}\,\mathbf{N}^{(i,k)}(\mathbf{B},\mathbf{S},\mathbf{Q})\ ; \tag{4}$$

- and the internal parameters have to fulfil evolution equations

$$\dot{\mathbf{Q}}(i,k) = \lambda^{(i,k)}\,\mathbf{M}^{(i,k)}(\mathbf{B},\mathbf{S},\mathbf{Q})\ ; \tag{5}$$

These assumptions are fulfilled by a wide class of models in continuum physics e.g. models of pure mechanics, thermomechanics or magnetoelectrodynamics. If an irreversible process for any independent variable $\mathbf{E}^{(i,k)}$ is out of considertaion, $\Phi_\lambda^{(i,k)} = \lambda^{(i,k)}$ can be chosen such that (3) , (4) and (5) give zero values for the variables of irreversibility. Some of the above notations (especially the notation of temperature as independent variable) are in conflict to the conventional notations of thermodynamics and material theories, but they are very usueful for the numerical treatment as it can seen below. Keeping the history constant, i.e. the internal variables constant, we obtain from (2) and (4) by time differentation

$$\dot{\mathbf{S}}^{(i,k)} = \sum_{l=0}^{N_u}\sum_{j=0}^{M_l} \frac{\partial^2\Psi}{\partial\mathbf{E}_R^{(i,k)}\partial\mathbf{E}_R^{(j,l)}} : \left[\dot{\mathbf{E}}_R^{(j,l)} - \lambda^{(j,l)}\,\mathbf{N}^{(j,l)}(\mathbf{B},\mathbf{S},\mathbf{Q})\right] \tag{6}$$

Where the double-dot product $\mathbf{A}:\mathbf{B}$ of tensors

$$\mathbf{A} = A^{I\cdots,J,K,\cdots,L}\,\frac{\partial}{\partial X^I}\otimes\cdots\otimes\frac{\partial}{\partial X^L}$$

and

$$\mathbf{B} = B_{K\cdots,L,M,\cdots,N}\,dX^K\otimes\cdots\otimes dX^N$$

is defined by

$$\mathbf{A}:\mathbf{B} = [A^{I\cdots,J,K,\cdots,L}\,B_{K\cdots,L,M,\cdots,N}]\,\frac{\partial}{\partial X^I}\otimes\cdots\otimes\frac{\partial}{\partial X^J}\otimes dX^M\otimes\cdots\otimes dX^N\ . \tag{7}$$

Thus we are well motivated to formulate the following constitutive hypothesis, containing the above assumptions:

[1]Often the load criterion is given as Kuhn-Tucker condition: $\lambda \geq 0$, $F \leq 0$, $\lambda\,F = 0$; which are equivalent to the equation $|\lambda + F|^3 + F^3 - \lambda^3 = 0$ (see also [5])

[2]wich can be obtained as a consequence of a principle of the maximum of dissipation of the irreversible parts ([8])

(HC) There are given smooth functions

$$\mathcal{F}_s^{(i,k)} = \mathcal{F}_s^{(i,k)}(\mathbf{E},\dot{\mathbf{E}},\mathbf{S},\mathbf{Q}) \;, \qquad \mathcal{G}_s^{(i,k)} = \mathcal{G}_s^{(i,k)}(\mathbf{E},\lambda,\mathbf{S},\mathbf{Q}) \;,$$

$$\mathcal{F}_q^{(i,k)} = \mathcal{F}_q^{(i,k)}(\mathbf{E},\dot{\mathbf{E}},\mathbf{S},\mathbf{Q}) \;, \qquad \mathcal{G}_q^{(i,k)} = \mathcal{G}_q^{(i,k)}(\mathbf{E},\lambda,\mathbf{S},\mathbf{Q}) \quad \text{and}$$

$$\Phi_\lambda^{(i,k)} = \Phi_\lambda^{(i,k)}(\mathbf{B},\lambda^{i,k},\mathbf{S},\mathbf{Q}) \;, \quad (k=0,\cdots,N_u\;;\; i=0,\cdots,M_k)$$

such that

- the evolution equations of dependent variables

$$\dot{\mathbf{S}}^{(i,k)} = \mathcal{F}_s^{(i,k)}(\mathbf{E},\dot{\mathbf{E}},\mathbf{S},\mathbf{Q}) + \mathcal{G}_s^{(i,k)}(\mathbf{E},\lambda,\mathbf{S},\mathbf{Q})$$

- the evolution equations of internal variables

$$\dot{\mathbf{Q}}^{(i,k)} = \mathcal{F}_q^{(i,k)}(\mathbf{E},\mathbf{S},\mathbf{Q}) + \mathcal{G}_q^{(i,k)}(\mathbf{E},\lambda,\mathbf{S},\mathbf{Q})$$

- and state (loading) criterion $\quad \Phi_\lambda^{(i,k)}(\mathbf{B},\lambda^{i,k},\mathbf{S},\mathbf{Q}) = 0$

are fulfilled.

1.3 The initial boundary value problem

Now we are able to formulate a mathematical model, which is physically well motivated and allows a unified numerical treatment. With the aim of a compact description we introduce the functions

$$\mathcal{A} = \left(\mathcal{A}^{(0)},\cdots,\mathcal{A}^{(N_u)}\right)^T \quad \text{with} \quad \mathcal{A}^{(k)} = \left(\mathcal{A}^{(0,k)},\cdots,\mathcal{A}^{(M_k,k)}\right) . \tag{8}$$

in the first instance for $\mathcal{A} = \mathcal{F}_s, \mathcal{F}_q, \mathcal{G}_s, \mathcal{G}_q, \Phi_\lambda$. Furthermore let

$$\mathbf{V} = (\mathbf{V}^{(0)},\cdots,\mathbf{V}^{(N_u)})^T \in \mathcal{V}_0 = \mathcal{V}_0^{(0)} \times \cdots \times \mathcal{V}_0^{(N_u)}$$

$$\Pi = (\Pi^{(0)},\cdots,\Pi^{(N_u)})^T \;, \quad \mathcal{L} = (\mathcal{L}^{(0)},\cdots,\mathcal{L}^{(N_u)})^T \in \mathcal{V}_0^* \;;$$

which gives

$$\left\langle \Pi(\dot{\mathbf{U}},\mathbf{U},\dot{\mathbf{S}},\mathbf{S}) \;,\; \mathbf{V}\right\rangle = \sum_{k=0}^{N_u} \left\langle \Pi^{(k)}(\dot{\mathbf{U}}^{(k)},\mathbf{U}^{(k)},\dot{\mathbf{S}}^{(k)},\mathbf{S}^{(k)}) \;,\; \mathbf{V}^{(k)}\right\rangle \tag{9}$$

Thus, we consider coupled field problems in a material convective description, modelled as initial boundary value problems in the following form:

For a given generalized load function $\mathcal{L}(X,t)$, find
$\mathbf{U} = \mathbf{U}(X,t)$, $\mathbf{S} = \mathbf{S}(X,t)$, $\mathbf{Q} = \mathbf{Q}(X,t)$, $\lambda = \lambda(X,t)$;
such that for all (quasi-) time parameters $t \in [0,T]$

- the weakly formulated field equations:

$$\text{(I)} \quad \left\langle \Pi(\dot{\mathbf{U}},\mathbf{U},\dot{\mathbf{S}},\mathbf{S}),\mathbf{V}\right\rangle = \langle \mathcal{L},\mathbf{V}\rangle \qquad \forall \mathbf{V} \in \mathcal{V}_0$$

- the geometric relations:

$$\text{(II)} \quad \mathbf{E} = \mathbf{E}(\mathbf{U}) \; ; \qquad \dot{\mathbf{E}} = \sum_{k=0}^{N_u} \sum_{i=0}^{M_k} \frac{\partial \mathcal{E}}{\partial (\nabla^{(i)}\mathbf{U})^{(k)}} : \nabla^{(i)}\dot{\mathbf{U}}^{(k)} \qquad in\ \mathcal{B}_0$$

- the constitutive evolution equations:

$$\text{(III)} \quad \Phi_s(\dot{\mathbf{E}},\mathbf{E},\lambda,\dot{\mathbf{E}},\mathbf{S},\mathbf{Q}) = \dot{\mathbf{S}} - \mathcal{F}_s(\dot{\mathbf{E}},\mathbf{E},\mathbf{S},\mathbf{Q}) - \mathcal{G}_s(\mathbf{E},\lambda,\mathbf{S},\mathbf{Q}) = 0 \qquad in\ \mathcal{B}_0$$

$$\text{(IV)} \quad \Phi_q(\dot{\mathbf{E}},\mathbf{E},\lambda,\mathbf{S},\dot{\mathbf{Q}},\mathbf{Q}) = \dot{\mathbf{Q}} - \mathcal{F}_q(\dot{\mathbf{E}},\mathbf{E},\mathbf{S},\mathbf{Q}) - \mathcal{G}_q(\mathbf{E},\lambda,\mathbf{S},\mathbf{Q}) = 0 \qquad in\ \mathcal{B}_0$$

- the loading criterion:

$$\text{(V)} \quad \Phi_\lambda(\mathbf{E},\lambda,\mathbf{S},\mathbf{Q}) = 0 \qquad in\ \mathcal{B}_0$$

- the initial conditions:

$$(VI) \quad \mathbf{U}(X,0) = \mathbf{U}_0 \; , \; \mathbf{S}(X,0) = \mathbf{S}_0 \; , \; \mathbf{Q}(X,0) = \mathbf{Q}_0 \qquad in\ \mathcal{B}_0$$

are fulfilled. (10)

2 Numerical Solution

2.1 Implicit time integration

If a (not necessarily equidistant) partition $t_0 = 0, t_1, \cdots t_N = T$ of the time interval is given and if the solutions $\mathbf{U}^n = \mathbf{U}(t_n)$, $\mathbf{S}^n = \mathbf{S}(t_n)$, $\mathbf{Q}^n = \mathbf{Q}(t_n)$ and $\lambda^n = \lambda(t_n)$ at time $t = t_n$ are known, an incremental solution strategy leads to the problem of finding the solutions $\mathbf{U}^{n+1} = \mathbf{U}(t_{n+1})$, $\mathbf{S}^{n+1} = \mathbf{S}(t_{n+1})$, $\mathbf{Q}^{n+1} = \mathbf{Q}(t_{n+1})$ and $\lambda^{n+1} = \lambda(t_{n+1})$ at time $t = t_{n+1}$ Using a full implicit time integration scheme we replace the derivatives

$\dot{\mathbf{U}}$, $\dot{\lambda}$, $\dot{\mathbf{S}}$ and $\dot{\mathbf{Q}}$ by their backward differences:

$$\dot{\mathbf{U}}(t_{n+1}) \approx \mathbf{U}_t^{n+1} = \frac{\mathbf{U}(t_{n+1}) - \mathbf{U}(t_n)}{t_{n+1} - t_n} \; , \quad \dot{\lambda}(t_{n+1}) \approx \lambda_t^{n+1} = \frac{\lambda(t_{n+1}) - \lambda(t_n)}{t_{n+1} - t_n} \; , \quad \text{etc..}$$

This defines the functionals

$$\left.\begin{aligned} < \Pi^n(\mathbf{U}^{n+1}, \mathbf{S}^{n+1}), \mathbf{V} > &:= \; < \; \Pi(\mathbf{U}_t^{n+1}, \mathbf{U}^{n+1}, \mathbf{S}_t^{n+1}, \mathbf{S}^{n+1}, \mathcal{L}^{n+1}, \mathbf{V}) \, , \, \mathbf{V} > \\ \Phi_s^n(\mathbf{E}^{n+1}, \lambda^{n+1}, \mathbf{S}^{n+1}, \mathbf{Q}^{n+1}) &= \mathbf{S}_t^{n+1} - \mathcal{F}_s(\mathbf{E}_t^{n+1}, \mathbf{E}^{n+1}, \mathbf{S}^{n+1}, \mathbf{Q}^{n+1}) \\ &\quad - \mathcal{G}_s(\mathbf{E}_t^{n+1}, \lambda^{n+1}, \mathbf{S}^{n+1}, \mathbf{Q}^{n+1}) \\ \Phi_q^n(\mathbf{E}^{n+1}, \lambda^{n+1}, \mathbf{S}^{n+1}, \mathbf{Q}^{n+1}) &= \mathbf{S}_t^{n+1} - \mathcal{F}_q(\mathbf{E}_t^{n+1}, \mathbf{E}^{n+1}, \mathbf{S}^{n+1}, \mathbf{Q}^{n+1}) \\ &\quad - \mathcal{G}_q(\mathbf{E}_t^{n+1}, \lambda^{n+1}, \mathbf{S}^{n+1}, \mathbf{Q}^{n+1}) \end{aligned}\right\} \quad (11)$$

and the problem (10) leads approximately to a sequence of nonlinear problems of equations:

$$\boxed{\begin{array}{lll} \text{(I)} & \langle \Pi^n(\mathbf{U}^{n+1}, \mathbf{S}^{n+1}) \, , \, \mathbf{V} \rangle = \langle \mathcal{L}^{n+1} \, , \, \mathbf{V} \rangle & \forall \mathbf{V} \in \mathcal{V}_0 \\ \text{(II)} & \mathbf{E}^{n+1} = \mathbf{E}^n + \sum\limits_{k=0}^{N_u} \sum\limits_{i=0}^{M_k} [\frac{\partial \mathcal{E}^n}{\partial (\nabla^{(i)} \mathbf{U})^{(k)}}] : \nabla^{(i)} [(\mathbf{U}^{(k)})^{n+1} - (\mathbf{U}^{(k)})^n] & in\ \mathcal{B}_0 \\ \text{(III)} & \Phi_s^n(\mathbf{E}^{n+1}, , \lambda^{n+1}, \mathbf{S}^{n+1}, \mathbf{Q}^{n+1}) = 0 & in\ \mathcal{B}_0 \\ \text{(IV)} & \Phi_q^n(\mathbf{E}^{n+1}, \lambda^{n+1}, \mathbf{S}^{n+1}, \mathbf{Q}^{n+1}) = 0 & in\ \mathcal{B}_0 \\ \text{(V)} & \Phi_\lambda(\mathbf{E}^{n+1}, \lambda^{n+1}, \mathbf{S}^{n+1}, \mathbf{Q}^{n+1}) = 0 & in\ \mathcal{B}_0 \\ (VI) & \mathbf{U}(X, t_n) = \mathbf{U}^n \; , \; \mathbf{S}(X, t_n) = \mathbf{S}^n \; , \; \mathbf{Q}(X, t_n) = \mathbf{Q}^n & in\ \mathcal{B}_0 \end{array}} \quad (12)$$

2.2 Finite Element Discretization

Let us now introcuce a standard finite element triangularization (see, e.g. [4]) $\mathcal{B}_0^h$ for the body manifold $\mathcal{B}_0$, i.e.

$$\overline{\mathcal{B}_0^h} = \bigcup_{e=1}^{\mathcal{N}_{el}} \overline{\Omega}_e \; ; \quad \Omega_s \cap \Omega_r = \emptyset \;\; (s \neq r)$$

with $\mathcal{N}_{el}$ elements defined on the element geometry Ω_e. We consider the finite element approximations

$$\begin{aligned} \mathbf{U}_h(X,t) &= \sum_{I=1}^{\mathcal{N}} [\mathbf{N}_I(X)] \, \mathbf{U}_I(t) = [\mathbf{N}(X)] \, \underline{\mathbf{U}}(t) \in \mathcal{V}_h \subset \mathcal{V} \\ &\text{and} \\ \mathbf{V}_h(X,t) &= \sum_{J=1}^{\mathcal{N}} [\mathbf{N}_J(X)] \, \mathbf{V}_J(t) = [\mathbf{N}(X)] \, \underline{V}(t) \in \mathcal{V}_{0h} \subset \mathcal{V}_0 \end{aligned} \quad (13)$$

of the placement fields $\mathbf{U}$ and test functions $\mathbf{V}$ which have local restrictions

$$\mathbf{U}_{h|\Omega_e} = \mathbf{U}_{(e)}(X,t) = [\mathbf{N}_{(e)}(X)]\,\underline{\mathbf{U}}_{(e)}(t) = \sum_{k=0}^{N_u} [\mathbf{N}_{(e)}^{(k)}(X)]\,\underline{\mathbf{U}}_{(e)}^{(k)}(t)$$

and (14)

$$\mathbf{V}_{h|\Omega_e} = \mathbf{V}_{(e)}(X,t) = [\mathbf{N}_{(e)}(X)]\,\underline{V}_{(e)}(t) = \sum_{k=0}^{N_u} [\mathbf{N}_{(e)}^{(k)}(X)]\,\underline{V}_{(e)}^{(k)}(t)\,.$$

respectively. Here the $(1 \times \mathcal{N})$ matrices $\mathbf{N}$ and the $(1 \times \mathcal{N}_e)$ matrices $\mathbf{N}_{(e)}$ are constructed from the finite element basis functions and their element restrictions (the so called shape functions), where $\mathcal{N}_e$ is the local number of degrees of freedom in Ω_e and $\mathcal{N}$ is the global number of degrees of freedom. Furthermore $[\mathbf{N}_{(e)}^{(k)}(\mathbf{X})]$ are the matrix blocks of $[\mathbf{N}_{(e)}(X)]$ connected with the subvectors $\underline{\mathbf{U}}^{(k)}$. The local $\mathcal{N}_e$-dimensional element vectors $\underline{\mathbf{U}}_{(e)}$ and the $\mathcal{N}$-dimensional global node vector $\underline{\mathbf{U}}$ are connected by the Boolean connectivity $(\mathcal{N}_e \times \mathcal{N})$-matrices $\mathcal{A}_{(e)}$ with

$$\underline{\mathbf{U}}_{(e)} = \mathcal{A}_{(e)}\,\underline{\mathbf{U}}. \tag{15}$$

After the well known finite element procedure we obtain from the first equation in (12) a nonlinear system of algebraic equations

$$\underline{\mathcal{P}}^n(\underline{\mathbf{U}}^{n+1}, \mathcal{S}^{n+1}) = \underline{\mathcal{L}}^{n+1} \tag{16}$$

with the column vectors $\underline{\mathcal{P}}^n$ and $\underline{\mathcal{L}}^{n+1}$ of length $\mathcal{N}$, defined by

$$\underline{\mathcal{P}}^n = (< \Pi^n, \mathbf{N}_1 >, \cdots, < \Pi^n, \mathbf{N}_{\mathcal{N}} >)^T\,, \quad \underline{\mathcal{L}}^n = (< \mathcal{L}^n, \mathbf{N}_1 >, \cdots, < \mathcal{L}^n, \mathbf{N}_{\mathcal{N}} >)^T \tag{17}$$

Using a quadratur formula over the elements Ω_e (with integration weights $\omega^{(e,\alpha)}$ and integration points $X_{(e,\alpha)}$) this leads to

$$\begin{aligned} \underline{\mathcal{P}}_h^n(\underline{\mathbf{U}}^{n+1}, \underline{\mathcal{S}}^{n+1}) &= \sum_{e=1}^{\mathcal{N}_{el}} [\mathcal{A}_{(e)}]^T\, \underline{\mathcal{P}}_{h(e)}^n(\underline{\mathbf{U}}_{(e)}^{n+1}, \mathcal{S}^{n+1}) \\ &\approx \sum_{e=1}^{\mathcal{N}_{el}} [\mathcal{A}_{(e)}]^T \sum_{\alpha=1}^{\mathcal{N}_{int}} \omega^{(e,\alpha)}\, \underline{\mathcal{P}}_{(e,\alpha)}^n(\underline{\mathbf{U}}_{(e)}^{n+1}, \mathcal{S}_{(e,\alpha)}^{n+1}) \end{aligned} \tag{18}$$

and

$$\underline{\mathcal{L}}^{n+1} = \sum_{e=1}^{\mathcal{N}_{el}} [\mathcal{A}_{(e)}]^T\, \underline{\mathcal{L}}_{(e)}^{n+1} \approx \sum_{e=1}^{\mathcal{N}_{el}} [\mathcal{A}_{(e)}]^T \sum_{\alpha=1}^{\mathcal{N}_{int}} \omega^{(e,\alpha)}\, \underline{\mathcal{L}}_{(e,\alpha)}^{n+1} \tag{19}$$

respectively. Because the computing of the node vectors needs only the values of the dependent variables in the integration points, we can replace the system (12) by its approximate problem

Find the element node vectors $\underline{U}_{(e)}$ and the values $\mathbf{S}_{(e,\alpha)} = \mathbf{S}^{n+1}_{(e,\alpha)}$, $\mathbf{Q}_{(e,\alpha)} = \mathbf{Q}^{n+1}_{(e,\alpha)}$ and $\lambda_{(e,\alpha)} = \lambda^{n+1}_{(e,\alpha)}$, such that for all elements Ω_e and in each integration point $X_{e,\alpha}$ $(e = 1, \cdots, \mathcal{N}_{el}\ ;\ \alpha = 1, \cdots, \mathcal{N}_{int})$, the following equation are fulfilled

$$
\begin{aligned}
&(I) && \sum_{e=1}^{\mathcal{N}_{el}} [\mathcal{A}_{(e)}]^T \sum_{\alpha=1}^{\mathcal{N}_{int}} \omega^{(e,\alpha)} \left[\underline{\mathcal{P}}^n_{(e,\alpha)}(\underline{U}_{(e)}, \mathbf{S}_{(e,\alpha)}) - \underline{\mathcal{L}}^{n+1}_{(e,\alpha)} \right] = 0 \\
&(II) && \mathbf{E}_{(e,\alpha)} = \mathbf{E}^n_{(e,\alpha)} + [\underline{\mathcal{E}}_{(e,\alpha)}(\underline{U}_{(e)})]\ (\underline{U}_{(e)} - \underline{U}^n_{(e)}) \\
&(III) && \Phi^n_s(\mathbf{E}_{(e,\alpha)}, \lambda_{(e,\alpha)}, \mathbf{S}_{(e,\alpha)}, \mathbf{Q}_{(e,\alpha)}) = 0 \\
&(IV) && \Phi^n_q(\mathbf{E}_{(e,\alpha)}, \lambda_{(e,\alpha)}, \mathbf{S}_{(e,\alpha)}, \mathbf{Q}_{(e,\alpha)}) = 0 \\
&(V) && \Phi_\lambda(\mathbf{E}_{(e,\alpha)}, \lambda_{(e,\alpha)}, \mathbf{S}_{(e,\alpha)}, \mathbf{Q}_{(e,\alpha)}) = 0 \\
&(VI) && \mathbf{U}_{(e)}(t_n) = \mathbf{U}^n_{(e)}\ ,\ \mathbf{S}(X_{e,\alpha}, t_n) = \mathbf{S}^n_{e,\alpha}\ ,\ \mathbf{Q}(X_{e,\alpha}, t_n) = \mathbf{Q}^n_{e,\alpha}
\end{aligned}
\tag{20}
$$

In the second equation the matrix $[\underline{\mathcal{E}}_{(e,\alpha)}] = (\underline{\mathcal{E}}^{(0)}, \cdots, \underline{\mathcal{E}}^{(N_u)})^T$ with $\underline{\mathcal{E}}^{(k)} = (\underline{\mathcal{E}}^{(0,k)}, \cdots, \underline{\mathcal{E}}^{(M_k,k)})^T$ is formed by the submatrices

$$
\underline{\mathcal{E}}^{(i,k)}(\underline{U}_{(e)}) = \sum_{k=0}^{N_u} \sum_{i=0}^{M_k} [\frac{\partial \mathcal{E}^n}{\partial(\nabla^{(i)} \mathbf{U}(k))}]\ \nabla^{(i)}[(\mathbf{N}^{(k)}_{(e)})(X_{e,\alpha})]\ \underline{U}^{(k)} .
$$

2.3 Consistent linearization

In order to linearize (20) we use a nested linearization method, beginning with a Newton-Raphson-iteration for the equation (20.I). Since the functions $\underline{\mathcal{P}}^n_{e,\alpha}$ depends not only on the node points, the Frechet-derivative *is not* the Jacobian-matrix of partial derivatives in the element nodes. An analysis via the *Implicit Function Theorem* ([8]) leads to

$$
[\underline{\mathcal{P}}^n_{(e,\alpha)}]'(\underline{U}_{(e)}, S_{(e,\alpha)}) = \frac{\partial \underline{\mathcal{P}}^n_{(e,\alpha)}}{\partial \underline{U}_{(e)}} + \frac{\partial \underline{\mathcal{P}}^n_{(e,\alpha)}}{\partial \mathbf{S}_{(e,\alpha)}} \frac{\partial \mathbf{S}_{(e,\alpha)}}{\partial \mathbf{E}_{(e,\alpha)}} [\underline{\mathcal{E}}_{(e,\alpha)}(\underline{U}_{(e)})] \tag{21}
$$

with the generalized *consistent tangent modul* $\dfrac{\partial \mathbf{S}_{(e,\alpha)}}{\partial \mathbf{E}_{(e,\alpha)}}$. It is computable ([7]) by the formula

$$
\frac{\partial \mathbf{S}_{(e,\alpha)}}{\partial \mathbf{E}_{(e,\alpha)}} = \mathcal{B}^{-1}_{SS}\ \mathcal{B}_{SE} \tag{22}
$$

whith the matrices

$$
\mathcal{B}_{ab} = \mathcal{A}_{ab} - \mathcal{A}_{a\lambda}\ [\mathcal{A}^{-1}_{\lambda\lambda}]\ \mathcal{A}_{\lambda b}\ , \qquad a, b = \mathbf{S}, \mathbf{Q}, \lambda . \tag{23}
$$

and

$$\mathcal{A}_{ab} = \frac{\partial \phi_a}{\partial b} - \frac{\partial \phi_a}{\partial \mathbf{Q}} [\frac{\partial \phi_Q}{\partial \mathbf{Q}}]^{-1} \frac{\partial \phi_Q}{\partial \phi_b} \quad , \qquad a,b = \mathbf{S},\mathbf{Q},\lambda.quad \tag{24}$$

(i) We are now in a position to realize *one* Newton-Raphson iteration

$$[(\underline{\mathcal{P}}^n)'(\underline{\mathbf{U}},\underline{\mathbf{S}})]\ (\check{\underline{\mathbf{U}}} - \underline{\mathbf{U}}) = -\underline{\mathcal{P}}^n(\underline{\mathbf{U}},\underline{\mathbf{S}}) + \underline{\mathcal{L}}^{(n+1)} \tag{25}$$

to compute the new iterate $\check{\underline{\mathbf{U}}}$, where the global stiffness system is assembled by

$$[(\underline{\mathcal{P}}^n)'(\underline{\mathbf{U}},\underline{\mathbf{S}})] = \sum_{e=1}^{\mathcal{N}_{el}} [\mathcal{A}_{(e)}]^T \left(\sum_{\alpha=1}^{\mathcal{N}_{int}} \omega^{(e,\alpha)} [\underline{\mathcal{P}}^n_{(e,\alpha)}]'(\underline{\mathbf{U}}_{(e)}, S_{(e,\alpha)}) \right) \mathcal{A}_{(e)}$$

and (26)

$$\underline{\mathcal{P}}^n(\underline{\mathbf{U}},\underline{\mathbf{S}}) - \underline{\mathcal{L}}^{(n+1)} = \sum_{e=1}^{\mathcal{N}_{el}} [\mathcal{A}_{(e)}]^T \sum_{\alpha=1}^{\mathcal{N}_{int}} \omega^{(e,\alpha)} \left[\underline{\mathcal{P}}^n_{(e,\alpha)}(\underline{\mathbf{U}}_{(e)}, \mathbf{S}_{(e,\alpha)}) - \underline{\mathcal{L}}^{(n+1)}_{(e,\alpha)} \right] .$$

(ii) Then we can compute one iterated independent variableswith the help of (20.II) by

$$\check{\mathbf{E}}_{(e,\alpha)} = \mathbf{E}^n_{(e,\alpha)} + [\underline{\mathcal{E}}_{(e,\alpha)}(\underline{\mathbf{U}}_{(e)})]\ (\check{\underline{\mathbf{U}}}(e) - \underline{\mathbf{U}}(e) \tag{27}$$

and we obtain the following nonlinear system of equations in each integration point

$$\begin{aligned} \Phi^n_s(\check{\mathbf{E}}_{(e,\alpha)}, \check{\lambda}_{(e,\alpha)}, \check{\mathbf{S}}_{(e,\alpha)}, \check{\mathbf{Q}}_{(e,\alpha)}) &= 0 \\ \Phi_\lambda(\check{\mathbf{E}}_{(e,\alpha)}, \check{\lambda}_{(e,\alpha)}, \check{\mathbf{S}}_{(e,\alpha)}, \check{\mathbf{Q}}_{(e,\alpha)}) &= 0 \\ \Phi^n_q(\check{\mathbf{E}}_{(e,\alpha)}, \check{\lambda}_{(e,\alpha)}, \check{\mathbf{S}}_{(e,\alpha)}, \check{\mathbf{Q}}_{(e,\alpha)}) &= 0 . \end{aligned} \tag{28}$$

(iii) Before linearizing of the system (29) we compute an improved initial solution by a *predictor step*. For that purpose we choose $\lambda_{(e,\alpha)} = \mathbf{O}$ and find the trial solutions $\mathbf{S}^{tr}_{(e,\alpha)}$ and $\mathbf{S}^{tr}_{(e,\alpha)}$ of the system

$$\begin{aligned} \Phi^n_s(\check{\mathbf{E}}_{(e,\alpha)}, \mathbf{O}, \mathbf{S}^{tr}_{(e,\alpha)}, \mathbf{Q}^{tr}_{(e,\alpha)}) &= 0 \\ \Phi^n_q(\check{\mathbf{E}}_{(e,\alpha)}, \mathbf{O}, \mathbf{S}^{tr}_{(e,\alpha)}, \mathbf{Q}^{tr}_{(e,\alpha)}) &= 0 . \end{aligned}$$

If the trial solutions fulfil the equation

$$\Phi_\lambda(\check{\mathbf{E}}_{(e,\alpha)}, \mathbf{O}, \mathbf{S}^{tr}_{(e,\alpha)}, \mathbf{Q}^{tr}_{(e,\alpha)}) = 0 , \tag{29}$$

they are the solutions of (29) too.

- If the trial solutions are not the solution of (29), then we start the *corrector* with the initial solutions $\widehat{\mathbf{S}}_{(e,\alpha)} = \mathbf{S}^{tr}{}_{(e,\alpha)}$, $\overline{\mathbf{Q}}_{(e,\alpha)} = \mathbf{Q}^{tr}{}_{(e,\alpha)}$ and $\widetilde{\lambda}_{(e,\alpha)} = \mathbf{O}$.

(iv) In the first corrector step we execute *one* Newton-Raphson iteration, replacing $\mathbf{S}_{(e,\alpha)} := \widehat{\mathbf{S}}_{(e,\alpha)}$ and computing the new solution $\widehat{\mathbf{S}}_{(e,\alpha)}$ by.

$$\widehat{\mathbf{S}}_{(e,\alpha)} = [(\Phi^n_s)'(\check{\mathbf{E}}_{(e,\alpha)}, \widetilde{\lambda}_{(e,\alpha)}, \mathbf{S}_{(e,\alpha)}, \overline{\mathbf{Q}}_{(e,\alpha)})]^{-1} \mathbf{S}_{(e,\alpha)} - \Phi^n_s(\check{\mathbf{E}}_{(e,\alpha)}, \widetilde{\lambda}_{(e,\alpha)}, \mathbf{S}_{(e,\alpha)}, \overline{\mathbf{Q}}_{(e,\alpha)}) . \tag{30}$$

The derivative in (30) is given by

$$[\Phi_s^n(\check{\mathbf{E}}_{(e,\alpha)}, \lambda_{(e,\alpha)}, \mathbf{S}_{(e,\alpha)}, \mathbf{Q}_{(e,\alpha)})] = \mathcal{B}_{SS}(\check{\mathbf{E}}_{(e,\alpha)}, \lambda_{(e,\alpha)}, \mathbf{S}_{(e,\alpha)}, \mathbf{Q}_{(e,\alpha)} \tag{31}$$

with $\mathcal{B}_{SS}$ from (23) .

(v) In the next substep we replace $\lambda_{(e,\alpha)} := \widetilde{\lambda}_{(e,\alpha)}$ and compute the next iterative with the help of

$$\widetilde{\lambda}_{(e,\alpha)} = [(\Phi_\lambda^n)'(\check{\mathbf{E}}_{(e,\alpha)}, \lambda_{(e,\alpha)}, \mathbf{S}_{(e,\alpha)}, \mathbf{Q}_{(e,\alpha)})]^{-1}\lambda_{(e,\alpha)} - \Phi_\lambda^n(\check{\mathbf{E}}_{(e,\alpha)}, \lambda_{(e,\alpha)}, \mathbf{S}_{(e,\alpha)}, \mathbf{Q}_{(e,\alpha)}) \;, \tag{32}$$

where the derivative in (32) is given by

$$[\Phi_s^n(\check{\mathbf{E}}_{(e,\alpha)}, \lambda_{(e,\alpha)}, \mathbf{S}_{(e,\alpha)}, \mathbf{Q}_{(e,\alpha)})] = \mathcal{A}_{\lambda\lambda}(\check{\mathbf{E}}_{(e,\alpha)}, \lambda_{(e,\alpha)}, \mathbf{S}_{(e,\alpha)}, \mathbf{Q}_{(e,\alpha)} \tag{33}$$

with $\mathcal{A}_{\lambda\lambda}$ from (24) .

(vi) The most inner iteration begins with the replacement $\mathbf{Q}_{(e,\alpha)} := \overline{\mathbf{Q}}_{(e,\alpha)}$ and computes the next solution by

$$\overline{\mathbf{Q}}_{(e,\alpha)} = [(\Phi_\lambda^n)'(\check{\mathbf{E}}_{(e,\alpha)}, \lambda_{(e,\alpha)}, \mathbf{S}_{(e,\alpha)}, \mathbf{Q}_{(e,\alpha)})]^{-1}\mathbf{Q}_{(e,\alpha)} - \Phi_\lambda^n(\check{\mathbf{E}}_{(e,\alpha)}, \lambda_{(e,\alpha)}, \mathbf{S}_{(e,\alpha)}, \mathbf{Q}_{(e,\alpha)}) \;. \tag{34}$$

- and is to execute up to the fulfilment of a given tolerance criterion.
- With the last solution of (34) we go to (32) and repeat the procedure up to the fulfilment of an tolerance for $\widetilde{\lambda}_{(e,\alpha)}$.
- With this solution we go to (30) and repeat the procedure up to the fulfilment of an tolerance for $\widehat{\mathbf{S}}_{(e,\alpha)}$.
- The steps (iii) - (vii) are to execute for all elements and all integration points giving the values for generating the element stiffness system and for assambling the global (linearized) stiffness system, which allowes us to continue with the step (i).

In order to clarify this process, a structogram of the above algorithm is given in the appendix

2.4 Example: Thermoplasticity

Let's now demonstrate our formalism at an example of thermoplasticity. By a standard procedure the field equations are obtained from the equations of the balance of mass, of momentum, of moment of momentum and of energy. For the normal stresses $\mathcal{S}_n$, prescribed at the boundary part Γ_S , the heat flux $\mathcal{Q}_n$, prescribt at boundary part Γ_Q and for given volume forces $\mathbf{b}$ and internal heat sources $\mathbf{R}$ the field equations are given by the weak formulation of the equilibrium subject to mechanical loads

$$\int\limits_{\mathcal{B}_0} \mathcal{S} : \mathbf{E}'(\mathbf{U};\mathbf{V})\, d\mathcal{B}_0 = \int\limits_{\mathcal{B}_0} \mathbf{b}^T\, \mathbf{V}\, d\mathcal{B}_0 + \int\limits_{\Gamma_S} \mathcal{S}_n \cdot \mathbf{V} d\Gamma_S \qquad \forall \mathbf{V} \in \mathcal{V} \subset [W^{1,2}(\mathcal{B}_0)]^3 \tag{35}$$

and the weak formulation of the energy balance

$$\int_{\mathcal{B}_0} \left(\dot{\mathcal{W}}_{th}\, \widehat{T} + \mathcal{Q}\, \nabla \widehat{T} \right)\, d\mathcal{B}_0 = \int_{\mathcal{B}_0} R\, \widehat{\widehat{T}} d\mathcal{B}_0 + \int_{\Gamma_Q} \mathcal{Q}_n \cdot \widehat{T} d\Gamma_Q \quad \forall T \in \mathcal{V}_1 \subset W^{1,2}(\mathcal{B}_0)\,. \quad (36)$$

In these equations $\mathbf{U}$ and T denotes the classical displacement vector and the absolute temperature. The dependent variables $\mathbf{S}$, $\mathcal{W}_{th}$ and $\mathcal{Q}$ are the second Piola-Kirchhoff stress tensor, the thermal energy[3] and the thermal flux. The function $\mathbf{E}'$ is defined by the equation

$$\mathbf{E}'(\mathbf{U};\mathbf{V}) = \frac{1}{2}[\nabla\mathbf{V} + \nabla\mathbf{V}^T + (\nabla\mathbf{V})^T\nabla\mathbf{U} + (\nabla\mathbf{U})^T\nabla\mathbf{V}]\,. \quad (37)$$

We assume the existence of a free Helmholz energy $\Psi = \Psi(\mathbf{E}_{el}, T, \mathbf{Q})$, which is a function of the elastic part of the Green deformation tensor $\mathbf{E}$, the temperature T and the ordered set $\mathbf{Q}$ of plastic hardening parameters. Furthermore let be given a temperature dependent flow function $F = F(\mathbf{E}, \mathcal{S}, T, \mathbf{Q})$. Using the normality rule

$$\dot{\mathbf{E}}_{pl} = \lambda \mathbf{N}(\mathbf{E}, \mathcal{S}, T, \mathbf{Q}) = \frac{\partial F(\mathbf{E}, \mathcal{S}, T, \mathbf{Q})}{\partial \mathcal{S}}\,,$$

we consider the following evolution equations of dependent variables:

$$\begin{aligned}
\dot{\mathcal{S}} &= \frac{\partial^2 \Psi}{\partial \mathbf{E}^{el}\, \partial \mathbf{E}^{el}}\, \dot{\mathbf{E}} + \frac{\partial^2 \Psi}{\partial \mathbf{E}^{el}\, \partial \mathbf{T}}\, \dot{T} - \lambda\, \frac{\partial^2 \Psi}{\partial \mathbf{E}^{el}\, \partial \mathbf{E}^{el}}\, \mathbf{N} \\
\dot{\mathcal{W}}_{th} &= -\frac{\partial^2 \Psi}{\partial T\, \partial T}\, \dot{T} - T \frac{\partial^2 \Psi}{\partial \mathbf{E}^{el}\, \partial \mathbf{T}}\, \dot{\mathbf{E}} + \lambda\, [T \frac{\partial^2 \Psi}{\partial \mathbf{E}^{el}\, \partial \mathbf{T}} + \frac{\partial \Psi}{\partial \mathbf{E}^{el}}]\, \mathbf{N} \qquad (38) \\
\mathcal{Q} &= \kappa\, \nabla T \qquad \text{(Fourier law of heat condition)}\,.
\end{aligned}$$

To complete the initial boundary value problem of thermo-plasticity we consider a hardening law

$$\dot{\mathbf{Q}} = \lambda \mathbf{M}(\mathbf{E}, \mathcal{S}, T, \mathbf{Q})\,, \quad (39)$$

the loading criterion

$$|\lambda + F|^3 + F^3 - \lambda^3 = 0\,, \quad (40)$$

the consistence condition

$$\lambda > 0 \implies \dot{F} = 0\,, \quad (41)$$

and the initial conditions

$$\begin{aligned}
\mathbf{U}(X,0) &= \mathbf{U}_0\,, \quad T(X,0) = T_0\,, \quad \mathcal{S}(X,0) = \mathcal{S}_0 \\
\mathcal{Q}(X,0) &= \mathbf{Q}_0\,, \quad \mathcal{W}_{th}(X,0) = \mathcal{W}_0\,, \quad \mathbf{Q}(X,0) = \mathbf{Q}_0
\end{aligned} \quad (42)$$

[3] We define the thermal energy as difference between the internal energy and the mechanical energy

To obtain a formulation in the context of our formalism we transfer the notations in following sense

$$
\begin{array}{lll}
\mathbf{U}^{(0)} := \mathbf{U} \ , & \mathbf{U}^{(1)} := T \ , & \\
\mathbf{E}^{(0,0)} := \mathcal{E}^{(0,0)}(\nabla \mathbf{U}) \ , & \mathbf{E}^{(0,1)} := \mathcal{E}^{(0,1)}(\nabla^0 T) = T \ , & \mathbf{E}^{(1,1)} := \mathcal{E}^{(1,1)}(\nabla T) = \nabla T \\
\mathbf{S}^{(0,0)} := \mathcal{S} \ , & \mathbf{S}^{(0,1)} := \mathcal{W}_{th} \ , & \mathbf{S}^{(1,1)} := \mathcal{Q}
\end{array}
\tag{43}
$$

with

$$
\mathcal{E}^{(0,0)}(\nabla \mathbf{U}) = \frac{1}{2}[\nabla \mathbf{U} + \nabla \mathbf{U}^T + (\nabla \mathbf{U})^T \nabla \mathbf{U}] \ . \tag{44}
$$

The equations (39) can be transformed in the form of (10.III), as follows

$$
\begin{pmatrix} \dot{\mathcal{S}} \\ \dot{\mathcal{W}}_{th} \\ \dot{\mathcal{Q}} \end{pmatrix} - \begin{pmatrix} -\frac{\partial^2 \Psi}{\partial \mathbf{E}^{el}\, \partial \mathbf{E}^{el}} \dot{\mathbf{E}} + \frac{\partial^2 \Psi}{\partial \mathbf{E}^{el}\, \partial \mathbf{T}} \dot{T} \\ -\frac{\partial^2 \Psi}{\partial T\, \partial T} \dot{T} - T \frac{\partial^2 \Psi}{\partial \mathbf{E}^{el}\, \partial \mathbf{T}} \dot{\mathbf{E}} \\ \kappa \nabla T \end{pmatrix} - \lambda \begin{pmatrix} -\frac{\partial^2 \Psi}{\partial \mathbf{E}^{el}\, \partial \mathbf{E}^{el}} \mathbf{N} \\ T \frac{\partial^2 \Psi}{\partial \mathbf{E}^{el} \partial T} \frac{\partial \Psi}{\partial \mathbf{E}^{el}} \mathbf{N} \\ \mathbf{O} \end{pmatrix} = \begin{pmatrix} \mathbf{O} \\ \mathbf{O} \\ \mathbf{O} \end{pmatrix} \tag{45}
$$

A more detailed representation including more examples is still in preparation ([8]).

References

[1] In A. K. Miller, editor, *Unified Constitutive Equations for Creep and Plasticity.*. EASP, Barking, 1987.

[2] In H. Bergander, editor, *Inelastische Deformationsgesetze in Standartformulierung*, Wissenschaftliche Schriftenreihe der TU Karl-Marx-Stadt. TU Karl-Marx-Stadt, 5/1990.

[3] A. Bertram. Materielle Systeme mit inneren Zwangsbedingungen. Dissertation, TU Berlin, 1980.

[4] P. Ciarlet. *The Finite Element Method for Elliptic Problems.* Nort-Holland Publ., Amsterdam, 1978.

[5] O. L. Mangasarian. Equivalence of the Complementarity Problem to a System of Nonlinear Equations. *SIAM J. Appl. Math.*, 31(1): 89–92, 1976.

[6] G. A. Maugin. The Method of Virtual Power in Continuum Mechanics: Application to Coupled Fields. *Acta Mechanica*, 35(1-2): 1–70, 1980.

[7] A. Tamme. Initial-Boundary-Value Problems of the Elastic-Plastic Flow Theory for Reduced Stress-/Deformations States and Their Numerical Solution by means of Nested Linearization. *submitted for publication*, 1991.

[8] A. Tamme. in preparation. 1992.

Appendix A

Structogram for the linearization of the field equations

INITIAL VALUES: $\mathbf{U} := \mathbf{U}^0$, $\mathbf{E} := \mathbf{E}^0$, $\mathbf{S} := \mathbf{S}^0$, $\mathbf{Q} := \mathbf{Q}^0$, $[\mathcal{P}']$

Loop over all time levels $n = 0, \cdots, N-1$

- $\widehat{\mathbf{U}} := \mathbf{U}$
  - $\mathbf{U} := \widehat{\mathbf{U}}$
  - $\widehat{\mathbf{U}} := \mathbf{U} - [\mathcal{P}'](\mathbf{U},\mathbf{S})]^{-1} \; (\mathcal{P}(\mathbf{U},\mathbf{S}) + \mathcal{L}^{n+1}) \;\; \rightarrow \;\; \widehat{\mathbf{U}}$
  - $[\mathcal{P}]' := \mathbf{O}$
  - Loop over all elements
    - $\mathbf{U}_e := \mathcal{A}_e \, \mathbf{U}$, $\mathcal{P}_e := \mathbf{O}$
    - Loop over all integration points
      - INPUT: e , α , $\mathbf{U}_e$, $\mathbf{E}_{(e,\alpha)}$, $\mathbf{S}_{(e,\alpha)}, \mathbf{Q}_{(e,\alpha)}$

        PREDICTOR - CORRECTOR (RETURN-) ALGORITHM

        OUTPUT: e , α $\widehat{\mathbf{U}}$, $\mathbf{E}_{(e,\alpha)}$, $\mathbf{S}_{(e,\alpha)}, \mathbf{Q}_{(e,\alpha)}$, $\widehat{\mathbf{E}}_{(e,\alpha)}$, $\widehat{\mathbf{S}}_{(e,\alpha)}$, $\widehat{\mathbf{Q}}_{(e,\alpha)}$, $\mathcal{P}_{(e,\alpha)}$
      - $\mathcal{P}_e' := \mathcal{P}_e' + \omega^{(e,\alpha)} \mathcal{P}_{e,\alpha}'$
    - $\widehat{\mathcal{P}}' = \widehat{\mathcal{P}}' + \mathcal{A}_e^T \mathcal{P}_e' \mathcal{A}_e$
  - $\|\mathcal{P}^n(\widehat{\mathbf{U}},\widehat{\mathbf{S}})\| < \epsilon_u$
- $\mathbf{U}^{n+1} := \widehat{\mathbf{U}}$, $\mathbf{E}^{n+1} : \widehat{\mathbf{E}}$, $\mathbf{S}^{n+1} := \widehat{\mathbf{S}}$, $\mathbf{Q}^{n+1} := \widehat{\mathbf{Q}}$

Appendix B

Structogram of the generalized return algorithm

INPUT: e , α $\mathbf{U}$, $\mathbf{E}^n$, $\mathbf{S}^n$, λ^n , $\mathbf{Q}^n$, $\mathbf{U}$

$[\mathcal{P}^n] := \mathbf{O}$, $\mathbf{E} = \mathbf{E}^n + [\mathcal{E}(\mathbf{U})]\,\mathbf{U}$

Solve $\left[\begin{array}{c} \Phi^n_s(\mathbf{E}, 0, \mathbf{S}^{tr}, \mathbf{Q}^{tr}) \\ \Phi^n_q(\mathbf{E}, 0, \mathbf{S}^{tr}, \mathbf{Q}^{tr} \end{array}\right] = \mathbf{O} \implies \mathbf{S}^{tr}, \mathbf{Q}^{tr}$

$\widehat{\mathbf{S}} := \mathbf{S}^{tr}$, $\widetilde{\lambda} := \mathbf{O}$, $\overline{\mathbf{Q}} := \mathbf{Q}^{tr}$

$\Phi^n_\lambda(\mathbf{E}, \widetilde{\lambda}, \widehat{\mathbf{S}}^{tr}, \overline{\mathbf{Q}}^{tr}) \neq 0$?

YES

$\mathbf{S} := \widehat{\mathbf{S}}$

$\widehat{\mathbf{S}} := \mathbf{S} - \left[\mathcal{B}_{SS}(\mathbf{E}, \widetilde{\lambda}, \mathbf{S}, \overline{\mathbf{Q}})\right]^{-1} : \Phi_S(\mathbf{E}, \widetilde{\lambda}, \mathbf{S}, \overline{\mathbf{Q}})$

$\lambda := \widetilde{\lambda}$

$\widetilde{\lambda} := \lambda - \left[\mathcal{A}_{\lambda\lambda}(\mathbf{E}, \lambda, \widehat{\mathbf{S}}, \overline{\mathbf{Q}})\right] : \Phi_\lambda(\mathbf{E}, \lambda, \widehat{\mathbf{S}}, \overline{\mathbf{Q}})$

$\mathbf{Q} := \overline{\mathbf{Q}}$

$\overline{\mathbf{Q}} := \mathbf{Q} - \dfrac{\partial \Phi_q(\mathbf{E}, \widetilde{\lambda}, \mathbf{S}, \overline{\mathbf{Q}})}{\partial Q} : \Phi_\lambda(\mathbf{E}, \widetilde{\lambda}, \widehat{\mathbf{S}}, \mathbf{Q})$

$\|\Phi_q(\mathbf{E}, \widetilde{\lambda}, \widehat{\mathbf{S}}, \overline{\mathbf{Q}})\| < \epsilon_q$

$\|\Phi_q(\mathbf{E}, \widetilde{\lambda}, \widehat{\mathbf{S}}, \widetilde{\mathbf{Q}})\| < \epsilon_\lambda$

$\|\Phi'_\lambda(\mathbf{E}, \widehat{\lambda}, \widehat{\mathbf{S}}, \widehat{\mathbf{Q}})\| < \epsilon_S$

NO

OUTPUT: e , α $\mathbf{U}$, $\mathbf{E}^n$, $\mathbf{S}^n$, λ^n , $\mathbf{Q}^n$; $\widehat{\mathbf{S}}$, $\widetilde{\lambda}$, $\overline{\mathbf{Q}}$

On the Thermodynamics of Rate-Independent Plasticity as an Asymptotic Limit of Viscoplasticity for Slow Processes

P. Haupt, M. Kamlah, Ch. Tsakmakis

Institute of Mechanics, University of Kassel
Mönchebergstr. 7, 3500 Kassel, Germany

Summary

The thermodynamics of internal variables, formulated by Coleman and Gurtin [1], provides a general frame for constitutive theories of viscoelasticity. This thermodynamic background has been extended to viscoplasticity by Kratochvil and Dillon [5]. On the basis of their argumentations we study here the rate-independent thermoplasticity as an asymptotic limit for slow processes or for the limit case of a vanishing viscosity. We notice, that it is not necessary to postulate the existence of this limit: Actually, it follows from the properties of the considered material model. Furthermore, the evolution equation for the plastic strain is obtained from the limit process.

1. Thermodynamics with Internal Variables

To represent thermomechanical effects the basic relations of continuum mechanics must be supplemented by the **energy balance** (1st Law),

$$\dot{e} = \frac{1}{\rho}\mathbf{T}\cdot\dot{\mathbf{E}} - \frac{1}{\rho}\operatorname{div}\mathbf{q} + r \quad , \tag{1}$$

$\left((\,)^{\cdot}\right.$: material time derivative of $\left.(\,)\right)$ and the **entropy inequality** (2nd Law):

$$\Theta\dot{s} \geq -\frac{1}{\rho}\operatorname{div}\mathbf{q} + r + \frac{1}{\rho\Theta}\mathbf{q}\cdot\mathbf{g} \tag{2}$$

Eliminating the heat flux and supply with eq. (1) we obtain the formulation

$$-\dot{e} + \frac{1}{\rho}\mathbf{T}\cdot\dot{\mathbf{E}} + \Theta\dot{s} - \frac{1}{\rho\Theta}\mathbf{q}\cdot\mathbf{g} \geq 0 \tag{3}$$

of the entropy inequality, which is widely used in continuum mechanics to restrict the explicit form of constitutive equations. In the above formulae, e is the specific energy, ρ the mass density, $\mathbf{T}$ the stress tensor, $\mathbf{E}$ the strain tensor, $\mathbf{q}$ the heat flux vector, r the heat supply per unit mass, Θ the absolute temperature, s the specific entropy and $\mathbf{g} = \operatorname{grad}\Theta$ the temperature gradient.

Throughout this paper no distinction will be made between different kinds of stress and strain tensors, spatial and material coordinates and similar fundamental aspects. Thus, all equations may only apply in the context of geometric linearization.

D. Besdo, E. Stein (Eds.)
Finite Inelastic Deformations –
Theory and Applications
IUTAM Symposium Hannover/Germany 1991

However, it is possible to employ an appropriate notation, which makes all formulations meaningful for arbitrarily large deformations. In this context we refer to the concept of dual variables and derivatives, developed in [2].

As a basic assumption, we postulate the additive decomposition of the total strain rate tensor $\dot{\mathbf{E}}$ into elastic and inelastic parts:

$$\dot{\mathbf{E}} = \dot{\mathbf{E}}_e + \dot{\mathbf{E}}_i \tag{4}$$

Accordingly, the stress power decomposes, i.e.,

$$w = \frac{1}{\rho}\mathbf{T}\cdot\dot{\mathbf{E}} = \frac{1}{\rho}\mathbf{T}\cdot\dot{\mathbf{E}}_e + \frac{1}{\rho}\mathbf{T}\cdot\dot{\mathbf{E}}_i = w_e + w_i \quad . \tag{5}$$

The theory of internal variables as formulated by Coleman and Gurtin [1] is a generalization of thermoelasticity in the sense that a set of additional variables is introduced to represent the inelastic material behavior. These may be assumed to be the inelastic strain tensor $\mathbf{E}_i$ and a further set of quantities $q_1, q_2, \ldots, q_n$ to be specified in the particular context.

On the basis of the specific internal energy e two other thermodynamic quantities are defined, namely the **free energy** and the **free enthalpy**, respectively:

$$\psi = e - \Theta s \quad , \quad g = \frac{1}{\rho}\mathbf{T}\cdot\mathbf{E}_e - \psi$$

It is important to recognize that the free enthalpy is defined on the basis of the elastic part $\mathbf{E}_e$ of the total strain. For the free energy the following general constitutive equation may be postulated:

$$\psi = \hat{\psi}\left(\mathbf{E}_e, \Theta, q_1, \ldots, q_n\right) \tag{6}$$

Here, it is assumed that the free energy depends on the total strain $\mathbf{E}$, the temperature Θ and the internal variables $\mathbf{E}_i, q_1, \ldots, q_n$; however, the dependence on $\mathbf{E}$ and $\mathbf{E}_i$ is restricted to the dependence on the difference $\mathbf{E} - \mathbf{E}_i = \mathbf{E}_e$ [5]. Standard arguments suggest that the free energy (6) represents a potential for the stress tensor and the entropy:

$$\frac{1}{\rho}\mathbf{T} = \frac{\partial\hat{\psi}}{\partial\mathbf{E}_e} \quad , \quad s = -\frac{\partial\hat{\psi}}{\partial\Theta} \quad .$$

A quite natural constitutive assumption for the free enthalpy, corresponding to eq. (6) is related to the stress-space formulation. In this representation the free enthalpy is the potential of the elastic strain and the entropy:

$$g = \hat{g}\left(\mathbf{T}, \Theta, q_1, \ldots, q_n\right) \quad , \quad \frac{1}{\rho}\mathbf{E}_e = \frac{\partial\hat{g}}{\partial\mathbf{T}} \quad , \quad s = \frac{\partial\hat{g}}{\partial\Theta} \tag{7}$$

Thereby, the entropy inequality reduces to the internal **dissipation inequality,**

$$\frac{1}{\rho}\mathbf{T}\cdot\dot{\mathbf{E}}_i + \sum_{k=1}^{n}\frac{\partial\hat{g}}{\partial q_k}\dot{q}_k - \frac{1}{\rho\Theta}\mathbf{q}\cdot\mathbf{g} \geq 0 \quad , \tag{8}$$

and the energy balance (1) implies

$$\Theta\dot{s} = -\frac{1}{\rho}\operatorname{div}\mathbf{q} + r + \frac{1}{\rho}\mathbf{T}\cdot\dot{\mathbf{E}}_i + \sum_{k=1}^{n}\frac{\partial\hat{g}}{\partial q_k}\dot{q}_k \quad . \tag{9}$$

In passing we note that calculating the time derivative $\dot{s}$ of the specific entropy from the potential g and introducing the specific heat at constant stress,

$$c_s\left(\mathbf{T}, \Theta, q_1, \ldots, q_n\right) = \Theta \frac{\partial^2 \hat{g}}{\partial \Theta^2} \quad , \tag{10}$$

we arrive at the **equation of heat conduction** (see also [3]):

$$c_s \dot{\Theta} + \Theta \frac{\partial^2 \hat{g}}{\partial \Theta\, \partial \mathbf{T}} \cdot \dot{\mathbf{T}} = -\frac{1}{\rho} \operatorname{div} \mathbf{q} + r + \frac{1}{\rho} \mathbf{T} \cdot \dot{\mathbf{E}}_i + \sum_{k=1}^{n} \frac{\partial}{\partial q_k} (g - \Theta s) \dot{q}_k \tag{11}$$

This equation illustrates the fact that (apart from the thermoelastic coupling on the left-hand side) the change of temperature is caused by the local heating, the inelastic stress power and, finally, the change of the internal variables expressed in terms of the enthalpy $h = g - \Theta s$.

The above constitutive equations are completed by evolution equations for the inelastic strain rate and the other internal variables:

$$\dot{\mathbf{E}}_i(t) = \mathbf{F}\left(\mathbf{T}, \Theta, q_1, \ldots, q_n\right) \quad , \quad \dot{q}_k(t) = f_k\left(\mathbf{T}, \Theta, q_1, \ldots, q_n\right) \tag{12}$$

The whole set of evolution equations (12) together with a constitutive equation for the heat flux vector $\mathbf{q}$ must satisfy the dissipation inequality (8).

2. Rate-Dependent Viscoplasticity

The thermodynamics of internal variables can be utilized for a consistent representation of viscoelastic as well as viscoplastic behavior. For the following it suffices to assume a special case of viscoplasticity, namely a linear thermoelasticity relation

$$\dot{\mathbf{T}}(t) = \mathbb{C}\left[\dot{\mathbf{E}}_e(t)\right] + \mathbf{C}\dot{\Theta}(t) \quad , \tag{13}$$

an associated flow rule

$$\dot{\mathbf{E}}_i(t) = \frac{1}{\eta} \left\langle f(\mathbf{T}, \mathbf{X}, \Theta) \right\rangle \frac{\partial f}{\partial \mathbf{T}} \quad , \tag{14}$$

and the Armstrong-Frederick hardening model*) $\left(|\dot{\mathbf{E}}_i| = \sqrt{\tfrac{2}{3} \operatorname{tr}(\dot{\mathbf{E}}_i)^2} \right)$

$$\dot{\mathbf{X}} = c\dot{\mathbf{E}}_i - b|\dot{\mathbf{E}}_i|\,\mathbf{X} \quad . \tag{15}$$

Here, $\mathbb{C}$, $\mathbf{C}$ are elasticity tensors, and η, c and b are material constants. The yield function f is a scalar-valued material function. The condition

$$f(\mathbf{T}, \mathbf{X}, \Theta) = 0 \tag{16}$$

defines a surface in the stress space, the **static yield surface.** A special case would be the v. Mises yield function ($\mathbf{T}^D$: deviator of $\mathbf{T}$, σ_o: static yield stress)

---

*) The Armstrong-Frederick equation (15) has been investigated intensively in [7] and can be interpreted as the first term in a far more general representation of the kinematic hardening behavior. It is compatible with the entropy inequality; in a recent paper [4] we demonstrate how the hardening model (15) can be modified to incorporate the detailed hardening and softening behavior, which can be observed in the context of cyclic strain processes with different amplitudes.

$$f(\mathbf{T}, \mathbf{X}, \Theta) = \frac{1}{2}\,\mathrm{tr}(\mathbf{T}^D - \mathbf{X}^D)^2 - \frac{1}{3}\sigma_o^2(\Theta) \quad .$$

In this case $\mathbf{X}$ is the center of the static yield surface. Usually, $\mathbf{X}$ is called the **back stress** tensor. In the flow rule (14) the brackets $\langle\ \rangle$ denote the following definition:

$$\langle x \rangle = \begin{cases} x \text{ for } x \geq 0 \\ 0 \text{ for } x < 0 \end{cases} \tag{17}$$

Thus, the static yield surface (16) forms the boundary of the elastic region, where the inelastic strain $\mathbf{E}_i$ is constant in time. Inelastic strains are produced for stresses outside the static yield surface, i.e. for positive values of the yield function f. If a positive value of f is interpreted as a distance with respect to the static yield surface, the inelastic strain rate is proportional to that distance and the parameter η corresponds to the **viscosity** of the material. Finally, the center of the static yield surface may move according to the evolution equation (15).

Obviously, the above constitutive model of a viscoplastic material is a special case of the general theory of internal variables. To recognize this, we identify the back stress tensor $\mathbf{X}$ with the internal variables $q_1, \ldots, q_n$ and insert eq. (14) into eq. (15). Clearly, the elasticity relation (13) corresponds to a special assumption for the free enthalpy g; the entropy relation is omitted. It is not needed in the sequel.

For the following considerations it is not important that the associated flow rule has been assumed: In eq. (14) the gradient of the yield function may be replaced by a tensor valued material function $\mathbf{G} = \mathbf{G}(\mathbf{T}, \mathbf{X}, \Theta)$. The special choice of an associated flow rule is a simple possibility to guarantee the stability property of the material model, expressed by the condition (25) (see below).

3. A basic identity for the yield function

Now, let a strain and temperature process

$$\Lambda(t) = \{\mathbf{E}(t)\ ,\ \Theta(t)\} \tag{18}$$

be given and initial conditions for $\mathbf{E}_i$, $\mathbf{T}$ and $\mathbf{X}$ be prescribed such, that the static yield condition is satisfied for $t=t_o$. We assume that the set of eqs. (4), (13), (14) and (15) determine a unique solution $\mathbf{E}_i(t)$, $\mathbf{T}(t)$ and $\mathbf{X}(t)$. Accordingly, the yield function f is also determined as a function of time, i.e.,

$$f(t) = f\big(\mathbf{T}(t), \mathbf{X}(t), \Theta(t)\big) \quad .$$

The relation between plasticity and viscoplasticity can be studied, if we calculate the total time derivative $\dot{f}(t)$ [5], [8]. Observing the chain rule and inserting the decomposition (4) and the constitutive assumptions (13)-(15), we arrive at the identity

$$\dot{f}(t) + \frac{1}{\eta}K(t)\,\langle f(t)\rangle = \mathbf{\Sigma}(t)\cdot\dot{\Lambda}(t) \quad , \tag{19}$$

where the following definitions have been introduced:

$$K(t) = \frac{\partial f}{\partial \mathbf{T}}\cdot \mathbf{C}\Big[\frac{\partial f}{\partial \mathbf{T}}\Big] - c\,\frac{\partial f}{\partial \mathbf{T}}\cdot\frac{\partial f}{\partial \mathbf{X}} + b\Big(\mathbf{X}\cdot\frac{\partial f}{\partial \mathbf{X}}\Big)\Big|\frac{\partial f}{\partial \mathbf{T}}\Big| \tag{20}$$

$$\mathbf{\Sigma}(t) = \Big\{\mathbf{C}\Big[\frac{\partial f}{\partial \mathbf{T}}\Big]\ ,\ \mathbf{C}\cdot\frac{\partial f}{\partial \mathbf{T}} + \frac{\partial f}{\partial \Theta}\Big\} \tag{21}$$

$$\mathbf{\Sigma}(t) \cdot \dot{\mathbf{\Lambda}}(t) = \mathbf{C}\left[\frac{\partial f}{\partial \mathbf{T}}\right] \cdot \dot{\mathbf{E}}(t) + \left(\mathbf{C} \cdot \frac{\partial f}{\partial \mathbf{T}} + \frac{\partial f}{\partial \Theta}\right)\dot{\Theta}(t) \tag{22}$$

It is important to note that K and $\mathbf{\Sigma}$ depend on the current state only. As a restriction on the regularity of the constitutive functions we assume K and $\mathbf{\Sigma}$ to be finite. For inelastic processes ($f > 0$) we have

$$\dot{f}(t) + \frac{1}{\eta} K(t)\, f(t) = \mathbf{\Sigma}(t) \cdot \dot{\mathbf{\Lambda}}(t) , \quad t \geq t_o \tag{23}$$

Another formulation of this identity is given by

$$f(t) = \int_{t_o}^{t} e^{-\frac{1}{\eta}\int_{\tau}^{t} K(\sigma)\, d\sigma}\, \mathbf{\Sigma}(\tau) \cdot \dot{\mathbf{\Lambda}}(\tau)\, d\tau , \tag{24}$$

which is the solution of the differential equation (23) for the initial condition $f(t_o) = 0$. This equation may be understood as a first integral of the system of differential equations (4), (13)-(15). The identities (23) and (24) are known from the paper of Kratochvil and Dillon [5]. They lead to quite interesting consequences which will be analyzed in the sequel. These consequences are based on the assumption that K(t) is strictly positive, i.e. ,

$$K(t) \geq K_o > 0 . \tag{25}$$

In fact, this assumption corresponds to stability properties of the material. It imposes a restriction on the system of constitutive equations and can be shown to be met by eq. (14) and (15) in combination with a v. Mises yield function. The identity (23) or, equivalently, eq. (24) relates the evolution of the yield function to the given strain and temperature process.

As a first observation we recognize a **relaxation property**: A static continuation of an arbitrary process implies $\dot{\mathbf{\Lambda}} = \{\mathbf{0}, 0\}$ and, because of the stability condition (25), f(t) tends to zero. Consequently, both the inelastic strain rate $\dot{\mathbf{E}}_i$ and the rate $\dot{\mathbf{X}}$ of the back stress according to the evolution equation (22) tend to zero. Thus, an equilibrium solution of the system of differential equations is approached during a relaxation process at constant strain and temperature. As a characteristic feature of viscoplasticity this equilibrium state depends on the foregoing process history: This is due to the particular structure of the flow rule (14) and the evolution equation (15) for the back stress $\mathbf{X}$.

We may guess, that in the case of inelastic processes corresponding to very small strain and temperature rates the values of f(t) will be very close to zero. The same might be true for finite rates and very small values of the viscosity η (see eq. (23). Indeed, these speculations can be corroborated. Moreover, it can be seen that in these cases the flow rule (21) of viscoplasticity tends asymptotically to the flow rule of rate-independent thermoplasticity.

4. Slow processes

On the basis of the generalized vector $\mathbf{\Sigma}$, defined by eq. (21) we define the kernel function

$$\mathbf{\Gamma}(t, \tau) = \left\{ e^{-\frac{1}{\eta}\int_{\tau}^{t} K(\sigma)\, d\sigma} \right\} \mathbf{\Sigma}(\tau) \tag{26}$$

and we rewrite eq. (24) as

$$f(t) = \int_{t_o}^{t} \boldsymbol{\Gamma}(t, \tau) \cdot \dot{\boldsymbol{\Lambda}}(\tau)\, d\tau \quad . \tag{27}$$

Integration by parts and the substitution τ = t-s leads to

$$f(t) = \boldsymbol{\Gamma}(t, t_o) \cdot \left[\boldsymbol{\Lambda}(t) - \boldsymbol{\Lambda}(t_o)\right] + \int_0^{t-t_o} \frac{\partial}{\partial s} \boldsymbol{\Gamma}(t, t-s) \cdot \boldsymbol{\Lambda}_d^t(s)\, ds \ , \tag{28}$$

where

$$\boldsymbol{\Lambda}_d^t(s) = \boldsymbol{\Lambda}(t-s) - \boldsymbol{\Lambda}(t) = \left\{\mathbf{E}(t-s) - \mathbf{E}(t),\ \Theta(t-s) - \Theta(t)\right\} \tag{29}$$

is the relative difference history. Following Coleman and Noll [6], the **retardation of the strain-temperature history** is introduced by

$$\boldsymbol{\Lambda}_{d(\alpha)}^t(s) = \boldsymbol{\Lambda}_d^t(\alpha s) = \boldsymbol{\Lambda}(t - \alpha s) - \boldsymbol{\Lambda}(t) \ , \tag{30}$$

with $0 < \alpha < 1$. According to the retardation the starting point t_o is replaced by

$$t_{o(\alpha)} = t - \frac{t - t_o}{\alpha} \ .$$

Inserting this into eq. (28) we get

$$f_{(\alpha)}(t) = \boldsymbol{\Gamma}_{(\alpha)}(t, t_{o(\alpha)}) \cdot \left[\boldsymbol{\Lambda}(t) - \boldsymbol{\Lambda}(t_o)\right] + \int_0^{t-t_{o(\alpha)}} \frac{\partial}{\partial s} \boldsymbol{\Gamma}_{(\alpha)}(t, t-s) \cdot \boldsymbol{\Lambda}_d^t(\alpha s)\, ds \ , \tag{31}$$

where the subscripts (α) indicate the dependence on the retardation factor α, which is induced by the solution of the constitutive differential equations (4), (13)-(15) as a response to the retarded strain-temperature process. $f_{(\alpha)}(t)$ is a solution for the-retarded form of the differential eq. (23), which was won from the identity (19) for inelastic processes. Consequently, eq. (31) is only valid for $0 < \alpha < 1$, if the process remains inelastic for all retardations, i.e. $f_{(\alpha)}(t) > 0$ for $\alpha \neq 0$. A corresponding criterion will be derived later.

Now, we are interested to investigate the limit $\alpha \to 0$, which implies $t_{o(\alpha)} \to -\infty$. Looking at eqs. (25) and (26) we infer

$$\lim_{\alpha \to 0} \boldsymbol{\Gamma}_{(\alpha)}(t, t_{o(\alpha)}) = \{\mathbf{0}, 0\}$$

as well as

$$\lim_{\alpha \to 0} \frac{1}{\alpha} \boldsymbol{\Gamma}_{(\alpha)}(t, t_{o(\alpha)}) = \{\mathbf{0}, 0\} \quad . \tag{32}$$

In the sense of the fading memory theory, outlined in [6], we replace the retarded difference history by its Taylor-approximation:

$$\boldsymbol{\Lambda}_d^t(\alpha s) = -\alpha s\, \dot{\boldsymbol{\Lambda}}(t) + O(\alpha^2 s^2) \tag{33}$$

With (32) and (33) eq. (31) then reduces to

$$\lim_{\alpha \to 0} \frac{1}{\alpha} f_{(\alpha)}(t) = -\lim_{\alpha \to 0} \left\{ \int_0^{t-t_{o(\alpha)}} s \frac{\partial}{\partial s} \boldsymbol{\Gamma}_{(\alpha)}(t, t-s)\, ds \right\} \cdot \dot{\boldsymbol{\Lambda}}(t) \ . \tag{34}$$

From the definition (26) of $\boldsymbol{\Gamma}$ we conclude, that the right-hand side of eq. (34) is finite; thus, we have the asymptotic relation

$$f_{(\alpha)}(t) = O(\alpha) \quad ; \tag{35}$$

in view of eq. (23) this implies also

$$\dot{f}_{(\alpha)}(t) = O(\alpha) \ . \tag{36}$$

We now investigate the term

$$-\lim_{\alpha\to 0}\left\{\int_0^{t-t_{o(\alpha)}} s\frac{\partial}{\partial s}\boldsymbol{\Gamma}_{(\alpha)}(t, t-s)\,ds\right\} = \lim_{\alpha\to 0}\left\{\int_0^{t-t_{o(\alpha)}}\boldsymbol{\Gamma}_{(\alpha)}(t, t-s)\,ds\right\} : \tag{37}$$

From eq. (26) we conclude

$$\int_0^{t-t_{o(\alpha)}}\boldsymbol{\Gamma}_{(\alpha)}(t, t-s)\,ds = \int_0^{t-t_{o(\alpha)}} e^{-\frac{1}{\eta}\int_0^s K_{(\alpha)}(t-\sigma)\,d\sigma}\,\boldsymbol{\Sigma}_{(\alpha)}(t-s)\,ds =$$

$$= \int_0^{t-t_{o(\alpha)}} e^{-\frac{1}{\eta}K_{(\alpha)}(t)\,s}\left\{e^{-\frac{1}{\eta}\int_0^s\left[K_{(\alpha)}(t-\sigma)-K_{(\alpha)}(t)\right]d\sigma}\,\boldsymbol{\Sigma}_{(\alpha)}(t-s)\right\}ds .$$

Integration by parts leads to

$$\int_0^{t-t_{o(\alpha)}}\boldsymbol{\Gamma}_{(\alpha)}(t, t-s)\,ds = \frac{\eta}{K_{(\alpha)}(t)}\boldsymbol{\Sigma}_{(\alpha)}(t) - \frac{\eta}{K_{(\alpha)}(t)}\,e^{-\frac{1}{\eta}\int_0^{t-t_{o(\alpha)}}K_{(\alpha)}(t-\sigma)\,d\sigma}\,\boldsymbol{\Sigma}_{(\alpha)}(t_{o(\alpha)}) +$$

$$+\frac{\eta}{K_{(\alpha)}(t)}\int_0^{t-t_{o(\alpha)}} e^{-\frac{1}{\eta}\int_0^s K_{(\alpha)}(t-\sigma)\,d\sigma}\,\frac{\partial}{\partial s}\boldsymbol{\Sigma}_{(\alpha)}(t-s)\,ds \ +$$

$$-\frac{1}{K_{(\alpha)}(t)}\int_0^{t-t_{o(\alpha)}} e^{-\frac{1}{\eta}\int_0^s K_{(\alpha)}(t-\sigma)\,d\sigma}\left[K_{(\alpha)}(t-s)-K_{(\alpha)}(t)\right]\boldsymbol{\Sigma}_{(\alpha)}(t-s)\,ds. \tag{38}$$

It is a technical matter to prove the asymptotic relations

$$\frac{\partial}{\partial s}\boldsymbol{\Sigma}_{(\alpha)}(t-s) = O(\alpha) \ , \tag{39}$$

$$K_{(\alpha)}(t-s) - K_{(\alpha)}(t) = O(\alpha) \quad . \tag{40}$$

Thus, equations (34), (37), (38), (39) and (40) lead to

$$\lim_{\alpha\to 0}\frac{1}{\alpha}f_{(\alpha)}(t) = \lim_{\alpha\to 0}\frac{\eta}{K_{(\alpha)}(t)}\boldsymbol{\Sigma}_{(\alpha)}(t)\cdot\dot{\boldsymbol{\Lambda}}(t) \tag{41}$$

for processes, which are inelastic for all redardations, i.e. $\frac{1}{\alpha}f_{(\alpha)}(t)>0$ for $\alpha\neq 0$. Recalling the flow rule (14), we have for such processes

$$\frac{1}{\alpha}\dot{\mathbf{E}}_{i(\alpha)}(t) = \left\{\frac{1}{\alpha\eta}f_{(\alpha)}(t)\right\}\frac{\partial f_{(\alpha)}}{\partial\mathbf{T}} \quad , \tag{42}$$

and with the definition

$$\lim_{\alpha\to 0} \frac{1}{\alpha} \dot{\mathbf{E}}_{i(\alpha)}(t) = \dot{\mathbf{E}}_p(t) \tag{43}$$

we arrive at the result

$$\dot{\mathbf{E}}_p(t) = \left\{ \frac{1}{K_{(0)}(t)} \boldsymbol{\Sigma}_{(0)}(t) \cdot \dot{\boldsymbol{\Lambda}}(t) \right\} \frac{\partial f_{(0)}}{\partial \mathbf{T}} \quad , \tag{44}$$

or in view of eqs. (20)-(22), omitting the index (0):

$$\dot{\mathbf{E}}_p(t) = \frac{\mathbb{C}\left[\frac{\partial f}{\partial \mathbf{T}}\right] \cdot \dot{\mathbf{E}}(t) + \left(\mathbf{C} \cdot \frac{\partial f}{\partial \mathbf{T}} + \frac{\partial f}{\partial \Theta}\right) \dot{\Theta}(t)}{\frac{\partial f}{\partial \mathbf{T}} \cdot \mathbb{C}\left[\frac{\partial f}{\partial \mathbf{T}}\right] - c\, \frac{\partial f}{\partial \mathbf{T}} \cdot \frac{\partial f}{\partial \mathbf{X}} + b \left(\mathbf{X} \cdot \frac{\partial f}{\partial \mathbf{X}}\right) \left|\frac{\partial f}{\partial \mathbf{T}}\right|} \frac{\partial f}{\partial \mathbf{T}} \tag{45}$$

This equation is recognized as the flow rule of rate-independent plasticity, which follows from the consistency condition $\dot{f}(t) = 0$ in the context of $f(t) = 0$ and the case of loading.

5. Loading criterion

Finally, we have to specify a criterion for such processes, which remain inelastic during retardation. In case $f_{(\alpha)}(t)$ becomes negative as α approaches zero, the underlying differential eq. (23) is no longer relevant. In this case we simply have an elastic process (see eq. (14)), i.e.,

$$\lim_{\alpha\to 0} \frac{1}{\alpha} \dot{\mathbf{E}}_{i(\alpha)}(t) = \mathbf{0} \quad . \tag{46}$$

According to the definition (30) of the retardation of the prescribed strain-temperatur process (18) we find

$$\dot{\mathbf{E}}(t) = -\frac{1}{\alpha} \frac{d}{ds} \mathbf{E}_d^t(\alpha s)\Big|_{s=0} = \frac{1}{\alpha} \dot{\mathbf{E}}_{(\alpha)}(t) = \frac{1}{\alpha}\left(\dot{\mathbf{E}}_{e(\alpha)}(t) + \dot{\mathbf{E}}_{i(\alpha)}(t) \right) \quad , \tag{47}$$

where in the last step the additive decomposition (4) has been applied to the retarded strain history. From the elasticity law (13) we get for the retarded process

$$\dot{\mathbf{T}}_{(\alpha)}(t) = \mathbb{C}\left[\dot{\mathbf{E}}_{e(\alpha)}(t)\right] + \alpha \mathbf{C}\, \dot{\Theta}(t) \qquad \left(\dot{\Theta}_{(\alpha)}(t) = \alpha\, \dot{\Theta}(t) \right) \quad . \tag{48}$$

With the help of eqs. (42), (47) and (48) we obtain from eq. (41) for inelastic processes

$$\lim_{\alpha\to 0} \frac{1}{\alpha} f_{(\alpha)}(t) = \lim_{\alpha\to 0} \frac{\eta}{K_{(\alpha)}(t)} \left[\frac{\partial f_{(\alpha)}}{\partial \mathbf{T}} \cdot \frac{1}{\alpha} \dot{\mathbf{T}}_{(\alpha)}(t) + \frac{\partial f_{(\alpha)}}{\partial \Theta} \dot{\Theta}(t) + \right.$$

$$\left. + \frac{f_{(\alpha)}(t)}{\alpha \eta} \mathbb{C}\left[\frac{\partial f_{(\alpha)}}{\partial \mathbf{T}} \right] \cdot \frac{\partial f_{(\alpha)}}{\partial \mathbf{T}} \right] \quad . \tag{49}$$

This can be rewritten as

$$\lim_{\alpha\to 0} \frac{f_{(\alpha)}(t)}{\alpha}\left(1 - \frac{1}{K_{(\alpha)}(t)}\,\mathbf{C}\left[\frac{\partial f_{(\alpha)}}{\partial \mathbf{T}}\right]\cdot\frac{\partial f_{(\alpha)}}{\partial \mathbf{T}}\right) =$$

$$= \lim_{\alpha\to 0} \frac{\eta}{K_{(\alpha)}(t)}\left[\frac{\partial f_{(\alpha)}}{\partial \mathbf{T}}\cdot\frac{1}{\alpha}\dot{\mathbf{T}}_{(\alpha)}(t) + \frac{\partial f_{(\alpha)}}{\partial \Theta}\,\dot{\Theta}(t)\right] \quad . \qquad (50)$$

Looking at the definition (22), i.e.

$$K(t) = \frac{\partial f}{\partial \mathbf{T}}\cdot\mathbf{C}\left[\frac{\partial f}{\partial \mathbf{T}}\right] + \overline{K}(t) \quad , \quad \overline{K}(t) = -c\,\frac{\partial f}{\partial \mathbf{T}}\cdot\frac{\partial f}{\partial \mathbf{X}} + b\left(\mathbf{X}\cdot\frac{\partial f}{\partial \mathbf{X}}\right)\left|\frac{\partial f}{\partial \mathbf{T}}\right| \quad , \qquad (51)$$

we see, that the property (25) together with the assumption

$$\overline{K}(t) \geq 0 \quad , \qquad (52)$$

implies

$$1 - \frac{1}{K_{(\alpha)}(t)}\,\mathbf{C}\left[\frac{\partial f_{(\alpha)}}{\partial \mathbf{T}}\right]\cdot\frac{\partial f_{(\alpha)}}{\partial \mathbf{T}} \geq 0 \quad , \qquad (53)$$

since the elasticity tensor $\mathbf{C}$ is positive definite. Now, if a process remains inelastic, i.e. $f_{(\alpha)}(t) > 0$, then eq. (50) yields

$$\lim_{\alpha\to 0} \frac{\partial f_{(\alpha)}}{\partial \mathbf{T}}\cdot\frac{1}{\alpha}\dot{\mathbf{T}}_{(\alpha)}(t) + \frac{\partial f_{(\alpha)}}{\partial \Theta}\,\dot{\Theta}(t) \geq 0 \quad . \qquad (54)$$

In this relation, the equality holds, if $\overline{K}(t)=0$. The special case of ideal plasticity has this property. If $\overline{K}(t)>0$, the expression (53) is positive and then the left-hand side of (54) is positive, if and only if the limit of $\frac{1}{\alpha} f_{(\alpha)}(t)$ is positive. In this sense the inequality (54) coincides with the loading condition of rate-independent plasticity, where the equality sign corresponds to the case of neutral loading.

6. Vanishing viscosity

In order to investigate the limit case of a vanishing viscosity η we look at eq. (24), which holds for inelastic processes and is equivalent to

$$f(t) = \int_0^{t-t_o} e^{-\frac{1}{\eta}\int_0^s K(t-\sigma)d\sigma}\,\boldsymbol{\Sigma}(t-\sigma)\cdot\dot{\boldsymbol{\Lambda}}(t-s)\,ds$$

$$= \int_0^{t-t_o} e^{-\frac{1}{\eta}K(t)\,s}\; e^{-\frac{1}{\eta}\int_0^s [K(t-\sigma)-K(t)]d\sigma}\,\boldsymbol{\Sigma}(t-s)\cdot\dot{\boldsymbol{\Lambda}}(t-s)\,ds \qquad (55)$$

Again, integration by parts yields

$$\frac{1}{\eta}f(t) = \frac{1}{K(t)}\boldsymbol{\Sigma}(t)\cdot\dot{\boldsymbol{\Lambda}}(t) - e^{-\frac{1}{\eta}\int_0^{t-t_o} K(t-\sigma)d\sigma}\,\boldsymbol{\Sigma}(t_o)\cdot\dot{\boldsymbol{\Lambda}}(t_o) +$$

$$- \frac{1}{K(t)}\int_0^{t-t_o} e^{-\frac{1}{\eta}\int_0^s K(t-\sigma)d\sigma}\left[\dot{\boldsymbol{\Sigma}}(t-s)\cdot\dot{\boldsymbol{\Lambda}}(t-s) + \boldsymbol{\Sigma}(t-s)\cdot\ddot{\boldsymbol{\Lambda}}(t-s)\right]ds +$$

$$- \frac{1}{K(t)}\int_0^{t-t_o}\frac{1}{\eta}\,e^{-\frac{1}{\eta}\int_0^s K(t-\sigma)d\sigma}\left[K(t-s) - K(t)\right]\boldsymbol{\Sigma}(t-s)\cdot\dot{\boldsymbol{\Lambda}}(t-s)\,ds \,.$$

Technical proofs corroborate the presumption that for the limit case of vanishing viscosity all terms on the right-hand side vanish, apart from the first one. Thus, we arrive at

$$\lim_{\eta \to 0} \frac{1}{\eta} f(t) = \frac{1}{K(t)} \mathbf{\Sigma}(t) \cdot \dot{\mathbf{\Lambda}}(t) \quad . \tag{56}$$

With the definition

$$\dot{\mathbf{E}}_p(t) = \lim_{\eta \to 0} \dot{\mathbf{E}}_i(t) = \lim_{\eta \to 0} \frac{1}{\eta} f(t) \frac{\partial f}{\partial \mathbf{T}} \quad , \tag{57}$$

we again obtain the flow rule of rate-independent plasticity, namely eq. (45). In view of the loading criterion, the arguments of section 5 remain valid.

7. Conclusions

Generalizing and extending arguments of Kratochvil and Dillon [5] it has been shown, that the theory of rate-independent plasticity can be embedded into the "classical" theory of internal variables as formulated by Coleman and Gurtin [1]. In this context the relaxation property of the yield function of viscoplasticity is most important: A static continuation at constant strain and temperature leads to a relaxation process which tends to a stable equilibrium state, where the state of stress lies on the static yield surface. For sufficiently slow strain and temperature processes the flow rule of rate-independent plasticity originates from the flow rule of viscoplasticity as an asymptotic limit. Furthermore the loading condition is satisfied in the limit. The same result is obtained for finite strain and temperature rates, if the viscosity tends to zero. The thermodynamic interpretation of these results leads to the conclusion that rate-independent elastoplastic deformation processes occur within a certain neighborhood of stable equilibrium states. However, these equilibrium states depend on the foregoing history of the thermodynamic process.

References

1. Coleman, B.D.; Gurtin, M.E.: Thermodynamics with Internal State Variables. **The Journal of Chemical Physics 47** (1967), 597 - 613
2. Haupt, P.; Tsakmakis, Ch.: On the Application of Dual Variables in Continuum Mechanics, **J. Cont. Mech. Thermodyn. 1** (1989) 165 - 196
3. Lehmann, Th.: On Thermodynamically-Consistent Constitutive Laws in Plasticity and Viscoplasticity. **Archive of Mechanics 40** (1988), 415 - 431
4. Haupt, P.; Kamlah, M.; Tsakmakis, Ch.: Continuous Representation of Hardening Properties in Cyclic Plasticity. Submitted for Publication
5. Kratochvil, J.; Dillon, O.W.: Thermodynamics of Elastic-Plastic Materials as a Theory with Internal State Variables. **J. Appl. Phys. 40** (1969), 3207 - 3218
6. Coleman, B.D.; Noll, W.: An Approximation Theorem for Functionals with Applications in Continuum Mechanics. **Arch. Rat. Mech. Anal. 6** (1960), 355 - 370
7. Chaboche, J.L.; Rousselier, G.: On the Plastic and Viscoplastic Constitutive Equations. Parts I and II. **J. Pressure Vessel Technol. 105** (1983), 153-164
8. Travnicek, L.; Kratochvil, J.: On the rate-independent limit of viscoplastic constitutive equations. **Arch. Mech. 32** (1980), 101-110

Incremental Evolution of Anisotropic Hardening

J.P. CORDEBOIS [1][2] and M. BOUCHER [1]

[1] Laboratoire de Mécanique et Technologie, E.N.S. Cachan/C.N.R.S./ Université Paris 6, 61 avenue du Président Wilson-94235 Cachan cedex-France.
[2] C.N.A.M. 292 rue Saint Martin-75141 Paris cedex 03-France

I- Introduction

Predictive numerical computations or simulations of forming processes required by industry must be more and more efficient. A lot of works contibuted to their improvement It was given with the help of various tools (theoretical, numerical, computational,...) but unfortunately not with the same degree of elaboration. In such a case, the tool which is in late give the level of the result. It seems that the constitutive equation of plasticity is not developped as much as other aspects like large deformation or numerical methods. In order to progress, it is necessary to take into consideration some more physical aspects as the description of the elastoplastic behaviour.

The aim of this work is to give a significant improvement of the models used in predictive calculations. At present, the more used laws are the standard materials laws associated with the von Mises or Hill criterion. Such laws cannot lead to realistic description of induced anisotropy. In a general way, a plasticity criterion expressed by $\mathcal{F}(\mathbb{T}, a_i..., \mathbb{X}_j...) = 0$ where $\mathcal{F}$ is a numerical function of a tensorial variable $\mathbb{T}$ and hardening parameters, either scalar a_i or tensorial $\mathbb{X}_j$ leads to a particular kind of strain hardening directly connected to the shape of $\mathcal{F}$.The yield surfaces can be interpreted as sections by the hyperplanes $(a_i..., \mathbb{X}_j...)$ = constant of the hypersurface implicitly defined in the space $(\mathbb{T}, a_i.., \mathbb{X}_j..)$ by $\mathcal{F}(\mathbb{T}, a_i.., \mathbb{X}_j..) = 0$. Strain hardening can evolve only within a determined family through the evolutions of the parameters which are given by the way of differential equations. The analytic functions $\mathcal{F}$ usually employed are too simple to interpret the complexity of the phenomenon. At the opposite, the proposed approach is based on a function $\mathcal{F}$ given as the solution of a differential equation using known strain hardening functions.

II- Finite deformations framework

For each particle p of a solid which is in motion with respect to a referential system $\mathcal{R}$, we associate the corotational referential frame $\underline{\mathcal{R}}(p)$, [1]. The isometric operator $\mathbb{Q}(p,t)$ allows a transport from $\underline{\mathcal{R}}(p)$ to $\mathcal{R}$. For example, if $\mathbb{T}$ is the Cauchy stress tensor, we note $\underline{\mathbb{T}}$ the transported operator by $\mathbb{Q}^{ad}$:

D. Besdo, E. Stein (Eds.)
Finite Inelastic Deformations -
Theory and Applications
IUTAM Symposium Hannover/Germany 1991

$$\underline{T} = Q^{ad} \circ T \circ Q$$

The particular derivative with respect to $\underline{\mathcal{R}}$ is objective. Its transport on $\mathcal{R}$ remains objective and is the Jaumann derivative :

$$T^J = Q \circ \dot{\underline{T}} \circ Q^{ad}$$

Since the particular derivative with respect to the corotational frame is objective, we define, in this space, strain tensors by time integration of quantities wich represent strain velocities. For example, we note, [1] :

$$\underline{\varepsilon} = \int_0^t \underline{D}(\tau)\, d\tau$$

where D is the strain rate tensor. So, $\underline{D}$ is the particular derivative of $\dot{\underline{\varepsilon}}$:

$$\underline{D} = \dot{\underline{\varepsilon}}$$

The use of the Hooke's operator $\mathfrak{K}$ allows the calculation of the elastic strain velocity tensor :

$$\dot{\underline{\varepsilon}}_e = \underline{\mathfrak{K}}^{-1}(\dot{\underline{T}})$$

Then, we can define the plasic strain velocity tensor by :

$$\dot{\underline{\varepsilon}}_p = \dot{\underline{\varepsilon}} - \dot{\underline{\varepsilon}}_e \qquad \text{or} \qquad \underline{\varepsilon}_p = \underline{\varepsilon} - \underline{\varepsilon}_e$$

Afterwards, we shall suppose that $\underline{\mathfrak{K}}$ is time independent. So :

$$\underline{\varepsilon}_e(t) = \int_0^t \dot{\underline{\varepsilon}}_e(\tau)\, d\tau = \underline{\mathfrak{K}}^{-1}(\underline{T}(t))$$

With respect to $\mathcal{R}$, we define the following operators :

$$\varepsilon = Q \circ \underline{\varepsilon} \circ Q^{ad} \qquad D = Q \circ \dot{\underline{\varepsilon}} \circ Q^{ad}$$

$$\varepsilon_p = Q \circ \underline{\varepsilon}_p \circ Q^{ad} \qquad D_p = Q \circ \dot{\underline{\varepsilon}}_p \circ Q^{ad}$$

$$\varepsilon_e = Q \circ \underline{\varepsilon}_e \circ Q^{ad} \qquad D_e = Q \circ \dot{\underline{\varepsilon}}_e \circ Q^{ad}$$

It can be remarked that :

$$D = \varepsilon^J, \qquad D_p = \varepsilon_p^J, \qquad D_e = \varepsilon_e^J, \qquad D = D_e + D_p$$

and $\varepsilon_e^J = \mathfrak{K}^{-1}(T^J)$ which shows that the elastic law leads to a relation between two Jaumann derivatives.

III- Intrinsic representation of the loading path

An intrinsic characterisation of symmetric second order tensors is proposed with the help of the following expression:

$$T = \mathfrak{I}_T\, O \circ T^{\circ} \circ O^{ad}$$

where

- $\mathfrak{I}_T = \sqrt{tr(T \circ T)}$ is the intensity of T.

- O gives the orientation of T with respect to the material by means of the principal directions.

- T' represents the type of T (uniaxial tensor, isotropic tensor ...).

In a first step, the model is applied to thin metal sheet forming and then, it is convenient to use a plane stress condition.

In this case, a single scalar parameter α, $\alpha \in [0,\pi[$ is necessary to give the position of T. α is the angle between the principal direction associated with the eigenvalue of greatest magnitude and a material direction which has been previously chosen. For example, in the case of an initial orthtropic material, we can choose the transported dircection by Q of one of the initial orthotropic directions.

The two eigenvalues T'_1 and T'_2 of T' are defined by : $T'_i = \frac{T_i}{\mathfrak{I}_T}$.

and satisfy: $T'^2_1 + T'^2_2 = 1$

So, there is an angle Ψ which represents the type of T':

$$T'_1 = \cos\Psi \quad T'_2 = \sin\Psi \qquad \text{with} \qquad \Psi \in [-\pi/4, \pi/4] \cup]3\pi/4, 5\pi/4]$$

For the general case, the intrinsic representation of the loading path is possible and can be found in [2].

IV- Modelling of induced anisotropy

Considering the plane stresses hypothesis, we describe the evolution of the yield surface in the three dimensionnal space ($\underline{T}_{11}$, $\underline{T}_{22}$, $\sqrt{2}\underline{T}_{12}$).

1- Hypothesis

Experimental results [3], lead to neglect, in a first step, the influence of the parameters α and Ψ.This hypothesis allows to build a model which wants to be the most simple as possible. As a consequence, it will not be able to predict all industrial applications [6].

2- Strain hardening functions

The observations made during the previous tests lead to introduce three kinds of strain hardening (figure 1) :

- direct strain hardening defined by the scalar σ_d whose rate, $\dot{\sigma}_d$, gives the yield surface evolution along its normal N* at the loading point m^*: $\dot{\sigma}_d = tr(\dot{\underline{T}}^* \circ N^*)$.

- indirect strain hardening defined by the scalar σ_i whose rate, $\dot{\sigma}_i$, gives the evolution of the yield surface at m_i , in the direction N_i, opposite to N*.In an incremental way, we have : $d\sigma_i(\overline{\varepsilon}_p) = tr(d\underline{T}_i \circ N_i(\overline{\varepsilon}_p))$

with $d\underline{T}_i = m_i.(\overline{\varepsilon}_p)\, m_i.(\overline{\varepsilon}_p + d\overline{\varepsilon}_p)$ and $\overline{\varepsilon}_p = \int_0^t \sqrt{tr(\dot{\underline{\varepsilon}}_p \circ \dot{\underline{\varepsilon}}_p)}\, d\tau$

- cross effect or latent strain hardening defined by the scalar σ_ℓ whose rate, $\dot{\sigma}_\ell$, gives the evolution of the yield surface at every point m_ℓ of $\Gamma_{\ell m^*}$ in the direction N_ℓ, perpendicular to N^*.In an incremental way, we have :

$$d\sigma_\ell(\overline{\varepsilon}_p) = tr\,(d\underline{T}_\ell \circ N_\ell(\overline{\varepsilon}_p)) \quad \text{with} \quad d\underline{T}_\ell = m_\ell.(\overline{\varepsilon}_p)\, m_\ell.(\overline{\varepsilon}_p + d\overline{\varepsilon}_p).$$

Without experimental indication on the repartition of the latent strain hardening along $\Gamma_{\ell m^*}$, we suppose it is uniform.

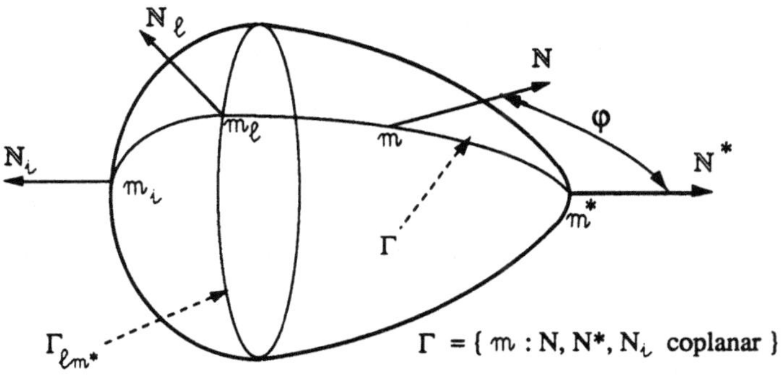

Figure 1 : characteristic points

In conformity with the previous hypothesis, the functions g_d, g_i, and g_ℓ are independent on α and Ψ. So, they are known from a tensile test. For example, in the case of an aluminium alloy, we obtain [7] (figure 2) :

These different strain rate hardenings can be expressed by functions of $\bar{\varepsilon}_p$ and $\dot{\bar{\varepsilon}}_p$ in the following manner:

$$\dot{\sigma}_d = g'_d(\overline{\varepsilon}_p)\,\dot{\overline{\varepsilon}}_p \qquad \dot{\sigma}_i = g'_i(\overline{\varepsilon}_p)\,\dot{\overline{\varepsilon}}_p \qquad \dot{\sigma}_\ell = g'_\ell(\overline{\varepsilon}_p)\,\dot{\overline{\varepsilon}}_p$$

3- Evolution of the yield surface

A loading path for which $\dot{\overline{\varepsilon}}_p$ is always strictly positive is called a monotonous plastic loading path. At the opposite, we shall talk about intermittent plastic loading paths.

The experimental results [5], allow putting three principal phenomena forward : distortion, cross effect and, in case of an intermittent loading path, come back effect.

3-1- Monotonous plastic loading path

The three strain hardening functions determine the motion of the points m^*, m_i and m_ℓ. For any point m of the surface, the evolution along the normal

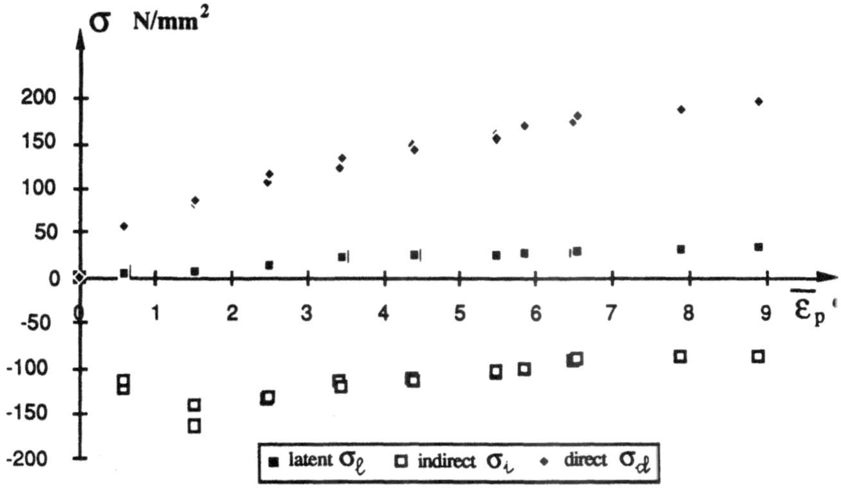

Figure 2 : generative strain hardening functions

N is given by a strain hardening rate, $\dot{\sigma}(m,\overline{\varepsilon}_p)$, which is a combination of two among the three particular strain hardenings. If the point m is between m^* and one of the points m_ℓ , then it is a combination of direct strain hardening and latent strain hardening. The coefficients of this combination depend on the angle φ between the normal N at m and the normal N* at the loading point m^*: $\cos\varphi = tr(N \circ N^*)$, (figure 1). Of course, if m is between m_i and one of the points m_ℓ, a combination of indirect strain hardening and latent strain hardening is used. This procedure can be justified, in a qualitative way, by means of the system of slip planes which is at the root of the theory of plasticity.

$\dot{\sigma}(m,\overline{\varepsilon}_p)$ can be expressed in the form :

$$\dot{\sigma}(m,\overline{\varepsilon}_p) = \dot{\sigma}_d(\overline{\varepsilon}_p)\, f_d(\varphi) + \dot{\sigma}_i(\overline{\varepsilon}_p)\, f_i(\varphi) + \dot{\sigma}_\ell(\overline{\varepsilon}_p)\, f_\ell(\varphi)$$

where f_d, f_i and f_ℓ are functions of φ through cos φ and $|\sin\varphi|$. In the case of the tested aluminium alloy, we have the following expressions :

$$f_d(\varphi) = (<\cos\varphi>^+)^2 \qquad f_i(\varphi) = (<\cos\varphi>^-)^2 \qquad f_\ell(\varphi) = \sin^2\varphi$$

3-2- Intermittent plastic loading path

After a first monotonous plastic loading path for which the end is a_1, we make an elastic loading till the point c_1 ,followed by a second monotonous plastic loading path starting from c_1. At the beginning of this second plastic loading path, the evolution of the surface, at the point a_1, has a velocity along N_{a_1} whose sign is opposed to the one of the velocity of a_1 along N_{a_1} at the end of the first monotous plastic loading. This is the come back effect for point a_1 (figure 3).

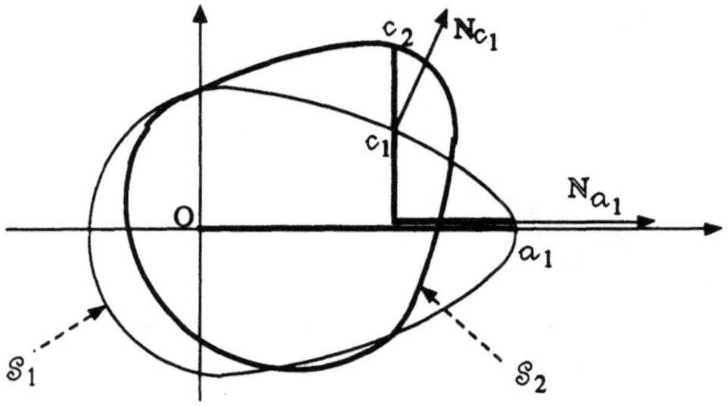

Figure 3 : intermittent plastic loading path - come back effect

Figure (4) shows, in thick line, the yield surface obtained at the end of the first plastic loading path, and also at the beginning of the second plastic loading path. In thin line is the yield surface which would be obtained from a direct loading path ended at point c. In both cases, if the loading goes on from c_1, both yield surfaces tend to a common limit. In other words, points a, b, d, ℓ and respectively a', b', d', ℓ' tend to common points.

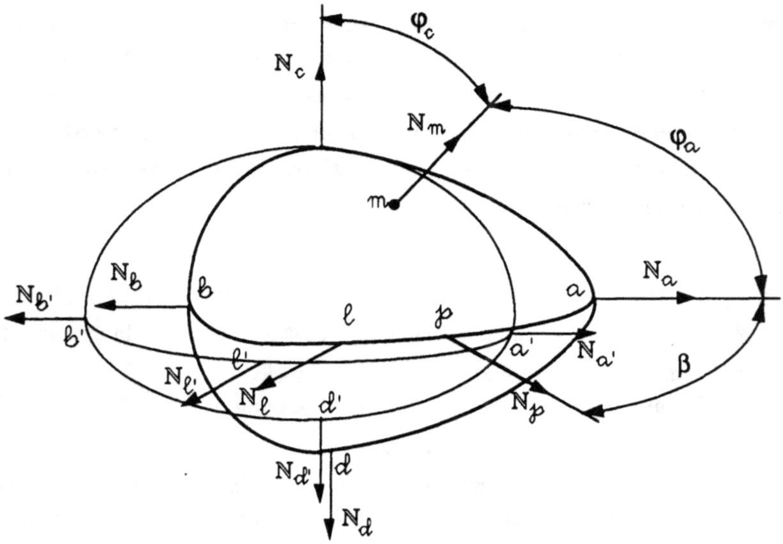

Figure 4 : perpendicular loading paths

For perpendicular plasic loading paths, the evolution, during and after the come back effect, of the current point m is given with the help of the motions of the particular points c, a, b , d and ℓ which are described in detail in [2] (figure 4). We introduced points p such that the normals N_p are perpendicular to N_c and coplanar with N_c and N_m. The velocity of point p is given by the following expression :

$$\dot{\sigma}(p,\overline{\varepsilon}_p) = f_i\,(\pi-\beta)\,\dot{\sigma}(a,\overline{\varepsilon}_p) + f_d\,(\pi-\beta)\,\dot{\sigma}(b,\overline{\varepsilon}_p) + f_\ell\,(\pi-\beta)\,\dot{\sigma}(\ell,\overline{\varepsilon}_p)$$

and for point m :

$$\dot{\sigma}(m,\overline{\varepsilon}_p) = \dot{\sigma}(c,\overline{\varepsilon}_p)\,f_d\,(\varphi_c) + \dot{\sigma}(d,\overline{\varepsilon}_p)\,f_i\,(\varphi_c) + \dot{\sigma}(p,\overline{\varepsilon}_p)\,f_\ell\,(\varphi_c)$$

When the both plastic loading paths are no longer perpendicular, we determine the evolution of point m in the same way. But now, $\dot{\sigma}(m,\overline{\varepsilon}_p)$ is depending of the angle φ^* defined by $\cos\varphi^* = \mathrm{tr}\,(N_a \circ N_c)$. All details about that can be found in [2] as well as the extension to the case of general stresses.

IV- Conclusion

The laid out model, which has the most simple formulation as possible, allows to take into account the principal phenomena observed during the tests :

- distortion
- cross effect
- come back effect.

In the case of some industrial applications, this formulation is not sufficient to lead to realistic simulation. So, it is necessary to revise some hypothesis and then, to consider that generative functions of strain hardening can be dependent of the whole loading path's characteristic parameters. At the moment, tests are made in view to extract this dependence [7]. Introduction of the type of the loading path, perhaps the orientation, in generative functions must be sufficient to improve, in a significant way, the accuracy of the model without changing the foundations used in its construction.

The convexity is an essential point in the theory of the standard laws associated with the yield criteria, as it has been shown before. Till a level of loading, the convexity of the initial surface is sufficient to assure, by continuity, the convexity of the actual surface, and that without any special condition. The convexity of the actual surface for any level of loading requires that the generative functions of strain hardening and the functions of distribution f_d, f_i and f_ℓ must satisfy conditions.

Introduction of this model in industry codes needs a mathematical description of surfaces which don't take up a too large storage It is the aim of a work which is in process [8].

REFERENCES

1 P. Ladevèze, Sur la théorie de la plasticité en grandes déformations, Rapport interne L.M.T. n°9, 1980

2 M. Boucher and J.P. Cordebois, Evolution incrémentale de l'anisotropie induite, Rapport interne L.M.T. n°117, 1991

3 J.P. Cordebois and M. Boucher, Evolution des surfaces seuil en chargements complexes: aspects théorique et expérimental, Rapport IRSID, RE 90308 PPM/QC.

4 B. Halphen and Q.S. Nguyen, Sur les matériaux standards généralisés, Journal de mécanique, Vol.14, n°1, 1975.

5 M. Rousset, Surface seuil de plasticité. Détermination automatique et modélisation, Thèse de Docteur Ingénieur, Université Paris 6, 1985.

6 H. Le Gac, Evolution de la contrainte d'écoulement et de l'anisotropie d'une tôle mince lors d'une sollicitation en expension, Rapport IRSID, PE 3876 HLG/OS.

7 P. Cayla and M. Boucher, A theoretical and experimental approach of strain induced anisotropy, I.D.D.R.G, Working group meeting, Gothenburg, 1990.

8 M. Jamal, P. Gilormini, N. Dahan and P. Rougée, A model of distortion of the plane stress yield surface in sheet forming simulations, MECAMAT 91, Internationnal seminar on large plastic deformations.

A New Constitutive Formulation for Finite Elastoplastic Deformation

F. Ellyin and Z. Xia
Department of Mechanical Engineering
University of Alberta
Edmonton, Alberta, Canada T6G 2G8

ABSTRACT

In this paper we present a constitutive model for finite elastoplastic deformations which incorporate two novel features, viz. a strain hardening law applicable to complex load histories, and an objective rate measure based on the spin of an instantaneously orthogonal triad of material unit vectors. The proposed formulation is used to predict the large deformation response of a thin-walled cylinder subjected to cyclic torsional loading.

1. INTRODUCTION

Plastic flow of metals is a rather complex process involving various forms of hardening. For example, apart from the well known Bauschinger effect, other phenomena such as strain-amplitude effect, cross-effect in non-proportional loading, induced anisotropy, erasure of certain memory property, have been reported in the literature, e.g. see Ellyin and Xia (1989). The classical isotropic and kinematic hardening rule cannot described the above mentioned behaviour; consequently, a number of hardening models have been proposed, see for example Ellyin and Xia (1989) for a list of references. However, almost all of these models are formulated within the assumption of a small (infinitesimal) strain range.

There are situations in which plastic strains or even both elastic and plastic strains are very large (finite). Examples of this type are drawing, extrusion, rolling, sheet metal forming, and deformation from impact loading. In extending a small deformation theory to finite deformations, the analytical requirement of material objectivity has to be satisfied. That is, an objective derivative of the stress tensor should be adopted. The most popular rate measure used in the past has been the Jaumann rate. However, Dienes (1979) showed that the Jaumann rate furnishes an unrealistic oscillatory response for a hypoelastic material in

D. Besdo, E. Stein (Eds.)
Finite Inelastic Deformations –
Theory and Applications
IUTAM Symposium Hannover/Germany 1991

simple shear deformation. Nagtegaal and de Jang (1981) found a similar response for rigid-plastic materials with a kinematic hardening rule.

A number of co-rotational rates have been proposed in an effort to correct the anomalous behaviour of the Jaumann rate in simple shear, among them, Dienes (1979), Lee et al. (1983), Dafalias (1983), Loret (1983), Paulun and Pecherski (1985). While these rate measures remedy the unwanted oscillatory stress response in simple shear, they unfortunately suffer from other drawbacks. Among these are the need to perform an eigen-analysis or a polar decomposition of the deformation gradient, etc.

Recently Xia and Ellyin (1991) have proposed a new objective stress rate based on the spin of a material triad that momentarily coincides with the principal stress axes. A comparative study of various alternative stress rates indicates the superior performance of the proposed rate on physical bases, general applicability and ease of implementation.

In this paper we present a constitutive model for finite elastoplastic deformations which incorporates two novel features, viz. a strain hardening model based on our previous rate-independent model (Ellyin and Xia, 1989), and the aforementioned stress-rate measure. The constitutive model is then used to predict the large deformation response of a thin-walled cylinder subject to cyclic torsional loading under deformation or stress controlled modes.

2. THEORETICAL FORMULATION

Let $\underline{D} = d_{ij}\underline{g}^i \otimes \underline{g}^j$, $\underline{W} = w_{ij}\underline{g}^i \otimes \underline{g}^j$, $\underline{\tau} = \tau_{ij}\underline{g}^i \otimes \underline{g}^j$ and $\underline{H} = \eta_{ij}\underline{G}^i \otimes \underline{G}^j$ be the spatial strain rate, the spin, the Kirchoff stress and the Lagrangian strain tensors, where $\underline{g}^i$ and $\underline{G}^i$ are the contravariant base vectors of the deformed and undeformed configurations, respectively. The material derivative of the Lagrangian strain $\dot{\underline{H}} = \dot{\eta}_{ij}\underline{G}^i \otimes \underline{G}^j$ and the spatial strain rate, $\underline{D} = d_{ij}\underline{g}^i \otimes \underline{g}^j$, have the same covariant components (but are defined in different configurations) i.e.

$$d_{ij} = \dot{\eta}_{ij}. \tag{2.1}$$

In this model three hypersurfaces are introduced, namely, a yield surface, a stress memory surface (both defined in the Kirchhoff stress space) and a strain memory surface which is specified in the Lagrangian strain space. The yield surface specifies the locus of the elastic regime at a given stress state and is represented by

$$\phi_y = f_y(\underline{\tau},\underline{\alpha}) - q^2 = 3/2 \, tr(\underline{\bar{S}}^2) - q^2 = 0, \tag{2.2}$$

where $\underline{\bar{S}}$ is the deviatoric part of $\underline{\bar{\tau}} = \underline{\tau}-\underline{\alpha}$. The $\underline{\alpha}$ and q represent centre and radius of the yield surface. Their evolution rules will be discussed in sections 2.1 and 2.2.

The stress memory surface is related to the maximum equivalent stress level experienced by the material during its previous history, i.e.

$$\phi_m^\tau = f_m^\tau(\underline{\tau}) - Q^2(\tau_{eq,max}) = 3/2 \, tr(\underline{S}^2) - Q^2(\tau_{eq,max}) = 0, \tag{2.3}$$

where $\underline{S}$ is the deviatoric part of $\underline{\tau}$ and $\tau_{eq} = [3/2 \, tr(\underline{S}^2)]^{1/2}$.

The strain memory surface is introduced to describe additional hardening due to the loading path history (cross effect). It is defined as,

$$\phi_m^H = f_m^H(\underline{H}) - C^2(\eta_{eq,max}) = 2/3 \, tr(\underline{\Gamma}^2) - C^2(\eta_{eq,max}) = 0, \tag{2.4}$$

where $\underline{\Gamma}$ is the deviatoric part of $\underline{H}$ and $\eta_{eq} = [2/3 \, tr(\underline{\Gamma}^2)]^{1/2}$.

It is assumed that the spatial strain rate, $\underline{D}$, can be decomposed into elastic and plastic parts:

$$\underline{D} = \underline{D}^e + \underline{D}^p = \frac{1}{E}[(1+\nu)\overset{*}{\underline{\tau}} - \nu(tr\overset{*}{\underline{\tau}})\underline{1}] + \frac{9c}{4q^2}(\frac{1}{E_t} - \frac{1}{E})tr(\overset{*}{\underline{\tau}}\underline{\bar{S}})\underline{\bar{S}}, \tag{2.5}$$

where $\overset{*}{\underline{\tau}}$ is an objective rate of the Kirchhoff stress tensor, to be discussed in section 2.3, E and ν are elastic modulus and Poisson ratio, E_t is the slope for the uniaxial Kirchhoff stress-logarithmic strain curve and c = 1 for $tr(\overset{*}{\underline{\tau}}\underline{\bar{S}}) > 0$ and $3/2 \, tr(\underline{\bar{S}}^2) - q^2 = 0$, and c = 0 for $tr(\overset{*}{\underline{\tau}}\underline{\bar{S}}) < 0$ or $3/2 \, tr(\underline{\bar{S}}^2) - q^2 < 0$. The inverse form of eq. (2.5) can be obtained after algebraic manipulation as,

$$\overset{*}{\underline{\tau}} = \frac{E}{1+\nu}\left[\underline{D} + \frac{\nu}{1-2\nu}(tr\underline{D})\underline{1} - \left(\frac{9c}{2q^2}\right)\cdot\frac{E-E_t}{3E+(2\nu-1)E_t}tr(\underline{D}\underline{\bar{S}})\underline{\bar{S}}\right]. \tag{2.6}$$

2.1 Description of Transient Hardening

The transient hardening behaviour of material is described by an evolution rule for both q and E_t,

$$\frac{\partial q}{\partial \ell_p} = \beta(q_{st} - q), \qquad \frac{\partial E_t}{\partial \ell_p} = \beta(E_{t,st} - E_t), \tag{2.7}$$

where $\ell_p = \int\{2/3\ \mathrm{tr}[(\mathbb{D}^p)^2]\}^{1/2}$ is the accumulated plastic strain length and the subscript "st" refers to the stable value. The above rule is physically equivalent to a uniaxial stress-strain curve which evovles from a virgin state to a saturated one, see Fig. 1.

The effect of strain amplitude and the loading path on the hardening are described by

$$q_{st} = (1 + kF)(A\eta_{eq,max} + B), \tag{2.8}$$

where k, A, B are material constants, and

$$F = A_{sw}/A_{mem}, \tag{2.9}$$

is an internal variable measuring the extent of the cross-effect. A_{mem} is the projected area of the strain memory surface, eq. (2.4) on the strain Π-plane and A_{sw} is the area of a polygon connecting all vertices of the projection of the strain vector, see Fig. 2. The value of F is bounded $0 \leq F \leq 1$, and $F = 0$ when the loading path is proportional.

Two types of plastic loading are distinguished by the introduction of the stress memory surface (see Fig. 3a). The first is called monotonic loading (ML) in which the stress memory surface expands with the movement of the yield surface and the two surface remain tangent to each other (path AB, Fig. 3a). The second type is termed reloading (RL). After an elastic unloading (path BC, Fig. 3a) plastic deformation takes place again. But in this case, the current loading point is inside the stress memory surface (path CD, Fig. 3a). When the current stress point touches the memory surface and moves it through the expansion, we incorporate the latter stage into the ML case.

In the ML case, the tangent modulus, E_t, can be calculated from the uniaxial stress-strain curve corresponding to a current accumulated plastic length, ℓ_p, viz.

$$E_t = E_t(\tau_{eq}, \ell_p). \tag{2.10}$$

In the RL case, it is obvious that the current equivalent stress, τ_{eq}, is no longer a proper variable to calculate the E_t. We can find a ratio in the Kirchhoff stress space (see Fig. 3a)

$$r = \delta_1/\delta_2, \tag{2.11}$$

where δ_1 is the distance measured from the loading point D to a corresponding point on the stress memory surface, E, whose exterior normal is parallel to that at the point D, and δ_2 is the distance measured from the point of the onset of the plastic flow, C, to the point D. For every ratio r, a corresponding point D′ on the uniaxial unloading branch can be found (Fig. 3b). It is often assumed that the uniaxial unloading branch can be approximately described by a uniaxial curve magnified by a factor of two. Thus, a corresponding point D″ on the uniaxial curve can be found (Fig. 3b). The tangent modulus is then calculated from eq. (2.10).

2.2 Evolution Rule for the Movement of the Yield Surface

Different evolution rule is adopted for the two types of plastic loading. For the ML case, a Ziegler type rule is used, i.e.

$$\overset{*}{\underset{\sim}{\alpha}} = \dot{\mu}(\underset{\sim}{\tau} - \underset{\sim}{\alpha}). \tag{2.12}$$

For the RL case, it has assumed that $\underset{\sim}{\alpha}$ moves in the direction of the vector connecting the loading point D, to a point on the stress memory surface, E, whose exterior normal is parallel to the normal at the point D (see Fig. 3a). This is expressed by

$$\overset{*}{\underset{\sim}{\alpha}} = \dot{\mu}(\underset{\sim}{\tau}^m - \underset{\sim}{\tau}). \tag{2.13}$$

By using the consistency condition $\dot{\phi}_y = 0$ and one of the equations (2.12) or (2.13), $\dot{\mu}$ is completely determined as

$$\dot{\mu} = \frac{3tr(\bar{\underset{\sim}{S}}\overset{*}{\tau}) - 2q\dot{q}}{3tr(\bar{\underset{\sim}{S}}\bar{\tau})} \qquad \textit{for ML case,} \tag{2.14}$$

$$\dot{\mu} = \frac{3tr(\bar{\underset{\sim}{S}}\overset{*}{\tau}) - 2q\dot{q}}{3tr[\bar{\underset{\sim}{S}}(\underset{\sim}{\tau}^m - \underset{\sim}{\tau})]} \qquad \textit{for RL case.} \tag{2.15}$$

2.3 A Co-Rotational Rate Measure

The proposed objective stress rate, Xia and Ellyin (1991), is expressed by

$$\overset{*}{\underset{\sim}{\tau}} = \dot{\underset{\sim}{\tau}} + \underset{\sim}{\tau}\hat{\underset{\sim}{W}} - \hat{\underset{\sim}{W}}\underset{\sim}{\tau}, \tag{2.16}$$

where a dot represents the material rate and

$$\hat{\underset{\sim}{W}} = \underset{\sim}{W} + (\underset{\sim}{D}\underset{\sim}{\tau} - \underset{\sim}{\tau}\underset{\sim}{D})/\tau_{eq}. \tag{2.17}$$

The new stress rate, based on the spin of a material triad that coincides momentarily with the principal stress axes, accounts for the general stress state in a weighted average sense that emphasizes the contributions of the dominant principal stress.

3. RESPONSE OF A THIN-WALLED CYCLINDRICAL TUBE SUBJECTED TO CYCLIC TORSIONAL LOADING

Two types of applied cyclic loading are considered. Following Harren et al. the first type is termed constrained shear, i.e. deformation of the thin-walled tube is controlled by

$$r = R, \quad \theta = \Theta + \psi(t)Z, \quad z = Z, \tag{3.1}$$

where $\psi(t) = \psi_a \cos\omega t$, R, Θ, Z and r, θ, z are cylindrical polar coordinates of material points in the undeformed and deformed configurations, respectively. The second type is termed unconstrained shear, for which only a pair of cyclic shear stress, $\tau = \tau_a \cos\omega t$, is applied at two ends of the tube and there is no constraint in the longitudinal and radial directions.

3.1 Constrained Cyclic Shear

From eqs. (3.1) and (2.17) we get

$$\underset{\sim}{D} = \begin{bmatrix} 0 & 0 & 0 \\ 0 & 0 & \frac{r}{2}\dot{\psi} \\ 0 & \frac{r}{2}\dot{\psi} & 0 \end{bmatrix} \qquad \hat{\underset{\sim}{W}} = \begin{bmatrix} 0 & -z\dot{\psi} & 0 \\ z\dot{\psi} & 0 & \frac{r}{2}\dot{\psi} \\ 0 & -\frac{r}{2}\dot{\psi} & 0 \end{bmatrix} + \frac{r\dot{\psi}}{2\tau_{eq}} \begin{bmatrix} 0 & -\tau_{13} & -\tau_{12} \\ \tau_{13} & 0 & -(\tau_{22}-\tau_{33}) \\ \tau_{12} & (\tau_{22}-\tau_{33}) & 0 \end{bmatrix} \tag{3.2}$$

Note that the above matrices are already expressed in the physical components of the tensors in the deformed cylinderical polar coordinate configuration. All tensors to appear in the following derivation will be expressed in the physical components without further explanation. By using eqs. (2.6), (2.12-15) it is found that $\tau_{22} = -\tau_{33}$, $\alpha_{22} = -\alpha_{33}$ and $\tau_{11} = \tau_{12} = \tau_{13} = \alpha_{11} = \alpha_{12} = \alpha_{13} = 0$, when loading from a virgin state. There are four differential equations for τ_{22}, τ_{23}, α_{22}, α_{23} with respect to variable ψ. They will not be produced here due to space limitation.

Let us now consider a thin-walled tube with mean radius R = 1 subject to a cyclic twist with $\psi = 1 \cdot \cos\omega t$. The material constants used in the numerical calculation are: $E = 10^5$ MPa, $\nu = 0.25$, $q_{st} = q_o = 207$ MPa, $E_{t,st} = E_t = 310$ MPa and $\beta = k = 0$. The numerical results are shown in Fig. 4. It is seen that the stress response (shear stress τ_{23}, circumferential stress τ_{22}, and longitudinal stress τ_{33}) reaches a stabilized state after first two cycles. It is found that the equivalent stress reaches its maximum value after one and a quarter cycles at point 6 (Fig. 4a). Thereafter, the yield surface moves inside the stress memory surface and a stable cyclic response is obtained due to the evolution rule for the RL case, eq. (2.15). The predicted result using the Ziegler rule (2.14) without distinguishing between the monotonic and reloading cases is also shown in Fig. 4c. It can be seen that the stress response does not reach a stable state, especially for τ_{22}. Furthermore, the stress components do not show reversal when the deformation ψ is reversed.

3.2 Uncontrained Cyclic Shear

The following deformation mode is assumed for this stress-controlled cyclic loading, $\tau = \tau_a \cos\omega t$:

$$\begin{cases} r = f[R, \psi(t)] \\ \theta = \Theta + \psi(t)Z \\ z = \lambda[\psi(t)]Z. \end{cases} \tag{3.7}$$

It can be shown that,

$$\underline{D} = \begin{bmatrix} \dot{h}/h & 0 & 0 \\ 0 & & \\ 0 & \frac{\dot{r}}{r} & \frac{r\dot{\psi}}{2\lambda} \\ 0 & \frac{r\dot{\psi}}{2\lambda} & \frac{\dot{\lambda}}{\lambda} \end{bmatrix} \qquad \hat{\underline{W}} = \begin{bmatrix} 0 & -\frac{z\dot{\psi}}{\lambda} & 0 \\ \frac{z\dot{\psi}}{\lambda} & 0 & \frac{r\dot{\psi}}{2\lambda} + \frac{1}{\sqrt{3}}(\frac{\dot{r}}{r} - \frac{\dot{\lambda}}{\lambda}) \\ 0 & \frac{-r\dot{\psi}}{2\lambda} - \frac{1}{\sqrt{3}}(\frac{\dot{r}}{r} - \frac{\dot{\lambda}}{\lambda}) & 0 \end{bmatrix} \tag{3.8}$$

where h is the thickness of the thin-walled tube measured in the deformed configuration. Substituting (3.8) into eqs. (2.5) and (2.12-15) we find that $\alpha_{22} = -\alpha_{33}$, $\alpha_{11} = \alpha_{12} = \alpha_{13} = 0$, h = constant and $\dot{r}/r = -\dot{\lambda}/\lambda$, when loading from a virgin state. As before, there are four differential equations for r, ψ, α_{22}, and α_{23} with respect to variable τ.

The numerical results are shown in Fig. 5. The tube is twisted up to τ_{max} = 207.6 MPa and the corresponding ψ is increased to about 1 (Fig. 5a, OAB). It is noted that the mean radius, r, decreases while the length of the tube z increases at the same proportion (Fig. 5b, OAB) thus keeping the tube deformation isochoric. Upon the unloading, BCD, a singularity is observed in the governing differential equations when the reversed shear stress reaches a value about τ = -201.2 MPa. To avoid this singularity, we integrate the equations with respect to the variable ψ instead of τ. From Fig. 5a,b it can be seen that an instability occurs when unloading to a certain limit value. In the same figure the predicted results using the classical kinematic hardening model (Ziegler's rule) are also plotted. For this unconstrained shear case, the predicted results of the two different hardening rules do not differ significantly.

4. CONCLUSIONS

A constitutive model is formulated for finite elasto-plastic deformation. The model is based on two recent contributions by the authors, i.e. a hardening law applicable to complex loading paths and histories, and an objective stress-rate measure which provides physically more realistic predictions for both constrained and unconstrained shear problems. The proposed model is applied to the finite deformation of a thin-walled cylindrical tube subjected to cyclic torsional loading (deformation-or stress-controlled). The predicted response appears to be fairly satisfactory, thus indicating the usefulness of the constitutive model for application to other engineering problems.

ACKNOWLEDGEMENT

We are indebted to D.J. Steigmann for his valuable contribution throughout this investigation. We also gratefully acknowledge support provided by the Natural Sciences and Engineering Research Council of Canada.

REFERENCES

1. Dafalias, Y.F., "Corotational Rates for Kinematic Hardening at Large Plastic Deformations", ASME Journal of Applied Mechanics, Vol. 50, 1983, pp. 561-565.

2. Dienes, J.K., "On the Analysis of Rotation and Stress Rate in Deforming Bodies", Acta Mechanics, Vol. 32, 1979, pp. 217-232.

3. Ellyin, F. and Xia, Z., "A Rate-Independent Constitutive Model for Transient Nonproportional Loading", Journal of the Mechanics and Physics of Solids, Vol. 37, 1989, pp. 71-91.

4. Lee, E.H., Mallett, R.L. and Wertheimer, T.B., "Stress Analysis for Anisotroic Hardening in Finite Deformation Plasticity", ASME Journal of Applied Mechanics, Vol. 50, 1983, pp. 554-569.

5. Loret, B., "On the Effects of Plastic Rotation in the Finite Deformation for Anisotropic Elastoplastic Materials", Mechanics of Materials, Vol. 2, 1983, pp. 287-304.

6. Nagtegaal, J.C. and de Jong, J.E., "Some Aspects of Non-Isotropic Work-Hardening in Finite Strain Plasticity", Proc. of the Workshop on Plasticity of Metals at Finite Strain: Theory, Experiment and Computation, Lee, E.L. and Mallett, R.L. (eds.), pp. 65-102.

7. Paulun, J.E. and Pecherski, R.B., "Study of Corotational Rates for Kinematic Hardening in Finite Deformation Plasticity", Arch. Mech., Vol. 37, 1985, pp. 661-667.

8. Xia, Z. and Ellyin, F., "A Stress Rate Measure for Finite Elastic Plastic Deformations", (to appear), 1991.

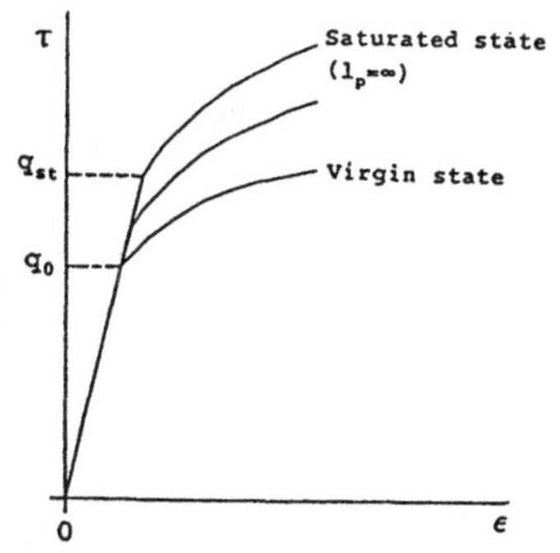

Fig. 1 Uniaxial curve evolving from a virgin state to a saturated state with the increasing ℓ_p.

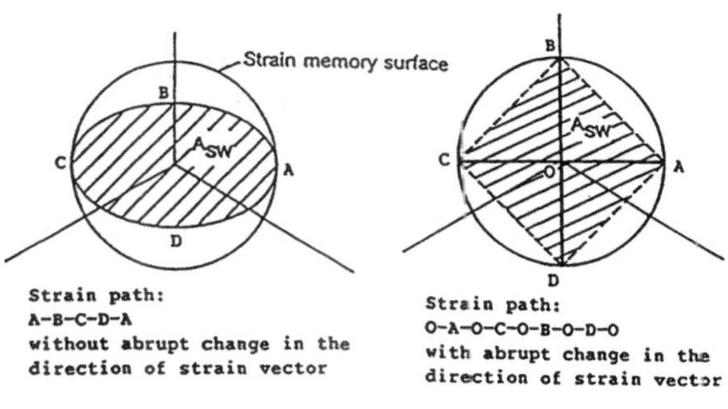

Fig. 2 Projection of the strain memory surface and the area A_{sw} for two types of strain loading paths

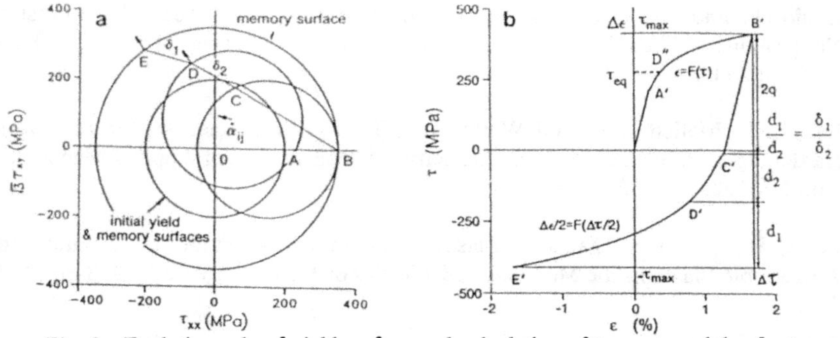

Fig. 3 Evolution rule of yield surface and calculation of tangent modulus for two types of plastic loading

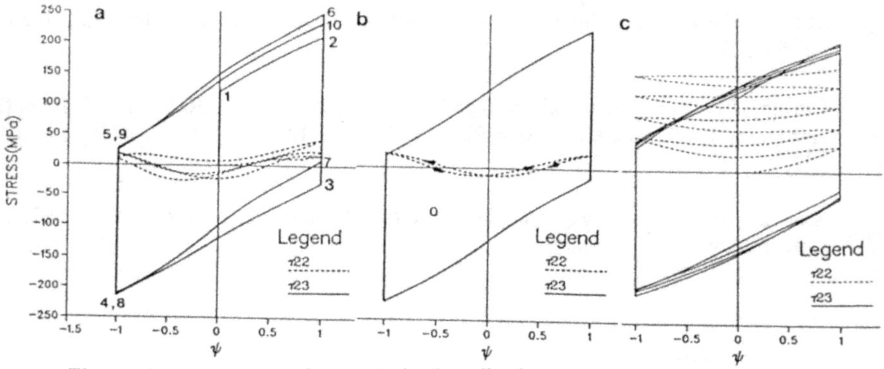

Fig. 4 Stress responses for constrained cyclic shear
(a) first two cycles by the present model (b) stabilized response by the present model
(c) result by only using Ziegler's hardening rule

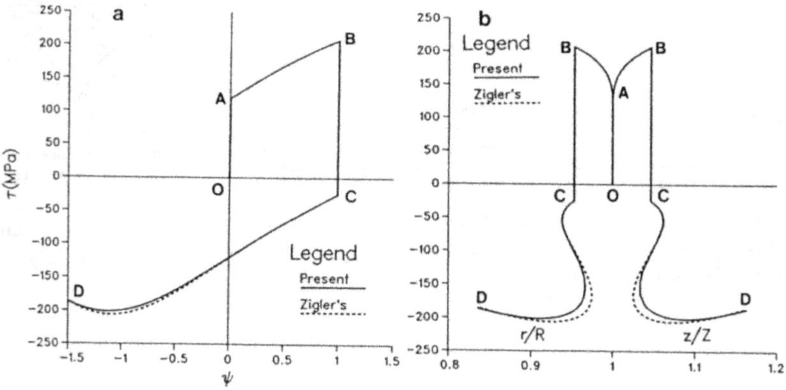

Fig. 5 Deformation responses for unconstrained cyclic shear
(a) rotation ψ versus shear stress τ (b) r/R versus τ and z/Z versus τ

The Influence of Texture Development on the Plastic Behaviour of Polycrystals

Dieter **Besdo** and Michael **Müller***)

1. Motivation

A usual theory of continuous media is based on the assumption that the material is mainly homogeneous and that isotropy is its natural state. In reality a technical medium is an assembly of grains or crystallites with different orientations of their lattice each of them being highly anisotropic. Isotropy is - if it is observed globally - the result of a random distribution of the different orientations. A very realistic model of these polycrystals has to consider the grains and their orientations.

Since such a detailled consideration has to be applied in every Gauß-point of a FE-net finally, the mathematical model must be as simple as possible but simultaneously rigorous enough. The Taylor-theory of texture development seemes to be a possible compromise if it is improved towards the special necessities of a theory of rate independent plasticity. Also a rate dependent model was introduced by Asaro et.al. [1].

2. Texture Model

2.1 The Taylor-Theory [2] in an Appropriate Version

Every grain has a lattice orientation with axes x, y, and z (see Fig. 1). Its glide directions $\vec{s}_i$ and the normals of the appropriate glide planes $\vec{m}_i$ are fixed with respect to this system. The grain (number may be α) has an arbitrary orienta-

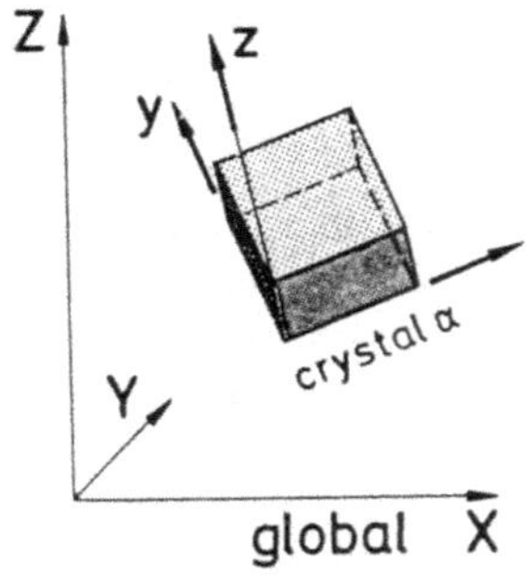

Fig. 1: *Global and local coordinate system*

*) Prof. Dr.-Ing. D. Besdo and Dipl.-Ing. M. Müller, Institut für Mechanik Universität Hannover, Appelstraße 11, W-3000 Hannover 1

D. Besdo, E. Stein (Eds.)
Finite Inelastic Deformations -
Theory and Applications
IUTAM Symposium Hannover/Germany 1991

tion in the global reference X,Y,Z-frame what can be expressed by a variable orthogonal tensor $\underline{\underline{A}}_\alpha$.

The velocity gradient $(\nabla \otimes \vec{v})^{\mathrm{T}}$ of grain α is the result of at least five superimposed glide-motions with activity factors $\dot{\gamma}_i(\alpha)$ each satisfying div $\vec{v} = 0$ and of a lattice spin $\underline{\underline{\Omega}}_\alpha$:

$$(\nabla \otimes \vec{v}(\alpha))^{\mathrm{T}} = \sum_i \dot{\gamma}_i(\alpha)\ \underline{\underline{A}}_\alpha \cdot (\vec{s}_i \otimes \vec{m}_i) \cdot \underline{\underline{A}}_\alpha^{\mathrm{T}} + \underline{\underline{\Omega}}_\alpha . \tag{1}$$

The basic idea of the Taylor-theory is that these gradients are identical with the global one being observed in experiments with polycrystals:

$$(\nabla \otimes \vec{v}(\alpha)) = (\nabla \otimes \vec{v}_{\mathrm{E}}). \tag{2}$$

The parameters $\dot{\gamma}_i$ are determined by a comparison of the symmetric part of eq. (1) taking into account that the kinematical assumptions (1) and (2) are due to an upper bound theorem. Hence, the internal power of each grain has to be minimized by an appropriate choice of the five active glide systems. Finally also the lattice spins $\underline{\underline{\Omega}}_\alpha$ are determined.

Often applications of this theory are restricted to equal shear stresses $\tau_i(\alpha)$ in each glide system i of one crystallite α. The development of the texture is then predicted precise enough but not the plastic behaviour. Hence, improvements were necessary. The general relation for the τ_i has to be denoted in terms of a functional ($\pm \rightarrow$ different directions):

$$\tau_i^{\pm} = \tau_0 + \int f_i^{\pm}(\dot{\gamma}_i, |\dot{\gamma}_i|, \tau_i^{\pm}, \dot{\varepsilon}_e)\ \mathrm{d}t, \tag{3}$$

including also latent and local kinematical hardening. Yield loci calculated on the basis of such a hardening rule were published by Canova et.al. [3]. None of these yield surfaces shows kinematical hardening.

2.2 Global Kinematical Hardening

A kinematical hardening of single glidings is almost not recognized by experimentalists. Hence, the kinematical hardening observed for polycrystals (s. [4]) must be due to other effects. One of them can be that there exists a particular interaction between the different crystallites averaging out their states of stresses. This can be expressed in the following manner:

The (deviatoric) stress state of the grain α is a superposition of the original plastic part $\underline{\underline{\sigma}}_\alpha$ due to active glide-situation and an additve part $\Delta\underline{\underline{\sigma}}_\alpha$ due to the tendency of the grains to equalize their stress-states. This tendency is assumed to be a (bi-)linear function of the difference between the average state $\underline{\underline{\sigma}}_{\mathrm{M}}$ for all crystallites and the stress-state of the considered grain $\underline{\underline{\sigma}}_\alpha$, of the existing value of $\Delta\underline{\underline{\sigma}}_\alpha$, of the strain rates $\dot{\underline{\underline{\varepsilon}}}$, and of the equivalent strain rate $\dot{\varepsilon}_e$ which can be fitted by an appropriate choice of c_1, c_2, and c_3:

$$\Delta\overset{\nabla}{\underline{\underline{\sigma}}}_\alpha = \dot{\varepsilon}_e\ [\ c_1\ (\underline{\underline{\sigma}}_{\mathrm{M}} - \underline{\underline{\sigma}}_\alpha) - c_2\ \Delta\underline{\underline{\sigma}}_\alpha\ ] + c_3\ \dot{\underline{\underline{\varepsilon}}} . \tag{4}$$

Here ∇ denotes a rate relative to the lattice and Δ one relative to a translatorically moving frame:

$$\Delta\overset{\nabla}{\underline{\underline{\sigma}}}_\alpha = \Delta\overset{\Delta}{\underline{\underline{\sigma}}}_\alpha - \underline{\underline{\Omega}}_\alpha \cdot \Delta\underline{\underline{\sigma}}_\alpha + \Delta\underline{\underline{\sigma}}_\alpha \cdot \underline{\underline{\Omega}}_\alpha . \tag{5}$$

2.3 Determination of Global Yield Loci

If, after unloading, the system is loaded by (e.g. two) global stresses it could be assumed that these stresses are also active inside each of the grains. So a yield locus can be determined for every grain from Schmid's law. The question is how to find an appropriate global locus.

As it is well-known from theories including kinematical hardening (c.f. [5]), the position and the shape of the yield locus can depend on the backstresses, in special cases it can die out completely. This can now happen for the crystallites, too. Hence, the growth of the appropriate stresses has to be limited anyhow.

In a five-dimensional stress-space the yield surface consists of several hyperplanes each corresponding to one slip system. Figure 2 shows a three-dimensional projection of a local yield surface schematically. The excentricity $\Delta\underline{\underline{\sigma}}_{\alpha}$ causes a displacement of the intersecting plane from p_1 to p_2. The evaluated two-dimensional yield loci are their cross sections c_1 and c_2. As easily can be seen, $\Delta\underline{\underline{\sigma}}_{\alpha}$ gives rise to kinematical hardening. Additionally, it changes the shape of the local yield surface, the more, since also different sets of slip systems can be involved.

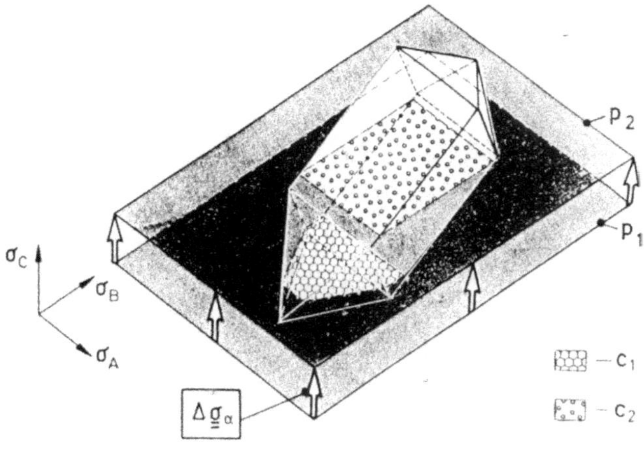

Fig. 2: *Schematic illustration of the influence of $\Delta\underline{\underline{\sigma}}_{\alpha}$ on the local yield locus*

Nevertheless, the calculation of the equivalent global yield locus is not very easy but needs several additional assumptions:

A first method uses the innermost local yield loci. The result is then comparable to the idea of Tresca's material law. As it is well-known this law is not really optimal for finite plastic deformations. On the other hand, the often-applied Huber-Mises-Hencky-criterion can be interpreted as the result of an appropriate averaging procedure.

The question is then: What is ment by "averaging of a yield locus"? Two possible procedures of averaging are:

a) Averaging of the centres combined with radial averaging

In a first step the centres of the yield loci are computed and averaged. This step is due to the fact that the stresses in the single grains are never identical with those applied from outside, hence, a direct application of outer stresses to describe the local situation would not make any sense. After this, the distances on parallel lines through the different local loci from the local centre to the locus line are linearly averaged. This procedure is very simple but has no rigorous foundament.

b) Averaging of the centres combined with some offset-simulation

The first step is identical with that of the first method. The second part is looked upon like an experiment starting from the new over-all centre to which all the other centres are shifted. Every streight line through this centre is now considered as a radial loading path (c.f. Fig. 3). At each moment when a yield locus is reached, the appropriate elastic constant reduces. An assumption is that this reduction is equal for every grain. So a curve of an elastic-plastic transition is approximated. At any strain offset $\Delta\varepsilon$ the equivalent yield locus is reached. This method is comparable to methods being applied by lots of experimentalists.

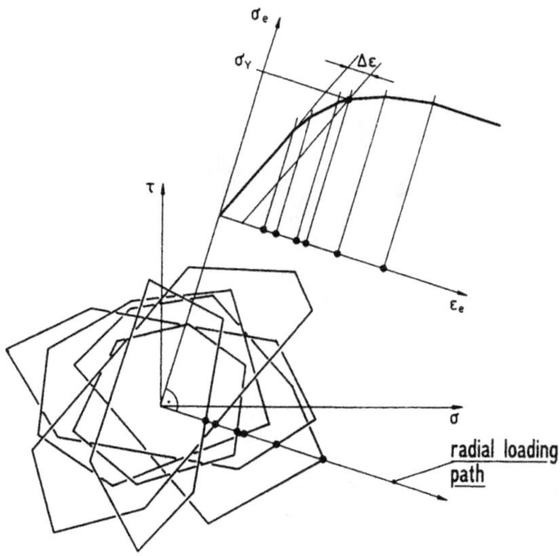

Fig. 3: *Simulation of radial loading experiments to obtain stress-strain curves. The offset $\Delta\varepsilon$ determines the global yield locus.*

As a great advantage, comparisons of these methods showed that the differences are not important if the strain offset $\Delta\varepsilon$ is chosen as a relatively high value. Hence, the use of the simple method can be justified.

The averaging scheme is also influenced by the simulation of grain interactions determined by eqs. (4) and (5). The effect of this artifical approach was tested with particular interest. It seemes to be a necessary improvement of the original theory if a significant kinematical hardening has to be expected.

3. Discretisation of the ODF

3.1 Reasons for a Discretisation

Usual applications of the Taylor-theory do not discretize the orientations but use orientation distribution functions (ODF) for Euler- or Kardan-angles denoting the angular positions of the different crystallites (c.f. [6]). This is no longer the best idea, since so many local effects have to be considered that a pre-calculation of changes of the ODF in a more or less closed version is completely impossible now.

On the other hand, these improved theories are necessary to establish a meaningful theory of plastic polycrystals. Hence, the only aim could be to minimize the number of the necessary "representative" orientations by an optimal choice of initial positions. Canova et.al. [3] e.g. used 800 grains with randomly distributed initial orientations.

Additionally it is necessary to represent all effects which are known from detailled computations with ODF, by the discretized scheme, too. Especially the fact that several grains tend to this, others to that orientation under the influence of an over-all tension should be described as precise as possible.

3.2 Initial Isotropy

If no starting distribution of the orientations and the deformational states is given, the best idea is to assume an inital globally-isotropic state. That means that if normal stresses are applied the response must not be shear and vice-versa, furtheron, initial global yield loci must have a shape like those due to the Huber-Mises-Hencky-criterion etc.

In order to ensure this, groups of 8 orientations were considered. A good collection needs 128 representative orientations. Several assemblies of local initial yield loci and their averaged (global) figures are shown in Fig. 4. The method how to find out these special orientations, will later be described in a detailled thesis.

3.3 Isotropy Test

Before these 128 representative initial orientations can be used for numerical simulations, an additional test is necessary to check whether really an inital isotropy is garanteed and not only an orthotropy. In order to check this, arbi-

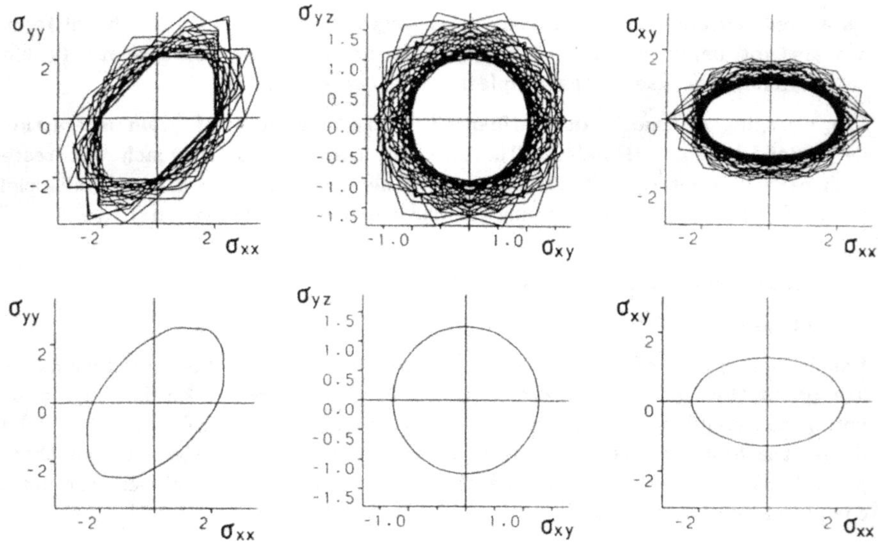

Fig. 4: *Initial yield loci* (ε_e = 0)

trary rotations were applied to the complete collection of 128 representatives. After this, the resultant average yield loci were computed and tested with respect to deviations from an isotropic behaviour. This test ended positive, so the computations simulating technical processes are justified.

4. Numerical Results

4.1 Material Parameters

It has been assumed that each of the 24 critical stresses $\tau_i^\pm$ consists of three additive parts according to the relation

$$\tau_i^\pm = \tau_{\alpha i} \pm \tau_{\beta i} \pm \tau_{\gamma i} \tag{6}$$

where the initial value of $\tau_{\gamma i}$ is τ_0 whereas the other initial values vanish. The three parts denote the following effects:

kinematical hardening of one glide system: $\dot{\tau}_{\alpha i} = a_\alpha \dot{\gamma}_i - b_\alpha \tau_{\alpha i} |\dot{\gamma}_i|$, (7a)

two-sided hardening of one glide system: $\dot{\tau}_{\beta i} = (a_\beta - b_\beta \tau_{\beta i}) |\dot{\gamma}_i|$, (7b)

isotropic hardening: $\dot{\tau}_{\gamma i} = (a_\gamma - b_\gamma \tau_{\gamma i}) \dot{\varepsilon}_e$. (7c)

The nine dimensionless material parameters $a_{...}$, $b_{...}$, and $c_{...}$ ($c_{...}$ defined by eq. (4)) applied in the numerical tests besides τ_0 = 1.0 , were:

a_α = 0.3 , b_α = 1.0 , c_1 = 1.0 ,

a_β = 1.0 , b_β = 1.2 , c_2 = 12.0 ,

a_γ = 1.5 , b_γ = 1.0 , c_3 = 13.0 .

4.2 Tension Test Applied to f.c.c. Crystals

A first group of numerical tests simulates a tension test of a f.c.c. polycrystal in Z-direction. After an equivalent strain of 1, the local yield loci differ visuably, but at ε_e = 3, the two possible end-positions of the monocrystals differing only by rotations around the axis of tension are almost reached. The global σ_{xx},σ_{yy}-yield-locus lookes like a mixture of the Tresca- and the Huber-Mises-Hencky-one (see Fig. 5).

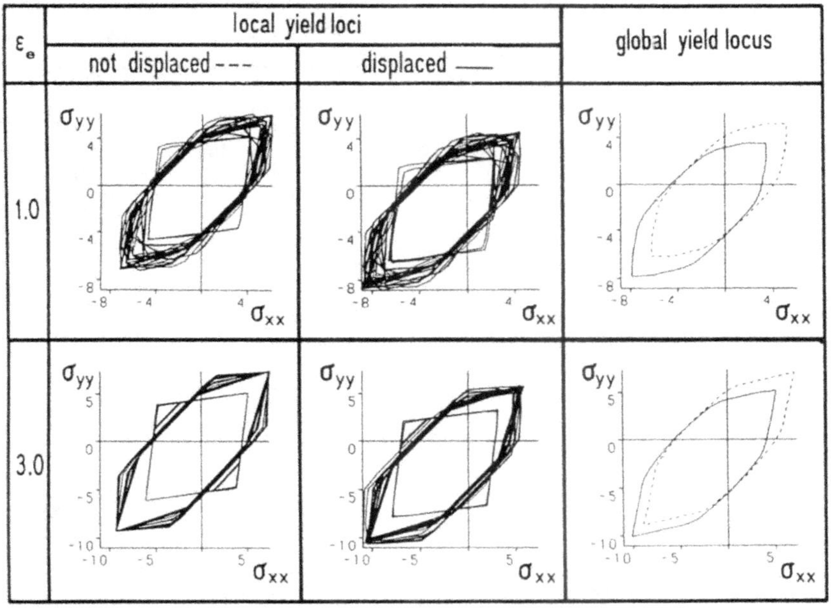

Fig. 5: *Tension test: Yield loci*

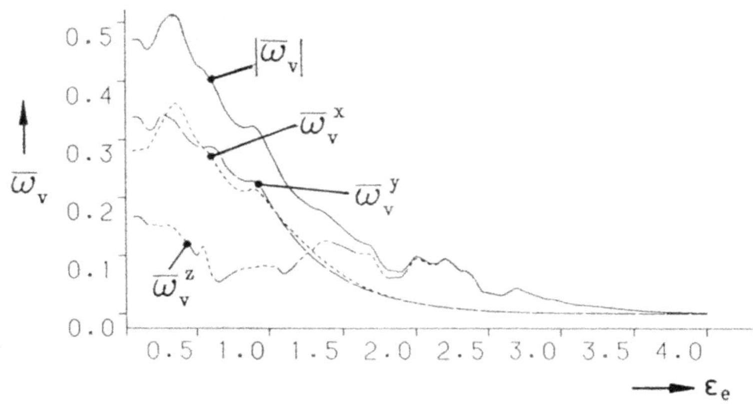

Fig. 6: *Tension test: Variance of lattice rotation in global coordinates*

In Fig. 5, also the influence of the effect described by eqs. (4), (5) is pointed out. Additionally it might be mentioned, that almost no average lattice rotation was computed, but decreasing variances of the rotations could be recognized (Fig. 6). Similar results were found for a compression-test.

4.3 Influence of the Loading Path

In next series of numerical tests (Figs. 7 and 8) the influence of the loading history is demonstrated. The loading path contains two parts of plane straining within the *X,Y*-plane in opposite sequence:

- Simple shear according to $\partial v_X / \partial Y \neq 0$, final value of $\Delta\varepsilon_e$ is 1.6,
- "Pure stretching": Tension in *Y*- and compression in *X*-direction, $\Delta\varepsilon_e = 3$.

The final yield loci depend very much on the history (s. Fig. 7).

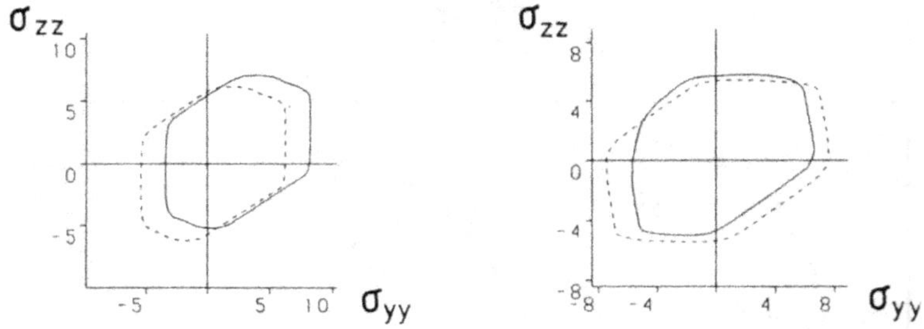

1. shear
2. pure stretching

1. pure stretching
2. shear

Fig. 7: *Influence of loading history on global yield loci.*
(—— displaced y.l., - - - not displaced y.l.)

The average lattice rotations presented in Fig. 8 computed for these processes show two effects:

- Average values of ω_Z do almost not exist during a phase of pure stretching. This is true also after shearing.
- Within a phase of shear, ω_Z does not die out completely. This might be due to the guessed evolutional laws of the crystallites.

4.4 Simulation of a Rolling Process for f.c.c. and b.c.c. Crystals

A last numerical test had to check the influence of the crystal type (f.c.c. or b.c.c.) on the final equivalent yield locus after a three-axial test simulating a rolling procedure denoted by:

$$\varepsilon_{XX} = -\lambda, \qquad \varepsilon_{YY} = 0.7\,\lambda, \qquad \varepsilon_{ZZ} = 0.3\,\lambda.$$

In Fig. 9, first the initial cube and the final sheet are plotted, afterwards the final yield loci for σ_{XX} and σ_{YY}. The crystal type has a remarkable effect.

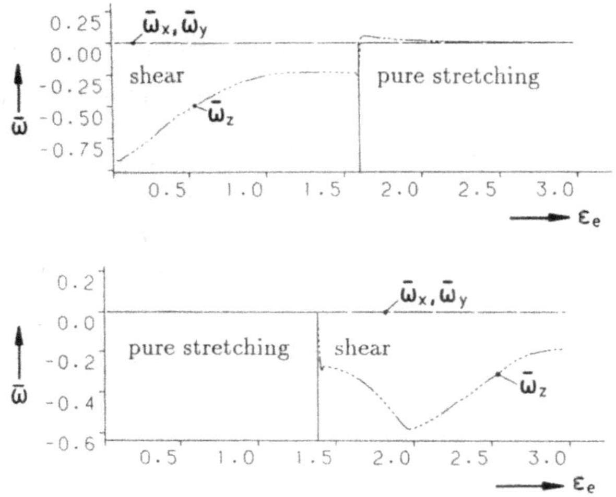

Fig. 8: *Influence of loading history on mean lattice rotation in global coordinates*

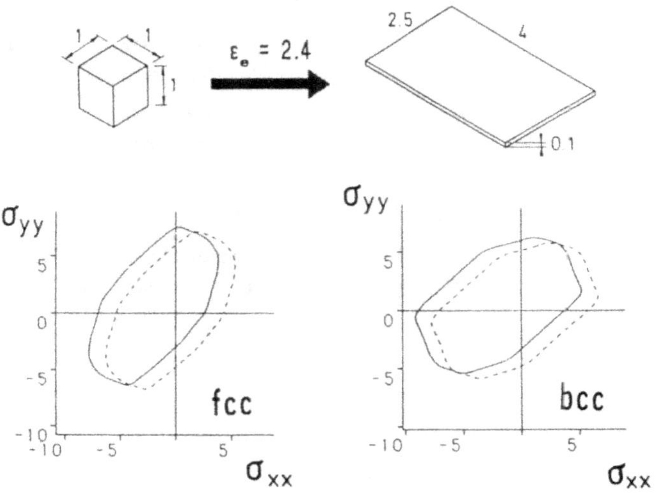

Fig. 9: *Rollling process: Influence of crystal type (fcc/bcc) on global yield loci (——— displaced yield loci, - - - - not displaced yield loci)*

5. Conclusions

* A modern theory of plasticity should take into account the influence of the inner structure. For large deformations an improved Taylor-theory seemes to be adequate. The average yield loci calculated by two different methods show isotropic and kinematical hardening as well as shape changes.

* The kinematical hardening is due to a local kinematical hardening but also to the interaction between the numerous grains denoted here by a special collection of "representatives". The results are influenced significantly by the loading history and by the type of the crystals (f.c.c. or b.c.c.).

* It might be possible that the average lattice rotation is comparable to the plastic spin appearing in several modern theories of continua (s. e.g. Dafalias et.al. [7]).

References:

[1] ASARO, R.J., and A. NEEDLEMAN: Texture Development and Strain Hardening in Rate Dependent Polycrystals. *Acta metall.* 33 (1985) 923-953

[2] TAYLOR, G.I.: Plastic Strain in Metals. (28th May-Lecture of the Inst. of Metals). In: *J. of the Inst. of Metals* 62 (1938) 307-325

[3] CANOVA, G.R., U.F. KOCKS, C.N. TOMÉ, and J.J. JONAS: The Yield Surface of Textured Polycrystals. *J. Mech. Phys. Solids* 33 (1985) 371-397

[4] KREISSIG, R., and. J. SCHINDLER: Some Experimental Results on Yield Condition in Plane Stress State. *Acta Mechanica* 65 (1986) 169-179

[5] BESDO, D.: Constitutive laws for metal forming purposes in stress- and strain-space representation. In: *FE-Simulation of 3-D Sheet Metal Forming Processes in Automotive Industry.* VDI Berichte 894 (1991) 1-15

[6] BUNGE, H.-J.: *Texture Analysis in Material Science. Butterworths* 1982

[7] DAFALIAS, Y.F., and H.-W. CHO: Verification of the Plastic Spin Concept in Viscoplasticity. In: Akhtar S. Khan, Masataka Tokuda: *Advances in Plasticity* 1989

Elastic Invariants for Crystal Defects: A Comparison with Classical Continuum Theory of Dislocations

C. DAVINI

Istituto di Meccanica Teorica ed Applicata, Universita' di Udine (Italy)

Summary
Defect measures of the theory of DAVINI & PARRY are compared with those of the classical continuum theory of dislocations described by BILBY and NOLL. A short discussion of the latter is also given.

1. Introduction

In the early 50' s an idea of KONDO led to the development of a kind of geometrical approach to the modelling of defects in crystals that is described in classical articles by ESHELBY [1], BILBY [2] and KRÖNER [3]. In that theory BURGERS' vectors and the dislocation density tensor play a central role as descriptors of dislocations providing, in particular, a nice interpretation for certain geometrical quantities that occur in the theory of differentiable manifolds. The idea has been stretched somewhat by various authors and other descriptors have been proposed on the basis of analogy with geometry, although their identification is not always as convincing as it is for dislocations. The result is a rather abstract theory where geometry, rather than mechanics, seems to be the main source of inspiration.

In a series of papers DAVINI [4][5] and DAVINI & PARRY [6][7] have shown that the subject can be addressed from a different point of view starting from the rather general assumptions that any defect descriptor be additive over subregions of the crystal and stay invariant under a special class of deformations that DAVINI calls *elastic*. The latter assumption is in fact a generalization of TAYLOR' s view on the role of defects in crystal plasticity. When these ingredients are cooked up appropriately, one obtains defect descriptors quite systematically. Among them, the BURGERS' integrals of the continuum theory of dislocations and others that have less evident correspondence in that theory, but allow for expressive molecular interpretation. In fact, the analysis in [7] provides a finite functional basis for the defect descriptors obtained in this way and shows that the prescription of the elements in the basis determines the state of the crystal

D. Besdo, E. Stein (Eds.)
Finite Inelastic Deformations -
Theory and Applications
IUTAM Symposium Hannover/Germany 1991

at each point to within a locally elastic deformation (that is, in a sense slightly weaker than TAYLOR' s), and that states corresponding to the same descriptors are to be connected with one another by combination of elastic deformations and rearrangements or slips of the types encountered in the theories of crystal plasticity.

It is the emerging of the slip mechanism from abstract ideas that strengthens the belief that this may be a useful and general approach to crystalline defects. Therefore, a comparison with other approaches is an interesting exercise that can help a judgement on their qualities.

Here I try to make clear the connections with the geometric theory of defects described in the classical reviews [1]-[3] and [8], Sect. 34. The comparison reveals points in common and a few important differences. In particular, I have found some difficulties of both kinematical and mechanical nature in interpreting the geometric theory in the general context of finite kinematics. Finite and infinitesimal cases are often treated together, but the distinction between the two is not always marked out clearly. In the sequel I discuss this in some detail.

2. Elastic invariants, slips and rearrangements

In the kinematical model for defective crystals proposed by DAVINI [4] the macroscopic state of the body is described by three linearly independent vectors $\mathbf{d}_a$ (*lattice vectors*) and a mass density ρ assigned over some region $\mathcal{B}$ of the physical space. Roughly, the model reflects observations at two distinct levels of accuracy. So, $\mathcal{B}$ is the region macroscopically occupied by the body; ρ describes the gross distribution of mass; and the $\mathbf{d}_a$ describe the (average) microscopic order of the matter at each point. The fields $\mathbf{d}_a(\cdot)$ $\rho(\cdot)$ are assumed to vary smoothly over macroscopic distances, but reflect in some way the presence of distributed defects at submicroscopic level. From a mechanical point of view one expects that the knowledge of these fields around each point is all that is needed to evaluate the mechanical response: in the molecular theory of elasticity of CAUCHY, for instance, only the value of $\mathbf{d}_a(\mathbf{x})$ at $\mathbf{x}$ is of importance; in more refined theories one might require to know $\mathbf{d}_a(\mathbf{x})$, $\rho(\mathbf{x})$ and their gradients at $\mathbf{x}$ up to a certain order. DAVINI assumes that $\mathbf{d}_a(\mathbf{x})$ and $\rho(\mathbf{x})$, possibly supplemented by suitable descriptors of defects at $\mathbf{x}$, are important. So, in particular, no role is assigned to the macroscopic deformation or to the notion of a reference configuration, and in this fundamental sense the material is not simple. Also, it is primarily for constitutive reasons that, in the absence of a natural notion of defectiveness in a model where the basic fields vary smoothly, a good definition of what is meant by defectiveness of the crystal is crucial.

Ignoring density for brevity, let us denote by $\Sigma = \left\{ \mathbf{d}_a(\cdot),\ \mathcal{B} \right\}$ the present state of the crystal. A change of state $\Sigma \longrightarrow \Sigma^* = \left\{ \mathbf{d}_a^*(\cdot),\ \mathcal{B}^* \right\}$, here regarded as a *deformation* of the crystal, is then the assignment of a mapping $x^*(\cdot)$ that transfer material points $\mathbf{x} \in \mathcal{B}$ to $\mathbf{x}^* = x^*(\mathbf{x})$, and new lattice vector fields $\mathbf{d}_a^*(\cdot)$ over $\mathcal{B}^* = x^*(\mathcal{B})$. *A priori*, there is no reason to suppose any connection between the new fields $\mathbf{d}_a^*(\cdot)$ and $x^*(\cdot)$, since they reflect observations at different scales. So, kinematics encompasses both the deformations that obey the *CAUCHY-BORN' s rule* of the classical molecular theories of elasticity:

$$\mathbf{d}_a^*(\mathbf{x}^*) = (\nabla\mathbf{x}^*)\ \mathbf{d}_a(\mathbf{x}) \quad , \quad \text{with} \quad \nabla\mathbf{x}^* \equiv (\partial\ \mathbf{x}^*/\partial\ \mathbf{x}) \quad , \tag{1}$$

that DAVINI calls *elastic deformations*, and many others the role of which in the theory is to be decided.

Deviations from CAUCHY-BORN' s rule are commonly observed when anelastic processes occur and by common accord, after TAYLOR [9], the evolution of what we call defects in the crystal is held responsible for the discrepancy between the macroscopic deformation and the behavior of the average lattice. Accordingly, it is assumed by implication that there is no evolution of defects whenever the lattice vectors transform according to (1).

On this basis, a notion of defectiveness can be established by adding the requests that defect descriptors are additive over subregions of $\mathcal{B}$ and invariant under (1). The former assumption corresponds to the way common defects are counted and summed up; the latter reflects TAYLOR' s view on the role of them in crystal plasticity. In particular, a general theorem states that additive quantities are to be integral forms defined over submanifolds of $\mathcal{B}$. DAVINI calls them *elastic invariants* and proposes to identify the assignment of all the elastic invariants over the parts of $\mathcal{B}$ with that of a given distribution of defects in the crystal. Thus, a full characterization of the elastic invariants becomes important in order to establish what is meant by *defectiveness* in the context and to describe it in specific situations. DAVINI characterizes in [5] all the invariants depending on up to the first gradient of the $\mathbf{d}_a$; DAVINI & PARRY in [7] show that there are in fact infinitely many invariants, although a finite number of them is sufficient for a full characterization of all others.

To give an example, a basis for the elastic invariants up to the first order is

$$\oint_c \mathbf{d}^a \cdot d\mathbf{x}, \quad \int_{\mathfrak{v}} s^{ab}\, d\mathfrak{v}, \quad \int_{\mathfrak{v}} n\, d\mathfrak{v} \tag{2}$$

with c any circuit in $\mathcal{B}$, $\mathfrak{v}$ any three-dimensional part of $\mathcal{B}$ and

$$\mathbf{d}^a \cdot \mathbf{d}_b = \delta^a{}_b, \quad s^{ab} \equiv \nabla\times\mathbf{d}^a \cdot \mathbf{d}^b, \quad n \equiv \mathbf{d}^1 \cdot \mathbf{d}^2 \times \mathbf{d}^3. \tag{3}$$

For brevity, it is enough to say that $(2)_1$ are the BURGERS' integrals of the classical theory of dislocations; the remaining integrals are new, but they also allow for nice molecular interpretation for s^{ab} and n are the lattice components of the dislocation density tensor and the number of lattice cells per unit volume, respectively, cf. [4].

Associated with the elastic invariants one finds densities (like the vectors $\mathbf{b}^a = \nabla\times\mathbf{d}^a$ associated with $(2)_1$ through STOKES' theorem) that are assumed to describe defectiveness locally. Giving these density fields is equivalent to the assignment of the corresponding invariants over the parts of $\mathcal{B}$. In particular, it is shown in [7] that the list

$$\mathbf{b}^a,\ s^{ab},\ n,\ \nabla(s^{ab}/n)\times\mathbf{d}^c,\ \nabla(\nabla(s^{ab}/n)\cdot\mathbf{d}_c)\times\mathbf{d}^\delta \tag{4}$$

is a functional basis for densities of all orders. Said differently, (4) gives a full characterization of defectiveness at points of $\mathcal{B}$ if we accept the proposal to identify defectiveness with the assignment of the whole set of the elastic invariants. So these quantities, or suitable combinations of them, should be included in the constitutive list together with the lattice vectors.

As a matter of fact, the analysis of DAVINI & PARRY qualifies for giving more fundamental evidence of the exhaustiveness of the list (4). They offer a complete description of the classes of states corresponding to the same defect densities and show that, if Σ and Σ' are two distinct states in one class, then they can be elastically mapped onto one another locally at each point. It turns out that these states must show some kind of uniformity along surfaces or planes, and are connectable to one another through elastic deformations and rearrangements or slips along those planes. The analysis is rather elaborate and the reader is referred to [6] [7] for details.

Again, to give an example in the simplest instance, consider the case of "perfect crystals", *i.e.*, bodies in the states $\Sigma = \left\{ \mathbf{d}_a(\cdot),\ \mathcal{B} \right\}$ and $\Sigma' = \left\{ \mathbf{d}_a'(\cdot),\ \mathcal{B} \right\}$, with $\mathbf{d}^a = \nabla\psi^a$ and $\mathbf{d}^{a\prime} = \nabla\xi^a$ respectively, so that there are no dislocations: $\mathbf{b}^a = s^{ab} = 0$. Here the ψ^a and ξ^a, $a = 1,2,3$, are functionally independent potentials since the $\mathbf{d}^a$ and $\mathbf{d}^{a\prime}$ are independent by assumption.

Then, if $\underline{\psi}$ and $\underline{\xi}$ denote the vectors with Cartesian components (ψ^{α}) and (ξ^{α}) and elastic deformations $\Sigma \longrightarrow \Sigma^{*}$ and $\Sigma' \longrightarrow \Sigma'^{*}$ are defined through the mappings $\mathbf{x} \longrightarrow \underline{\psi}(\mathbf{x})$ and $\mathbf{x} \longrightarrow \underline{\xi}(\mathbf{x})$, respectively, one calculates from (1) and (3) that Σ^{*} and Σ'^{*} are homogeneous states with unit orthogonal lattice vectors defined over two regions of equal volume and different shape. So they can be reconstructed from each other by simply rearranging the material from point $\underline{\psi}(\mathbf{x})$ to $\underline{\xi}(\mathbf{x})$ in a deformation that leaves unchanged the lattice vectors at each point. Seen the homogeneity of the states Σ^{*} and Σ'^{*}, it is difficult to see how this operation could change defectiveness in any reasonable sense, and this is the basis for concluding that Σ and Σ' are equally defective also in a less abstract sense, cf. [7].

3. Review of the classical theory of dislocations

The classical theory of dislocations adopts a quite different point of view. The subject is standard, so I shall only sketch it hereafter.

Elaborating old ideas on distortions in solids, a continuously defective body is regarded as the result of a process where infinitely small pieces of an ordinary continuum in a natural configuration (denoted by κ) are pre-deformed and patched together in space so as the field $\mathbf{K}(\cdot)$ of the local deformation gradient from κ to be smooth over the placement $\mathcal{B}$ of the body. The theory envisages then what is now called a *materially uniform simple body with inhomogeneities* [8]. The assignment of the placement and of the collection of local configurations $\Sigma = \left\{ \mathbf{K}(\cdot),\ \mathcal{B} \right\}$ defines a *state* of the body, *i.e.*, a condition of the body one is able to get all mechanical information about. The collection of local natural configurations with respect to the present one, $\mathbf{K}^{-1}(\cdot)$, is what NOLL calls a *uniform reference* for the body.

To remind the peculiarity of the situation, let us concentrate on a few mechanical issues. Let $\mathfrak{W}_{\kappa} = \mathfrak{W}_{\kappa}(\mathbf{H})$ be the stored energy density with respect to the natural configuration κ and consider a local deformation $\mathbf{H} = \mathbf{F}\ \mathbf{K}(\mathbf{x})$ at $\mathbf{x} \in \mathcal{B}$. Then, the CAUCHY stress is given by

$$\mathbf{T}(\mathbf{x}) = (\det\ \mathbf{H})^{-1} \left. \frac{\partial\ \mathfrak{W}_{\kappa}}{\partial\ \mathbf{H}}\ \mathbf{H}^{T} \right|_{\mathbf{H}\ =\ \mathbf{F}\ \mathbf{K}(\mathbf{x})} = \hat{\mathbf{T}}(\mathbf{F};\mathbf{x}) \tag{5}$$

and the response function, when using Σ as a reference, is different at different points, generally.

Another simple problem is to find the equilibrium configuration when the body is free of external forces and constraints. This amounts to solving the variational problem

$$\int_{\mathcal{B}} (\det \mathbf{K}(\mathbf{x}))^{-1} \mathfrak{W}_{\kappa}(\mathbf{F}(\mathbf{x})\mathbf{K}(\mathbf{x}))\, d\omega = \min \tag{6}$$

with $\mathbf{F} = \nabla \mathbf{x}^*$ and $\mathbf{x}^* = \mathbf{x}^*(\mathbf{x})$ any transplacement of the body. Now, if $\mathbf{K}^{-1}(\cdot)$ happens to be a gradient of some diffeomorphism $\chi(\cdot)$ of $\mathcal{B}$, $\mathbf{x}^* = \chi(\mathbf{x})$ is a minimizer, provided $\mathfrak{W}_{\kappa}(1) \le \mathfrak{W}_{\kappa}(\mathbf{H})\ \forall\, \mathbf{H}$, and $\mathbf{T}(\mathbf{x}) \equiv 0$. If it does not, the problem is less trivial: $\mathbf{T}(\mathbf{x}) \not\equiv \mathbf{0}$ and the body is self-stressed, in general. In the former instance, the response function measured from the configuration defined by $\chi(\cdot)$ turns to be independent of $\mathbf{x}$ and we say that the body is homogeneous. In the latter, the body is inhomogeneous and distortions remained trapped in Σ are responsible for the internal state of stress.

So there is no real peculiarity here, but the prejudice that one can split the contribution of distortions from that of the external forces, cf. [1], provides the motivation for searching appropriate descriptors of the distortions. Thus, attention is addressed to the geometrical structure $\mathbf{K}(\cdot)$ endows $\mathcal{B}$ with, rather than on $\mathbf{K}(\cdot)$ itself, which would be a more straightforward way to address the problem. I shall discuss some weaker points connected with this view later.

$\mathbf{K}(\cdot)$ endows $\mathcal{B}$ with a twofold structure. First, there is a *material metric* defined by assigning any two material vectors $d\mathbf{x}$, $d\mathbf{y}$ at a point $\mathbf{x}$ in the state Σ the inner product of their images $d\mathbf{X}$, $d\mathbf{Y}$ in the natural configuration, *e.g.* $d\mathbf{x} = \mathbf{K}^{-1}(\mathbf{x})\, d\mathbf{X}$:

$$d\mathbf{X} \cdot d\mathbf{Y} = d\mathbf{x} \cdot \mathbf{B}^{-1}(\mathbf{x})\, d\mathbf{y} \quad \text{with} \quad \mathbf{B}^{-1} \equiv \mathbf{K}^{-T}\mathbf{K}^{-1} \ . \tag{7}$$

The tensor $g_{ij} = (\mathbf{B}^{-1})_{ij}$ of the Cartesian components of $\mathbf{B}^{-1}$ is the metric tensor associated with the metric. Note that it is $\mathbf{B}^{-1}$ which can be measured in the usual gedanken experiment of cutting portions of the body and letting them release. $\mathbf{B}^{-1}$ may or may not define a Euclidean metric on $\mathcal{B}$. Said it in mechanical terms, it does so if and only if

$$\exists\, \chi : \mathbf{B}^{-1} = \mathbf{F}^T \mathbf{F} \quad \text{with} \quad \mathbf{F} = \nabla\chi \ . \tag{8}$$

In this case $\mathbf{B}^{-1}$ is called *compatible* because the distortions can be released by deforming the body by χ. It is known that (8) holds if and only if the curvature tensor based on $\mathbf{B}^{-1}$ vanishes. By this reason the two rank tensor $\mathbf{S}$ associated with the curvature tensor $\mathring{R}_{hijk}$:

$$S^{rs} \equiv \frac{1}{4}\, \varepsilon^{rhi}\, \varepsilon^{sjk}\, \mathring{R}_{hijk} \ , \tag{9}$$

with ε^{sjk} RICCI symbols, is called the *incompatibility tensor*.

Secondly, it is natural to assume that two material vectors at distinct points of $\mathcal{B}$ are to be equivalent in some sense, if they are the images of the same vector in the natural configuration. In geometrical terms, this view defines a parallelism notion, or *material connection*, over $\mathcal{B}$ whose coefficients are

$$\Gamma_{jk.}^{\ \ \ l} \equiv K^{l}_{.r} \, K^{-1r}_{\ \ .j,k} \tag{10}$$

It follows that the body is homogeneous if and only if the CARTAN torsion

$$T^{rs} \equiv \varepsilon^{rjk} \, \Gamma_{jk.}^{\ \ \ s} \tag{11}$$

vanishes. $\mathbb{T} = (T^{sl})$ is indeed the *inhomogeneity tensor* introduced by NOLL. Notice that **B** and Γ define independent structures on $\mathcal{B}$, although there are compatibility conditions between the two that are discussed in some detail in [8].

4. Some difficulties

Use of the above kinematics is sometimes convoluted. So, I find in [1] some confusion between $\mathbf{C} = \mathbf{K}^T \mathbf{K}$ and $\mathbf{B} = \mathbf{K}\,\mathbf{K}^T$ when imposing compatibility. Remarks raised by BLUME [10] indicate that the use of either of them can make substantial difference in treating compatibility and indeterminacies that arise in the kinematics of finite deformations. I have not checked whether such a confusion is present in other treatments of the subject, so this is perhaps a minor point not worth to emphasize. It is instead common that attention is primarily addressed to the incompatibility and the inhomogeneity tensors, with no attempt to discuss how they restrict **B** and Γ. So, the assessment of their kinematical exhaustiveness on less abstract basis is missing. Another fundamental question one might ask is the following: " Are **B** and Γ sufficient to describe the mechanical response of the material when, say, the stored energy function of the material with respect to the natural configuration is known?".

To answer this question, let $\mathfrak{W}_K = \hat{\mathfrak{W}}_K(\mathbf{C})$ be the reduced form of the strain energy function. Then, from $\mathbf{C} = \mathbf{R}^T\mathbf{B}\,\mathbf{R}$ and (5), it follows that the stress in state Σ is given by

$$\mathbf{T}(\mathbf{x}) = (\det \mathbf{V})^{-1}\, \mathbf{V}\, \mathbf{R}\, \frac{\partial \hat{\mathfrak{W}}_K}{\partial \mathbf{C}} (\mathbf{R}^T\, \mathbf{B}\, \mathbf{R})\, \mathbf{R}^T\, \mathbf{V} \ , \tag{12}$$

where **V** and **R** are the left stretch and rotation of **K**, respectively. Note that **R** is not determined by **B** since $\mathbf{K}^{-1}$ is not integrable in general. Thus, **B** alone is not sufficient to determine **T**.

If Γ is also known, one can easily prove from the integration of (10) and the definition of **B** that **B** and Γ give **K** to within a rotation to the right, so that $\mathbf{K}'(\mathbf{x}) = \mathbf{K}(\mathbf{x})\,\mathbf{Q}_\circ$, $\mathbf{Q}_\circ \in$ Orth, corresponds to the same pair **B**, Γ as **K**(**x**). Thus,

$$\mathbf{T}'(\mathbf{x}) = (\det \mathbf{V})^{-1}\, \mathbf{V}\, \mathbf{R}\, \mathbf{Q}_\circ \frac{\partial \hat{\mathfrak{W}}_K}{\partial \mathbf{C}} (\mathbf{Q}_\circ^{\ T}\, \mathbf{R}^T\mathbf{B}\, \mathbf{R}\, \mathbf{Q}_\circ)\, \mathbf{Q}_\circ^{\ T}\, \mathbf{R}^T\, \mathbf{V} \tag{13}$$

is different from $\mathbf{T}(\mathbf{x})$ and depends on $\mathbf{Q}_\circ$. We conclude that $\mathbf{B}$, Γ are not sufficient to determine the internal stress, contrarily to the expectation of the theory. Of course, this is not so in the infinitesimal theory and KRÖNER has given elegant results in this respect, cf. [1].

We notice from (13) and (12) that $\mathbf{T}'(\mathbf{x}) = \mathbf{T}(\mathbf{x})$ if the material is isotropic. So, $\mathbf{B}$ is enough in this case. Moreover, if $\mathbf{B}^{-1}$ is compatible, cf. (8), the state Σ^* obtained from Σ by means of the deformation χ has a uniform reference $\mathbf{K}^*(\mathbf{x}^*) = \mathbf{F}(\mathbf{x})\mathbf{K}(\mathbf{x}) \in \mathrm{Orth}$ for all $\mathbf{x}^* \in \chi(\mathcal{B})$. Then, the response to a deformation $\mathbf{H}$ from Σ^*

$$\hat{\mathfrak{B}}_{K_*}(\mathbf{K}^{*T}(\mathbf{x}^*)\ \mathbf{H}^T\mathbf{H}\ \mathbf{K}^*(\mathbf{x}^*)) = \hat{\mathfrak{B}}_{K}(\mathbf{H}^T\mathbf{H}) \tag{14}$$

is the same at all $\mathbf{x}^*$ and the body is homogeneous. So, $S^{rs} = 0$ implies homogeneity, independently of Γ.

More generally, two states over $\mathcal{B}$ whose uniform references differ by a rotation field (to the right) taking values in the symmetry group of the material pertain to the same material metric and are mechanically indistinguishable, although they have different inhomogeneity tensors. This points out that material symmetry has to play a non-subsidiary role in a theory of inhomogeneities. CAPRIZ & DAVINI brought the question up in [11], but they restricted to the characterization of the inhomogeneity equivalence classes that correspond to homogeneous changes of the uniform reference, avoiding the more difficult problem to find which types of inhomogeneities are mechanically detectable in a material with a given material symmetry group. For instance, I cannot figure out which inhomogeneities are possible in materially uniform simple fluids. The problem is of some interest, but it has been overlooked as far as I know.

5. Comparison and final remarks

The differences between the two approaches above are numerous. As was already observed, DAVINI & PARRY describe a non simple material where the characterization of defectiveness serves primarily constitutive purposes, because accounting for defect measures is regarded as a correction at the constitutive level of, say, CAUCHY' s theory of perfect crystals. In the classical theory, on the contrary, the scope of characterizing defects through T^{sr}, S^{sr} is less clear, being these measures unable to enter the constitutive equations of a material which is assumed to be simple from the outset.

At the kinematical level differences are also relevant. In DAVINI & PARRY the lattice vectors have a direct meaning since they are observable measures of the microscopic order in any current state, whereas in the classical theory we always need a reference configuration in the background. A formal comparison is possible if we write

$$\mathbf{K}(\mathbf{x}) = \mathbf{d}_a(\mathbf{x}) \otimes \mathbf{e}^a \quad , \tag{15}$$

with $\left\{ \mathbf{e}^a \right\}$ some Cartesian basis chosen in κ. Observed that $\mathbb{T} = \nabla\times\mathbf{d}^a \otimes \mathbf{d}_a$, it turns out from (15) and (3) that

$$\mathfrak{n} = \det \mathbf{K}^{-1} \,, \quad \mathfrak{b}^a = \mathbb{T}\, \mathbf{K}^{-T} \mathbf{e}^a \quad \text{and} \quad \mathfrak{s}^{ab} = \mathbf{e}^b \cdot \mathbf{K}^{-1} \mathbb{T}\, \mathbf{K}^{-T} \mathbf{e}^a \,. \tag{16}$$

Notice that the indeterminacy of $\mathbf{K}$ for given $\mathbf{B}$ and Γ reflects into a corresponding arbitrariness in $\mathfrak{b}^a$ and $\mathfrak{s}^{ab}$ that is absent in DAVINI & PARRY' s theory. Notice also, from (16), that we need both $\mathbf{K}$ and $\mathbb{T}$ to reconstruct the defect measures in (16). An example in [7], Appendix 2 shows that $\mathfrak{n}$, $\mathfrak{b}^a$ and $\mathfrak{s}^{ab}$ are not a functional basis of the defect densities. Thus, use of $\mathbb{T}$ only does not give a complete description of defects and one has to consider also $\mathbb{S}$, at least.

If $\mathbb{S}$ is calculated from (9) and the standard definition of the curvature tensor in terms of g_{ij}, it is found that some components of $\mathbb{S}$ contain the symmetric part of the $\nabla\mathbf{d}^a$ which is related to elastic changes of the $\mathbf{d}^a$, cf. [7], rather than to the notion of defectiveness as it emerges from elastic invariance arguments. It follows that $\mathbb{S}$ is hardly comparable with the list of defect measures (4), if one remains with finite kinematics. If one switches to the infinitesimal case, however, it can be shown that

$$\mathfrak{b}^a = \mathfrak{s}^{ab}\, \mathbf{e}_b, \quad \mathfrak{s}^{ab}/\mathfrak{n} = \mathfrak{s}^{ab} \quad \text{and} \quad \nabla(\mathfrak{s}^{ab}/\mathfrak{n}) \times \mathbf{d}^c = \nabla(\mathfrak{s}^{ab}) \times \mathbf{e}^c \,. \tag{17}$$

and, to the same order of approximation,

$$\mathbb{S}^{rs} = \delta_{ab} \left[\nabla\mathfrak{b}^a \times \mathbf{d}^b \right]^{(rs)} \quad . \tag{18}$$

Thus, it is found that

$$\mathbb{S}^{rs} = \delta_{ab}\, \delta^{(r}_{\;\cdot c} \left[\nabla(\mathfrak{s}^{ac}) \times \mathbf{e}^b \right]^{s)} \quad , \tag{19}$$

which shows the relation of $\mathbb{S}$ with some of the measures in (4).

Apart from this, however, very little can be said in the general case. It is worth noticing that $\mathbb{S}$ contains at most the second derivative of the $\mathbf{d}^a$. Therefore, since the measures $\nabla(\nabla(\mathfrak{s}^{ab}/\mathfrak{n})\cdot\mathbf{d}_c)\times\mathbf{d}^\delta$ involve the third derivatives of the $\mathbf{d}^a$, they cannot have any place in the classical theory. PARRY has shown me an example proving that including these measures in (4) is essential for the validity of Theorem 3, case *b* of [7]. As the following proof of completeness is based on that theorem, one should conclude that the pair $\mathbb{S}$, $\mathbb{T}$

does not give a complete description of defectiveness in the classical theory, if we stay with that proof. Of course, one might find a better proof to go around the difficulty, but we have been unable to do so.

Acknowledgement: This work has been supported by the Italian Ministry of Education. The final version of the paper has been prepared during a visit with prof. I. MÜLLER at the TU of Berlin.

References

1. ESHELBY, J.D., The continuum theory of lattice defects, *Solid State Physics* **3** (1956), p. 79-144.

2. BILBY, B.A., Continuous distributions of dislocations, In: *Progress in Solid Mechanics*, Vol. I (I.E. SNEDDON ed.), North HOLLAND Publishing Co., Amsterdam, 1960.

3. KRÖNER, E., Continuum theory of defects, in: *Physics of defects*, (R. BALLIAN et al., eds.), North-Holland Publishing Co., 1981.

4. DAVINI, C., A proposal for a continuum theory of defective crystals, *Arch. Rational Mech. Anal.* **96** (1986), 295 -317.

5. DAVINI C., Elastic invariants in crystal theory, In: *Material instabilities in continuum mechanics and related mathematical problems* (J.M. Ball, ed.), Clarendon Press, Oxford, 1988.

6. DAVINI, C. and G. P. PARRY, On defect-preserving deformations in crystals, *Int. J. Plasticity* **5** (1989), 337-369.

7. DAVINI, C., and G.P. PARRY, A complete list of invariants for defective crystals, *Proc. R. Soc. London* A **432** (1991), 341-365.

8. TRUESDELL, C., and W. NOLL, *The non-linear field theories of mechanics*, Handbuch der Physik Band III/3, Springer-Verlag, Berlin Heidelberg New York, 1965.

9. TAYLOR, G.I., The mechanism of plastic deformation of crystals, Parts I and II, *Proc. Roy. Soc.* A **145** (1934), 362-387, 388-404.

10. BLUME, J. A., Compatibility conditions for the left Cauchy-Green strain field, *Int. J. Elasticity* **21** (1989), 271-308.

11. CAPRIZ, G., and C. DAVINI, A geometrical classification of inhomogeneities in continua with dislocations, *Meccanica* **19** (1984), 1-6.

Analysis of the Influence of Non-Schmid and Thermal Effects on Adiabatic Shear Band Localization in Elastic-Plastic Single Crystals

M.K. DUSZEK-PERZYNA and P. PERZYNA

Institute of Fundamental Technological Research
Polish Academy of Science, Warsaw, Poland

Summary
The main objective of the paper is the investigation of shear band localization criteria for finite elastic-plastic deformation of single crystal subjected to adiabatic process. The next objective is to focus attention on temperature dependent plastic behaviour of single crystal considered. A non-Schmid constitutive model is developed within the thermodynamic framework of the rate type covariance constitutive structure, i.e., it is invariant with respect to diffeomorphism. Thermomechanical couplings are investigated and a method is developed which allows us to use the standard bifurcation procedure in the examination of the adiabatic shear band localization. It has been shown that by incorporating the thermomechanical effects and the spatial covariance effects into a non-Schmid constitutive law of elastic-plastic single crystal, the plastic hardening modulus h_{crit} at the inception of localization is in fact small but positive. It has been also proved that this thermomechanical theory of single crystals can describe the misalignment of the shear bands from the active slip systems in the crystal's matrix. The influence of non-Schmid, thermomechanical coupling and covariance effects on adiabatic shear band localization criteria is investigated. Numerical estimations of particular effects are given and comparison with available experimental results are presented.

1 Introduction

Experimental observations of the shear band localization in single crystals have been performed by Chang and Asaro [2] and Spitzig [7]. These observations have shown that the plastic hardening rate h_{crit} at the inception of localization is small but positive and the direction of the macroscopic shear band is misaligned by several degrees from the active slip systems in the crystal matrix.

To describe both observed experimentally facts, namely that the shear band localization may occur for small positive value of the plastic hardening rate h_{crit} and that the formed shear bands are misaligned with the active slip systems in the crystal's matrix by several degrees we intend to consider the synergetic effects generated by simultaneous incorporation in the non-Schmid constitutive model of single crystals the spacial covariance effects and the thermomechanical couplings.

The importance of non-Schmid effect in the explanation of localized plastic deformation in single crystals has been firstly noticed by Asaro and Rice [1], while the incorporation of the thermomechanical and covariance effects into a constitutive law of

D. Besdo, E. Stein (Eds.)
Finite Inelastic Deformations -
Theory and Applications
IUTAM Symposium Hannover/Germany 1991

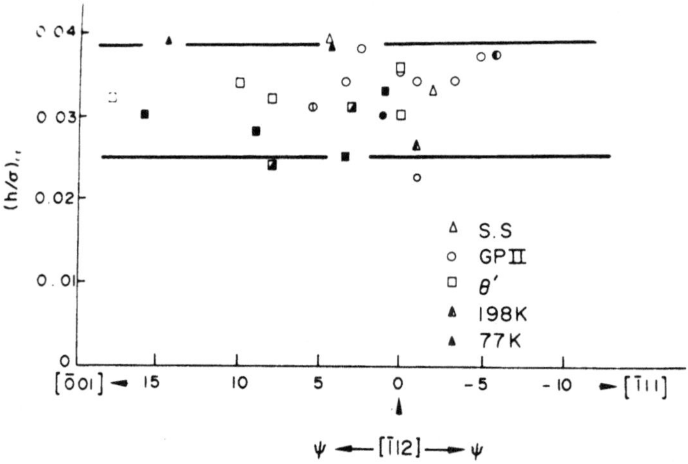

Figure 1: Critical ratio $(h/\sigma)_{crit}$ versus ψ, the angle between the tensile axis and $[\bar{1}12]$ for various ageing treatments (After Chang and Asaro [2]).

elastic–plastic model has been given by Duszek–Perzyna and Perzyna [3,5].

In chapter 2 the experimental observations of localizations of plastic deformations in single crystals are discussed. Heuristic considerations of the different localization modes and physical motivation for the present research are presented. Chapter 3 is devoted to the development of a non-Schmid constitutive model within the thermodynamic framework of the rate type covariance constitutive structure. The Lie derivative is used to define objective rates. Thermomechanical couplings are investigated. In chapter 4 the macroscopic shear band formation during adiabatic process for single slip is studied. For some simplified cases the criteria for adiabatic shear band localization are obtained in exact analytical form. The influence of non–Schmid effect and two important thermal effects, namely thermal expansion and thermal plastic softening on the criteria of localization is investigated. The similar influence of spatial covariance effects (which arise from the difference between the Lie derivative and the material derivative of the Kirchhoff stress tensor) is also examined.

Numerical estimations of the main effects obtained and comparison with available experimental results are presented in chapter 5.

It has been found that the predicted by theory critical value of strain hardening rate h_{crit} as well as the misalignment obtained for the shear bands are in accord with recent experimental observations (cf. Chang and Asaro [2] for Al–Cu single crystals and Spitzig [7] for Fe–Ti–Mn single crystals).

Chapter 5 is also focused on a discussion of the results obtained. At the closure of the paper final comments are summarized.

2 Experimental and physical motivations

Chang and Asaro [2] have studied Al – 2.8 wt% Cu single crystals aged to contain GP zones, θ' and θ precipitates. Uniaxial tension tests (for specimens with a 2.5 cm gage length and a square cross section of approx. 4 × 4 mm) were carried out at room temperature (298 K) and at 77 and 198 K. Macroscopic shear bands soon developed within the

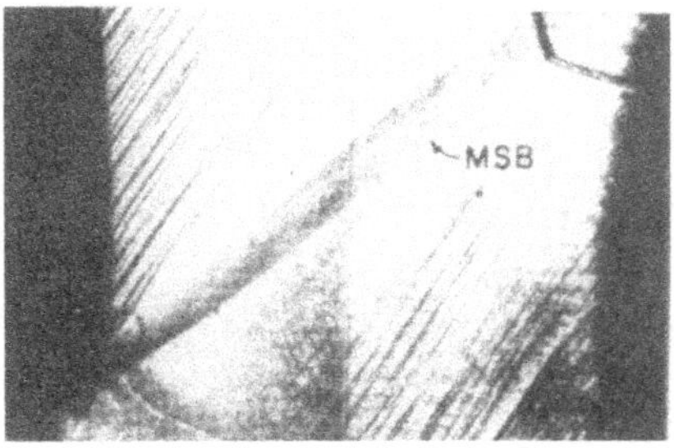

Figure 2: Coarse slip bands (CSB) and macroscopic shear bands (MSB) in a θ' strengthened crystal tested at 298 K (After Chang and Asaro [2]).

diffusely necked region. The onset of macroscopic shear band localization corresponded to the attainment of a critically low value, of the ratio of slip plane strain–hardening rate, h, to current tensile stress σ, i.e. $(h/\sigma)_{crit}$. Figure 1 shows some obtained data for the ratio $(h/\sigma)_{crit}$. The bands were not aligned with active slip systems in the crystal's matrix but were misaligned by several degrees. Figure 2 shows how the material planes of the macroscopic shear bands were inclined by small angle to adjacent slip traces. The amount of rotation was estimated by Chang and Asaro [2] to be 4 – 5 degrees.

In Fig. 3 the macroscopic shearing in θ strengthened crystals is shown. After necking (and load maximum) macroscopic localized shear sets in at nearly 45 degrees to the tensile axis on the material plane of maximum stress. In all cases fractures took place within the macroscopically localized bands (cf. Fig.3). The fractures were ductile and involved void and crack initiation where shear bands intersected the free surfaces and crack growth along the bands.

It is noteworthy that Chang and Asaro [2] found no indication of strain softening at the initiation of coarse slip bands or macroscopic shear bands, or even in fully developed macroscopic shear bands.

Spitzig [7] performed uniaxial tension tests of nitrogenated Fe – 0.19 Ti – 0.39 Mn single crystals (for sheet tension specimens with a gage length of about 1.5 cm and cross–section of about 2.5 by 0.9 mm) at temperature between 77 and 473 K. He has observed that after necking begins, macroscopic localized shear bands form, with subsequent deformation occurring by localized shearing within a shear band under a decreasing load. Final separation occurred by a ductile rupture mechanism along the shear band. The onset of the localized shear bands was estimated for the ratio of the strain hardening rate to current tensile stress $(h/\sigma)_{crit}$ to be between 0.03 and 0.06. Analysis of the localized shear bands in Fig.4 indicates that the angle between the primary slip traces and the localized shear bands is about 5 degrees.

From the analysis of experimental investigations of localized shearing in single crystal performed by Chang and Asaro [2] and Spitzig [7] during the uniaxial tests it follows that in the first stage of the process a crystal specimen undergoes uniform extension in single slip. At the moment when the load–engineering strain trajectory reaches its maximum, a crystal specimen exhibits slight amount of very diffuse necking. At this stage of the process the thermomechanical coupling effects begin to play crucial role. With continued extension macroscopic, adiabatic shear bands have soon developed within the diffusely

Figure 3: Macroscopic shearing in θ strengthened crystals. After necking (and load maximum) macroscopic localized shear sets in at nearly 45 degrees to tensile axis on the material plane of maximum shear (After Chang and Asaro [2]).

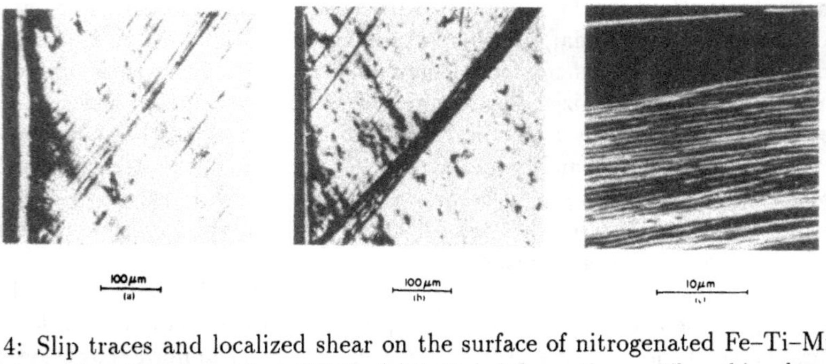

Figure 4: Slip traces and localized shear on the surface of nitrogenated Fe–Ti–Mn crystal of orientation D deformed at 295 K: (a) initial deformation until necking began, (b) subsequent deformation after removal of neck and localized shear bands from initial deformation, (c) slip traces within localized shear band in (b) (After Spitzig, [7]).

necked region. It has been observed that at the inception of the macroscopic, adiabatic shear band, the critical value of the strain hardening rate $h_{crit} = \frac{\partial \tau}{\partial \gamma}$ is small but positive and that the macroscopic shear bands are not aligned with the active slip systems in the crystal's matrix but are misaligned by angle δ_b. To describe these experimentally observed facts, we intend to consider synergetic effects resulting from taking into account spatial covariance effects and thermomechanical coupling together with non–Schmid constitutive law.

3 Constitutive model of crystal

3.1 Constitutive postulates

For rate independent behaviour of single crystals we adopt the non–Schmid law. Namely, a slip initiates on system α when the Schmid resolved shear stress $\tau^{(\alpha)}$ satisfies the relation

$$\tau^{(\alpha)} - \boldsymbol{\alpha} : \boldsymbol{\tau} = \tau_c^{(\alpha)}, \tag{1}$$

where $\boldsymbol{\alpha}$ denotes the symmetric tensor of non–Schmid effects. The work hardening – softening law is postulated in the form

$$\tau_c^{(\alpha)} = \tau_c^{(\alpha)}(\gamma, \vartheta), \tag{2}$$

$\gamma = \sum_{\alpha=1}^{n} \gamma^{(\alpha)}$, $\gamma^{(\alpha)}$ denotes the shearing on slip system α and ϑ is absolute temperature.

3.2 Thermodynamic restrictions

Introducing the following assumptions: (i) conservation of mass, (ii) balance of momentum, (iii) balance of moment of momentum, (iv) balance of energy, (v) entropy production inequality, as well as two fundamental postulates: (i) existence of the free energy function $\hat{\psi}$ and (ii) the axiom of entropy production, we get the results as follows

$$\boldsymbol{\tau} = \rho_{Ref} \frac{\partial \hat{\psi}}{\partial \mathbf{e}}, \qquad \eta = -\frac{\partial \hat{\psi}}{\partial \vartheta}, \tag{3}$$

$$\vartheta \hat{i} - \frac{1}{\rho \vartheta} \mathbf{q} \cdot \operatorname{grad} \vartheta \geq 0, \tag{4}$$

where $\boldsymbol{\tau}$ is the Kirchhoff stress, $\mathbf{e}$ is the Eulerian strain tensor, η entropy, $\mathbf{q}$ denotes the heat flux and

$$\vartheta \hat{i} = -\sum_{\alpha=1}^{n} \frac{\partial \hat{\psi}}{\partial \gamma^{(\alpha)}} \dot{\gamma}^{(\alpha)} \tag{5}$$

denotes the rate of the internal dissipation.

Introduction the denotation

$$\chi \tau^{(\alpha)} = -\rho \frac{\partial \hat{\psi}}{\partial \gamma^{(\alpha)}} \tag{6}$$

leads to

$$\vartheta \hat{i} = \chi \sum_{\alpha=1}^{n} \frac{1}{\rho} \tau^{(\alpha)} \dot{\gamma}^{(\alpha)}. \tag{7}$$

The coefficient χ is determined from the equation

$$\chi \rho_{Ref} \frac{\partial \hat{\psi}}{\partial \mathbf{e}} : \mathbf{N}^{(\alpha)} = -\rho \frac{\partial \hat{\psi}}{\partial \gamma^{(\alpha)}} \tag{8}$$

where $\mathbf{N}^{(\alpha)}$ is the symmetric part of the tensor $\mathbf{s}^{(\alpha)}\mathbf{m}^{(\alpha)}$, and $\mathbf{s}^{(\alpha)}$, $\mathbf{m}^{(\alpha)}$ denote the slip direction vector and the slip plane normal vector, respectively.

3.3 Rate type constitutive relation

Assuming fixed prior history and operating on the stress relation $(3)_1$ with the Lie derivative we obtain

$$(\overset{el}{L_{\boldsymbol{v}}\boldsymbol{\tau}}) = \mathcal{L}^e : \mathbf{d}^e - \theta \mathbf{Z} \dot{\vartheta}, \tag{9}$$

where

$$(\overset{el}{L_{\boldsymbol{v}}\boldsymbol{\tau}})^{ij} = \dot{\tau}^{ij} - \tau^{ik}(d^e_{lk} + w^e_{lk})g^{lj} - \tau^{kj}(d^e_{lk} + w^e_{lk})g^{il}, \tag{10}$$

$$\mathcal{L}^e = \rho_{Ref} \frac{\partial^2 \hat{\psi}}{\partial \mathbf{e}^2}, \qquad \theta \mathbf{Z} = -\rho_{Ref} \frac{\partial^2 \hat{\psi}}{\partial \mathbf{e} \partial \vartheta}, \tag{11}$$

and θ denotes the thermal expansion coefficient.
Denoting by

$$\mathbf{b}^{(\alpha)} = (\mathbf{N}^{(\alpha)} + \mathbf{W}^{(\alpha)}) \cdot \boldsymbol{\tau} + \boldsymbol{\tau} \cdot (\mathbf{N}^{(\alpha)} - \mathbf{W}^{(\alpha)}) \tag{12}$$

where $\mathbf{W}^{(\alpha)}$ is the antysymmetric part of the tensor $\mathbf{s}^{(\alpha)}\mathbf{m}^{(\alpha)}$, we finally have the resulting constitutive law of the rate type in the form

$$L_{\boldsymbol{v}}\boldsymbol{\tau} = \mathcal{L}^e : \mathbf{d} - \sum_{\alpha=1}^{n} (\mathcal{L}^e : \mathbf{N}^{(\alpha)} + \mathbf{b}^{(\alpha)}) \dot{\gamma}^{(\alpha)} - \theta \mathbf{Z} \dot{\vartheta}. \tag{13}$$

3.4 Thermomechanical couplings

Substituting the rate of free energy function $\dot{\psi}$ into the energy balance equation and taking advantage of the results (7) gives

$$\rho \vartheta \dot{\eta} = -\mathrm{div}\mathbf{q} + \rho \vartheta i. \tag{14}$$

Operating on the entropy relation $(3)_2$ with the Lie derivative and substituting the result together with the Fourier constitutive law for the heat vector field

$$\mathbf{q} = -k \mathrm{grad} \vartheta, \tag{15}$$

where k is the conductivity coefficient, into (14) we obtain the heat conduction equation in the form as follows

$$\rho c_p \dot{\vartheta} = \mathrm{div}(k \mathrm{grad} \vartheta) + \vartheta \frac{\rho}{\rho_{Ref}} \frac{\partial \boldsymbol{\tau}}{\partial \vartheta} : \mathbf{d} + \rho \vartheta \hat{i} + \rho \vartheta \sum_{\alpha=1}^{n} \frac{\partial^2 \hat{\psi}}{\partial \vartheta \partial \gamma^{(\alpha)}} \dot{\gamma}^{(\alpha)}, \tag{16}$$

where

$$c_p = -\vartheta \frac{\partial^2 \hat{\psi}}{\partial \vartheta^2} \tag{17}$$

denotes the specific heat.

The therm $\rho\vartheta\hat{i} = \chi\sum_{\alpha=1}^{n}\tau^{(\alpha)}\dot{\gamma}^{(\alpha)}$ on the right hand side of Eq. (16) represents the internal heating generated by the rate of internal dissipation. This term describes the main contribution to the thermomechanical coupling phenomena.

It is noteworthy that in the description of thermomechanical couplings (Section 3.4) and adiabatic process (Section 3.5) for single crystal we take advantage of similar procedure as has been developed for polycrystalline elastic–plastic solids by Duszek and Perzyna [3].

3.5 Adiabatic process

It is proved that for the adiabatic process the evolution equation for temperature (16) can be simplified to the form

$$\rho c_p\dot{\vartheta} = \chi\sum_{\alpha=1}^{n}\tau^{(\alpha)}\dot{\gamma}^{(\alpha)}, \tag{18}$$

where the coefficient χ is determined by Eq.(8).

4 Macroscopic shear band formation

4.1 Adiabatic process for single slip

Differentiation of (1) written for single slip with respect to time gives

$$\dot{\gamma} = \frac{1}{h}(\dot{\tau} + \boldsymbol{\alpha} : \dot{\boldsymbol{\tau}}) + \pi\dot{\vartheta} \tag{19}$$

provided $\boldsymbol{\alpha}$ is assumed constant in time and

$$h = \frac{\partial\tau_c}{\partial\gamma}, \qquad \pi = -\frac{1}{h}\frac{\partial\tau_c}{\partial\vartheta}. \tag{20}$$

Operating on the relation $\tau = \mathbf{s}\cdot\boldsymbol{\tau}\cdot\mathbf{m}$, defining the Schmid resolved shear stress for single slip, with the Lie derivative, we obtain

$$\dot{\tau} = (\overset{el}{L_{\boldsymbol{v}}}\boldsymbol{\tau}) : \mathbf{N} + 2\boldsymbol{\tau}\cdot(\mathbf{N}+\mathbf{W}) : \mathbf{g}\cdot\mathbf{d}^e. \tag{21}$$

Replacing in (19) $\dot{\tau}$ by (21) and $\dot{\boldsymbol{\tau}} : \boldsymbol{\alpha}$ by

$$\begin{aligned}\dot{\boldsymbol{\tau}} : \boldsymbol{\alpha} &= (\overline{\boldsymbol{\tau} : \boldsymbol{\alpha}}) = (\overset{el}{L_{\boldsymbol{v}}})\,(\boldsymbol{\tau} : \boldsymbol{\alpha}) \\ &= (\overset{el}{L_{\boldsymbol{v}}}\boldsymbol{\tau}) : \boldsymbol{\alpha} - \boldsymbol{\tau} : [(\mathbf{d}^e + \mathbf{w}^e)\cdot\boldsymbol{\alpha} + \boldsymbol{\alpha}\cdot(\mathbf{d}^e + \mathbf{w}^e)] \\ &\approx (\overset{el}{L_{\boldsymbol{v}}}\boldsymbol{\tau}) : \boldsymbol{\alpha} \approx (\mathcal{L}^e : \mathbf{d}^e) : \boldsymbol{\alpha}\end{aligned} \tag{22}$$

(provided $\boldsymbol{\alpha}$ is assumed to be of the order of $\tau/\mathcal{L}$) we obtain

$$\dot{\gamma} = \frac{(\mathbf{Q}^* + \mathcal{L}^e : \boldsymbol{\alpha}) : \mathbf{d}}{h - \Pi\boldsymbol{\tau} : \mathbf{N} + (\mathbf{Q}^* + \frac{\Theta}{\mu}\boldsymbol{\tau} : \mathbf{NZ} + \mathcal{L}^e : \boldsymbol{\alpha}) : \mathbf{N}}, \tag{23}$$

where

$$\mathbf{Q}^* = \mathcal{L}^e : \mathbf{N} + \mathbf{b}, \qquad \Theta = \frac{\chi\theta}{\rho c_p}\mu, \qquad \Pi = \frac{\chi}{\rho c_p} h\pi. \tag{24}$$

Eliminating $\dot{\gamma}$ and $\dot{\vartheta}$ from a set of equations (13), (18) and (24) we obtain the fundamental rate equations in the form

$$L_{\boldsymbol{v}}\boldsymbol{\tau} = \mathbb{L}^* : \mathbf{d} \tag{25}$$

where

$$\mathbb{L}^* = \left[\mathcal{L}^e - \frac{(\mathbf{Q}^* + \frac{\Theta}{\mu}\tau\mathbf{Z} + \mathcal{L}^e : \boldsymbol{\alpha})\mathbf{Q}^*}{h - \Pi\tau + (\mathbf{Q}^* + \frac{\Theta}{\mu}\tau\mathbf{Z} + \mathcal{L}^e : \boldsymbol{\alpha}) : \mathbf{N}}\right]. \tag{26}$$

It is noteworthy that the scalar coefficients Θ and Π describe the thermal expansion and thermal plastic softening effects, respectively, while the tensor $\mathbf{Q}^*$ takes account of the covariance effects (μ denotes the elastic shear modulus).
Assuming $\boldsymbol{\alpha} = 0$ (the Schmid law is applied) then the fundamental rate equation (25) reduces to

$$L_{\boldsymbol{v}}\boldsymbol{\tau} = \mathbb{L} : \mathbf{d}, \tag{27}$$

where

$$\mathbb{L} = \left[\mathcal{L}^e - \frac{(\mathbf{Q}^* + \frac{\Theta}{\mu}\tau\mathbf{Z})\mathbf{Q}^*}{h - \Pi\tau + (\mathbf{Q}^* + \frac{\Theta}{\mu}\tau\mathbf{Z}) : \mathbf{N}}\right], \tag{28}$$

discussed first by Duszek–Perzyna and Perzyna [4].
Using the Zaremba–Jaumann rate instead of the Lie derivative and assuming an isothermal process then Eqs. (25)–(26) reduce to those obtained first by Asaro and Rice [1].

4.2 Shear band localization criteria

Taking advantage of the fundamental constitutive rate equation (25) and applying standard bifurcation method (cf. Rice [5] and Rudnicki and Rice [6]) as well as perturbation procedure developed by Asaro and Rice [1] we obtain the critical hardening rate at the onset of localization and the orientation of shear band as follows (for the perturbation about $\mathbf{n} = \mathbf{m}$)

$$\begin{aligned} h_{crit} &= \Pi\tau + \frac{1}{4}(\Theta\frac{\tau}{\mu}\mathbf{Z} : \mathcal{N} + 2\tau\mathbf{s} + \boldsymbol{\alpha} : \mathcal{M}\cdot\mathbf{s})\cdot(\mathbf{s}\cdot\mathcal{M}\cdot\mathbf{s})^{-1}\cdot(\Theta\frac{\tau}{\mu}\mathbf{Z} : \mathcal{N} \\ &\quad + 2\tau\mathbf{s} + \mathbf{s}\cdot\mathcal{M} : \boldsymbol{\alpha}) + O(\alpha\tau, \tau^2/\mathcal{L}^e, \alpha^3\mathcal{L}^e), \\ \mathbf{n} &= \mathbf{m} + \frac{1}{2}(\mathbf{s}\cdot\mathcal{M}\cdot\mathbf{s})^{-1}\cdot\left[\left(\frac{\Theta}{\mu}\mathbf{Z} : \mathcal{N} + 2\mathbf{s}\right)\tau + \mathbf{s}\cdot\mathcal{M} : \boldsymbol{\alpha}\right] + O\left(\alpha^2, \tau/\mathcal{L}^e\right), \end{aligned} \tag{29}$$

where

$$\begin{aligned} \mathcal{M} &= \mathcal{L}^e - (\mathcal{L}^e\cdot\mathbf{m})\cdot(\mathbf{m}\cdot\mathcal{L}^e\cdot\mathbf{m})^{-1}\cdot(\mathbf{m}\cdot\mathcal{L}^e), \\ \mathcal{N} &= \mathbf{s}\mathbf{1} - \mathbf{m}(\mathbf{m}\cdot\mathcal{L}^e\cdot\mathbf{m})^{-1}\cdot(\mathbf{m}\cdot\mathcal{L}^e\cdot\mathbf{m}). \end{aligned} \tag{30}$$

To obtain the direct analytical results we shall introduce the following simplifications:
(i) Let assume (cf. Duszek–Perzyna and Perzyna [4])

$$\mathbf{Z} = \mathbf{s}\cdot(\mathbf{s}\cdot\mathcal{L}^e). \tag{31}$$

(ii) Let us restrict our considerations to the linear, isotropic and homogeneous elastic properties of the crystal, i.e.

$$\overset{e}{\mathcal{L}}{}^{abcd} = \tau^{bd} g^{ac} + \mu (g^{ac} g^{bd} + g^{ad} g^{bc}) + \lambda g^{ab} g^{cd}, \tag{32}$$

where the constants μ and λ are the Lamé moduli.
Taking advantage of the simplifications (31) and (32) we obtain

$$\begin{aligned} h_{crit} &= \Pi\tau + \frac{1}{4\nu\mu}\left[2\Theta\tau\nu + \tau + \mu(2\nu - 1)\alpha_{zz} + 2\mu\nu\alpha_{ss}\right]^2 + \mu\alpha_{zs}^2, \\ \mathbf{n} &= \mathbf{m} + \frac{1}{2}\Theta\frac{\tau}{\mu}\mathbf{s} + \frac{1}{4}\frac{\tau}{\mu}\frac{1}{\nu}\mathbf{s} + \alpha_{zs}\mathbf{z} + \frac{1}{4\nu}\left[(2\nu - 1)\alpha_{zz} + 2\nu\alpha_{ss}\right]\mathbf{s}, \end{aligned} \tag{33}$$

where ν denotes Poisson's ratio.
Assuming particular case when $\alpha_{sz} = \alpha_{zs} = 1/2\alpha$, $\alpha_{zz} = \alpha_{ss} = 0$ we have

$$\begin{aligned} \left(\frac{h}{\tau}\right)_{crit} &= \Pi + \frac{\tau}{\mu}\left(\Theta^2\nu + \Theta + \frac{1}{4\nu}\right) + \frac{1}{4}\alpha^2\frac{\mu}{\tau}, \\ \mathbf{n} &= \mathbf{m} + \frac{1}{4\nu}\frac{\tau}{\mu}(2\Theta\nu + 1)\,\mathbf{s} + \frac{1}{2}\alpha\mathbf{z}. \end{aligned} \tag{34}$$

The results (34) have sufficiently simple form to be used in quantitative analysis as well as in numerical estimations of the influence of the effects considered on shear band localization criteria.

5 Numerical estimations and comparison with experimental results

5.1 Numerical estimations

We shall make the numerical computations for Al–Cu single crystals. Some particular material parameters are taken from Chang and Asaro [2] experimental data.

Aluminum–copper single crystal tested at 298 K:

$$\left(\frac{\tau}{\mu}\right)_{crit} = 7.5 \times 10^{-3}, \qquad \frac{\partial \tau_{crit}}{\partial \vartheta} = -0.06 \times 10^6 \text{ Pa/K}.$$

It has been assumed that $\alpha = 0.015$. Then we have

$$\begin{aligned} \left(\frac{h}{\tau}\right)_{crit} &= 0.0235 + 0.0004 + 0.0018 + 0.0025 + 0.0075 \\ &= 0.0282 + 0.0075 = 0.0357, \\ \mathbf{n} &= \mathbf{m} + (0.0009 + 0.0025)\mathbf{s} + 0.0075\mathbf{z} \end{aligned}$$

We can approximate

$$\tan\delta_b = 0.0082 \quad \Rightarrow \quad \delta_b = 30'$$

5.2 Comparison with experimental results

It seems that the comparison of the theoretical prediction estimated for the critical hardening rate $(h/\tau)_{crit}$ with the Chang and Asaro [2] experimental results for aluminum–copper single crystal (c.f. Fig. 1) is extremely good.

To compare the theoretical predictions for the misalignment of the direction of the macroscopic shear band from the active slip systems in the crystal matrix we have first take into account that the theoretical estimations concerned the angle δ_b i.e. at the inception of the shear band localization while the experimental observations of the misalignment performed by Chang and Asaro [2] for aluminum–copper single crystals and by Spitzig [7] for nitrogenated Fe–Ti–Mn crystals have been detected at the end of the process considered. This interpretation of the experimental observations helps us to drawn conclusion that the description of the misalignment by incorporating the synergetic effects resulting from simultaneous consideration of the spatial covariance effects and thermomechanical couplings (as has been estimated in Section 5.1) is realistic and looks promising.

6 Discussion of the results

Four different effects namely non–Schmid, thermal expansion, thermal plastic softening and spatial covariance which can affect the shear band localization conditions have been investigated. It has been shown that non–Schmid, thermal expansion and spatial covariance effects have influence on the critical hardening rate $(h/\tau)_{crit}$ as well as on the direction of the shear band **n**, while thermal plastic softening affects only the critical hardening rate $(h/\tau)_{crit}$. The contribution of the non–Schmid effect to the critical hardening rate $(h/\tau)_{crit}$ is of the order 10^{-3} and is comparable to the contribution implied by cooperative phenomena generated by thermal expansion and spatial covariance effects.

The main contribution to the critical hardening rate $(h/\tau)_{crit}$ is implied by thermal plastic softening represented by the term Π, which is of the order of 10^{-2}.

The misalignment of the direction of the macroscopic shear bands from the active slip systems in the crystal's matrix is affected by three phenomena, namely by non–Schmid, thermal expansion and spatial covariance effects. The last term in Eq. $(34)_2$ which represents the non–Schmid effect has dominated influence on the direction of the shear band and is 2 times larger than a sum of the thermal expansion and spatial covariance terms. It has been found that obtained value of the angle δ_b (for aluminum–copper single crystals) is small but sufficiently distinct to explain experimentally observed misalignment. To fit experimental observations of h_{crit} in the range of 5×10^{-4} to $5 \times 10^{-3}\mu$ Asaro and Rice [1] estimated the parameter α in the range from 0.04 to 0.14. However their estimations were obtained under very strong assumption that h_{crit} is affected only by the non–Schmid effect.

If we consider situation in which h_{crit} can be influenced by such effects as non–Schmid, thermal expansion, thermal plastic softening and spatial covariance then final estimation for the non–Schmid parameter α has to be smaller (in our computation in the range of 0.01 to 0.02) than that obtained by Asaro and Rice [1].

References

1. Asaro, R.J. and Rice, J.R., 1977, Strain localization in ductile single crystals. *J. Mech. Phys. Solids*, **25**, 309–338.

2. Chang, Y.W. and Asaro, R.J., 1981, An experimental study of shear localization in aluminum–copper single crystals. *Acta Metall.*, **29**, 241–257.

3. Duszek–Perzyna, M.K. and Perzyna, P., 1991, The localization of plastic deformation in thermoplastic solids. *Int. J. Solids Structures*, **27**, 1419–1443.

4. Duszek–Perzyna, M.K. and Perzyna, P., 1991, Adiabatic shear band localization in elastic–plastic single crystals. *Int. J. Solids Structures*, (submitted for publication).

5. Rice, J.R., 1977, The localization of plastic deformation. In *Theoretical and Applied Mechanics*, ed. W.T.Koiter, N.–Holland, Amsterdam, pp. 207–220.

6. Rudnicki, J.W. and Rice, J.R., 1975, Conditions for the localization of deformation in pressure–sensitive dilatant materials. *J. Mech. Phys. Solids*, **23**, 371–394.

7. Spitzig, W.A., 1981, Deformation behaviour of nitrogenated Fe–Ti–Mn and Fe–Ti single crystals. *Acta Metall.*, **29**, 1359–1377.

Physical and Theoretical Aspects of Large Plastic Deformations Involving Shear Banding

Ryszard B. PĘCHERSKI
Institute of Fundamental Technological Research,
Polish Academy of Sciences, Warsaw

Summary

Shear banding is related with large plastic deformations produced by mechanisms of crystallographic slip and/or twinning and micro–shear. Plane deformations of elasto-plastic material accounting for micro–shear bands are considered. Micro–shear bands produce the non–coaxiality between principal directions of the tensors of stress and rate of plastic deformations. Relations to known vertex–type plasticity models are discussed.

Introduction

Large plastic deformations of polycrystalline metallic materials, in particular under highly constrained conditions such as rolling or in the case of ductile fracture, while coalescence of voids occurs, can give rise to profuse flow localization into shear bands. They appear on different scales: grain–scale shear bands, sometimes called *micro–shear bands*, which extend through a few grains; and sample–scale shear bands, often called *macroscopic shear bands*, which extend through the entire sample. The so–called "shear banding" is related with plastic deformations produced by competing mechanisms of crystallographic slip and/or twinning and micro–shear. The study will be confined to plane deformations of polycrystalline metallic materials at moderate temperature and quasi static loading conditions. According to DÈVE and ASARO [1], in heavily deformed metals, shear localization progressively replaces the current deformation mode, *viz.* slip and/or twinning. This change in deformation mode contributes to the development of strain induced anisotropy and modifies material properties. Therefore, formulation of a model of large plastic deformations accounting for micro–shear bands seems to be important for adequate constitutive description of inelastic behaviour of metallic materials. The model proposed shows that micro–shear bands produce the non–coaxiality between principal directions of stress tensor and those of rate of plastic deformations tensor. Relations to known vertex–type plasticity models are discussed.

D. Besdo, E. Stein (Eds.)
Finite Inelastic Deformations –
Theory and Applications
IUTAM Symposium Hannover/Germany 1991

Basic physical mechanisms

The early observations of microscopic and macroscopic shear band formation in different metals and alloys are reported by GIL SEVILLANO et al. [2]. Usually, the following terminology is used (cf. [1], [2] and KORBEL and MARTIN [3]):

– ***Microband***; *a long thin band, parallel to crystallographic slip planes, often formed across the width of the grain. (The distinct markings on the surface produced by the microbands are called sometimes* ***coarse slip bands****).*

– ***Micro–shear band***; *a long and thin sheet–like region of concentrated plastic shear crossing grain boundaries without deviation and having a characteristic orientation with respect to the directions of principal strain. It bears very large shear strains and lies in a position that is usually non–crystallographic. This means that micro–shear bands are not parallel with a particular slip plane of the crystallites they intersect.*

As it was stressed in DUGGAN et al. [4] and HATHERLY and MALIN [5] a particular micro–shear band operates only once and develops rapidly to its full width. The micro-shear bands, once formed, do not contribute further to the deformation process. Thus, it appears that the successive generations of active micro–shear bands competing with the mechanisms of crystallographic slip and/or twinning are responsible for the process of plastic flow. Also the observations of DUBOIS et al. [6], concerning initiation and evolution of micro–shear bands in polycrystalline sheet of pure copper subjected to rolling, confirm the sequential character of the active micro–shear bands as carriers of large plastic strains. Application of the fiducial grid technique and scanning electron microscopy made possible to observe simultaneously traces of slip lines and coarse slip bands, as well as, the grid and some grain boundaries sheared off by active micro–shear bands initiated at different stages of deformation process. The angle between the observed micro–shear bands and the rolling axis was nearly 35°. The authors in [2] argue that at least during part of the deformation the strain can be entirely or almost entirely concentrated in the shear bands. On the other hand, YEUNG and DUGGAN [7] studied the question of the contribution that active micro–shear bands make to accommodate the total imposed strain on the example of rolling of 70/30 brass slab. Applying the geometrical model of shear banding developed in [2] and introducing the fraction of the plastic deformation increment carried by active micro–shear bands they have found experimentally that this fraction changes very irregularly in deformation and in different locations along the specimen. The mean value of the fraction parameter lies between 0.6 and 0.8. According to [5] the shear bands formed in rolling are usually inclined by ± 35° to the rolling plane and are orthogonal to the specimen lateral face, although there can be considerable deviations from this value within the 15° to 50° range.

The problem of specifying the angle is complicated by the difficulty of distinguishing the most recently formed shear bands from those that were formed earlier and subsequently rotated with material towards the rolling plane. As it is reported in [5], in some cases, two sets of bands intersecting each other were observed, whereas in other cases only one set is present, or in some parts of the structure there is one set and in other part there are two sets. In the case of fcc metals of low stacking fault energy, two sets of shear bands form an array of rhomboidal prisms with axes parallel to the transverse direction. Shear bands of each sign exist together in any particular material volume. In the fcc metals of medium or high stacking fault energy as well as in bcc metals only one set of shear bands forms in a particular volume and neighbouring volumes develop bands of opposite sign. The volume concerned usually consists of several grains. Similar shear band systems were observed also in plane–strain tension and compression tests. ANAND and SPITZIG [8] conducted experiments on a maraging steel and examined with use of light microscopy the polished and etched cross–sections taken from the middle of the tension and compression specimens deformed to various values of strain. Narrow shear bands were first detected in both tension and compression specimens when they were deformed to a strain just beyond that corresponding to diffuse necking in plane–strain tension test. These bands were approximately 1 to 2 μm wide. The inclination of shear bands to the tensile axis was $\pm(38 \pm 2)^\circ$ whereas the inclination of shear bands to the compression axis was $\pm(55 \pm 2)^\circ$. This angle did not appear to change with increasing strain. Physical nature of shear banding is not completely explained. In particular, mechanisms of initiation and propagation of micro–shear bands across grain boundaries is a little known process. Nevertheless, the reported observations can make the basis to formulate a tentative model of large plastic deformations accounting for micro–shear bands.

Physical motivation to formulate simplified phenomenological model

The physical constraint on any continuum mechanics approach to metal plasticity, i.e. the physical dimension of the smallest representative volume element (r.v.e.) of crystalline material for which it is possible to define significant overall measures of stress and strain during plastic deformation was thoroughly discussed by HILL [9] and HAVNER [10, 11]. The minimum physical dimension of the r.v.e. at the *microscopic level* can be, accordingly, of the order 1 μm, i.e. $> 10^3$ lattice spacings. The linear dimension of the r.v.e. corresponding to the *macroscopic level* is often assumed of the order of 1 mm, for moderately fine–grained metals, [11]. This discussion is valid under the general assumption that the dominant mechanism of inelastic behaviour is crystallographic slip. In such a case the theory describing kinematics and constitutive structure of finite elastic–plastic deformations of crystalline solids is well established and the transition between the microscopic and macroscopic levels is well understood (cf. e.g. HAVNER [10], [11] and HILL [12], [13], as well as, MANDEL [14], HILL and RICE [15] and PETRYK [16]). In particular, relations between macro–measures of stress, strain and plastic work are related with the volume averages of their micro–measures. It has also been shown that certain structural features of the constitutive relations, as the normality rule or certain constitutive inequalities are transmitted upwards through a hierarchy of observational levels unchanged, irrespective of the heterogeneity, no matter what its origin (HILL [12]). The aforementioned discussion on the proper choice of the r.v.e. corresponding to the *microscopic level* and the *macroscopic level* substantiate the assumptions of the averaging theorem (HILL [13]). It is assumed that within the macroscopic r.v.e. (macro–element) micro–stresses are self–equilibriated whereas micro-velocities are continuous and piecewise continuously differentiable throughout the volume. In the simple case also the displacements and velocities given on the surface of macro–element are prescribed such that they would be compatible with uniform fields within the element. If these assumptions hold and macroscopic nominal stress and deformation gradient are the averages $\{\mathbf{n}\}$ and $\{\mathbf{f}\}$ of the respective microscopic fields over the r.v.e. in the reference configuration then the following basic averaging theorem can be proved, $\{\mathbf{f}\,\mathbf{n}\} = \{\mathbf{n}\}\,\{\mathbf{f}\}$.

The situation changes when additional mechanism of micro–shear banding is considered. Micro–shear bands extend over the distances of several grain diameters, produce large shear strains and form often clusters or packets of the width of ~ 1÷10 μm. Due to this, the assumption of the r.v.e. corresponding to the *microscopic level* is not adequate. The minimum physical dimension of the r.v.e. that could be appropriate for a valid continuum mechanics description of plastic deformation accounting for micro–shear bands should be probably of order ~ 10÷100 μm (depending on the grain diameter). This would correspond to the *mezoscopic level.* However, even at the *mezoscopic level* the assumptions of the averaging theorem are not fulfilled. This is due to the dynamic character of the micro–shear bands and high values of shear strains they carry. The redistribution of residual stresses inside the elementary volume takes place when the geometry changes with the propagation of micro–shear bands. In such a case the system is far from equilibrium and the known averaging procedures loose their validity (cf. STOLZ [17] for similar discussion concerning propagating damaged zones). The elaboration of suitable averaging procedures and precise transition between mezoscopic and macroscopic levels accounting for micro–shear banding is a challenging task and needs further studies. In face of these difficulties a simplified elastic–plastic material model accounting for macroscopic effects of active micro–shear bands is proposed.

Basic equations

The known assumption of the local, relaxed, isoclinic configurations, results in the unique decomposition of the deformation gradient $\mathbf{F} = \mathbf{EP}$ into the elastic component, $\mathbf{E}$, and the plastic one, $\mathbf{P}$ (cf. e.g. MANDEL [14]). The simple model of small elastic and finite plastic deformations with Huber–Mises yield criterion

$$f = \bar{\tau} - \kappa(\bar{\gamma}) = 0\,, \quad \text{where} \quad \bar{\tau} = \left(\tfrac{1}{2}\,\tau' : \tau'\right)^{1/2}, \qquad \bar{\gamma} = \left(2\,\mathbf{D}^p : \mathbf{D}^p\right)^{1/2} \tag{1}$$

makes a basis to incorporate new effects of active micro–shear bands. The polar decomposition $\mathbf{E} = \mathbf{V}^e\mathbf{R}^s$, with $\mathbf{V}^e = 1 + \varepsilon$, where $\|\,\varepsilon\,\| << 1$, leads to

$$\mathbf{D} = \mathbf{D}^e + \mathbf{D}^p\,, \quad \mathbf{D}^e = \overset{\circ}{\varepsilon} = \dot{\varepsilon} + \varepsilon\boldsymbol{\Omega} - \boldsymbol{\Omega}\varepsilon, \quad \mathbf{D}^p = \mathbf{R}^s\{\dot{\mathbf{P}}\mathbf{P}^{-1}\}_s(\mathbf{R}^s)^{-1} \tag{2}$$

$$\mathbf{W} = \boldsymbol{\Omega} + \mathbf{W}^p\,, \quad \boldsymbol{\Omega} = \dot{\mathbf{R}}^s(\mathbf{R}^s)^{-1}, \quad \mathbf{W}^p = \mathbf{R}^s\{\dot{\mathbf{P}}\mathbf{P}^{-1}\}_a(\mathbf{R}^s)^{-1}, \tag{3}$$

where $\mathbf{R}^s$ is the rotation of structure arising from elastic deformations and the geometric constrains imposed by boundary and compatibility conditions. The J_2 flow theory is assumed for the case when the sole mechanism of crystallographic multiple slip is responsible for plastic deformation

$$\mathbf{D}^p_S = <\lambda>\mathbf{N} = \frac{\sqrt{2}}{2}\,\dot{\bar{\gamma}}_S\,\mathbf{N}\,, \qquad \dot{\bar{\gamma}}_S = \sqrt{2}<\lambda>\,, \qquad \mathbf{N} = \frac{\tau'}{\sqrt{2}\,\bar{\tau}}\,, \tag{4}$$

$$\mathbf{D} = \mathbf{D}^e + \mathbf{D}^p_S = \left(\mathscr{L}^{-1} + \frac{1}{2\,h}\,\mathbf{N}\otimes\mathbf{N}\right) : \overset{\circ}{\tau}\,, \quad \text{for } \mathbf{D}^p = \mathbf{D}^p_S\,, \tag{5}$$

where $h = \partial\kappa/\partial\bar{\gamma}$ denotes plastic hardening modulus, $\lambda = \dfrac{\overset{\circ}{\tau}\!:\!\mathbf{N}}{2\,h}$ is the loading index, $\dot{\bar{\gamma}}_S$, $\mathscr{L}$, τ' represent, respectively, effective shear strain rate, tensor of elastic moduli and deviator of the Kirchhoff stress. The bracket $<\cdot>$ gives $<\lambda> = \lambda$ if $\lambda > 0$ and $<\lambda> = 0$ if $\lambda \leq 0$. In general,

$$\overset{\circ}{\tau} = \dot{\tau} - \boldsymbol{\Omega}\,\tau + \tau\,\boldsymbol{\Omega}\,, \qquad \boldsymbol{\Omega} = \mathbf{W} - \mathbf{W}^p\,, \tag{6}$$

provided $\mathbf{W}^p \neq \mathbf{0}$. In case of the J_2 flow theory, with isotropic hardening, $\mathbf{W}^p = \mathbf{0}$.

Phenomenological model of plastic flow involving shear banding

To capture the gross effects of active micro–shear bands, consider the mode of plane deformation at hand and assume that the contribution of the mechanisms of crystallographic multiple slip is represented by the flow rule $(4)_1$, whereas the contribution of active micro–shear bands to rate of plastic deformation is idealized by means of a double shear system with the directions of shear $\mathbf{n}_1^{(i)}$ and the normal to the shear plane $\mathbf{n}_2^{(i)}$, $i = 1, 2$, cf. Fig. 1. The double shear system can be oriented according to the fixed principal axes of strain, (e. g. $\mathbf{i}_1$ may correspond to the direction of rolling, extrusion, tension or compression). In such a case the angles φ_1, φ_2 can be taken as constants approximating the aforementioned characteristic orientation of micro–shear bands. It can be supposed that separate yield surfaces that intersect at a vertex are related with active micro–shear bands. An elastic–plastic material model accounting for the double shear system with pertinent yield planes intersecting at the yield–locus (1) is studied elsewhere, PĘCHERSKI and TVERGAARD [18]. The conjecture is made here, following the works of BUDIANSKY [19], RUDNICKI and RICE [20] and STÖREN and RICE [21], that for continued plastic flow with small deviations from proportional loading the hypoelastic form of J_2 deformation theory can be used to model in a simple way incremental plastic response from a local yield–locus vertex. It is then justified to relate the unit vectors of the double shear system $(\mathbf{n}_1^{(i)}, \mathbf{n}_2^{(i)})$ with the principal directions of applied stress tensor, in particular, with the unit vectors $(\mathbf{e}_1, \mathbf{e}_2)$ describing the orientation of the plane of maximum shear stress.

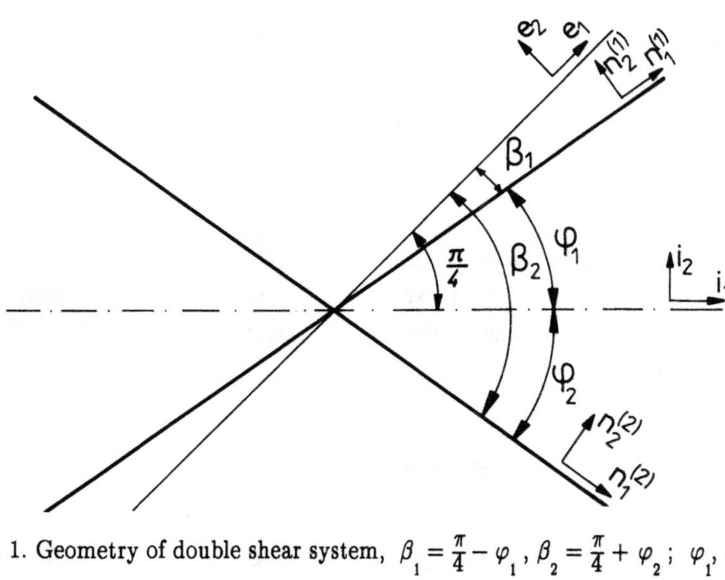

Fig. 1. Geometry of double shear system, $\beta_1 = \frac{\pi}{4} - \varphi_1$, $\beta_2 = \frac{\pi}{4} + \varphi_2$; $\varphi_1, \varphi_2 \in (0, \frac{\pi}{2})$.

According to Fig. 1. the following geometrical relations hold

$$\mathbf{e}_1 = \cos\beta_i\, \mathbf{n}_1^{(i)} + \sin\beta_i\, \mathbf{n}_2^{(i)} \qquad \mathbf{n}_1^{(i)} = \cos\beta_i\, \mathbf{e}_1 - \sin\beta_i\, \mathbf{e}_2$$
$$\mathbf{e}_2 = -\sin\beta_i\, \mathbf{n}_1^{(i)} + \cos\beta_i\, \mathbf{n}_2^{(i)} \qquad \mathbf{n}_2^{(i)} = \sin\beta_i\, \mathbf{e}_1 + \cos\beta_i\, \mathbf{e}_2\,, \tag{7}$$

i = 1, 2, no summation over repeated indices is assumed. The rate of plastic deformations and plastic spin produced by active micro–shear bands is assumed in the form

$$\mathbf{D}^{\mathrm{p}}_{\mathrm{MS}} = \sum_{i=1}^{2} \dot{\gamma}^{(i)}_{\mathrm{MS}} (\mathbf{n}_1^{(i)} \otimes \mathbf{n}_2^{(i)})_{\mathrm{s}}, \qquad \mathbf{W}^{\mathrm{p}}_{\mathrm{MS}} = \sum_{i=1}^{2} \dot{\gamma}^{(i)}_{\mathrm{MS}} (\mathbf{n}_1^{(i)} \otimes \mathbf{n}_2^{(i)})_{\mathrm{a}}\,, \tag{8}$$

where $\dot{\gamma}^{(i)}_{\mathrm{MS}}$ is the rate of plastic shearing along the ith direction of shear. Equations (8) can be transformed with use of $(7)_2$ into the following form (cf. ZBIB [22], [23], where formally similar relations but in a different physical context were obtained)

$$\mathbf{D}^{\mathrm{p}}_{\mathrm{MS}} = \sum_{i=1}^{2} \dot{\gamma}^{(i)}_{\mathrm{MS}} \, [\cos2\beta_i \, (\mathbf{e}_1 \otimes \mathbf{e}_2)_{\mathrm{s}} + \tfrac{1}{2} \sin2\beta_i \, (\mathbf{e}_1 \otimes \mathbf{e}_1 - \mathbf{e}_2 \otimes \mathbf{e}_2)],$$
$$\mathbf{W}^{\mathrm{p}}_{\mathrm{MS}} = \sum_{i=1}^{2} \dot{\gamma}^{(i)}_{\mathrm{MS}} \, (\mathbf{e}_1 \otimes \mathbf{e}_2)_{\mathrm{a}}\,, \tag{9}$$

or, equivalently, by means of the inclination angles φ_1 and φ_2, (cf. Fig. 1),

$$\mathbf{D}^{\mathrm{p}}_{\mathrm{MS}} = \dot{\gamma}_{\mathrm{MS}} \, (\mathbf{e}_1 \otimes \mathbf{e}_2)_{\mathrm{s}} + \tfrac{1}{2} \, \dot{\epsilon} \, (\mathbf{e}_1 \otimes \mathbf{e}_1 - \mathbf{e}_2 \otimes \mathbf{e}_2)\,, \tag{10}$$

where

$$\dot{\gamma}_{\mathrm{MS}} = \dot{\gamma}^{(1)}_{\mathrm{MS}} \sin2\varphi_1 - \dot{\gamma}^{(2)}_{\mathrm{MS}} \sin2\varphi_2 \,, \dot{\epsilon} = \dot{\gamma}^{(1)}_{\mathrm{MS}} \cos2\varphi_1 + \dot{\gamma}^{(2)}_{\mathrm{MS}} \cos2\varphi_2\,, \tag{11}$$

represent, respectively, the rate of plastic shearing in the direction of the maximum shear stress and the axial strain rate component in the direction $\mathbf{e}_1$. Assuming that the contribution of the mechanisms of crystallographic multiple slip and micro–shear allows for the additive decomposition of the rate of plastic deformations,

$$\mathbf{D}^{\mathrm{p}} = \mathbf{D}^{\mathrm{p}}_{\mathrm{S}} + \mathbf{D}^{\mathrm{p}}_{\mathrm{MS}}\,, \tag{12}$$

and taking into account that $\tau' = 2\bar{\tau}(\mathbf{e}_1 \otimes \mathbf{e}_2)_S$, for plane deformations, the relation for rate of plastic deformations reads

$$\mathbf{D}^p = \frac{\sqrt{2}}{2}(\dot{\bar{\gamma}}_S + \dot{\gamma}_{MS})\,\mathbf{N} + \tfrac{1}{2}\,\dot{\epsilon}\,\mathbf{T}\,. \tag{13}$$

The tensor $\mathbf{T} = \mathbf{e}_1 \otimes \mathbf{e}_1 - \mathbf{e}_2 \otimes \mathbf{e}_2$ is orthogonal to the normal $\mathbf{N}$. Accordingly, the following relation for the rate of plastic deformations is obtained, that can model the behaviour at a vertex produced by active micro–shear bands in the same spirit as in the constitutive model for fissured rock masses by RUDNICKI and RICE [20],

$$\mathbf{D}^p = \frac{\sqrt{2}}{2}\,\dot{\gamma}^*\,\mathbf{N} + \tfrac{1}{2}\,\dot{\epsilon}\,\Lambda\,[\overset{\circ}{\tau}{}' - (\overset{\circ}{\tau} : \mathbf{N})\,\mathbf{N}]\,, \tag{14}$$

where

$$\dot{\gamma}^* = \dot{\bar{\gamma}}_S + \dot{\gamma}_{MS}, \quad \mathbf{T} = \Lambda\,[\overset{\circ}{\tau}{}' - (\overset{\circ}{\tau} : \mathbf{N})\,\mathbf{N}]\,, \tag{15}$$

and Λ is a normalization factor. The rate $\dot{\bar{\gamma}}^* = \sqrt{2} < \lambda >$ contributes to the change of the radius of the Huber–Mises yield locus. The following *fractions of the rate of plastic shearing carried by the active micro–shear bands*, $f_{MS}^{(1)}$, $f_{MS}^{(2)}$, are introduced, respectively, to eliminate the unknown rates, $\dot{\bar{\gamma}}_{MS}^{(1)}$, $\dot{\bar{\gamma}}_{MS}^{(2)}$,

$$\dot{\gamma}_{MS}^{(1)} \sin 2\varphi_1 = f_{MS}^{(1)}\,\dot{\gamma}^*, \qquad -\dot{\gamma}_{MS}^{(2)} \sin 2\varphi_2 = f_{MS}^{(2)}\,\dot{\gamma}^*, \tag{16}$$

where due to $(11)_1$ and for $\dot{\gamma}^* > 0$ the following constraints hold

$$\frac{\dot{\bar{\gamma}}_S}{\dot{\gamma}^*} + f_{MS}^{(1)} + f_{MS}^{(2)} = 1\,, \quad f_{MS}^{(1)} + f_{MS}^{(2)} \in [0, 1], \quad f_{MS}^{(1)}, f_{MS}^{(2)} \in [0, 1] \tag{17}$$

Furthermore, it is assumed that for $\dot{\gamma}^* = 0$, $f_{MS}^{(1)} = f_{MS}^{(2)} = 0$. This assumption is based on the observation that the micro–shear bands can be active only in the case of continued plastic flow, i.e. when loading condition is fulfilled.

Consequently, the total rate of plastic deformations (14) takes form

$$\mathbf{D}^p = \frac{1}{2\,h}(\overset{\circ}{\tau} : \mathbf{N})\,\mathbf{N} + \frac{1}{2h_1}\left[f^{(1)}_{MS}\ \mathrm{ctan}2\varphi_1 - f^{(2)}_{MS}\ \mathrm{ctan}2\varphi_2\right]\left[\overset{\circ}{\tau}{}' - (\overset{\circ}{\tau} : \mathbf{N})\,\mathbf{N}\right], \quad (18)$$

where Λ is assumed as, $\Lambda = 1/(\dot{\gamma}^* h_1)$,for $\dot{\gamma}^* > 0$, and h_1 is a hardening modulus in the direction orthogonal to $\mathbf{N}$. In particular, for often observed symmetry of the double shear system with respect to the direction i_1, $\varphi_1 = \varphi_2 = \varphi \in (0, \frac{\pi}{2})$

$$\mathbf{D}^p = \frac{1}{2\,h}(\overset{\circ}{\tau} : \mathbf{N})\,\mathbf{N} + \frac{1}{2h_1}\mathrm{ctan}2\varphi\,(f^{(1)}_{MS} - f^{(2)}_{MS})\left[\overset{\circ}{\tau}{}' - (\overset{\circ}{\tau} : \mathbf{N})\,\mathbf{N}\right]. \quad (19)$$

The second term in (18) and (19), denoted with $\mathbf{D}_T$, is responsible for the non–coaxiality between the principal directions of the stress tensor and those of the plastic rate of deformation tensor.

If $\lambda \leq 0$, $\dot{\gamma}^* = 0$ and $\mathbf{D}^p = 0$. Observe that $\mathbf{D}_T = 0$ if $f^{(1)}_{MS} = f^{(2)}_{MS}$ or $\varphi = \pi/4$. The following special cases can be considered:

(i) for $\varphi < \pi/4$, and $f^{(1)}_{MS} \gtrless f^{(2)}_{MS}$ $\quad \mathbf{D}^p = \frac{1}{2\,h}(\overset{\circ}{\tau} : \mathbf{N})\,\mathbf{N} \pm \mathbf{D}_T$

(ii) for $\varphi > \pi/4$, and $f^{(1)}_{MS} \gtrless f^{(2)}_{MS}$ $\quad \mathbf{D}^p = \frac{1}{2\,h}(\overset{\circ}{\tau} : \mathbf{N})\,\mathbf{N} \mp \mathbf{D}_T$.

To eliminate in $(10)_2$ the orientation tensor $(\mathbf{e}_1 \otimes \mathbf{e}_2)_a$ the following relation is derived with use of (13)

$$(\mathbf{e}_1 \otimes \mathbf{e}_2)_s\,\mathbf{D}^p - \mathbf{D}^p\,(\mathbf{e}_1 \otimes \mathbf{e}_2)_s = -(\dot{\gamma}^{(1)}_{MS}\cos2\varphi_1 + \dot{\gamma}^{(2)}_{MS}\cos2\varphi_2)\,(\mathbf{e}_1 \otimes \mathbf{e}_2)_a\,, \quad (20)$$

Hence, the equation for plastic spin reads

$$\mathbf{W}^p = \frac{\dfrac{f^{(2)}_{MS}}{\sin2\varphi_1} - \dfrac{f^{(1)}_{MS}}{\sin2\varphi_2}}{2\ \bar{\tau}(f^{(1)}_{MS}\,\mathrm{ctan}2\varphi_1 - f^{(2)}_{MS}\,\mathrm{ctan}2\varphi_2)}(\tau\,\mathbf{D}^p - \mathbf{D}^p\tau)\,, \quad (21)$$

or equivalently, applying (18),

$$\mathbf{W}^{p} = \frac{\dfrac{f^{(2)}_{MS}}{\sin 2\varphi_1} - \dfrac{f^{(1)}_{MS}}{\sin 2\varphi_2}}{4\,\bar{\tau}\,h_1}\,(\mathring{\tau}\,\tau - \tau\,\mathring{\tau}). \tag{22}$$

For $\varphi_1 = \varphi_2 = \varphi$ the relations (21) and (22) take form

$$\mathbf{W}^{p} = \frac{-1}{2\,\bar{\tau}\,\cos 2\varphi}\,(\tau\,\mathbf{D}^{p} - \mathbf{D}^{p}\tau) = \frac{f^{(2)}_{MS} - f^{(1)}_{MS}}{4\,\bar{\tau}\,h_1\,\sin 2\varphi}\,(\mathring{\tau}\,\tau - \tau\,\mathring{\tau}). \tag{23}$$

The fractions $f^{(1)}_{MS}$, $f^{(2)}_{MS}$ that appear in discussed relations can be taken in the first approximation as constants that fulfill the conditions (17). On the other hand, certain microscopic models lead to the conjecture that shear banding can contribute to the rate of plastic deformations as a sequence of generations of active micro–shear bands governed by logistic equation (Verhulst equation), taken from the population dynamics, PĘCHERSKI [24]. The problem of 'unconstrained' large–strain shear has been solved numerically in [24] to study the effect of micro–shear bands on axial elongation, called the 'Swift effect'. It appears that evolution of micro–shear bands can produce irregular, non–monotonic dependence of the axial elongation on shear strain.

Concluding remarks

The developed J_2 deformation theory accounting for micro–shear bands can be applied to give a new physical insight into J_2 corner theory of plasticity studied by CHRISTOFFERSEN and HUTCHINSON [25]. In the J_2 corner theory yield surface vertex is modeled as a stress space hypercone with the axis τ'. The J_2 corner theory is devised to coincide with J_2 deformation theory for nearly proportional stress increments, corresponding to 'total loading range', and provide a smooth transition to elastic unloading for stress increments directed along or within the corner of the yield surface. The hypoelastic J_2 deformation theory accounting for micro–shear bands, given in (18) can be assumed in the total loading cone. In such a case the yield surface vertex is controlled by the development of active micro–shear bands. This problem is studied in a separate paper, [26]. Further experimental and theoretical studies are necessary to formulate a theory valid for an arbitrary three–dimensional case.

Acknowledgement

This work was supported by the Committee for Research (KBN), Poland, under the project No. 3 1112 91 01.

References

1. H. E. DÈVE, R. J. ASARO: The development of plastic failure modes in crystalline materials: shear bands in fcc polycrystals, Metall. Trans., **20A**, 59–593, 1989.

2. J. GIL SEVILLANO, P. van HOUTTE, E. AERNOUDT: Large strain work hardening and textures, Progress in Materials Science, **25**, 69–412, Pergamon Press, 1981.

3. A. KORBEL, P. MARTIN: Microstructural events of macroscopic strain localization in prestrained tensile specimens, Acta. Metall. **36**, 2575, 1988

4. B. J. DUGGAN, M. HATHERLY, W. B. HUTCHINSON, P. T. WAKEFIELD: Deformation structures and textures in cold–rolled 70:30 brass, Metal Sci., **12**, 343–351, 1978.

5. M. HATHERLY, A. S. MALIN: Shear bands in deformed metals, Scripta Metall., **18**, 449–454, 1984.

6. Ph. DUBOIS, M. GASPERINI, C. REY, A. ZAOUI: Crystallographic analysis of shear bands initiation and propagation in pure metals. Part II. Initiation and propagation of shear bands in pure ductile rolled polycrystals, Arch. Mech. **40**, 35–40, 1988.

7. W. Y. YEUNG, B. J. DUGGAN: On the plastic strain carried by shear bands in cold–rolled α–brass, Scripta Metall. **21**, 485–490, 1987.

8. L. ANAND, W. A. SPITZIG: Initiation of localized shear bands in plane strain, J. Mech. Phys. Solids, **28**, 113–128, 1980.

9. R. HILL: The mechanics of quasi–static plastic deformation in metals, in Surveys in Mechanics, 7–31, G. K. Batchelor, R. M. Davies (eds.), Cambridge, 1956.

10. K. HAVNER: On the mechanics of crystalline solids, J. Mech. Phys. Solids, **21**, 383–394, 1973.

11. K. HAVNER: Aspects of theoretical plasticity at finite deformation and large pressure, ZAMP, **25**, 765–781, 1974.

12. R. HILL: The essential structure of constitutive laws for metal composites and polycrystals, J. Mech. Phys. Solids, **15**, 779–95, 1967.

13. R. HILL: On constitutive macro–variables for heterogeneous solids at finite strain, Proc. R. Soc. Lond. A **326**, 131–147, 1972.

14. J. MANDEL: Plasticité classique et viscoplasticité, C.I.S.M., Udine, Springer–Verlag, 1972.

15. R. HILL, J. R. RICE: Constitutive analysis of elastic–plastic crystals at arbitrary strain, J. Mech. Phys. Solids, **20**, 401–413, 1972.

16. H. PETRYK: On constitutive inequalities and bifurcation in elastic–plastic solids with a yield–surface vertex, J. Mech. Phys. Solids, **37**, 265–291, 1989.

17. C. STOLZ: On relationship between micro and macro scales for particular cases of nonlinear behaviour of heterogeneous media, Proc. of IUTAM/ICM Symposium on Yielding, Damage and Failure of Anisotropic Solids, J.–P. Boehler ed., Mechanical Engineering Publications, London, 1990.

18. R. B. PĘCHERSKI, V. TVERGAARD: An elastic–plastic material model accounting for micro shear bands and effects on the development of macroscopic shear localization – under preparation.

19. B. BUDIANSKY: A reassessment of deformation theories of plasticity, J. Appl. Mech., **26**, 259–264, 1959.

20. J. W. RUDNICKI, J. R. RICE: Conditions for the localization of deformation in pressure–sensitive dilatant materials, J. Mech. Phys. Solids, **23**, 371–394, 1975.

21 S. STÖREN, J. R. RICE: Localized necking in thin sheets, J. Mech. Phys. Solids, **23**, 421–441, 1975.

22. H. M. ZBIB: On the mechanics of large inelastic deformations: noncoaxiality, axial effects in torsion and localization, Acta Mech. **87**, 179–196, 1991.

23. H. M. ZBIB: On the concept of noncoaxiality and large material rotations: A double–slip system, Mech. Res. Comm. (in print).

24. R. B. PĘCHERSKI: A model of plastic flow with an account of micro–shear banding, ZAMM, 1992, T 339 (in print).

25. J. CHRISTOFFERSEN, J. W. HUTCHINSON: A class of phenomenological corner theories of plasticity, J. Mech. Phys. Solids, **27**, 465–487, 1979.

26. R. B. PĘCHERSKI: A study of phenomenological corner theory accounting for the effects of micro–shear bands – under preparation.

Shear Localization in Thermo-Elasto-Viscoplastic Plane Strain Blocks

YOSHIHIRO TOMITA* AND TETSUYA NAKAO**

* Faculty of Engineering, Kobe University, Nada, Kobe, Japan
** Graduate Student, Kobe University

SUMMARY

Thermocoupled flow localization of plane strain blocks, obeying the elasto-viscoplastic constitutive equation with strain-gradient-dependent flow stress, has been analyzed. The effects of the strain gradient, finite element mesh size and the size of the specimen on flow localization have been clarified.

INTRODUCTION

Flow localizations frequently accompany the large plastic straining of materials. These are often recognized as prefracturing processes of ductile materials and have already been investigated extensively. However, most of the investigations employed the classical constitutive equation, and hence the flow localization zones were found to be dependent essentially upon the size of the finite elements. Although this is usually attributed to the ill-posedness of the problem, it must be a material property which may not depend upon the mesh size employed. To resolve the problem, considerable efforts have been made to develop a suitable regularization method by introducing the strain rate sensitivity[1,2], locally-enriched shape function[3,4] and a nonlocal theory[5-7].

In a previous paper[8], the authors have presented the flow localization analysis of tension block using the thermo-elasto-viscoplastic constitutive equation. In this paper, a thermo-elasto-viscoplastic version of the rate-dependent constitutive equation[8] is generalized by introducing the strain gradient

D. Besdo, E. Stein (Eds.)
Finite Inelastic Deformations –
Theory and Applications
IUTAM Symposium Hannover/Germany 1991

into the flow stress. The numerical results of full plane strain finite element simulation of velocity and temperature fields of the block subjected to shear and tension are presented. The influence of the strain gradient in the flow stress, the finite element mesh size and the size of the specimens on the shear band-type flow localization is discussed.

CONSTITUTIVE EQUATION

It is appropriate to show the constitutive equation employed in this investigation. As discussed by Tomita et al.[8], the thermo-elastic constitutive equation and the constitutive equation for the viscoplastic strain rate give the following relation for the the Jaumann rate of Kirchhoff stress $\overset{\nabla}{S}_{ij}$, deformation rate tensor d_{kl}, and the rate of change in temperature $\dot{T}$:

$$\overset{\nabla}{S}_{ij}=L'_{ijkl}\,d_{kl}-\dot{\bar{\varepsilon}}^{vp}P_{ij}-\dot{T}\,\beta_{ij}, \qquad (1)$$

where

$$p_{ij}=\frac{3\sigma'_{ij}}{2\bar{\sigma}}, \quad P_{ij}=D^e_{ijkl}p_{kl}, \quad \beta_{ij}=\frac{\alpha E}{1-2\nu}\delta_{ij},$$

$$L'_{ijkl}=\frac{H_s}{H_s+3G}[D^e_{ijkl}+\frac{3G}{H_s}\{\frac{1}{3}\delta_{ij}\delta_{kl}\frac{2G(1+\nu)}{1-2\nu}+\frac{3G}{\bar{\sigma}^2}\sigma'_{ij}\sigma'_{kl}\}], \qquad (2)$$

D^e_{ijkl} is the elastic stiffness tensor, α is the thermal expansion coefficient, E is the elastic modulus, ν is Poisson's ratio, G is the elastic shear modulus, $\bar{\varepsilon}^{vp}$ and $\dot{\bar{\varepsilon}}^{vp}$ are representative viscoplastic strain and its rate, respectively, and $H_s=\bar{\sigma}/\bar{\varepsilon}^{vp}$.

According to Zbib and Aifantis [7], the above-mentioned thermo-elasto-viscoplastic constitutive equation is generalized by introducing higher strain gradient terms such that

$$\bar{\sigma}=\bar{\bar{\sigma}}(\bar{\varepsilon}^{vp},\dot{\bar{\varepsilon}}^{vp},\ T)-c\nabla^2\bar{\varepsilon}^{vp}, \qquad (3)$$

where $\bar{\bar{\sigma}}$ is representative stress. ∇^2 is the Laplacian operator. The second term of Eq.(3) represents the nonlocal effects. The

Laplacian operator is introduced to describe the isotropic nonlocal effects. C is assumed to be constant. Introduction of Eqs.(3) into Eq.(2) yields the nonlocal generalization of the thermo-elasto-viscoplastic constitutive equation. Since the present constitutive equation is an elasto-viscoplastic one, the problem associated with the condition of loading and unloading will not arise. In order to increase the length of the time steps for stable computation, a forward gradient method[9] is applied.

METHOD OF ANALYSIS

The numerical procedure utilizes the finite element method for velocity and temperature fields[8] along with the constitutive equation derived above. Ninety-five percent of the viscoplastic work is assumed to transform into heat. In most of the calculations described here, representative stress $\bar{\bar{\sigma}}$, absolute temperature T, representative viscoplastic strain $\bar{\varepsilon}^{vp}$, and representative viscoplastic strain rate $\dot{\bar{\varepsilon}}^{vp}$ are related by

$$\bar{\bar{\sigma}} = \tau_c\{K(T)/K(T_0)\}\{2.287+1.986\tanh(31.16\bar{\varepsilon}^{vp})\}(\dot{\bar{\varepsilon}}^{vp}/\dot{\varepsilon}_y)^m \qquad (4)$$
$$K(T)=0.002145T^2-2.632T+991.3$$

where τ_c and $\dot{\varepsilon}_y$ are material constants. The constitutive equation of the temperature and strain rate independent limit well describes the deformation behavior of the polycrystal. Material and computational parameters are as follows: $\nu=0.3$; $\dot{\varepsilon}_y=0.001/s$; $\tau_c=60.84MPa$; $C_0=0.005$; $C=\gamma C_0$; $E=1000\tau_c$; $m=0.005$; $T_0=T_{air}=296.15K$; $\alpha=11.6*10^{-6}/K$; thermal conductivity, $\lambda=59W/(mK)$; heat transfer coefficient, $\lambda_{air}= 0.1256*10^5W/(m^2K)$; density, $\rho=7.87*10^3Kg/m^3$; $\dot{U}=(\dot{u}/L)/\dot{\varepsilon}_y=1,10^6$; and $W_0=0.032mm$.

PLANE STRAIN DEFORMATION

In order to understand the features of plane strain deformation behavior, the thermo-viscoplastic shear and tensile deformation of blocks has been investigated. Fig.1(a) shows the computational model and finite element discretization of shear deformation. Due to the one-dimensional feature of the present problem, a line finite element has been employed and the Laplacian ∇^2 has been approximately estimated by the representative strain of elements

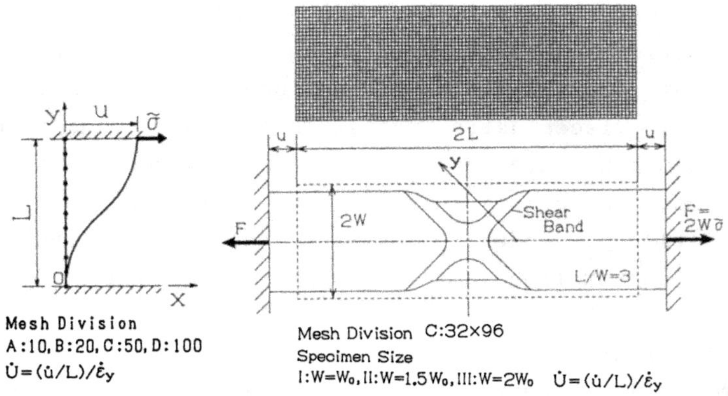

(a) Shear of blocks (b) Tension of blocks

Fig.1 Computational model and finite element discretization

located in the neighborhood at the previous incremental step.

Before discussing the main problem, the validity of the present method has to be checked numerically by comparison between the analytical solution of Zbib and Aifantis[7] and the present author's solution. The constitutive equation employed is rigid-plastic softening type one expressed by $\tau(\gamma)=\tau_m-a(\gamma-\gamma_m)^2-c\,\nabla^2\gamma$ with material constants τ_m, a and γ_m. Zbib and Aifantis[7] obtained the shear strain distribution $\gamma=\gamma_m+\sqrt{\delta}\,[3\,\mathrm{sech}^2(\sqrt{2a/c}\,\delta^{1/4}(y-\bar{y})/2)-1]$ and band width $B=2.924\sqrt{c/2a}\;\delta^{-1/4}$.

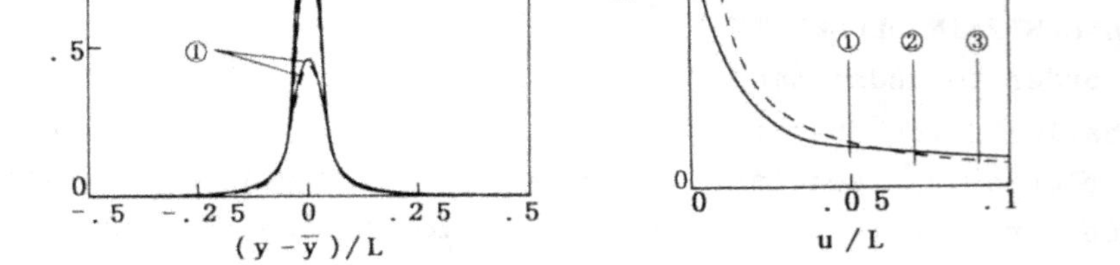

(a)Shear strain distribution (b)Evolution of shear-band width

Fig.2 Comparison betweeen the analytical results and present results.

In γ and B, $\bar{y}$ is the center of shear band and $\delta=(\gamma_m-\gamma_1)^2$ with homogeneous strain γ_1 corresponding to global deformation.

Fig.2 shows the shear strain distribution and the evolution of shear band width. According to the numerical results, although the width of shear band and the distribution of shear strain are influenced by the size of the specimen at the first stage of deformation, they are similar to those of Zbib and Aifantis[7] as the shear band width reduces to less than 20% of the specimen size. Beyond a certain level of maximum strain, (i.e. 100%), however, analytical prediction of shear strain exhibits steeper distribution. Accordingly, the present method will be restrictively employed for the problem with the above-mentioned limit of maximum strain.

Here, we discuss the thermo-elasto-viscoplastic shear deformation of blocks shown in Fig.1(a). The effect of the strain gradient term on the flow localization has been investigated initially with four different C values, namely, C=0, C_0, $4C_0$, $20C_0$, in Eq.(3). The normalized end displacement rate was set at $\dot{U}$=1,10^6. As predicted from analytical investigation[7], the nonlocal effect stabilizes the deformation, and considerably postpones the growth of the shear bands accompanied with a rapid drop in force. The softening due to temperature rise by irreversible work done causes significant flow localization. Although, a proper value of C in Eq.(3) should be determined by some other method, henceforth, the value of C will be set at 0 or C_0. The former stands for the local case and the latter represents the nonlocal case.

Fig.3 shows the distribution of shear strain for the four different mesh sizes shown in Fig.1(a). The notations shown in Fig.3 are introduced to specify the computational conditions. The first letter of the code identifies the mesh division. The two numbers in the notation indicate the logarithmic value of the nominal deformation rate, $\log_{10}\dot{U}$, and the value $\gamma=C/C_0$, respectively. A quite strong mesh dependency is observed in local

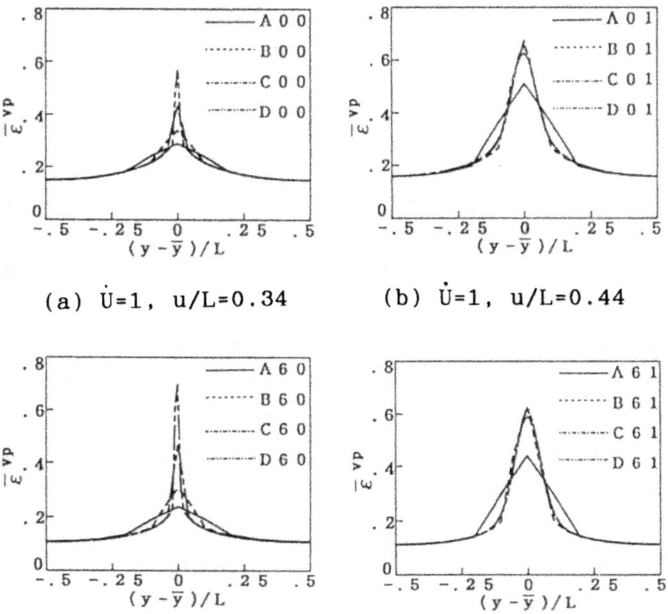

(a) $\dot{U}$=1, u/L=0.34 (b) $\dot{U}$=1, u/L=0.44

(c) $\dot{U}=10^6$, u/L=0.26 (d) $\dot{U}=10^6$, u/L=0.34

Fig.3 Representative strain distribution in the shear band

case shown in (a) and (c). In other words, refinement of the mesh size causes strong flow localization and eventually the mesh size sets the width of the shear band. With regard to the nonlocal case depicted by (b) and (d), the convergence of the solution is quite good for the models having a sufficient number of elements. Furthermore, the results imply the suitable finite element mesh size to have a converged solution so that an equivalent mesh size of B in Fig.1(a) will be employed.

Thermo-elasto-viscoplastic analysis is performed for the plane strain blocks subjected to tension at both ends. Due to the symmetry of the deformation, 1/4th of the block with the finite element discretization shown in Fig.1(b) is investigated. Each quadrilateral shown in the figure consists of four crossed constant-strain triangular elements. Heat generated by viscoplastic work is assumed to be discharged through convection to air at the side surface. The remaining surfaces are assumed to be adiabatic boundaries. The gradient term of representative strain for an element was approximately estimated by the numerical differentiation based on the values of representative strain of neighboring elements. In order to clearly specify the computational results, the four-part notations shown in Fig.4 will be employed henceforth. Note that the first letter of the code in the left-hand column indicates the specimen size and the second stands for the mesh division shown in Fig.1(b). The two numbers in the notation denote the logarithmic value of the nominal deformation rate, $\log_{10}\dot{U}$, and the value $\gamma = C/C_0$, respectively.

Since the general features of the deformation behavior which depend on the magnitude of C in Eq.(3) and the mesh size are similar to those of shear, the size effect on the shear localization is discussed here. Figure 4 shows the force $\tilde{\sigma}$ vs displacement u. Figure 5 shows the deformed meshes and representative strain contours. In order to directly compare the results, the sizes of II and III are normalized to that of I. For the case with $\dot{U}$=1, both patterns of meshes and strain contours for the local case are almost identical. This implies a strong size-dependent feature of the flow localization behavior. As can be seen in the nonlocal case, more intense localizations appear

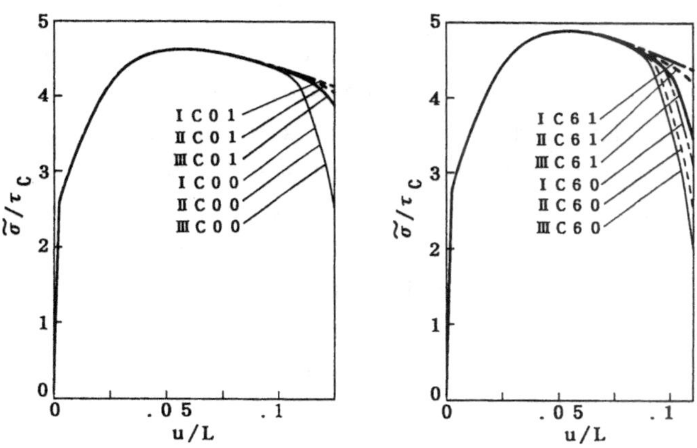

(a) $\dot{U}$=1 (b) $\dot{U}=10^6$

Fig.4 Force $\tilde{\sigma}$ vs. displacement u

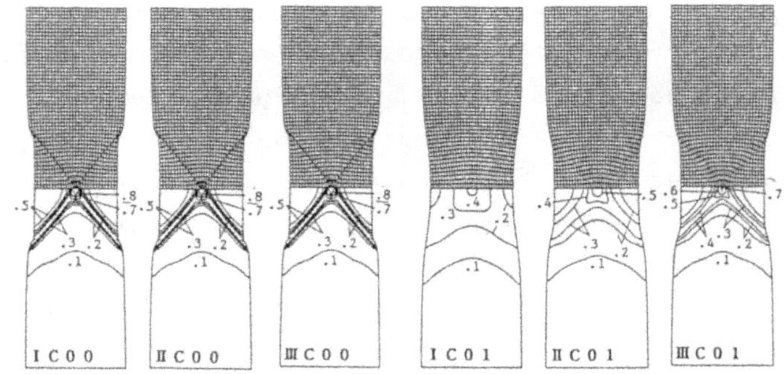

(a)C=0, $\dot{U}$=1, u/L=0.125 (b)C=C_O, $\dot{U}$=1, u/L=0.125

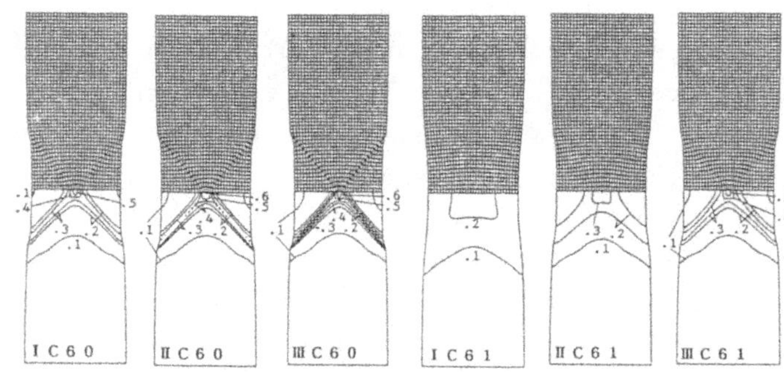

(c)C=0, $\dot{U}=10^6$, u/L=0.11 (d)C=C_O, $\dot{U}=10^6$, u/L=0.11

Fig.5 Deformed mashes and representative strain $\bar{\varepsilon}^{vp}$ contours

as the size of the specimen increases. This can be more clearly seen in Fig.6 which shows the representative strain distribution across the shear band. The thickness of shear bands is almost independent of the size of the specimen in the nonlocal cases, whereas strong dependence is observed in the local case. Of course, as discussed by Tomita et al.[8], the thermal effect contributes to the size-dependent flow localizations. Therefore, in the case with $\dot{U}=10^6$ where the thermal effect is active, more localized deformation is equally observed in Figs.4-6 as the specimen size increases.

It should be pointed out that the difference in the magnitude of the representative strain in Fig.6(b),(d) for the same displacement u/L is attributed to the almost identical thickness

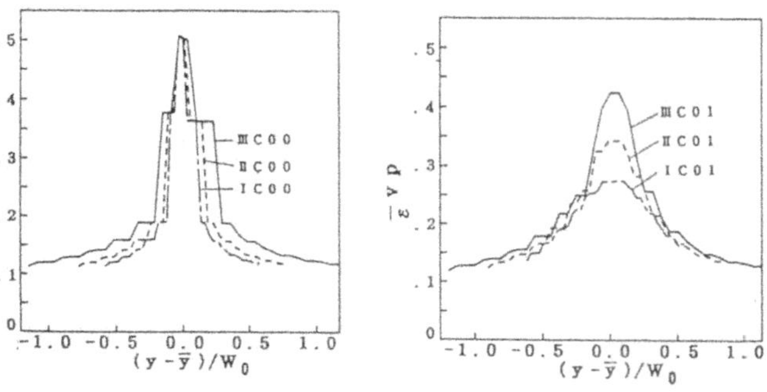

(a)C=0, $\dot{U}$=1, u/L=0.125 (b)C=C_0, $\dot{U}$=1, u/L=0.125

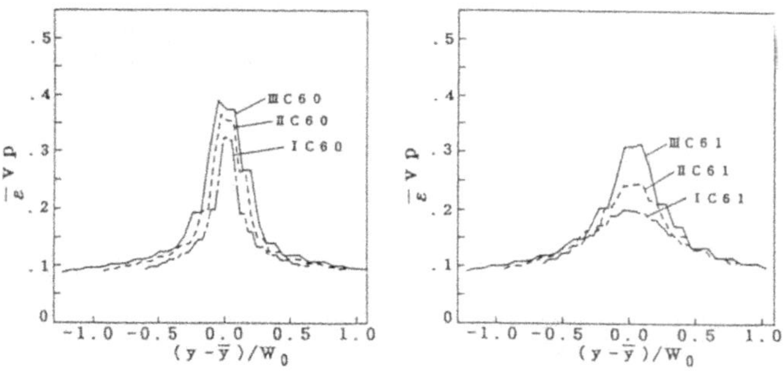

(c)C=0, $\dot{U}=10^6$, u/L=0.11 (d)C=C_0, $\dot{U}=10^6$, u/L=0.11

Fig.6 Representative strain $\bar{\varepsilon}^{vp}$ distribution in the shear band

of the shear band. When the shear band develops sufficiently with the same thickness for the specimens having different sizes, the displacement arising from the shear localization with almost similar inclination should be independent of the size. A more intense localization is required to attain the same displacement u/L for larger blocks. As discussed in this paper, the results presented above may be slightly affected by the size-dependent nature of the Laplacian operator in flow stress.

CONCLUSIONS

A thermo-elasto-viscoplastic version of the constitutive equation was generalized by introducing the strain gradient into the flow stress. The numerical results obtained for shear and tensile deformation clarified the influence of the strain gradient in the flow stress, the finite element mesh size and the specimen size on the shear band-type flow localization.

ACKNOWLEDGMENTS

The present work has been supported in part by the Grant-in-Aid for Scientific Research, Japanese Ministry of Education. Their financial support is gratefully acknowledged.

REFERENCES

1. Needleman,A.;Material rate dependence and mesh sensitivity in localization problems, Comp. Meth. Appl. Mech. Engng., 67-1,(1988)69-85.

2. Loret,B. and Prevost,J.H.; Dynamic strain localization in elasto-(visco) plastic solids, Comp. Meth. Appl. Mech. Engng., 83-3,(1990)247-273.

3. Ortiz,M., Leroy,Y., and Needleman,A.;A finite element method for localized failure analysis, Comp. Meth. Appl. Mech. Engng., 17-12,(1987) 323-334.

4. Fish,J., and Belytschko,T.; A general finite element procedure for problems with high gradient, Computers & Structures,35-4,(1990)309-319.

5. Bazant,Z.P., Lin,F.B. and Pijaudier-Cabot,G:Yield limit degradation: Nonlocal continuum with local strain. Proc. Int. Conf. on Computational Plasticity. Barcelona. ed. by Onate, E. et al., University of Swansea. (1987)1757-1779.

6. Aifantis,E.C.: On the microstructural origin of certain inelastic models,Trans. ASME, J. Eng. Mat. Tech. 106.(1984) 326-330

7. Zbib,H.M., and Aifantis,E,C.; On the localization and post localization behavior of plastic deformation. I. On the initiation of shear bands, Res Mechanica 23 (1988) 261-277.

8. Tomita,Y., Shindo,A., and Sasayama, T.; Plane strain tension of thermo-elasto-viscoplastic blocks, Int. J. Mech. Sci., 32-7 (1990) 613-622.

9. Peirce,D., Shih,C.F., and Needleman,A.; A tangent modulus method for rate dependent solids. Comp. Struct, 18 (1984)875.

On Stability of Time-Independent Materials at Finite Strain

H. PETRYK

Institute of Fundamental Technological Research,
Polish Academy of Sciences, Warsaw, Poland

Summary

Stability of plastic deformation at the level of a homogeneous material element imbedded in a deforming continuum is discussed. Distinction is made between instability of equilibrium and instability of uniform straining of the element. Under the general assumption that constitutive rate equations admit a potential, pointwise energy criteria of instability are formulated and related to propagation of acceleration waves and to incipient localization of deformation. For incrementally nonlinear materials they are shown to differ in general from the familiar condition of ellipticity loss. Existence of a continuous range of shear band bifurcations is demonstrated. In a numerical example, a finite strain version of the Christoffersen-Hutchinson model of material behaviour at a yield-surface vertex is used.

Introduction

We are concerned here with stability of quasi-static, isothermal deformation of an incrementally nonlinear continuum, at the level of a material element and at unrestricted strains. The restriction is made that the constitutive law does not contain a natural time and, in the rate form, admits a velocity-gradient potential. No spatial discontinuities in displacements or velocities are considered: fracture, slip discontinuities or cavitation are thus excluded from consideration. A certain class of macroscopic instabilities, primarily in metals, is intended to be so modeled. Physical aspects of plastic instabilities, especially at the level of dislocation motion, are not discussed here.

The concept of an "intrinsic" stability or instability of the material in the plastic range was propounded in the mechanics of continua in various contexts, but no general agreement about its precise meaning has been worked out yet. Definitions of Drucker's type [1] implicitly involve an arbitrary choice of the measure of strain (or stress-rate) when geometric changes are not disregarded [2]. In turn, the measure-invariant requirement based on comparison of the internal and external work for a virtual *uniform* deformation increment involves an arbitrary specification of the interaction rule between the configuration of a material element and the tractions on the element boundary applied by its environment [3]. An analogous difficulty is met if material instability is understood in the sense of "intrinsic eigenstates" associated with constitutive branching [4].

D. Besdo, E. Stein (Eds.)
Finite Inelastic Deformations –
Theory and Applications
IUTAM Symposium Hannover/Germany 1991

In another approach, used recently in many investigations of shear band formation, instability of the material is understood in the sense of possible bifurcation in a planar band in an infinite homogeneous continuum [5]. Admissibility of quasi-static bifurcation in a band is strictly related to vanishing speed of dynamic propagation of acceleration waves [6] in the *deforming* material. If the constitutive domain of the currently active "tangent" moduli in strain-rate space coincides with the half-space (classical elasto-plasticity) then the onset of instability in the above sense can be determined from the familiar condition of (strong) ellipticity loss, as in elastic or incrementally linear solids [7][6]. Moreover, uniform post-critical deformations exhibit then still another kind of instability, namely, an intrinsic instability of equilibrium in the energy sense (*op.cit.*) and also in a dynamic sense [8], related to imaginary speed of wave propagation with respect to the *non-deforming* material.

For an incrementally nonlinear response, predicted e.g. by micromechanical studies [9][10] and regarded as more realistic, the situation is more complex since the critical points associated with shear band bifurcation or loss of stability of equilibrium or ellipticity loss for tangent moduli need not coincide with each other in general. The question thus arises, not fully explored yet, what is the correct extension of the condition of ellipticity loss to an incrementally nonlinear case. That question is discussed here by starting from the concept of local instability of a quasi-static deformation process in the energy sense [11]. Further results presented below are extracted from the recent work [12] but illustrated on another numerical example.

Following Hill [13], constitutive rate equations will be expressed in terms of the (forward) rates of the deformation gradient $\mathbf{F}$ and the nominal stress $\mathbf{N}$ as

$$\dot{N}_{ij} = \frac{\partial U}{\partial \dot{F}_{ji}} = C_{ijkl}(\dot{\mathbf{F}}, \mathcal{H})\dot{F}_{lk}, \qquad U = U(\dot{\mathbf{F}}, \mathcal{H}) = \frac{1}{2}\dot{N}_{ij}(\dot{\mathbf{F}}, \mathcal{H})\dot{F}_{ji}, \tag{1}$$

where the subscripts (obeying the summation convention) denote tensor components on a fixed rectangular basis, a dot over a symbol denotes right-hand differentiation by a time-like parameter t,

$$C_{ijkl} = \frac{\partial^2 U}{\partial \dot{F}_{ji} \partial \dot{F}_{lk}} = C_{klij} \tag{2}$$

are the instantaneous "stiffness" moduli dependent (piecewise-continuously) on the direction of the strain-rate, and $\mathcal{H}$ indicates the influence of the deformation history prior to the current instant; the symbolic parameter $\mathcal{H}$ will be omitted below for simplicity. U is a positively homogeneous of degree two and continuously differentiable function of $\dot{\mathbf{F}}$, being arbitrary otherwise. Constitutive laws for hyperelastic solids or for classical elastoplastic solids obeying the normality flow rule are included in (1) as particular cases.

Pointwise conditions sufficient for instability in the energy sense

Consider a material element which constitutes a small part of a quasi-statically deforming continuum. We are concerned with stability of a fundamental deformation process $\mathbf{F}^0(t)$ in the element; the process is regarded as known and spatially homogeneous in the limit as the element size tends to zero. Perturbations of the deformation increment are considered which preserve continuity of the material and are *non-uniform inside and vanish outside the element.* For stability of the fundamental process in the energy sense it is required [11] that among all virtual deformations of that type the work of deformation in the element, evaluated to *second* order, is minimized by the uniform mode. It can be shown by appealing to Graves' [14] theorem that the above minimum property is necessarily violated if

$$\mathcal{E}(\dot{\mathbf{F}}^0, \dot{\mathbf{F}}) < 0 \qquad \text{for some } \dot{F}_{ji} = \dot{F}^0_{ji} + g_j n_i, \tag{3}$$

where $\mathcal{E}$ is the Weierstrass function associated with the constitutive potential $U(\dot{\mathbf{F}})$:

$$\mathcal{E}(\dot{\mathbf{F}}^0, \dot{\mathbf{F}}) = U(\dot{\mathbf{F}}) - U(\dot{\mathbf{F}}^0) - \dot{N}^0_{ij}(\dot{F}_{ji} - \dot{F}^0_{ji}), \qquad \dot{N}^0_{ij} = \frac{\partial U}{\partial \dot{F}_{ji}}(\dot{\mathbf{F}}^0). \tag{4}$$

A direct derivation which provides a mechanical interpretation to (3) as the condition that incipient localization of deformation in a band becomes energetically preferable to uniform deformation is given in [12]. Other interpretations of (3) as a condition *sufficient for instability of uniform straining with the rate* $\dot{\mathbf{F}}^0$ will be discussed in the next section.

A *possible* critical stage associated with (3) is that $\mathcal{E}(\dot{\mathbf{F}}^0, \dot{\mathbf{F}})$ vanishes non-trivially for some $\dot{F}_{ji} = \dot{F}^0_{ji} + g_j n_i$ while (3) is still not satisfied; obviously, this corresponds to $\partial\mathcal{E}/\partial g_j = 0$ at such $(\dot{\mathbf{F}}^0, \dot{\mathbf{F}})$. At that instant we thus have [11]

$$(\dot{N}_{ij}(\dot{\mathbf{F}}) - \dot{N}^0_{ij}) n_i = 0 \qquad \text{for some } \dot{F}_{ji} = \dot{F}^0_{ji} + g_j n_i \tag{5}$$

which are just the statical and kinematical first-order compatibility conditions on a stationary discontinuity (of normal n_i) in a dynamic problem [6] or on a boundary of a bifurcation band in a quasi-static problem [5].

We shall assume below that the tangent moduli $C^0_{ijkl} = C_{ijkl}(\dot{\mathbf{F}}^0)$ are well defined whenever they appear in the considerations. If the material is incrementally linear ($\mathbf{C}$ independent of $\dot{\mathbf{F}}$) then (3) is equivalent to

$$(C^0_{ijkl} n_i n_k) g_j g_l < 0 \qquad \text{for some } n_i, g_j \tag{6}$$

and (5) reduces to

$$\det(C^0_{ijkl} n_i n_k) = 0. \tag{7}$$

For the classical elastic-plastic solids obeying normality, (3) is still equivalent to (6) provided the moduli C^0_{ijkl} correspond to the loading branch. However, if the material is incrementally nonlinear then in general (3) may hold even if C^0_{ijkl} satisfy the strong ellipticity condition, i.e. if (6) and (7) are both excluded. In turn, (6) implies (3) as can be seen e.g. from the Taylor series expansion of $\mathcal{E}(\dot{\mathbf{F}}^0, \dot{\mathbf{F}})$ at $\dot{\mathbf{F}} = \dot{\mathbf{F}}^0$. The condition (6) may be regarded as a specification of (3) for the *linearized* constitutive law.

If the work argument leading to (3) is applied to a material element imbedded in a *non-deforming* continuum ($\dot{\mathbf{F}}^0 = \mathbf{0}$) then it leads to the following condition *sufficient for instability of equilibrium*:

$$U(\dot{\mathbf{F}}) < 0 \qquad \text{for some } \dot{F}_{ji} = g_j n_i. \tag{8}$$

The instability is meant here in the following energy sense: (8) implies that energy can be extracted from the material element without perturbing the equilibrium *state* of the surrounding continuum. Interpretation of a typical critical stage associated with (8) is analogous to (5) with the restriction that now the material at one side of the discontinuity surface (or outside the band) is *non-deforming*. It can be seen [11] that for the classical elastoplastic solids with normality (and obviously also for incrementally linear materials), the condition (8) is equivalent to (3) and also to (6) for C^0_{ijkl} taken from the loading branch, but in a general nonlinear case we only have the implication (8) $\Longrightarrow$ (3).

Interpretations of instability

To understand better what kind of material instability is predicted by the above energy criteria, their connection with propagation of acceleration waves and formation of shear bands is now briefly discussed. A more detailed study is presented in [12] where the proofs of the statements quoted below are given.

For the speed of dynamic propagation of all acceleration waves with respect to the *non-deforming* material to be real it is necessary and sufficient that (8) is not satisfied. The speed is always real and nonzero if and only if

$$U(\dot{\mathbf{F}}) > 0 \qquad \text{for every nonzero } \dot{F}_{ji} = g_j n_i. \tag{9}$$

The inequality (8) may also be interpreted as the condition sufficient for initiation of *dynamic localization of deformation* in a planar band, at a fixed value of the external loading parameter. Namely, if (8) holds then, in the first approximation, a dynamic growth of an arbitrarily small disturbance within a band can take place under constant loading along the (non-deforming) band boundaries.

Non-fulfillment of (3) is necessary (but not sufficient) for the speed of dynamic propagation of all acceleration waves into the material *deforming* with the rate $\dot{\mathbf{F}}^0$ to be real. A general connection between (3) and instability in a dynamic sense is less clear than in the case when (8) holds. However, if (9) is satisfied then uniform straining with $\dot{\mathbf{F}}^0 \neq \mathbf{0}$ satisfying (3) should be regarded as unstable in the *quasi-static* sense. For, it can be proved, also if n_i is fixed in advance, that validity of (5), and hence existence of admissible *stationary discontinuities*, is implied by (3) so long as (9) holds. This means that quasi-static *bifurcation* in a band can take place (in the conventional formulation of the problem for an infinite continuum) at *every* point on a deformation path $\mathbf{F}^0(t)$ along which (3) is satisfied simultaneously with (9). In turn, at a given instant from that range the bifurcation can take place for a continuous "fan" of band orientations; an illustration is given below.

It can be seen that in a general case of an incrementally nonlinear material, loss of ellipticity for the *tangent* moduli, in the form (6) or (7), is not necessary for localization of deformation to be initiated, neither in dynamic nor quasi-static way. Validity of the criterion of ellipticity loss as a *critical* condition for shear band formation can be, however, re-established for material models which satisfy a constitutive inequality recently derived from micromechanical considerations [15]. The inequality is satisfied automatically for classical elastic-plastic solids with normality, and also for the material model examined below *provided* the tangent moduli coincide with "total loading" moduli.

Example

The general results are now illustrated on an incrementally nonlinear model of time-independent material behaviour at a yield-surface vertex, of the form proposed in [16] for polycrystalline metals as the so-called J_2 corner theory of plasticity. More essential material parameters adopted in the calculations are: a power-law hardening exponent $n = 0.1$, an external angle of the (rather sharp) yield-surface corner $\beta_c = 135^\circ$, an angle of total loading cone $\theta_0 = 0$ (thoroughly nonlinear plastic response). The material is assumed to be practically incompressible, with the Poisson ratio for elastic strains taken as 0.499. The transition function is described by eq. (2.46) in [16] with $m = 3$, the Zaremba-Jaumann flux of Kirchhoff stress is substituted for a stress-rate, and the deformation theory moduli in the hyperelastic version are used as total loading moduli.

A particular family of non-proportional fundamental deformation paths $\mathbf{F}^0(t)$ is considered (Fig. 1). Along each path, plane-strain compression up to a fixed engineering strain $F^0_{11} - 1 = -0.25$ is followed by further compression accompanied by additional shear F^0_{21} which increases with a constant ratio $\dot{F}^0_{21}/|\dot{F}^0_{11}|$ from an interval $[0, 0.5]$. In a manner described in more detail in [12], for several such paths the range has been

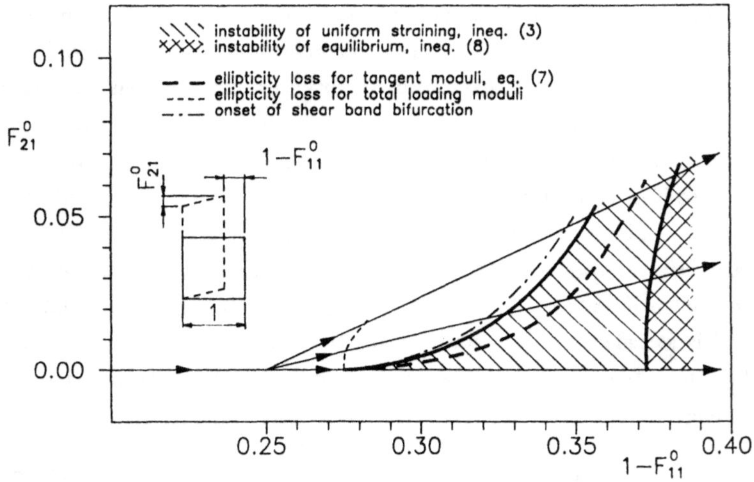

Fig. 1. Instability ranges on non-proportional deformation paths for an incrementally nonlinear material obeying the J_2 corner theory of plasticity

calculated where the inequalities (3), (6) or (8) are satisfied. To complete the picture, two further curves have also been determined. It is pointed out that due to strong path- and parameter-sensitivity, the results shown in Fig. 1 are only valid for the specified family of deformation paths for the assumed model; it is not discussed here whether the quantitative predictions are realistic or not. For any path with $\dot{F}^0_{21} \neq 0$, the order of events is the following. First, the limit of ellipticity is reached for the total loading moduli; this corresponds to the onset of some "latent" instability which *would* manifest itself *if* the direction of the fundamental deformation rate $\dot{\mathbf{F}}^0$ were abruptly changed to satisfy (6). For $\dot{\mathbf{F}}^0$ as prescribed above, the critical condition (7) is met at a later stage which, however, is preceded by two other critical points. The second critical point on the path is found when $\mathcal{E}(\dot{\mathbf{F}}^0, \dot{\mathbf{F}})$ with $\dot{F}_{ji} = \dot{F}^0_{ji} + g_j n_i$ has for the first time a nontrivial stationarity point (in fact, a local minimum point was searched numerically), which corresponds to (5) and hence to the onset of shear band bifurcation. However, such bifurcations are not energetically preferable (in the sense discussed above) to the fundamental continuation of deformation unless the third critical point is crossed and the range of validity of (3) is entered. Finally, the range of unstable equilibrium is reached and uncontrolled dynamic localization of deformation in a shear band becomes possible. It can be seen that fulfillment of the condition (7) does not correspond to the onset of instability in any sense discussed here. However, one can still relate (7) to a critical deformation stage when a *finite* growth of an *infinitesimal* initial imperfection becomes possible without bifurcation (cf. [5][17]).

In the limiting case $\dot{F}^0_{21} = 0$ (plane-strain compression), all the critical points coincide except the one related to loss of stability of equilibrium. In that particular case, to be examined from now on, the actual tangent moduli coincide with the total loading moduli, and (3) reduces to (6). The general result that (3) and (9) satisfied simultaneously along a path imply existence of a *continuous* range of shear band bifurcations is visualized in Fig. 2 where a domain of admissible band angles (in the current configuration) is indicated. The solid line represents band normals n_i satisfying (7), while *any* band orientation from the shaded area, at the state attained at a certain compressive strain, satisfies the conditions (6), (3) *and* the bifurcation condition (5). A point on the broken line inside this area corresponds to an absolute minimum of $\mathcal{E}(\dot{\mathbf{F}}^0, \dot{\mathbf{F}})$ among all $\dot{F}_{ji} = \dot{F}^0_{ji} + g_j n_i$ at a fixed state. That bifurcation band may be regarded as energetically most preferable since $\mathcal{E}(\dot{\mathbf{F}}^0, \dot{\mathbf{F}})$ is proportional to the second-order work difference due to

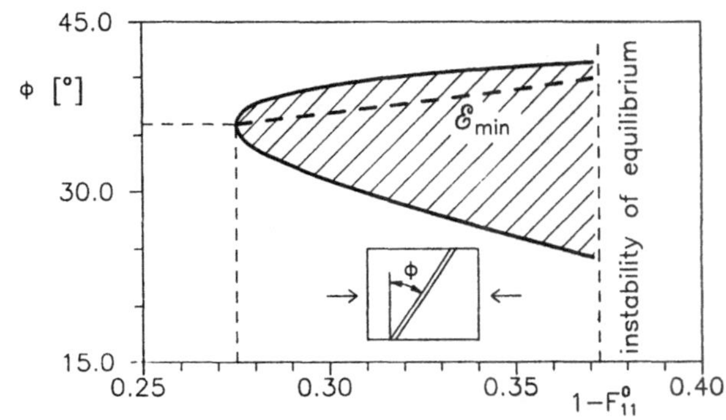

Fig. 2. Continuous range of shear band bifurcations in plane strain compression

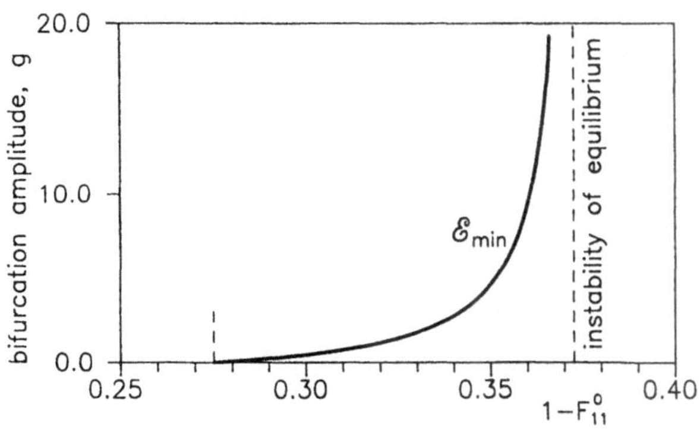

Fig. 3. Magnitude of superimposed shear rate in the energetically most preferable bifurcation band

superposition of the bifurcation mode in a vanishingly thin band [12]. The bifurcation amplitude $g = (g_j g_j n_i n_i)^{1/2}/|\dot{F}^0_{11}|$ for that band grows from 0 at the instant of ellipticity loss (we recall that the plastic response is here thoroughly nonlinear) and tends to infinity when the critical strain corresponding to transition from the range (9) to (8) is approached (Fig. 3). Shear band bifurcations are possible also beyond that instant but appear to be not of much interest in view of the loss of stability of equilibrium.

Acknowledgement. This work was supported by the Committee for Research (KBN), Poland under the project No. 3 1082 91 01.

References

1. Drucker, D.C.: On the postulate of stability of material in the mechanics of continua. *J. Méc.* **3** (1964) 235-249.

2. Hill, R.: On constitutive inequalities for simple materials. *J. Mech. Phys. Solids* **16** (1968) 229-242, 315-322.

3. Hill, R.: Aspects of invariance in solids mechanics. *Advances in Applied Mechanics, Vol.* **18**, pp.1-75, Acad. Press, New York 1978.

4. Hill, R.: On intrinsic eigenstates in plasticity with generalized variables. *Math. Proc. Camb. Phil. Soc.* **93** (1983) 177-189.

5. Rice, J.R.: The localization of plastic deformation. *Theoretical and Applied Mechanics* (W.T. Koiter, ed.), pp.207-220, North- Holland, Amsterdam 1977.

6. Hill, R.: Acceleration waves in solids. *J. Mech. Phys. Solids* **10** (1962) 1-16.

7. Hadamard, J.: *Leçons sur la Propagation des Ondes et les Equations de l'Hydrodynamique,* Hermann, Paris 1903.

8. Mandel, J.: Conditions de stabilité et postulat de Drucker. *Rheology and Soil Mechanics* (J. Kravtchenko, ed.), pp.58-67, Springer, Berlin 1966.

9. Hill, R.: The essential structure of constitutive laws for metal composites and polycrystals. *J. Mech. Phys. Solids* **15** (1967) 79-95.

10. Hutchinson, J.W.: Elastic-plastic behaviour of polycrystalline metals and composites. *Proc. Roy. Soc. Lond.* **A 319** (1970) 247-272.

11. Petryk, H.: On energy criteria of plastic instability. *Plastic Instability, Proc. Considère Memorial,* pp.215-226. Ecole Nat. Ponts Chauss., Paris 1985.

12. Petryk, H.: Material instability and strain-rate discontinuities in incrementally nonlinear continua. *J. Mech. Phys. Solids* (to appear).

13. Hill, R.: Some basic principles in the mechanics of solids without a natural time. *J. Mech. Phys. Solids* **7** (1959) 209-225.

14. Graves, L.M.: The Weierstrass condition for multiple integral variation problems. *Duke Math. J.* **5** (1939) 656-660.

15. Petryk, H.: On constitutive inequalities and bifurcation in elastic-plastic solids with a yield-surface vertex. *J. Mech. Phys. Solids* **37** (1989) 265-291.

16. Christoffersen, J.; Hutchinson, W: A class of phenomenological corner theories of plasticity. *J. Mech. Phys. Solids* **27** (1979) 465-487.

17. Hutchinson, J.W.; Tvergaard, V.: Shear band formation in plane strain. *Int. J. Solids Structures* **17** (1981) 451-470.

Extension of Anisotropic Hardening Elastic-Plastic Theory to Finite Strains based on Visco-Elastic Finite Strain Theory

J. HUÉTINK
Department of Mechanical Engineering
University of Twente, the Netherlands

Summary

Based on the observations of visco-elastic theory, an extension of kinematic hardening to finite strains is found showing that a formulation based on only one convective rate may not be sufficient. The back-stress should be considered as the assembly of two components (or modes), one for which the lower- convective rate is proportional to the rate of deformation and another component for which the upper-convective rate is proportional to the rate of deformation. A smooth transition from elastic to plastic deformation including a proper description of the Bauschinger effect after load reversals is formulated using a two yield surface model. The additional yield surface is used in order to determine the history and state dependent hardening rate. Large strain cyclic shear experiments on sheet metal specimens show effects which can be predicted with the proposed formulation.

Introduction

The oscillating stresses found by Nagtegaal and de Jong [8] in generalization of kinematic hardening to finite strains are not peculiar to anisotropy. Atluri [1] found these oscillations also in some isotropic hypo-elastic material models. He showed that the anomalies disappear if the Zaremba-Jaumann-Noll rigid body derivative is replaced by either the upper- convective (Truesdell or Oldroyd) rate or by the lower-convective (Cotter- Rivlin) rate, whereas mixed convective rates appear physically unacceptable.

In the current paper further investigations of finite strain isotropic elasticity in relation to anisotropic plasticity are presented. The principles of visco-elastic theories are more or less similar to finite strain plasticity; The Leonov-model [6] for visco-elastic materials is a parallel assembly of isotropic elastic and isotropic viscous fractions. It has the same structure as the Besseling-(fraction)-model [2] which for small strains contains kinematic hardening as a special case. However, in the Leonov-model finite elastic strains are taken into account.

D. Besdo, E. Stein (Eds.)
Finite Inelastic Deformations –
Theory and Applications
IUTAM Symposium Hannover/Germany 1991

Finite strain elasticity

The stresses in isotropic elastic materials are related to the free energy by

$$\underline{\sigma} = 2\rho \left(\frac{\partial H}{\partial I_1} \underline{B} + \frac{\partial H}{\partial I_2} \underline{B}^{-1} + \frac{\partial H}{\partial I_3} (\det \underline{B})\, \underline{I} \right) \tag{1}$$

where ρ is the mass density, $\underline{B}$ is the left Cauchy Green or finger tensor, and H is the free energy depending on the invariants (I_1, I_2, I_3) of $\underline{B}$.

$$I_1 = \operatorname{tr} \underline{B} \quad , \quad I_2 = \operatorname{tr} \underline{B}^{-1} \quad , \quad I_3 = \det \underline{B}$$

The finger tensor is related to the deformation tensor $\underline{F}$ by

$$\underline{B} = \underline{F} \cdot \underline{F}^T \tag{2}$$

The deformation tensor expresses the transition of a line element from the reference state to the current state

$$d\tilde{x} = \underline{F} \cdot d\tilde{X} \tag{3}$$

hence $$\underline{F} = \left(\frac{\partial \tilde{x}}{\partial \tilde{X}}\right) \tag{4}$$

For the sake of convenience we consider two simple neo Hookean types of material satisfying the simplified expression

$$\underline{\sigma}_1 = G_1 \, [\underline{B} - \underline{I}] \tag{5}$$

$$\underline{\sigma}_2 = G_2 \, [\underline{I} - \underline{B}^{-1}] \tag{6}$$

The stress in formula (1) can be regarded as a (quasi) linear combination of these simple forms and an additional pressure.

In order to determine the rate of change of $\underline{\sigma}$ we first derive

$$\dot{\underline{B}} = (\tilde{v} \overleftarrow{\nabla}) \cdot \underline{B} + \underline{B} \cdot (\overrightarrow{\nabla} \tilde{v}) = \underline{\Omega} \cdot \underline{B} - \underline{B} \cdot \underline{\Omega} + \underline{D} \cdot \underline{B} + \underline{B} \cdot \underline{D} \tag{7}$$

$$\dot{\underline{B}}^{-1} = -(\tilde{v} \overleftarrow{\nabla}) \cdot \underline{B}^{-1} - \underline{B}^{-1} \cdot (\overrightarrow{\nabla} \tilde{v}) = \underline{\Omega} \cdot \underline{B}^{-1} - \underline{B}^{-1} \cdot \underline{\Omega} - \underline{D} \cdot \underline{B}^{-1} - \underline{B}^{-1} \cdot \underline{D} \tag{8}$$

Here the velocity gradient is subdivided into a symmetric part and an anti symmetric part

$$\underline{D} = \frac{1}{2} (\tilde{v} \overleftarrow{\nabla} + \overrightarrow{\nabla} \tilde{v}) \tag{9}$$

and

$$\underline{\Omega} = \frac{1}{2} (\tilde{v} \overleftarrow{\nabla} - \overrightarrow{\nabla} \tilde{v}) \tag{10}$$

The tensors $\underline{D}$ and $\underline{\Omega}$ are known as the rate of deformation tensor and the spintensor respectively.

The rate of change of the stresses can now be written as

$$\dot{\underline{\sigma}}_1 - \underline{\Omega}\cdot\underline{\sigma}_1 + \underline{\sigma}_1\cdot\underline{\Omega} - \underline{D}\cdot\underline{\sigma}_1 - \underline{\sigma}_1\cdot\underline{D} = 2G_1\ \underline{D} \tag{11}$$

$$\dot{\underline{\sigma}}_2 - \underline{\Omega}\cdot\underline{\sigma}_2 + \underline{\sigma}_2\cdot\underline{\Omega} + \underline{D}\cdot\underline{\sigma}_2 + \underline{\sigma}_2\cdot\underline{D} = 2G_2\ \underline{D} \tag{12}$$

The left hand side of (11) is known as the upper convective (Truesdell or Oldroyd) rate of $\underline{\sigma}_1$. The left hand side of (12) is known as the lower convective (Cotter- Rivlin) rate of $\underline{\sigma}_2$. From eq.(11) and (12) it is obvious that the physically acceptable convective rates found by Atluri [1] correspond to thermodynamically reversible finite strain (isotropic) elasticity. The unacceptable (mixed) convective rates correspond to irreversible hypoelasticity. Consequently the back-stress resulting from the convective rates proposed by Atluri can, for a constant hardening rate, be regarded as a reversible isotropic elastic stress which is superimposed on an isotropic non hardening plastic stress. Similar models are used in visco-elasticity.

We now consider an elastic material satisfying the relation

$$\underline{\sigma} = G\ [\underline{B} - \underline{B}^{-1}] \tag{13}$$

This equation can be regarded as the sum of (5) and (6) with $G_1=G_2=G$. However, the rate of change of the stress cannot be expressed by the sum of (11) and (12); the sum of (11) and (12) results into the Zaremba- Jaumann-Noll rigid body derivative which cannot express reversible elastic deformation. This leads to the conclusion that the stress rate in rather simple reversible elasticity cannot be expressed by only one of the corotational convective rates. Hence it is to be expected that the rate of change of the back-stress in a kinematic hardening model expressed in only one of the acceptable convective rates may not be sufficient.

Anisotropic hardening

Based on the observations of visco-elastic theory, an extension of kinematic hardening to finite strains is proposed. The back-stress should (at least) be considered as the assembly of two components (or modes), one for which the lower- convective rate is proportional to the rate of deformation and another component for which the upper-convective rate is proportional to the rate of deformation;

$$\underline{\alpha} = \psi\ \underline{\alpha}_1 + (1-\psi)\ \underline{\alpha}_2 \tag{14}$$

$$\dot{\underline{\alpha}}_1 - \underline{\Omega}\cdot\underline{\alpha}_1 + \underline{\alpha}_1\cdot\underline{\Omega} - \underline{D}^P\cdot\underline{\alpha}_1 - \underline{\alpha}_1\cdot\underline{D}^P = h_1\ \mathbb{K}_1 : \underline{D}^P \tag{15}$$

$$\dot{\underline{\alpha}}_2 - \underline{\Omega}\cdot\underline{\alpha}_2 + \underline{\alpha}_2\cdot\underline{\Omega} + \underline{D}^P\cdot\underline{\alpha}_2 + \underline{\alpha}_2\cdot\underline{D}^P = h_2\ \mathbb{K}_2 : \underline{D}^P \tag{16}$$

The proportionality factors h_1 and h_2 represent the hardening rate and are generally depending on the deformation history and the state of stress, and may be different for each mode. This assembly cannot be replaced by one simple rate equation because the convective rates do not satisfy simple multiplication and additive rules. The hardening rate and the weight factor ψ must experimentally be established.

The fourth or de tensors $\mathbb{K}_1$ and $\mathbb{K}_2$ depend on the state of (back) stress and can be for instance be derived from Ziegler's hardening rule [10].

In accordance with visco elastic theory and as an extension of the small strain fraction model we assumed that the tensors $\mathbb{K}_i$ are both equal to the the shear modulus times the (fourth order) unit tensor $\mathbb{I}$. The hardening rate parameters h_i are assumed to be equal. For a fixed value of this hardening rate an extension of the classical kinematic hardening model is obtained. Figure 1 shows predictions of cyclic shear with h=0.01.

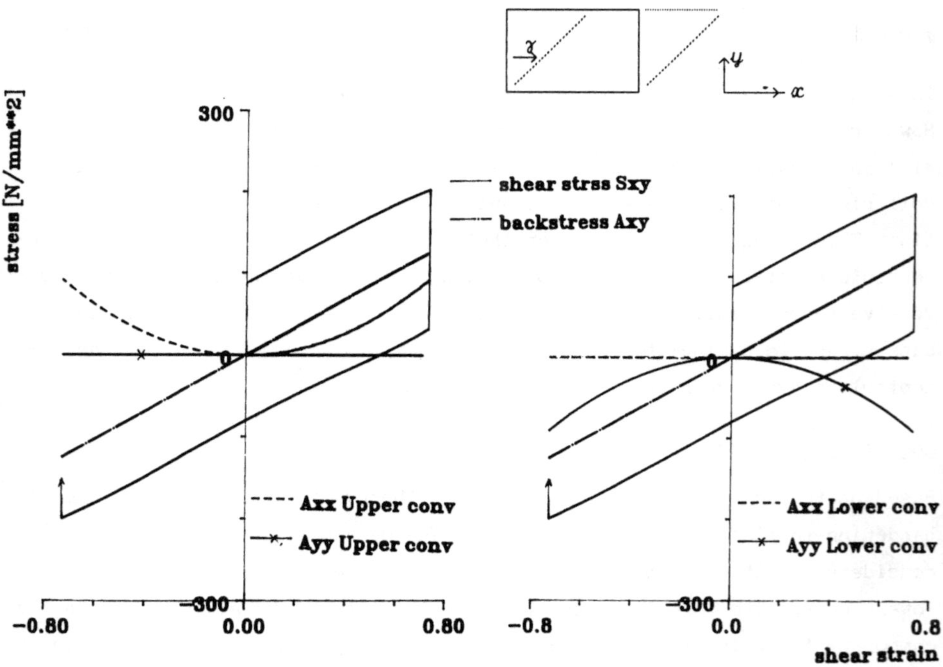

Figure 1. Large cyclic shear kinematic hardening model, Back-stress bases on upper and lower convective rate.

In order to obtain a more appropriate description of a real hardening behaviour including the Bauschinger effect, the hardening rate is determined by introducing a fictive nontranslating (isotropic) yield surface. This fictive yield surface corresponds to the maximum stress ever reached during the deformation history (see fig. 2). When the stress tensor σ approaches the fictive yield surface, the hardening rate tends to the value obtained from an uniaxial (equivalent) stress-strain curve, and the outer surface expands accordingly.

In literature, this fictive yield surface is called memory surface (Ellyin), [3,4] and limit surface (Krieg) [5]. The hypothesis now is that the rate of hardening depends on the distance d_y between the actual stress point σ (lying on the translating kinematic yield surface) and a stress point σ_{fict} on the fictive isotropic yield surface in the direction of the normal at the yield surface, this is illustrated by figure 2. The fictive yield surface (re)-starts growing when the kinematic yield surface touches the isotropic yield surface.

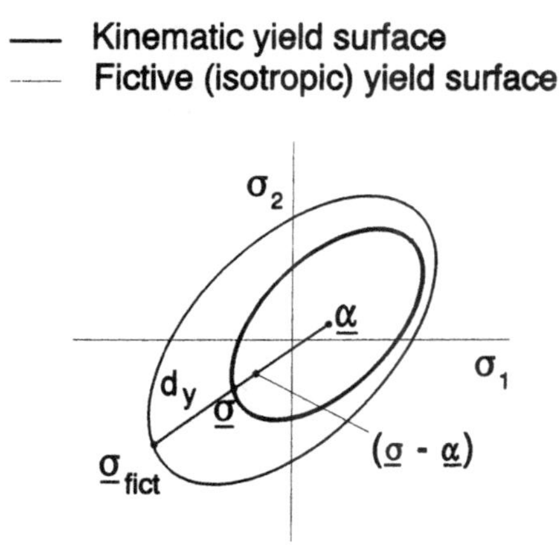

Figure 2. Double yield surface model

The resulting constitutive equations can in the usual way be written as:

$$\dot{\underline{\sigma}} - \underline{\Omega}\cdot\underline{\sigma} + \underline{\sigma}\cdot\underline{\Omega} = [\mathbb{E} - (1-h)\,\mathbb{Y}\,]:\underline{D} \qquad (17)$$

with the elasticity tensor $\mathbb{E}$ and the

$$\text{yield tensor} \quad \mathbb{Y}(\underline{\sigma},\underline{\alpha}) = \frac{\mathbb{E}: \dfrac{\partial\varphi}{\partial\sigma}\,\dfrac{\partial\varphi}{\partial\sigma} : \mathbb{E}}{\dfrac{\partial\varphi}{\partial\sigma} : \mathbb{E} : \dfrac{\partial\varphi}{\partial\sigma}} \qquad (18)$$

The hardening rate coefficient h_1 and h_2 are assumed to be equal and depending on the distance d_y between the inner (Kinematic) surface and the outer (Isotropic or Memory) surface;

$$h = f(d_y)\, h_{max} + (1-f(d_y))\, h_{iso}, \qquad 0 \le f(d_y) \le 1 \tag{19}$$

where h_{max} is the hardening rate coefficient just after load reversal and h_{iso} is the value related to the stress level of the outer surface.
The yield surfaces and hardening rate are material dependent. We considered:

- (Inner) Yield criterion $\varphi(\underline{\sigma},\underline{\alpha},\ldots) = J_2(\underline{\sigma}-\underline{\alpha}) - \frac{2}{3}\sigma_0^2 = 0$ (20)
- (Outer) Yield criterion $\Phi(\underline{\sigma},\sigma_y(\varepsilon)) = J_2(\underline{\sigma}) - \frac{2}{3}\sigma_y^{\,2} = 0$ (21)
- uniaxial stress strain relation $\sigma_y(\varepsilon) = C_1(\varepsilon + \varepsilon_0)^n + C_2\,\varepsilon$ (22)
- Shift rule tensors $\mathbb{K}_1, \mathbb{K}_2(\underline{\sigma},\underline{\alpha}_1,\underline{\alpha}_2,h) = h\ G\ \mathbb{I}$ (23)
- Anisotropic hardening rate function $f(d_y) = \left(\frac{d_y}{d_0}\right)^p$ (24)

<u>Experimental verification by cyclic shear experiments</u>

To verify the proposed hardening model, a test facility is build in the laboratory of the University. The original test was invented by Miyauchi [7] he tried to reinforce his theory of simple and pure shear. Nevertheless the test is appropriate to achieve moderate shear strains during alternating loading. The principle of this test is displayed in figure 3. The sheet is clamped in the hatched part of figure 3, manufactured from metal files. Tests revealed that negligible slip occurs between the files and the specimen.

EXPERIMENTS of CYCLIC SHEAR

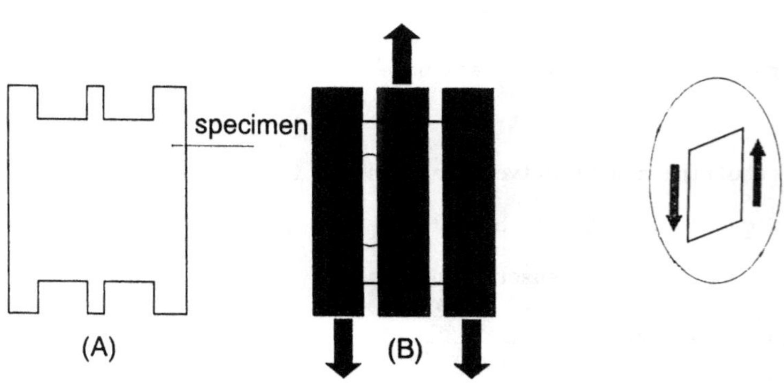

Figure 3 Shear test experiment on sheet metal.

The sheet parts can be subjected to an alternating displacement. The great advantage of this test is the possibility to achieve relative high strains within sheet metal.
Note that in most experiments found in literature a tube is used as specimen. Consequence is that the the specimens must be manufactured out of massive material, instead of sheet metal. Although a lot of different stress states can be achieved, only small strains are possible because buckling restricts the tube deformation. Besides, the sheet specimens used are relatively cheep compared with tubular specimens.

The response of the specimen is displayed by figure 4. The results can be compared with the response of the proposed hardening model and the agreement is quite satisfactory.

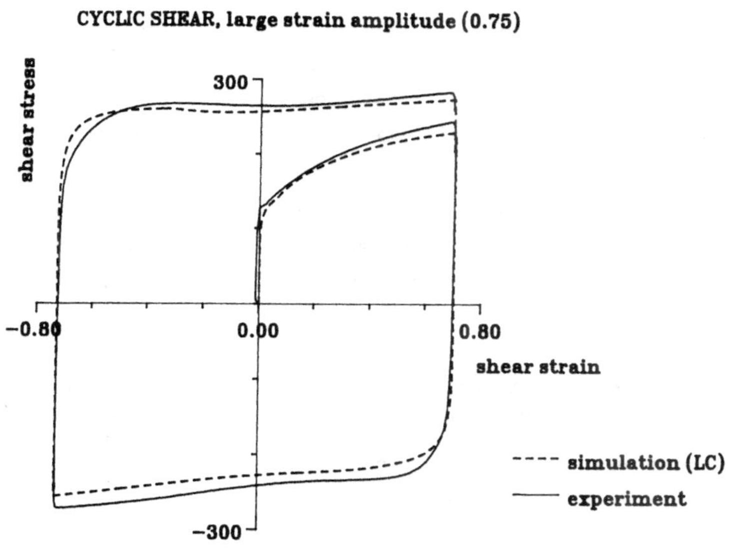

Figure 4 prediction of stress-strain path compared with a cyclic test

The material parameter are obtained from a uniaxial tension test. The stress strain curve resulting from this test is approximated by a Nadai like formula (22). The size of the kinematic yield surface is assumed to be fixed to the level of initial yielding. The value of ε_0 is determined from (22) by requiring that the initial yield stress is found for ε=0. The value for p in equation (24) is determined from a cyclic shear test with a small amplitude (0.05). The weight factor for the upper en lower convective contribution to the back-stress is determined by a trial and error fit of the cyclic shear test shown in figure 4. It appeared that the best fit was found for $\psi = 0$ which means that for the considered material the lower convective rate only contributes to the back-stress. More details concerning the experiments and material data will be published (P.T.Vreede) [9].

Concluding remarks

The presented anisotropic hardening model in which the back-stress is an assembly of two components, –one for which the upper convective rate is proportional to the rate of plastic deformation and one for which the lower convective rate is proportional to the rate of deformation–, shows satisfactory agreement with a cyclic shear experiment with a shear strain amplitude up to 0.75. Particularly remarkable is the observed alternately decreasing and increasing hardening rate which is well predicted.
For the sheet metal material used the best agreement was obtained when the contribution of the upper convective rate component vanishes. Further experimental investigations on other materials are required to investigate whether this observation holds more generally.
Considerable longitudinal and/or transverse normal stresses or strains are predicted. The experimental setup must be modified to measure these effects, and to determine more appropriate material tensors in the shift rule.

Acknowledgment

The author thanks Mr. P.T. Vreede and Mr. G. Rekers for their contribution to the development of the material model and the simulation programme, Mr. P.E van Riesen and Mr. N. van Vliet for carrying out the experiments and Mr. M. Louwes for the development of the graphical presentation.

References

1. Atluri , S.N.; On constitutive relations at finite strain: hypo-elasticity and elasto-plasticity with isotropic and kinematic hardening. comp. meth. appl. mech. 43 (1984) 137-171.
2. Besseling, J.F.; A theory of elastic, plastic and creep deformation of an initially isotropic material. J. Appl. Mech. 25 (1958), 529-536.
3. Ellyin, F.; An Anisotropic Hardening rule for Elastoplastic Solids based on Experimental observations. J. appl. mech. 56 (1989), 499-507.
4. Ellyin, F. & Xia, Z.; A rate independent constitutive model for transient non-proportional loading. J. mech .phys. solids. 37, no 1 (1989) 71-91.
5. Krieg, R.D.; A practical two surface plasticity theory. J. appl. mech. (1975) 641-646.
6. Leonov, A.I.; Nonequilybrium thermodynamics and rheology of viscoelastic polymer media. Rheologica Acta 15 (1976) 85-98

7. Miyauchi, K.; Deformation behavior of sheet metals in simple shear. Proc. IDDRG meeting 1985, Amsterdam.

8. Nagtegaal, J.C. & de Jong, J.E. Some aspects of non-isotropic workhardening in finite strain plasticity. in E.H. Lee and R.L. Malett, eds. Plasticity of metals at finite strain, Stanford (1982), 65-102.

9. Vreede, P.T.; Finite element formulation for three dimensional sheet metal forming. ph.d. dissertation, University of Twente, Netherlands (to be published 1992).

10. Ziegler, H.; A modification of Prager's hardening rule. Quaterly Appl. Math. Provedence, 17 (1959), 55 - 65.

A Formulation of Anisotropic Plasticity at Large Strains

W. Gambin

Institute of Fundamental Technological Research,
Polish Academy of Sciences, Warsaw, Poland

Summary

A description of an appearance and development of a plastic anisotropy in strongly deformed elastic-plastic materials is proposed within a modified continuum plasticity approach. The polycrystalline structure of the materials and crystallographic texture development are assumed as the main reason of the anisotropy. A continuous model of the elastic-plastic material with a texture is introduced. The basic idea is the following: to consider continuum of lattice frames attached to a material particle instead of a finite number of grains in a small volume of the material. The above concept leads to the notion of a "textured material body" $\mathcal{C}$, which may be embodied into a six-dimensional space which is the Cartesian product of the Euclidean point-space and the Riemanian orientation-space. A motion of $\mathcal{C}$ determines texture changes. The behaviour of the model is described on two levels. On the local (micro-) level, an evolution equation for the orientation distribution function and constitutive relations for the single crystal plasticity are used. Here, all considered fields are functions of particle positions, lattice orientations and time. On the global (macro-) level, the virtual work principle (VWP) in the extended (six-dimensional) physical space is assumed. Due to the local constitutive relations, the VWP may be expressed in terms of a velocity field only. It leads to the finite element formulation of elastic-plastic analysis of textured materials. Because the form of VWP is the same as in the classical continuum plasticity, one can adapt the standard FEM procedures to obtain an adequate numerical code.

Introduction

An appearance and development of a plastic anisotropy in strongly deformed elastic-plastic materials can not be described within the classical continuum plasticity. A new approach, which take into account a polycrystalline structure of the materials is necessary. It is a known fact, that plastic deformations of

D. Besdo, E. Stein (Eds.)
Finite Inelastic Deformations -
Theory and Applications
IUTAM Symposium Hannover/Germany 1991

polycrystals are accompanied by the formation of preferred lattice orientations. The phenomenon of a crystallographic texture development observed on the micro-level is noticed as changes of the macroscopic anisotropy on the maco-level. In the paper, a continuous model of the elastic-plastic material with a texture is proposed. The model neglects other microscopic effects like those connected with a shape of grains and their distribution. To explain the basic idea, consider a small volume of the polycrystalline material with a finite, large number of grains. Such an element of the material will be modelled by continuum of lattice frames attached to one material particle. This continuum is prescribed at each particle of the considered material body. Following this idea, a notion of a "textured material body" $\mathcal{B}$ is introduced. Elements of $\mathcal{B}$ are pairs composed of one material particle and one lattice frame. A motion of $\mathcal{B}$ may be investigated in an "extended physical space" $\mathcal{E}^3\times\mathcal{F}^3$. Elements of $\mathcal{E}^3$ are identified with points of the physical space, elements of $\mathcal{F}^3$ - with triads of the Euler angles. All fields prescribed on $\mathcal{B}$ may be regarded as given on a certain configuration of $\mathcal{B}$ in $\mathcal{E}^3\times\mathcal{F}^3$, and then they may be considered as functions of $x_k \in \mathcal{E}^3$, $\varphi_\alpha \in \mathcal{F}^3$ and time t. One can prescribe a certain measure on $\mathcal{B}$ with a density $f(x_k,\varphi_k,t)$ called the orientation distribution function (ODF). For the case of uniform texture, when $f = f(\varphi_\alpha,t)$, the introduced ODF has the same meaning as that used in the standard texture analysis. The evolution equation for $f = f(x_k,\varphi_\alpha,t)$ has been derived in [1].

Below, following the paper [1], kinematics of $\mathcal{B}$ is described. Next, the model the elastic-plastic material with texture based on the concept of the textured material body is introduced. A constitutive behaviour of the model is defined for a single lattice frame attached to an uniformly deformed material element, and the appropriate constitutive relations are taken from a modified single crystal plasticity proposed in [2]. An overall behaviour of the model is governed by the virtual work principle (VWP). A total virtual work is a sum of virtual works for material elements with single lattice frames. The sum is taken with a weight prescribed by a current ODF. The VWP is given in the incremental form which may be directly applied in the finite element analysis. It is very important, that the VWP and the final constitutive relations have the same form as those in the classical isotropic plasticity at large strains (see f.c. [3]). It enables to adapt the standard FEM procedures for numerical analysis of large deformations of elastic-plastic materials with texture.

Kinematics of textured material body

The notion of the extended physical space $\mathcal{E}^3 \times \mathcal{F}^3$ was introduced and described in [1]. Recall, that $\mathcal{E}^3$ is the Euclidean point-space in which one can introduce a Cartesian system of coordinates $\{x_1, x_2, x_3\}$. $\mathcal{F}^3$ is the Riemanian orientation-space with the metric (see f.c. [4])

$$ds^2 = (d\varphi_1)^2 + (d\varphi_2)^2 + (d\varphi_3)^2 + 2 \cos \varphi_2 \, d\varphi_1 d\varphi_3 \tag{1}$$

and the volume element

$$dg = 1/(8\pi^2) \sin \varphi_2 \, d\varphi_1 d\varphi_2 d\varphi_3 \tag{2}$$

where $\{\varphi_1, \varphi_2, \varphi_3\}$ are triads of the Euler angles, usually denoted by $\{\varphi_1, \Phi, \varphi_2\}$. In the curvilinear system of coordinates induced by the metric (1), one can consider contravariant fields of the Euler angle change rates

$$R^\alpha = \{\dot{\varphi}_1, \dot{\varphi}_2, \dot{\varphi}_3\} \tag{3}$$

Elements of $\mathcal{E}^3 \times \mathcal{F}^3$ are parameterized by six coordinates $\{x_1, x_2, x_3; \varphi_1, \varphi_2, \varphi_3\}$. Because a metric in $\mathcal{E}^3 \times \mathcal{F}^3$ is not defined, the extended physical space may be regarded as a three--parametric Euclidean point-space.

The textured material body $\mathcal{C}$ is defined as a pair $(\mathcal{B}, \mathcal{F})$, where $\mathcal{B}$ is the material body (commonly assumed as a three-dimensional manifold homeomorphic with open subsets of $\mathcal{E}^3$), and $\mathcal{F}$ is a "texture support" (a three-dimensional manifold homeomorphic with $\mathcal{F}^3$). Pairs (P,R), composed of a particle $P \in \mathcal{B}$ and a triad of orthonormal vectors $R \in \mathcal{F}$, are point-elements of $\mathcal{C}$. Then, $\mathcal{C}$ may be considered as six-dimensional manifold homeomorphic with those open subsets of $\mathcal{E}^3 \times \mathcal{F}^3$ which contain the whole space $\mathcal{F}^3$. For a fixed configuration of $\mathcal{E}$ in $\mathcal{E}^3\times\mathcal{F}^3$, assumed as a reference one, the localized lattice frame $R_P = (P,R)$ is labeled by six parameters $\{X_K, \Phi_A, t\}$, where X_K is a reference position, and Φ_A – a reference orientation. Motion of the textured material body is described by one-parameter family of transformations

$$x_k = x_k(X_K, \Phi_A, t) \tag{4}$$

$$\varphi_\alpha = \varphi_\alpha(X_K, \Phi_A, t) \tag{5}$$

for $t \geq t_0$. In the above, the rule (4) describes a trajectory of R_P with the reference position X_K and the reference orientation Φ_A, and the rule (5) gives its current orientation. Moreover, the Jacobian

$$J_{x\varphi} \equiv \left| \begin{bmatrix} \dfrac{\partial x_i}{\partial X_K} & \dfrac{\partial \varphi_\alpha}{\partial X_K} \\ \dfrac{\partial x_i}{\partial \Phi_A} & \dfrac{\partial \varphi_\alpha}{\partial \Phi_A} \end{bmatrix} \right| \tag{6}$$

should be always positive.

Orientation distribution function

Consider R_P with a current position x_k and an arbitrary orientation. Probability that this orientation belongs to a certain interval $\{\varphi_1 + d\varphi_1, \varphi_2 + d\varphi_2, \varphi_3 + d\varphi_3\}$ is denoted by $f(x_k, \varphi_\alpha, t)$ and called the orientation distribution function (ODF). The knowledge of the ODF is useful in the standard texture analysis and it is essential in the considered model. As it was mentioned, the ODF may be also introduced as a density of a certain measure on $\mathcal{E}$, as it has been done in [1]. The conservation law for this measure leads to the following evolution equation for ODF (see [1]):

$$\frac{\partial f}{\partial t} + \frac{1}{\sin\varphi_2}\frac{\partial}{\partial \varphi_\alpha}(R^\alpha f \sin\varphi_2) + \frac{\partial}{\partial x_i}(v^i f) = 0 \tag{7}$$

where $v^i = v^i(x_k, \varphi_\alpha, t)$ is a velocity field, and R^α is defined by the rule (3). For an uniform texture, i.e. when $f = f(\varphi_\alpha, t)$, and for negligible changes of a material volume ($v^i{}_{,i} = 0$), the last term in the rule (7) vanishes, and the relation reduces to that proposed by Clement in the paper [5].

Local constitutive relations

As it was mentioned, instead of a real crystalline grain, a single lattice frame attached to an uniformly deformed material element is considered. Recall, that the geometry of a single crystal in a fixed reference configuration is described by a set of slip systems $\{s_i^{(\alpha)}, m_i^{(\alpha)}\}$, where $s_i^{(\alpha)}$ is a slip direction, $m_i^{(\alpha)}$ - a normal to the slip plane and $\alpha = 1,2,...,M$. The slip systems are rigidly connected with the lattice frame and they rotate commonly, under the rotation tensor R^*_{ij}, to the current configuration. Actual slip vectors are denoted by $s_i^{*(\alpha)}$ and $m_i^{*(\alpha)}$. Increments of the tensor R^*_{ij}, one can calculate from the relation:

$$\dot{R}^*_{ij} = \omega^*_{ik} R^*_{kj} \tag{8}$$

where

$$\omega^*_{ij} = \omega_{ij} - \omega^P_{ij} \tag{9}$$

is the lattice spin, ω_{ij} - the total spin, and ω^P_{ij} - the plastic spin. For ω^P_{ij}, an additional constitutive relation should be given. Further, the following quantities will be used (see f.c. [6]):

$$P^{(\alpha)}_{ij} = \frac{1}{2}(s_i^{(\alpha)} m_j^{(\alpha)} + m_i^{(\alpha)} s_j^{(\alpha)}) \tag{10}$$

$$P^{*(\alpha)}_{ij} = \frac{1}{2}(s_i^{*(\alpha)} m_j^{*(\alpha)} + m_i^{*(\alpha)} s_j^{*(\alpha)}) \tag{11}$$

$$W^{*(\alpha)}_{ij} = \frac{1}{2}(s_i^{*(\alpha)} m_j^{*(\alpha)} - m_i^{*(\alpha)} s_j^{*(\alpha)}) \tag{12}$$

$$\lambda^{(\alpha)}_{ij} = \mathcal{L}_{ijkl} P^{*(\alpha)}_{kl} + \beta^{(\alpha)}_{ij} \tag{13}$$

$$\beta^{(\alpha)}_{ij} = W^{*(\alpha)}_{ik} \tau_{kj} - \tau_{ik} W^{*(\alpha)}_{kj} \tag{14}$$

where $\mathcal{L}_{ijkl}$ are components of the elastic moduli tensor, and

$\tau_{ij} = (\rho/\rho_o)\sigma_{ij}$ is the Kirchhoff stress tensor (ρ/ρ_o is the ratio of a current mass density to a reference one, and σ_{ij} - the Cauchy stress). Moreover, denote by $h_{\alpha\beta}$ - the matrix of hardening moduli ($\alpha,\beta = 1,2,\ldots,M$), by $\tau^{(\alpha)} = \sigma_{ij} m_i^{*(\alpha)} s_j^{*(\alpha)}$ - a resolved shear stress and by k_c and $\tau_c^{(\alpha)}$ - its initial and current critical value, respectively. Finally, let us introduce residual shear stresses on each of the slip systems - $a^{(\alpha)}$.

The constitutive relations for elastic-plastic crystals, proposed in the paper [2], assume a nonlinear, smooth yield condition with the associated flow rule. A degree of the nonlinearity is determined by a parameter $n = 1,2,3,\ldots$ assumed to be a material constant. Both the isotropic and kinematic hardening on slip systems are considered. The final relations have the same form as the Prandtl-Reuss equations in the case of the isotropic plasticity at large strains [3]. This fact enables to adapt the standard FEM procedures in the finite element analysis of elastic-plastic crystals. Recall the results of the paper [2]. Let $\overset{\nabla}{\tau}_{ij}$ - will be the Jaumann derivative of τ_{ij}, and d_{ij} - the total strain rate. The incremental constitutive relation has the following form (compare with the rule (20) in [3]):

$$\overset{\nabla}{\tau}_{ij} = \left(\mathcal{L}_{ijkl} - \frac{\mathcal{F}_{ij}\mathcal{F}_{kl}}{\mathcal{F}_{pq}\mathcal{G}_{pq} + h_o} \right) d_{kl} \tag{15}$$

where

$$\mathcal{F}_{ij} = \sum_{\alpha=1}^{M} \frac{\lambda_{ij}^{(\alpha)}}{\tau_c^{(\alpha)}} \left(\frac{\tau^{(\alpha)}}{\tau_c^{(\alpha)}} \right)^{2n-1} \tag{16}$$

$$\mathcal{G}_{ij} = \sum_{\alpha=1}^{M} \frac{P_{ij}^{*(\alpha)}}{\tau_c^{(\alpha)}} \left(\frac{\tau^{(\alpha)}}{\tau_c^{(\alpha)}} \right)^{2n-1} \tag{17}$$

$$h_o = \sum_{\alpha=1}^{M}\sum_{\beta=1}^{M} \frac{h_{\alpha\beta}}{\tau_c^{(\alpha)}\tau_c^{(\beta)}} \left(\frac{\tau_c^{(\alpha)}}{\tau_c^{(\alpha)}}\right)^{2n} \left(\frac{|\tau^{(\beta)}|}{\tau_c^{(\beta)}}\right)^{2n-1} +$$

$$+ \frac{1}{M}\sum_{\alpha=1}^{M}\sum_{\beta=1}^{M}\sum_{\delta=1}^{M} \left(\frac{h_{\beta\delta}}{\tau_c^{(\beta)}\tau_c^{(\delta)}} - \frac{h_{\alpha\delta}}{\tau_c^{(\alpha)}\tau_c^{(\delta)}} \right) \left(2\,\frac{\tau_c^{(\beta)}}{\tau_c^{(\alpha)}} P_{ij}^{(\beta)} P_{ij}^{(\alpha)}\right)^{2n} \left(\frac{|\tau^{(\delta)}|}{\tau_c^{(\delta)}}\right)^{2n-1} \tag{18}$$

for the isotropic hardening approach, and

$$\mathcal{F}_{ij} = \sum_{\alpha=1}^{M} \frac{\lambda_{ij}^{(\alpha)}}{k_c^{(\alpha)}} \left(\frac{\tau^{(\alpha)} - a^{(\alpha)}}{k_c^{(\alpha)}}\right)^{2n-1} \tag{19}$$

$$\mathcal{G}_{ij} = \sum_{\alpha=1}^{M} \frac{P_{ij}^{*(\alpha)}}{k_c^{(\alpha)}} \left(\frac{\tau^{(\alpha)} - a^{(\alpha)}}{k_c^{(\alpha)}}\right)^{2n-1} \tag{20}$$

$$h_o = \sum_{\alpha=1}^{M}\sum_{\beta=1}^{M} \frac{h_{\alpha\beta}}{k_c^{(\alpha)}k_c^{(\beta)}} \left(\frac{\tau^{(\alpha)} - a^{(\alpha)}}{k_c^{(\alpha)}}\right)^{2n-1} \left(\frac{\tau^{(\beta)} - a^{(\beta)}}{k_c^{(\beta)}}\right)^{2n-1} \tag{21}$$

for the kinematic hardening approach.

The constitutive relation for the plastic spin has the form:

$$\omega_{ij}^{P} = \frac{\mathcal{F}_{kl}\, d_{kl}}{\mathcal{F}_{pq}\mathcal{G}_{pq} + h_o}\, \mathcal{H}_{ij} \tag{22}$$

where

$$\mathcal{H}_{ij} = \sum_{\alpha=1}^{M} \frac{W_{ij}^{*(\alpha)}}{\tau_c^{(\alpha)}} \left(\frac{\tau^{(\alpha)}}{\tau_c^{(\alpha)}}\right)^{2n-1} \tag{23}$$

or

$$\mathcal{H}_{ij} = \sum_{\alpha=1}^{M} \frac{W_{ij}^{*(\alpha)}}{k_c^{(\alpha)}} \left(\frac{\tau^{(\alpha)} - a^{(\alpha)}}{k_c^{(\alpha)}}\right)^{2n-1} \tag{24}$$

The yield condition is as follows

$$\sum_{\alpha=1}^{M}\left(\frac{\tau^{(\alpha)}}{\tau_c^{(\alpha)}}\right)^{2n}-\frac{1}{M}\sum_{\alpha=1}^{M}\sum_{\beta=1}^{M}\left(2\frac{\tau_c^{(\beta)}}{\tau_c^{(\alpha)}}P_{ij}^{(\beta)}P_{ij}^{(\alpha)}\right)^{2n}=0 \qquad (25)$$

or

$$\sum_{\alpha=1}^{M}\left(\frac{\tau^{(\alpha)}-a^{(\alpha)}}{k_c^{(\alpha)}}\right)^{2n}-\frac{1}{M}\sum_{\alpha=1}^{M}\sum_{\beta=1}^{M}\left(2\frac{k_c^{(\beta)}}{k_c^{(\alpha)}}P_{ij}^{(\beta)}P_{ij}^{(\alpha)}\right)^{2n}=0 \qquad (26)$$

and the appropriate hardening laws take the following forms:

$$\dot{\tau}_c^{(\alpha)}=\frac{\mathcal{F}_{kl}\,d_{kl}}{\mathcal{F}_{pq}\mathcal{G}_{pq}+h_o}\sum_{\beta=1}^{M}\frac{h_{\alpha\beta}}{\tau_c^{(\beta)}}\left(\frac{|\tau^{(\beta)}|}{\tau_c^{(\beta)}}\right)^{2n-1} \qquad (27)$$

and

$$\dot{a}^{(\alpha)}=\frac{\mathcal{F}_{kl}\,d_{kl}}{\mathcal{F}_{pq}\mathcal{G}_{pq}+h_o}\sum_{\beta=1}^{M}\frac{h_{\alpha\beta}}{k_c^{(\beta)}}\left(\frac{\tau^{(\alpha)}-a^{(\alpha)}}{k_c^{(\alpha)}}\right)^{2n-1} \qquad (28)$$

Global virtal work principle

The presented approach does not use neither global equations of motion nor global constitutive relations. The only relation which describes a behaviour of the body $\mathcal{C}$ on the global level is the virtual work principle extended on the space $\mathcal{E}^3\times\mathcal{F}^3$. Its incremental form is the following (compare with the rule (5) in [3]):

$$\int_{\mathcal{F}^3}\left\{\int_V f\left[\overset{\nabla}{\tau}_{ij}\,\delta d_{ij}-\frac{1}{2}\,\sigma_{ij}\,\delta(2\,d_{ik}d_{kj}-v_{k,i}v_{k,j})\right]dV\right\}dg=$$

$$=\int_{\mathcal{F}^3}\left(\int_S f\,\dot{p}_i\delta v_i\,dS\right)dg+\int_{\mathcal{F}^3}\left(\int_V f\,\dot{b}_i\delta v_i\,dV\right)dg \qquad (29)$$

where f is a current orientation distribution function,and dg is defined by (2). Moreover, $\dot{p}_i$ and $\dot{b}_i$ are nominal surface and body force intensity rates per unit current area and unit current volume, respectively. The unknown fields: f, σ_{ij}, d_{ij} and v_i, are functions of x_k, φ_α and t. The field $\dot{p}_i$ and the non-inertial part of $\dot{b}_i$ may be assumed as functions of x_k and t only.

Finite element formulation

The principle (29) may be used to formulate an Eulerian finite element method. Owing the local constitutive relations the all unknown fields may be expressed by $v_i(x_k,\varphi_\alpha,t)$. Introducing modified finite elements, one can approximate the velocity field by a set of shape functions $N_i(x_k,\varphi_\alpha)$. Notice, that in the space $\mathcal{F}^3$, one can apply trigonometric series and assume only few integration points for each of the Euler angles. Denote by $\{\dot{\delta}\}$ - the vector of rates of nodal degrees of freedom, by [C] - the incremental constitutive matrix and let $[B_{ij}] = 1/2\ ([N_i]_{,j} + [N_j]_{,i})$. Then, a set of finite element equations is the following (compare with the rules (9-10) in [3]):

$$\left[\int_{\mathcal{F}^3}\left(\int_V f\ [B]^T[C][B]\ dV\right) dg + [k_s]\right]\{\dot{\delta}\} = \int_{\mathcal{F}^3}\left(\int_S f\ [N]^T\{\dot{p}\}\ dS\right) dg + \int_{\mathcal{F}^3}\left(\int_V f\ [N]^T\{\dot{b}\}\ dS\right) dg \tag{30}$$

where

$$[k_s] = \int_{\mathcal{F}^3}\left[\int_V f\ \left([N_k]_{,i}^T\sigma_{ij}[N_k]_{,j} - 2[B_{ki}]^T\sigma_{ij}[B_{kj}]\right)dV\right]dg \tag{31}$$

is the initial stress stiffness. In the above the function f is calculated from the rule (5), and increments of the Euler angles - from the rules (8-9) and the following ones:

$$\dot{\varphi}_1 = -\frac{\sin\varphi_1}{\sin\varphi_2}\omega^*_{23} + \frac{\cos\varphi_3}{\sin\varphi_2}\omega^*_{13} \tag{32}$$

$$\dot{\varphi}_2 = -\cos\varphi_3\ \omega^*_{23} - \sin\varphi_3\ \omega^*_{13} \tag{33}$$

$$\dot{\varphi}_3 = -\dot{\varphi}_1\cos\varphi_2 - \omega^*_{12} \tag{34}$$

The presented formulation was applied to an analysis of a radially extended circular disc made of a material with an initial texture. The elastic-ideally plastic, f.c.c. metal sheet with a rolling texture close to the {001}<110> component was assumed. The problem was solved analytically. assuming only three degree of freedom of the circular element (a radial extension rate, a thickness change rate and a rate of change of harmonic amplitude of deviation from the radial extension). Despite such a rough approximation of the velocity field, the results are quite reasonable: one can observe characteristic "ears" of the deformed disc on the rolling and transversal direction.

References

1. Gambin, W.: Phenomenological model of deformation texture development. Proc. Int. Sem. Large Plastic Deformations, MECAMAT'91, Fontainebleau 1991.

2. Gambin, W.: Refined analysis of elastic-plastic crystals. Submitted to Int. J. Solids Structures.

3. McMeeking, R. M. and Rice, J. R.: Finite-element formulation for problems of large elastic-plastic deformation. Int. J. Solids Structures 11 (1975), 601-616.

4. Synge, J. L. and Schild, A.: Tensor calculus. Toronto: University Toronto Press, 1959.

5. Clement, A.: Prediction of deformation texture using a physical principle of conservation. Mat. Sci. Engng 55 (1982), 203-210.

6. Asaro, R. J.: Crystal plasticity. J.Appl. Mech. 50 (1983), 921-924.

The Intrinsic Lagrangian Metric and Stress Variables

P. ROUGEE
Laboratoire de Mécanique et Technologie
Cachan (France)

1- Introduction

Strain and stress, and their rates, are fundamental concepts. Their modeling largely depends on basic kinematic choices that need to be well defined and uncontested. As shown in [1], to which we refer also for its extensive bibliography, this is not the case in large deformation theory. Even in the simplest case without any microstructure that we consider in this paper, many insufficiencies, inadequacies and inconsistencies remain, which are not sufficiently recognized.

Beside the mathematical model of space that is the chosen space-frame, the Lagrangian approaches ([1-5]...) build a mathematical model for the material body and its properties, which is the natural place to treat material properties that strain, stress and constitutive laws are [4]. In particular, they always furnish the following expected simple general scheme: *a "point"* m *representative of the current local metric state, which moves in a known space* M *modeling the "manifold" of all the possible local metric states.* From this starting point, the (*true, thus uniquely defined)* time derivative $\dot{m}$, element of the velocity space Tm tangent to M at point m, models the strain-rate, and the associated force θ, the stress in this case, is in the space dual to Tm possibly identified with Tm by means of an existing scalar product.

Unfortunately, the tools used until now in this good way are defective. First, they depend generally on a particular reference configuration (r.c.) (which, being a time independent representative for the material body, is taken as a support for the announced *material mathematical model*) and on a particular strain-measure (s.m.) e which is chosen to be the metric state variable m. Now it is known that there are many possibilities for the choice of e, any one being intrinsic: the small deformation theory physical concept of strain measure cannot be extended to large deformations, and s.m.-independent properties are needed [4]. Second, and important, the chosen variables for m are always tensorial variables *belonging thus always to linear spaces unable to take into account the non linearity of large deformation processes.*

On the opposite, the Eulerian approach ([6-7] ...) is not the best way, because it treats material properties through the spatial position of the body and, consequently, in the mathematical model of space. Nevertheless, two of its basic tools are uncontested: the strain-rate D, symmetrical part of the Eulerian velocity field gradient, and, to write constitutive laws for a

D. Besdo, E. Stein (Eds.)
Finite Inelastic Deformations –
Theory and Applications
IUTAM Symposium Hannover/Germany 1991

time-independent local material element, the Kirchhoff stress $\tau = \rho^{-1}\sigma$ where ρ is the current mass density and σ is the Cauchy stress. In particular, eigen elements of D and τ describe essential properties of strain-rate and stress. Nevertheless, D does not appear, as it should be, as an uncontested time derivative of an uncontested metric variable. Moreover, many questions arise about the choice for an objective derivative for τ or σ ([8-10]...). The known attempts to solve these problems are more or less explicitly based upon the choice of a push-pull rule between the current configuration (c.c.) and the material body. For convective transport rules, the problem is brought back to the foregoing unsatisfying Lagrangian form. For rules associated with a particular rotation, the associated integration of D leads to cumulated tensorial strain variables depending on the strain process, which thus may not be the expected metric variable.

Obviously, the main question that arises from this review, and to which we propose an answer in this paper, is: *what is the good choice for the metric variable* m, *leading to an as good strain-rate* $\dot{m}$ *and an as good associated stress* θ *that* D *and* τ *are*? Or, more precisely: *what is the space* M *of all possible local metric states (the "six-dimension configurational space" evoked in [11]) and what is its (probably nonlinear!) geometry*? We think that other problems will be clarified when this first one will be solved, and that, if it is not solved, the study of large deformation processes, which are long travels on M, will stay as hazardous as were deep-sea travels before knowing that the Earth is a spherical surface.

Clearly, the metric variables e used in Lagrangian approaches are only various *characterizations* of the wanted m. Each of them is the projection (or set of coordinates) of m in some particular and non intrinsic *linear chart of the manifold* M. And thus, *despite they are largely used*, the $\dot{e}$ and , consequently, their associated "stress" variables K_e defined as usual by

$$\mathcal{P} = -\rho^{-1}\sigma : D = -\tau : D = -\rho_1^{-1} K_e : \dot{e}$$

where $\mathcal{P}$ is the specific internal stress power and ρ_1 is the mass density in the r.c., *may only be very bad representatives for strain-rate and stress*. Exactly as the time derivatives of its surfacic coordinates fail to describe the velocity of a point moving on a surface. As examples, the eigen axes of the classical Piola-Kirchhoff *stress* tensor K, associated with the Green-Lagrange strain measure ε, are the *strain* eigen axes in the case of an hydrostatic pressure, and those of $\dot{\varepsilon}$ are also the strain eigen axes when D is spherical. More generally,*there is no direct Lagrangian way today to exhibit the principal axes and values for strain-rate and stress!*

2 - Some basic elements and notations.

For two linear spaces G and F, let L(G,F) denote the linear space of the linear maps of G in F. We note L(F) for L(F,F), F* for the dual space L(F,R) , $< , >_F$ for the associated duality, A* for the adjunct of $A \in L(G,F)$ (defined by $A^* \in L(F^*,G^*)$ and $\langle A^*f,u\rangle_G = \langle f,Au\rangle_F$ for any $u \in G$ and $f \in F^*$), $L_S(F,F^*)$ for the subset of $A \in L(F,F^*)$ such that $\langle Au,v\rangle_F$ is a symmetrical bilinear form, and $L_S^+(F,F^*)$ for the open subset of $A \in L_s(F,F^*)$ such that $\langle Au,u\rangle_F$ is positive-definite.

We note $\mathscr{E}$ the chosen space-frame (set of spatial points), E its associated linear space and $g \in L_s^+(E,E^*)$ the metric tensor defining the Eulerian structure of the space-frame. The scalar product in E is $u.v = <gu,v>_E$. To any $A \in L(E)$ is associated $A^T = g^{-1}A^*g \in L(E)$ (such as $A^Tv.u = v.Au$ for any u and v in E). The space L(E) is canonically a Euclidean space, with the scalar product $A : B = Tr(A^TB)$. Any $A \in L(E)$ is symmetrical if $A^T=A$ and skew-symmetrical if $A^T=-A$. We note $L_S(E)$ the subspace of symmetrical elements A. The algebraic structure of this six-dimensional linear space $L_S(E)$ (i.e., the well known algebra of symmetrical Euclidian second order tensors, making use of trace, determinant, scalar invariants, real eigen values, orthogonal eigen vectors...) will be called the L_S-structure.

We note $\mathfrak{B}$ the three-dimensional manifold representative of the material body, X the considered point of $\mathfrak{B}$ and $\mathfrak{t}$ (for $T_X\mathfrak{B}$) the tangent linear space to $\mathfrak{B}$ at point X. This space $\mathfrak{t}$ is the classical support of the "material simple element in X". It models the space of the little material segments at X. It is the basic space of our Lagrangian local study. Its only additional intrinsic (physical and time-independent) structure is a "mass volume form " [1].

Let x be the current place of X in $\mathscr{E}$ and $a = dx/dX \in L(\mathfrak{t},E)$ be the current spatial placement of the vicinity of X. This a is the basic tool for push-pull rules between matter and space models. In Eulerian approaches it is used from $\mathfrak{t}$ to E to see the material body in space. In Lagrangian approaches it is used from E to $\mathfrak{t}$ *to see the space on the material body*. Tensors D, σ and τ are elements on the Eulerian side. They are in $L_S(E)$.

We develop our theory with the above geometrical tools only, without the help of any reference configuration and any coordinate system. But we also, for convenience, translate it with these classical but non intrinsic tools. We thus consider a particular time-independent r.c., for which we note x_1 the position of X and a_1 the local placement. Let E_1 be the space E when it is associated with the point x_1. Obviously, E_1 plays the Lagrangian role of $\mathfrak{t}$, but with an additional metric tensor $g_1 = g$ (the metric of the chosen r.c.). Putting $F = aa_1^{-1}$, the gradient of the displacement transformation from the r.c. to the c.c., we introduce the dilatation tensor $C = F^TF$, the Green-Lagrange and Almansi-Lagrange s.m. $\varepsilon = 1/2(C - 1_E)$ and $\varepsilon' = 1/2(1_E - C^{-1})$, and their associated stresses $K = \rho_1 F^{-1} \tau F^{-T}$ (second Piola-Khirchhoff) and $K' = \rho_1 F^T \tau F$. These elements, and also the space $L_S(E_1)$ in which they are, are really dependent on the metric of the r.c. (because $F^T = g_1^{-1}F^*g$). We also consider a particular Lagrangian coordinate system (L.c.s.) and its associated local bases in $\mathfrak{t}$, E (the convected basis) and E_1. Note that with this, tensors on the linear space $\mathfrak{t}$ have only one set of components, with the suitable variance, while an nth order tensor on the Euclidean spaces E or E_1 has 2^n sets with all the possible variances.

3 - Two intrinsic approaches.

Carrying by a the space metric tensor g in $\mathfrak{t}$, as done in [12-13], the current local metric properties that we attempt to *model* may be *characterized* by the time-dependent metric tensor

$\gamma = a^* g\, a \in \bar{\Gamma}^+$ with $\bar{\Gamma} = L_S(t,t^*)$, obtained by writing that two elements of t have the same scalar product as their positions in E. Taking, for convenience, $\bar{m} = 1/2\gamma$ as the metric variable, the associated stress variable is the element $\underline{\theta}$ of the space $\underline{\Gamma}=\bar{\Gamma}^* = L_S(t^*,t)$ defined by $\mathcal{P} = -\operatorname{Tr}(\underline{\theta}\, \dot{\bar{m}})$, and we have the convective transport rule

$$\dot{\bar{m}} = a^* D a \qquad \text{and} \qquad \underline{\theta} = a^{-1} \tau\, a^{-*} \tag{1}$$

Making use of a r.c., putting $\gamma_1 = a_1^* g a_1$ and (without any loss of generality) identifying t to the linear space support of E_1, we obtain $\dot{\varepsilon} = \gamma_1^{-1}\, \dot{\bar{m}}$ and $K/\rho_1 = \underline{\theta}\, \gamma_1$. This shows that the usual associated Lagrangian variables $\{\varepsilon, K\}$ differ from our $\{\bar{m}, \underline{\theta}\}$ variables only through constant (but r.c.-metric dependent) maps. Making use of a L.c.s., we have $\gamma_{ij} = g_{ij}$, $\dot{\bar{m}}_{ij} = \dot{\varepsilon}_{ij} = D_{ij}$, $\bar{\theta}^{ij} = K^{ij}/\rho_1 = \tau^{ij}$, which show that *an indirect way to work with the intrinsic Lagrangian tools* $\{\bar{m}, \underline{\theta}\}$, *which have only one set of components, consists in considering exclusively only one well chosen among the four sets of components of the* $\{\varepsilon, K\}$ [14] *or* $\{D, \tau\}$ *Eulerian tensors*. Note that doing this, to call g_{ij} "the space metric tensor variable" is a misuse: g is a physical constant and the variations of the g_{ij}'s reflect only those of the convected basis.

Although it is unused, if not unknown, *there does exist a second such intrinsic approach, in some way symmetrical of the first one*. In fact, the manifold $\mathfrak{B}$ is represented locally not only by t, modeling the small material segments issued from X, *but also by its dual space* t^* *modeling the small material slices at* X (the slice $s \in t^*$ is the set of material segments $dx \in t$ such that $0 < \langle s, dx \rangle_t < 1$). Thus, *the metric tensor of* t^*, i.e., $\gamma^{-1} \in \underline{\Gamma}^+$ or, for convenience, $\underline{m} = -1/2\gamma^{-1} \in \underline{\Gamma}^-$, *may also be taken to characterize the metric state*. While γ gives the lengthes and angles of material segments, γ^{-1} gives the thicknesses and dihedral angles of slices (usually more useful in technology). Noting $\bar{\theta}$ the associated stress, lying in $\bar{\Gamma}$, we obtain the inverse convective transport rule

$$\dot{\underline{m}} = a^{-1} D a^{-*} \qquad \text{and} \qquad \bar{\theta} = a^* \tau\, a \tag{2}$$

and, with a r.c. and a L.c.s., the relations $\dot{\varepsilon}' = \dot{\underline{m}} \gamma_1$, $K'/\rho_1 = \gamma_1^{-1}\, \bar{\theta}$, involving {e', K'} instead of {e , K}, and $\gamma^{-1\, ij} = g^{ij}$, $\dot{\underline{m}}^{ij} = \dot{\varepsilon}'^{ij} = D^{ij}$, $\bar{\theta}_{ij} = K'_{ij} = \tau_{ij}$, which invert the variance of the components used.

Both metric variables $\bar{m}$ and $\underline{m}$ are r.c. and s.m. independent, but any of them may be our expected m because they lie in $\bar{\Gamma}^+$ and $\underline{\Gamma}^-$, which are linear manifolds. We are still working in linear charts of M, whose only advantage is to be intrinsic and to have been obtained by two very symmetrical approaches. Note also that $\bar{m}$, $\underline{m}$, $\dot{\bar{m}}$, $\dot{\underline{m}}$, $\bar{\theta}$ and $\underline{\theta}$, lying in spaces $\bar{\Gamma}$ and $\underline{\Gamma}$, *cannot have eigen elements*. First this prevents the building of another intrinsic metric variable from $\bar{m}$ or $\underline{m}$, as is done for the strain-measures when starting from a particular r.c.. Second, this does not give a Lagrangian way to exhibit the eigen elements of strain rate and sress.

4 - The space M (Fig. 1)

Our thesis is that the two intrinsic charts (i.c.) $\bar{c}$: $\bar{m} \to m$ and $\underline{c}$: $\underline{m} \to m$ *are equally qualified* (a symmetry which, seen between Green and Almansi s.m., is clearly emphasized by Hill in [4]), *and that, equally treated, they implicitly define the geometry of the space* M. *More precisely, this geometry is the set of properties which are consistently expressed by these two charts and which thus are invariable by the change*

$$c = \bar{c} \circ \underline{c}^{-1} : \quad \bar{m} \to \underline{m} = c(\bar{m}) = -1/4\ \bar{m}^{-1} . \tag{3}$$

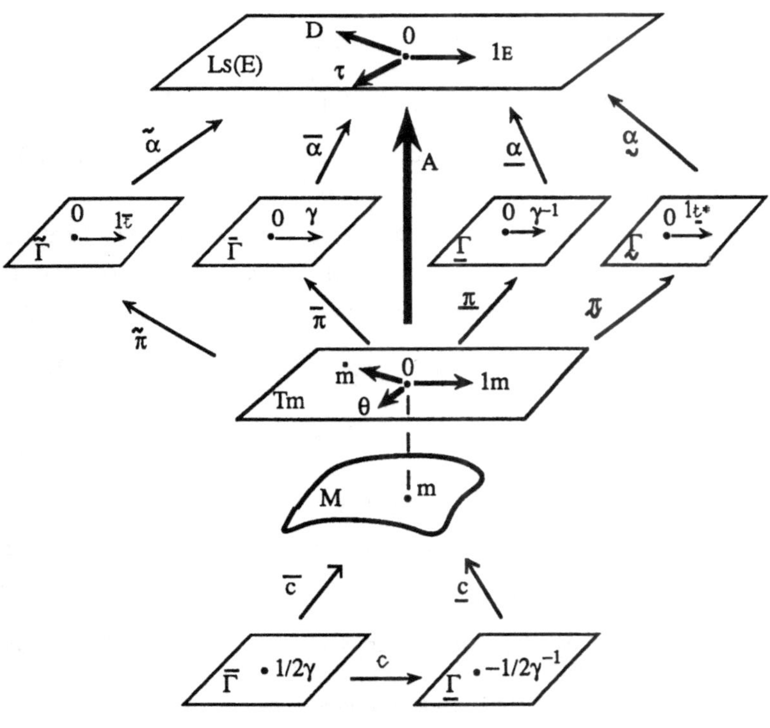

Fig. 1 - Intrinsic charts of M. Maps p and spatial convective transport A of M

As an example, $\bar{\Gamma}^{+}$ and $\underline{\Gamma}^{-}$ are both open subspaces of linear spaces, and thus each i.c. proposes this particular mathematical structure for M. But since the change c is a nonlinear map, these two propositions are inconsistent: the two linear structures obtained on M by carrying the linear structures of $\bar{\Gamma}^+$ and $\underline{\Gamma}^-$ by the i.c. are different, and elements of M which are linearly related for one are not for the other. As predicted, M is thus not a linear manifold. By contrast, c being a C^∞ map which conserves the C^∞-manifold structure, M *is a (six-dimensional)* C^∞-*manifold.* This gives *an upper and a lower bound for the mathematical structure of* M.

Let Tm be the linear space tangent to the manifold M at point m. Clearly, Tm is devoted to be the space of all the strain-rates $\dot{m}$ and all the small changes of metric dm to which the local material element in the metric state m may be submitted. Thus, looking at the Eulerian approach and at the small displacement theory, a L_S-stucture is strongly expected for Tm.

First, the derivatives of the maps $\bar{c}$ and $\underline{c}$ give two linear representations of Tm on the linear spaces $\bar{\Gamma}$ and $\underline{\Gamma}$. Putting

$$\bar{\pi} = \bar{c}'(\bar{m}) \in L(\bar{\Gamma},Tm)\ ,\ \ \underline{\pi} = \underline{c}'(\underline{m}) \in L(\underline{\Gamma},Tm)\ ,\ \ \pi = c'(\bar{m}) = \bar{\pi}\,\underline{\pi}^{-1} \in L(\bar{\Gamma},\underline{\Gamma}), \tag{4}$$

noting $\bar{\mu}$ and $\underline{\mu}$ the representatives of $\mu \in$ Tm, and computing π by (3), we obtain

$$\mu = \bar{\pi}\bar{\mu} = \underline{\pi}\,\underline{\mu} \quad \text{and} \quad \underline{\mu} = \pi\,\bar{\mu} = \gamma^{-1}\bar{\mu}\,\gamma^{-1} \tag{5}$$

The last in relations (5) is the characteristic condition for two elements $\bar{\mu}$ and $\underline{\mu}$ in $\bar{\Gamma}$ and $\underline{\Gamma}$, respectively, to be the representatives of an element μ of Tm.

Second, making a symmetrical use of the two i.c.'s, let $\bar{t}$ and $\underline{t}^*$ be the Euclidean spaces obtained by endowing t and t^* with the metric tensors γ and γ^{-1} respectively (i.e., the spaces of material segments and of material slices *in their current metric state*), and let $\tilde{\pi}$ and $\underset{\sim}{\pi}$ be the two linear representations of Tm on the linear spaces $\tilde{\Gamma} = L_S(\bar{t})$ and $\underset{\sim}{\Gamma} = L_S(\underline{t}^*)$ defined by

$$\tilde{\mu} = \tilde{\pi}\,(\mu) = \gamma^{-1}\bar{\mu} \quad \text{and} \quad \underset{\sim}{\mu} = \underset{\sim}{\pi}\,(\mu) = \underline{\mu}\,\gamma \tag{6}$$

The two m-dependent spaces $\tilde{\Gamma}$ and $\underset{\sim}{\Gamma}$ have the expected L_S-stucture, and this structure may be carried on Tm by $\tilde{\pi}$ and $\underset{\sim}{\pi}$. By this way, we obtain the same L_S-stucture only if the change of representation $\pi' = \tilde{\pi}\,\underset{\sim}{\pi}^{-1}$ is a L_S-isomorphism. Now this last property is satisfied because, first, $\pi'(\tilde{\mu}) = \underset{\sim}{\mu} = \gamma\tilde{\mu}\,\gamma^{-1}$ and, second, γ is an isometric map of $\tilde{\Gamma}$ on $\underset{\sim}{\Gamma}$. This proves that Tm *is a* L_S*-space, isomorphic to the* L_S*-spaces* $\tilde{\Gamma}$ *and* $\underset{\sim}{\Gamma}$ *by* $\tilde{\pi}$ *and* $\underset{\sim}{\pi}$. *Its elements may be identified with the classical second order Euclidean symmetrical tensors on the* m*(thus time)-dependent Euclidian space* $\tilde{\Gamma}$.The geometry that this structure of Tm induces on M has been studied in [15]. Let us only remark that, a L_S-space being a Euclidian space, M is a *Riemannian manifold*, in particular, and let us give some expressions of its first quadratic form (we denote ":" the scalar product in the L_S-space Tm as in L_S(E)):

$$ds^2 = dm : dm = \tilde{dm} : \tilde{dm} = \mathrm{Tr}\,(\gamma^{-1}\, d\bar{m}\, \gamma^{-1}\, d\bar{m}) = \mathrm{Tr}\,(d\bar{m}\, d\underline{m})$$

$$= -1/4\ \mathrm{Tr}\,(C^{-1}dC\, C^{-1}dC) = \mathrm{Tr}\,(C^{-1}d\varepsilon C^{-1}d\varepsilon) = d\varepsilon : d\varepsilon' \tag{7}$$

Making use of a L.c.s., a particular basis is introduced in each of the four spaces $\bar{\Gamma}$, $\underline{\Gamma}$, $\tilde{\Gamma}$, $\underset{\sim}{\Gamma}$ and, carried by the four representations ($\bar{\pi}$, $\underline{\pi}$, $\tilde{\pi}$, $\underset{\sim}{\pi}$), this gives four bases in Tm and thus four sets of components for an element of Tm. It is easily seen that these four sets present all the possible variances and that the indices of these components are moved downwards and upwards by means of the components γ_{ij} and $(\gamma^{-1})^{ij}$ of γ and γ^{-1}. Note also that, denoting by 1_m the unit element of Tm (such that $\tilde{\pi}\,(1_m) = 1_t$), we have $(1_m)_{ij} = \gamma_{ij}$ and $(1_m)^{ij} = (\gamma^{-1})^{ij}$ (and also $(1_m)^i{}_j = (1_m)_j{}^i = \delta^i_j$). We have also the two sets of coordinates $m_{ij} = \bar{m}_{ij} = 1/2\,\gamma_{ij}$ and $m^{ij} = \underline{m}^{ij} = -1/2\,\gamma^{-1ij}$ for the point m of M. *Despite their similar components, it is important not to confuse the geometric elements* γ, γ^{-1}, m *and* 1_m (Fig. 1).

5 - Lagrangian intrinsic strain rate, stress and stress rate.

Starting from the kinematic variable $m \in M$ modeling the metric state, the metric-rate or strain-rate is modeled by $\dot{m} \in \mathrm{Tm}$, whose set of projections in the linear spaces $\bar{\Gamma}$, $\underline{\Gamma}$, $\tilde{\Gamma}$, $\underset{\sim}{\Gamma}$ is ($\dot{\bar{m}}$, $\dot{\underline{m}}$, $1/2\,\gamma^{-1}\dot{\gamma}$, $1/2\,\dot{\gamma}\gamma^{-1}$). Using a particular r.c., identifying t and E_1 by means of a_1, and identifying $E_1{}^*$ and E_1 by means of g_1, this set is ($\dot{\varepsilon}$, $\dot{\varepsilon}'$, $1/2\, C^{-1}\dot{C}$, $1/2\dot{C}C^{-1}$).

It may be seen from (1) and (2) that the elements $\bar{\theta}$ and $\underline{\theta}$ defined above satisfy the last of relations (5). *They are thus the representatives in the i.c. of an element θ of Tm.which, again from* (1)*and* (2), *satisfy*

$$-\ \theta : \dot{m} \ = -\,\tau : D \ = \mathcal{P} \tag{8}$$

and which is thus the intrinsic Lagrangian stress variable associated with the metric variable m. Note that, in section 3, the stress tensor associated to the representative of m in one of the i.c. is the representative of θ in the second i.c..

To model the material stress-rate we have to take the time derivative of the stress θ, which is an element of the time-dependent linear space Tm. For this, the covariant derivative ∇ associated with the Riemannian geometry of M is needed. Its expression may be obtained from the above first form by classical derivations [16]. For example this gives in the first i.c.:

$$\nabla/\overline{dt}\ \theta = d/dt\,\bar{\theta} + 2(\dot{\bar{m}}\ \underline{m}\,\bar{\theta} + \bar{\theta}\,\underline{m}\,\dot{\bar{m}})\ ,\quad \widetilde{\nabla/dt}\theta = d/dt\,\tilde{\theta} + \tilde{\dot{m}}\tilde{\theta} - \tilde{\theta}\tilde{\dot{m}} = (d/dt\,\tilde{\theta})_{s\text{-}m} \tag{9}$$

where the "s-m" index indicates the symmetric part in $L_S(\bar{t})$. The translation of the last expression in a L.c.s. gives the following relation (writing γ^{ij} for $(\gamma^{-1})^{ij}$):

$$(\nabla/dt\ \theta)^i_j = 1/2[\ d/dt\ (\theta^i_j\ ) +\ \gamma^{ik}\ d/dt\ (\theta^l_k\ )\ \gamma_{lj}\ ] \qquad (10)$$

6 - Consistency with the Eulerian approach.

We have obtained four linear representations of Tm, by the four maps $\bar{\pi}$, $\underline{\pi}$, $\tilde{\pi}$, $\underset{\sim}{\pi}$ in the four linear spaces $\bar{\Gamma}$, $\underline{\Gamma}$, $\tilde{\Gamma}$, $\underset{\sim}{\Gamma}$. When m is the metric induced by a local placement a, these four spaces may themselves be linearly represented in the spatial space $L_S(E)$ by four classical convective transports $\bar{\alpha}$, $\underline{\alpha}$, $\tilde{\alpha}$, $\underset{\sim}{\alpha}$ associated with a:

$$\bar{\alpha}(\bar{\mu}) = g^{-1}a^{-*}\bar{\mu}\,a^{-1} \quad , \underline{\alpha}(\underline{\mu}) = a\,\underline{\mu}\,a^{*}g \quad , \quad \tilde{\alpha}(\tilde{\mu})) = a\,\tilde{\mu}\,a^{-1} , \underset{\sim}{\alpha}(\underset{\sim}{\mu}) = g^{-1}a^{-*}\underset{\sim}{\mu}\,a^{*}g$$

Note that identifying $\mathfrak{t}$ and E_1 by means of a_1, and identifying $E_1{}^*$ and E_1 by means of g_1, when using a particular r.c., the four spaces Γ are identified with $L_S(E_1)$ and the maps α are the four classical convective transports between $L_S(E)$ and $L_S(E_1)$ built by using F and F^{-T} alternatively .

We thus obtain by this way (Fig. 1) four linear representations of Tm on $L_S(E)$, $\bar{\alpha}\,\bar{\pi}$, $\underline{\alpha}\,\underline{\pi}$, $\tilde{\alpha}\,\tilde{\pi}$, $\underset{\sim}{\alpha}\,\tilde{\pi}$, which are intrinsic generalizations of the four possible classical convective transports for Euclidian symmetrical second order tensors, *but which are equal* , as is easily seen. Denoting by A their common value (Fig. 1), *we have thus only one placement of* Tm *in* $L_S(E)$ *associated with the local placement a of* $\mathfrak{t}$ *in* E:

$$\mu \in Tm \ \rightarrow\ \boldsymbol{\mu} = A(\mu) = g^{-1}a^{-*}\,\bar{\mu}\,a^{-1} = a\,\underline{\mu}\,a^{*}g = a\tilde{\mu}a^{-1} = g^{-1}a^{-*}\,\underset{\sim}{\mu}\,a^{*}g \in L_S(E) \qquad (11)$$

We may say that *this* A *is the intrinsic convective transport rule. It is a* L_S*-isomorphism*, because both $\tilde{\alpha}$ and $\tilde{\pi}$ are L_S-isomorphisms (a is an isometry of $\mathfrak{t}$ on E). *It thus conserves eigen elements:* eigen values of μ and $\boldsymbol{\mu} = A(\mu)$ are equal, and eigen vectors of μ (identified with eigen vectors of $\tilde{\mu}$ in $\mathfrak{t}$) and of $\boldsymbol{\mu}$ are homologous by a. By acting on $\dot{m}$ and θ it gives

$$A(\dot{m}) = D \qquad \text{and} \qquad A(\theta) = \tau \qquad (12)$$

and *this proves that as models for strain-rate and stress,* $\dot{m}$ *and* θ, *and in particular their eigen elements, are totally consistent with the uncontested Eulerian tools* D *and* τ. Also, *the first of these relations gives the wanted interpretation of D as the representative of the (true) derivative of the metric variable* m. Note that, A being an isometry, we have $(\dot{m}:\dot{m})^{1/2} = (D:D)^{1/2}$ and thus *the length of the trajectory of* m *on* M *during a deformation process is equal to the classical scalar cumulated strain.* This set of properties indicates that *the insufficiencies of the classical Lagrangian tools listed in section 1 disappear with our intrinsic tools.*

Making use of a L.c.s.and of the associated convected basis in E, the four sets of components of $\mu \in Tm$ and of $\boldsymbol{\mu} = A(\mu) \in L_S(E)$ are equal. *Working with the components of the Eulerian and Euclidian tensors* D*and* τ *in a convected basis is thus an indirect way to work with our Lagrangian tensors* $\dot{m}$ *and* θ, *at least for algebraic computations* [17].

By A acting on the covariant derivative of θ, we obtain after a short derivation

$$A(\nabla/dt\,\theta) = \dot{\tau} + \tau\,\Omega - \Omega\,\tau = \tau^J \tag{13}$$

where Ω is the skew-symmetrical part of the gradient of the Eulerian velocity field. *The convected image by A of our covariant derivative on M is thus the classical Jaumann derivative.*

7 - Conclusion

First, we have defined the kinematic variable $m \in M$ which, independently from its position and thus from its orientation in a particular space-frame, and also independently from any r.c. or s.m., *models* and not simply characterizes the current local metric state of continuous media. Exactly as a point on a sphere models a point on the Earth surface, while its longitude and latitude only characterize it. This implies that the geometry of the chosen space M expresses some intrinsic property. Its nonlinearity is the basis of the well known nonlinearities of the kinematics of large deformations. This variable m is closely related to the metric tensor γ classically used (directly or through the Green-Lagrange s.m.). But, and this is the key point of this paper, γ is only one of two equally possible intrinsic linear representations of M. The second one is the metric tensor of material slices γ^{-1} (undirectly, the Almansi-Lagrange s. m.), and the geometry of M is obtained by an equal treatment of these two representations.

Second, starting from this well elucidated kinematic model, we have defined the Lagrangian strain-rate and its associated stress variable, and we have given the material derivative rule for such elements as the stress tensor, furnishing a strong Lagrangian interpretation of the Eulerian Jaumann derivative. *By this way, all insufficiencies of the Lagrangian approach and all their inconsistencies with the Eulerian approach listed section 1 have disappeared.* Moreover, *a so great formal similitude is obtained for the two approaches that they have not to be fundamentally distinguished any longer.* As examples, the constitutive laws for a perfect gas and for an isotropic hypoelastic solid are

$$\tau = 1_E \qquad \text{and} \qquad t^J = \lambda\,\mathrm{Tr}\,D\,1_E + 2\mu\,D \qquad \text{(Eul.)}$$

$$\theta = 1_m \qquad \text{and} \qquad \nabla/dt\,\theta = \lambda\,\mathrm{Tr}\,\dot{m}\,1_m + 2\mu\,\dot{m} \qquad \text{(Lag.)}$$

Further developments and examples may be found in [15] and [18].

References

[1] NAGHDI P.M., "A critical review of the state of finite plasticity", Journal of Appl. Math.and Phys., Vol. 41, 315-394, 1990

[2] GREEN A.E. and NAGHDI P.M., "A General Theory of an Elastic Plastic Continuum", Arch. Rat. Mech. Analysis 18, 251-281, 1965

[3] HILL R., "On constitutive inequalities for simple material. I and II", J. Mech. Phys. Solids", Vol. 16, 229-242 and 315-322, 1968

[4] HILL R., "Aspects of invariance in solid mechanics", Adv. Appl. Mech., Vol. 18, 1-75, 1978

[5] PERIC Dj., OWEN D.R.J., "A model for finite strain elastoplasticity", Proc. Sd Int. Conf. on Comput. Plasticity, Barcelonne, OWEN, HINTON, ONATE editors, Pineridge Press, Swansea U.K., 1989

[6] LEE E.H., "Elastic-plastic deformation at finite strains", J. Appl. Mech. 36, 1-6, 1969

[7] LUBARDA V.A. and LEE E.H., "A correct definition of elastic and plastic deformation and its comput. significance" J. Appl. Mech. 48, 35-40, 1981

[8] DIENES J.K., "On the analysis of rotation and stress rate in deforming bodies" Acta Mech. 53, 217-232, 1979

[9] LEE E.H., MALLET R.L. and WERTHEIMER T.B., "Stress analysis for anisotropic for anisotropic hardening in finite def. plast.", J. Appl. Mech., 50, 554-560, 1983

[10] DOGUI A. and SIDOROFF F., "Large strain formulation of anisotropic elasto-plasticity for metal forming". In "Computational methods for predicting materials processing defects", M. Predeleanu Ed., Elsevier Science, Amsterdam, 1987

[11] KLEIBER M. and RANIECKI B., "Elastic-Plastic Materials at Finite Strains", in "Plasticity today" Ed A. Sawczuk, 1983

[12] NOLL W., "A new Mathematical Theory of Simple Material", Arch. Rat. Mech. Anal., Vol. 48, 1-50, 1972

[13] ROUGEE P., "Formulation Lagrangienne Intrinsèque en Mécanique des Milieux Continus", Journal de mécanique, Vol. 18, n°1, 7-32, 1980

[14] SIMO J.C., "A framework for finite strain elastoplasticity based on max. plastic dissipation and on multiplicative decomposition. Part 1", Comp. Meth. in Appl.Mech and Eng., 66, 199-219, 1988

[15] ROUGEE P., 1990, "A new Lagrangian intrinsic approach of continuous media in large deformation", Eur. J. Mech. A/ Solids , 10, n° 1, 15-39, 1991

[16] PHAM MAU QUAN, "Introduction à la géométrie des variétés différentielles", Dunod Ed. Paris, 1968

[17] LEHMANN Th., "On the balance of energy and entropy at inelastic deformations of solid bodies", Eur. J. Mech., A/Solids, 8, n° 3, 235-251, 1989

[18] ROUGEE P., "Kinematics of finite deformations"(to appear)

Constitutive Equation and Mixed Variational Principle of Hyperelasticity Using Rotationless Strain

O. WATANABE
Institute of Engineering Mechanics, University of Tsukuba
Tsukuba, Ibaraki 305, JAPAN

Summary
The present paper proposes a new constitutive equation for rubber-like material through the strain energy function using strain measure called 'rotationless strain'. This strain is defined as an accumulated strain obtained by integrating strain rate excluding from the stretching tensor the effects of rotation entering through the polar decomposition of the deformation gradient tensor. The proposed constitutive equation takes the form of a nonlinear stress-strain relation, and include an isotropic linear relation for metal alloy as a special case, where the objective stress rate is taken as the Green and Naghdi rate in both of the constitutive equations. The validity of the present constitutive modeling is confirmed by good agreements with the previous experimental results. Particularly, the present model cam predict accurately the simple shear response, which can not be be explained by the previous model. The present paper is also concerned with the theoretical treatment of volume constraint arising from nearly incompressible response of materials. By applying the method of Lagrange multipliers with respect to the internal work related to volumetric change, derived are three-field Hu-Washizu and two-field Hellinger-Reissner variational principles. The mixed variational principle proposed here is proved to hold exactly the conditions of equilibrium in rate form, which has been bypassed in the previous works.

Deviatoric and Mean Parts of Internal Work Rate

For the theoretical development, we use the inner product between the second-order tensors, which retains the following properties:

$$(\mathbf{A}\cdot\mathbf{B}):\mathbf{C} = \mathbf{A}:(\mathbf{C}\cdot\mathbf{B}^t) = \mathbf{B}:(\mathbf{A}^t\cdot\mathbf{C}), \qquad \mathbf{A}:\mathbf{B} = \mathbf{A}^t:\mathbf{B}^t, \tag{1}$$

where the notations $(\cdot)$, $(:)$ and $(\ )^t$ indicate dot product, inner product and transpose, respectively. From Eq.(1), we have, for any **A** and **B**,

$$\mathbf{A}:\mathbf{B} = \mathbf{A}_s:\mathbf{B}_s + \mathbf{A}_a:\mathbf{B}_a \ , \tag{2}$$

where the notations $(\ )_s$ and $(\ )_a$ indicate the symmetric and anti-symmetric part of the tensor in the parenthesis, respectively, defined by

$$\mathbf{A}_s \equiv \frac{1}{2}(\mathbf{A} + \mathbf{A}^t), \qquad \mathbf{A}_a \equiv \frac{1}{2}(\mathbf{A} - \mathbf{A}^t)\ . \tag{3}$$

The initial and current configurations are denoted by C_o and C, respectively. According to the polar decomposition theorem of the deformation gradient tensor **F**, we have (see Truesdell and Noll [1], p.52)

$$\mathbf{F} = \mathbf{R}\cdot\mathbf{U}; \qquad \mathbf{R}^{-1} = \mathbf{R}^t, \quad \mathbf{U} = \mathbf{U}^t, \tag{4}$$

where **U** is the symmetric deformation tensor, and **R** is the orthogonal rotation tensor, and the

D. Besdo, E. Stein (Eds.)
Finite Inelastic Deformations -
Theory and Applications
IUTAM Symposium Hannover/Germany 1991

notation $(\ )^{-1}$ indicate inverse. The velocity gradient tensor $\mathbf{L}$ is calculated by

$$\mathbf{L} = \dot{\mathbf{F}}\cdot\mathbf{F}^{-1} = \Omega + \mathbf{R}\cdot(\dot{\mathbf{U}}\cdot\mathbf{U}^{-1})\cdot\mathbf{R}^t; \qquad \Omega \equiv \dot{\mathbf{R}}\cdot\mathbf{R}^t. \tag{5}$$

The notation $(\dot{\ })$ indicates the material time derivative. The tensor $\mathbf{L}$ is decomposed into the symmetric stretching tensor $\mathbf{D}$ and the anti-symmetric spin tensor $\mathbf{W}$ as

$$\mathbf{D} \equiv \mathbf{L}_s = \mathbf{R}\cdot(\dot{\mathbf{U}}\cdot\mathbf{U}^{-1})_s\cdot\mathbf{R}^t, \qquad \mathbf{W} \equiv \mathbf{L}_a = \Omega + \mathbf{R}\cdot(\dot{\mathbf{U}}\cdot\mathbf{U}^{-1})_a\cdot\mathbf{R}^t. \tag{6}$$

The Cauchy true stress tensor $\mathbf{T}$ and the stretching tensor $\mathbf{D}$ can be decomposed into the deviatoric and mean parts as follows:

$$\mathbf{T} = \mathbf{T}' + T_m\mathbf{I}, \qquad \mathbf{D} = \mathbf{D}' + D_m\mathbf{I}, \tag{7}$$

where $\mathbf{I}$ is the identity tensor, the notations $(\ )'$ and $(\ )_m$ indicate the deviatoric and mean components, respectively. The mean parts T_m and D_m in Eq.(7) are defined by

$$T_m \equiv \frac{1}{3}(\mathbf{T}:\mathbf{I}), \qquad D_m \equiv \frac{1}{3}(\mathbf{D}:\mathbf{I}), \tag{8}$$

The deviatoric parts $\mathbf{T}'$ and $\mathbf{D}'$ are defined by substituting Eq.(8) into Eq.(7). By definition, the inner products of the deviatoric tensors $\mathbf{T}'$ and $\mathbf{D}'$ with the identity tensor $\mathbf{I}$ are reduced to zero as

$$\mathbf{T}':\mathbf{I} = 0, \qquad \mathbf{D}':\mathbf{I} = 0. \tag{9}$$

The internal work rate $\dot{W}_o$ per unit volume of the initial configuration C_o is obtained by the following formula: ([1], p.301)

$$\dot{W}_o = J\,\mathbf{T}:\mathbf{D}, \tag{10}$$

where the volume change J and its time derivative are given by the following: ([1], p.50, p.132)

$$J = \det \mathbf{F} = \det \mathbf{U}, \qquad \frac{\dot{J}}{J} = \mathbf{D}:\mathbf{I} = 3D_m. \tag{11}$$

Substituting Eq.(7) into Eq.(10) yields the following relation uncoupled with respect to the deviatoric and mean parts by using Eq.(9):

$$\dot{W}_o = \dot{W}_o' + \dot{W}_{om}; \qquad \dot{W}_o' = J\,\mathbf{T}':\mathbf{D}', \quad \dot{W}_{om} = 3J\,T_mD_m. \tag{12}$$

Hyperelasticity Based on Rotationless Strain

The strain rate $\dot{E}_R$ on the intermediate configuration C_R rotated back from the current configuration C by the rotation $\mathbf{R}$ is given by the orthogonal transformation:

$$\dot{\mathbf{E}}_R \equiv \mathbf{R}^t\cdot\mathbf{D}\cdot\mathbf{R} = (\dot{\mathbf{U}}\cdot\mathbf{U}^{-1})_s. \tag{13}$$

where Eq.$(6)_1$ is used in deriving Eq.(13). Since $\dot{E}_R$ is expressed as a function of the pure stretch $\mathbf{U}$ and excludes the effects of the rotation $\mathbf{R}$, it may be labeled as 'rotationless strain', as is suggested by Stören and Rice [2]. The accumulated measure of deformations is obtained by integrating the entire history of the rate $\dot{E}_R$ up to the present time t as

$$\mathbf{E}_R = \int_0^t \dot{\mathbf{E}}_R dt = \int_0^t (\dot{\mathbf{U}}\cdot\mathbf{U}^{-1})_s\, dt\,. \tag{14}$$

As is indicated by Stören and Rice [2], $\mathbf{E}_R$ is path-dependent and does not posses the property of a strain measure required by Hill [3]. It is noted that $\mathbf{E}_R$ is defined on the state C_R rotated back from the current configuration with the present rotation $\mathbf{R}(t)$. The deviatoric and mean parts of the strain rate in the intermediate configuration C_R are defined by

$$\dot{E}'_R \equiv \mathbf{R}^t \cdot \mathbf{D}' \cdot \mathbf{R}, \qquad \dot{E}_{Rm} \equiv \mathbf{R}^t \cdot (D_m \mathbf{I}) \cdot \mathbf{R} = D_m \mathbf{I} \equiv \dot{E}_{Rm} \mathbf{I} \,. \tag{15}$$

In 1961, Flory [4] introduced the following modified deformation gradient tensor, based on which the recent hyperelasticity analysis is performed by Hibbitt, Karlsson and Sorensen [5]:

$$\overline{\mathbf{F}} \equiv J^{-1/3} \mathbf{F} \,. \tag{16}$$

The polar-decomposition of the tensor $\overline{\mathbf{F}}$ and its components are related to the following:

$$\overline{\mathbf{F}} \equiv \overline{\mathbf{R}} \cdot \overline{\mathbf{U}}; \qquad \overline{\mathbf{R}} = \mathbf{R}, \quad \overline{\mathbf{U}} = J^{-1/3} \mathbf{U} \,. \tag{17}$$

The configurations defined by Eqs.(4) and (16) are illustrated in Fig.1. The intermediate configuration mapping from the initial state C_o by the tensor $(J^{1/3}\mathbf{I})$ is denoted by $\overline{C}$. From Eqs.(17) and $(11)_2$, we calculate the stretch rate of $\overline{\mathbf{U}}$ as follows:

$$\dot{\overline{\mathbf{U}}} \cdot \overline{\mathbf{U}}^{-1} = \dot{\mathbf{U}} \cdot \mathbf{U}^{-1} - D_m \mathbf{I} \,. \tag{18}$$

The stretching tensor $\overline{\mathbf{D}}$ defined by Eq.$(6)_1$ using $\overline{\mathbf{F}}$ is related to,

$$\overline{\mathbf{D}} \equiv \overline{\mathbf{R}} \cdot \left(\dot{\overline{\mathbf{U}}} \cdot \overline{\mathbf{U}}^{-1} \right) \cdot \overline{\mathbf{R}}^t = \mathbf{R} \cdot \left(\dot{\mathbf{U}} \cdot \mathbf{U}^{-1} - D_m \mathbf{I} \right) \cdot \mathbf{R}^t = \mathbf{D}' \,. \tag{19}$$

This relation indicates that the strain rate $\overline{\mathbf{D}}$ measured from the configuration $\overline{C}$ is identical with the deviatoric strain rate $\mathbf{D}'$ measured from the configuration C_o. Substituting Eq.(19) into Eq.$(15)_1$ and integrating over the entire history up to the present time yields the deviatoric rotationless strain,

$$\mathbf{E}'_R \equiv \int_0^t \dot{\mathbf{E}}'_R dt = \int_0^t (\dot{\overline{\mathbf{U}}} \cdot \overline{\mathbf{U}}^{-1})_s \, dt. \tag{20}$$

From Eq.(20), the strain $\mathbf{E}_R$' is also expressed as a function of the stretch $\overline{\mathbf{U}}$ $(=J^{-1/3}\mathbf{U})$ and exclude effects of the rotation $\mathbf{R}$. On the other hand, the mean part of the total strain, which is obtained by substituting Eq.$(11)_2$ in Eq.(15) and integrating the entire history, is reduced to the path-independent logarithmic strain of J:

$$E_{Rm} = \int_0^t D_m \, dt = \int_0^t \frac{1}{3} \frac{\dot{J}}{J} \, dt = \frac{1}{3} \ln J \,. \tag{21}$$

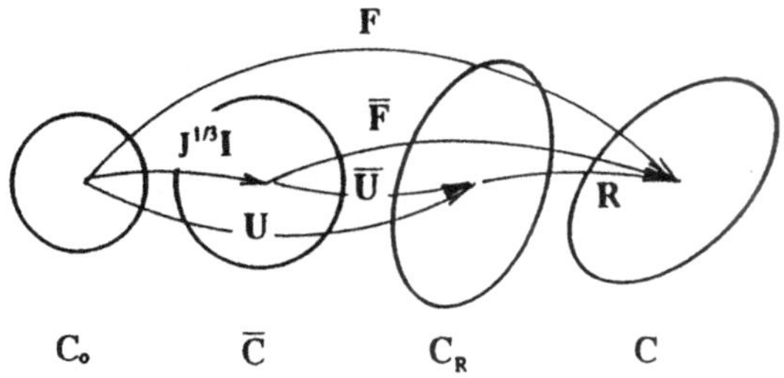

Fig.1 Intermediate configurations defined by Flory's deformation gradient tensor

As for the deviatoric stress T_R' and the mean stress T_{Rm} in the state C_R, we define the followings:

$$T_R' \equiv J R^t \cdot T' \cdot R, \qquad T_{Rm} \equiv J R^t \cdot (T_m I) \cdot R = J T_m I \equiv T_{Rm} I. \tag{22}$$

From stresses of Eq.(22) and strain rates of Eq.(15) and using the inner product of tensors of Eq.(1), we find that the following conjugate relations hold:

$$\dot{W}_o' = T_R' : \dot{E}_R', \qquad \dot{W}_{om} = 3 T_{Rm} \dot{E}_{Rm}. \tag{23}$$

If the strain energy functions are introduced a priori as a function of the utilized strains, the stress can be obtained by differentiating the strain energy functions of Eq.(23) with the strains as follows:

$$T_R' = \frac{\partial W_o'}{\partial E_R'}, \qquad 3T_{Rm} = \frac{\partial W_{om}}{\partial E_{Rm}}. \tag{24}$$

Since the deviatoric strain energy function W_o' is a scalar regardless of the employed coordinate system, it may be convenient to expand W_o' by the strain invariants of the first, second and third kinds I_1', I_2' and I_3' as follows:

$$I_1' = E_R' : I, \qquad I_2' = \frac{1}{2}(E_R')^2 : I, \qquad I_3' = \frac{1}{3}(E_R')^3 : I. \tag{25}$$

From Eqs.(15) and $(25)_1$, the first strain invariant I'_1 is calculated as follows:

$$I_1' = \left(\int_0^t R^t \cdot D' \cdot R \, dt \right) : I = \left\{ \int_0^t \left(R^t \cdot D' \cdot R \right) : I \right\} dt = 0. \tag{26}$$

Thus, the deviatoric strain energy W_o' is expressed in terms of I_2' and I_3', and the mean strain energy W_{om} in terms of E_{Rm} i.e., J as

$$W_o' = W_o'(I_2', I_3'), \qquad W_{om} = W_{om}(J). \tag{27}$$

Differentiation of the strain invariants I_2' and I_3' in Eq.(25) with E_R' yields

$$\frac{\partial I_2'}{\partial E_R'} = E_R', \qquad \frac{\partial I_3'}{\partial E_R'} = (E_R')^2. \tag{28}$$

Substituting Eq.(28) into $\text{Eq.}(24)_1$ yields the deviatoric constitutive equation as

$$T_R' = a_1 E_R' + a_2 (E_R')^2; \qquad a_1 \equiv \frac{\partial W_o'}{\partial I_2'}, \quad a_2 \equiv \frac{\partial W_o'}{\partial I_3'}, \tag{29}$$

where the coefficients a_1 and a_2 are expressed as a function of strain invariants. Since the stress T_R'=O, for the strain E_R'=O from $\text{Eq.}(29)_1$, there exists no restriction for the coefficients a_1 and a_2. The nonlinear relation $(29)_1$ of stress and strain for rubber-like material is regarded as a natural extension of the generalized hypoelasticity, which has been proposed by Green and Naghdi [6] and Green and McInnis [7] to describe elastic-plastic behaviors of metal alloys. The constitutive equation for the mean part can be obtained by Eqs.$(24)_2$, $(27)_2$ and (21) as

$$3T_{Rm} = 3JT_m = \frac{\partial W_{om}}{\partial J} \frac{\partial J}{\partial E_{Rm}} = 3J \frac{\partial W_{om}}{\partial J}. \tag{30}$$

If we expand W_{om} in the quadratic form of (ln J) using the bulk modulus κ_o, then the constitutive equation for the mean part is given by using Eq.(30) as

$$W_{om} = \frac{1}{2} \kappa_o (\ln J)^2; \qquad T_m = \frac{1}{J} \kappa_o (\ln J). \tag{31}$$

Comparison with Experimental Results

We shall apply the present constitutive equation for the typical loading problems of simple shear, simple elongation and uniform 2-dimensional extension. Treloar [8] carried out the systematic experiment using the same material for these three problems. Assuming the material is perfectly incompressible, the deviatoric strain energy function W_o' is expanded as follows:

$$W_o' = \sum_{s=1} \sum_{s=1} C_{st}(I_2')^s (I_3')^t = \sum_{s=1} C_{s0}(I_2')^s + \sum_{t=1} C_{0(2t)}(I_3')^{2t}. \tag{32}$$

Fig.2(a) shows the comparison of simple shear problem. Such a strain occurs, for example, in the base-rubber to absorb shear deformation due to earthquake. The figure also includes the theoretical results of Treloar based on neo Hooke's law of hyperelasticity of Green, which cannot predict the experimental fact that the shear stress responses behind shear deformation. Fig.2(b) shows the comparison of simple elongation, where the present constitutive modeling can predict the experimental results at large strain range. The details of the theoretical treatment and the determination method of material constants may be found in Watanabe[9].

Rate Constitutive Equation and Its Rate Potential

We derive the constitutive equations in rate form from Eq.(24). By taking the material time derivative of $\text{Eq.}(24)_1$, we have the following deviatoric constitutive relation:

$$\dot{\mathbf{T}}_R' = \mathbf{A}(\mathbf{E}_R') : \dot{\mathbf{E}}_R'; \qquad \mathbf{A}(\mathbf{E}_R') = \frac{\partial^2 W_o'}{\partial {\mathbf{E}_R'}^2}, \tag{33}$$

where $\mathbf{A}(\mathbf{E}_R')$ is the coefficient tensor of the fourth order.

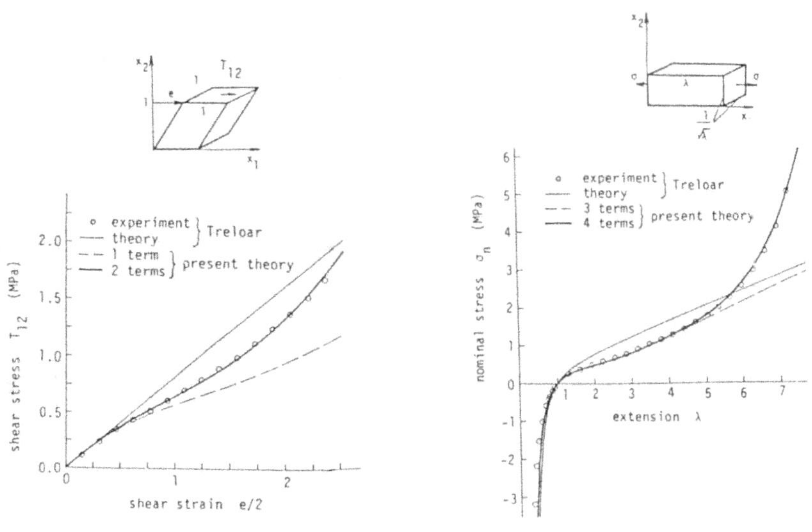

(a) Simple shear (b) Simple elongation

Fig.2 Comparison of the previous experimental and theoretical results

We introduce the following strain e'_R on the current state C corresponding to the strain E'_R on the intermediate state C_R from the orthogonal transformation of **R** as

$$\mathbf{e}'_R \equiv \mathbf{R}\cdot\mathbf{E}'_R\cdot\mathbf{R}^t. \tag{34}$$

For the deviatoric part, we have the relation of stress rates from Eq.(22)$_1$ as

$$\dot{\mathbf{T}}'_R = J\mathbf{R}^t\cdot\mathring{\mathbf{T}}'_R\cdot\mathbf{R}; \qquad \mathring{\mathbf{T}}'_R \equiv \dot{\mathbf{T}}' - \boldsymbol{\Omega}\cdot\mathbf{T}' + \mathbf{T}'\cdot\boldsymbol{\Omega} + (\mathrm{tr}\ \mathbf{D})\mathbf{T}'. \tag{35}$$

If the material is isotropic, we can prove the following relation: (see [1], p.23 or Watanabe [10])

$$\mathbf{R}\cdot\{\mathbf{A}(\mathbf{E}'_R) : \dot{\mathbf{E}}'_R\}\cdot\mathbf{R}^t = \mathbf{A}(\mathbf{e}'_R) : \mathbf{D}'. \tag{36}$$

Then, we have the deviatoric constitutive equation in rate form from Eqs.(35)$_1$, (36) and (33)$_1$:

$$\mathring{\mathbf{T}}'_R = \frac{1}{J}\mathbf{R}\cdot\dot{\mathbf{T}}'_R\cdot\mathbf{R}^t = \frac{1}{J}\mathbf{A}(\mathbf{e}'_R) : \mathbf{D}'. \tag{37}$$

The rate potential U'_R relating $\mathring{T}'_R$ to **D'** is calculated by Eq.(37) as

$$\mathring{\mathbf{T}}'_R \equiv \frac{\partial U'_R}{\partial \mathbf{D}'}; \qquad U'_R = \frac{1}{2J}\mathbf{A}(\mathbf{e}'_R):\mathbf{D}':\mathbf{D}'. \tag{38}$$

The mean part of the constitutive equation in rate form is derived from Eq.(24) as

$$3(JT_m)^\bullet = \frac{\partial^2 W'_o}{\partial E_{Rm}{}^2}\dot{E}_{Rm}. \tag{39}$$

Using D_m and Eqs.(39), (30) and (31) yield the mean part of the constitutive equation,

$$\mathring{T}_{Rm} = \frac{1}{J}3\kappa_o D_m; \qquad \mathring{T}_{Rm} \equiv \dot{T}_m + 3D_m T_m. \tag{40}$$

The rate potential U_{Rm} relating $\mathring{T}_{Rm}$ to D_m is calculated by Eq.(40) as,

$$\mathring{T}_{Rm} \equiv \frac{\partial U_{Rm}}{\partial D_m}; \qquad U_{Rm} = \frac{1}{2}3\mathring{T}_{Rm}D_m = \frac{1}{2J}\kappa_o D_m{}^2. \tag{41}$$

The complementary energy U^c_{Rm} is obtained through Legendre transformation based on Eq.(41) as,

$$U^c_{Rm} = 3\mathring{T}_{Rm}D_m - U_{Rm} = \frac{J}{2\kappa_o}\mathring{T}_{Rm}{}^2. \tag{42}$$

<u>Principle of Stationary Potential Energy</u>

The linear and angular momentum rate balances referred to the current configuration C are expressed by (see, for example, Watanabe [11])

$$\nabla\cdot(\mathring{\mathbf{T}}_T + \mathbf{T}\cdot\mathbf{L}^t) + \rho\,\dot{\bar{\mathbf{a}}} = 0, \qquad (\mathring{\mathbf{T}}_T)^t_a = 0, \tag{43}$$

where ∇ is the gradient operator, ρ is the mass density, $\dot{\bar{\mathbf{a}}}$ is the rate of effective force including inertia force, and $\mathring{T}_T$ is the Truesdell stress rate expressed by ([1], p.133)

$$\mathring{\mathbf{T}}_T \equiv \dot{\mathbf{T}} - \mathbf{L}\cdot\mathbf{T} - \mathbf{T}\cdot\mathbf{L}^t + (\mathrm{tr}\ \mathbf{D})\mathbf{T}. \tag{44}$$

We shall choose the arbitrary set of virtual velocity δv such that the displacement condition on the geometrical boundary S_v are not violated. The principle of virtual work can be obtained by multiplying the linear and angular rate equilibrium equations by the velocity field

$\delta\mathbf{v}$ and the spin field $\delta\mathbf{W}$ arising from $\delta\mathbf{v}$, respectively, and integrating over the entire volume as

$$\int\left[\left\{\nabla\cdot(\overset{\circ}{\mathbf{T}}_T + \mathbf{T}\cdot\mathbf{L}^t) + \rho\,\dot{\bar{\mathbf{a}}}\right\}\cdot\delta\mathbf{v} + (\overset{\circ}{\mathbf{T}}_T)^t_a:\delta\mathbf{W}\right] dV = 0. \tag{45}$$

This principle is valid regardless of the employed constitutive equation. Since the first term in Eq.(45) can be converted to

$$\left\{\nabla\cdot(\overset{\circ}{\mathbf{T}}_T + \mathbf{T}\cdot\mathbf{L}^t)\right\}\cdot\delta\mathbf{v} = \nabla\cdot\left\{(\overset{\circ}{\mathbf{T}}_T + \mathbf{T}\cdot\mathbf{L}^t)\cdot\delta\mathbf{v}\right\} - (\overset{\circ}{\mathbf{T}}_T + \mathbf{T}\cdot\mathbf{L}^t):\nabla\delta\mathbf{v}, \tag{46}$$

and using the divergence theorem and the property of inner product of Eq.(2), then the principle of virtual work of Eq.(45) can be reduced to the following principle of stationary potential energy, where the spin tensor $\mathbf{W}$ is eliminated:

$$\Pi^P(\mathbf{v}) = \int\left\{U_T + \frac{1}{2}\mathbf{T}:(\nabla\mathbf{v}\cdot\mathbf{v}\nabla)\right\}dV - \int\rho\dot{\bar{\mathbf{a}}}\cdot\mathbf{v}\,dV - \int\dot{\bar{\mathbf{t}}}\cdot\mathbf{v}\,dS_t\,, \tag{47}$$

where dV is the volume element, dS_t is the surface element on the traction boundary. The term U_T expresses the internal work rate related to the Truesdell stress rate $\overset{\circ}{\mathbf{T}}_T$, and its variation is given by

$$\delta U_T = \overset{\circ}{\mathbf{T}}_T:\delta\mathbf{D}\ . \tag{48}$$

Therefore, it is necessary to express δU_T by the employed constitutive equation. The tensors $\mathbf{D}$, $\mathbf{W}$ and $\mathbf{L}$ are expressed by the velocity field $\mathbf{v}$ and the gradient operator ∇ as follows:

$$\mathbf{D} = (\mathbf{v}\nabla)_s\,, \qquad \mathbf{W} = (\mathbf{v}\nabla)_a\,, \qquad \mathbf{L} = (\mathbf{v}\nabla). \tag{49}$$

Since the stretching tensor $\mathbf{D}$ consists of the deviatoric and mean components, as shown in Eq.$(7)_2$, the work rate δU_T also should be decomposed into the following two terms:

$$\delta U_T = \delta U_T' + \delta U_{Tm}\,, \tag{50}$$

where

$$\delta U_T' = \overset{\circ}{\mathbf{T}}{}_T':\delta\mathbf{D}'\,, \qquad \delta U_{Tm} = 3\overset{\circ}{T}_{Tm}\delta D_m\ . \tag{51}$$

The stress rates $\overset{\circ}{\mathbf{T}}{}_T'$ and $\overset{\circ}{T}_{Tm}$ are defined by the definition of Eq.$(7)_2$ as follows:

$$\overset{\circ}{T}_{Tm} \equiv \frac{1}{3}(\overset{\circ}{\mathbf{T}}_T:\mathbf{I}), \qquad \overset{\circ}{\mathbf{T}}{}_T' \equiv \overset{\circ}{\mathbf{T}}_T - \overset{\circ}{T}_{Tm}\mathbf{I} = \dot{\mathbf{T}}' - (\mathbf{L}\cdot\mathbf{T})' - (\mathbf{T}\cdot\mathbf{L}^t)' + (\mathrm{tr}\,\mathbf{D})\,\mathbf{T}'\,. \tag{52}$$

Using Eqs.(6) and (7) and property of inner product of Eq.(2), we have the relations

$$(\mathbf{L}\cdot\mathbf{T}):\mathbf{I} = (\mathbf{T}\cdot\mathbf{L}^t):\mathbf{I} = \mathbf{T}':\mathbf{D}' + 3T_mD_m\ . \tag{53}$$

The deviatoric parts of $(\mathbf{L}\cdot\mathbf{T})'$ and $(\mathbf{T}\cdot\mathbf{L}^t)'$ in Eq.(51) are calculated as follows:

$$(\mathbf{L}\cdot\mathbf{T})' = (\mathbf{L}\cdot\mathbf{T}) - \frac{1}{3}(\mathbf{L}\cdot\mathbf{T}):\mathbf{I} = \mathbf{L}'\cdot\mathbf{T}' + D_m\mathbf{T}' + T_m\mathbf{L}' - \frac{1}{3}(\mathbf{T}':\mathbf{D}')\mathbf{I}, \tag{54}$$

$$(\mathbf{T}\cdot\mathbf{L}^t)' = (\mathbf{T}\cdot\mathbf{L}^t) - \frac{1}{3}(\mathbf{T}\cdot\mathbf{L}^t):\mathbf{I} = \mathbf{T}'\cdot\mathbf{L}^{t\prime} + D_m\mathbf{T}' + T_m\mathbf{L}^{t\prime} - \frac{1}{3}(\mathbf{T}':\mathbf{D}')\mathbf{I}. \tag{55}$$

Finally, the stress rates $\overset{\circ}{\mathbf{T}}{}_T'$ and $\overset{\circ}{T}_{Tm}$ are calculated by substituting Eqs.(53)-(55) into Eq.(52) as

$$\overset{\circ}{T}_{Tm} = \dot{T}_m - \frac{2}{3}(\mathbf{T}':\mathbf{D}') - T_mD_m\,, \qquad \overset{\circ}{\mathbf{T}}{}_T' = \overset{\circ}{\bar{\mathbf{T}}}{}_T' - 2D_m\mathbf{T}' - 2T_m\mathbf{D}' + \frac{2}{3}(\mathbf{T}':\mathbf{D}')\mathbf{I}, \tag{56}$$

where

$$\overset{\circ}{\mathbf{T}}{}'_T \equiv \dot{\mathbf{T}}' - \mathbf{L}'\cdot\mathbf{T}' - \mathbf{T}'\cdot\mathbf{L}'^t + (\mathrm{tr}\,\mathbf{D})\mathbf{T}'; \qquad \mathbf{L}' \equiv \mathbf{L} - D_m\mathbf{I} = \mathbf{D}' + \mathbf{W}. \tag{57}$$

The difference between the stress rates $\overset{\circ}{\mathbf{T}}{}'_T$ and $\overset{\circ}{\mathbf{T}}{}'_R$ are given from Eqs.$(56)_2$ and $(35)_2$ as

$$\overset{\circ}{\mathbf{T}}{}'_T = \overset{\circ}{\mathbf{T}}{}'_R - (\mathbf{L}' - \Omega)\cdot\mathbf{T}' - \mathbf{T}'\cdot(\mathbf{L}'^t - \Omega) - 2D_m\mathbf{T}' - 2T_m\mathbf{D}' + \frac{2}{3}(\mathbf{T}':\mathbf{D}')\,\mathbf{I}. \tag{58}$$

Since the work due to the last term of the right-hand side of Eq.(58) becomes $2/3(\mathbf{T}':\mathbf{D}')(\mathbf{I}:\delta\mathbf{D}')=0$, the rate potential U_T' defined in Eq.$(51)_1$ is derived from Eqs.(58) and $(38)_2$ as follows

$$\delta U_T' = \delta U_R' + \delta U_{Rr}' - \mathbf{T}':\delta(\mathbf{D}'^2) - 2D_m(\mathbf{T}':\delta\mathbf{D}') - 2T_m(\mathbf{D}':\delta\mathbf{D}'), \tag{59}$$

where

$$\delta U_{Rr}' \equiv -\{(\mathbf{W} - \Omega)\cdot\mathbf{T}' + \mathbf{T}'\cdot(\mathbf{W} - \Omega)\}:\delta\mathbf{D}'. \tag{60}$$

The rate potential U_{Tm} defined in Eq.$(51)_2$ is derived from Eqs.(40) and (41) as follows:

$$\delta U_{Tm} \equiv \delta U_{Rm} - 2(\mathbf{T}':\mathbf{D}')\delta D_m - 6T_m\delta(D_m)^2. \tag{61}$$

Thus, the principle of stationary potential energy is given by

$$\Pi^P(\mathbf{v}) = \int [\, U_R'(\mathbf{D}') + U_{Rm}(D_m) + U_{Re}(\mathbf{D}', D_m) + \frac{1}{2}\mathbf{T}:(\nabla\mathbf{v}\cdot\mathbf{v}\nabla)\,]dV - \int \rho\dot{\bar{\mathbf{a}}}\cdot\mathbf{v}dV - \int \dot{\bar{\mathbf{t}}}\cdot\mathbf{v}dS_t\,, \tag{62}$$

where

$$U_{Re}'(\mathbf{D}', D_m) \equiv U_{Rr}' - \mathbf{T}':(\mathbf{D}')^2 - T_m(\mathbf{D}':\mathbf{D}') - 6T_m(D_m)^2 - 2(\mathbf{T}':\mathbf{D}')D_m\,. \tag{63}$$

Since the term U'_{Rr} can be expressed as a potential, it should be treated as a load vector in the finite element analysis. The subsidiary conditions imposed on this principle are as follows:

$$D_m = \frac{1}{3}(\mathbf{v}\nabla):\mathbf{I} \equiv (\mathbf{v}\nabla)_m\,,\quad \mathbf{D}' = (\mathbf{v}\nabla)_s - (\mathbf{v}\nabla)_m\mathbf{I}:\ \text{ in } V, \tag{64}$$

$$\mathbf{v} = 0:\quad \text{on } S_v\,. \tag{65}$$

Hu-Washizu Variational Principle

The principle of stationary potential energy can be generalized by the introduction of the Lagrange multiplier method to yield a family of variational principles such as Hu-Washizu as well as Hellinger-Reissner types. If we employ the total Lagrangian approach referred to the initial configuration, we introduce the following inner product between the compatible condition and stress rate into the principle of stationary potential energy in order to obtain the Hu-Washizu principle: (see Watanabe [12])

$$-\int \dot{\mathbf{T}}_{(2)}:\left(\dot{\mathbf{E}} + \dot{\mathbf{S}} - \mathbf{F}^t\cdot\dot{\mathbf{u}}\nabla_o\right) dV_o\,, \tag{66}$$

where $\mathbf{T}_{(2)}$ is the second Piola-Kirchhoff stress tensor defined by $(J\mathbf{F}^{-1}\cdot\mathbf{T}\cdot\mathbf{F}^{-t})$, $\mathbf{u}$ is the displacement, ∇_o is the gradient operator, $\dot{\mathbf{E}}$ and $\dot{\mathbf{S}}$ are the symmetric Green-Lagrange strain rate and the anti-symmetric spin tensor on the initial state C_o, respectively. If we employ the updated Lagrangian approach referred to the current configuration, Eq.(66) is equivalently transformed to the following:

$$-\int \overset{\circ}{T}_T:\left(D + W - v\nabla\right) dV. \tag{67}$$

Among the work of Eq.(67), we choose the following term related to the volumetric change as well as the Legendre transformation of Eq.(42):

$$-\int 3\overset{\circ}{T}_{Rm}\left\{D_m - (v\nabla)_m\right\} dV. \tag{68}$$

Thus we have Hu-Washizu principle as follows:

$$\Pi^{HW}(v, \overset{\circ}{T}_{Rm}, D_m) = \int [\, U_R'(D') + U_{Rm}(D_m) + U_{Re}(D', (v\nabla)_m) - 3\overset{\circ}{T}_{Rm}\left\{D_m - (v\nabla)_m\right\}$$
$$+ \frac{1}{2}T:(\nabla v \cdot v\nabla)\,]dV - \int \rho\dot{\bar{a}}\cdot v\,dV - \int \dot{\bar{t}}\cdot v\,dS_t - \int \dot{t}\cdot(v - \bar{v})dS_v\,, \tag{69}$$

where the independent variables are v, $\overset{\circ}{T}_{Rm}$ and D_m, and the subsidiary condition for this principle is Eq.$(64)_2$. According to the rule of the calculus of variation, Eq.(69) is equivalent to

$$\delta\Pi^{HW} = \int \Bigg[\left\{\frac{\partial U_R'}{\partial D'} - (L' - \Omega)\cdot T' - T'\cdot(L'^{t} - \Omega) - 2D_m T' - 2T_m D' + \frac{2}{3}(T':D')I\right\}:\delta D'$$
$$+ 3\left\{\overset{\circ}{T}_{Rm} - \frac{2}{3}(T':D') - 4T_m D_m\right\}\delta(v\nabla)_m - T:(\nabla v \cdot \delta v\nabla) - \left(3\overset{\circ}{T}_{Rm} - \frac{\partial U_{Rm}}{\partial D_m}\right)\delta D_m$$
$$- 3\left(D_m - (v\nabla)_m\right)\delta\overset{\circ}{T}_{Rm}\Bigg]dV - \int \rho\dot{\bar{a}}\cdot\delta v\,dV - \int \dot{\bar{t}}\cdot\delta v\,dS_t - \int \left\{\delta\dot{t}\cdot(v - \bar{v}) + \dot{t}\cdot\delta v\right\}dS_v\,. \tag{70}$$

The obtained stationary conditions are as follows: from δv, we find that the rate linear momentum balance is attained by tracing back Eqs.(56) and (46); from δD_m, we get the mean part of the constitutive equation of Eq.$(40)_1$; from $\delta\overset{\circ}{T}_{Rm}$, we get the relation of strain rate and velocity field in mean part of Eq.$(61)_1$.

Hellinger-Reissner Variational Principle

We now consider certain special case of the Hu-Washizu principle to reduce the Hellinger-Reissner principle. If the condition of Eq.(42) is satisfied a priori, one can eliminate the strain rate D_m as a variable from Eq.(69) and derive the following rate functional for the Hellinger-Reissner principle:

$$\Pi^{HR}(v, \overset{\circ}{T}_{Rm}) = \int [\, U_R'(D') + U_{Re}(D', (v\nabla)_m) - \frac{J}{2\kappa_o}(\overset{\circ}{T}_{Rm})^2 + 3\overset{\circ}{T}_{Rm}(v\nabla)_m$$
$$+ \frac{1}{2}T:(\nabla v \cdot v\nabla)\,]dV - \int \rho\dot{\bar{a}}\cdot v\,dV - \int \dot{\bar{t}}\cdot v\,dS_t - \int \dot{t}\cdot(v - \bar{v})dS_v\,, \tag{71}$$

where the subsidiary condition for this principle is Eq.$(64)_2$. Since the bulk modulus κ_o appears in denominator, we can expect to perform the stable numerical calculation. The variation of the functional of Eq.(71) is given by

$$\delta\Pi^{HR} = \int \Big[\Big\{ \frac{\partial U_R'}{\partial \mathbf{D}'} - (\mathbf{L}'-\Omega)\cdot\mathbf{T}' - \mathbf{T}'\cdot(\mathbf{L}'^{t}-\Omega) - 2D_m\mathbf{T}' - 2T_m\mathbf{D}' + \frac{2}{3}(\mathbf{T}':\mathbf{D}')\mathbf{I} \Big\} : \delta\mathbf{D}'$$
$$+ 3\Big\{ \mathring{T}_{Rm} - \frac{2}{3}(\mathbf{T}':\mathbf{D}') - 4T_mD_m \Big\} \delta(\mathbf{v}\nabla)_m - \mathbf{T}:(\nabla\mathbf{v}\cdot\delta\mathbf{v}\nabla) - 3\Big\{ \frac{J}{3\kappa_o}\mathring{T}_{Rm} - (\mathbf{v}\nabla)_m \Big\} \delta\mathring{T}_{Rm} \Big] dV$$
$$- \int \rho\dot{\bar{\mathbf{a}}}\cdot\delta\mathbf{v}\, dV - \int \dot{\bar{\mathbf{t}}}\cdot\delta\mathbf{v}\, dS_t - \int \{ \delta\dot{\mathbf{t}}\cdot(\mathbf{v} - \bar{\mathbf{v}}) + \dot{\mathbf{t}}\cdot\delta\mathbf{v} \} dS_v . \tag{72}$$

The obtained stationary conditions are as follows: from $\delta\mathbf{v}$, we get the rate linear momentum balance; and from $\delta\mathring{T}_{Rm}$, we have the mean part constitutive equation $(40)_1$.

Acknowledgement

The author expresses his sincere appreciation for the financial support of the Japanese Ministry of Education to attend this IUTAM Symposium.

References

1. Truesdell, C.; Noll, W.: *The nonlinear field theories of mechanics*. Handbuch der Physik. III/3 Springer, (1965)

2. Stören, S.; Rice, J. R.:Localized necking in thin sheets. J. Mech. Phys. Solids. 23 (1975) 421-441.

3. Hill, R.: On Constitutive inequalities for simple materials-I. J. Mech. Phys. Solids. 16 (1968) 229-242.

4. Flory, P. J.: Thermodynamic relations for high elastic materials. Trans. Faraday Soc. 57- 5 (1961) 829-838.

5. Hibbitt, Karlsson and Sorensen, Inc.:ABAQUS theory manual. Ver.4-8 (1989) 5.11.1-6

6. Green, A. E.; Naghdi, P. M.: A general theory of an elastic-plastic continuum. Arch. Rational Mech. Anal.18 (1965) 251-281.

7. Green, A. E.; McInnis, B. C.: Generalized hypo-elasticity. Proc. Roy. Soc. Edinburgh A 57 (1967) 220-230.

8. Treloar, L. R. G.:*The physics of rubber elasticity*. 2nd ed. Oxford Press (1958)

9. Watanabe, O.: Hyperelasticity theory based on rotationless strain (constitutive equation for incompressible material). Trans. JSME. A-57-533 (1991) 154-161, to appear in JSME Int. J.

10. Watanabe, O.: Stability of solutions for hyperelasticity based on rotationless strain. submitted to Trans. JSME (1991).

11. Watanabe, O.: Renewal and change of frame in finite deformation. JSME Int. J. I-34-3 (1991) 287-296.

12. Watanabe, O.: Variational principles of elastoplasticity in finite deformation. JSME Int. J. I-33-3 (1990) 480-489.

Numerical Analysis and Computational Aspects

State Updates and State-Transfer Operators in Computational Plasticity

A. M. Cuitiño and M. Ortiz

Division of Engineering

Brown University

Providence, RI 02912

USA

Summary

Several aspects of computational plasticity pertaining to the formulation and implementation of implicit state updates are considered. We review a recently proposed method [1] for automatically extending small-strain state-update algorithms and their correspondent consistent tangents into the finite deformation range within the framework of multiplicative plasticity. The procedure, when it applies, operates at the level of kinematics and, hence, can be implemented once and for all independently of the material-specific details of the constitutive model. Next, we address the problem of formulating state-transfer operators in the context of adaptive meshing calculations. A recently proposed method [2] for defining variationally consistent transfer operators is summarized. As a new example of application, we develop a procedure for analyzing a growing crack which enables an arbitrary fixed mesh to be kept continuously focused about the crack tip. This in turn permits the accurate resolution of the evolving near-tip fields as the crack advances.

1. Introduction

Implicit finite element methods applied to elastoplastic solids can prove advantageous in applications in which the stress and deformation paths remain nearly proportional. Under these conditions, implicit methods can significantly speed up the calculations by enabling the use of larger time steps than otherwise permitted by their explicit counterparts. Implicit methods which seek to exactly satisfy equilibrium, compatibility and some suitable state-update rule at every load increment were formalized by Hughes and Taylor [3] for the viscoplastic solid, and by Simó and Taylor [4] for the rate-independent solid. A Newton-Raphson solution of the resulting incremental nonlinear equations requires the formulation of so-called 'consistent tangents' [3,4] obtained by direct linearization of the state-update algorithm.

For small strains, it is straightforward to compute the consistent tangents for a number of widely used constitutive relations, such as J_2-flow theory, a variety of pressure-sensitive formulations, including the Drucker-Prager and Gurson models [5], and others. In the finite deformation range, however, the need to linearize the finite kinematics renders the analytical computation of the consistent tangents exceedingly cumbersome. In a recent paper [1] we have shown, however, that for a broad class of models it is possible to extend small-strain implicit updates into the finite deformation range within the framework of multiplicative plasticity by operating entirely at the level of kinematics. Consequently,

D. Besdo, E. Stein (Eds.)
Finite Inelastic Deformations –
Theory and Applications
IUTAM Symposium Hannover/Germany 1991

the method of extension can be implemented once and for all independently of the material-specific details of the constitutive description. In addition, the finite consistent tangents follow from their small-strain counterparts by recourse to purely kinematical material-independent manipulations. In Section 2 we describe the salient aspects of this method in broad outline.

In Section 3, we turn our attention to the problem of defining state updates in situations where the meshes supporting the initial and final states differ. Inevitably, at some point during the update the state fields need to be transferred from mesh to mesh. Such state transfer operators have often been defined in an *ad-hoc* manner. By contrast, a recently proposed method [2] furnishes a prescription for constructing variationally consistent transfer operators directly from the Hu-Washizu principle. As a new example of application of state-transfer operators, we address the problem of analyzing a crack growing in a ductile solid. Conventional methods of analysis typically involve a crack propagating through a fine square mesh. Methods of this type are unable to keep the mesh sharply focused about the crack tip, and thus afford limited resolution of the near-tip fields. The procedure developed here overcomes this limitation. In essence, the method allows any convenient mesh, such as a radially graded fan-like grid, to be continuously focused on the crack tip. The mesh is simply translated along with the crack tip and, therefore, the analysis proceeds as if on a *fixed* mesh. The only remeshing capability required for the implementation of the method consists of some suitable transfer operator for extrapolating the state fields to a shifted mesh. The performance of the numerical methods presented in this paper is illustrated by an application involving the quasistatic growth of a mathematically sharp crack through a finitely deforming plastic solid obeying J_2-flow theory.

2. Finite extension of small-strain updates.

We begin by outlining a material-independent method for extending state-update algorithms from small-strain plasticity to finite plasticity with multiplicative kinematics. Further details of the method can be found in [1].

2.1. General framework.

Consider a solid initially occupying a reference configuration B_0, and a process of incremental loading whereby the deformation mapping over B_0 changes from $\boldsymbol{\phi}_n$, at time t_n, to $\boldsymbol{\phi}_{n+1} = \boldsymbol{\phi}_n + \mathbf{u}$, at time $t_{n+1} = t_n + \Delta t$. We enforce equilibrium at time t_{n+1} weakly by recourse to the principle of virtual work:

$$\int_{B_0} \mathbf{P}_{n+1} : \nabla_0 \delta\boldsymbol{\phi}\, dV_0 - \int_{B_0} \mathbf{B}_{n+1} \cdot \delta\boldsymbol{\phi}\, dV_0 - \int_{\partial B_{0\tau}} \bar{\mathbf{T}}_{n+1} \cdot \delta\boldsymbol{\phi}\, dS_0 = 0 \tag{1}$$

where $\mathbf{P}_{n+1}$ denotes the first Piola-Kirchhoff stress field at time t_{n+1}, $\mathbf{B}_{n+1}$ and $\bar{\mathbf{T}}_{n+1}$ are the corresponding body forces and boundary tractions, respectively, $\delta\boldsymbol{\phi}$ is an admissible virtual displacement field, and ∇_0 denotes the material gradient. Assume for now that we have determined a rule to update the stress field of the general form

$$\mathbf{P}_{n+1} = \hat{\mathbf{P}}(\mathbf{F}_{n+1};\ state\ at\ t_n, \Delta t) \tag{2}$$

where the deformation gradients

$$\mathbf{F}_{n+1} = \nabla_0 \boldsymbol{\phi}_{n+1} \tag{3}$$

are assumed given. Inserting (3) and (2) into (1) one obtains

$$\int_{B_0} \hat{\mathbf{P}}(\nabla_0 \boldsymbol{\phi}_{n+1};\ state\ at\ t_n, \Delta t) : \nabla_0 \delta\boldsymbol{\phi}\, dV_0 - \int_{B_0} \mathbf{B}_{n+1} \cdot \delta\boldsymbol{\phi}\, dV_0 - \int_{\partial B_{0\tau}} \bar{\mathbf{T}}_{n+1} \cdot \delta\boldsymbol{\phi}\, dS_0 = 0 \tag{4}$$

which defines a set of nonlinear equations which can be solved for the updated deformation mapping $\boldsymbol{\phi}_{n+1}$. If one chooses to solve (4) by means of a Newton-Raphson iteration, the linearized problem for the incremental displacements takes the form

$$\int_{B_0} \nabla_0 \delta\boldsymbol{\phi} : \hat{\mathbf{K}}_{n+1} : \nabla_0 \mathbf{u}\, dV_0 - \ forcing\ terms = 0 \tag{5}$$

where the 'consistent tangents' $\hat{\mathbf{K}}_{n+1}$ follow by linearization of the stress update (2) as

$$\hat{\mathbf{K}}_{n+1} \equiv \frac{\partial \hat{\mathbf{P}}}{\partial \mathbf{F}_{n+1}}(\mathbf{F}_{n+1};\ state\ at\ t_n, \Delta t) \tag{6}$$

Alternatively, the linearized virtual work principle (5) can be expressed in spatial form as

$$\int_{B_0} \nabla \delta\boldsymbol{\phi} : \hat{\mathbf{k}}_{n+1} : \nabla \mathbf{u}\, dV_0 - \ forcing\ terms = 0 \tag{7}$$

where ∇ is the spatial gradient operator and the spatial consistent tangents $\mathbf{k}$ are defined as

$$\hat{k}_{ijkl} = \hat{K}_{iJkL} F_{jJ} F_{lL} \tag{8}$$

Whereas for small strains the computation of the consistent tangents reduces to a straightforward exercise for many commonly used models, in the finite deformation range the calculations are considerably more cumbersome. Next we summarize a method given in [1] which provides a means of extending small-strain updates and consistent tangents to the finite deformation range which operates strictly at the level of kinematics.

2.2. Small-strain formulation.

We begin by outlining the computational formulation of small-strain plasticity which we seek to extend into the finite deformation range. Further details and properties of the small-strain theory adopted here can be found elsewhere [6,7]. As in the preceding section, we concern ourselves with the incremental analysis of a solid, now presumed to undergo small deformations. Under these conditions, the distinction between deformed and reference configurations becomes spurious, the first Piola-Kirchhoff stress tensor reduces to the Cauchy stress tensor $\boldsymbol{\sigma}$, and the deformation is fully described by the small-strain tensor $\boldsymbol{\epsilon}$.

The equations governing the small-strain constitutive behavior of an elastoplastic solid can be formulated and discretized as follows:

Constitutive:		**Incremental:**	
$\boldsymbol{\epsilon} = \boldsymbol{\epsilon}^e + \boldsymbol{\epsilon}^p$	$\longleftrightarrow$	$\boldsymbol{\epsilon}_{n+1} = \boldsymbol{\epsilon}^e_{n+1} + \boldsymbol{\epsilon}^p_{n+1}$	(9)
$\dot{\boldsymbol{\epsilon}}^p = \dot{\epsilon}^p\, \mathbf{r}(\boldsymbol{\sigma}, \mathbf{q})$	$\longleftrightarrow$	$\boldsymbol{\epsilon}^p_{n+1} = \boldsymbol{\epsilon}^p_n + \triangle\epsilon^p\, \mathbf{r}_{n+1}$	(10)
$\boldsymbol{\sigma} = \boldsymbol{\sigma}(\boldsymbol{\epsilon}^e)$	$\longleftrightarrow$	$\boldsymbol{\sigma}_{n+1} = \boldsymbol{\sigma}(\boldsymbol{\epsilon}^e_{n+1})$	(11)
$\dot{\mathbf{q}} = \dot{\epsilon}^p\, \mathbf{h}(\boldsymbol{\sigma}, \mathbf{q})$	$\longleftrightarrow$	$\mathbf{q}_{n+1} = \mathbf{q}_n + \triangle\epsilon^p\, \mathbf{h}_{n+1}$	(12)
$\dot{\epsilon}^p = \frac{1}{\eta}\phi(\boldsymbol{\sigma}, \mathbf{q})$	$\longleftrightarrow$	$\triangle\epsilon^p = \frac{\Delta t}{\eta}\phi_{n+1}$	(13)

where $\boldsymbol{\epsilon}^p$ is the plastic strain tensor, $\mathbf{q}$ is some suitable (model dependent) set of internal variables, $\mathbf{h}$ are the hardening moduli, ϵ^p is an effective plastic strain, ϕ is an effective overstress, and η is a scalar viscosity parameter. Eq. (9) represents the conventional additive decomposition of strain into elastic and plastic components; eq. (10) is the flow rule; eq. (11) is the elastic response, which we allow to be nonlinear; and eqs. (12) and (13) are the hardening and viscosity laws, respectively. We note for later reference that eqs. (9) and (10) can be combined to give

$$\boldsymbol{\epsilon}^e_{n+1} = \boldsymbol{\epsilon}^{e*}_{n+1} - \triangle\epsilon^p\, \mathbf{r}_{n+1} \tag{14}$$

where the elastic predictor

$$\boldsymbol{\epsilon}^{e*}_{n+1} = \boldsymbol{\epsilon}_{n+1} - \boldsymbol{\epsilon}^p_n \tag{15}$$

gives the elastic strain which would result from a purely elastic step.

As is evident from the discretization of the constitutive equations adopted in eqs. (9-13), we shall confine our attention to the backward-Euler or fully-implicit update algorithm. Several popular schemes, such as the radial return [8-12], and operator-splitting and return-mapping algorithms [13-15] , are closely related to the backward-Euler method. Furthermore, the backward-Euler method possesses some unique properties which set it apart from all other update rules, such as: it satisfies the yield criterion identically in the rate-independent limit; it is unconditionally stable in the sense of defining a contractive (resolvent) mapping [13]; it preserves the symmetries of the constitutive relations, in the sense of giving rise to symmetric consistent tangents whenever the flow rule is associative [7]; for rate-independent associative plasticity, it maximizes the incremental work of deformation, and satisfies the constitutive equations exactly along maximal strain paths for a restricted class of hardening rules [7].

As in the finite deformation case, the discretized eqs. (9-13) implicitly determine the updated value of the stress tensor in the form

$$\boldsymbol{\sigma}_{n+1} = \hat{\boldsymbol{\sigma}}(\boldsymbol{\epsilon}_{n+1};\ state\ at\ t_n, \Delta t) \tag{16}$$

The appropriate definition of consistent tangent is, then, [3,4]

$$\hat{\mathbf{D}}_{n+1} \equiv \frac{\partial \hat{\boldsymbol{\sigma}}}{\partial \boldsymbol{\epsilon}_{n+1}}(\boldsymbol{\epsilon}_{n+1};\ state\ at\ t_n, \Delta t) \tag{17}$$

and the corresponding linearized virtual work equation is identical to (7) with $\hat{\mathbf{k}}$ replaced by $\hat{\mathbf{D}}$.

2.3. Finite deformation formulation

Next we adopt a standard formulation of finite deformation plasticity based on multiplicative kinematics. Details on the historical background and physical basis of the theory may be found elsewhere (e. g. [16]). The governing constitutive equations and their discretized counterparts can be written as:

Constitutive:		**Incremental:**	
$\mathbf{F} = \mathbf{F}^e\mathbf{F}^p$	$\longleftrightarrow$	$\mathbf{F}_{n+1} = \mathbf{F}^e_{n+1}\mathbf{F}^p_{n+1}$	(18)
$\dot{\mathbf{F}}^p\mathbf{F}^{p-1} = \dot{\epsilon}^p\,\bar{\mathbf{R}}(\bar{\mathbf{S}}, \bar{\mathbf{Q}})$	$\longleftrightarrow$	$\mathbf{F}^p_{n+1} = \exp\left(\Delta\epsilon^p\,\bar{\mathbf{R}}_{n+1}\right)\mathbf{F}^p_n$	(19)
$\bar{\mathbf{S}} = \bar{\mathbf{S}}(\frac{1}{2}\log\bar{\mathbf{C}}^e)$	$\longleftrightarrow$	$\bar{\mathbf{S}}_{n+1} = \bar{\mathbf{S}}(\frac{1}{2}\log\bar{\mathbf{C}}^e_{n+1})$	(20)
$\dot{\bar{\mathbf{Q}}} = \dot{\epsilon}^p\,\bar{\mathbf{H}}(\bar{\mathbf{S}}, \bar{\mathbf{Q}})$	$\longleftrightarrow$	$\bar{\mathbf{Q}}_{n+1} = \bar{\mathbf{Q}}_n + \Delta\epsilon^p\,\bar{\mathbf{H}}_{n+1}$	(21)
$\dot{\epsilon}^p = \frac{1}{\eta}\bar{\phi}(\bar{\mathbf{S}}, \bar{\mathbf{Q}})$	$\longleftrightarrow$	$\Delta\epsilon^p = \frac{\Delta t}{\eta}\bar{\phi}_{n+1}$	(22)

Eq. (18) embodies the assumption that the deformation gradients $\mathbf{F}$ admit a multiplicative decomposition into elastic and plastic parts, $\mathbf{F}^e$ and $\mathbf{F}^p$, respectively. The plastic part $\mathbf{F}^p$ of the deformation gradient field defines an additional (generally incompatible) configuration $\bar{B}_t$, or intermediate configuration. We shall use an overbar to identify fields defined over the intermediate configuration.

Eq. (19) is the plastic flow rule which, as stated here, determines $\dot{\mathbf{F}}^p$ fully. The manner in which the flow rule is discretized is an essential part of the method of extension, and consists of taking the plastic flow direction $\bar{\mathbf{R}}$ to be constant throughout the increment and equal to its final value $\bar{\mathbf{R}}_{n+1}$. This reduces the flow rule to a system of linear equations for $\mathbf{F}^p$ with initial conditions $\mathbf{F}^p_n$, the exact solution of which is given by the exponential mapping in (19).

Without loss of generality, we express the elastic response (20) in terms of logarithmic elastic strains $(1/2)\log\bar{\mathbf{C}}^e$, where

$$\bar{\mathbf{C}}^e = \mathbf{F}^{eT}\mathbf{F}^e \tag{23}$$

is the elastic right Cauchy-Green deformation tensor. The stress measure $\bar{\mathbf{S}}$ in (20) is the second Piola-Kirchhoff stress tensor relative to the intermediate configuration, and is defined as

$$\bar{\mathbf{S}} = \mathbf{F}^{e-1}\boldsymbol{\tau}\mathbf{F}^{e-T} \tag{24}$$

in terms of the Kirchhoff stress tensor $\boldsymbol{\tau}$. The symbol $\bar{\mathbf{Q}}$ denotes some suitable set of internal variables referred to the intermediate configuration, with $\bar{\mathbf{H}}$ the corresponding hardening moduli. Finally, ϵ^p is an effective plastic strain, $\bar{\phi}$ is an effective overstress, and η a viscosity parameter. We remark that, by expressing the flow rule, elastic response and hardening laws on the intermediate configuration, the formulation automatically satisfies material frame indifference.

Eqs. (18-22) define a system of nonlinear equations which, for given $\mathbf{F}_{n+1}$, can be solved for the updated state variables $\bar{\mathbf{S}}_{n+1}$, $\bar{\mathbf{Q}}_{n+1}$, $\bar{\mathbf{F}}^p_{n+1}$, as well as for $\triangle\epsilon^p$. As in the small-strain case, it is possible to reduce the system to a single equation for $\triangle\epsilon^p$, which can be solved by recourse to a local Newton-Raphson iteration. The consistent tangents (6) or (8) follow by linearization of (18-22), which entails considerable difficulty in the general case. Next we show, however, that in many cases update algorithms and tangents implemented in small strains can be readily extended to the finite deformation range in a manner consistent with the constitutive and algorithmic framework expressed in (18-22).

2.4. The method of extension.

The key step in establishing a formal connection between the finite deformation formulation (18-22) and its small-strain counterpart (9-13) is as follows. Inserting the discretized flow rule (19) into (23), the elastic right Cauchy Green deformation tensor can be written as

$$\bar{\mathbf{C}}^e_{n+1} = \exp\big(-\triangle\epsilon^p\,\bar{\mathbf{R}}^T_{n+1}\big)\bar{\mathbf{C}}^{e*}_{n+1}\exp\big(-\triangle\epsilon^p\,\bar{\mathbf{R}}_{n+1}\big) \tag{25}$$

Here,

$$\bar{\mathbf{C}}^{e*}_{n+1} = \mathbf{F}^{p-T}_n\mathbf{C}_{n+1}\mathbf{F}^{p-1}_n \tag{26}$$

with $\mathbf{C}_{n+1} = \mathbf{F}^T_{n+1}\mathbf{F}_{n+1}$, plays the role of an elastic predictor for the elastic deformation. But, if $\bar{\mathbf{R}}_{n+1}$ commutes with $\bar{\mathbf{C}}^{e*}_{n+1}$, taking logarithms in eq. (25) gives

$$\frac{1}{2}\log\bar{\mathbf{C}}^e_{n+1} = \frac{1}{2}\log\bar{\mathbf{C}}^{e*}_{n+1} - \triangle\epsilon^p\,\mathrm{sym}\bar{\mathbf{R}}_{n+1} \tag{27}$$

which is identical to the small-strain kinematic relation (14). A consequence of this correspondence is that the relations which determine $\bar{\mathbf{S}}_{n+1}$, $\bar{\mathbf{Q}}_{n+1}$ and $\triangle\epsilon^p$ in terms of the predictor strains $(1/2)\log\bar{\mathbf{C}}^{e*}_{n+1}$ are *identical* to those arising in small-strains from eqs. (9-13), provided that the identification of variables listed in Box 1 is effected.

Box 1. Correspondence of Variables

Small strains:		Finite deformations:
ϵ^e_{n+1}	$\longleftrightarrow$	$\frac{1}{2}\log\bar{\mathbf{C}}^e_{n+1}$
ϵ^{e*}_{n+1}	$\longleftrightarrow$	$\frac{1}{2}\log\bar{\mathbf{C}}^{e*}_{n+1}$
σ_{n+1}	$\longleftrightarrow$	$\bar{\mathbf{S}}_{n+1}$
$\mathbf{q}_{n+1}$	$\longleftrightarrow$	$\bar{\mathbf{Q}}_{n+1}$
$\triangle\epsilon^p$	$\longleftrightarrow$	$\triangle\epsilon^p$
$\mathbf{r}_{n+1}$	$\longleftrightarrow$	$\text{sym}\bar{\mathbf{R}}_{n+1}$

The full finite deformation update comprises three steps:

i) *Preprocessor step.* Compute the predictor logarithmic elastic strains $(1/2)\log\bar{\mathbf{C}}^{e*}_{n+1}$ from the given updated deformation gradients $\mathbf{F}_{n+1}$ as in (26). Set ϵ^{e*}_{n+1} equal to $(1/2)\log\bar{\mathbf{C}}^{e*}_{n+1}$. Identify σ_n with $\bar{\mathbf{S}}_n$ and $\mathbf{q}_n$ with $\bar{\mathbf{Q}}_n$.

ii) *Small-strain update.* Effect a small-strain update driven by ϵ^{e*}_{n+1} with initial conditions σ_n and $\mathbf{q}_n$, to compute σ_{n+1}, $\mathbf{q}_{n+1}$ and $\triangle\epsilon^p$.

iii) *Postprocessor step.* Identify $\bar{\mathbf{S}}_{n+1}$ with σ_{n+1} and $\bar{\mathbf{Q}}_{n+1}$ with $\mathbf{q}_{n+1}$, and compute $\mathbf{F}^p_{n+1}$ by the exponential mapping (19).

It bears emphasis that the steps preceding and following the small-strain update are purely kinematical in nature, and, hence, material independent. Thus, when it applies, the above procedure provides a material-independent prescription for extending small-strain updates into the finite deformation range within the framework of multiplicative plasticity.

Next we turn our attention to the conditions under which the method of extension applies. The key assumption leading to (27) is that the plastic flow direction $\bar{\mathbf{R}}_{n+1}$ commutes with the predictor $\bar{\mathbf{C}}^{e*}_{n+1}$. Because of the symmetry of $\bar{\mathbf{C}}^{e*}_{n+1}$, this immediately requires that $\bar{\mathbf{R}}_{n+1}$ be likewise symmetric, i. e., that the plastic spin vanish, and that $\bar{\mathbf{R}}_{n+1}$ have the same principal directions as $\bar{\mathbf{C}}^{e*}_{n+1}$. Equivalently, $\mathbf{r}_{n+1}$ must commute with ϵ^{e*}_{n+1} in the small-strain formulation. But, by (14), $\mathbf{r}_{n+1}$ commutes with ϵ^{e*}_{n+1} if and only if $\mathbf{r}_{n+1}$ commutes with ϵ^e_{n+1}. By inserting the elastic relation (11) into the flow direction $\mathbf{r}(\sigma, \mathbf{q})$, this can be thought of as a function $\mathbf{r}(\epsilon^e, \mathbf{q})$ of the elastic strains ϵ^e. The sought condition is then that the function $\mathbf{r}(\epsilon^e, \mathbf{q})$ have the same principal directions as ϵ^e, for all values of its arguments. A general representation of a tensor-valued function satisfying this requirement is [18]

$$\mathbf{r}(\epsilon^e) = \lambda_1(\epsilon^e)\mathbf{I} + \lambda_2(\epsilon^e)\epsilon^e + \lambda_3(\epsilon^e)\epsilon^{e2} \tag{28}$$

where the scalar functions $\{\lambda_1, \lambda_2, \lambda_3\}$ of $\boldsymbol{\epsilon}^e$ need not be isotropic. Thus, the direction of plastic flow $\mathbf{r}$ regarded as a function of $(\boldsymbol{\epsilon}^e, \mathbf{q})$, while restricted to be of the form (28), can still possess some measure of anisotropy. It bears emphasis that these requirements place no restrictions on the elastic relation (11), which can be specified arbitrarily. In particular, the elastic response can be anisotropic.

Next we show how the consistent tangents (6), or in spatial form (8), can be recovered from their small-strain counterpart (17) by means of material-independent kinematical manipulations. To simplify writing, we drop the subindex $n+1$ from the expressions with the understanding that all quantities, unless otherwise stated, correspond to time t_{n+1}. A lengthy but straightforward calculation gives the following expression for the finite deformation spatial tangents:

$$\hat{k}_{ijkl} = \hat{d}_{ijkl} - \tau_{nj} A_{inkl} - \tau_{ni} A_{jnkl} + \delta_{ik} \tau_{lj} \tag{29}$$

where $\boldsymbol{\tau}$ is the Kirchhoff stress tensor and one writes

$$\hat{d}_{ijkl} = F^e_{iI} F^e_{jJ} \hat{D}_{IJMN} L_{MNKL} (F^{e*}_{lL} F^{e*}_{kK} + F^{e*}_{kL} F^{e*}_{lK}); \tag{30}$$

$$\begin{aligned} A_{inkl} &= F^e_{iI} F^{-1}_{Bn} \frac{\partial F^p_{IB}}{\partial F_{kD}} F_{lD} \\ &= F^e_{iI} F^{-1}_{Bn} E_{IBRS} H_{RSMN} L_{MNKL} (F^{e*}_{lL} F^{e*}_{kK} + F^{e*}_{kL} F^{e*}_{lK}); \end{aligned} \tag{31}$$

$$F^{e*}_{iJ} = F_{iA} (F^{p-1}_n)_{AJ}; \tag{32}$$

$$L_{MNKL} = \frac{\partial (\log \sqrt{\bar{\mathbf{C}}^e})_{MN}}{\partial \bar{\mathbf{C}}^e_{KL}}; \tag{33}$$

$$E_{IBRS} = \frac{\partial F^p_{IB}}{\partial (\Delta \boldsymbol{\epsilon}^p)_{RS}} = \frac{\partial [\exp(\Delta \boldsymbol{\epsilon}^p)]_{IJ}}{\partial \Delta \boldsymbol{\epsilon}^p_{RS}} (F^p_n)_{JB}; \tag{34}$$

$$H_{RSMN} = \frac{\partial \Delta \boldsymbol{\epsilon}^p_{RS}}{\partial \boldsymbol{\epsilon}^{e*}_{MN}} = \frac{1}{2} (\delta_{RM} \delta_{SN} + \delta_{RN} \delta_{SM}) - C^e_{RSAB} \hat{D}_{ABMN}; \tag{35}$$

Here, as before, $\hat{\mathbf{D}}$ represent the small-strain tangent (17), which is the only material-dependent quantity in the above expressions, $\mathbf{F}^{e*}$ may be thought of as an elastic predictor for $\mathbf{F}^e$, and

$$\mathbf{C}^e = \frac{\partial \boldsymbol{\epsilon}^e}{\partial \boldsymbol{\sigma}} \tag{36}$$

are the tangent elastic compliances. The geometrical tensors $\mathbf{L}$ and $\mathbf{E}$ follow by linearization of the logarithmic and exponential mappings, respectively. These linearizations have been given Simó and Taylor [17] in the general case.

2.5. Discussion.

The method outlined above (see [1] for a full account) permits small-strain plasticity models and updates to be extended into the finite deformation range within the framework of multiplicative plasticity. The method operates at the level of kinematics and, consequently, is independent of the material-specific details of the constitutive description. For the method of extension to apply, the small-strain model is subject to certain restrictions on the plastic flow direction, which is required to be of the form (28). This representation allows for some measure of anisotropy in the plastic flow direction regarded as a function of the elastic strains. In addition, the elastic response can be chosen arbitrarily and, in particular, it can be fully anisotropic. Then, even if the direction of plastic flow were chosen to be an isotropic function of the elastic strains, which is a special case of (28), it still would be an anisotropic function of the stresses. In this case, the plastic anisotropy is directly inherited from that of the elastic response, a proposition not without physical appeal. The finite deformation model obtained by the method of extension is one in which the plastic spin is zero. This is appropriate as a constitutive assumption in circumstances where texturing effects are not important. In conclusion, the method proposed in [1], when applicable, simplifies the implementation of finite deformation plasticity models, by allowing state updates and consistent tangents to be formulated in small strains. An example of application illustrative of the performance of the method is given in Section 3.3.

3. State-transfer operators with application to crack growth.

Even modest remeshing capabilities can add considerable versatility to a finite element code. In this section we develop a method for analyzing a growing crack which requires the implementation of a simple state-transfer operator. The method then allows a *fixed* mesh to be continuously focused about the propagating crack tip, which facilitates the detailed resolution of the evolving near-tip fields.

3.1. Variationally consistent state-transfer operators.

Here we consider the problem of formulating finite element solution procedures for history-dependent materials on an evolving mesh. In contrast to linear elastic solids, for which the displacements can be re-calculated readily as successive mesh refinements are introduced, the history-dependent nature of plastic solids necessitates the transfer of the state variables from mesh to mesh. As shown in [2], the Hu-Washizu variational principle provides a prescription for constructing state-transfer operators. This prescription is uniquely determined once the state variable interpolation is defined. Some relevant results from [2] are summarized in the sequel.

For the formulation of finite element procedures it proves convenient to rephrase the field equations governing the motion of the continuum in weak form, which leads to the integral statements

$$\begin{aligned} &\int_{B_0} \mathbf{P} : \nabla_0 \delta\boldsymbol{\phi} dV_0 - \int_{B_0} \rho_0 \mathbf{B} \cdot \delta\boldsymbol{\phi} dV_0 - \int_{\partial B_{0\tau}} \bar{\mathbf{T}} \cdot \delta\boldsymbol{\phi} dS_0 = 0 \\ &- \int_{B_0} (\mathbf{F} - \nabla_0 \boldsymbol{\phi}) : \delta\mathbf{P} dV_0 - \int_{\partial B_{0u}} (\boldsymbol{\phi} - \bar{\boldsymbol{\phi}}) \cdot \delta\mathbf{P} \cdot \mathbf{N} = 0 \end{aligned} \tag{37}$$

where $\delta\boldsymbol{\phi}$ and $\delta\mathbf{P}$ are arbitrary variations of the deformation mapping and stress field. For a plastically deforming solid, the constitutive response is path dependent. Envision, therefore, a process of step-by-step integration of the governing equations of the type formulated in Section 2. Enforcing eqs. (37) at time t_{n+1} gives

$$\begin{aligned} \int_{B_0} \mathbf{P}_{n+1} : \nabla_0 \delta\boldsymbol{\phi} dV_0 - \int_{B_0} \rho_0 \mathbf{B}_{n+1} \cdot \delta\boldsymbol{\phi} dV_0 - \int_{\Gamma_{\tau 0}} \bar{\mathbf{T}}_{n+1} \cdot \delta\boldsymbol{\phi} dS_0 = 0 \\ - \int_{B_0} (\mathbf{F}_{n+1} - \nabla_0 \boldsymbol{\phi}_{n+1}) : \delta\mathbf{P} dV_0 - \int_{\Gamma_{u0}} (\boldsymbol{\phi}_{n+1} - \bar{\boldsymbol{\phi}}_{n+1}) \cdot \delta\mathbf{P} \cdot \mathbf{N} = 0 \end{aligned} \tag{38}$$

For convenience, we collect all state variables $\{\mathbf{F}, \mathbf{F}^p, \mathbf{P}, \bar{\mathbf{Q}}\}$ under the symbol $\boldsymbol{\Lambda}$. Assume that a state-update algorithm has been defined, giving the updated state $\boldsymbol{\Lambda}_{n+1}$ as a function of the initial conditions $\boldsymbol{\Lambda}_n$, the updated value of the deformation gradients $\mathbf{F}_{n+1}$ and Δt. In particular, we write $\mathbf{P}_{n+1}$ as in (2). As in the case of equilibrium and compatibility, we shall endeavor to satisfy (2) in its weak form

$$\int_{B_0} [\hat{\mathbf{P}}(\mathbf{F}_{n+1}; \boldsymbol{\Lambda}_n, \Delta t) - \mathbf{P}_{n+1}] : \delta\mathbf{F} dV_0 = 0 \tag{39}$$

In the special case of finite elasticity, (39) of all the field equations can be jointly derived as the Euler equations of the Hu-Washizu functional specialized for time t_{n+1}.

Eqs. (38) and (39) are now taken as a basis for consistently deriving the finite element equations in the presence of mesh refinement. To this end, assume that B_0 is partitioned into finite element meshes $\mathcal{M}_n$ at times t_n, $n = 0, 1, 2, \ldots$. For a given mesh, the following spatial interpolation is introduced for the deformation mapping and the state variables

$$\boldsymbol{\phi}(\mathbf{X}) = \sum_a \mathbf{x}_a N_a(\mathbf{X}), \qquad \boldsymbol{\Lambda}(\mathbf{X}) = \sum_r \sum_\alpha \boldsymbol{\Lambda}_\alpha^r M_\alpha^r(\mathbf{X}) \tag{40}$$

where N_a are the conforming interpolation functions for the deformation mapping, M_α^r are the element interpolation functions for the state variables, $\mathbf{x}_a$ the spatial coordinates of the nodes and $\boldsymbol{\Lambda}_\alpha^r$ are the state degrees of freedom for element r. Because the state variables enter eqs. (38) and (39) undifferentiated, the shape functions M_α^r need not be continuous across element boundaries. The displacement shape functions N_a, on the other hand, are subject to the usual requirement of C^0-continuity.

For simplicity, we shall assume that the displacement boundary conditions are identically satisfied over ∂B_{0u}, so that the second term in (38-b) vanishes identically. Substitution of (40) into (38) and (39) gives the set of discretized equations

$$\begin{aligned} &\sum_r \mathbf{L}_{n+1}^{rT} \mathbf{P}_{n+1}^r - \mathbf{f}_{n+1} = 0 \\ &\mathbf{M}_{n+1}^r \mathbf{F}_{n+1}^r - \mathbf{L}_{n+1}^r \mathbf{x}_{n+1} = 0 \\ &\hat{\mathbf{P}}^r(\mathbf{F}_{n+1}^r; \boldsymbol{\Lambda}_n^r, \Delta t) - \mathbf{M}_{n+1}^r \mathbf{P}_{n+1}^r = 0 \end{aligned} \tag{41}$$

where one writes

$$
\begin{aligned}
L^r_{iI\alpha\, jb} &= \int_{\Omega^r_0} \delta_{ij} N_{b,I} M^r_\alpha dV_0 \\
f_{ia} &= \sum_r \left[\int_{\Omega^r_0} \rho_0 B_i N_a dV_0 + \int_{\partial\Omega^r_{0\tau}} \bar{T}_i N_a dS_0 \right] \\
M^r_{iI\alpha\, jJ\beta} &= \int_{\Omega^r_0} \delta_{ij}\delta_{IJ} M^r_\alpha M^r_\beta dV_0 \\
\hat{P}^r_{iI\alpha}(\mathbf{F}^r; \mathbf{\Lambda}^r_n, \Delta t) &= \int_{\Omega^e_0} \hat{P}_{iI}(\mathbf{F}^r; \mathbf{\Lambda}^r_n, \Delta t) M^r_\alpha dV_0
\end{aligned}
\tag{42}
$$

In writing (41) in matrix form, the groups of indeces of the type $\{ia\}$ and $\{iI\alpha\}$ have been redefined as single indeces in the usual way. Thus, $L^r_{jJ\beta\, ia}$ and $M^r_{iI\alpha\, jJ\beta}$ are mapped into matrices $\mathbf{L}^r$ and $\mathbf{M}^r$, respectively, and f_{ia}, $F^r_{iI\alpha}$ and $P^r_{iI\alpha}$ into linear arrays denoted $\mathbf{f}$, $\mathbf{F}^r$ and $\mathbf{P}^r$, respectively. The superindex r in the latter arrays refers to element r in the mesh and differentiates them from the symbols $\mathbf{F}$ and $\mathbf{P}$ employed for the deformation gradients and the stress field, respectively. The subindeces n and $n+1$ in (41) refer to the time at which the various arrays are sampled. Arrays such as $\mathbf{L}^r$ vary from t_n to t_{n+1} as a consequence of the change of mesh from $\mathcal{M}_n$ to $\mathcal{M}_{n+1}$. It should be carefully noted that in (41-c), the initial state $\mathbf{\Lambda}^r_n$ is interpolated on mesh $\mathcal{M}_n$, whereas all the remaining variables are interpolated on mesh $\mathcal{M}_{n+1}$.

Eqs. (41-b) and (41-c) are decoupled on an element by element basis. These equations can be simplified further by recourse to a lumping scheme. To this end, we start by computing all integrals in (42) by numerical quadrature based on element quadrature points and weights $\{\boldsymbol{\xi}^r_\alpha\}$ and $\{w^r_\alpha\}$, respectively. Next, we normalize the element shape functions M^r_α to satisfy

$$
M^r_\alpha(\boldsymbol{\xi}^r_\beta) = \delta_{\alpha\beta} \tag{43}
$$

Then, the array $\mathbf{\Lambda}^r_\alpha$ in (40-b) represents the value of the state variables $\mathbf{\Lambda}$ at quadrature point α in element r. Furthermore, eqs. (41-b) and (41-c) decouple on a quadrature point by quadrature point basis to become

$$
\begin{aligned}
\mathbf{F}^r_{n+1} &= \mathbf{L}^r_{n+1}\mathbf{x}_{n+1} \\
\mathbf{P}^r_{n+1} &= \hat{\mathbf{P}}^r(\mathbf{F}^r_{n+1}; \mathrm{T}\mathbf{\Lambda}^r_n, \Delta t)
\end{aligned}
\tag{44}
$$

where the lumped $\mathbf{L}^r$-array is redefined as

$$
L^r_{iI\alpha jb} = \delta_{ij} N_{b,I}(\boldsymbol{\xi}^r_\alpha) \tag{45}
$$

and the state 'transfer operator' T between meshes $\mathcal{M}_n$ and $\mathcal{M}_{n+1}$ is defined as

$$
\left(\mathrm{T}\mathbf{\Lambda}^r_n\right)_\alpha = \sum_\beta \mathbf{\Lambda}^r_{\beta,n} M^r_{\beta,n}(\boldsymbol{\xi}^r_{\alpha,n+1}) \tag{46}
$$

Eq. (44-a) simply states that the deformation gradients at the quadrature points are computed by direct differentiation of the displacement field. Likewise, in the case of an elastic solid for which the dependence of $\mathbf{P}^r_{n+1}$ on the initial conditions $(\mathbf{\Lambda}^r_n, \Delta t)$ is lost, eq. (44-b) simply states that the stresses at the quadrature points are computed directly from the corresponding deformation gradients by means of the stress-strain relations. By contrast, in the case of a history-dependent solid eqs. (45) and (46) convey non-trivial information and constitute the central result of this section. Thus, eqs. (45) and (46) provide a precise rule for updating the state variables in the presence of mesh refinement. The content of eqs. (45) and (46) may be stated by saying that, to effect a state-update between two different meshes, the initial state fields $\mathbf{\Lambda}_n(\mathbf{X}) = \sum_\beta \mathbf{\Lambda}^r_{\beta,n} M^r_{\beta,n}(\mathbf{X})$ must first be 'transferred' to the new mesh by sampling them at the new quadrature points $\boldsymbol{\xi}^r_{\alpha,n+1}$. This defines the transfer operator T. Then, a regular state-update is effected at the quadrature points of the new mesh. The mesh transfer operator T for displacements can be arbitrarily defined as

$$\left(\mathrm{T}\mathbf{x}_n\right)_{ia} = \sum_b x_{ib,n} N_{b,n}(\mathbf{X}_{a,n+1}) \tag{47}$$

Thus, to obtain the nodal displacements of $\mathrm{T}\mathbf{x}_n$ in $\mathcal{M}_{n+1}$, one simply samples the interpolated deformation mapping $\boldsymbol{\phi}_n(\mathbf{X}) = \sum_a \mathbf{x}_n N_{a,n}(\mathbf{X})$ on $\mathcal{M}_n$ at the nodal points of $\mathcal{M}_{n+1}$. The precise nature of this operation is not critical since it is only used to define an initial guess for the equilibrium iteration. Possible ways in which the above stress updates can be integrated into a Newton-Raphson iterations are discussed in [2]. In general, the quality of the transferred state fields depends critically on the accuracy of the state variable interpolation within the elements. If, as commonly happens when four-node isoparametric elementsare employed, extrapolation from the quadrature points defines a jagged variation of the state fields across elements, the state update picks up the noise and the solution becomes corrupted. We have based all our calculations on six-noded triangular elements with three quadrature points, resulting in a linear representation of the state variables within the element. The fact that the displacements are quadratic and the strains are linear within an element seems to minimize the amount of noise in the state variables. In addition, triangular meshes can be conveniently constructed by a triangulation algorithm. These desirable properties render the six-noded element particularly attractive for adaptive meshing calculations involving history-dependent solids.

3.2. Application to the analysis of growing cracks.

As an application of the methods outlined in the preceding sections, consider the problem of a crack growing through a ductile solid under the action of remote loads. Assume that small-scale yielding conditions prevail, so that three well-differentiated length scales can be identified: a macroscopic length scale commensurate with the overall geometrical dimensions of the solid, including the size of the crack; an intermediate region surrounding the crack tip wherein a linear elastic K-field dominates; and an innermost region where large plastic deformations take place. The objective of the calculations is to characterize the deformation fields which develop in the near-tip region as the crack begins to propagate through the solid. Under these conditions, the crack can be idealized as being semi-infinite in extent and as growing under the action of a remotely applied K-field. We let the crack lie within the negative X_1-axis, with its tip at the origin, Fig. 1, and assume that the loads are symmetric about the X_1-axis and that the crack grows within its plane.

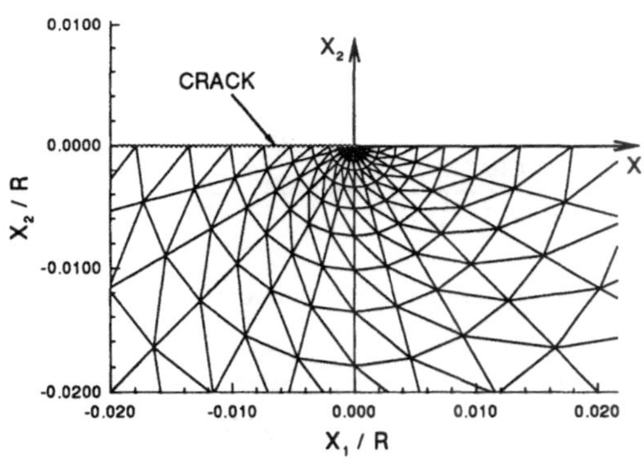

Fig. 1. Finite element mesh near the crack tip.

Because the near-tip fields of interest are often of a singular nature, and because their angular variations are a sought outcome of the analysis, a convenient mesh geometry takes the form of a fan-like grid centered at the tip with radial grading spanning several orders of magnitude in spatial resolution, Fig. 1. If the crack is simply allowed to run through a stationary mesh of this type, however, the mesh soon becomes out of focus and the fine resolution of the near-tip fields is lost. Evidently, a means of avoiding this difficulty is to *translate* the whole mesh so as to follow the crack tip. Conceptually, this shifting operation can be broken down into the steps shown in Fig. 2. Assume that the solution is known for some configuration of the crack, Fig. 2a. Imagine that the crack tip now advances by a distance Δa, resulting in the configuration of Fig. 2b. Evidently, now the solid is not in equilibrium. Before engaging in an equilibrium iteration, we wish to refocus the mesh about the new tip. To accomplish this, we introduce a *new* mesh which is shifted from the preceding one by Δa in the positive X_1 direction, and we extrapolate the state fields to the new mesh by means of a suitable state-transfer operator. We also shift the remotely applied K-field so as to be centered on the new tip. After re-focusing the mesh and the loads, we proceed to equilibrate the solid. The net result of these operations is the equilibrium configuration of the solid containing a slightly longer crack.

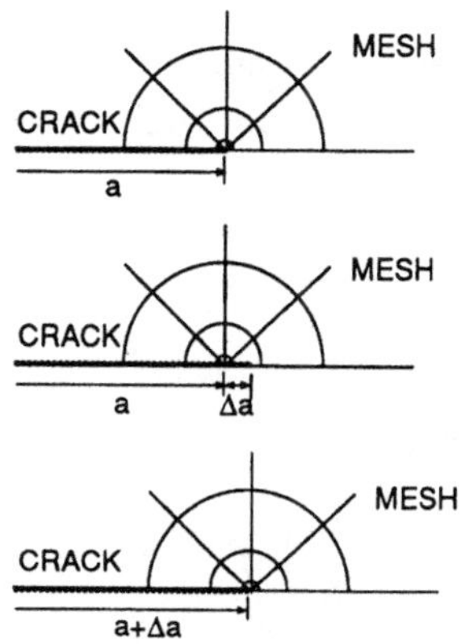

Fig. 2. Schematic representation of the crack advance algorithm.

It bears emphasis that, because successive meshes are geometrically identical, the analysis, save for the repeated shifting of the state fields, proceeds as if on a fixed mesh. In particular, the number of nodes and elements, as well as the nodal coordinates and element connectivity, remain unchanged throughout the calculations. Consequently, the only remeshing capability necessitated by the method is the state-transfer operator.

3.3. Numerical example.

We consider the case of a ductile solid obeying J_2-flow theory of plasticity with power-law hardening and rate-sensitivity. A detailed formulation of the model is given in [1]. The remote K-field is parametrized by the applied stress-intensity factor K or, alternatively, by the remote value of the J-integral, namely, $J = K^2(1-\nu^2)/E$, where E and ν are the Young's modulus and Poisson's ratio of the solid, respectively. The crack is first loaded monotonically up to the critical value K_c (respectively, J_c) of the applied stress intensity factor (respectively, applied J) for crack growth initiation. During this stage of the loading the crack blunts visibly as a result of the finite plastic deformations which accumulate in the vicinity of the tip. Subsequently, crack growth is governed by the J-resistance curve of the solid $J_R(\Delta a)$ [19]. Experimental measurements [19,20] and numerical simulations [20,21] suggest that, during extended crack growth, the crack-tip opening angle (CTOA) remains nearly constant while the remote J increases linearly with Δa.

The finite element mesh employed in the calculations is shown in Fig. 1. The analysis can be reduced to the lower half plane $X_2 \leq 0$ by virtue of the symmetries of the problem. The remote boundary is semi-circular of radius R. The mesh comprises 1039 nodal points and 492 six-noded triangular elements. The smallest element size is of the order of $10^{-4}R$. The material parameters adopted in the calculations are $E/\sigma_0 = 200$, $\nu = 0.3$, $n = 10$, $m = 200$, $\epsilon_0 = \sigma_0/E = 1/200$ and $\dot{\epsilon}_0 = 1$, where n is the hardening exponent, m is the rate sensitivity exponent, ϵ_0 is a reference strain, and $\dot{\epsilon}_0$ is a reference strain rate. Displacements corresponding to a plane-strain K-field are prescribed on the remote boundary. During the blunting phase, the value of the applied J is increased up to a critical value $J_c/\sigma_0 R = 0.00455$ for crack growth initiation at the constant rate $\dot{J}/\dot{\epsilon}_0\sigma_0 R = 0.00455$ in 20 increments. Subsequently, the crack is assumed to grow at a constant rate $\dot{a}/\dot{\epsilon}_0 R = 0.1$ and J to increase linearly with a according to the relation $\dot{J}/\dot{\epsilon}_0\sigma_0 R = \dot{a}/\dot{\epsilon}_0 R$. The length of the crack is increased by $0.01R$ in 20 steps.

The calculations are based on the state-update method discussed in Section 2. During the blunting phase, the stress paths at all material points are roughly proportional and the global Newton-Raphson iteration typically converges in 5 iterations or less to within a relative error of 10^{-9} in the residual. Convergence is asymptotically quadratic. The stress-update procedure handles deformation increments of the order of several times the yield strain quite effectively. During crack growth, the local stress paths become highly non-proportional and the Newton-Raphson iteration has to be combined with line searches to facilitate convergence. The average number of global equilibrium iterations then rises to 10, with an asymptotically quadratic convergence rate still maintained.

The state-transfer between shifted meshes is effected by the transfer operator defined in Section 3.1. The following variant of the implementation of the transfer operator was found useful as a means of ensuring that the transferred state variables be compatible with the constitutive relations. The deformation gradients $\mathbf{F}$, their elastic part $\mathbf{F}^e$, and the internal variables $\bar{\mathbf{Q}}$ are transferred according to (46). The plastic part of the deformations gradients is then computed as $\mathbf{F}^p = \mathbf{F}^{e-1}\mathbf{F}$, which ensures that the relation (18) is satisfied after the transfer. It should also be carefully noted that we transfer the deformation gradients $\mathbf{F}$ directly, as opposed to computing them by differentiation of the deformation mapping transferred by (47). This second method is not desirable since differentiation greatly amplifies the errors incurred in transferring the nodal displacements.

Fig. 3 displays the computed relation between CTOA and crack advance. Also included for reference is the variation in applied J. In this figure, the extent of crack advance is normalized by $J_c E/(1-\nu^2)\sigma_0^2$, which is of the order of the plastic zone size. Prior to crack growth initiation, the CTOA is a monotonically increasing function of the applied

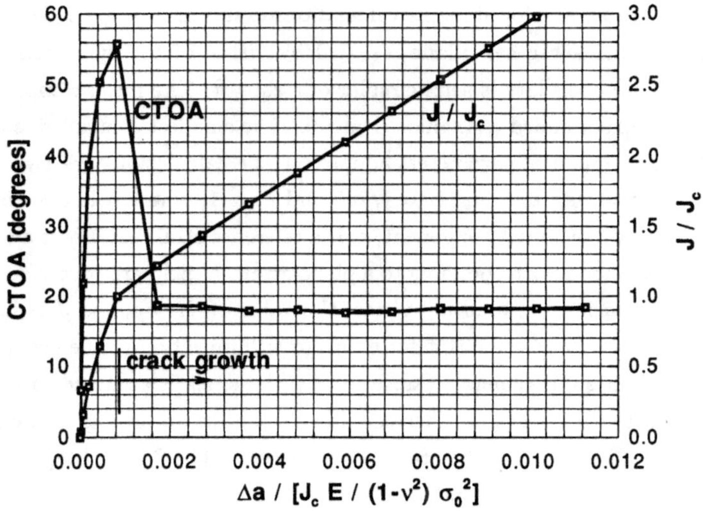

Fig. 3. Variation of crack-tip opening angle (CTOA) and applied J with crack advance.

J. The onset of crack growth is clearly marked by an sharp reduction in the CTOA. For an applied J increasing linearly with Δa, the CTOA is observed to drop to a constant level during subsequent extension. Thus, the crack advances at constant CTOA for constant $\dot{J}/\dot{a}$. The constant value of the CTOA maintained during growth is a function of the rate $\dot{J}/\dot{a}$ at which J is increased with crack advance. This relation can be deduced by repeating the present calculations for different values of $\dot{J}/\dot{a}$.

Fig. 4 shows the crack opening profile at various stages of deformation and growth. As expected, the crack initially blunts under increasingly load. Following crack growth initiation a rather abrupt transition to a V-shaped opening profile is observed. The constancy of the CTOA with crack advance is clearly apparent in this figure. A typical deformed mesh after some crack extension is shown in Fig. 5. It bears emphasis that the mesh remains constantly focused about the tip throughout the calculations.

Finally, the evolution of the near-tip distribution of effective plastic strain with crack advance is shown in Fig. 6. Each plot depicts isocontours corresponding to a fixed level of effective plastic strain at every other of the 20 steps taken to propagate the crack. A general expansion of the plastic zone with increasing applied load is evident from the contours corresponding to the smaller values of effective plastic strain, which lie on the outer reaches of the plastic zone. Contours for higher values of effective plastic strain represent conditions prevailing closer to the tip. From these latter contours a contraction is observed in the plastic zone ahead of the tip under the conditions of

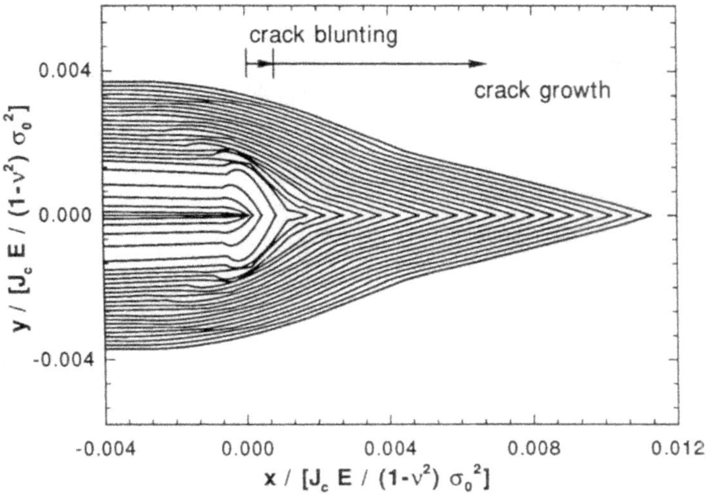

Fig. 4. Variation of crack opening profile with crack advance.

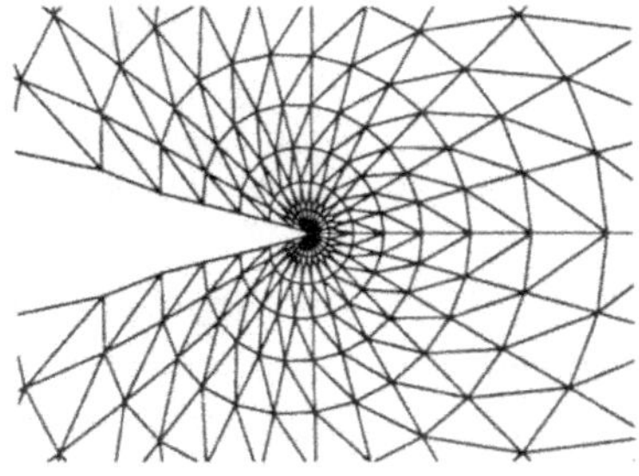

Fig. 5. Typical deformed mesh near the crack tip.

the analysis. The early stages of development of a wake in the distribution of effective plastic strain can also be discerned.

3.4. Discussion

A finite element method has been developed which enables a mesh of a fixed geometry to be kept continuously focused on the tip of a semi-infinite crack growing within its plane. This is accomplished by translating the mesh so as to follow the crack tip advance. The method requires the implementation of a state-transfer operator to extrapolate the state variables between shifted meshes, but the calculations otherwise proceed as if on a fixed

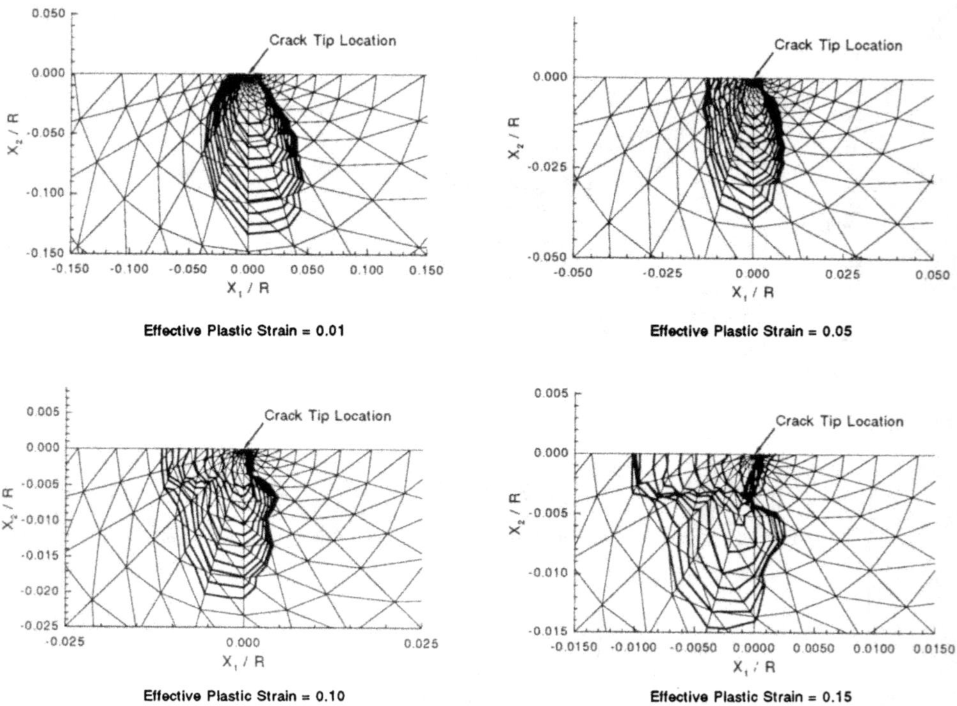

Fig. 6. Variation of isocontours of effective plastic strain with crack advance.

mesh. Our numerical examples demonstrate the ability of the method to accurately resolve the near-tip fields as they evolve with crack advance. While the method, in its present form, is restricted to planar semi-infinite cracks, it can be applied under conditions of dynamic propagation. In this case, the appropriate remote field to be prescribed is the known linear elastic solution for a dynamically growing semi-infinite crack [22,23].

Acknowledgements

The support of the U. S. Office of Naval Research through grant N00014-90-J-1758 is gratefully acknowledged.

References.

1. Cuitiño, A. M., and Ortiz, M., A material-independent method for extending stress update algorithms form small-strain plasticity to finite plasticity with multiplicative kinematics, submitted for publication.
2. Ortiz, M., and Quigley, J. J., IV, Adaptive mesh refinement in strain localization problems, *Comput. Methods Appl. Mech. Engrg.*, **90** (1991) 781-804.

3. Hughes, T. J. R., and Taylor, R. L., Unconditionally stable algorithm for quasi-static elasto/visco-plastic finite element analysis, *Comp. Struct.*, **8** (1978) 169-173.
4. Simó, J. C., and Taylor, R. L., Consistent tangent operators for rate independent elasto-plasticity, *Comput. Methods. Appl. Mech. Engrg.*, **48** (1985) 101-118.
5. Aravas, N., On the numerical integration of a class of pressure-dependent plasticity models, *Int. J. Numer. Methods. Eng.*, **24** (1987) 1395-1416.
6. Ortiz, M., and Popov E. P., Accuracy and stability of integration algorithms for elastoplastic constitutive relations, *Int. J. Numer. Methods. Eng.*, **21** (1985) 1561-1576.
7. Ortiz, M., and Martin, J. B., Symmetry-preserving return mapping algorithms and incremental extremal paths: A unification of concepts, *Int. J. Numer. Methods. Eng.*, **28** (1989) 1839-1853.
8. Wilkins, M. L., Calculation of elastoplastic flow, *Methods of Computational Physics*, **3**, Alder, B., *et al.*, (eds.), Academic Press (1964).
9. Krieg, R. D., and Key, S. W., Implementation of a time dependent plasticity theory into structural computer programs, *Constitutive Equations in Viscoplasticity: Computational and Engineering Aspects*, Stricklin, J. A., and Saczlski, K. J., (eds.), ADM **20**, ASME (1976).
10. Krieg, R. D., and Krieg, D. B., Accuracies of numerical solution methods for the elasto-perfectly plastic models, *J. Pressure Vessel Tech., ASME*, **99** (1977) 510-515.
11. Schreyer, H. L., Kulak, R. L., and Kramer, J. M., Accurate numerical solutions for elastoplastic models, *J. Pressure Vessel Tech., ASME*, **101** (1979) 226-234.
12. Nagtegaal, J. C., On the implementation of inelastic constitutive equations with special reference to large deformation problems, *Comput. Methods. Appl. Mech. Engrg.*, **33** (1982) 469-484.
13. Ortiz, M., Pinsky, P. M., and Taylor, R. L., Operator split methods for the numerical solution of the elastoplastic dynamic problem, *Comput. Methods. Appl. Mech. Engrg.*, **39** (1983) 137-157.
14. Pinsky, P. M., Ortiz, M., and Taylor, R. L., Operator split methods for the numerical solution of the finite deformation elastoplastic dynamic problem, *Comp. Struct.*, **13**, (1983) 345-359.
15. Simó, J. C., and Ortiz, M., A unified approach to finite deformation elastoplastic analysis based on the use of hyperelastic constitutive equations, *Comput. Methods. Appl. Mech. Engrg.*, **49** (1985) 221-245.
16. Moran, B., Ortiz, M., and Shih C. F., Formulation of implicit finite element methods for multiplicative finite deformation plasticity, *Int. J. Numer. Methods. Eng.*, **29** (1990) 483-514.

17. Simó, J. C., and Taylor, R. L., Quasi incompressible finite elasticity in principal stretches: Continuum basis and numerical algorithms, *Comput. Methods. Appl. Mech. Engrg.*, **85** (1991) 273-310.
18. Blume, J. A., private communication.
19. Kanninen, M. F. and Popelar, C. H. *Advanced Fracture Mechanics.* Oxford University Press (1985).
20. Shih, C. F., deLorenzi, H. G. and Andrews, W. R. Studies on crack initiation and stable crack growth, *Elastic-Plastic Fracture,* ASTM STP **668**, Landes, J. D., Begley, J. A. and Clarke G. A., (eds.), Amer. Soc. Testing Mater., (1979) 65-120.
21. de Konig, A. U., A contribution to the analysis of quasi-static crack growth fracture, Proc. 4th Intl. Conference on Fracture, Vol. 3A, Pergamon (1978) 25-31.
22. Freund, L. B., Crack propagation in an elastic solid subjected to general loading - I. Constant rate of extension, *J. Mech. Phys. Solids*, **20** (1972) 129-140.
23. Freund, L. B., Crack propagation in an elastic solid subjected to general loading - II. Nonuniform rate of extension, *J. Mech. Phys. Solids*, **20** (1972) 121-152.

Recent Advances in the Numerical Analysis and Simulation of Thermoplasticity at Finite Strains

J. C. SIMO & F. ARMERO

Division of Applied Mechanics,
Department of Mechanical Engineering,
Stanford University.

Abstract

Novel numerical analysis aspects involved in the algorithmic treatment of a recently proposed formulation of thermoplasticity at finite strains are discussed. Two key aspects in these new developments are: **i.** The formulation of a new class of return mapping algorithms for multiplicative decomposition which inherit without modification all the features of the standard algorithms restricted to the infinitesimal theory; **ii.** The development of a new class of staggered coupled algorithms for coupled problems which retain the key property of unconditional stability. The work reported on herein summarizes some of the recent developments at Stanford in the area of inelastic coupled thermomechanical problems at finite strains.

§1. Introduction.

In a recent paper, SIMO & MIEHE [1991] have presented a formulation of multiplicative thermoplasticity at finite strains characterized by two distinctive features: **i.** First, the model explicitly identifies that part of the configurational entropy arising at the micro-level as a result of dislocation and lattice defect motion by means of an internal variable called the *plastic entropy*. **ii.** Second, for associative plasticity the assumption of maximum dissipation uniquely defines the flow rule and specifies the evolution of the plastic entropy in terms of the derivative of the flow criterion with respect to the temperature field. In Section **2** we summarize the salient features of this approach and review those aspects needed for our subsequent numerical analysis of the thermomechanically coupled nonlinear initial boundary value problem.

In Section **3** we address the integration of the constrained local problem of evolution defined by the flow rule. In a finite element context this problem is typically solved at the quadrature points for prescribed strain increments by means of the so-called local return mapping algorithms. These schemes go back to the work of MOREAU [1975] and have been extensively studied within the context of the infinitesimal theory; see e.g. SIMO, KENNEDY & TAYLOR [1988] and SIMO [1991a]. Remarkably, within the framework described in Section **3**, the classical return mapping schemes of the infinitesimal theory carry over to the finite strain regime *without modification* but with an additional simplification: The

[1] Invited lecture presented by the first author at the IUTAM symposium in Hannover on large inelastic deformations. This paper is dedicated to the late Professor Th. Lehmann whose many contributions have had a lasting impact in the field.

D. Besdo, E. Stein (Eds.)
Finite Inelastic Deformations -
Theory and Applications
IUTAM Symposium Hannover/Germany 1991

return mapping is formulated in fixed principal (Eulerian) axis as defined by the trial elastic step. A key aspect of this approach is the use of an exponential approximation to the flow rule which results in algorithms that exactly preserve plastic volume for pressure insensitive yield criteria.

In Section **4** we discuss the application to coupled thermoplasticity of a new class of staggered algorithms proposed in ARMERO & SIMO [1991] which retain the key property of *unconditional stability*. The design of these methods exploits the observation that staggered algorithms for coupled problems can be viewed as fractional steps methods, in the sense of YANENKO [1971], arising from an operator split of the coupled problem of evolution. This view is advocated in SIMO & MIEHE [1991] and, somewhat surprisingly, has not been systematically exploited in the literature on coupled problems. The key result is a split of the coupled problem into a *mechanical phase* at constant entropy, followed by a heat conduction phase at fixed configuration. This split preserves the formal contractivity property of the thermoplastic problem (relative to certain a priori stability estimates) and leads naturally to the formulation of unconditionally B-stable coupled algorithms. We conclude this overview of the recent developments with two representative simulations which, in addition, illustrate the good performance in localization problems exhibited by a new class of nonlinear mixed finite element methods recently proposed in SIMO & ARMERO [1991].

§2. Associative Thermoplasticity at Finite Strains.

Our interest is a formulation of thermoplasticity at finite strains in which the thermoelastic domain E possesses the same functional form as in the infinitesimal theory, now defined in the current configuration in terms of true stresses. Following SIMO & MIEHE [1991], we outline the basic structure of a theory in which the evolution of plastic flow obeys the thermomechanical version of the principle of maximum dissipation.

2.1. Basic Assumptions. Free Energy and Plastic Entropy.

i. *Thermoelastic Domain.* Consider for simplicity a *smooth* yield criterion $\phi(\boldsymbol{\tau}, q, \Theta)$, assumed to be convex, depending on the Kirchhoff stress tensor $\boldsymbol{\tau}$, a single (stress-like) scalar hardening variable q and the absolute temperature Θ. The thermoelastic domain E is the *convex* set

$$\mathsf{E} := \left\{ (\boldsymbol{\tau}, q, \Theta) \in \mathsf{S} \times \mathsf{R} \times \mathsf{R}_{+} : \quad \phi(\boldsymbol{\tau}, q, \Theta) \leq 0 \right\}. \tag{2.1}$$

We remark that the principle of objectivity restricts the possible forms of ϕ in (2.1) to *isotropic functions.* Consequently, a classical formulation of the yield criterion in terms of true stresses necessarily imposes the restriction to isotropy of the resulting theory.

ii. *Multiplicative decomposition.* The generalization to finite strains of the additive decomposition used in the infinitesimal theory is motivated by the structure of the single crystal models of metal plasticity, and takes the form of the local multiplicative decomposition

$$\boldsymbol{F}(X,t) = \boldsymbol{F}^{e}(X,t)\boldsymbol{F}^{p}(X,t) \quad \forall (X,t) \in \Omega \times \mathsf{I}, \tag{2.2}$$

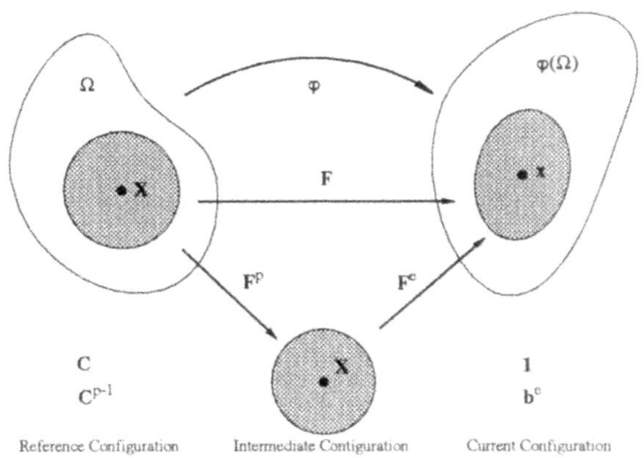

FIGURE 2.1. Illustration of the local multiplicative decomposition of the deformation gradient, and the basic strain tensors.

where $\boldsymbol{F}(X,t) := D\varphi(X,t)$ denotes the deformation gradient of the motion $\varphi(\cdot,t)$ at $X \in \Omega$. From a micromechanical point of view $\boldsymbol{F}^p$ is an internal variable related to the amount of dislocation flow through the crystal lattice. From a phenomenological standpoint $\boldsymbol{F}^{e-1}$ defines the local, stress-free, unloaded configuration.

iii. *Free energy and plastic entropy.* The elastic response of the material is characterized by a free energy ψ assumed to depend locally on the deformation gradient through the elastic part $\boldsymbol{F}^e$. From a micromechanical standpoint, $\boldsymbol{F}^e$ gives the distortion of the crystal lattice. Consistent with the restriction to isotropy it is further assumed that ψ is independent of the orientation of the intermediate configuration. Frame invariance then implies the functional form

$$\psi = \tilde{\psi}(\boldsymbol{b}^e, \xi, \Theta), \quad \text{where} \quad \boldsymbol{b}^e := \boldsymbol{F}^e \boldsymbol{F}^{e\,T}. \tag{2.3}$$

where ξ is the internal variable conjugate to q and $\boldsymbol{b}^e$ is the elastic left Cauchy-Green tensor. Our final assumption is the *additive* decomposition $\eta = \eta^e + \eta^p$ of the total entropy η of the thermoplastic body into and elastic part η^e and a plastic (*configurational*) part η^p associated with the motion of dislocations and defects through the crystal lattice. Equivalently, in terms of the Legendre transformation,

$$\tilde{e}(\boldsymbol{b}^e, \xi, \eta^e) = \tilde{\psi}(\boldsymbol{b}^e, \xi, \Theta) + \eta^e \Theta \quad \text{where} \quad \eta^e := \eta - \eta^p. \tag{2.4}$$

It remains to specify *local* constitutive equations and evolution equations for the internal variables describing plastic flow; in particular, for the configurational entropy η^p.

2.2. Dissipation Inequality and Associative Evolution Equations.

The local dissipation function $\mathcal{D}$ per unit of reference volume, associated with a material point $X \in \Omega$, is defined as the difference between the total power (i.e., the stress power plus the thermal power) and the local rate of change of the internal energy. Assuming the relation $\boldsymbol{q} \cdot \nabla \Theta \geq 0$, where $\boldsymbol{q}$ is the (spatial) Kirchhoff heat flux and $\nabla \Theta$ denotes the gradient relative to spatial coordinates, the Clausius-Duhem form of the second law yields

$$\mathcal{D}:= \boldsymbol{\tau} \cdot \boldsymbol{d} + \Theta\, \dot{\eta} - \frac{d}{dt}\tilde{e}(\boldsymbol{b}^e, \xi, \eta^e) \geq 0 \quad \forall t \in \mathsf{I}, \tag{2.5}$$

where $\boldsymbol{d} = \text{sym}[\boldsymbol{l}]$ is the rate of deformation tensor and $\boldsymbol{l} := \dot{\boldsymbol{F}}\boldsymbol{F}^{-1}$ is the spatial velocity gradient.

i. *Reduced dissipation inequality: Constitutive equations.* The rate of change $\frac{d}{dt}\tilde{e}$ is computed by exploiting the identity $\boldsymbol{b}^e = \boldsymbol{F}\boldsymbol{C}^{p-1}\boldsymbol{F}^T$, which follows from the multiplicative decomposition (2.2), and identifies $\boldsymbol{C}^p := \boldsymbol{F}^{pT}\boldsymbol{F}^p$ as the plastic right Cauchy-Green tensor. Time differentiation of this expression gives

$$\dot{\boldsymbol{b}}^e = \boldsymbol{l}\boldsymbol{b}^e + \boldsymbol{b}^e\boldsymbol{l}^T + \boldsymbol{\mathcal{L}}_{\boldsymbol{v}}\boldsymbol{b}^e \quad \text{with} \quad \boldsymbol{\mathcal{L}}_{\boldsymbol{v}}\boldsymbol{b}^e := \boldsymbol{F}\frac{\partial}{\partial t}[\boldsymbol{C}^{p-1}]\boldsymbol{F}^T, \tag{2.6}$$

where $\boldsymbol{\mathcal{L}}_{\boldsymbol{v}}\boldsymbol{b}^e$ is referred to as the *Lie derivative* of $\boldsymbol{b}^e$. Time differentiation of the internal energy function, use of the relation $(2.6)_1$ along with the fact $\boldsymbol{b}^e$ and $\partial_{\boldsymbol{b}^e}\tilde{e}$ commute (as a result of elastic isotropy), gives the following reduced form of inequality (2.5):

$$\begin{aligned} \mathcal{D} &= \left[\boldsymbol{\tau} - 2\partial_{\boldsymbol{b}^e}\tilde{e}\,\boldsymbol{b}^e\right] \cdot \boldsymbol{d} + \left[\Theta - \partial_{\eta^e}\tilde{e}\right] \cdot \dot{\eta} \\ &\quad + \Big[\left[2\partial_{\boldsymbol{b}^e}\tilde{e}\,\boldsymbol{b}^e\right] \cdot \left[-\tfrac{1}{2}(\boldsymbol{\mathcal{L}}_{\boldsymbol{v}}\boldsymbol{b}^e)\boldsymbol{b}^{e-1}\right] + \left[-\partial_{\xi}\tilde{e}\right]\dot{\xi} + \Theta\,\dot{\eta}^p\Big] \geq 0. \end{aligned} \tag{2.7}$$

Since (2.7) must hold for all admissible processes, a standard argument then gives the following constitutive equations:

$$\boxed{\boldsymbol{\tau} = 2\left[\partial_{\boldsymbol{b}^e}\tilde{e}\right]\boldsymbol{b}^e = 2\left[\partial_{\boldsymbol{b}^e}\tilde{\psi}\right]\boldsymbol{b}^e \quad \text{and} \quad \Theta = \partial_{\eta^e}\tilde{e} \iff \eta^e = -\partial_{\Theta}\tilde{\psi},} \tag{2.8}$$

along with the relation $q := -\partial_{\xi}\psi$. Inserting these relations into (2.7) gives the final reduced form of the dissipation inequality:

$$\mathcal{D} = \Theta\dot{\eta}^p + \mathcal{D}_{\text{mech}} \geq 0 \quad \text{where} \quad \mathcal{D}_{\text{mech}} := \boldsymbol{\tau} \cdot \left[-\tfrac{1}{2}(\boldsymbol{\mathcal{L}}_{\boldsymbol{v}}\boldsymbol{b}^e)\boldsymbol{b}^{e-1}\right] + q\,\dot{\xi}. \tag{2.9}$$

Observe that $\boldsymbol{\tau}$ and $\boldsymbol{b}^e$ commute as a result of the restriction to isotropy.

ii. *Associative Model.* The simplest model of plastic flow corresponds to the assumption of *maximum dissipation.* Accordingly, the actual state $(\boldsymbol{\tau}, q, \Theta) \in \mathsf{E}$ in the plastically deformed body at a given (fixed) configuration, with prescribed intermediate configuration and prescribed rates $\{\boldsymbol{\mathcal{L}}_{\boldsymbol{v}}\boldsymbol{b}^e, \dot{\xi}\}$, renders a maximum of the dissipation function $\mathcal{D}$. Thus, the inequality

$$[\boldsymbol{\tau} - \overset{*}{\boldsymbol{\tau}}] \cdot \left[-\tfrac{1}{2}(\boldsymbol{\mathcal{L}}_{\boldsymbol{v}}\boldsymbol{b}^e)\boldsymbol{b}^{e-1}\right] + [q - \overset{*}{q}]\dot{\xi} + [\Theta - \overset{*}{\Theta}]\dot{\eta}^p \geq 0, \tag{2.10}$$

holds for all *admissible* $(\overset{*}{\boldsymbol{\tau}}, \overset{*}{q}, \overset{*}{\Theta}) \in \mathsf{E}$. A standard result in convex analysis then shows that inequality (2.10) holds if and only if the coefficients $\{-\frac{1}{2}(\boldsymbol{\mathcal{L}}_{\boldsymbol{v}}\boldsymbol{b}^e)\boldsymbol{b}^{e-1}, \dot{\xi}, \dot{\eta}^p\}$ lie in the cone normal to $\partial\mathsf{E}$ at the point $(\boldsymbol{\tau}, q)$. In particular, if $\partial\mathsf{E}$ is defined by a smooth yield function as in (2.1), the evolution equations read

$$\boxed{\begin{array}{c}\left.\begin{aligned} -\tfrac{1}{2}\,\pounds_{\boldsymbol{v}}\boldsymbol{b}^e &= \gamma\,[\,\partial_{\boldsymbol{\tau}}\phi(\boldsymbol{\tau}, q, \Theta)\,]\,\boldsymbol{b}^e, \\ \dot{\xi} &= \gamma\,\partial_q\phi(\boldsymbol{\tau}, q, \Theta), \\ \dot{\eta}^p &= \gamma\,\partial_\Theta\phi(\boldsymbol{\tau}, q, \Theta), \end{aligned}\right\} \\ \gamma \geq 0, \quad \phi(\boldsymbol{\tau}, q, \Theta) \leq 0 \quad \text{and} \quad \gamma\,\phi(\boldsymbol{\tau}, q) = 0. \end{array}} \tag{2.11}$$

Despite its unconventional appearance the flow rule (2.11) possesses a number of interesting properties. In particular, it gives the correct evolution of plastic volume changes as the following observations show.

a. The total and elastic volume changes are given by $J := \det[\boldsymbol{F}] > 0$ and $J^e := (\det[\boldsymbol{b}^e])^{\frac{1}{2}} > 0$, respectively.

b. Let $J^p := \det[\boldsymbol{F}^p]$. The rate of plastic volume change predicted by the flow rule $(2.11)_1$ is given by the evolution equation

$$d(\log J^p)/dt = \gamma\,\mathrm{tr}\left[\partial_{\boldsymbol{\tau}}\phi\right], \tag{2.12}$$

which implies exact conservation of plastic volume for pressure insensitive yield conditions; i.e. $\dot{J}^p = 0 \iff \mathrm{tr}\left[\partial_{\boldsymbol{\tau}}\phi\right] = 0$.

Within the preceding thermomechanical framework, the simplest case corresponding to J_2–flow theory is obtained when ϕ is the mises yield criterion. Observe that objective rates or hypoelastic constitutive equation play no role in this development.

§3. Algorithmic Approximation to the Flow Rule.

Following SIMO [1991], we construct below an algorithmic approximation consistent with the associative flow rule derived above which is exact for incrementally elastic processes, independent of the specific form adopted for the stored energy function and exact plastic volume preserving for pressure insensitive plasticity models. Remarkably, the implementation of this algorithm takes a form essentially identical to that of the infinitesimal theory.

3.1. The Local Problem of Evolution. The Algorithmic Flow Rule.

Consider a typical time sub-interval $[t_n, t_{n+1}]$, an arbitrary material point $X \in \mathcal{B}$ and assume that the primary variables $\{\boldsymbol{F}_n, \Theta_n\}$ along with $\{\boldsymbol{b}^e_n, \xi_n, \eta^p_n\} \in \mathsf{S}_+ \times \mathsf{R}_+ \times \mathsf{R}$ which define the constitutive response of the material, are prescribed initial data at $X \in \mathcal{B}$. The objective is to construct an algorithmic approximation to the plastic flow evolution equations for incremental displacement and temperature fields $(\boldsymbol{u}_t, \vartheta)$ prescribed on $\varphi_n(\mathcal{B}) \times [t_n, t_{n+1}]$, where $\varphi_n(\cdot)$ is a give configuration at time t_n with temperature field $\Theta_n(\cdot)$ defined on $\mathcal{B}$. Denoting by $\boldsymbol{f}_t(\boldsymbol{x}_n) := \boldsymbol{I} + \nabla\boldsymbol{u}_t(\boldsymbol{x}_n)$ the relative deformation gradient at $\boldsymbol{x}_n = \varphi_n(X)$, the local evolution of plastic flow at is governed by the first order constrained system

$$\left.\begin{aligned} \dot{\boldsymbol{b}}_t^e &= \left[\boldsymbol{l}_t \boldsymbol{b}_t^e + \boldsymbol{b}_t^e \boldsymbol{l}_t^T\right] - 2\gamma_t \partial_{\boldsymbol{\tau}} \phi(\boldsymbol{\tau}_t, q_t, \Theta_t) \boldsymbol{b}_t^e, \\ \dot{\xi}_t &= \qquad\qquad \gamma_t \partial_q \phi(\boldsymbol{\tau}_t, q_t, \Theta_t), \\ \dot{\eta}_t^p &= \qquad\qquad \gamma_t \partial_\Theta \phi(\boldsymbol{\tau}_t, q_t, \Theta_t), \end{aligned}\right\} \tag{3.1}$$
$$\gamma_t \geq 0, \quad \phi(\boldsymbol{\tau}_t, q_t, \Theta_t) \leq 0 \quad \text{and} \quad \gamma_t \phi(\boldsymbol{\tau}_t, q_t, \Theta_t) = 0,$$

where $\boldsymbol{l}_t = \dot{\boldsymbol{f}}_t \boldsymbol{f}_t^{-1}$ is the spatial velocity gradient. The solution of (3.1) defines $\{\boldsymbol{b}_t^e, \xi_t, \eta_t^p\}$ at the point X within the interval of time $[t_n, t_{n+1}]$ for configuration and temperature prescribed by the relations $\varphi_t = \varphi_n + \boldsymbol{u}_t \circ \varphi_n$ and $\Theta_t = \Theta_n + \vartheta_t \circ \varphi_n$, respectively.

The key to the numerical integration (3.1) is the transformation of this system to a standard problem of evolution. To this end, define the following (pull-back) tensor fields on the configuration $\varphi_n(\mathcal{B})$:

$$\boldsymbol{b}_t^{e\,*} := \boldsymbol{f}_t^{-1} \boldsymbol{b}_t^e \boldsymbol{f}_t^{-T}, \quad \boldsymbol{c}_t := \boldsymbol{f}_t^T \boldsymbol{f} \quad \text{and} \quad \boldsymbol{n}_t := \boldsymbol{f}_t^T \partial_{\boldsymbol{\tau}} \phi(\boldsymbol{\tau}_t, q_t, \Theta_t) \boldsymbol{f}_t. \tag{3.2}$$

In geometric terms $\boldsymbol{n}_t$ is interpreted as the pull-back of the normal $\partial_{\boldsymbol{\tau}}\phi$ to the configuration $\varphi_n(\mathcal{B})$, with the tensor $\boldsymbol{c}_t = \boldsymbol{f}_t^T \boldsymbol{f}_t$ viewed as a Riemannian metric on $\varphi_n(\mathcal{B})$. A straightforward manipulation shows that the system $(3.1)_1$ is equivalent to

$$\frac{\partial}{\partial t} \boldsymbol{b}_t^{e\,*} = \left[-2\gamma \boldsymbol{c}_t^{-1} \boldsymbol{n}_t\right] \boldsymbol{b}_t^{e\,*}, \quad \text{with} \quad \boldsymbol{b}_t^{e\,*}\big|_{t=t_n} = \boldsymbol{b}_n^e. \tag{3.3}$$

The form of equation (3.3) suggests an exponential approximation within the interval $[t_n, t] \subset [t_n, t_{n+1}]$ with standard form $\tilde{\boldsymbol{b}}_t^{e\,*} = \exp\left[-2\Delta\gamma_t \boldsymbol{c}_t^{-1} \tilde{\boldsymbol{n}}_t\right] \boldsymbol{b}_n^e$, where $\Delta\gamma \geq 0$ is a Lagrange multiplier to be determined by enforcing consistency as described below, and a superposed $\tilde{}$ refers to the algorithmic approximation. Using $(3.2)_1$ this relation becomes

$$\tilde{\boldsymbol{b}}_t^e = \left(\boldsymbol{f}_t \exp\left[-2\Delta\gamma_t \boldsymbol{f}_t^{-1} \tilde{\boldsymbol{n}}_t\right] \boldsymbol{f}_t^{-1}\right) \boldsymbol{b}_t^{e\,\mathrm{tr}} \quad \text{where} \quad \boldsymbol{b}_t^{e\,\mathrm{tr}} = \boldsymbol{f}_t \boldsymbol{b}_n^e \boldsymbol{f}_t^T. \tag{3.4}$$

The well-known property $\boldsymbol{F} \exp[\boldsymbol{A}] \boldsymbol{F}^{-1} = \exp[\boldsymbol{F}\boldsymbol{A}\boldsymbol{F}^{-1}]$ finally yields the following algorithmic approximation to the flow rule:

$$\left.\begin{aligned} \tilde{\boldsymbol{b}}_t^e &= \exp\left[-2\Delta\gamma_t \partial_{\boldsymbol{\tau}} \phi(\tilde{\boldsymbol{\tau}}_t, \tilde{q}_t, \tilde{\Theta}_t)\right] \boldsymbol{b}_t^{e\,\mathrm{tr}}, \\ \tilde{\xi}_t &= \xi_n + \Delta\gamma_t \partial_q \phi(\tilde{\boldsymbol{\tau}}_t, \tilde{q}_t, \tilde{\Theta}_t), \\ \tilde{\eta}_t^p &= \eta_n^p + \Delta\gamma_t \partial_\Theta \phi(\tilde{\boldsymbol{\tau}}_t, \tilde{q}_t, \tilde{\Theta}_t) \end{aligned}\right\} \tag{3.5}$$
$$\Delta\gamma_t \geq 0, \quad \phi(\tilde{\boldsymbol{\tau}}_t, \tilde{q}_t, \tilde{\Theta}_t) \leq 0, \quad \Delta\gamma_t \phi(\tilde{\boldsymbol{\tau}}_t, \tilde{q}_t, \tilde{\Theta}_t) = 0,$$

where $(\tilde{\boldsymbol{\tau}}_t, \tilde{q}_t)$ are defined by the elastic stress-strain relations evaluated at $(\tilde{\boldsymbol{b}}_t^e, \tilde{\xi}_t, \tilde{\Theta}_t)$. A standard argument using a Taylor series expansion confirms that the product formula algorithm (3.5) is first order accurate. Unconditional stability follows from the properties

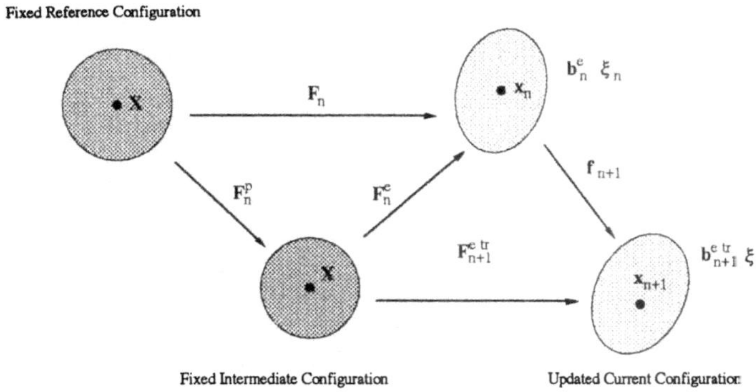

FIGURE 3.1. Interpretation of the trial state: The (local) reference and intermediate configurations remain fixed, while the current configuration is updated for prescribed relative deformation gradient.

of the backward Euler scheme and the exponential map. A geometric illustration of (3.5) is given in Figure 3.1.

The derivation of the algorithmic flow rule (3.5) is identical to the algorithmic treatment in SIMO [1988], with the backward–Euler scheme replaced here by an exponential approximation. Although both algorithms possess identical accuracy properties, the backward Euler method does not inherit the following property conservation properties which are identically satisfied by (3.5).

i. *Preservation of the constraint* $\det[\boldsymbol{b}^e_t] > 0$; a property which follows from the identity $\det\big[\exp[\boldsymbol{A}]\big] = \exp\big[\text{tr}[\boldsymbol{A}]\big]$, for any matrix $\boldsymbol{A} \in \mathsf{L}(3)$.

ii. *Exact conservation of the plastic volume.* For pressure insensitive yield criteria the plastic volume preserving property of the exact flow rule is preserved by the algorithmic approximation, since (3.5) implies $\text{tr}\big[\partial_{\boldsymbol{\tau}}\phi(\tilde{\boldsymbol{\tau}}_t, \tilde{q}_t, \tilde{\Theta}_t)\big] \equiv 0 \iff \tilde{J}^p_t = J^p_n$.

A modification of the backward Euler method which does inherit property **ii** above is given in SIMO & MIEHE [1991]. We remark that both the backward Euler and the exponential approximations become exact for incrementally elastic steps since $\Delta\gamma_t = 0$.

3.2. Closed–Form Implementation: Spectral Decomposition.

By isotropy the principal directions $\{\boldsymbol{n}^{(A)}_t\}_{A=1,2,3}$ of the Kirchhoff stress tensor $\boldsymbol{\tau}_t$ coincide with the principal directions of the elastic left Cauchy-Green tensor $\boldsymbol{b}^e_t$. Let $\{\beta_A\}$ and $\{\lambda^{e\,2}_A\}$ denote the respective set of principal values. The key to the implementation of the algorithm lies in the following observations:

i. Let $\boldsymbol{n}^{\text{tr}\,(A)}_t$ and $\lambda^{e\,\text{tr}}_A > 0$ denote the principal directions and the square root of the associated principal values of $\boldsymbol{b}^{e\,\text{tr}}_t$, respectively. Then the final principal directions *coincide* with the principal directions; i.e, $\boldsymbol{n}^{(A)}_t \equiv \boldsymbol{n}^{\text{tr}\,(A)}_t$ for $A = 1, 2, 3$.

ii. Let $\varepsilon^e_{t\,A} := \log[\lambda^e_{t\,A}]$ and $\varepsilon^{e\,\mathrm{tr}}_{t\,A} := \log[\lambda^{e\,\mathrm{tr}}_{t\,A}]$ denote the logarithmic stretches. Setting $\psi(\boldsymbol{b}^e, \xi, \Theta) = \hat{\psi}(\varepsilon^e_A, \xi, \Theta)$, the constitutive equations become

$$\beta_A = \partial_{\varepsilon^e_A}\hat{\psi}(\varepsilon^e_A, \xi_t, \Theta), \quad \eta^e = \partial_\Theta \hat{\psi}(\varepsilon^e_A, \xi_t, \Theta), \qquad q_t = -\partial_\xi \hat{\psi}(\varepsilon^e_t, \xi_t), \tag{3.6}$$

which possess a functional form identical to that of the infinitesimal theory.

iii. Set $\phi(\boldsymbol{\tau}, q, \Theta) = \hat{\phi}(\beta_1, \beta_2, \beta_3, q, \Theta)$. Then the algorithmic flow rule collapses to

$$\left.\begin{aligned} \varepsilon^e_{t\,A} &= \varepsilon^{e\,\mathrm{tr}}_{t\,A} - \Delta\gamma_t \partial_{\beta_A}\hat{\phi}(\boldsymbol{\beta}_t, q_t, \Theta_t), \\ \xi_t &= \xi_n \quad + \Delta\gamma_t \partial_q \hat{\phi}(\boldsymbol{\beta}_t, q_t, \Theta_t), \\ \eta^p_t &= \eta^p_n \quad + \Delta\gamma_t \partial_\Theta \hat{\phi}(\boldsymbol{\beta}_t, q_t, \Theta_t), \end{aligned}\right\} \tag{3.7}$$
$$\Delta\gamma_t \geq 0, \quad \hat{\phi}(\boldsymbol{\beta}_t, q_t, \Theta_t) \leq 0, \quad \Delta\gamma_t \hat{\phi}(\boldsymbol{\beta}_t, q_t, \Theta_t) = 0,$$

which along with (3.6) define a return mapping algorithm at *fixed principal axis* (defined by the trial state) identical to that of the infinitesimal theory.

The advantage of the preceding scheme is clear: It allows the extension to the finite strain regime of the classical return mapping algorithms of the infinitesimal theory without modification (see e.g., SIMO, KENNEDY & GOVINDJEE [1988]). Observe that the return map is now formulated in the principal (Eulerian) axis defined by the trial state, which can be computed in *closed form* directly from $\boldsymbol{b}^{e\,\mathrm{tr}}_t$. Finally, the linearization of the algorithm can be performed in closed form leading to algorithmic tangent moduli with form identical to those of the infinitesimal theory first proposed in SIMO & TAYLOR [1985]. We refer to SIMO [1991b] for a proof of the results quoted above and a detailed discussion of the actual implementation of the return mapping algorithm within the context of the purely mechanical theory.

§4. The IBVP: Unconditionally Stable Staggered Schemes.

By exploiting the recent results in ARMERO & SIMO [1991], we construct a product formula approximation to the IBVP of thermoplasticity which leads naturally to staggered algorithms possessing unconditional stability, in the sense described below. This stability result should be contrasted with the conditional stability property inherent to the standard staggered algorithms currently used.

4.1. Strong Form of the IBVP for Thermoplasticity.

The local or strong form of the IBVP for thermoplasticity is obtained by supplementing the constitutive equations is Section **2** with the balance of momentum and (reduced) balance of energy equations, written in a Lagrangian description as

$$\left.\begin{aligned} \dot{\boldsymbol{\varphi}} &= \boldsymbol{V} \\ \rho_0 \dot{\boldsymbol{V}} &= \mathrm{DIV}[\boldsymbol{P}] + \rho_0 \boldsymbol{f} \\ \Theta\dot{\eta} &= -\mathrm{DIV}[\boldsymbol{Q}] + \mathcal{D} + r \end{aligned}\right\} \quad \text{in} \quad \Omega \times \mathsf{I}, \tag{4.1a}$$

where $\boldsymbol{P}:=\boldsymbol{\tau}D\boldsymbol{\varphi}^{-T}$ and $\boldsymbol{Q}:=D\boldsymbol{\varphi}^{-1}\boldsymbol{q}$ denote the nominal stress tensor and the nominal heat flux vector, respectively, $\boldsymbol{f}$ is the body force, r is the heat source and $\rho_0>0$ is the reference density. For simplicity we assume Dirichlet boundary conditions of the form

$$\boldsymbol{\varphi}=\bar{\boldsymbol{\varphi}} \quad \text{and} \quad \Theta=\bar{\Theta}\,(\text{constant}) \quad \text{on} \quad \partial\Omega\times\mathsf{I}, \tag{4.1b}$$

along with the initial conditions $\boldsymbol{\varphi}(\cdot,0)=\boldsymbol{\varphi}_0(\cdot)$ and $\Theta(\cdot,0)=\Theta_0(\cdot)$ in Ω. The dynamics generated by problem (4.1) has the following property.

i. *A priori stability estimates.* Define functionals L and V of the dynamics by the expressions

$$\mathsf{V}:=\int_\Omega\left[e+\tfrac{1}{2}\rho_0|\boldsymbol{V}|^2-\bar{\Theta}\eta^e\right]d\Omega\;+\;\Pi_{\text{ext}}(\boldsymbol{\varphi}) \quad \text{and} \quad \mathsf{L}:=\mathsf{V}-\bar{\Theta}\int_\Omega\eta^p\,d\Omega, \tag{4.2}$$

where $\Pi_{\text{ext}}(\boldsymbol{\varphi})$ corresponds to the potential energy of the external loading (assumed conservative). In the context of thermoelasticity the functional V is known as the *canonical free energy function* which, according to a classical result of DUHEM [1911], defines a Lyapunov function for the dynamics; see ARMERO & SIMO [1991] and references therein. For thermoplasticity, the following result holds.

Time differentiation of $(4.2)_1$ and use of relation (2.5), along with the local equations (4.1), expression $\boldsymbol{\tau}\cdot\boldsymbol{d}=\boldsymbol{P}\cdot\text{GRAD}[\boldsymbol{V}]$ and integration by parts, yields the result

$$\frac{d}{dt}\mathsf{L}=\int_\Omega\left[(\Theta-\bar{\Theta})\dot{\eta}-\mathcal{D}\right]d\Omega=\bar{\Theta}\int_\Omega\left[\frac{\text{GRAD}[\Theta]\cdot\boldsymbol{Q}}{\Theta^2}-\frac{\mathcal{D}}{\Theta}\right]d\Omega\le 0, \tag{4.3}$$

where the last inequality is a consequence of the Clausius-Duhem form of the second law. Thus L is a *non-increasing function* along the dynamics generated by (4.1).

ii. *Thermoplastic softening.* A material with the property $\partial_\Theta\phi>0$ is said to exhibit thermoplastic softening (in the perfectly plastic range) since, for J_2–flow theory with flow stress $\sigma_Y(\Theta)$, this property corresponds to $\sigma'_Y(\Theta)<0$. If the material obeys the principle of maximum dissipation then

$$\dot{\eta}^p=\gamma\partial_\Theta\phi \quad \text{and} \quad \partial_\Theta\phi>0 \quad \Rightarrow \quad \Theta\dot{\eta}^p\ge 0 \quad \text{since} \quad \gamma\ge 0. \tag{4.4}$$

For a material exhibiting thermoplastic softening the property $\mathcal{D}\ge 0$ follows from the rather natural assumption that $\mathcal{D}_{\text{mech}}\ge 0$. A calculation similar to that in (4.3) then gives

$$\frac{d}{dt}\mathsf{V}=\int_\Omega\left[(\Theta-\bar{\Theta})\dot{\eta}^e-\mathcal{D}_{\text{mech}}\right]d\Omega=\bar{\Theta}\int_\Omega\left[\frac{\text{GRAD}[\Theta]\cdot\boldsymbol{Q}}{\Theta^2}-\frac{\mathcal{D}_{\text{mech}}}{\Theta}\right]d\Omega\le 0, \tag{4.5}$$

which shows that V is non-increasing along the dynamics. As in the infinitesimal theory of classical plasticity (see SIMO [1991]) an algorithm that inherits this property will be termed B-stable.

4.2. Operator split and Product Formula Algorithm.

Our goal is to design a time-stepping algorithm that allows the staggered solution of the coupled system (4.1) while retaining the property $\dot{\mathsf{V}} \leq 0$; i.e., B–stability. Algorithms of this type are fractional step methods associated with a *split* of the coupled problem (4.1). Unfortunately, the 'obvious' split defined by an isothermal phase followed by a heat conduction phase at fixed configurations results in the classical staggered algorithms which, even for thermoelasticity, are at best conditionally stable; see the analysis in ARMERO & SIMO [1991]. To circumvent this difficulty, consider the following alternative split:

Phase **1**. (Adiabatic) *Phase* **2**. (Heat Conduction)

$$\left.\begin{aligned} \dot{\varphi} &= \boldsymbol{V} \\ \rho_o \dot{\boldsymbol{V}} &= \mathrm{DIV}[\boldsymbol{P}] + \rho_0 \boldsymbol{f} \\ \Theta\, \dot{\eta}^e &= 0 \\ \dot{\eta}^p &= 0 \end{aligned}\right\} \qquad \left.\begin{aligned} \dot{\varphi} &= \boldsymbol{0} \\ \rho_o \dot{\boldsymbol{V}} &= \boldsymbol{0} \\ \Theta\, \dot{\eta}^e &= -\mathrm{DIV}[\boldsymbol{Q}] + \mathcal{D}_{\mathrm{mech}} + r \\ \dot{\eta}^p &= \gamma\, \partial_\Theta \phi \end{aligned}\right\} \tag{4.6}$$

Observe that the heat conduction phase takes place at fixed configuration. The local constitutive equations describing plastic flow, as summarized in Section **2**, are appended to both phases and assumed to hold pointwise. This implies that in the first phase the evolution equation for the plastic entropy is compatible with condition $\dot{\eta}^p = 0$ provided that

$$\phi(\boldsymbol{\tau}_t, q_t, \Theta_t) = \phi(\boldsymbol{\tau}_t, q_t, \Theta_n) \quad \text{in Adiabatic Phase} \tag{4.7}$$

This observation is in fact crucial to obtain a symmetric formulation in the actual implementation of the method. It can be easily shown that each of the sub-problems in this split inherits the formal contractivity properties (4.3) and (4.5) in the sense that

$$\frac{d}{dt}\mathsf{V}\Big|_{\text{Adiabatic Phase}} \leq 0 \qquad \text{and} \qquad \frac{d}{dt}\mathsf{V}\Big|_{\text{Heat Cond. Phase}} \leq 0. \tag{4.8}$$

Consequently, B–stable schemes for the coupled problem are constructed merely as the product of two B–stable algorithms, consistent with each of the sub-problems above. These unconditionally stable schemes should once more be contrasted with the standard staggered, conditionally stable, coupled algorithms based on an isothermal split of the problem. We refer to ARMERO & SIMO [1991] for a proof of this result and the statement of sharp stability bounds.

§5. Representative Numerical Simulations.

The formulation presented in the preceding Sections is illustrated below in two full three dimensional numerical simulations. The goal is to demonstrate the robustness of the overall finite element formulation in both static and dynamic analyses. All the calculations are performed with the enhanced an version of the general purpose finite element program FEAP developed by R.L. TAYLOR and J.C. SIMO.

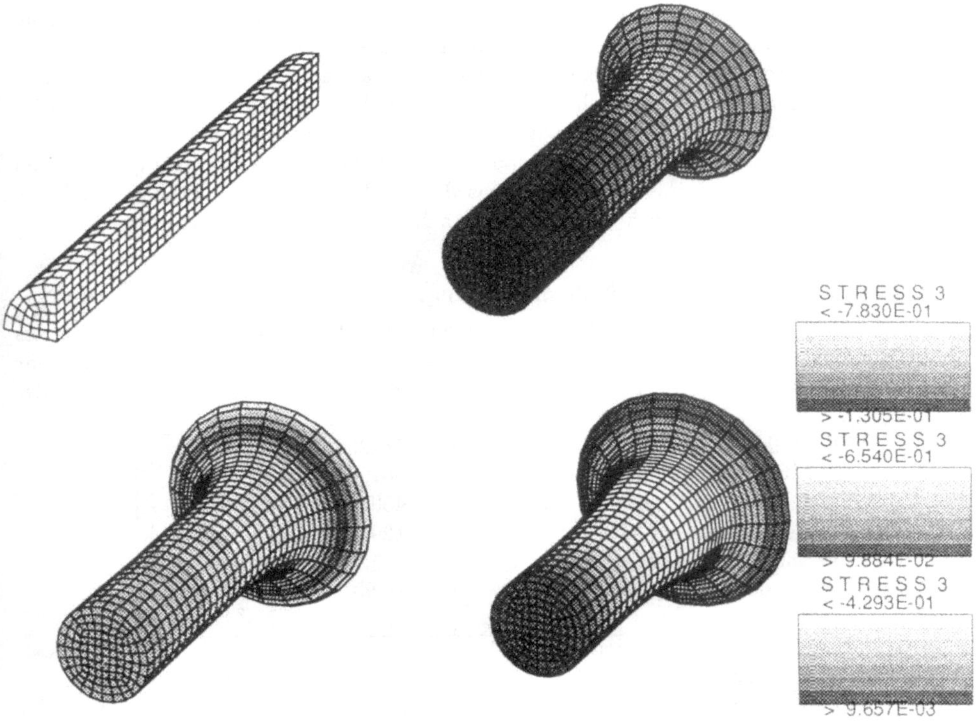

FIGURE 5.1. Three dimensional impact of a circular bar. Large mesh consisting of 972 Q1/P0 mixed finite elements. Sequence of deformed shapes corresponding to times $t = 20$, $t = 40$ and $t = 80$, obtained in 32 time time steps with the exact momentum-conserving algorithm and the exponential return map.

5.1. Three Dimensional Dynamic Impact of a Circular Bar.

The objective here is to assess the performance in nonlinear dynamic plasticity of the class of return mapping algorithms discussed in Section **2** together two different time stepping integrators: The exact momentum–conserving schemes of SIMO, TARNOW & WONG [1991] and the standard Newmark method. The model problem selected is the dynamic impact on a rigid frictionless wall of a three-dimensional bar, with length $L = 32.4mm$ and circular cross section with radius $R_0 = 3.2mm$, presented in HALLQUIST & BENSON [1988]. The initial velocity is $v_0 = 0.227mm/\mu s$ and the finite element mesh is described in Figure **5.1.**

The sequence of deformed configurations corresponding to times $t = 20s$ $t = 40s$ and

$t = 80$, computed with the exact momentum-conserving algorithm and the exponential return map, is shown in Figure 5.1. This simulation is successfully completed in 32 time steps. Despite this extremely large time step, the final shape of the specimen at time $t = 80s$ agrees remarkably well with the results reported in HALLQUIST & BENSON [1988]. The same simulation was attempted with the standard Newmark algorithm and the one-step exponential backward-Euler return map. The computation was successfully completed only when the number of time steps was increased to 64. In all these simulations a quadratic rate of asymptotic convergence was attained in a Newton iterative solution strategy.

5.2. Plane strain "adiabatic" shear banding.

We consider the plane strain tensile test of a rectangular bar, with material properties corresponding to J_2–theory. Localization of the deformation and nearly adiabatic shear banding arises as a result of thermal softening (see the table below). In the problem under

Bulk Modulus	κ	164206.	N/mm^2
Shear Modulus	μ	801938.	N/mm^2
Flow Stress	y_0	450.	N/mm^2
Linear Hardening	h	129.24	N/mm^2
Saturation Hardening	$y_{0,\infty}$	715.	N/mm^2
Hardening Exponent	δ	16.93	
Density	ρ	$7.8 \cdot 10^{-9}$	Ns^2/mm^4
Expansion Coefficient	β	$1 \cdot 10^{-5}$	K^{-1}
Conductivity	k	45.	N/sK
Specific Capacity	c_s	$0.46 \cdot 10^9$	mm^2/s^2K
Dissipation Factor	χ	0.9	
Flow Stress Softening	ω_0	0.002	K^{-1}
Hardening Softening	ω_h	0.002	K^{-1}

consideration, after a first state of essentially homogeneous deformation, thermally triggered shear bands arise at 45 degrees with the axial direction of loading.

Assuming plane strain conditions, a quarter of a specimen shown in Figure **5.2.a**, with 12.826 mm width and 53.334 mm length, is discretized with 200 4–node isoparametric quad based on the enhanced formulation as presented in SIMO & ARMERO [1991]. The bar is assumed to be insulated along its lateral face, while the temperature is kept constant to the reference value on the upper and lower faces. The simulation is performed by imposing the top displacement with $\dot{L}/L_o = 4 \cdot 10^4\, s^{-1}$, i.e., at very high strain rate. No initial imperfections are introduced; the final localized pattern of the deformation is completely triggered by the thermal field. Figures **5.3.b–c** show the distribution of the temperature and equivalent plastic strain, respectively, at a imposed top displacement of $\bar{u} = 5.0\, mm$. In particular, the shear bands appear remarkably well–resolved.

Acknowledgements. We are indebted to C. Miehe, N. Tarnow and K. Wong for numerous discussion. The support of NSF through Grant No. 2-DJA-792 with Stanford University is gratefully acknowledged.

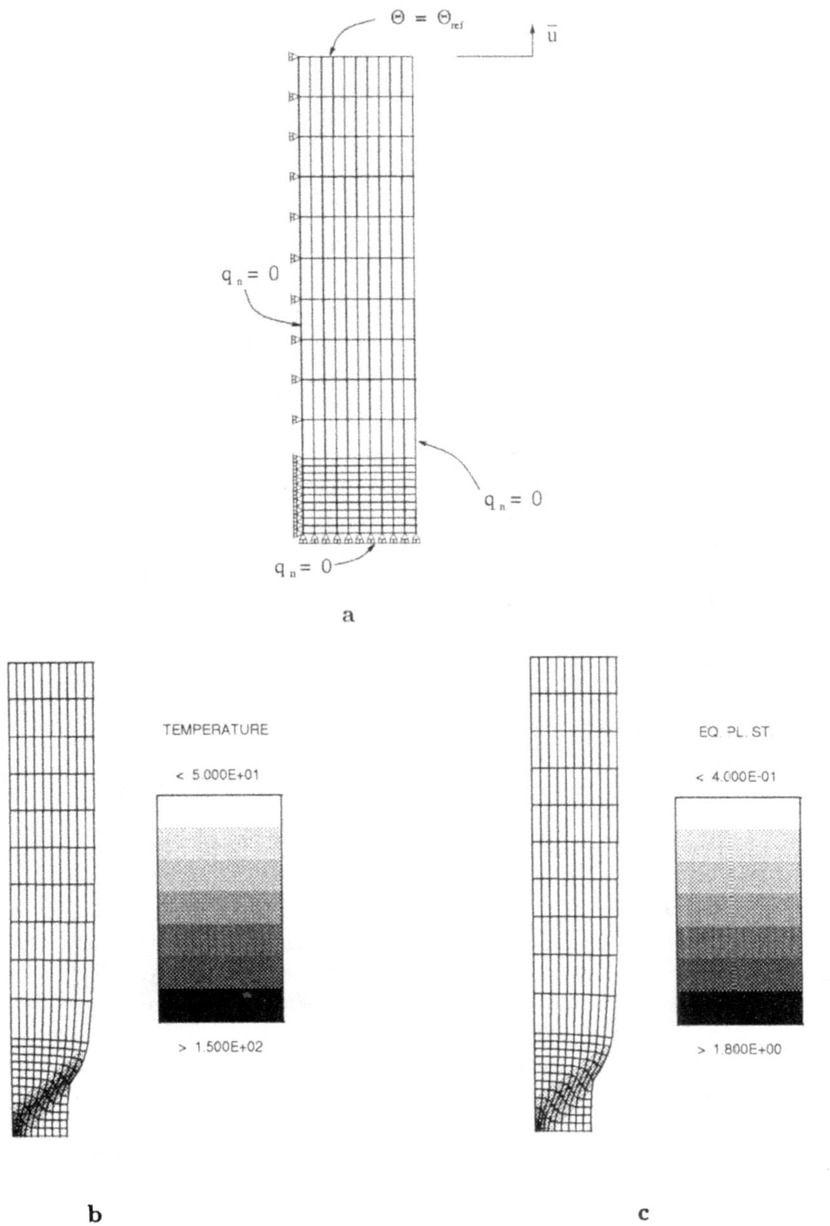

FIGURE 5.2 Plane strain "adiabatic" shear banding. a) Initial configuration and boundary conditions. b) Temperature and, c) Equivalent Plastic Strain distributions at $\bar{u} = 5.0\,mm$.

References

1. J.H. Argyris & J.St. Doltsinis [1981], "On the Natural Formulation and Analysis of Large Deformation Coupled Thermomechanical Problems", *Computer Methods in Applied Mechanics and Engineering*, **25**, 195–253.
2. F. Armero & J.C. Simo, [1991], "A New Unconditionally Stable Fractional Step Method for Nonlinear Coupled Thermomechanical Problems," SUDAM Report No. 91-5, Stanford University. To appear in *International J. Numerical Methods in Engineering.*
3. P. Duhem [1911] *Traite d'Energetique ou de Thermodynamique Generale,* Gauthier–Villars, Paris.
4. J.J. Moreau, [1977], "Evolution Problem Associated with a Moving Convex Set in a Hilbert Space", *Journal of Differential Equations,* Volume **26**, pp. 347.
5. J.C. Simo and R.L. Taylor, [1985], "Consistent Tangent Operators for Rate-independent Elastoplasticity", *Computer Methods in Applied Mechanics and Engineering* **48**, 101–118.
6. J.C. Simo, [1988a,b], "A Framework for Finite Strain Elastoplasticity Based on Maximum Plastic Dissipation and Multiplicative Decomposition: Part I. Continuum Formulation; Part II.; Computational Aspects", *Computer Methods in Applied Mechanics and Engineering* **66**, 199–219 and **68**, 1-31.
7. J.C. Simo, J.G.Kennedy and S.Govindjee, [1988], "General Return Mapping Algorithms for Multisurface Plasticity and Viscoplasticity," *International Journal of Numerical Methods in Engineering,* **26**, No. 2, 2161-2185.
8. J.C. Simo and R.L. Taylor, [1991], "Finite Elasticity in Principal stretches; Formulation and Finite element implementaion," *Computer Methods in Applied Mechanics and Engineering,* **85**, 273-310.
9. J.C. Simo and C.Miehe [1991], "Coupled Associative Thermoplasticity at finite strains. Formulation, numerical analysis and implementation, *Computer Methods in Applied Mechanics and Engineering,* (in press)
10. J.C. Simo, [1991a], "Nonlinear Stability of the Time Discrete Variational Problem in Nonlinear Heat Conduction and Elastoplasticity," *Computer Methods in Applied Mechanics and Engineering,* **88**, 111–131.
11. J.C. Simo, [1991b], "Algorithms for Multiplicative Plasticity which Preserve the Structure of the Classical Return Mappings of the Infinitesimal Thery," SUDAM Report No. 91-4. In press in *Computer Methods in Applied Mechanics and Engineering.*
12. J.C. Simo and F. Armero [1991], "Geometrically Nonlinear Enhanced Strain Methods and the Method of Incompatible Modes," In press in *International J. Numerical Methods in Engineering.*
13. J.C. Simo, N. Tarnow and K.W. Wong [1991], "Exact Energy-Momentum Conserving Algorithms and Symplectic Schemes for Nonlinear Dynamics, *Computer Methods in Applied Mechanics and Engineering.*
14. N.N. Yanenko, [1971] *The method of Fractional Steps,* Springer-Verlag, Berlin, Heidelberg, New-York.

Adaptive Finite Element Methods for Small Strain Elasto-Plasticity

Claes Johnson and Peter Hansbo

Mathematics Department
Chalmers University of Technology
412 96 Göteborg

Summary

We prove *a posteriori* error estimates for a finite element method for small strain elasto-plasticity based on Hencky's and Prandtl-Reuss' flow rules, and formulate corresponding adaptive methods. We also present the results of some numerical experiments.

1. Introduction

In this note we present an adaptive finite element method for elasto-plasticity modeled by Hencky's and Prandtl-Reuss' flow rules, in the case of small strains and displacements. The adaptive method is based on a *a posteriori* estimate for the L_2-error in stresses in terms of the residual of the finite element solution. The present work is on the one hand related to our earlier work on plasticity problems, ([1] - [4]), in particular the *a priori* error estimate of [5], and on the other hand to our earlier work on adaptive finite element methods for linear elliptic, parabolic and hyperbolic problems (see [5], [6]) and variational inequalities ([7]). As far as we know the *a posteriori* error estimates of this note are the first such estimates for plasticity problems. Extensions to large strain elasto-plasticity will be considered in future work.

An outline of this paper is as follows: In Section 2 we consider a two-dimensional model problem of Hencky type for perfect plasticity: the antiplane shear problem (cf. [8]). In Section 3 we extend the analysis to a corresponding Prandtl-Reuss model problem. In Section 4 we consider briefly the case of three-dimensional small strain elasto-plasticity with the von Mises yield condition. Finally, in Section 5 we present some numerical results for the model problem.

D. Besdo, E. Stein (Eds.)
Finite Inelastic Deformations -
Theory and Applications
IUTAM Symposium Hannover/Germany 1991

2. A Hencky model problem

We shall consider the following model problem of Hencky type (the antiplane shear problem, cf. [8]): Find $\sigma \in P_f$ such that

$$J(\sigma) \leq J(\tau) \qquad \forall \tau \in P_f, \tag{2.1}$$

where

$$J(\tau) = \frac{1}{2}\|\tau\|^2, \quad \|\tau\|^2 = \int_\Omega |\tau|^2 dx,$$

$$P_f = P \cap H_f,$$

$$P = \{\tau \in H: |\tau(x)| \leq 1 \quad \text{a.e. } x \in \Omega\},$$

$$H_f = \{\tau \in H: -\operatorname{div} \tau = f \text{ in } \Omega\},$$

$$H = [L_2(\Omega)]^2,$$

where Ω is a bounded domain in the plane and $f \in L_2(\Omega)$ is a given function. Here $P_f = P \cap H_f$ is the set of plastically admissible stresses P intersected with the set H_f of (shear) stresses τ satisfying the equilibrium equation $-\operatorname{div} \tau = f$ in Ω, and $J(\tau)$ is the complementary energy. We shall also consider the following regularized version of (2.1) with $\mu > 0$ a regularization parameter: Find $\sigma_\mu \in H_f$ such that

$$J_\mu(\sigma_\mu) \leq J_\mu(\tau) \qquad \forall \tau \in H_f, \tag{2.2}$$

where

$$J_\mu(\tau) = \frac{1}{2}\|\tau\|^2 + \frac{1}{2\mu}\|\tau - \pi\tau\|^2,$$

$$\pi\tau(x) = \begin{cases} \tau(x) & \text{if} \quad |\tau(x)| \le 1, \\ \frac{\tau(x)}{|\tau(x)|} & \text{if} \quad |\tau(x)| > 1. \end{cases}$$

Note that $\pi\tau(x)$ is the projection of $\tau(x)$ onto the unit disc $\{r \in \mathbb{R}^2: |r| \le 1\}$. Equivalently, the regularized problem (2.2) may be formulated as follows: Find $(\sigma_\mu, u_\mu) \in H \times V$, $V = H_0^1(\Omega)$, such that

$$\sigma_\mu + \frac{1}{\mu}(\sigma_\mu - \pi\sigma_\mu) = \nabla u_\mu \qquad \text{in } \Omega, \tag{2.3a}$$

$$-\text{div } \sigma_\mu = f \qquad \text{in } \Omega. \tag{2.3b}$$

Here u_μ is the displacement corresponding to the stress σ_μ (note that u_μ is zero on the boundary of Ω) via the Hencky flow rule (2.3a) expressing the total strain ∇u_μ as the sum of an elastic strain σ_μ and a viscoplastic strain $(\sigma_\mu - \pi\sigma_\mu)/\mu)$. For all $\mu > 0$ the problem (2.3) has a unique solution $(\sigma_\mu, u_\mu) \in H \times V$ with σ_μ satisfying (2.2). If $P_f \neq \emptyset$, then the problem (2.1) will have a unique solution $\sigma \in P_f$ and σ_μ tends weakly to σ in H as μ tends to zero. To be able to prove existence of a displacement u associated to the solution $\sigma \in P_f$ of (2.1), we need the following *safe load hypothesis*:

$$\exists \chi \in H_f \text{ such that for some } \delta > 0, \text{ we have } |\chi(x)| \le 1 - \delta, \text{ a.e. } x \in \Omega. \tag{2.4}$$

If (2.4) is satisfied then (see e.g. [1]) there is a constant C_1 independent of μ such that

$$\|\frac{1}{\mu}(\sigma_\mu - \pi\sigma_\mu)\|_{L_1(\Omega)} + \|\nabla u_\mu\|_{L_1(\Omega)} \le C_1. \tag{2.5}$$

From (2.5) follows that $\exists u \in BV(\Omega) \equiv \{v \in L_2(\Omega): \nabla v \in [M(\bar{\Omega})]^2\}$, where $M(\bar{\Omega})$ is the set of bounded measures on $\bar{\Omega}$, such that u_μ converges to u weak star in $BV(\Omega)$ and such that u is a displacement corresponding to the stress solution σ of (2.1). Roughly speaking, we will have the following relation between $\sigma \in P$ and u:

$$(\sigma, \tau - \sigma) - (\nabla u, \tau - \sigma) \geq 0 \qquad \forall \tau \in P, \tag{2.6a}$$

or

$$\sigma(x) = \pi \nabla u(x) \qquad \text{a.e. } x \in \Omega, \tag{2.6b}$$

a relation which needs to be precised if $u \in BV(\bar{\Omega})$ but $u \notin H_0^1(\Omega)$. To avoid the technical complications (see [9]) involved in defining (2.6) with $u \in BV(\Omega)$, we consider in this note the regularized problem (2.3) where the relation (2.3a) between σ_μ and u_μ is well defined with $(\sigma_\mu, u_\mu) \in H \times V$. With the safe load hypothesis (2.4) and the consequence (2.5) thereof satisfied, the *a posteriori* error estimate to be proved below will be essentially independent of μ (for small μ) and will not require a coupling of the mesh size h and the regularization parameter μ. Thus, the μ-regularization of this note is introduced only to simplify the statement and proof of the *a posteriori* error estimate and the actual value of μ (small) used in the computation will be insignificant (cf. [7], where the regularization is coupled to the mesh size). Notice further that if in (2.6) we have $|\nabla u(x)| \leq 1$ so that $\sigma(x) = \nabla u(x)$, then the material is *elastic* at x, while if $|\nabla u(x)| > 1$ the material is *plastic* at x. Correspondingly, we say for the regularized problem (2.3) that the material is elastic at x if $|\sigma(x)| \leq 1$ so that $\sigma_\mu(x) = \nabla u_\mu(x)$, and otherwise plastic.

Let now $V_h \subset H_0^1(\Omega)$ be the standard finite element space of continuous piecewise linear functions v on a triangulation $T_h = \{K\}$ of Ω into triangles K of diameter h_K, such that $v = 0$ on Γ. We consider the following discrete analog of (2.3): Find $(\sigma_{\mu h}, u_{\mu h}) \in H \times V_h$ such that

$$\sigma_{\mu h} + \frac{1}{\mu}(\sigma_{\mu h} - \pi\sigma_{\mu h}) = \nabla u_{\mu h} \quad \text{in } \Omega, \tag{2.7a}$$

$$(\sigma_{\mu h}, \nabla v) = (f, v) \qquad \forall v \in V_h. \tag{2.7b}$$

Letting $\mu \to 0$ in (2.7) with V_h fixed, it follows easily since $V_h \subset H_0^1(\Omega)$ is finite dimensional that $(\sigma_{\mu h}, u_{\mu h}) \to (\sigma_h, u_h) \in P \times V_h$ where the limit (σ_h, u_h) satisfies

$$\sigma_h = \pi \nabla u_h \qquad \text{in } \Omega, \tag{2.8a}$$

$$(\sigma_h, \nabla v) = (f,v) \qquad \forall v \in V_h, \tag{2.8b}$$

that is, the displacement $u_h \in V_h$ is characterized by the equation

$$(\pi \nabla u_h, \nabla v) = (f,v) \qquad \forall v \in V_h,$$

or

$$(a(u_h)\nabla u_h, \nabla v) = (f,v) \qquad \forall v \in V_h \tag{2.9}$$

where

$$a(u_h)(x) = \begin{cases} 1 & \text{if } |\nabla u_h(x) \le 1, \\ \frac{1}{|\nabla u_h(x)|} & \text{if } |\nabla u_h(x)| > 1. \end{cases}$$

In the numerical experiments below the problem (2.9) was solved iteratively as follows: Find $u_h^{(j)} \in V_h$, $j = 1,2,\ldots$ such that

$$(a(u_h^{(j-1)}) \nabla u_h^{(j)}, \nabla v) = (f,v) \qquad \forall v \in V_h, \tag{2.10}$$

where $a(u_h^{(0)}) \equiv 1$.

Let us now derive an a posteriori error estimate for the quantity $\|\sigma_\mu - \sigma_{\mu h}\|$. Using the monotonicity of $\tau \to \frac{1}{\mu}(\tau - \pi\tau)$ resulting from the convexity of $\tau \to \frac{1}{2\pi}\|\tau - \pi\tau\|^2$, we have by (2.3a) and (2.7a)

$$\|\sigma_\mu - \sigma_{\mu h}\|^2 \le (\sigma_\mu - \sigma_{\mu h}, \sigma_\mu - \sigma_{\mu h}) + (\tfrac{1}{\mu}(\sigma_\mu - \pi\sigma_\mu) - \tfrac{1}{\mu}(\sigma_{\mu h} - \pi\sigma_{\mu h}), \sigma_\mu - \sigma_{\mu h})$$

$$= (\sigma_\mu + \tfrac{1}{\mu}(\sigma_\mu - \pi\sigma_\mu), \sigma_\mu - \sigma_{\mu h}) - (\sigma_{\mu h} + \tfrac{1}{\mu}(\sigma_{\mu h} - \pi\sigma_{\mu h}), \sigma_\mu - \sigma_{\mu h})$$

$$= (\nabla u_\mu, \sigma_\mu - \sigma_{\mu h}) - (\nabla u_{\mu h}, \sigma_\mu - \sigma_{\mu h})$$

$$= (\nabla(u_\mu - u_{\mu h}), \sigma_\mu) - (\nabla(u_\mu - u_{\mu h}), \sigma_{\mu h})$$

$$= (u_\mu - u_{\mu h}, f) - (\nabla(u_\mu - u_{\mu h}), \sigma_{\mu h}),$$

where we also used (2.3b) in the last step. Thus by (2.7b), letting $I_h v \in V_h$ denote the standard nodal interpolant of $v \in H_0^1(\Omega)$, we have integrating by parts over each element $K \in T_h$ recalling that by (2.8a) $\sigma_{\mu h}$ is constant over each element K:

$$\|\sigma_\mu - \sigma_{\mu h}\|^2 \leq ((u_\mu - u_{\mu h}) - I_h(u_\mu - u_{\mu h}), f) \tag{2.11}$$

$$+ \sum_S \int_S [\sigma_{\mu h} \cdot n]((u_\mu - u_{\mu h}) - I_h(u_\mu - u_{\mu h}))ds,$$

where we sum over all sides S of the triangulation T_h and $[\sigma_{\mu h} \cdot n]$ is the jump in normal stress $\sigma_{\mu h} \cdot n$ across a side S with unit normal n.

To proceed and estimate the right hand side of (2.11), we now split Ω into an "elastic part" Ω^e and a "plastic part" Ω^p, depending on the relation between u_μ and σ_μ, and $u_{\mu h}$ and $\sigma_{\mu h}$, as follows:

$$\Omega^e = \Omega_\mu^e \cap \Omega_{\mu h}^e, \quad \Omega^p = \Omega \backslash \Omega^e,$$

where

$$\Omega_\mu^e = \{x \in \Omega: |\sigma_\mu(y)| \leq 1 \qquad \forall y \in K, \text{ where } x \in K\},$$

$$\Omega_{\mu h}^e = \{x \in \Omega: |\sigma_{\mu h}(y)| \leq 1 \qquad \forall y \in K, \text{ where } x \in K\}.$$

Clearly, by (2.3a) and (2.7a)

$$(\nabla u_\mu - \nabla u_{\mu h})(x) = (\sigma_\mu - \sigma_{\mu h})(x) \quad \text{if } x \in \Omega^e.$$

We can now estimate the right hand side of (2.11) as follows:

$$((u_\mu - u_{\mu h}) - I_h(u_\mu - u_{\mu h}), f)$$

$$\leq C\|hf\|_{L_2(\Omega^e)}\|\nabla(u_\mu - u_{\mu h}\|_{L_2(\Omega^e)}$$

$$+ C\|hf\|_{L_\infty(\Omega^p)}(\|\nabla u_\mu\|_{L_1(\Omega^p)} + \|\nabla u_{\mu h}\|_{L_1(\Omega^p)})$$

$$\leq C\|hf\|_{L_2(\Omega^e)}\|\sigma_\mu - \sigma_{\mu h}\|_{L_2(\Omega^e)}$$

$$+ C\|hf\|_{L_\infty(\Omega^p)}(C_1 + \|\nabla u_{\mu h}\|_{L_1(\Omega^p)}),$$

where $C_1 = \|\nabla u_\mu\|_{L_1(\Omega^p)}$, and we used the following interpolation error estimates (see [6])

$$|(f, v - I_h v)| \leq C\|hf\|_{L_p(K)}\|\nabla v\|_{L_q(K)}, \quad p^{-1} + q^{-1} = 1, \quad K \in T_h,$$

with C independent of v,f,h and K. Similarly, we have (see [6])

$$\sum_S \int_S [\sigma_{\mu h} \cdot n]((u_\mu - u_{\mu h}) - I_h(u_\mu - u_{\mu h}))ds$$

$$\leq C\|h\, D_h\, \sigma_{\mu h}\|_{L_2(\Omega^e)}\|\sigma_\mu - \sigma_{\mu h}\|_{L_2(\Omega^e)}$$

$$+ C\|h\, D_h\, \sigma_{\mu h}\|_{L_\infty(\Omega^p)}\, (C_1 + \|\nabla u_{\mu h}\|_{L_1(\Omega^p)}),$$

where

$$D_h\sigma_{\mu h} = \max_{\tau \subset \partial K} \frac{|[\sigma_{\mu h} \cdot n_\tau]|}{h_K},$$

where τ is a side of $K \in T_h$ with unit normal n_τ. Recalling (2.11) we thus have proved the following error estimate for the stresses.

Theorem 2.1. There is a constant C such that if $(\sigma_\mu, u_\mu) \in H \times V$ is the solution of (2.3) and $(\sigma_{\mu h}, u_{\mu h}) \in H \times V_h$ that of (2.8), then for $\mu > 0$ we have

$$\|\sigma_\mu - \sigma_{\mu h}\|_{L_2(\Omega)} \tag{2.12}$$

$$\leq C(\|h\, D_h \sigma_{\mu h}\|_{L_2(\Omega^e)} + \|hf\|_{L_2(\Omega^e)})$$

$$+ C(\|hD_h \sigma_{\mu h}\|_{L_\infty(\Omega^p)} \|hf\|_{L_\infty(\Omega^p)})^{\frac{1}{2}}(C_1 + \|\nabla u_{\mu h}\|_{L_1(\Omega^p)})^{\frac{1}{2}}$$

where $C_1 = \|\nabla u_\mu\|_{L_1(\Omega^p)}$.

We now comment on the error estimate (2.12), which is the main theoretical result of this note. First, (2.12) is not a full *a posteriori* type since Ω^e and Ω^p depend on (σ_μ, u_μ) and the constant $C_1 = \|\nabla u_\mu\|_{L_1(\Omega^p)}$ depends on u_μ. The dependence of Ω^e and Ω^p may be eliminated by replacing Ω^e and Ω^p by Ω at the expense of increasing the right hand side of (2.12). The constant C_1 depends on the constant δ in the safe load hypothesis. To get a full *a posteriori* error estimate in practice, we propose to replace Ω^e by $\Omega^e_{\mu h}$ and choose $\Omega^p = \Omega^p_{\mu h} \equiv \Omega \setminus \Omega^e_{\mu h}$, and further to replace C_1 by $\|\nabla u_{\mu h}\|_{L_1(\Omega^p_{\mu h})}$. This would lead to the following adaptive method for (2.8): Choose $h = h(x)$ so that

$$E(h, u_{\mu h}, \sigma_{\mu h}, f) = TOL, \tag{2.13}$$

where $TOL > 0$ is a given tolerance and

$$E(h, u_{\mu h}, \sigma_{\mu h}, f)$$

$$= C(\|hD_h\sigma_{\mu h}\|_{L_2(\Omega^e_{\mu h})} + \|hf\|_{L_2(\Omega^e_{\mu h})}) +$$

$$+ C(\|hD_h\sigma_{\mu h}\|_{L_\infty(\Omega^p_{\mu h})} + \|hf\|_{L_\infty(\Omega^p_{\mu h})})^{\frac{1}{2}}\|\nabla u_{\mu h}\|^{\frac{1}{2}}_{L_1(\Omega^p_{\mu h})}.$$

Further, we suppose h in (2.13) is chosen so as to minimize approximatively the number of elements K in T_h corresponding to equidistribution of the element contributions to E (cf. [6]).

3. A Prandtl-Reuss model problem

In this section we extend the above results to the following rate-dependent model problem for perfect plasticity based on the Prandtl-Reuss flow rule: For $\mu > 0$ find (σ_μ, u_μ): $I \to H \times V$ such that

$$\dot{\sigma}_\mu + \frac{1}{\mu}(\sigma_\mu - \pi\sigma_\mu) = \nabla u_\mu \qquad \text{in } \Omega \times I, \tag{3.1a}$$

$$-\operatorname{div} \sigma_\mu = f \qquad \text{in } \Omega \times I, \tag{3.1b}$$

$$\sigma_\mu(0) = \sigma^0, \tag{3.1c}$$

where $I = (0,T)$ is a "time interval", $\dot{\sigma} = \frac{\partial\sigma}{\partial t}$ is the derivative of σ: $I \to H$, and f: $I \to L_2(\Omega)$ is a given time dependent load and $\sigma^0 \in H$ a given initial stress. Again we consider a regularized problem with $\mu > 0$ small to be able to easily formulate the Prandtl-Reuss problem (3.1a) stating that the strain-rate ∇u_μ is the sum of elastic strain rate $\dot{\sigma}_\mu$ and the visco-plastic strain rate $(\sigma_\mu - \pi\sigma_\mu)/\mu$. For $\mu > 0$ the problem (3.1) admits a unique solution (σ_μ, u_μ) and under a safe load hypothesis corresponding to (2.4), (σ_μ, u_μ) will converge to a limit (σ,u): $I \to P \times BV(\bar{\Omega})$ satisfying

$$(\dot{\sigma}, \tau - \sigma) - (\nabla u, \tau - \sigma) \geq 0 \qquad \forall \tau \in P, \ t \in I, \tag{3.2a}$$

$$-\operatorname{div} \sigma = f \qquad \text{in } \Omega \times I, \tag{3.2b}$$

$$\sigma(0) = \sigma^0, \tag{3.2c}$$

where (3.2a) again has to be made precise if $u(t) \notin H_0^1(\Omega)$, cf. [9].

We now consider the following discrete analog of (3.1) based on space-discretization using the fem-space $V_h \subset H_0^1(\Omega)$ introduced above, and time-discretization with the backward Euler scheme on a subdivision $0 = t_0 < t_1 < \dots t_N = T$ of (0,T):
For $n = 1,2,\dots, N$, find $(\sigma_{\mu h}^n, u_{\mu h}^n) \in H \times V_h$ such that

$$\frac{1}{k_n}(\sigma_{\mu h}^n - \sigma_{\mu h}^{n-1}) + \frac{1}{\mu}(\sigma_{\mu h}^n - \pi\sigma_{\mu h}^n) = \nabla u_{\mu h}^n \quad \text{in } \Omega \times I, \tag{3.3a}$$

$$(\sigma_{\mu h}^n, \nabla v) = (f^n, v) \quad \forall v \in V_h, \tag{3.3b}$$

$$\sigma_{\mu h}^0 = \sigma^0, \tag{3.3c}$$

where $k_n = t_n - t_{n-1}$ and $f^n = f(\cdot, t_n)$. Letting $\mu \to 0$ in (3.3) as above we find that the corresponding limit $(\sigma_h^n, u_h^n) \in P \times V_h$ satisfies

$$\sigma_h^n = \pi(\sigma_h^{n-1} + k_n \nabla u_h^n) \qquad \text{in } \Omega, \tag{3.4a}$$

$$(\sigma_h^n, \nabla v) = (f^n, v) \qquad \forall v \in V_h, \tag{3.4b}$$

that is, $u_h^n \in V_h$ satisfies

(3.5) $$(a(u_h^n, \sigma_h^{n-1})(\sigma_h^{n-1} + k_n \nabla u_h^n), \nabla v) = (f^n, v) \qquad \forall v \in V_h, \tag{3.5}$$

where

$$a(u_h^n, \sigma_h^{n-1})(x) = \begin{cases} 1 & \text{if } |(\sigma_h^{n-1}+k_n\nabla u_h^n)(x)| \le 1, \\ |(\sigma_h^{n-1}+k_n\nabla u_h^n)(x)|^{-1} & \text{if } |(\sigma_h^{n-1}+k_n\nabla u_h^n)(x) > 1. \end{cases}$$

The discrete problem (3.5) may be solved by an iterative method analogous to (2.10).

The analog in the present case of the *a posteriori* error estimate (2.10) reads as follows:

$$\|\sigma_\mu - \sigma_{\mu h}\|_{L_\infty(L_2(\Omega))} \le E_h(T, h, u_{\mu h}, \sigma_{\mu h}, f) \tag{3.6}$$

where

$$E_h(T,h,u_{\mu h}, \sigma_{\mu h}, f) = C \int_0^T (\|hD_h\sigma_{\mu h}\|_{L_2(\Omega^e(t))} + \|hf\|_{L_2(\Omega^e(t))})dt$$

$$+ C \left(\int_0^T (\|hD_h\sigma_{\mu h}\|_{L_\infty(\Omega^p(t))} + \|hf\|_{L_\infty(\Omega^p(t))})^{\frac{1}{2}}\right.$$

$$\left.\times (\|\nabla u_\mu\|_{L_1(\Omega^p(t))} + \|\nabla u_{\mu h}\|_{L_1(\Omega^p(t))} dt\right)^{\frac{1}{2}}.$$

Here $(\sigma_{\mu h}, u_{\mu h})\colon I \to H \times V_h$ is defined by $(\sigma_{\mu h}(t), u_{\mu h}(t)) = (\sigma_{\mu h}^n, u_{\mu h})$ for $t \in (t_{n-1}, t_n]$.

The proof of (3.6) is analogous to (2.10) and we leave the details to the reader. Again we may formulate an adaptive version of (3.3) by seeking to satisfy

$$E_h(T,h, u_{\mu h}, \sigma_{\mu h}, f) = TOL,$$

with (approximately) as few degrees of freedom as possible.

4. Small strain plasticity.

In this section we indicate the extension of above results to the basic model for three-dimensional isotropic small strain elasto-plasticity based on the von Mises yield condition. In this case the corresponding visco-plastic Hencky problem reads (cf. (2.3)): Find $(\sigma_\mu, u_\mu) \in H \times V$ such that

$$A\,\sigma_\mu + \frac{1}{\mu}(\sigma_\mu^D - \pi\sigma_\mu^D) = \epsilon(u_\mu) \quad \text{in } \Omega, \tag{4.1a}$$

$$(\sigma_\mu, \epsilon(v)) = (f, v) \quad \forall v \in V, \tag{4.1b}$$

where $\Omega \subset \mathbb{R}^3$,

$$H = \{\tau = (\tau_{ij})_{i,j=1}^3 : \tau_{ij} = \tau_{ji} \in L_2(\Omega)\}, \quad V = [H_0^1(\Omega)]^3,$$

$$\tau_{ij}^D = \tau_{ij} - \frac{1}{3}\sum_{k=1}^{3} \tau_{kk}\,\delta_{ij},$$

$$(\tau, \epsilon(v)) = \sum_{i,j=1}^{3} \int_\Omega \tau_{ij}\,\epsilon_{ij}(v)dx, \quad (f,v) = \sum_{i=1}^{3}\int_\Omega f_i v_i dx,$$

$$\epsilon(v) = \{\epsilon_{ij}(v)\}_{i,j=1}^3, \quad \epsilon_{ij}(v) = \frac{1}{2}\{\frac{\partial v_i}{\partial x_j} + \frac{\partial v_j}{\partial x_i}\}, \quad v = (v_i)_{i=1}^3,$$

$$(A\tau)_{ij} = \alpha\,\tau_{ij}^D + \frac{\beta}{3}\sum_{k=1}^{3} \tau_{kk}\,\delta_{ij},$$

where α and β are positive constants ($\alpha = (2\mu)^{-1}$ and $\beta = (3\lambda + 2\mu)^{-1}$, where here μ and λ are the usual Lamé constants of the elastic material), and $\delta = \{\delta_{ij}\}$ is the usual Kronecker delta. Here τ^D is the stress deviatoric corresponding to the stress τ and $\epsilon(v)$ is the strain corresponding to the displacement v.

Letting $\mu \to 0$ in (4.1a) we obtain the following analog of (2.6b) which is (formally) the Hencky constitutive equation for perfect plasticity with the von Mises set of plastically admissible stresses

$$P = \{\tau \in H: \ |\tau^D(x)| \equiv (\sum_{i,j=1}^{3} \tau_{ij}^D\,\tau_{ij}^D)^{\frac{1}{2}} \leq 1 \quad \text{a.e. } x \in \Omega\},$$

$$\sigma = \pi(\frac{1}{\alpha}\,\epsilon(u)^D) + \frac{1}{3\beta}\operatorname{div} u\ \delta. \tag{4.2}$$

The finite element method for (4.1) with μ small thus may be formulated (cf. (2.9)): Find $u_h \in V_h \subset V$ such that

$$(\pi(\frac{1}{\alpha}\,\epsilon(u_h)^D),\,\epsilon(v)) + \frac{1}{3\beta}(\operatorname{div} u_h,\,\operatorname{div} v) = (f,v) \qquad \forall v \in V_h\,. \tag{4.3}$$

The corresponding discrete equation in the case of the Prandtl-Reuss flow rule reads as follows:

$$\sigma_h^n = \pi(\sigma_h^{n-1,D} + k_n\,\frac{1}{\alpha}\,\epsilon(u^n)^D) + \frac{1}{3}\sum_{k=1}^{3}\sigma_{h,kk}^{n-1} + k_n\,\frac{1}{3\beta}\operatorname{div} u_h^n\,. \tag{4.4}$$

The extensions of the *a posteriori* error estimates (2.12) and (3.6) to the present case will be presented elsewhere.

5. Numerical results

In this section we present the results of some numerical experiments with the finite element method (2.9) for the Hencky model problem (2.3) with μ small. The adaptivity is based on (2.13). In Fig. 5.1 we give the final mesh and elevations of u_h, $|\nabla u_h|$ and $|\sigma_h|$, in the case $\Omega = (0,1) \times (0,1)$, $f(x) = \lambda$ for $x \in \Omega$, with $\lambda = 2 + \sqrt{\pi}$. In Fig. 2 we give the corresponding results with $f(x) = f(x_1, x_2) = \lambda \sin \pi x_1 \sin \pi x_2$, $\lambda = 7.5$. In the first case the exact displacement u (with $\mu = 0$) is discontinuous with slip lines on the sides of Ω corresponding to concentrated plastic zones where $|\sigma(x)| = 1$, $|\nabla u(x)| = \infty$. The exact *limit load* in this case is $\lambda = 2 + \sqrt{\pi}$, see [8]. In the second example the displacement u appears to be continuous with $|\nabla u|$ bounded and the plastic zones now have positive area. In both cases the adaptive algorithm refines the mesh in the plastic zones as expected.

References

1. C. Johnson, Existence theorems for plasticity problems, J. Math. Pure et Appl. 55 (1976), 431-444.

2. C. Johnson, A mixed finite element for plasticity with hardening, Siam J. of Numer. Anal. 14 (1977), 575-584.

3. C. Johnson and R. Scott, A finite element method for problems in plasticity using discontinuous trial functions, Finite Element Analysis in Structural Mechanics, Proc. of the Europe-US Workshop Ruhr-Universität Bochum, Germany, 1980, Eds. W. Wunderlich, E. Stein and K.J. Bathe.

4. C. Johnson, Error estimates for a finite element method for perfect plasticity, Conf. "Problem i matematicci nella meccanica dei continui", Univ. of Trento, 1981.

5. C. Johnson, Adaptive finite element methods for convection and diffusion problems, Comp. Meth. Appl. Mech. Engrng 82 (1990), 301-322.

6. K. Eriksson and C. Johnson, Adaptive finite element methods for parabolic problems I: A linear model problem, Siam J. Numer. Anal. 28 (1991), 43 - 77.

7. C. Johnson, Adaptive finite element methods for the obstacle problem, Technical report, Math. Dept. Chalmers Univ. of Technology, 1991.

8. G. Strang, A minimax problem in plasticity theory, in Functional Analysis Methods in Numerical Analysis, ed. M.Z. Nashed, Springer Lecture Notes in Mathematics 701 (1979).

9. R. Temam, Mathematical Problems in Plasticity, Gauthier-Villar, 1985.

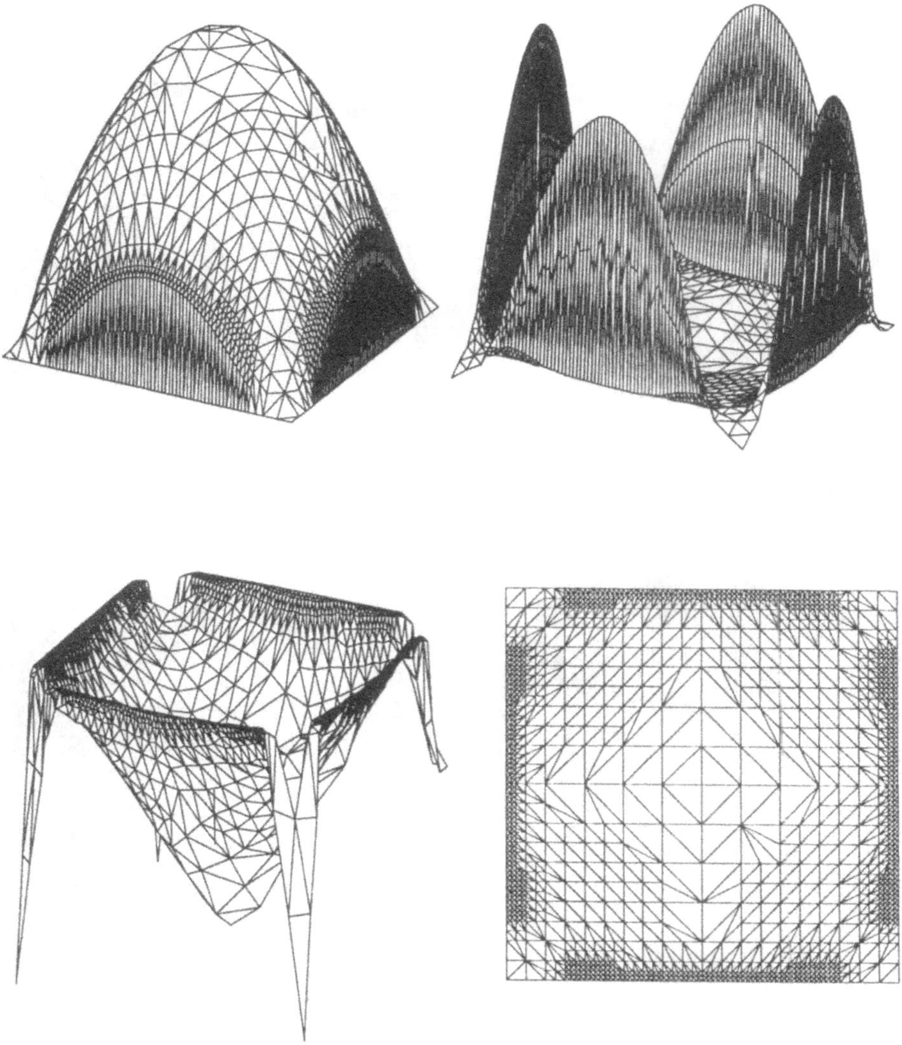

Fig. 5.1. Elevations of u_h, $|\nabla u_h|$, $|\sigma_h|$, and final mesh with $f(x) = 2 + \sqrt{\pi}$.

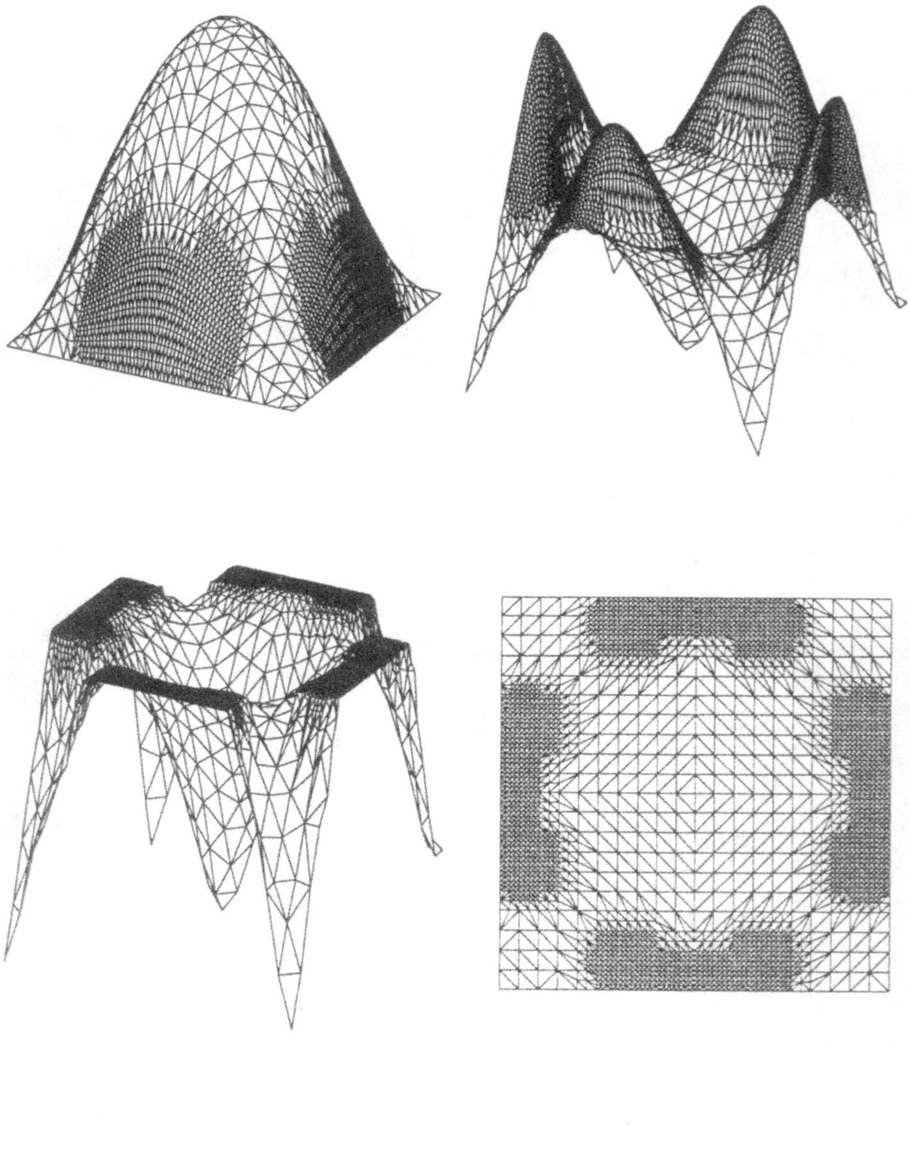

Fig. 5.2. Same as in Fig. 5.1 with $f(x) = 7.5 \sin \pi x_1 \sin \pi x_2$.

A Numerical Method for Solution of Nonsteady Elastoviscoplastic Problems at Large Strains

V.N.Kukudzanov

The Institute for Problems in Mechanics
USSR Academy of Sciences
Moscow, 117526, U.S.S.R.

SUMMARY

The new numerical approach to the solution of the finite deformation elastoplasticity is proposed. The equation of the motion are integrating by the natural approximation of the spatial derivatives for the curvilinear arbitrary mesh, and for the integration of the constitutive equations the splitting method is developed.
First the idea of the method is explained for the classical case of the Prandtl-Reiss equations and after the method is generalized for the finite deformation and nonclassical plasticity.
The moving adopted mesh for very large deformations of the media were used.
The model of damaged elastoplastic media with microdefects is also presented. The failure of material is considered as the process of the accumulation of the growing microvoids.
The developed approach was applied to this model and finally some numerical examples of dynamic problems were presented which illustrate the possibilities of the method.

1. GOVERNING EQUATIONS IN MOVING COORDINATES

The movement of continuous media is described in "Eulerian" or "Lagrangian" coordinates usually. But for the numerical calculations of nonsteady problems with large elastoplastic deformations the both methods are not sufficient, the accuracy of calculations is decreased then the deformations are increased [1-3]. The most effective way to overcome this difficulties is to use the moving coordinates or the moving

D. Besdo, E. Stein (Eds.)
Finite Inelastic Deformations -
Theory and Applications
IUTAM Symposium Hannover/Germany 1991

nets which are adopted to the obtained solution, and have to be found together the solution of the problem.
Let us introduce the coordinates η^i moving relatively to the frame of reference x^k

$$x^k = x^k(\eta^i, t), \quad \eta^i \in V, \; t \ge 0 \tag{1.1}$$

The motion of material particles ξ_k we shall describe in the coordinates η^i

$$\eta^i = \eta^i(\xi^k, t), \quad \xi_k \in V_0, t \ge 0 \tag{1.2}$$

It is clear that eq-s (1.1) and (1.2) describe particles motion at x^k.
The conservation law of the value **A** in the volume V_i has the following differential form of equation in coordinates η^i

$$\frac{\partial}{\partial t}\Delta \mathbf{A} + \frac{\partial}{\partial \eta^\alpha}\Delta(\mathbf{A}v^k_r + \mathbf{B}) = \Gamma^i_{\alpha\beta}\Delta(\mathbf{A}v^k_r + \mathbf{B}) + \mathbf{f}\rho, \quad \Delta = \left|\frac{\partial x^k}{\partial \eta^i}\right|$$

in balance laws of mass, momentum and energy, **A** equals ρ, ρv^i_r, $\rho\varepsilon$ and **B** equals 0 , $\sigma^{i\alpha}$, $(\sigma^{k\alpha} g_{k\beta} v^\beta - \kappa\, g^{\alpha\beta} \nabla_\beta T)$ respectively.
The control of coordinates η^i is realized by means of the variation of the knots velocities **w** of the moving net. This question is discussed in details in works [2,3].

2. CONSTITUTIVE EQUATIONS OF ELASTOPLASTICITY AND THEIR INTEGRATION

The theory of plastic flow based on the associated rule with the loading surface $F(\sigma,\kappa)=0$ can be obtained in the form

$$d\varepsilon_{ij} = (D^{-1}_{ijkl} + HF_{ij}F_{kl})d\sigma_{kl} = A^{-1}_{ijkl}d\sigma_{kl} \tag{2.1}$$

$$D^{-1}_{ijkl} = \frac{1}{2\mu}\left[\delta_{ij}\delta_{jl} - (1 - \frac{2\mu}{3k})\delta_{ij}\delta_{kl}F_{ij}\right] = \frac{\partial F}{\partial \sigma_{ij}}, \quad H = -\frac{\partial F}{\partial \kappa}\,\frac{\partial F}{\partial \sigma_{kl}}\sigma_{kl}$$

Equation (2.3) can be inverted

$$d\sigma_{ij}=[D_{ijkl}-(2\mu)^2 F_{ij}F_{kl}(H^{-1}+2\mu F_{mn}F_{mn})^{-1}]d\varepsilon_{kl}=A_{ijkl}d\varepsilon_{kl} \quad (2.2)$$

This hypoelastic form of equation gives purely differential relations between stress and strain tensors. The eq. (2.1) are often used in the numerical analysis of elastoplastic structures [7]. But this form requires the differentiation of the yield surface which decreases the accuracy of calculations. An another way which doesn't need differentiation but directly use the yield condition in original form can be suggested.
Using the Hook's law for elastic deformation we have

$$\frac{ds_{ij}}{dt} = 2\mu \frac{d\dot{\varepsilon}_{ij}}{dt} - 2\mu\Lambda s_{ij} \quad , \qquad s_{ij} = \sigma_{ij} - \frac{1}{3}\sigma_{ii} \quad (2.3)$$

Here Λ is unknown scalar parameter. Eq. (2.3) can be numerically integrated by means of the splitting method according to the physical processes [3]. On the first step (predictor) we assume on the right side of eq. $\varepsilon^p = 0$ hence $\varepsilon = \varepsilon^e$, and consequently stress deviator is determined on the base of the Hook's law

$$s^e_{ij} = s^n_{ij} + 2\mu\dot{\varepsilon}^{n+1/2}_{ij} \Delta t \quad (2.4)$$

where strain rate tensor $\dot{\varepsilon}^{n+1/2}$ is obtained by explicit scheme from the eq-s of motion.
On the second step (corrector) the total strain rate tensor $\dot{\varepsilon}=0$ is assumed, so the deformations are fixed and stress tensor is relaxed according to the plastic law. The eq. (2.4) is approximated by the second order scheme of the accuracy with the initial conditions obtained on the first step

$$s^{n+1}_{ij}=s^e_{ij} - \Lambda^{n+1}(s^{n+1}_{ij}+\Lambda^n s^e_{ij})\mu\Delta t \quad (2.5)$$

$$s^{n+1}_{ij}=s^e_{ij}(1- \Lambda^n\mu\Delta t)(1 + \Lambda^{n+1}\mu\Delta t)=\hat{s}_{ij}N^{n+1}$$

Substituting the founded value s^{n+1}_{ij} to the yield condition one determines exactly the "correction" factor N^{n+1}.

$$N^{n+1}= 1+\Lambda^{n+1}\mu\Delta t = \frac{K}{(\hat{s}_{ij}\hat{s}_{ij})^{1/2}} \, , \quad s^{n+1}_{ij}= \frac{K}{(\hat{s}_{ij}\hat{s}_{ij})^{1/2}}$$

This procedure has the second order of accuracy and was suggested in [1,6],it generalized the Willkins' correction rule [8] of the first order of accuracy.
Now the general yield condition for the time dependent plasticity taking into account the strain rate's and strain hardening effects is considered of the following form

$$S=K(\kappa,\tau\dot{I}_p)\ ,\qquad \dot{I}_p=(\dot{\varepsilon}^p_{ij}\dot{\varepsilon}^p_{ij})^{1/2}\ ,\qquad \dot{\kappa}=\dot{\lambda}\ q(S) \tag{2.6}$$

τ is time of relaxation , κ strain hardening parameter which is described by the evolution equation, $K(\kappa,\tau\dot{I}_p)$ and $q(S)$ are functions determined from the experimental data. The expression of the $\hat{s}_{ij}$ doesn't change , but Λ^{n+1} and κ^{n+1} should be found from eq-s (2.6)

$$S^{n+1} = (s^{n+1}_{ij}\,s^{n+1}_{ij})^{1/2} = K(\kappa^{n+1},\tau\Lambda^{n+1}S^{n+1})$$
$$\kappa^{n+1} = \kappa^n + \frac{1}{2}\,\Delta t(\Lambda^{n+1}q^{n+1}+\Lambda^n q^n\ )$$

These eq-s are nonlinear according to Λ^{n+1} and it should be linearized and solved iteratively. Expanding function $K(x,y)$ and $q(\kappa)$ at the point found from the predictor we obtain

$$N^{n+1}_{(i+1)} =\mu(S^e - K^e + K_y(N_{(i)}-1)S^e\,\frac{\tau}{\mu\Delta t} - K_x(N_{(i)}-1)q^e\mu) \tag{2.7}$$
$$[S^e(K^e_x q^e_x + K^e_y\,\frac{\tau}{\Delta t}\ ) + \frac{1}{2}K^e_x(q^e-q^e_*\ S^e) + \mu K^e-K_x(N_{(i)}-1)S^e\frac{\tau}{\Delta t}\ ]^{-1}-1$$

index (i) indicates number of iteration, indices (x,y) mean the derivatives of the functions $K(x,y)$, index (e) shows that arguments of functions are equal to the predictor values.
From this formula one can see that N^{n+1} depends on the relation $\tau/\Delta t$ only and then $\Delta t \gg \tau$ the solution transforms to the solution for the rate nonsensitive materials. The yield condition of the nonsensitive material satisfies exactly when $\tau = 0$. This property of the numerical scheme is very important. If the scheme doesn't possess this asymptotic property it gives incorrect results for $t \gg \tau$. For example the

scheme used by Malvern didn't have this asymptotic property for $t \gg \tau$ and therefore the results obtained in [9] were incorrect as shown in [10].
The suggested method effectively can be generalized to the finite deformations. The constitutive eq-s differ from small deformations in elastic law which is nonlinear and in equations of plastic flow where the material derivatives should be replaced to the objective which excluded the rigid body rotation of the particle (hyperplastic case).

3. ELASTOPLASTIC DAMAGE MODEL

The analysis of experimental data of the dynamic fracture of elastoplastic materials shows that it is connected with the nucleation and growth of microdefects. So it can be described as a process of the damage accumulation in frames of continuum mechanics until the fragmentation of the material will take place. For ductile materials this process has an isotropic character and consequently the microdefects can be considered as spherical microvoids.
The deformation of such materials has not only elastic and plastic parts but also a part due to the formation and growth of defects. The deformation gradient **F** is decomposed into three parts **F=EDP**, where **P** is viscoplastic, **D** - spall dilatation caused by defects and **E** - elastic deformation gradients. It is convenient to introduce the following tensors of deformations:

$$\varepsilon = (\mathbf{I} - \mathbf{F}^{-T}\mathbf{F}^{-1})/2, \qquad \varepsilon_e = (\mathbf{I} - \mathbf{E}^{-T}\mathbf{E}^{-1})/2,$$

$$\varepsilon_P = \frac{1}{2}\,\mathbf{E}(\mathbf{I} - \mathbf{P}^{-T}\mathbf{P}^{-1})\mathbf{E}^{-1}d^{-2}, \qquad \varepsilon_D = \frac{1}{2}\,(1 - d^{-1})\mathbf{E}^{-T}\mathbf{E}^{-1}$$

Expressing the elastic deformation gradient in the form $\mathbf{E}=\alpha\mathbf{I}+\mathbf{U}$ and taking into account that **U** is small, we find

$$2\varepsilon_e = (1 - \alpha^{-2})\mathbf{I} + 2\epsilon_e\alpha^{-1} + O(\mathbf{U}^2) \tag{3.1}$$

Introducing the decomposition tensor of velocity gradient **L** we can write the following relation

$$\mathbf{L} = \dot{\mathbf{F}}\mathbf{F}^{-1} = \dot{\mathbf{E}}\mathbf{E}^{-1} + \dot{d}d^{-1}\mathbf{I} + \mathbf{E}\ddot{\mathbf{P}}\mathbf{P}^{-1}\mathbf{E}^{-1} = \mathbf{L}_e + \mathbf{L}_D + \mathbf{L}_p \qquad (3.2)$$

The constitutive eq-s can be described in terms of the internal variables

$$\dot{\mathbf{L}}_p = \mathbf{L}_p(\mathbf{Q}_p,\mathbf{Q}_E) \qquad \dot{d} = d(\mathbf{Q}_p,\mathbf{Q}_E) \qquad \dot{\chi}_l = \chi_l(\mathbf{Q}_E,\mathbf{Q}_p). \qquad (3.3)$$

$\varepsilon_p, \varepsilon_D$ and structural parameters χ_m are denoted as $\mathbf{Q}_p$ and ε_e, T, grad T as $\mathbf{Q}_E$.

The state variables as free energy Ψ , stress tensor σ and heat flux $\mathbf{q}$ are assumed to be functions of $\mathbf{Q}_E$ and $\mathbf{Q}_p$

$$\Psi = U - \eta T = \Psi(\mathbf{Q}_E) \ , \ \sigma = \sigma(\mathbf{Q}_E) \ , \ \mathbf{q} = \mathbf{q}(\mathbf{Q}_E,\mathbf{Q}_p). \qquad (3.4)$$

The final restrictions to be imposed on the of constitutive equations are following from the thermodynamic Clausius-Duhem inequality.

$$\sigma:\mathfrak{L}_p + (\sigma_{ll}d^{-1} - \frac{\partial\Psi}{\partial T}\rho)\dot{d} - \rho\frac{\partial\Psi}{\partial\chi_l}\dot{\chi}_l - \mathbf{q}\ \text{grad}\ T\ T^{-1} \geq 0 \qquad (3.5)$$

The thermodynamic analysis provides the usual stress and entropy relations and the dissipative inequality.

$$\sigma:\mathfrak{L}_p + (\sigma_{ll}d^{-1} - \frac{\partial\Psi}{\partial T}\rho)\dot{d} - \rho\frac{\partial\Psi}{\partial\chi_l}\dot{\chi}_l - \mathbf{q}\ \text{grad}\ T\ T^{-1} \geq 0$$

$$\sigma = p(\alpha,T,d)\mathbf{I} + 2\mu(\alpha,T,d)\ \epsilon_e \ , \qquad \eta = \frac{\partial\Psi}{\partial T} \qquad (3.6)$$

Eq-s (3.6) are in fact Hook's law for porous media with functions $p(\alpha,T,d)$ and $\mu(\alpha,T,d)$ which are determined as the corresponding function for porous materials.

Now we shall use the principal of the minimal irreversible forces to determine the viscoplastic response of material, which satisfies (3.6). Assuming that the mutual influence of thermal and mechanical "flows" can be neglected, and dissipating function D_m has a form $D_m = D_1(\mathfrak{L}_p) + D_2(\dot{d})$ we find from the principle of minimal irreversible forces

$$\sigma = \frac{D_1(\mathbf{I}_p)}{\mathbf{I}_p} \mathfrak{L}_p \qquad S = (s^{ij}s_{ij})^{1/2} = D_1(\mathbf{I}_p) \ . \tag{3.7}$$

S is the second invariant of the Cauchy stress tensor.
For viscoplastic materials S is a function of the strain rate, which can be obtained from a lot of recently available quasistatic and dynamic experimental data (see [1]).
Assume that the generalization of Sokolovski-Malvern theory takes place, it follows

$$S = K_s(\chi_m, T, d) + \varphi(\mathbf{I}_p) \tag{3.8}$$

the function $K_s(\chi_m, T, d)$ is determined as an approximation of the static data for porous materials and $\varphi(\mathbf{I}_p)$ from the dynamic data. The expression for $\mathfrak{L}_p$ takes the form

$$\mathfrak{L}_p = \frac{\Phi(S_{II} - K_s(\chi_m, T, d))}{S_{II}} \sigma \ , \quad \Phi(x) = \varphi^{-1}(x) \ . \tag{3.9}$$

And for d we find the differential equation

$$\dot{d} = D_2(d)\left(\frac{\sigma_{ii}}{\rho d} - \frac{\partial \Psi}{\partial d}\right) . \tag{3.10}$$

Eq-s (3.10) includes the wide class of dependencies, particularly the experimental one suggested in [4].
There are different approaches in the literature to the study of the large deformations and the damage of elastoplastic materials [1,4,5]. We can't discuss details of the model in this paper, some of them were analyzed in our works [1,2].

4. ELASTOPLASTIC IMPACT OF PLATES AND CYLINDERS

The impact of a cylindrical projectile on a plate been studied. Both the plate and the projectile were aluminium with following parameters the initial density $\rho_0 = 2.65$ g/cm^3 , the share yield stress K_{s0}=0.077 GPa, the elastic moduli were $\lambda=725K_{s0}$, $\mu=370K_{s0}$, the velocity of longitudinal sonic wave C_0=6350 m/s. the constants in the kinetic fracture equations were taken from Ref.[4]. The calculations were performed for two velocities V_0=0.03 C_0 (Fig. 1a) and V_0 = 0.24 C_0 (Fig. 1b). The choice of the fracture criterion was based on critical microvoids volume $v_e=1-d=8\pi R^3 N$.

Figs. 1b,c show the fracture zones in which v 0.12 .

Figure 1a also illustrates the calculations for the interaction between the projectile of radius 5.7H_0 and the plate for the moment of time $t=5.78t_0$. In this case the impact was close to one-dimensional. The dashed line indicates the position of the spalling line obtained from the experimental data [4] . it is seen that the spalling line is well determined. There are also some essential differences as well associated with a multi-dimensional character of the interaction - the plate starts fracturing not of the simmetry axis and then propagates to the center (Fig. 1a). At a certain moment of time almost penny-shaped zone of damaged material is formed.

Figure 1b presents the data calculated for a projectile of radius 1.42 and an initial velocity of $V_0=0.24C_0$ for the moments of time $t=3.16t_0$. Fracture appears first in the projectile. The characteristic feature is the formation of the of the fracture zone in the target prior to the arrival of unloading waves from the rear surface (Fig. 1a). The zone of tensile stresses in which microvoids nucleated and accumulated appears as a result of superposition of two unloading waves - one from the free face surface of the plate and another from the projectile. The latter wave generated due to almost complete unloading of the projectile. The unloading of the projectile results in its separation from the plate. The damaged zone on the face side of the plate may be interpreted as the "face spallation".

The general trend is such that with increase of the initial velocity, the other conditions being the same, fracture zones shift from the face side of the plate surface towards the rear surface and the zones themselves are formed earlier. it is explained by intense and rapid unloading of the projectile. It should be noted that an increase in the plate thickness does not affect the moment of formation and position of fracture zones on the face side of the plate . This confirms the above explanation of the face side of the plate. These results agree with the experimental data.

The fracture of an elastic (glass) plate is illustrated in Fig.1c for time $t=6.43t_0$. The projectile material was steel, the initial velocity $V_0=0.12C_0$, where C_0 is the sound velocity in the steel. The target thickness was $H_1=0.625H_0$. The projectile position at the initial moment of time is indicated

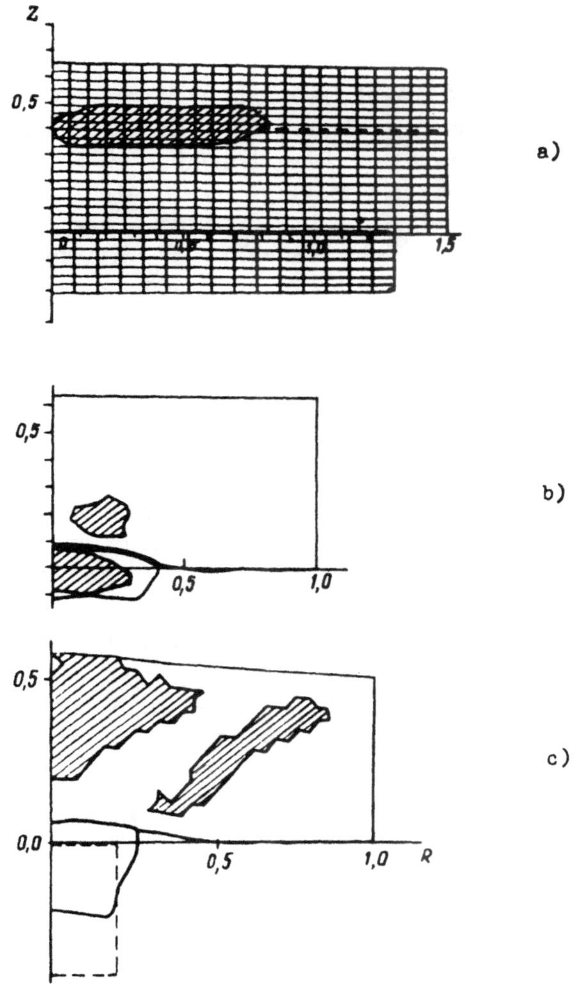

Fig.1

by the dashed line. The elastic target is strained under the action of the projectile without cratering and back ejection typical for elastic-plastic impact. The fracture zone of the elastic plate (dashed zones in Fig. 1c) has the shape of a return cone at the base of which the rear spallation occurs.

REFERENCES

1. Kukudzanov V.N.,Numerical modeling of dynamic processes of deformation and fracture in elastoplastic medium., Uspechi Mechaniki, 4,(1985),21-65.
2. Kukudzanov N.N.,Saparov K.I.,A numerical solution of non steady problems of the dynamics of elastoplastic medium by the method of moving coordinates.,Numerical methods in mechanics of solids, Computer Center Ac.Sci. USSR,Moscow,(1984), 85-103 3. Kovenia V.M., Janenko N.N.,Splitting method in gasdynamical problems., Novosibirsk, Nauka, 1981, 304.
4. Seaman L., Curran D.R. Shockey D.,Computational models for ductile and brittle fracture., Jour. Appl. Phys.,v.47, 1,(1977),4814-4826.
5. Davison L. Stevens A.L., Kipp M.E.,Theory of spall damage accumulation in ductile metals., Jour. Appl. Mech and Phys., v.25 , 1,(1977),11-29.6. Kukudzanov V.N.,Kudriashov Y.I.,Mixed problems of non stationary interaction between gas and solids.,IPM of Ac.Sci. USSR, Moscow,(1990),preprint 472,1-29.
7. O.C.Zienkevicz, The finite element method in engineering science., McGrow-Hill, London, (1971), 541.
8. M.Willkins ,Analysis of the elastoplastic flow., In book "The numrical methods in hydrodynamics"., Mir,Moscow,(1967), 232-263.
9. L.Malvern, The propagation of longitudional plastic wave with the influence of the strain rate., Jour.Appl.Mech.,N 2,(1951).
10. Kukudzanov V.N.,The asymptotic and numerical integration of equations of the dynamics of elastoplastic media with small parameter., Proceedings of BAIL IV. Bool Press, Dublin, (1986), 130-136.

A Model for Large Deformations of Elasto-Viscoplastic Solids at Finite Strain: Computational Issues

Djordje Perić and D.R.J. Owen

Department of Civil Engineering, University College of Swansea, Swansea SA2 8PP, Wales, U.K.

Abstract: *In the context of general isothermal processes, issues related to the constitutive modelling and computational treatment of large deformations of elasto-viscoplastic solids at finite strains are examined employing logarithmic stretches as strain measures. The material is assumed to remain isotropic throughout the process of deformation. A strain-energy function for isotropic elastic materials is utilized, which leads to a linear stress-strain relationship and constant and isotropic elastic modulus in material setting. A computationally effective scheme is proposed based on a fully implicit integration of the constitutive equations, which is exact for elastic processes. It is pointed out that within the proposed framework the small strain integration algorithms and corresponding consistent tangent operators automatically extend to the finite strain regime. The versatility and robustness of the method is illustrated through the numerical analysis of a structural problem typical for the metal forming applications, capturing all silent features of this complex problem and resulting in failure by the strain localisation.*

1. Introduction

The numerical simulation of large deformations of inelastic solids at finite strains is of great practical significance and has been intensively studied. Essentially, two approaches may be distinguished : rate-based (hypoelastic) formulations and those which rely on hyperelastic material characterization. By adopting *ab initio* non-objective variables within the rate-based formulation, the incremental objectivity of the numerical algorithm needs special attention, which is a non-trivial task (see Hughes [8] for details). Alternatively, if the hyperelastic characterization of the elastic response is employed, the need for integration of the rate form of the constitutive equation is avoided. This has been utilized in computational literature employing the rotationally invariant part of the deformation gradient in the rate constitutive equations based on the natural approach advocated by Argyris and co-workers [2], or employing the Green-Lagrange strains which offer a simple and general computational framework (Moran *et al.* [12] and Simo [26]). Some successes in formulating computational algorithms based on hyperelastic constitutive equations in the principal directions, mainly employing logarithmic strains, have already been achieved (see Eterović and Bathe [4], Perić *et al.* [15,17], Rolph and Bathe [24] and Weber and Anand [27]).

This work is concerned with the formulation and numerical implementation of a constitutive model for large deformations of elasto-viscoplastic solids at finite strains

D. Besdo, E. Stein (Eds.)
Finite Inelastic Deformations -
Theory and Applications
IUTAM Symposium Hannover/Germany 1991

when logarithmic stretches are employed as strain measures. This model is considered as a representative example of possible models in the principal axes and provides an extension of our previous work (see Perić *et al.* [15,17,18]) to rate dependent materials (see also [3,16,20]). The material is assumed to remain isotropic throughout the process of deformation. This necessitates a representation of the plastic flow based on the scalar internal variables.

By adopting the logarithmic strain as a kinematic variable and restricting attention to the materials which retain isotropy, a particularly simple model for large inelastic deformations at finite strains is recovered, based on the following observations:

(i) A strain energy function for isotropic elastic solids is utilized, which leads to linear stress-strain relationships in material and spatial settings, as well as to constant and isotropic elasticity tensor in material setting (see Perić [15]). This strain energy function is also in good agreement with experiments for a broad class of materials under moderately large deformations (Anand [1]).

(ii) The small strain form of the incompressibility constraint is automatically extended to the finite strain regime enabling the separation of distortional and dilatational deformations in a trivial manner. This allows for the automatic transfer of the standard stress-space based plasticity theories into the finite strain range.

An immediate consequence is that within the proposed model a finite strain theory effectively appears on the level of kinematics independently of the particular material characteristics of the constitutive model. Accordingly, a particularly simple computational framework is recovered which enables immediate transfer of the small strain integration algorithms and corresponding tangent operators into the finite strain range. Although it is not critical for the proposed approach, a restriction to small elastic strains simplifies treatment of the boundary value problem, as discussed in [15,17], where precise margins of validity for the introduced assumption are established. This situation is of considerable interest, especially in metal plasticity, and provides a basis for our large-scale computations.

2. Constitutive equations

2.1 Preliminaries

In this section, the essential notation and terminology which is already standard in contiuum mechanics is provided (see e.g. Marsden and Hughes [11] for complete account):

$\chi : \Omega \subset \mathbb{R}^{n_{\dim}} \times \mathbb{I} \rightarrow \mathbb{R}^{n_{\dim}}$ ... *deformation* from the *reference* configuration $\Omega \subset \mathbb{R}^{n_{\dim}}$, $1 \leq n_{\dim} \leq 3$ of the body $\mathcal{B}$ where $\mathbb{I} \in \mathbb{R}_+$ is the time interval of interest

$\boldsymbol{x} = \chi(\boldsymbol{X}, t)$ for each $\boldsymbol{X} \in \Omega$... *current* or *deformed* configuration

$\boldsymbol{E}_I(\boldsymbol{e}_i), (I, i = 1,2,3)$... bases in the reference (current) configuration

$\boldsymbol{F} = \frac{\partial \chi(\boldsymbol{X},t)}{\partial \boldsymbol{X}} \equiv \frac{\partial \boldsymbol{x}}{\partial \boldsymbol{X}}$... *deformation gradient*

$J = \det \boldsymbol{F}$... Jacobian of the deformation mapping

$\boldsymbol{F} = \boldsymbol{R}\boldsymbol{U} = \boldsymbol{V}\boldsymbol{R}$... polar decomposition of $\boldsymbol{F}$

$\boldsymbol{U}(\boldsymbol{V})$... right (left) *stretch* tensor

$\boldsymbol{R}$... *rotation* tensor

$$\lambda_{(I)} = \lambda_{(i)}(I, i = 1,2,3) \ldots \text{ \textit{principal stretches} – eigenvalues of } \boldsymbol{U}(\boldsymbol{V})$$
$$\boldsymbol{N}^{(I)}(\boldsymbol{n}^{(i)}), (I, i = 1,2,3) \ldots \text{ \textit{Lagrangian (Eulerian) triad} – eigenvectors of } \boldsymbol{U}(\boldsymbol{V})$$
$$\boldsymbol{C} = \boldsymbol{F}^T\boldsymbol{F} = \textstyle\sum_{I=1}^{3} \lambda_{(I)}^2 \boldsymbol{N}^{(I)} \otimes \boldsymbol{N}^{(I)} \ldots \text{ \textit{right Cauchy-Green} tensor}$$
$$\boldsymbol{b} = \boldsymbol{F}\boldsymbol{F}^T = \textstyle\sum_{i=1}^{3} \lambda_{(i)}^2 \boldsymbol{n}^{(i)} \otimes \boldsymbol{n}^{(i)} \ldots \text{ \textit{left Cauchy-Green} tensor}$$
$$\boldsymbol{E} = \ln \boldsymbol{U} = \textstyle\sum_{I=1}^{3} \ln(\lambda_{(I)}) \boldsymbol{N}^{(I)} \otimes \boldsymbol{N}^{(I)} \ldots \text{ material \textit{logarithmic} strain tensor}$$
$$\boldsymbol{e} = \ln \boldsymbol{V} = \textstyle\sum_{I=1}^{3} \ln(\lambda_{(i)}) \boldsymbol{n}^{(i)} \otimes \boldsymbol{n}^{(i)} \ldots \text{ spatial \textit{logarithmic} strain tensor}$$
$$\boldsymbol{v} = \boldsymbol{v}(\boldsymbol{x}, t) \ldots \text{ \textit{spatial velocity}}$$
$$\boldsymbol{l} = \nabla \boldsymbol{v}(\boldsymbol{x}) = \tfrac{\partial \boldsymbol{v}(\boldsymbol{x})}{\partial \boldsymbol{x}} \ldots \text{ \textit{velocity gradient}}$$
$$\boldsymbol{d} = \operatorname{sym} \boldsymbol{l} = (\boldsymbol{l} + \boldsymbol{l}^T) \ldots \text{ \textit{rate of deformation} tensor}$$
$$\boldsymbol{w} = \operatorname{asym} \boldsymbol{l} = (\boldsymbol{l} - \boldsymbol{l}^T) \ldots \text{ \textit{spin} tensor}$$
$$\boldsymbol{\sigma} \ldots \text{ \textit{Cauchy} stress tensor}$$
$$\boldsymbol{\tau} = J\boldsymbol{\sigma} \ldots \text{ \textit{Kirchhoff} stress tensor}$$
$$\rho_{\mathrm{Ref}}\ (\rho) \ldots \text{ reference (current) mass densities}$$
$$\dot{w} = \tfrac{1}{\rho_{\mathrm{Ref}}}(\boldsymbol{\tau} : \boldsymbol{d}) = \tfrac{1}{\rho}(\boldsymbol{\sigma} : \boldsymbol{d}) \ldots \text{ rate of stress work per unit mass}$$
$$\boldsymbol{S} = \boldsymbol{F}^{-1}\boldsymbol{\tau}\,\boldsymbol{F}^{-T} \ldots \text{ \textit{second Piola-Kirchhoff} stress tensor}$$
$$\boldsymbol{P} = \boldsymbol{\tau}\boldsymbol{F}^{-T} = \boldsymbol{F}\,\boldsymbol{S} \ldots \text{ \textit{first Piola-Kirchhoff} stress tensor}$$
$$\boldsymbol{T} = \boldsymbol{R}^T\boldsymbol{\tau}\,\boldsymbol{R} \ldots \text{ rotated stress tensor}$$

2.2 Isotropic elastic solids

We assume the existence of a strain energy function in the form of a scalar symmetric function of its principal stretches $W : \Omega \times \mathbb{R}^{n_{\mathrm{dim}}} \times \mathbb{I} \to \mathbb{R}$ in the form presented in Table 1. Standard procedure leads to the rate form of the hyperelastic constitutive equation $\boldsymbol{T} = \frac{\partial W(\boldsymbol{X}, \boldsymbol{E}^e, t)}{\partial \boldsymbol{E}^e}$ which is given in the material version as (see Perić [15] for details),

$$\dot{\boldsymbol{T}} = \mathsf{H}^e : \dot{\boldsymbol{E}}^e \tag{2.1}$$

where the stress tensor $\boldsymbol{T}$ and elasticity tensor H^e are given in Table 1. Note that H^e is a *constant* and *isotropic* fourth order tensor. Alternatively, in spatial description the rate form of the hyperelastic constitutive equation may be expressed as

$$\overset{\square}{\boldsymbol{\tau}} = J^e \mathsf{h}^e : \boldsymbol{d}^e \tag{2.2}$$

where

$$\overset{\square}{\boldsymbol{\tau}} = \dot{\boldsymbol{\tau}} - \boldsymbol{\omega}^{(R)e}\boldsymbol{\tau} + \boldsymbol{\tau}\boldsymbol{\omega}^{(R)e} \tag{2.3}$$

is the *Green-Naghdi rate* of Kirchhoff stress with $\boldsymbol{\omega}^{(R)e} = \dot{\boldsymbol{R}}^e \boldsymbol{R}^{e\,T}$ and h^e is a spatial elasticity tensor given in Table 1, which is in general *non-constant* and *anisotropic*.

2.3 Isotropic elasto-viscoplastic solids

Our approach to large deformations of elasto-viscoplastic solids at finite strains is based on the use of logarithmic strains as kinematic variables and is described in detail in Perić *et al.* [15,17,18] (see also [3,16,20]). We mention that logarithmic stretches as a 'convenient approximation in the constitutive description of plastic response' (Hill [7]) were recommended by Rice [23], and later confirmed by Hill [7], with an imposed interval of moderate stretches where the model is applicable. By adopting the logarithmic

TABLE 1.
Hyperelastic constitutive model

(i) Strain energy function

$$W^{(0)}(\lambda_{(1)},\lambda_{(2)},\lambda_{(3)}) = \mu\left((\ln\lambda_{(1)})^2 + (\ln\lambda_{(2)})^2 + (\ln\lambda_{(3)})^2\right) + \tfrac{1}{2}\lambda\,(\ln J)^2$$

(ii) Stress-strain relationships

- Material version

$$T_{IJ} = 2\mu E_{IJ} + \lambda(\operatorname{tr}\boldsymbol{E})\delta_{IJ}$$

- Spatial version

$$\tau_{ij} = 2\mu e_{ij} + \lambda(\operatorname{tr}\boldsymbol{e})\delta_{ij}$$

(iii) Elasticity tensors

- Material version

$$\mathsf{H}_{IJKL} = \mu\,(\delta_{IK}\delta_{JL} + \delta_{IL}\delta_{JK}) + \lambda\delta_{IJ}\delta_{KL}$$

- Spatial version

$$\mathsf{h}_{(ijkl)} = \left\{\begin{array}{ll} (J)^{-1}\,\widehat{\mathsf{h}}_{(ijkl)}\,, & k=l \\ (J)^{-1}\,\widehat{\mathsf{h}}_{(ijkl)}\dfrac{\ln(\lambda_{(l)}/\lambda_{(k)})}{\lambda_{(l)}^2-\lambda_{(k)}^2}2\lambda_{(k)}\lambda_{(l)}, & k\neq l \end{array}\right\}$$

(no summation over indices is implied) where

$$\widehat{\mathsf{h}}_{ijkl} = \mu\,(\delta_{ik}\delta_{jl} + \delta_{il}\delta_{jk}) + \lambda\delta_{ij}\delta_{kl}$$

---

strain as a kinematic variable the small strain form of the incompressibility constraint is automatically extended to the finite strain regime which enables the separation of distortional and dilatational deformations in a trivial manner. This allows for automatic transfer of the standard stress space based plasticity theories into the finite strain range. In this work we confine attention to large deformations of elasto-viscoplastic solids at finite strains which retain isotropy throughout the process of deformations, thus restricting inelastic strain rate to be coaxial with the stress tensor.

The viscoplastic constitutive model adopted here is based on the existence of the flow potential $\Phi(\boldsymbol{T}, A)$ (not necessarily in the stricter sence of Rice [22]; see also Halphen and Nguyen [5], Lemaitre and Chaboche [9], Lubliner [10] and Perić [17]) where $A : \Omega \times \mathbb{I} \rightarrow \mathbb{R}^{n_{int}}$ are the forces conjugate to the set of internal variables $\alpha \in \mathbb{R}^{n_{int}}$. Furthermore, we assume the existence of the elastic domain $\mathbb{K}$ as a closed convex set in the $(\boldsymbol{T}, A)$ space defined by the standard expression

$$\mathbb{K} = \left\{(\boldsymbol{T}, A) \in \mathbb{S} \times \mathbb{R}^{n_{int}} : F(\boldsymbol{T}, A) \leq 0\right\} \tag{2.4}$$

The general viscoplastic model which is presented in Table 2 and later further specified through the model problem, originates from the Perzyna's viscoplastic model based on the overstress function [21] and incorporates ideas from [16].

TABLE 2.
Constitutive model for elasto-viscoplasticity — Material version

(i) Additive decomposition of the rate of deformation tensor

$$\dot{\boldsymbol{E}} = \dot{\boldsymbol{E}}^e + \dot{\boldsymbol{E}}^{vp}$$

(ii) Elastic constitutive equation

$$\boldsymbol{T} = \frac{\partial W}{\partial \boldsymbol{E}^e}$$

(iii) Associative flow rule

$$\dot{\boldsymbol{E}}^{vp} = \gamma\langle \Phi(\boldsymbol{T}, A)\rangle \partial_{\boldsymbol{T}} F(\boldsymbol{T}, A)$$

(iv) Hardening law

$$\dot{A} = \gamma H(\boldsymbol{T}, A)$$

3. Numerical integration of constitutive equtions

From the computational standpoint the problem consists of formulating numerical integration procedures for updating the known state variables $\{\boldsymbol{F}_n^e, \boldsymbol{F}_n^p, \boldsymbol{T}_n, A_n\}$ at the configuration $\chi_n(\boldsymbol{X})$, to obtain $\{\boldsymbol{F}_{n+1}^e, \boldsymbol{F}_{n+1}^p, \boldsymbol{T}_{n+1}, A_{n+1}\}$ at the configuration $\chi_{n+1}(\boldsymbol{X})$, where the incremental displacement $\boldsymbol{u}(\boldsymbol{x}_n)$ is assumed given, and the updated configuration and updated deformation gradient can be evaluated through

$$\chi_{n+1} = \chi_n + \boldsymbol{u}, \qquad \boldsymbol{F}_{n+1} = \boldsymbol{F}_u \boldsymbol{F}_n \tag{3.1}$$

where

$$\boldsymbol{F}_u = \boldsymbol{1} + \operatorname{Grad} \boldsymbol{u} \tag{3.2}$$

These must be consistent with constitutive equations from Tables 1-2.

Particularly suitable for implementation are procedures connected with general operator split methodology, where the original problem of evolution is solved through composition, applying first the elastic and then the plastic algorithm.

- *Elastic predictor*

Calculating $\boldsymbol{F}_{n+1}$ from (3.1), and because

$$\boldsymbol{F}_{n+1}^{p\,\mathrm{trial}} = \boldsymbol{F}_n^p \tag{3.3}$$

we easily obtain

$$\boldsymbol{F}_{n+1}^{e\,\mathrm{trial}} = \boldsymbol{F}_{n+1}(\boldsymbol{F}_n^p)^{-1} \tag{3.4}$$

and with logarithmic stretches computed using the algorithm from [17] we evaluate the elastic trial stress. This leads to exact solution of the elastic problem governed by the strain-energy function (i) from Table 1.

- *Plastic corrector*

We solve the equations decribing inelastic response on the fixed intermediate configuration employing standard numerical procedures. We refer to Dutko *et al.* [3], Simo

BOX 1. *Finite strain extension of the stress integration procedure for elasto-viscoplasticity*

(i) Initialization

$$\chi_{n+1}^{(0)} = \chi_n\,, \qquad F_{n+1}^{(0)} = F_n$$

(ii) Update configuration

$$\chi_{n+1} = \chi_{n+1}^{(i)} + u$$

(iii) Compute current deformation gradient

$$F_{n+1} = F_u F_{n+1}^{(i)}\,, \qquad F_u = \mathbf{1} + \operatorname{Grad} u$$

(iv) Start elastic predictor phase

$$F_{n+1}^{p\,\text{trial}} = F_n^p\,, \qquad F_{n+1}^{e\,\text{trial}} = F_{n+1}(F_n^p)^{-1}$$

(v) Evaluate logarithmic strains

$$E_{n+1}^{e\,\text{trial}} = \tfrac{1}{2}\ln C_{n+1}^{e\,\text{trial}}\,, \qquad C_{n+1}^{e\,\text{trial}} = (F_{n+1}^{e\,\text{trial}})^T F_{n+1}^{e\,\text{trial}}$$

(vi) Perform the small strain update

GO TO BOX 2

(vii) Update intermediate configuration

$$V_{n+1}^e = R U_{n+1}^e R^T\,, \qquad F_{n+1}^p = (V_{n+1}^e)^{-1} F_{n+1}$$

and Hughes [25] and Zienkiewicz and Taylor [28] for discussion on the recent advances and details of implementation of the rate independent material models. Particular attention to the numerical integration of the viscoplastic problem of evolution has been given in the works of Ortiz *et al.* [13], Peirce *et al.* [14] and Zienkiewicz and Taylor [28]. Formulation and algorithmic aspects for the class of viscoplastic constitutive equations which are considered in this work are provided in Perić [16].

For convenience, the details of the algorithm relevant to the finite strain extension are summarized in Box 1. Observe that within the proposed framework, the finite strain theory operates only on the level of kinematics.

4. Numerical solution strategy - Model problem

4.1 Model problem

We consider a constitutive model for elasto-viscoplasticity based on a von Mises yield criterion and power law isotropic hardening represented by

$$F(T, K) = \mathcal{F}(T) - K\left(\overline{E}^{vp}\right) \tag{4.1}$$

$$\mathcal{F}(T) = \sqrt{J_2} \tag{4.2}$$

$$K = H_\circ \left(E_\circ + \overline{E}^{vp} \right)^n \tag{4.3}$$

where $H_\circ$, $E_\circ$ and n are material constants and $\overline{E}^{vp}$ is the equivalent viscoplastic strain†. Furthermore, we introduce the viscoplastic flow potential as

$$\Phi(\boldsymbol{T}, K) = \left(\frac{\mathcal{F}(\boldsymbol{T})}{K} \right)^N - 1 \tag{4.4}$$

where N is the material parameter representing the rate-sensitivity of the material. The evolution problem described by (4.1-4) is widely accepted as a description of rate-dependent deformations of solids and has a firm experimental basis. This has provided a motivation for its detailed study in following sections (see also Moran *et al.* [12], Peirce *et al.* [14] and Perić [16]).

4.2 Numerical solution strategy

Algorithmic treatment of the established problem of evolution relies on the operator split methodology (see e.g. Simo and Hughes [25]), which results in an elastic predictor / plastic corrector algorithm. In the elastic predictor phase (obtained by freezing of the plastic flow during the time interval $[t_n, t_{n+1}]$), values from the previous converged solution are used as the initial conditions to evaluate the ***elastic trial state***, i.e. exact elastic solution.

If the yield condition is violated the viscoplastic corrector phase is initiated. This is based on the Newton-Raphson procedure for the solution of of the nonlinear set of equations in $(\boldsymbol{T}_{n+1}, A_{n+1})$ (or equivalently in $(\boldsymbol{T}_{n+1}, \alpha_{n+1})$), arising from the discretized form of the viscoplastic problem of evolution (see Perić [16]).

Governing equations for one step of the Newton-Raphson iterative scheme are summarized in Box 2.

5. Numerical example

In this section a numerical example is presented to assess the accuracy and robustness of the constitutive model and the adopted numerical scheme. A full Newton-Raphson method is employed, with unsymmetric tangent stiffness arising from the non-associated frictional contact law as described in Perić and Owen [19]. Converegence of the finite element solution is established on the basis of the standard Euclidian norm of the out-of-balance forces. The use of consistent tangent moduli is shown to be of utmost importance for preserving the quadratic rate of convergence typical for the Newton-Raphson method.

EXAMPLE 5.1 ***Stretching of a circular thin sheet by a hemispherical punch.*** The geometry and material characteristics for this example are shown in Fig.5.1. The analysis is performed employing a 3-D formulation for the quarter of the mesh with appropriate boundary conditons. From a numerical point of view this problem necessitates full three-dimensional analysis with appropriate algorithmic treatment of the frictional contact problem (see [19] for details). To solve this problem we discretize the blank with

---

† For convenience, we shall later use notation $\alpha \equiv \overline{E}^{vp}$

BOX 2. *Iterative procedure for elasto-viscoplasticity*

(i) Initialize

$$k = 0, \quad \alpha_{n+1}^{(0)} = \alpha_n, \quad \boldsymbol{T}_{n+1}^{(0)} = \boldsymbol{T}_{n+1}^{\text{trial}}$$

(ii) Check yield condition and evaluate residuals

$$F_{n+1}^{(k)} = F(\boldsymbol{T}_{n+1}^{(k)}, K(\alpha_{n+1}^{(k)}))$$

$$\boldsymbol{R}_{n+1}^{(k)} = \left\{ \begin{array}{c} (\mathbf{H}^e)^{-1} : (\boldsymbol{T}_{n+1}^{(k)} - \boldsymbol{T}_{n+1}^{\text{trial}}) \\ (\alpha_{n+1}^{(k)} - \alpha_n) \end{array} \right\} + \gamma \Delta t \Phi_{n+1}^{(k)} \left\{ \begin{array}{c} \partial_{\boldsymbol{T}} F_{n+1}^{(k)} \\ -1 \end{array} \right\}$$

IF $\| \boldsymbol{R}_{n+1}^{(k)} \| < TOL_R$ THEN

Set $(\bullet)_{n+1} = (\bullet)_{n+1}^{(k)}$ then GO TO BOX 1

ELSE

(iii) Compute *algorithmic* modulus

$$\boldsymbol{\Xi}_{n+1}^{(k)-1} = \begin{bmatrix} \boldsymbol{\Xi}_{n+1}^{\boldsymbol{TT}(k)} & \boldsymbol{\Xi}_{n+1}^{\boldsymbol{T}\alpha(k)} \\ \boldsymbol{\Xi}_{n+1}^{\alpha \boldsymbol{T}(k)} & \Xi_{n+1}^{\alpha\alpha(k)} \end{bmatrix}^{-1}$$

where

$$\boldsymbol{\Xi}_{n+1}^{\boldsymbol{TT}(k)} = (\mathbf{H}^e)^{-1} + \gamma \Delta t \Phi_{n+1}^{(k)} \partial_{\boldsymbol{TT}}^2 \mathcal{F}_{n+1}^{(k)} + \gamma \Delta t \partial_{\mathcal{F}} \Phi_{n+1}^{(k)} \partial_{\boldsymbol{T}} \mathcal{F}_{n+1}^{(k)} \otimes \partial_{\boldsymbol{T}} \mathcal{F}_{n+1}^{(k)}$$

$$\boldsymbol{\Xi}_{n+1}^{\boldsymbol{T}\alpha(k)} = \gamma \Delta t \partial_\alpha \Phi_{n+1}^{(k)} \partial_{\boldsymbol{T}} \mathcal{F}_{n+1}^{(k)}$$

$$\boldsymbol{\Xi}_{n+1}^{\alpha \boldsymbol{T}(k)} = \gamma \Delta t \partial_{\mathcal{F}} \Phi_{n+1}^{(k)} \partial_{\boldsymbol{T}} \mathcal{F}_{n+1}^{(k)}$$

$$\Xi_{n+1}^{\alpha\alpha(k)} = 1 - \gamma \Delta t \partial_\alpha \Phi_{n+1}^{(k)}$$

(iv) Evaluate increment of stresses and internal variable

$$\left\{ \begin{array}{c} \Delta \boldsymbol{T}_{n+1}^{(k)} \\ \Delta \alpha_{n+1}^{(k)} \end{array} \right\} = -\boldsymbol{\Xi}_{n+1}^{(k)-1} : \boldsymbol{R}_{n+1}^{(k)}$$

(v) Update stresses and internal variable

$$\boldsymbol{\Sigma}^{(k+1)} = \left\{ \begin{array}{c} \boldsymbol{T}_{n+1}^{(k+1)} \\ \alpha_{n+1}^{(k+1)} \end{array} \right\} = \left\{ \begin{array}{c} \boldsymbol{T}_{n+1}^{(k)} \\ \alpha_{n+1}^{(k)} \end{array} \right\} + \left\{ \begin{array}{c} \Delta \boldsymbol{T}_{n+1}^{(k)} \\ \Delta \alpha_{n+1}^{(k)} \end{array} \right\} = \boldsymbol{\Sigma}_{n+1}^{(k)} + \Delta \boldsymbol{\Sigma}_{n+1}^{(k+1)}$$

Set $k = k + 1$ and GO TO (ii)

736 constant strain triangular finite elements, and use 2145 and 612 triangular flat elements to discretize the surfaces of the punch and die, respectively. Spatial discretization of the problem is depicted in Fig.5.2. Results are obtained for a coefficient of friction between tools and blank of $\nu_F = 0.30$. Fig.5.3 gives the deformed mesh for various punch displacements.

The punch force versus punch displacement diagram, presented in Fig.5.4, gives

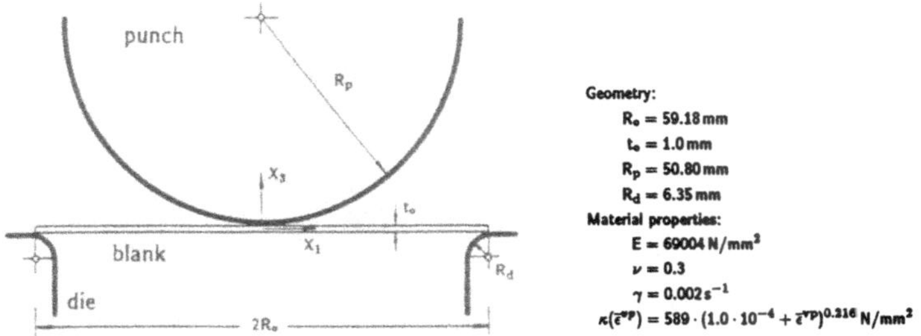

Fig.5.1: Stretching of a circular thin sheet by a hemispherical punch: Geometry and material characteristics.

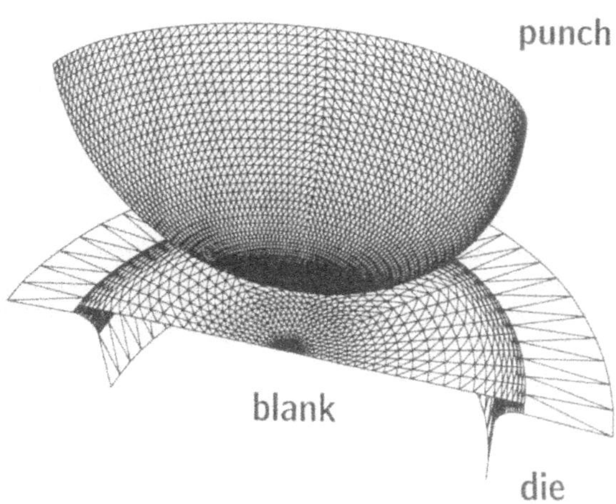

Fig.5.2: Stretching of a circular thin sheet by a hemispherical punch: Spatial finite element discretization of blank and punch and die surfaces.

comparison between results obtained by changing the loading program and the rate sensitivity of the material. The loading program was altered by varying the punch speed between $V_p = 0.5 - 500.0\,\mathrm{mm/s}$ increasing the speed by a factor of ten for every new test. Tests were performed for the rate sensitivity exponents $N = 10, 20, 50$ and 100. For comparison, the underlying rate independent solution based on the yield criterion (4.1-2) and hardening law (4.3) is also presented. The major influence of the rate of loading, demonstrated by the effect on the maximum value of the punch force, is observed for the high rate-sensitive material represented by N=10. As is expected, the maximum punch force increases with the rate of loading. We point out that for a very low punch speed of $V_p = 0.0005\,\mathrm{mm/s}$, the rate independent solution, characterized by

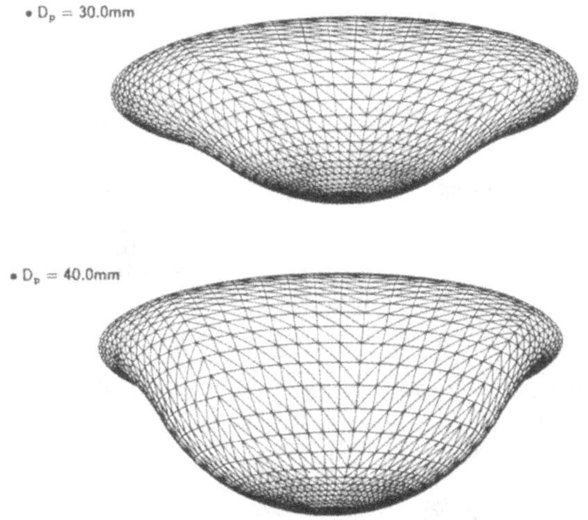

Fig.5.3: Stretching of a circular thin sheet by a hemispherical punch: Deformed finite element mesh at various stages of punch displacement.

the peak of the punch force followed by the falling branch, is recovered. This suggests failure by strain localisation and is illustrated in Fig.5.4. Only a minor difference in the overall force-displacement diagram is observed for the low rate-sensitive material characterized by N=100 over the whole range of the loading rates. A slight difference in the maximum punch force values in Fig.5.4(d) is associated with the influence of very high strain rates in the localisation region.

The distribution of true strain in radial direction is shown in Fig.5.5 for the punch speed $V_p = 0.5\,\text{mm/s}$ and the rate sensitivity exponents $N = 10, 20, 50\,\text{and}\,100$ for punch displacements $D_p = 30\,\text{and}\,40\,\text{mm}$. The underlying rate independent solution is also presented. For the rate independent solution we observe a typical localisation behaviour where strain accumulates in a narrow zone, reaching high levels and leading to failure. This trend is closely followed by the low rate-sensitive material represented by N=100, while the appearance of localisation and associated failure is less pronounced with the increase in rate sensitivity, and is completely averted in the higher rate-sensitive materials characterized by N=10.

A standard check of the residual forces based on the Euclidian norm was performed in this example with convergence tolerance $RTOL = 1.0 \times 10^{-5}$. To satisfy these convergence tolerances, 4-8 iterations were typically needed. The final configuration corresponding to a punch displacement $D_p = 40.0\,\text{mm}$ is attained with a total number of 100 increments for the rate sensitivity exponent $N = 10$, and varies between 104 and 126 increments for the $N = 100$ case. This is connected by an emerging localisation pattern of deformation. The rate of convergence of the Newton-Raphson method measured in residual norm exhibits a quadratic rate of asymptotic convergence whenever the solution is within the radius of convergence.

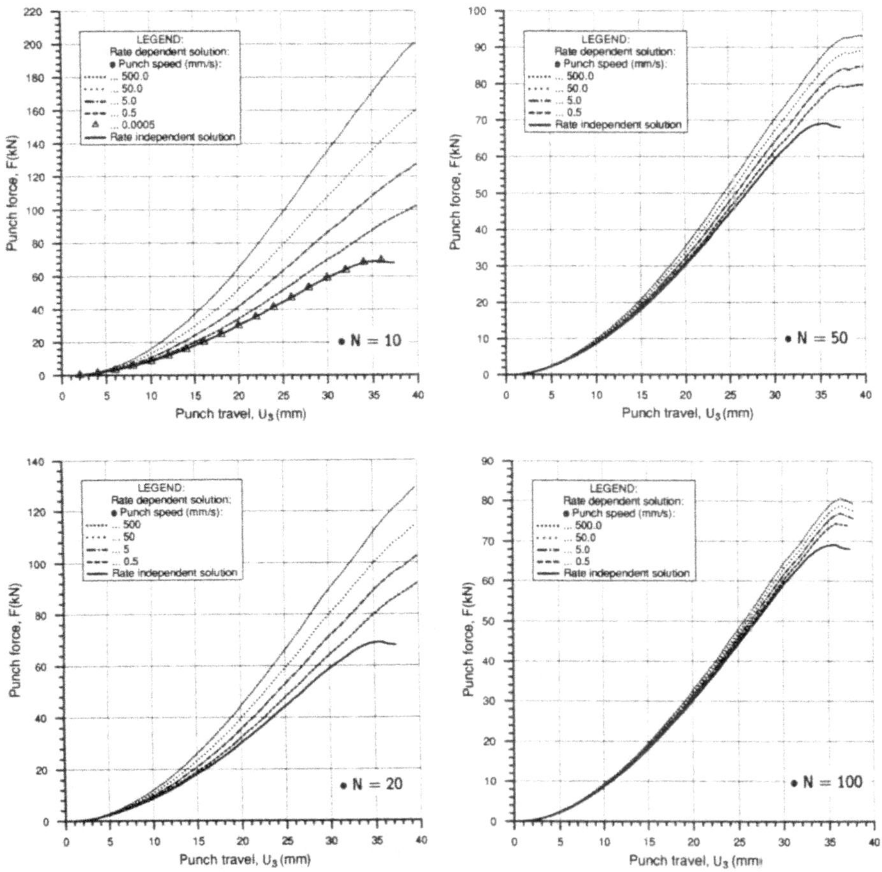

Fig.5.4: Stretching of a circular thin sheet by a hemispherical punch: Punch force versus punch displacement curves for various punch speeds and several values of the rate sensitivity exponent.

5. Concluding remarks

A model for large elasto-viscoplastic deformations at finite strains based on logarithmic strains has been proposed and applied to the large-scale computation of a structural problem typical for metal forming operations. Restriction to isotropic response provides analysis of the broad class of problems characterized by large elastc-viscoplastic deformations at finite strains under predominant monotonic and proportional loading conditions, and situations without significant texture development.

Due to intrinsic features of the logarithmic strain as a kinematic variable, formulation of the large strain elasto-viscoplasticity follows the formalism of the classical small

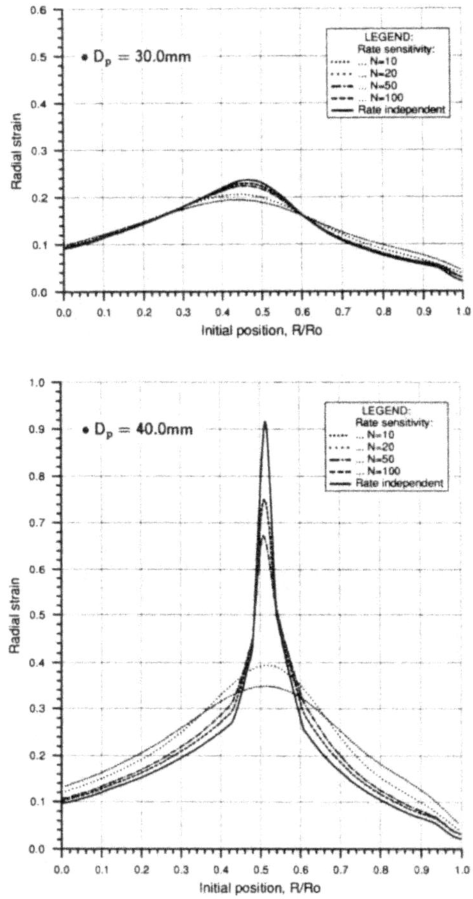

Fig.5.5: Stretching of a circular thin sheet by a hemispherical punch: Distribution of true strain in radial direction plotted over the initial configuration for punch speed V_p =0.5mm/s and several values of the rate sensitivity exponent at various stages of punch displacement.

strain plasticity theories. The finite strain extension effectively appears on the level of kinematics independently of the particular material characteristics of the constitutive model. Consequently, a particularly simple computational framework is recovered which enables immediate transfer of the small strain integration algorithms and corresponding tangent operators into the finite strain range.

The numerical algorithm for integration of differential equations of evolution arising in viscoplasticity developed in [16], which accomodates stiff regions of the viscoplastic problem of evolution, is, in this work, automatically extended to the large strain regime. Numerical results have been provided for large spectrum of material rate sensitivity parameters and various strain rates, with rate independent material behaviour as a limiting case. Both, the material rate sensitivity and the strain rate have proved to have a pronounced effect on the appearance of strain localisation and on the description of the failure mechanism.

Acknowledgment

This work was partially funded through the *BRITE* initiative of the *European Community* (Project P-2029) and this support is gratefully acknowledged.

References

[1] L.ANAND, 'Moderate deformations in extension-torsion of incompressible isotropic elastic materials', *J. Mech. Phys. Solids,* **34,** 293–304 (1986)

[2] J.H.ARGYRIS and M.KLEIBER, 'Incremental formulation in nonlinear mechanics and large strain elasto-plasticity — Natural approach. Part I and Part II', *Comp. Meth. Appl. Mech. Engng.*, **11,** 215–247 (1977) and **14,** 259–294 (1978)

[3] M.DUTKO, D.PERIĆ and D.R.J.OWEN, 'Universal anisotropic yield criterion based on superquadric functional representation. Part 1: Algorithmic issues and accuracy analysis', (submitted for publication) (1991)

[4] A.L.ETEROVIĆ and K.-J.BATHE, 'A hyperelastic-based large strain elasto-plastic constitutive formulation with combined isotropic-kinematic hardening using logarithmic stress and strain measures', *Int. J. Num. Meth. Engng.,* **30,** 1099–1114 (1990)

[5] B.HALPHEN and Q.S.NGUYEN, 'Sur les matériaux standards généralisés' , *J. Mécanique,* **14,** 39–63 (1975)

[6] R.HILL, 'On constitutive inequalities for simple materials. I and II', *J. Mech. Phys. Solids,* **16,** 229–242 and 315–322 (1968)

[7] R.HILL, 'Aspects of invariance in solid mechanics', *Adv. Appl. Mech.* **18,** 1–75 (1978)

[8] T.J.R.HUGHES, 'Numerical implementation of constitutive models: rate-independent deviatoric plasticity', In *Theoretical Foundations for Large Scale Computations of Nonlinear Material Behavior,* (Edited by S.Nemat-Nasser et al.), 29–57, Martinus Nijhoff Publishers, Dodrecht (1984)

[9] J.LEMAITRE and J.-L.CHABOCHE, *Mechanics of Solid Materials,* Cambridge University Press, Cambridge (1990)

[10] J.LUBLINER, *Plasticity Theory* , Macmillan, New York (1990)

[11] J.E.MARSDEN and T.J.R.HUGHES, *Mathematical Foundations of Elasticity*, Prentice-Hall, Englewood Cliffs, New Jersey (1983)

[12] B.MORAN, M.ORTIZ and C.F.SHIH, 'Formulation of implicit finite element methods for multiplicative finite deformation plasticity', *Int. J. Num. Meth. Engng.,* **29,** 483–514 (1990)

[13] M.ORTIZ, P.M.PINSKY and R.L.TAYLOR, 'Operator split methods for the numerical solution of the elastoplastic dynamic problem' , *Comp. Meth. Appl. Mech. Engng.,* **39,** 137–157 (1983)

[14] D.PEIRCE, C.F.SHIH and A.NEEDLEMAN, 'A tangent modulus method for rate dependent solids' , *Comput. Struct.,* **18,** 875–887 (1984)

[15] D.PERIĆ, 'On consistent stress rates in solid mechanics: computational implications', *Int. J. Num. Meth. Engng.,* (accepted for publication) (1991)

[16] D.PERIĆ, 'On a class of constitutive equations in viscoplasticity: formulation and computational issues', (submitted for publication) (1991)

[17] D.PERIĆ, D.R.J.OWEN and M.E.HONNOR, 'A model for finite strain elasto-plasticity based on logarithmic strains: computational issues', *Comp. Meth. Appl. Mech. Engng.,* (accepted for publication) (1991)

[18] D.PERIĆ, G.P.MITCHELL and D.R.J.OWEN, 'Some computational issues in strain localisation problems for rate independent solids', *Int. J. Num. Meth. Engng.,* **32** 79–101 (1991)

[19] D.PERIĆ and D.R.J.OWEN, 'Computational model for 3-D contact problems with friction based on the penalty method', *Int. J. Num. Meth. Engng.,* (accepted for publication) (1991)

[20] D.PERIĆ, D.R.J.OWEN and M.DUTKO, 'Universal anisotropic yield criterion based on superquadric functional representation. Part 2: Extension to large strain elastoplasticity with applications', (to appear) (1991)

[21] P.PERZYNA, 'Thermodynamic theory of viscoplasticity', In *Advances in Applied Mechanics,* **11**, 313–354, Academic Press, New York (1971)

[22] J.R.RICE, 'Inelastic constitutive relations for solids: An internal-variable theory and its aplication to metal plasticity' , *J. Mech. Phys. Solids* **19,** 433–455 (1971).

[23] J.R.RICE, 'Continuum mechanics and thermodynamics of plasticity in relation to microscale deformation mechanisms', In *Constitutive Equations in Plasticity,* (Edited by A.S.Argon), 23–75, MIT Press, Cambridge, Massachusetts (1975)

[24] W.D.ROLPH III and K.J.BATHE, 'On a large strain finite element formulation for elasto-plastic analysis', In *Constitutive Equations: Macro and Computational Aspects,* (Edited by K.J.Willam), Winter Annual Meeting, 131–147, ASME (1984)

[25] J.C.SIMO and T.J.R.HUGHES, 'General return mapping algorithms for rate-independent plasticity', In *Constitutive Laws for Engineering Materials: Theory and Applications,* (Edited by C.S.Desai et al.), 221–231, Elsevier Science Publ., New York (1987)

[26] J.C.SIMO, 'A framework for finite strain elastoplasticity based on maximum plastic dissipation and multiplicative decomposition: Part I. Continuum formulation and Part II. Computational aspects', *Comp. Meth. Appl. Mech. Engng.,* **66**, 199–219 and **68**, 1–31 (1988)

[27] G.WEBER and L.ANAND, 'Finite deformation constitutive equations and a time integration procedure for isotropic, hyperelastic-viscoplastic solids', *Comp. Meth. Appl. Mech. Engng.,* **79,** 173–202 (1990)

[28] O.C.ZIENKIEWICZ and R.L.TAYLOR, *The Finite Element Method. Volume 2: Solid and Fluid Mechanics, Dynamics and Non-linearity,* McGraw-Hill, Maidenhead, 4-th edition, (1991)

Finite Deformation Analysis of Inelastic Materials with Micro-Structure

RENE DE BORST
Delft University of Technology / TNO Building and Construction Research, Delft, The Netherlands

and

HANS-BERND MUHLHAUS
CSIRO Division of Geomechanics, Mt. Waverley, Australia

Summary

A large-strain formulation of an elasto-plastic Cosserat continuum is developed. The co-rotational rate of the Kirchhoff stress is defined using the Cosserat spin. For a two-dimensional configuration the finite element equations are elaborated in full detail. The example of an infinitely long layer loaded in simple shear is used to demonstrate the stabilizing effect of the additional terms of the Cosserat continuum.

Introduction

Ductile failure of metals and soil bodies is usually preceded by strain localization. At a microscale this phenomenon manifests itself by coalescence of micro-shearbands or nucleation of voids. At a macroscopic level strain localization is typically accompanied by a reduction of the load-carrying capacity. The simplest and most straightforward way of modelling this behaviour is by adopting a strain-softening model of rate-independent plasticity. In this case the critical condition for shear banding coincides with the condition for loss of ellipticity of the governing differential equations. The standard continuum theories commonly adopted to model localization, although appropriate for predicting the band orientation and the critical load level, leave the size of the localization zone unspecified. This observation is the reason for the frequently reported pathological dependence on the grid size of finite element solutions for strain-softening problems.

For transient conditions the ill-posedness of the governing set of differential equations can partially be remedied by introducing the effects of rate-dependence and inclusion of heat flow [1,2], which account for the physical phenomena that occur if the deformation is rapid enough for the system to react adiabatically.

In this contribution a general framework is presented for the description of a heterogeneously evolving microstructure under low deformation rates. It is assumed that microstructural effects manifest themselves at a macroscopic level, i.e. in a continuum theory, through higher-order gradients. This effect is accounted for by adopting the structure of Cosserats' continuum [3].

The higher-order terms enter the structure of classical plasticity theory in key places and modify it crucially by providing an internal length scale which allows to capture the characteristics of deformation patterns such as the thickness of shear bands. In contrast to conventional theories, higher-order plasticity theories do not necessarily suffer from ill-posedness of the field equations when a strain-localization zone develops. The latter property ensures meaningful numerical solutions of boundary-value problems where such phenomena are observed [5-8].

D. Besdo, E. Stein (Eds.)
Finite Inelastic Deformations –
Theory and Applications
IUTAM Symposium Hannover/Germany 1991

The present contribution is geared towards the numerical simulation of plastic instability phenomena in metals and soils under slow loading conditions. To model the large-deformation effect two alternatives that both fit within the framework of an Updated Lagrange Formulation have been considered for a Cosserat continuum. In the first approach the material spin is utilized for the definition of the co-rotational rates, while the second approach is based upon the use of the Cosserat spin [6]. The latter approach will be used in the numerical examples to be presented at the end of the paper and therefore, for the two-dimensional case, a detailed derivation is presented of the geometric terms that enter the tangential stiffness matrix.

For the inelastic effect a generalization of the J_2-flow theory for a Cosserat continuum will be used [4,5,7] and the algorithmic implementation will be according to the return-mapping method recently formulated for a Cosserat continuum [8]. The example of an infinitely long shear layer will be used to demonstrate the departure in results from the classical theory.

Basic kinematics in media with micro-structure

Most of the kinematic quantities that are used in the Cosserat continuum model also enter the description of a conventional continuum. With u_i the components of the displacement vector and x_j the components of the spatial coordinates of a material point, we for instance define

$$L_{ij}=\frac{\partial \dot{u}_i}{\partial x_j} \quad , \quad \dot{t}=1 \tag{1}$$

as the velocity gradient, t denoting the time,

$$W_{ij}=\frac{1}{2}\left(\frac{\partial \dot{u}_i}{\partial x_j}-\frac{\partial \dot{u}_j}{\partial x_i}\right) \tag{2}$$

as the spin tensor, and

$$D_{ij}=L_{ij}-W_{ij} \tag{3}$$

as the rate of deformation tensor or stretching. In addition we now introduce the tensors:

$$W^c_{ij}=-e_{ijk}\omega^c_k \, , \tag{4}$$

the Cosserat spin tensor, and

$$\gamma_{ij}=L_{ij}-W^c_{ij} \, , \tag{5}$$

the rate of the Cosserat relative deformation tensor. In eq. (4) ω^c_k are the components of the Cosserat rotation-rate vector, i.e. the rotation of the rigid triad attached to the material micro-elements. This vector can also be used to define the curvature-rates of the micro-elements of this continuum model:

$$\kappa_{ij}=\frac{\partial \omega^c_i}{\partial x_j} \, . \tag{6}$$

There is an obvious similarity between the kinematics of the Cosserat continuum and the kinematics of the phenomenological theories of single slip (for brief reviews, see e.g. [9,10]). In these theories the irreversible part of the objective tensor $W_{ij}-W^c_{ij}$ is interpreted as the average deviation of the rotations of the active slip systems from the spin of each infinitesimal volume element [10]. It should be mentioned

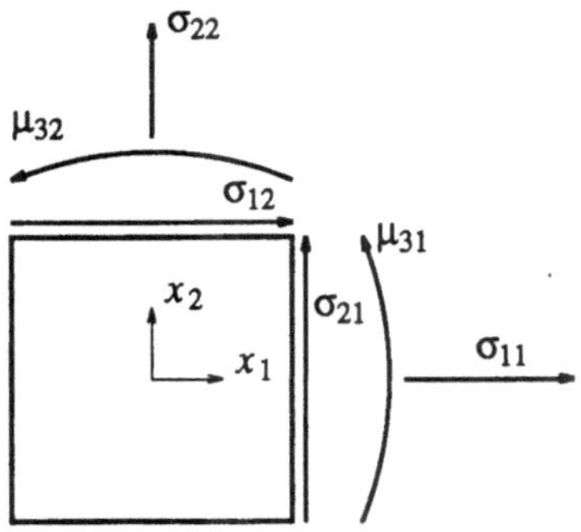

Fig. 1. Stress and couple-stress in a two-dimensional Cosserat continuum.

that Mandel [11] has introduced his plastic spin theory as a reduced Cosserat theory where the couple stresses are neglected. This assumption is not made here.

On the definition of co-rotational stress rates

When we denote the Cauchy stress by σ_{ij} (Figure 1) and adopt the common definition for the Kirchhoff stress $\tau_{ij} = (\det \mathbf{F})\sigma_{ij}$, two alternative definitions for the co-rotational rate of the Kirchhoff stress are possible. Conventionally, one employs the material spin to define:

$$\tau^{\mathrm{J}}_{ij} = \dot{\sigma}_{ij} + D_{kk}\sigma_{ij} - W_{ik}\sigma_{kj} + \sigma_{ik}W_{kj} \,, \tag{7}$$

where τ^{J}_{ij} designates the Jaumann rate of the Kirchhoff stress referred to the current configuration. In the present case, however, it seems more natural to use the Cosserat spin in the definition of the co-rotational rate, so that:

$$\tau^{\mathrm{c}}_{ij} = \dot{\sigma}_{ij} + D_{kk}\sigma_{ij} - W^{\mathrm{c}}_{ik}\sigma_{kj} + \sigma_{ik}W^{\mathrm{c}}_{kj} \,. \tag{8}$$

In the remainder of the paper we shall utilize the latter definition, since the ensuing expressions for the tangential stiffness matrix turn out to be more elegant and connect closer to the conventional formulation, e.g. [12]. Objective rates of the true couple-stress tensor μ_{ij} are defined analogously to (7)-(8).

The equilibrium rate equations

Translational equilibrium is most conveniently expressed as:

$$\sigma^{\mathrm{R}}_{ij,j} = 0 \,, \tag{9}$$

with σ^{R}_{ij} the rate of the first Piola-Kirchhoff stress tensor referred to the current configuration, that is

$$\sigma^{\mathrm{R}}_{ij} = \dot{\sigma}_{ij} + D_{kk}\sigma_{ij} - \sigma_{ik}L_{jk} \tag{10}$$

and the comma with the following index j denoting differentiation with respect to x_j. From rotational equilibrium of a small material element we have

$$\int_S \mu_{ji} n_i \mathrm{d}S + \int_V e_{ikj} x_i \sigma_{kl} n_l \mathrm{d}V = 0 \,, \tag{11}$$

with e_{ijk} designating the permutation symbol. After some algebra we obtain:

$$m_{ij,j} + e_{ijl}\tau_{lj} = 0 \,. \tag{12}$$

m_{ij} is a new quantity, which we shall name the first Piola-Kirchhoff couple-stress tensor. It is related to the true couple-stress μ_{ij} by

$$m_{ij} = (\det\mathbf{F})\mu_{ik}F^{-1}_{jk} \,. \tag{13}$$

Differentiating (12) and referring the result to the current configuration gives:

$$m^{\mathrm{R}}_{ij,j} + e_{ijl}(D_{kk}\sigma_{lj} + \dot{\sigma}_{lj}) = 0 \,, \tag{14}$$

where

$$m^{\mathrm{R}}_{ij} = \dot{\mu}_{ij} + D_{kk}\mu_{ij} - \mu_{ik}L_{jk} \,. \tag{15}$$

For planar deformations, the first Piola-Kirchhoff couple-stress rate tensor happens to be objective. This can be seen by substituting into (15) the identities

$$L_{jk} = D_{kj} - W_{kj} \quad \text{and} \quad W_{3k} = 0 \,, \tag{16}$$

and it turns out that for 2D m^{R}_{ij} equals the Jaumann rate of the "Kirchhoff" couple stress. In the present model we relate the Jaumann rate, that is eq. (15), to the rate of curvature eq. (6). This is somewhat inconsistent, since for the stress rate we shall employ definition (8), which is based on the Cosserat spin. This slight inconsistency is thought to be acceptable because the moments usually turn out to be small (perturbation-term type quantities). The advantage is that the present choice leads to an algebraically simple formulation.

From eqs. (9) and (14) we can, for planar deformations, derive the following expression for the weak form of the equilibrium equations:

$$\int_V \sigma^{\mathrm{R}}_{ij,j}\delta u_i \mathrm{d}V + \int_V [m^{\mathrm{R}}_{3j,j} + e_{3jl}(D_{kk}\sigma_{lj} + \dot{\sigma}_{lj})]\delta\omega_3 \mathrm{d}V = 0 \,. \tag{17}$$

Substituting the co-rotational rate (8) of the Kirchhoff stress, eq. (17) can be recasted in a format which closely connects to conventional updated Lagrangian formulations:

$$\int_V (\tau^{\mathrm{c}}_{ij} - \gamma_{ik}\sigma_{kj} + \sigma_{ik}\gamma_{jk})\delta\gamma_{ij}\mathrm{d}V + \int_V m^{\mathrm{R}}_{3k}\delta\kappa_{3k}\mathrm{d}V + \int_V L_{ik}\sigma_{kj}\delta F_{ij}\mathrm{d}V - \int_V L_{ik}(\sigma_{kj} - \sigma_{jk})\delta W^{\mathrm{c}}_{ij}\mathrm{d}V = 0 \,. \tag{18}$$

When Cosserat effects are absent, so that the second and the fourth terms in eq. (18) cancel, the Cosserat rate of deformation tensor γ_{ij} coincides with the stretching D_{ij} and there is no difference between both spin tensors W_{ij} and W^{c}_{ij}, the tangential stiffness expression of McMeeking and Rice [12] is recovered. The latter relation, namely $W_{ij} = W^{\mathrm{c}}_{ij}$, can be deduced from the moment equilibrium condition (11) and the constitutive relation (19) (see next section) if (a) the material is isotropic (in particular if there is no kinematic hardening), and (b) if $\mu_{ij} = 0$.

Matrix-vector formulation of the tangent stiffness

To arrive at a matrix-vector formulation that is suitable for implementation in a finite element environment, we first introduce the stress-strain relation:

$$\tau^{\mathrm{c}}_{ij} = D^{\tan}_{ijkl}\,\gamma_{kl} \,, \tag{19}$$

where $D^{\tan}_{ijkl}$ for example is the elasto-plastic tangent stiffness tensor. Defining δ_{ij} as the Kronecker symbol and considering the 'material part' of the tangent stiffness matrix first, we can rewrite the first two integrals as:

$$\int_V \gamma_{kl}(D^{\tan}_{ijkl} - \delta_{ik}\sigma_{lj} - \sigma_{il}\delta_{kj})\delta\gamma_{ij}\mathrm{d}V + \int_V m^{\mathrm{R}}_{3k}\delta\kappa_{3k}\mathrm{d}V = \int_V \delta\mathbf{u}^{\mathrm{T}}\mathbf{L}_L^{\mathrm{T}}(\mathbf{D}^{\tan} - \boldsymbol{\Sigma}_1)\mathbf{L}_L\dot{\mathbf{u}}\mathrm{d}V \,, \tag{20}$$

where

$$\mathbf{u}^{\mathrm{T}} = [u_1\,, u_2\,, \omega_3^c]\,, \tag{21}$$

$$\mathbf{L}_L^{\mathrm{T}} = \begin{bmatrix} \frac{\partial\cdot}{\partial x} & 0 & 0 & \frac{\partial\cdot}{\partial y} & 0 & 0 & 0 \\ 0 & \frac{\partial\cdot}{\partial y} & 0 & 0 & \frac{\partial\cdot}{\partial x} & 0 & 0 \\ 0 & 0 & 0 & +1 & -1 & \frac{\partial\cdot}{\partial x} & \frac{\partial\cdot}{\partial y} \end{bmatrix} \tag{22}$$

and

$$\boldsymbol{\Sigma}_1 = \begin{bmatrix} 2\sigma_{11} & 0 & 0 & \sigma_{12}+\sigma_{21} & 0 & 0 & 0 \\ 0 & 2\sigma_{22} & 0 & 0 & \sigma_{12}+\sigma_{21} & 0 & 0 \\ 0 & 0 & 0 & 0 & 0 & 0 & 0 \\ \sigma_{12} & \sigma_{12} & 0 & \sigma_{22} & \sigma_{11} & 0 & 0 \\ \sigma_{21} & \sigma_{21} & 0 & \sigma_{22} & \sigma_{11} & 0 & 0 \\ 0 & 0 & 0 & 0 & 0 & 0 & 0 \\ 0 & 0 & 0 & 0 & 0 & 0 & 0 \end{bmatrix}. \tag{23}$$

The first part of the geometric stiffness contribution is conventionally rewritten as

$$\int_V L_{ik}\sigma_{kj}\delta F_{ij}\mathrm{d}V = \int_V \delta\mathbf{u}^{\mathrm{T}}\mathbf{L}_{NL}^{\mathrm{T}}\boldsymbol{\Sigma}_2\mathbf{L}_{NL}\dot{\mathbf{u}}\mathrm{d}V \,, \tag{24}$$

where

$$\mathbf{L}_{NL}^{\mathrm{T}} = \begin{bmatrix} \frac{\partial\cdot}{\partial x} & \frac{\partial\cdot}{\partial y} & 0 & 0 \\ 0 & 0 & \frac{\partial\cdot}{\partial x} & \frac{\partial\cdot}{\partial y} \\ 0 & 0 & 0 & 0 \end{bmatrix} \tag{25}$$

and

$$\boldsymbol{\Sigma}_2 = \begin{bmatrix} \sigma_{11} & \sigma_{21} & 0 & 0 \\ \sigma_{12} & \sigma_{22} & 0 & 0 \\ 0 & 0 & \sigma_{11} & \sigma_{21} \\ 0 & 0 & \sigma_{12} & \sigma_{22} \end{bmatrix}. \tag{26}$$

The additional part of the geometric stiffness becomes:

$$\int_V L_{ik}(\sigma_{kj}-\sigma_{jk})\delta W^c_{ij}\mathrm{d}V=\int_V \delta\mathbf{u}^{\mathrm{T}}\mathbf{L}_R^{\mathrm{T}}\mathbf{L}_{NL}\dot{\mathbf{u}}\mathrm{d}V\ , \tag{27}$$

where

$$\mathbf{L}_R^{\mathrm{T}}=(\sigma_{21}-\sigma_{12})\begin{bmatrix}0&0&0&0\\0&0&0&0\\1&0&0&1\end{bmatrix}. \tag{28}$$

This term obviously cancels for a standard or Boltzmann continuum, where $\sigma_{12}=\sigma_{21}$.

Generalized plasticity theory

As in conventional plasticity we can assume that the strain rate can be decomposed in an additive fashion:

$$\boldsymbol{\gamma}=\boldsymbol{\gamma}^{\mathrm{e}}+\boldsymbol{\gamma}^{\mathrm{p}}\ . \tag{29}$$

In eq. (29) the elastic strain rate $\boldsymbol{\gamma}^{\mathrm{e}}$ is assumed to be linearly related to the co-rotational stress rate

$$\boldsymbol{\gamma}^{\mathrm{e}}=\mathbf{D}^{-1}\boldsymbol{\tau}^{\mathrm{c}}\ , \tag{30}$$

with:

$$\boldsymbol{\tau}^{\mathrm{c}}=[\ \tau^{\mathrm{c}}_{11}\ ,\ \tau^{\mathrm{c}}_{22}\ ,\ \tau^{\mathrm{c}}_{33}\ ,\ \tau^{\mathrm{c}}_{12}\ ,\ \tau^{\mathrm{c}}_{21}\ ,\ m^{\mathrm{R}}_{31}\ ,\ m^{\mathrm{R}}_{32}\ ]^{\mathrm{T}} \tag{31}$$

and

$$\boldsymbol{\gamma}=[\ \gamma_{11}\ ,\ \gamma_{22}\ ,\ \gamma_{33}\ ,\ \gamma_{12}\ ,\ \gamma_{21}\ ,\ \kappa_{31}\ ,\ \kappa_{32}\ ]^{\mathrm{T}}\ . \tag{32}$$

$\mathbf{D}$ is the elastic stiffness matrix:

$$\mathbf{D}=\begin{bmatrix}2Gc_1&2Gc_2&2Gc_2&0&0&0&0\\2Gc_2&2Gc_1&2Gc_2&0&0&0&0\\2Gc_2&2Gc_2&2Gc_1&0&0&0&0\\0&0&0&G+G^{\mathrm{c}}&G-G^{\mathrm{c}}&0&0\\0&0&0&G-G^{\mathrm{c}}&G+G^{\mathrm{c}}&0&0\\0&0&0&0&0&2Gl^2&0\\0&0&0&0&0&0&2Gl^2\end{bmatrix}, \tag{33}$$

with G the conventional shear modulus, $c_1=(1-\nu)/(1-2\nu)$ and $c_2=\nu/(1-2\nu)$, ν being Poisson's ratio. The parameter l is the internal length scale of the Cosserat continuum. It controls the size of the localization zone. G^{c} is the second additional material parameter, which completes the set of four material parameters (G, ν, l and G^{c}) that is needed to describe the elastic behaviour of a two-dimensional Cosserat medium completely.

The plastic strain rate $\boldsymbol{\gamma}^{\mathrm{p}}$ is assumed to be governed by the classical format of an associated flow rule

$$\boldsymbol{\gamma}^{\mathrm{p}}=\lambda\frac{\partial f}{\partial\boldsymbol{\sigma}}\ , \tag{34}$$

with λ the magnitude of the plastic flow and f the yield function. In the present computations a Von

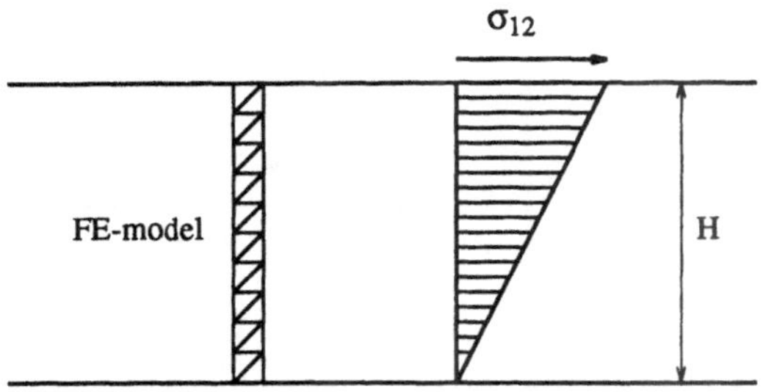

Fig. 2. Infinite shear layer: loading conditions and finite element grid.

Mises criterion has been used:

$$f=[\tfrac{3}{2}\boldsymbol{\sigma}^{\mathrm{T}}\mathbf{P}\boldsymbol{\sigma}]^{1/2}-\bar{\sigma}(\Gamma^{\mathrm{p}})\,. \tag{35}$$

For planar deformations $\mathbf{P}$ is given by:

$$\mathbf{P}=\begin{bmatrix} 2/3 & -1/3 & -1/3 & 0 & 0 & 0 & 0 \\ -1/3 & 2/3 & -1/3 & 0 & 0 & 0 & 0 \\ -1/3 & -1/3 & 2/3 & 0 & 0 & 0 & 0 \\ 0 & 0 & 0 & 2a_1 & 2a_2 & 0 & 0 \\ 0 & 0 & 0 & 2a_2 & 2a_1 & 0 & 0 \\ 0 & 0 & 0 & 0 & 0 & 2a_3 & 0 \\ 0 & 0 & 0 & 0 & 0 & 0 & 2a_3 \end{bmatrix}, \tag{36}$$

with a_1, a_2 and a_3 additional material constants. Based on micromechanical considerations Mühlhaus and Vardoulakis [4] and Mühlhaus [5] have proposed the values $a_1=3/4$, $a_2=-1/4$ and $a_3=1/8$. In the example calculations of the next section the values $a_1=1/4$, $a_2=1/4$ and $a_3=1/2$ have been adopted, since these values give rise to a particularly simple numerical algorithm [7]. Nevertheless, the basic characteristics are not significantly affected by the precise choice for this set of material parameters. To bring out the essential features of the model the present choice is permissible. When it is attempted to fit the material behaviour as closely as possible this is unlikely to be the case.

In eq. (35) the equivalent stress $\bar{\sigma}$ is a function of the hardening parameter Γ^{p}. Here, the strain-hardening hypothesis has been adopted to define this quantity, yielding

$$\dot{\Gamma}^{\mathrm{p}}=[\tfrac{2}{3}(\boldsymbol{\gamma}^{\mathrm{p}})^{\mathrm{T}}\mathbf{P}\boldsymbol{\gamma}^{\mathrm{p}}]^{1/2}\,. \tag{37}$$

For a Von Mises plasticity model we simply have [8]: $\dot{\Gamma}^{\mathrm{p}}=\lambda$.

Example calculations

In the example calculations we shall exclusively consider the infinite shear layer of Figure 2. The layer is modelled with 800 three-noded triangular elements. Because the problem is essentially one-dimensional, line elements could also have been used. Here, the 1D-situation has been enforced by adding linear

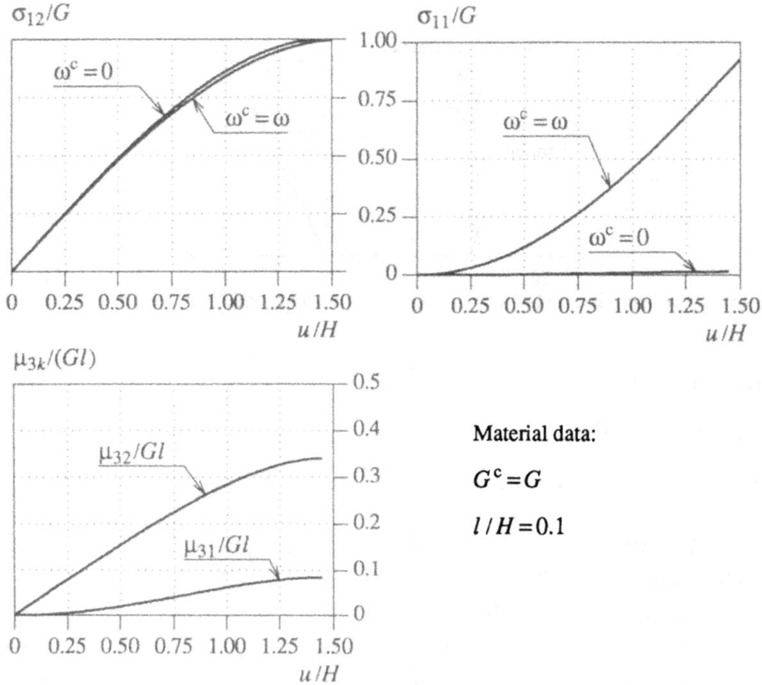

Fig. 3. Infinite shear layer: evolution of stress and couple-stress at the top of the layer assuming linear elasticity. Results for natural boundary conditions (ω^c free) and essential boundary conditions ($\omega^c=0$).

dependence relations for the horizontal translations and the rotations. For all nodes $u_y=0$ has been imposed. With respect to the rotational boundary conditions two different cases have been analysed: (i) natural boundary conditions (ω^c free), and (ii) essential boundary conditions ($\omega^c=0$).

Restricting attention to *elastic material behaviour* first, we note that the case with the natural boundary conditions is identical to the classical case, i.e. without Cosserat effects. Throughout the layer the analytical solution is given by:

$$\omega^c=\omega\,,\qquad \sigma_{12}/G=\sin(u/H)\,,\qquad \sigma_{11}/G=1-\cos(u/H)\,. \tag{38}$$

The numerical solution matches the analytical solution perfectly, thus showing that the implementation is correct at least for the limiting case of a conventional continuum. For the case with essential boundary conditions ($\omega^c=0$) differences with the classical case occur. Firstly, the deformations are no longer homogeneous over the height of the layer. Near the upper and lower boundaries a boundary layer develops with a thickness that is proportional to the internal length scale l. The evolution of the stresses and the couple-stresses at the boundaries of the layer has been plotted in Figure 3. It is remarkable that, in contrast to the classical case, the normal stress σ_{11} hardly develops. This is probably because of the evolution of the couple stresses μ_{31} and μ_{32}.

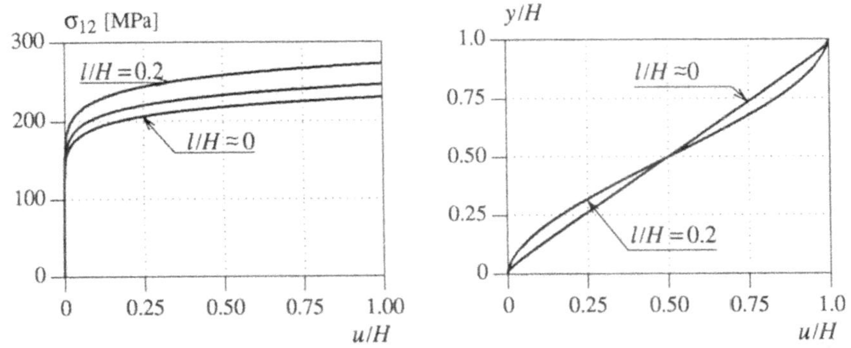

Fig. 4. Infinite shear layer: numerical results assuming elasto-plastic material properties for different ratios l/H. Left: shear stress σ_{12} *vs* horizontal displacement of the top of the layer u_{top}. Right: deformed profile over the height of the layer for $u_{top}/H=1$.

For the *elasto-plastic* calculations a power law has been adopted for the hardening behaviour:

$$\bar{\sigma}(\Gamma^p)=\sigma_0(1+\Gamma^p/\Gamma_0)^\alpha \tag{39}$$

where the following material data have been substituted: $\sigma_0=240\,\text{MPa}$, $\Gamma_0=10^{-3}$ and $\alpha=0.08$. The results of these calculations, all of which were performed for $\omega^c=0$ at both boundaries, are displayed in Fig. 4 for three different values of l/H, namely 0, 0.1 and 0.2. In the left part of Fig. 4 we clearly observe the stiffening influence of the addition of the Cosserat terms on the behaviour of the layer, while in the right part we see the development of boundary layers of high strain gradient near the boundaries.

To investigate whether adopting the structure of the Cosserats' continuum indeed preserves well-posedness of the governing set of differential equations the analysis for $l/H=0.1$ was repeated with a modified evolution law for $\bar{\sigma}$:

$$\bar{\sigma}(\Gamma^p)=\sigma_0[(1+\Gamma^p/\Gamma_0)^\alpha-\Gamma^p]\,, \tag{40}$$

so that a gradual transition of a hardening regime into a softening regime is incorporated. The results are shown in Fig. 5 for two different discretizations, namely for the standard 800 element mesh and for a mesh that is twice as coarse (400 elements). The differences between both analyses are not discernable, thus proving that the set of differential equations now remains well-posed after the onset of softening.

Concluding remarks

The use of a Cosserat continuum is very appealing when modelling phenomena in which large plastic strains occur locally, such as in necking and shear banding. Because of the presence of an internal length scale localization zones can develop while ellipticity of the governing set of differential equations is preserved. Accordingly, the mathematical problem remains well-posed, thus ensuring meaningful solutions when carrying out numerical simulations.

Nonetheless, the approach has some drawbacks. Firstly, it is unconventional, which makes its acceptance not straightforward. Secondly, there is the issue of how to obtain the additional material parameters. Here, one should bear in mind that the prime purpose of the use of the structure of the Cosserats'

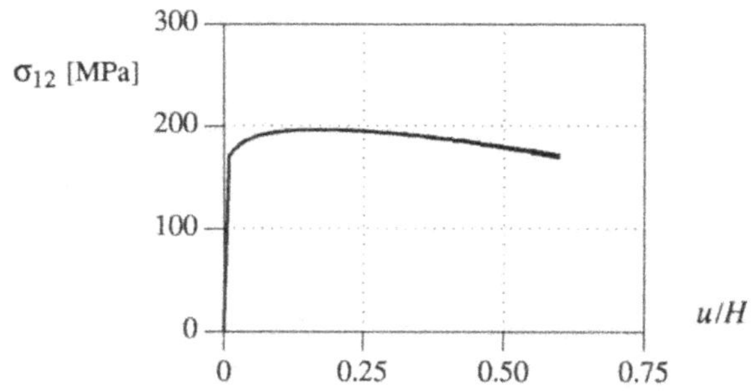

Fig. 5. Infinite shear layer: shear stress σ_{12} *vs* horizontal displacement of the top of the layer u_{top}. for elasto-plastic softening material behaviour according to eq. (40).

continuum is its proper behaviour during localization. Therefore, the additional material parameters, e.g., G^c, should be chosen such that in the pre-localization regime the differences with the conventional approach are as small as possible. During localization it is the l-parameter which sets the band width. The value of l can therefore be determined in a semi-inverse manner from measured band widths.

It is finally remarked that a consistent elasto-plastic tangent operator as formulated in [8] has been used throughout this study. Indeed, quadratic convergence was observed for all cases when Newton's method was used to solve the global set of equations.

References

[1] Lemonds, J.; Needleman, A.: Finite element analysis of shear localization in rate and temperature dependent solids. *Mechanics of Materials* 5 (1986) 339-361.

[2] Wriggers, P.; Miehe, C.; Kleiber, M.; Simo, J.C.: On the coupled thermo-mechanical treatment of necking problems via finite element methods. In: Owen, D.R.J.; Hinton, E.; Onate, E. (eds.): *Computational Plasticity: Models, Software and Applications*. Swansea: Pineridge Press, 1989, pp. 527-542.

[3] E. Cosserat and F. Cosserat, *Theórie des Corps Deformables*. Herman et fils, Paris, 1909.

[4] Mühlhaus, H.-B.; Vardoulakis, I.: The thickness of shear bands in granular materials. *Geótechnique* 37 (1987) 271-283.

[5] Mühlhaus, H.-B. Lamination phenomena in prestressed rock. In: Fairhurst, C. (ed.): *Rockbursts and Seismicity in Mines*. Balkema, Rotterdam and Brookfield, 1990, pp. 101-107.

[6] Mühlhaus, H.-B.: Application of Cosserat theory in numerical solutions of limit load problems. *Ingenieur-Archiv*, 59 (1989) 124-137.

[7] Mühlhaus, H.-B.; de Borst, R.; Aifantis, E.C. Constitutive models and numerical analyses for inelastic materials with microstructure. In: Beer, G.; Booker, J.R.; Carter, J.P. (eds.): *Computer Methods and Advances in Geomechanics*. Balkema, Rotterdam and Brookfield, 1991, pp. 377-386.

[8] de Borst, R.: Simulation of strain localisation: A reappraisal of the Cosserat continuum. *Eng. Comput.* 8 (1991).

[9] Dafalias, Y.: A missing link in the macroscopic constitutive formulations of large plastic deformation. In: Sawczuk, A.; Bianchi, G. (eds.): *Plasticity Today*. Elsevier Applied Science, London and New York, 1985, pp. 135-151.

[10] Aravas, N.; Aifantis, E.C.: On the geometry of slip and spin in finite plastic deformation. *Int. J. Plasticity* 7 (1991) 141-160.

[11] Mandel, J.: Equations constitutives et directeurs dans les milieux plastiques et visco-plastiques. *Int. J. Solids Structures* 9 (1973) 725-740.

[12] McMeeking, R.M.; Rice, J.R.: Finite-element formulations for problems of large elastic-plastic deformation. *Int. J. Solids Structures* 11 (1975) 601-616.

Localization Analysis in Finite Deformation Elasto-Plasticity

Paul Steinmann and Kaspar Willam
Institute of Mechanics, University of Karlsruhe, D-7500 Karlsruhe 1
CEAE-Department, University of Colorado at Boulder, CO 80309-0428

1 Introduction

A failure phenomenon which is frequently observed in laboratory experiments as well as in nature is the localization of deformations within narrow bands. Thereby, the underlying physical mechanisms vary with the particularly material under consideration. Large inelastic strains accumulate inside the localization zone and form a precursor to fracture. On the theoretical side, the micro-structure of grain particle interfaces is traditionally homogeneized into a continuum, allowing point-measures of strains and stresses to describe the response of a given configuration under prescribed load histories. Classical continuum theory signals the onset of localization via bifurcation at the constitutive level, i.e. the velocity gradient may develop a discontinuity across a singularity surface with orientation $\mathbf{N}$. Discontinuous bifurcation, which is synonymous with the loss of ellipticity of the governing equations, is reflected by a singularity of the acoustic tensor $\mathbf{Q}$. Localization analysis of constitutive models furnishes the critical directions $\mathbf{N}$ which are often in close agreement with the inclination of the observed failure bands. Nevertheless, continuum theory fails to describe the post-peak behaviour since discontinuities rather than steep gradients are embedded in the solution of the localization problem. This leads to severe mesh-dependence of numerical computations. Thus, the validity of numerical solutions of localization problems is very difficult to assess once localization has fully developed. Hence it is of utmost importance that the finite element formulation is capable of reflecting the singularity of the acoustic tensor not only at the element level but also throughout the spatial discretization. In this study we will extend the weak localization test [8][9] to finite deformations and we will verify the localization capturing abilities of single elements. Several variations of low order four noded elements will be studied within the finite deformation theory of hyperelastic-plastic solids. These findings will lead to the formulation of a singularized finite element which is designed to capture localization in the weak sense.

2 Localized Failure at the Constitutive Level

In the following, localization will be studied in the form of discontinuous bifurcation when hyperelastic-plastic solids are subjected to finite inelastic deformations. To this end the velocity field $\mathbf{v} = \dot{\mathbf{x}}$ and the deformation gradient field $\mathbf{F} = \mathbf{I} + \operatorname{Grad}\mathbf{u}$ remain continuous, while the material velocity gradient $\operatorname{Grad}\mathbf{v} = \dot{\mathbf{F}}$ may exhibit a finite discontinuity across a singularity surface which is defined by the normal $\mathbf{N}$ in the reference configuration. Maxwell's compatibility conditions require that the jump of the material velocity gradient

[1]To be presented at the IUTAM-Symposium on Finite Inelastic Deformations, August 19 to 23, 1991, Hanover, Germany

D. Besdo, E. Stein (Eds.)
Finite Inelastic Deformations -
Theory and Applications
IUTAM Symposium Hannover/Germany 1991

must be a rank one tensor of the form

$$[[\dot{\mathbf{F}}]] = \dot{\gamma}\ \mathbf{m} \otimes \mathbf{N} \tag{1}$$

Analogous to the wave propagation argument, the vector field $\mathbf{m}$ describes the polarization, and the localization amplitude $\dot{\gamma}$ is related to the wave speed. The first tangential material operator $\mathcal{L}_1$, which governs the differential behaviour of the comparison solid associated with plastic loading, relates the the conjugate rate of the first Piola stress tensor to the material velocity gradient, $\dot{\boldsymbol{\Sigma}}_1^t = \mathcal{L}_1 : \dot{\mathbf{F}}$. The balance of linear momentum across the discontinuity surface leads to the localization condition $[[\dot{\mathbf{T}}]] = \dot{\gamma}\ \mathbf{Q}\cdot\mathbf{m} \doteq \mathbf{0}$, or the corresponding argument of zero wave speed respectively, $\rho_o c^2 = \mathbf{m}\cdot\mathbf{Q}\cdot\mathbf{m} \doteq 0$. Thereby, the acoustic tensor is defined as $\mathbf{Q} = \mathbf{N}\cdot\mathcal{L}_1\cdot\mathbf{N}$. The equivalent statement in the spatial setting furnishes the localization condition in terms of $\dot{\gamma}\ \mathbf{q}\cdot\mathbf{m} \doteq \mathbf{0}$, or $\rho c^2 = \mathbf{m}\cdot\mathbf{q}\cdot\mathbf{m} \doteq 0$, with the spatial acoustic tensor being defined as $\mathbf{q} = \mathbf{n}\cdot\mathcal{E}_1\cdot\mathbf{n}$. Here $\mathcal{E}_1 = J^{-1}\mathbf{F}\cdot\mathcal{L}_1\cdot\mathbf{F}$ designates the push-forward operation of the tangential material operator $\mathcal{L}_1$, and $\mathbf{n} = JdAda^{-1}\mathbf{N}\cdot\mathbf{F}^{-1}$ denotes the surface normal in the spatial configuration. Localization initiates, when the zero wave speed condition is satisfied by some directions $\mathbf{N}$ and $\mathbf{m}$. In the following the hyperelastic format of the finite strain J_2-plasticity theory by SIMO [4] is adopted to study localization. The basic kinematic assumption for the extension of the classical von Mises J_2-plasticity to finite strains is based on the multiplicative split of the deformation gradient into elastic and inelastic parts, $\mathbf{F} = \mathbf{F}_e\cdot\mathbf{F}_p$. The hyperelastic response is governed by the stored energy function $W = W(\hat{\mathbf{b}}_e) = \frac{1}{2}\mu[\hat{I}_e - 3] + U(J)$ which characterizes a compressible extension of Neo-Hookean behaviour. Thereby $\hat{\mathbf{b}}_e = J^{-\frac{2}{3}}\mathbf{F}_e\cdot\mathbf{F}_e^t$ denotes the volume preserving part of the elastic Finger deformation tensor. The Kirchhoff stress is related to $\hat{\mathbf{b}}_e$, J and $\mathbf{g}$ with the aid of the Doyle-Ericksen formula $\boldsymbol{\tau} = \mu\mathrm{dev}[\hat{\mathbf{b}}_e] + JU'(J)\mathbf{g}^{-1}$. The yield condition is expressed in terms of the deviatoric part of the Kirchhoff stress tensor $\Phi(\mathbf{g},\boldsymbol{\tau},q) = \sqrt{\mathrm{dev}[\boldsymbol{\tau}]:\mathrm{dev}[\boldsymbol{\tau}]} - \sqrt{\frac{2}{3}}Y(q)$. The plastic flow rule is stated in terms of $\mathbf{F}\cdot\dot{\mathbf{C}}_p^{-1}\cdot\mathbf{F}^t = -\dot{\lambda}\frac{2}{3}\hat{I}_e\frac{\mathrm{dev}[\boldsymbol{\tau}]}{|\mathrm{dev}[\boldsymbol{\tau}]|}$, where the Lagrangian strain measure of plastic deformation is defined as $\mathbf{C}_p = \mathbf{F}_p^t\cdot\mathbf{F}_p = \mathbf{F}^t\cdot\mathbf{b}_e^{-1}\cdot\mathbf{F}$. The description of the finite hyperelastic-plastic model is closed by the definition of the evolution law, $Y = Y(q)$ which describes e.g. nonlinear isotropic hardening or softening behaviour in terms of the equivalent plastic strain rate $\dot{q} = \sqrt{\frac{2}{3}}\dot{\lambda}$.

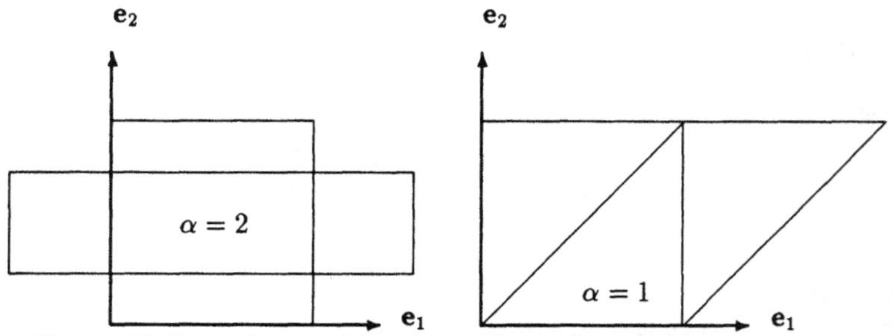

Fig. 1 : Uniform States of Pure and Simple Shear Deformation

For illustration, two finite deformation problems of pure and simple shear in Fig. 1 will be studied. Since both shear deformations are incompressible, it is not necessary to specify the volumetric part of the stored energy function $U(J = 1) = 0$ in this context. The elastic shear modulus $\mu = 30000/2.6$ ksi characterizes the elastic response and the yield limit $Y = 36$ ksi controls the plastic response which is governed here by ideally plastic flow. The deformation gradient of simple shear has the following representation $\mathbf{F} = \mathbf{I} + \alpha \mathbf{e}_1 \otimes \mathbf{e}_2$ in terms of inertial background co-ordinates. For pure shear the deformation gradient reads $\mathbf{F} = \alpha \mathbf{e}_1 \otimes \mathbf{e}_1 + \alpha^{-1} \mathbf{e}_2 \otimes \mathbf{e}_2$. The spatial acoustic tensor $\mathbf{q} = \mathbf{n} \cdot [\mathcal{E}_2^a + \boldsymbol{\sigma} \otimes \mathbf{I}] \cdot \mathbf{n}$ is analysed for $\alpha = 0.5$ in the case of pure shear and $\alpha = 1.0$ in the case of simple shear. To this end $\det \mathbf{q}$ is computed for a variation $-\pi/2 \leq \vartheta \leq \pi/2$, where the spatial normal of the discontinuity surface is expessed as $\mathbf{n}^t = (\cos\vartheta, \sin\vartheta)$. Thereby, $\mathcal{E}_2^a$ denotes the algorithmic counterpart of the second Eulerian material operator $\mathcal{E}_2$ conjugate to $\mathcal{L}_2$ defined in the next section. The algorithmic tangent operator is consistent with the fully implicit backward integration algorithm and takes into account the influence of finite loadsteps. The total deformation is obtained in $n_{incr} = 100, 10$ and 2 increments. The sensitivity of the algorithmic acoustic tensor with regard to the load step size is clearly visible in Fig. 4. The minima of $\det \mathbf{q}$, i.e. the critical directions $\mathbf{n}$, are shifting and the maxima $\det \mathbf{q}$ are diminishing as the load step increases. The geometric contributions $\boldsymbol{\sigma} \otimes \mathbf{I}$ are responsible for the lack of symmetry of the maxima of $\det \mathbf{q}$ with respect to $\vartheta = 0$. If these geometric stiffness terms were neclected, the acoustic tensor analysis in the spatial configuration would degenerate entirely to that of the linearized kinematic theory. Since in elasto-plasticity the magnitude of the elastic stress is restricted by the yield condition it was observed by SOBH [7], that the geometric stiffness terms have a minor effect on the onset of localization when compared with the tangential material model.

3 Weak Localization Element Test

Integration of the jump of the material velocity gradient renders the difference of the velocity field at the positive and negative side of the discontinuity, $\mathbf{v}^+ - \mathbf{v}^- = \dot{\gamma}\,[\mathbf{m} \otimes \mathbf{N}] \cdot \bar{\mathbf{X}}$, where $\bar{\mathbf{X}}$ denotes the material co-ordinate originating at the discontinuity surface. Accordingly, the bifurcated velocity field is defined as $\mathbf{v}^b = \mathbf{v} + \dot{\gamma}\,\Lambda(\mathbf{N} \cdot \bar{\mathbf{X}})[\mathbf{m} \otimes \mathbf{N}] \cdot \bar{\mathbf{X}}$. The jump distribution function $\Lambda(s)$ in Fig. 2 is defined as $\Lambda(s) = \begin{cases} \lambda & \text{if } s > 0 \\ \lambda - 1 & \text{if } s < 0 \end{cases}$

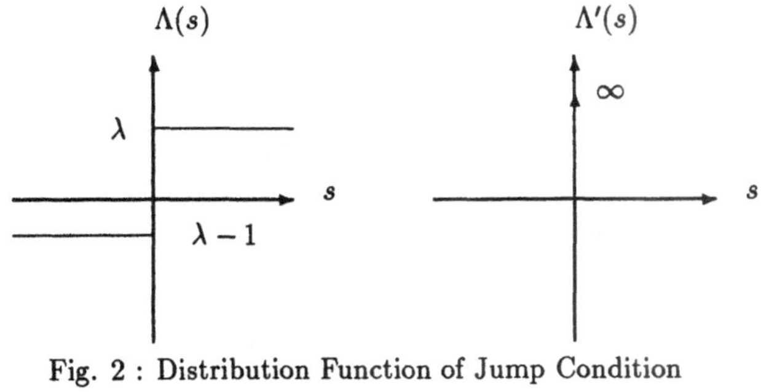

Fig. 2 : Distribution Function of Jump Condition

The derivation of the *"Weak Localization Test"* follows the development by STEINMANN & WILLAM [8][9], which was restricted however to infinitesimal motions. In the finite element setting we assume that a singularity of the acoustic tensor is detected at the centroid of the finite element. This indicates formation of a discontinuity surface normal to the vector $\mathbf{N}$ which divides the element area A into sub-areas A^+ and A^- at the positive and negative side of the discontinuity. Without loss of generality we may adopt $\lambda = 0.5$ and add the bifurcated velocity field to the standard finite element approximation $\mathbf{v}_h^b = \mathbf{v}_h + \dot{\gamma}\,\Lambda[\mathbf{m}\otimes\mathbf{N}]\cdot\bar{\mathbf{X}}$ where $\bar{\mathbf{X}} = [\mathbf{X}-\mathbf{X}_0]$ and $\mathbf{X}_0$ coincides with the element centroid. The resulting nodal velocities $\dot{\mathbf{d}}^b = \dot{\mathbf{d}} + \dot{\mathbf{d}}_{loc}$ correspond to those of discontinuous bifurcation within the element domain. However, since discontinuous velocity fields are usually not admitted within the C^0-continous element expansions, the discontinuity is distributed by the element shape functions and the gradient operator respectively. This leads to the smeared form of the bifurcated velocity gradient which is energetically equivalent to the formation of the localized discontinuity $\dot{\mathbf{F}}_{dis}^{h,b} = \mathbf{G}\cdot\dot{\mathbf{d}} + \mathbf{G}\cdot\dot{\mathbf{d}}_{loc} = \mathbf{G}\cdot\dot{\mathbf{d}} + [[\dot{\mathbf{F}}]]_{dis}$. Here $\mathbf{G}$ designates the finite element variation of the velocity gradient associated with the nodal velocities. The quadratic format of the localization condition furnishes a point-wise statement in terms of the jump of the material velocity gradient.

$$\rho_o c^2 = [\mathbf{m}\otimes\mathbf{N}] : \mathcal{L}_1 : [\mathbf{m}\otimes\mathbf{N}] = \dot{\gamma}^{-2}[[\dot{\mathbf{F}}]] : \mathcal{L}_1 : [[\dot{\mathbf{F}}]] = 0 \tag{2}$$

The spatial format of the localization condition $\rho c^2 = [\mathbf{m}\otimes\mathbf{n}] : \mathcal{E}_1 : [\mathbf{m}\otimes\mathbf{n}] = \dot{\gamma}^{-2}[[\mathbf{l}]] : \mathcal{E}_1 : [[\mathbf{l}]] = 0$ involves the spatial velocity gradient $\mathbf{l}$ rather than the material velocity gradient $\dot{\mathbf{F}}$. Finite element approximation of the discontinuous strain rate within the element domain $[[\dot{\mathbf{F}}]]_{dis} = \mathbf{G}\cdot\dot{\mathbf{d}}_{loc}$ violates pointwise localization, i.e. $[[\dot{\mathbf{F}}]]_{dis} : \mathcal{L}_1 : [[\dot{\mathbf{F}}]]_{dis} \neq 0$. However, the element description may capture discontinuous bifurcation in the integral sense. In that case $[[\dot{\mathbf{F}}]]_{dis}$ has to satisfy the localization condition over the element domain which results in the *"weak form"* of localization in the reference configuration.

$$\mathrm{d}^2 W_{loc} = \int_{V^e} \rho_o c^2 \mathrm{d}V = \dot{\gamma}^{-2}\int_{V^e} [[\dot{\mathbf{F}}]]_{dis} : \mathcal{L}_1 : [[\dot{\mathbf{F}}]]_{dis}\mathrm{d}V \doteq 0 \tag{3}$$

Reformulating this condition in terms of the second Piola stress rate and the rate of Green strain, $\dot{\boldsymbol{\Sigma}}_2 = \mathcal{L}_2 : \dot{\mathbf{E}}$, the first tangential material operator $\mathcal{L}_1 = \mathbf{F}\cdot\mathcal{L}_2\cdot\mathbf{F} + \boldsymbol{\Sigma}_2\otimes\mathbf{I}$ reveals the well-known initial stress effect in the kernel of the finite element tangent stiffness $\mathbf{k}_T = \int_{V^e}\mathbf{G}\cdot[\mathbf{F}\cdot\mathcal{L}_2\cdot\mathbf{F} + \boldsymbol{\Sigma}_2\otimes\mathbf{I}]\cdot\mathbf{G}\mathrm{d}V$. Hence the weak localization condition above may be alternatively expressed as

$$\mathrm{d}^2 W_{loc} = \dot{\gamma}^{-2}\int_{V^e} [[\dot{\mathbf{F}}]]_{dis} : [\mathbf{F}\cdot\mathcal{L}_2\cdot\mathbf{F} + \boldsymbol{\Sigma}_2\otimes\mathbf{I}] : [[\dot{\mathbf{F}}]]_{dis}\mathrm{d}V \tag{4}$$

Push-forward operation yields the corresponding formulation in the spatial description $\mathrm{d}^2 W_{loc} = \dot{\gamma}^{-2}\int_{v^e}[[\mathbf{l}]]_{dis} : [\mathcal{E}_2 + \boldsymbol{\sigma}\otimes\mathbf{I}] : [[\mathbf{l}]]_{dis}\mathrm{d}v$. Here the spatial form of the second tangential material operator $\mathcal{E}_2 = J^{-1}[\mathbf{F}\otimes\mathbf{F}] : \mathcal{L}_2 : [\mathbf{F}\otimes\mathbf{F}]$ relates the spatial velocity gradient to the objective Truesdell stress rate, i.e. $\overset{T}{\boldsymbol{\sigma}} = \mathcal{E}_2 : \mathbf{l}$. On the finite element level, the weak localization criterion reduces to the zero energy criterion associated with the quadratic form of the nodal velocities $\dot{\mathbf{d}}_{loc}$, such that

$$\mathrm{d}^2 W_{loc} = \dot{\mathbf{d}}_{loc}^t\,\mathbf{k}_T\,\dot{\mathbf{d}}_{loc} \doteq 0 \tag{5}$$

This weak localization test determines whether the singularity of the acoustic tensor, $\det\mathbf{Q} = 0$, may be reproduced at the element level. This test is satisfied identically if the localization kinematics associated with the zero energy criterion are embedded into the finite element discretization.

4 Mixed Variational Principles

In this section the underlying variational principles are outlined for the different element formulations considered. Stabilized elements based on the γ-projection method advocated by LIU, BELYTSCHKO & CHEN[2] are formulated within the standard Hu-Washizu principle which defines a three-field functional, where the displacement field $\mathbf{u}$, the displacement gradient $\mathbf{H} = \mathbf{F} - \mathbf{I}$, and the first Piola-Kirchhoff stress tensor $\boldsymbol{\Sigma}_1$ are the independent variables. For conservative loading the potential energy of the external forces is $\Pi_{ext}(\mathbf{u}) = -\int_{\mathcal{B}_0} \rho_0 \mathbf{B} \cdot \mathbf{u} \, dV - \int_{\Gamma} \mathbf{T} \cdot \mathbf{u} \, d\Gamma$ which renders the Hu-Washizu functional to

$$\Pi = \Pi(\mathbf{u}, \mathbf{H}, \boldsymbol{\Sigma}_1) = \int_{\mathcal{B}_0} W(\mathbf{H}) + \boldsymbol{\Sigma}_1^t : [\mathrm{Grad}\mathbf{u} - \mathbf{H}] \, dV + \Pi_{ext}(\mathbf{u}) \tag{6}$$

Stationarity conditions furnish the balance equations of motion, the kinematic compatibility equations and the constitutive relations. Pushing forward all expressions to the spatial configuration renders the Eulerian description of the three statements.

$$\delta\Pi = \int_{\mathcal{B}_0} \mathrm{grad}^t \delta\mathbf{u} : \boldsymbol{\tau} \, dV + \Pi_{ext}(\delta\mathbf{u}) \tag{7}$$

$$+ \int_{\mathcal{B}_0} \delta\boldsymbol{\tau}^t : [\mathrm{grad}\mathbf{u} - \mathbf{h}] \, dV \tag{8}$$

$$+ \int_{\mathcal{B}_0} \delta\mathbf{h}^t : [2\partial_{\mathbf{g}} W - \boldsymbol{\tau}] \, dV = 0 \tag{9}$$

Thereby $\mathbf{h} = \mathbf{H} \cdot \mathbf{F}^{-1}$ denotes the spatial displacement gradient. The spatial metric tensor $\mathbf{g} = \mathbf{F}^{-t} \cdot \mathbf{C} \cdot \mathbf{F}^{-1}$ results from push-forward operation of the right Cauchy-Green tensor, and the Kirchhoff stress is related to the nominal stress by $\boldsymbol{\tau} = \mathbf{F} \cdot \boldsymbol{\Sigma}_1$. In deviatoric plasticity the isochoric nature of the plastic flow leads to overconstrained displacement formulations. The isoparametric bilinear quadrilateral with uniform volume and pressure approximations as proposed e.g. by SIMO, TAYLOR & PISTER [6] is based on a modified Hu-Washizu principle. Exploiting the basic kinematic split of the deformation gradient $\mathbf{F} = J^{\frac{1}{3}} \hat{\mathbf{F}}$ into a volume preserving $\hat{\mathbf{F}} = J^{-\frac{1}{3}} \mathbf{F}$ and a volumetric part $J^{\frac{1}{3}} \mathbf{I}$ introduces an independent dilatation field θ, so that the deformation gradient may be modified into $\bar{\mathbf{F}} = \theta^{\frac{1}{3}} \hat{\mathbf{F}}$. Accordingly the modified right Cauchy-Green tensor is defined as $\bar{\mathbf{C}} = \theta^{\frac{2}{3}} \hat{\mathbf{C}} = \theta^{\frac{2}{3}} J^{-\frac{2}{3}} \mathbf{F}^t \cdot \mathbf{F}$. Requiring that $J = \theta$ the pressure field p enters the variational principle as Lagrange parameter.

$$\bar{\Pi} = \bar{\Pi}(\mathbf{u}, \theta, p) = \int_{\mathcal{B}_0} W(\bar{\mathbf{C}}) + p[J - \theta] \, dV + \Pi_{ext}(\mathbf{u}) \tag{10}$$

Defining the modified metric tensor $\bar{\mathbf{g}} = \theta^{\frac{2}{3}} \hat{\mathbf{g}} = \theta^{\frac{2}{3}} J^{-\frac{2}{3}} \mathbf{g}$ and the deviator of a spatial tensor as $\mathrm{dev}[\cdot] = [\cdot] - \frac{1}{3}([\cdot] : \mathbf{g})\mathbf{g}^{-1}$, tangential linearization yields the Euler-Lagrange equations in the current configuration.

$$\delta\bar{\Pi} = \int_{\mathcal{B}_0} \mathrm{grad}^t \delta\mathbf{u} : [\mathrm{dev}[J^{-\frac{2}{3}} 2\partial_{\hat{\mathbf{g}}} W] + pJ\mathbf{g}^{-1}] \, dV + \Pi_{ext}(\delta\mathbf{u}) \tag{11}$$

$$+ \int_{\mathcal{B}_0} \delta p [J - \theta] \, dV \tag{12}$$

$$+ \int_{\mathcal{B}_0} \delta\theta [\frac{2}{3\theta} \partial_{\mathbf{g}} W : \mathbf{g} - p] \, dV = 0 \tag{13}$$

Very recently SIMO & ARMERO [5] proposed an improved bending formulation which relies on the re-parametrization of the displacement gradient into a compatible and an enhanced part, $\mathbf{H} = \mathrm{Grad}\mathbf{u} + \tilde{\mathbf{H}}$. The enhanced three-field variational formulation follows from

$$\tilde{\Pi} = \tilde{\Pi}(\mathbf{u}, \mathrm{Grad}\mathbf{u} + \tilde{\mathbf{H}}, \boldsymbol{\Sigma}_1) = \int_{\mathcal{B}_0} W(\mathrm{Grad}\mathbf{u} + \tilde{\mathbf{H}}) - \boldsymbol{\Sigma}_1^t : \tilde{\mathbf{H}} \, dV + \Pi_{ext}(\mathbf{u}) \tag{14}$$

The first variation of the re-parametrized functional renders the stationarity conditions for the problem at hand. Requiring L_2-orthogonality between the nominal stress tensor $\boldsymbol{\Sigma}_1$ and the enhanced displacement gradient $\tilde{\mathbf{H}}$ leads to a remarkably simple format, where the spatial description is obtained via changing variables $\tilde{\mathbf{h}} = \tilde{\mathbf{H}} \cdot \mathbf{F}^{-1}$, $\tilde{\text{grad}}[\cdot] = \text{Grad}[\cdot] \cdot \mathbf{F}^{-1}$ and $\partial_{\tilde{\mathbf{g}}}[\cdot] = \mathbf{F} \cdot \partial_{\mathbf{C}}[\cdot] \cdot \mathbf{F}^t$ with $\mathbf{F} = \mathbf{I} + \text{Grad}\mathbf{u} + \tilde{\mathbf{H}}$.

$$\delta\tilde{\Pi} = \int_{\mathcal{B}_0} \tilde{\text{grad}}^t \delta\mathbf{u} : 2\partial_{\tilde{\mathbf{g}}} W \, \mathrm{d}V + \Pi_{ext}(\delta\mathbf{u}) \tag{15}$$

$$+ \int_{\mathcal{B}_0} \delta\tilde{\mathbf{h}}^t : 2\partial_{\tilde{\mathbf{g}}} W \, \mathrm{d}V = 0 \tag{16}$$

5 QS Singularized Element Formulation

Within the geometrically linearized theory a singularization strategy was proposed by STEINMANN & WILLAM [9] which fully satisfies the weak localization test. Thereby, the weak form of localization was introduced directly into the element formulation by chosing appropriately enhanced strain fields, such that the element stiffness is singularized for the eigenmode $\dot{\mathbf{d}}_{loc}$. The extension to the finite deformation case is straight-forward. For the sake of argument we adopt the **Q4**-element as reference stiffness and denote the singularized element by **QS**. In the continuum formulation the bifurcated velocity is expressed in terms of the jump distribution function $\Lambda(\mathbf{N} \cdot \bar{\mathbf{X}})$ as $\mathbf{v}^b = \mathbf{v} + \dot{\gamma}\,[\mathbf{m} \otimes \mathbf{N}] \cdot \bar{\mathbf{X}}_{loc}$; $\bar{\mathbf{X}}_{loc} = \bar{\mathbf{X}}\Lambda(\mathbf{N} \cdot \bar{\mathbf{X}})$. Introducing finite element approximations of the velocity field, the discretized counterpart reads

$$\mathbf{v}_h^b = \sum_I \boldsymbol{\omega}^I(\mathbf{X})\dot{\mathbf{d}}^I + \dot{\gamma}\,[\mathbf{m} \otimes \mathbf{N}] \cdot \sum_I \boldsymbol{\omega}^I(\mathbf{X})\bar{\mathbf{X}}_{loc}^I \tag{17}$$

where $\boldsymbol{\omega}^I$ denote the shape functions associated with node I and $\dot{\gamma}$ designates the internal d.o.f. defining the jump condition. In this case the jump distribution function is interpolated by smooth approximations resulting in a continuous representation of the discontinuity within the element domain. The discretized material velocity gradient follows from Eq. 17

$$\dot{\mathbf{F}}_h^b = \sum_I \dot{\mathbf{d}}^I \otimes \text{Grad}\boldsymbol{\omega}^I(\mathbf{X}) + \dot{\gamma}\,[\mathbf{m} \otimes \mathbf{N}] \cdot \sum_I \bar{\mathbf{X}}_{loc}^I \otimes \text{Grad}\boldsymbol{\omega}^I(\mathbf{X}) \tag{18}$$

As shown earlier, the nodal velocity vector corresponding to discontinuous bifurcation is specified by the rank one tensor within the element domain, $\dot{\mathbf{d}}_{loc}^I = \dot{\gamma}\,[\mathbf{m} \otimes \mathbf{N}] \cdot \bar{\mathbf{X}}_{loc}^I$, such that the discretized version of the jump of the material velocity gradient is distributed to $\dot{\mathbf{d}}_{loc}$ by the discrete gradient operator.

$$\dot{\mathbf{F}}_h^b = \sum_I \dot{\mathbf{d}}^I \otimes \text{Grad}\boldsymbol{\omega}^I(\mathbf{X}) + \sum_I \dot{\mathbf{d}}_{loc}^I \otimes \text{Grad}\boldsymbol{\omega}^I(\mathbf{X}) \tag{19}$$

Introduction of this enhanced deformation gradient into the re-parametrized Hu-Washizu principle leads to the desired singularity of the element stiffness. Accordingly, the weak localization test $\mathrm{d}^2 W_{loc}$ is satisfied identically. It remains to relate the **QS** formulation to the proposed enhancement of NACAR, NEEDLEMAN & ORTIZ [3]. Their **QLOC** strategy is recovered if the jump distribution function $\Lambda(\mathbf{N} \cdot \bar{\mathbf{X}})$ is introduced directly into the definition of the velocity field without interpolation by the shape functions. Then the velocity gradient results in

$$\dot{\mathbf{F}}_h^b = \sum_I \dot{\mathbf{d}}^I \otimes \text{Grad}\boldsymbol{\omega}^I(\mathbf{X}) + \dot{\gamma}\,\Lambda(\mathbf{N} \cdot \bar{\mathbf{X}})[\mathbf{m} \otimes \mathbf{N}] \tag{20}$$

with $\sum_I \bar{\mathbf{X}}^I \otimes \mathrm{Grad}\boldsymbol{\omega}^I(\mathbf{X}) = \mathbf{I}$, and with the internal d.o.f. defining the amplitude $\dot{\gamma}$ of the jump condition. It can be shown that the direct introduction of the jump distribution function is not sufficient to improve the localization performance of the underlying isoparametric element formulation for associated, perfectly plastic flow. Moreover, it is questionable if the discontinuous velocity field is integrated appropriatly by standard Gauss quadrature. These difficulties are circumvented if the jump condition is distributed over the element domain and approximated by smooth interpolation.

6 Element Localization Tests

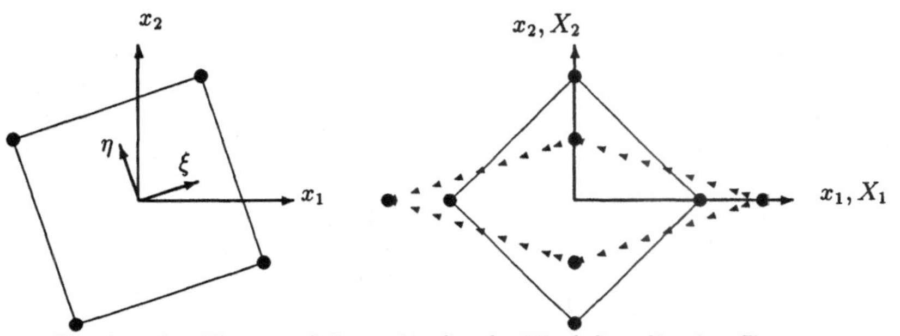

Fig. 3a, 3b : Element Orientation for the Weak Localization Test

Under the assumption of uniform state variables within the element domain several element formulations are investigated based on the aforementioned variational principles. To this end the co-ordinates of the element nodes in the spatial setting $\bar{\mathbf{x}}^I$ are assumed to form a square domain. The element $\xi - \eta$ co-ordinate sytem is rotated with respect to the fixed $x_1 - x_2$ co-ordinate system, i.e. the spatial nodal co-ordinates of the rotated element are $\bar{\mathbf{x}}^I_R = \mathbf{R} \cdot \bar{\mathbf{x}}^I$ with $\mathbf{R}$ the rotation matrix, see Fig. 3a. For a prescribed deformation gradient the material nodal co-ordinates corresponding to the rotated element geometry are $\bar{\mathbf{X}}^I_R = \mathbf{F}^{-1} \cdot \bar{\mathbf{x}}^I_R$ and the nodal displacements follow as $\mathbf{d}^I = \bar{\mathbf{x}}^I_R - \bar{\mathbf{X}}^I_R$. The natural $\xi - \eta$ co-ordinate system is rotated between $-\pi/2 \le \vartheta \le \pi/2$. For every increment $\Delta\vartheta$ the spatial and the reference geometry are computed, see e.g. Fig. 3b for the case of pure shear and $\vartheta = \pi/4$. The resulting nodal displacements are applied in one load step and the weak localization test is used to study the localization performance of different element formulations. For comparison, we examine the family of four node elements based on standard displacement expansions, i.e. the bilinear isoparametric **Q4**-element, the one-point underintegrated **QR4**-element and the assembly of four constant strain, or rather constant displacement gradient, triangles **CST4**. In addition we study the family of stabilized quadrilaterals advocated by LIU, BELYTSCHKO & CHEN [2] which were extended to finite deformations within the framework of the Hu-Washizu principle. Even if the formulation relies on the use of the γ-projection method, the interpretation as a stabilized one-point integrated method is lost as soon as the material operator varies within the element domain, see BELYTSCHKO & BINDEMAN [1]. By choosing appropriate parameters four different stabilization strategies were studied which were identified originally by the labels **OI**, **QBI**, **FIB** and **OB**. Alternatively, we consider the mixed model advocated by SIMO, TAYLOR & PISTER [6] for incompressible analysis based on the modified Hu-Washizu principle with uniform pressure and dilatation fields **Q1/P0**, which is an extension of the well-known $\bar{\mathbf{B}}$ approach by HUGHES. Very recently SIMO & ARMERO [5] extented the incompatible **QM6** element of TAYLOR, BERESFORD & WILSON within the framework of the re-parametrized Hu-Washizu principle to finite deformation problems which is labelled here by **Q1/E4**. Special enrichement strategies to capture discontinuous bifurcation on the element level were advocated by NACAR, NEEDLEMAN & ORTIZ [3] within the rate form of the principle of virtual work. As an equivalent rate-insensitive formulation in the spirit of the finite hyperelastic-plastic formulation above, the **QLOC**-element is compared with the finite deformation version of the **QS**-element proposed by STEINMANN & WILLAM [9]. The results of these different element formulations are depicted in Fig. 5 and Fig. 6. It turns out that both, the **QR4**- and the **QS**-elements do satisfy identically the weak localization test, while the **QLOC**-formulation reveals no differences to the results of the **Q4**-element with respect to the localization capturing ability. The largest improvement in the d^2W_{loc} test of the 'stabilized' elements is obtained by the **OI**-formulation. Even more dramatic is the reduction of d^2W_{loc} for those element formulations which effectively treat the incompressible constraint. The **Q1/P0** improves the representation of discontinuous bifurcation, but it is not able to satisfy the d^2W_{loc} test in general. In the past, the assembly of four triangles **CST4** has been considered particularly suitable for capturing localization in J_2-computations. Since this assembly is enriched by additional internal element edges, it satisfies the weak localization test for more orientations then the unsatisfactory elements, but it does not capture localization in general. The enhanced **Q1/E4**-element with four internal d.o.f. demonstrates the most interesting performance. For certain orientations the condensation of the internal d.o.f. fails due to a singularity of the partitioned substiffness operator. These singularities are reflected in Fig. 6 by the vertical discontinuities. Aside from these computational difficulties, however, the overall localization performance of this bending enrichment is very impressive.

7 Conclusion

The main thrust of this contribution has been to develop an element test which assesses the localization performance of arbitrary finite elements. The weak localization test is an element test of the localization energy required to form a discontinuity surface assuming that uniform conditions in the element domain prevail at incipient localization. It is remarkable that special element enhancements, which were originally conceived to improve the poor bending behaviour of low order expansions, did capture localization within a single element domain. Aside from the numerically difficulties with condensing the internal d.o.f. for specific orientations of the element geometry with respect to the discontinuity surface, the localization performance of these elements is astounding. Further enrichement to enforces satisfaction of the weak localization test is here proposed and will be pursued as a promising approach to extract those eigenmodes which are assoziated with discontinuous bifurcation and which lend themselves to regularise softening formulations within a single element domain in terms of constant fracture energy.

References

[1] T. BELYTSCHKO & L.P. BINDEMAN, [1991], "Assumed Strain Stabilization of the 4-Node Quadrilateral with 1-Point Quadrature for Nonlinear Problems", *Comp. Meth. Appl. Mech. Engr.*, Vol.88, pp.311-340.

[2] **W.K. Liu, T. Belytschko & J. Chen**, [1988], "Nonlinear Versions of Flexurally Superconvergent Elements", *Comp. Meth. Appl. Mech. Engr.*, Vol. 71, pp. 241-258.

[3] **A. Nacar, A. Needleman & M. Ortiz**, [1989], "A Finite Element Method for Analyzing Localization in Rate Dependent Solids at Finite Strain", *Comp. Meth. Appl. Mech. Engr.*, Vol. 73, pp. 235-258.

[4] **J.C. Simo**, [1988], "A Framework for Finite Strain Elastoplasticity Based on the Maximum Plastic Dissipation and the Multiplicative Decomposition: Part I", *Comp. Meth. Appl. Mech. Engr.*, Vol. 66, pp. 199-219.

[5] **J.C. Simo & F. Armero**, [1991], "Geometrically Nonlinear Enhanced Strain Mixed Method and the Method of Incompatible Modes", *Int. J. Num. Meth. Engr.*, to appear in 1991.

[6] **J.C. Simo, R.L. Taylor & K.S. Pister**, [1985], "Variational and Projection Methods for the Volume Constraint in Finite Deformation Elasto-Plasticity", *Comp. Meth. Appl. Mech. Engr.*, Vol. 51, pp. 177-208.

[7] **N. Sobh**, [1987], "Bifurcation Studies of Elasto-Plastic Operators", Ph.D. Dissertation at the University of Colorado, Boulder.

[8] **P. Steinmann & K. Willam**, [1991], "Finite Elements for Capturing Localized Failure", *Arch. Appl. Mech.*, Vol. 61, pp. 259-275.

[9] **P. Steinmann & K. Willam**, [1991], "Finite Elements Performance in Localized Failure Computations", *Comp. Meth. Appl. Mech. Engr.*, Vol. 85, p 23.

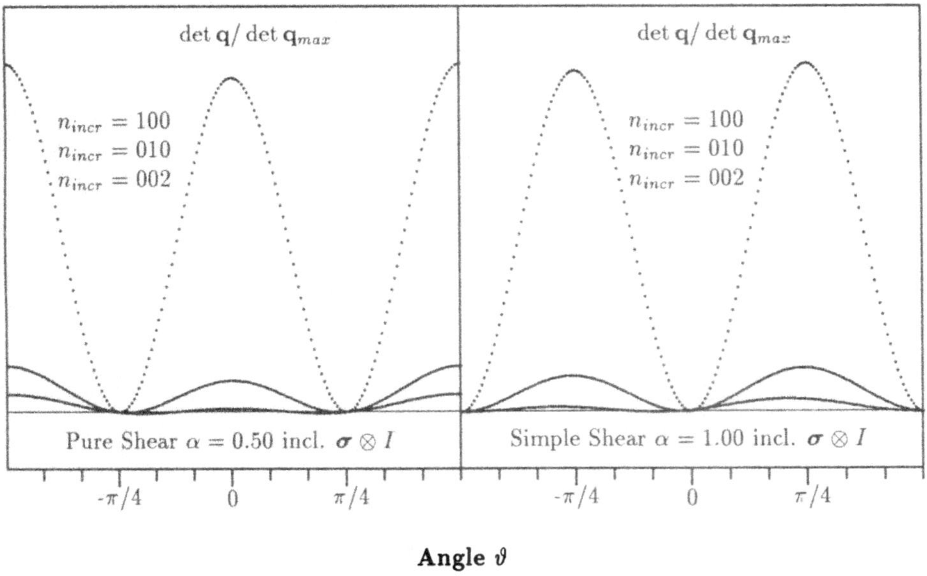

Fig. 4 : Determinant of Algorithmic Acoustic Tensor

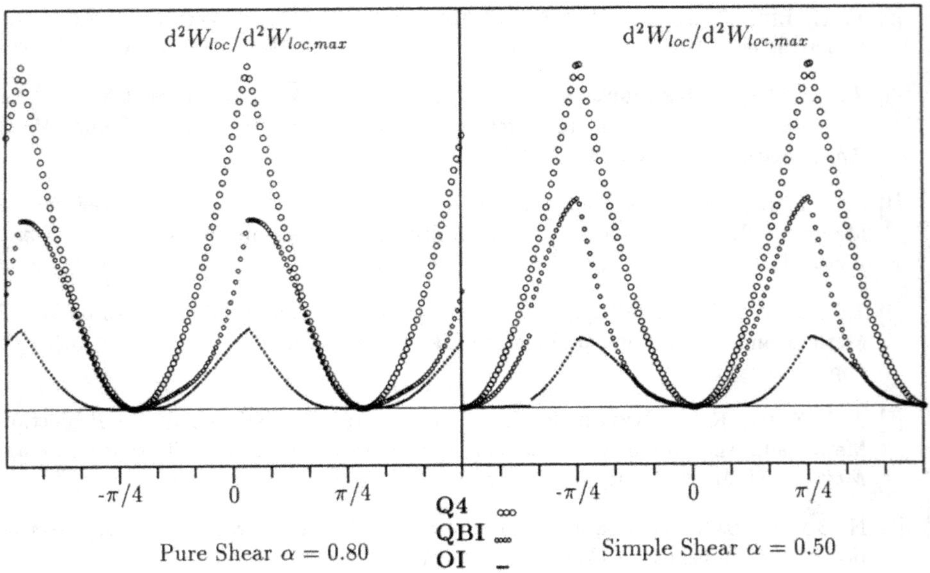

Fig. 5 : Weak Localization Test for the **Q4-**, **QBI-**, **OI-**Element

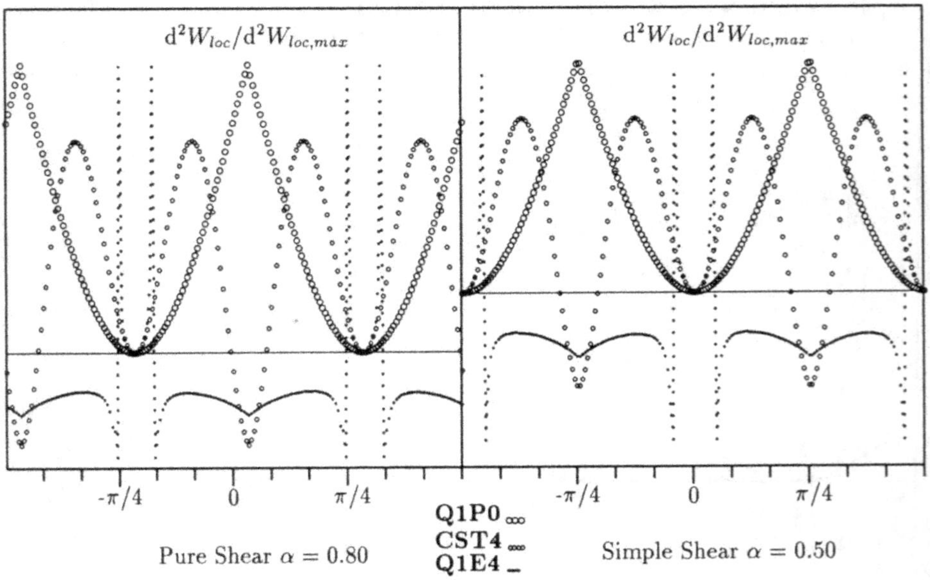

Fig. 6 : Weak Localization Test for the **Q1P0-**, **CST4-**, **Q1E4-**Element

On the Treatment of Contact Contraints within Coupled Thermomechanical Analysis

P. Wriggers* & C. Miehe**

* Institut für Mechanik, Technische Hochschule Darmstadt, Hochschulstr. 1, D-6100 Darmstadt, Germany

** Institut für Baumechanik und Numerische Mechanik, Universität Hannover, Appelstr. 9a, D-3000 Hannover 1, Germany

1 Introduction

In engineering problems like shrink fitting, deep drawing or hot rolling the effects of heat generation and heat transfer have to be considered during the process. Since many of these simulations involve contact problems with dry friction it is of interest to develop contact laws which are able to predict besides normal and tangential stresses also the amount of heat generated due to friction and the heat transfer across the contact surface.

During the last years a series of publication was devoted to the derivation of thermomechanical interface models that take into account the micro–geometrical shape of macroscopically smooth surfaces, see e.g. MADHUSUDANA & FLETCHER (1986). The associated theoretical studies consider a statistical characterization of the surface profile and of the related thermomechanical phenomena and usually pre-define the shape and the mechanical behaviour of the asperities. The derived constitutive relations for the thermomechanical contact interface include also effects due to the finishing treatments of the surface layer which involve different characteristics from the bulk material. Hence the effective yield pressure (determined by hardness tests) depends on the depth of the plastic zone, see SONG & YOVANOVICH (1987).

Connected to these sophisticated interface models for the heat transfer within the contact area are constitutive relations for the mechanical contact which are also based on statistical models, see e.g. the overview paper by ODEN & MARTINS (1985). Here we have to distinguish between constitutive equations for the normal contact and the tangential contact. For the normal contact statistical models have been stated in KRAGELSKY et. al. (1982). For the tangential contact which splits into the stick and slip part several constitutive relations have been developed, see also KRAGELSKY (1982) for a review.

Numerical formulations of contact problems within the finite element method have been given in many papers. The standard approach is to model the contact surface as an idealized smooth interface which then leads to a condition describing the non–penetration by a pure geometrical constraint. Within this assumption several approaches are possible for the variational formulation of contact problems in case of finite deformations. The so called Lagrangian multiplier technique enforces the contact constraints exactly, see e.g. HUGHES et. al. (1976) or NOUR–OMID & WRIGGERS (1986). A consistent formulation for implicit computations using penalty methods is given in e.g. WRIGGERS

D. Besdo, E. Stein (Eds.)
Finite Inelastic Deformations -
Theory and Applications
IUTAM Symposium Hannover/Germany 1991

& SIMO (1985) based on a discrete model discussed in HALLQUIST (1979). Recently also constitutive equations within the contact interface have been considered in finite element approximations, see e.g. WRIGGERS et. al. (1990).

Finite element formulations for the frictional contact have been discussed e.g. in ODEN & MARTINS (1985), CHAUDHARY & BATHE (1986), WRIGGERS et. al. (1990) or LAURSEN & SIMO (1991). Finite element models which include thermomechanical coupling for contact problems have been presented by e.g. CURNIER (1982) for lubricated contact, by ZAVARISE et. al. (1992) for large deformation frictionless contact and for small deformations by JOHANNSON & KLARBRING (1992) including frictional heating.

In this paper we will focus on a finite element formulation of a sophisticated interface model that includes frictional contact, heat exchange in the contact area and frictional heating. For this purpose we give (i) a *continuous formulation of the local contact geometry* based on a parametrization of a *predefined (moving) master surface*, where we restrict our consideration to two dimensional problems. This approach has similarities with the formulation proposed by LAURSEN & SIMO (1991) for the rigid obstacle problem. We postulate (ii) *constitutive equations for the contact stress, the contact heat flux and the frictional dissipation* on the contact surface. These equations have model character but exhibit essential features of known experimental and micro-geometrical investigations of contact interfaces. After having formulated (iii) the *structure of the initial-boundary-value problem of finite strain thermoelasticity combined with thermomechanical contact* we discuss (iv) two main algorithmic aspects: Firstly we discuss the *solution algorithm for the coupled global equations* based on an operatorsplit of the global thermomechanical evolution operator. This strategy was recently introduced in the context of finite strain thermoplasticity without contact constraints by SIMO & MIEHE (1991) and leads to an algorithmic decoupling of the coupled problem within the time step by preserving the symmetry properties of the subproblems. Secondly we consider an extension of the *return map algorithm* proposed by WRIGGERS (1987) for the update of the contact stress and frictional dissipation. Finally we discuss (v) the *spatial discretization of the contact surfaces* and show as an interesting aspect the transition of the continuous contact geometry to the discrete contact geometry of the three node contact element used e.g. in WRIGGERS et. al. (1990). Several numerical examples show the performance and the accuracy of the derived finite numerical formulation.

2 Contact Geometry

Let $\mathcal{B}^\alpha$, $\alpha = 1, 2$, denote *two bodies of interest* and $\boldsymbol{\varphi}_t^\alpha : \mathcal{B}^\alpha \to \mathbb{R}^2$ the associated deformation maps at time $t \in \mathbb{R}_+$. $\boldsymbol{\varphi}_t^\alpha$ maps points $\mathbf{X}^\alpha \in \mathcal{B}^\alpha$ of the reference configuration onto points $\mathbf{x}^\alpha = \boldsymbol{\varphi}_t^\alpha(\mathbf{X}^\alpha)$ of the current configuration. Assume that both bodies come into contact. Motivated by micromechanical investigations of contact problems we view *mechanical contact as a (microscopical) penetration of the current mathematical boundaries* $\boldsymbol{\varphi}_t^\alpha(\Gamma_c^\alpha)$ where $\Gamma_c^\alpha \subset \partial\mathcal{B}^\alpha$ are *possible* contact surfaces of the bodies $\mathcal{B}^\alpha$, see Figure 3.1 for an illustration of this idea. In what follows we denote $\boldsymbol{\varphi}_t^1(\Gamma_c^1)$ as the current *slave surface* which penetrates in the case of contact into the current *master surface* $\boldsymbol{\varphi}_t^2(\Gamma_c^2)$. The latter one plays within our formulation of the contact geometry the role of a (moving) reference surface. We parametrizise the master surface in its reference and current configuration by a *natural parameter* ξ (for 2–D problem), i.e. we consider material curves $\xi \mapsto \mathbf{X}^2 = \hat{\mathbf{X}}^2(\xi) \subset \Gamma_c^2$ and $\xi \mapsto \mathbf{x}^2 = \hat{\mathbf{x}}_t^2(\xi) \subset \varphi_t^2(\Gamma_c^2)$. Then the *local deformation gradient of the master surface* is given by $\mathbf{F}_t^2 := \frac{l}{L}\,\mathbf{e}_T^2 \otimes \mathbf{E}_T^2$ based on the tangent vectors $l\,\mathbf{e}_T^2 := \hat{\mathbf{x}}_{t,\xi}^2(\xi)$ and $L\,\mathbf{E}_T^2 := \hat{\mathbf{X}}_{,\xi}^2(\xi)$ with $\| \mathbf{e}_T^2 \| = \| \mathbf{E}_T^2 \| \equiv 1$.

2.1 Penetration. As the first relevant function for the contact geometry we define a *penetration function* on the current slave surface $\varphi_t^1(\Gamma_c^1)$ by setting

$$g_N = \begin{cases} \| \mathbf{x}^1 - \hat{\mathbf{x}}_t^2(\bar{\xi}) \| & \text{for } [\mathbf{x}^1 - \hat{\mathbf{x}}_t^2(\bar{\xi})] \cdot \bar{\mathbf{e}}_N^2 < 0 \\ 0 & \text{otherwise} \end{cases} \tag{2.1}$$

Here $\bar{\xi}$ is the *minimizer of the distance function* $\hat{d}^1(\xi) = \| \mathbf{x}^1 - \hat{\mathbf{x}}_t^2(\xi) \|$ for a *given slave point* $\mathbf{x}^1$, i.e. the current *master point* $\hat{\mathbf{x}}_t^2(\bar{\xi})$ is the orthogonal projection of a given slave point $\mathbf{x}^1$ onto the current master surface $\varphi_t^2(\Gamma_c^2)$. $\bar{\mathbf{e}}_N^2 := \mathbf{E}_3 \times \bar{\mathbf{e}}_T^2$ denotes the outward unit normal on the current master surface at the master point. The penetration function gives two informations: Firstly it serves as a *local contact check*, i.e. we set: *contact* $\Leftrightarrow$ $g_N > 0$. Secondly it enters for $g_N > 0$ as a *local kinematical variable* the constitutive function for the contact pressure, see section 3.1 below. By taking the time derivative of (2.1) one obtains in the case of contact the rate of penetration

$$-\dot{g}_N = [\mathbf{v}_t^1 - \hat{\mathbf{v}}_t^2(\bar{\xi})] \cdot \bar{\mathbf{e}}_N^2 \tag{2.2}$$

for given spatial velocities $\mathbf{v}_t^1$ and $\hat{\mathbf{v}}_t^2(\bar{\xi})$ at the slave and master points.

2.2 Tangential Relative Velocity and Tangential Relative Slip. The tangential relative slip between two bodies is related to the change of the solution point $\bar{\xi}$ of the minimal distance problem. Conseqently we define as the second important kinematical function (in view to problems with friction) a *tangential relative velocity function* on the current slave surface $\varphi_t^1(\Gamma_c^1)$ by setting

$$\dot{g}_T := \dot{\bar{\xi}}\, l \tag{2.3}$$

(2.3) determines *per definition* the evolution of the *tangential slip* g_T which enters as a *local kinematical variable* the constitutive function for the contact tangential stress, see section 3.2. The rate $\dot{\bar{\xi}}$ in (2.3) of the solution point $\bar{\xi}$ can be computed by taking the time derivative of the necessary condition $\hat{d}_{,\xi}^1(\bar{\xi}) = 0$ for the minimal distance problem. We obtain

$$\kappa\, \dot{\bar{\xi}}\, l = [\mathbf{v}_t^1 - \hat{\mathbf{v}}_t^2(\bar{\xi})] \cdot \bar{\mathbf{e}}_T^2 - \frac{g_N}{l}\, \hat{\mathbf{v}}_{t,\xi}^2(\bar{\xi}) \cdot \bar{\mathbf{e}}_N^2 \tag{2.4}$$

with the factor $\kappa := 1 + \frac{g_N}{l^2}\, \hat{\mathbf{x}}_{t,\xi\xi}^2(\bar{\xi}) \cdot \bar{\mathbf{e}}_N^2$ associated with the curvature of the master surface. Thus the evolution $\dot{g}_T$ in (2.3) is only in the case of *perfect contact*, i.e. for $g_N = 0$, given by the *projection* of the spatial velocities $\mathbf{v}_t^1$ and $\hat{\mathbf{v}}_t^2(\bar{\xi})$ evaluated at the slave and master points *onto the tangential direction of the master surface at the master point.* For a penetration $g_N > 0$ we have to take into account the second term in (2.4) and the scaling factor κ, both consequences of the time dependence of $\bar{\mathbf{e}}_T^2$.

2.3 Remark: Frame Invariance. Consider the vectors $\mathbf{g}_N := -g_N \bar{\mathbf{e}}_N^2$ and $\mathbf{g}_T := \dot{\bar{\xi}}\, l\, \bar{\mathbf{e}}_T^2$ normal and tangential to the master surface at the master point. Clearly these vectors transform objectively under rigid body motions superposed onto the spatial configuration $(\mathbf{x}_t^2)^+ = \mathbf{Q}(t)\,\mathbf{x}_t^2 + \mathbf{c}(t)\ \forall\ \mathbf{Q}(t) \in SO(2)$ where $SO(2)$ is the proper orthogonal group. Example: $(\mathbf{g}_N)^+ = \mathbf{Q}\,\mathbf{g}_N$ is consequence of $(\bar{\mathbf{e}}^2_N)^+ = \mathbf{Q}\,\bar{\mathbf{e}}^2_N$. Observe furthermore that the (a priori objective) Lie derivative of the tangential vector $\mathbf{g}_T$ has the representation $\mathcal{L}_v\,\mathbf{g}_T = \mathbf{F}^2_t\{\frac{d}{dt}[\mathbf{F}^{2-1}_t(\mathbf{g}_T)]\} = \dot{\bar{\xi}}\,l\,\bar{\mathbf{e}}_T$ based on the deformation gradient $\mathbf{F}^2_t$ of the master surface defined above. Thus (2.3) represents an evolution equation for the *objective rate* $\mathcal{L}_v\,\mathbf{g}_T$ of the tangential vector introduced above.

3 Constitutive Modelling on the Contact Surface

When the micromechanical behaviour of the contact area is studied different phenomena have to be considered as well for the mechanical as for the thermal interface description. In this paper we use constitutive models which have been derived based on micromechanical observations of physical contact surfaces. These models can be related to formulations relative to mathematical contact surfaces by an averaging process as symbolically indicated in Figure 3.1. Goal of this section is to formulate *local* constitutive equations for the *pressure, tangential stress and normal (outward) heat flux* on the slave surface at the slave point $\mathbf{x}^1$ relative to the bases $\{\bar{\mathbf{e}}^2_T, \bar{\mathbf{e}}^2_N\}$ acting on body $\mathcal{B}^1$, see general structure in (4.7).

3.1 Contact Normal Stress. It is well known that the contact pressure is related to the approach of the physical surfaces which come into contact, i.e. the penetration of the mathematical surfaces results from the deformation of the micro–asperities, see Figure 3.1. Based on experimental investigations KRAGELSKY et. al. (1982) formulated the following *nonlinear elastic constitutive equation* for the contact pressure

$$p_N = c_N\,(g_N)^m \tag{3.1}$$

in terms of the penetration g_N defined in (2.1). Here c_N and m are material parameters which have to be determined by experiments.

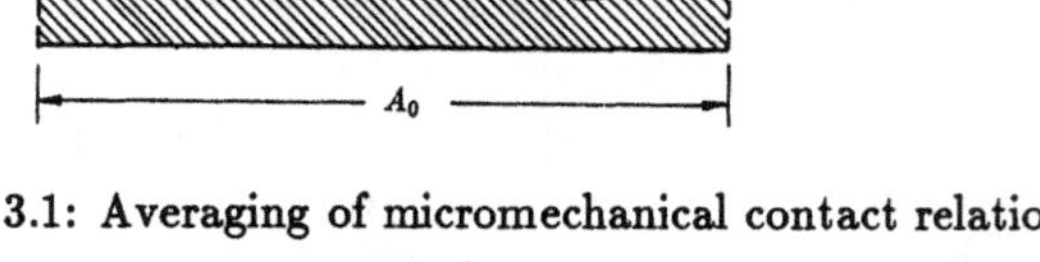

Figure 3.1: Averaging of micromechanical contact relations.

3.2 Contact Tangential Stress and Plastic Tangential Slip. In this paper we restrict our consideration to classical Coulomb–type friction problems. Coulomb's constitutive equation can formally be formulated in the framework of elastoplastitity. This has been considered by several authors, e.g. MICHALOWSKI, MROZ (1978) or CURNIER (1984) and within a finite element formulation by WRIGGERS (1987). The key idea of this approach is a *split of the tangential slip* g_T *into an elastic part* g^e_T (micro displacement describing the stick behaviour) *and a plastic part* g_T^p (frictional slip), see equation $(3.2)_2$ below. We assume a *linear elastic constitutive equation for the tangential contact stress*

$$t_T = c_T \, g_T^e \qquad \text{with} \qquad g_T^e := g_T - g_T^p \tag{3.2}$$

where c_T is a material parameter. The plastic tangential slip g_T^p is governed by a constitutive evolution equation which can be formally derived by using standard concepts of elastoplasticity theory. Let

$$\mathcal{D}^p := t_T \cdot \dot{g}_T^p \geq 0 \tag{3.3}$$

be the *dissipation due to the plastic slip.* Now consider an elastic domain $\mathbb{E}_t := \{ t_T \in \mathbb{R} \mid \hat{f}_p(t_T) \leq 0 \}$ in the space of the contact tangential stress. Here

$$\hat{f}_p(t_T) = |\, t_T \,| - \mu \, p_N \leq 0 \tag{3.4}$$

is the *plastic slip criterion function* for given contact pressure p_N with material parameter μ. Holding p_N fixed one obtains from the so called maximum dissipation principle, $(t_T - \overset{*}{t}_T) \cdot \dot{g}_T^p \geq 0 \;\forall\; \overset{*}{t}_T \in \mathbb{E}_t$, the *constitutive evolution equation for the plastic slip*

$$\dot{g}_T^p = \lambda \, \frac{\partial \hat{f}_p(t_T)}{\partial t_T} = \lambda \, \mathrm{sign}(t_T) \tag{3.5}$$

(normality rule for fixed contact pressure) along with the loading–unloading conditions in Kuhn–Tucker form

$$\lambda \geq 0, \quad \hat{f}_p(t_T) \leq 0, \quad \lambda \, \hat{f}_p(t_T) = 0 \,. \tag{3.6}$$

which determine the plastic parameter λ.

3.3 Contact Heat Flux. Different techniques for the computation of the thermal contact resistance, taking into account the dependence on various parameters have been proposed, see e.g. MADHUSUDANA & FLETCHER (1986). We assume the following structure for the *constitutive equation for the heat flux* q_N (outflux through slave surface)

$$q_N = \hat{q}_N(\Theta^1, \Theta^2, p_N) = \hat{h}(\Theta^1, \Theta^2, p_N)(\Theta^1 - \Theta^2) \,. \tag{3.7}$$

The *contact heat transfer coefficient* $h = \hat{h}(\Theta^1, \Theta^2, p_N)$ depends on the current temperatures Θ^1 and Θ^2 at the slave and master points, $\mathbf{x}^1$ and $\hat{\mathbf{x}}_t^2(\bar{\xi})$ respectively, as well as on the contact pressure p_N defined in (3.1). The resistance against heat transfer is mainly due to the low percentage of physical surface area which is really in contact. The presence of a reduced set of spots surrounded by micro-cavities characterizes the contact zone, hence heat exchange is possible by heat conduction through the *spots* (S), heat conduction through the *gas* (G) contained in the cavities and *radiation* (R) between micro–cavity surfaces. These mechanisms act in parallel which induces the additive decomposition

$$\hat{h} = \hat{h}_S + \hat{h}_G + \hat{h}_R \,. \tag{3.8}$$

of the contact heat transfer function. Based on statistical micromechanical models different relations for the gas and spot conductance have been developed, see e.g. SONG,

YOVANOVICH (1987). In this paper we use a simplyfied relation which neglects radiation effects and gas conductance but takes into account pressure dependency of the contact conductance

$$h_S = h_{S_o} \left[\frac{p_N}{H}\right]^{\epsilon} \qquad \text{and} \qquad h_G = h_R = 0 \tag{3.9}$$

based on the material parameters h_{S_o}, H, ϵ. This model is valid for high pressures. A more sophisticated constitutive model is implemented in ZAVARISE et. al. (1992) for the frictionless case.

4 Initial–Boundary–Value Problem

In the previous sections we have discussed the contact geometry and the constitutive equations associated with contact interface. Let us now formulate the initial–boundary–value problem for nonlinear thermoelasticity combined with thermomechanical (frictional) contact. In order to obtain a compact structure we introduce locally at the material point $\mathbf{X}^\alpha \in \mathcal{B}^\alpha$ and time $t \in \mathbb{R}_+$ a vector of *primary variables* $\mathcal{S}^\alpha(\mathbf{X}^\alpha, t) := \{\ \varphi^\alpha(\mathbf{X}^\alpha,t),\ \mathbf{V}^\alpha(\mathbf{X}^\alpha,t),\ \Theta^\alpha(\mathbf{X}^\alpha,t)\}$ (which contains the configuration φ^α, the material velocity $\mathbf{V}^\alpha$ and the material temperature Θ^α) and locally at points $\mathbf{X}^\alpha \in \Gamma_c^\alpha$ of the contact surfaces a vector of *history variables* $\mathcal{I}(\mathbf{X}^\alpha,t) := \{\ g_T^p(\mathbf{X}^\alpha,t)\}$ (which contains the plastic tangential slip g_T^p). Internal variables in the domains $\mathcal{B}^\alpha$ do not appear since we restrict our presentation to thermoelastic constitutive response. The thermomechanical initial–boundary–value problem is governed by the *local field equations* in the domains $\mathcal{B}^\alpha$

$$\begin{array}{rclcl} \frac{\partial}{\partial t}\varphi^\alpha & = & \mathbf{V}^\alpha & + & \mathbf{0} \\ \frac{\partial}{\partial t}\mathbf{V}^\alpha & = & \frac{1}{\rho_0^\alpha}\mathrm{DIV}\,\mathbf{P}^\alpha & + & \mathbf{0} \\ \frac{\partial}{\partial t}\Theta^\alpha & = & 0 & - & \frac{1}{c^\alpha}\mathrm{DIV}\,\mathbf{Q}^\alpha + \frac{1}{c^\alpha}\mathrm{S}^\alpha \end{array} \tag{4.1}$$

which represent the definition of the material velocity, the balance of linear momentum and the balance of internal energy in form of the temperature evolution equation. $\mathbf{P}^\alpha$ is the first Piola–Kirchhoff stress tensor, $\mathbf{Q}^\alpha$ denotes the material heat flux vector and S^α is a heat source which describes in the framework of thermoelasticity the Gough–Joule coupling effect. The field equation (4.1) form a coupled first order evolution system for the primary variable vector $\mathcal{S}^\alpha$ introduced above, i.e.

$$\frac{\partial}{\partial t}\mathcal{S}^\alpha = \hat{\mathbf{A}}^\alpha(\mathcal{S}^\alpha)\,, \tag{4.2}$$

where the nonlinear evolution operator $\hat{\mathbf{A}}$ represents the right hand side of (4.1) by taking into account the dependency of the thermoelastic *constitutive equations in the domains* $\mathcal{B}^\alpha$

$$\begin{aligned} \mathbf{P}^\alpha &= \hat{\mathbf{P}}^\alpha(\mathcal{S}^\alpha) \\ \mathbf{Q}^\alpha &= \hat{\mathbf{Q}}^\alpha(\mathcal{S}^\alpha) \\ \mathrm{S}^\alpha &= \hat{\mathrm{S}}^\alpha(\mathcal{S}^\alpha) \end{aligned} \tag{4.3}$$

on $\mathcal{S}^\alpha$. Next the *initial conditions* for the primary and history variables for a time interval $[t_n, t_{n+1}] \in \mathbb{R}_+$ of interest are given by

$$\mathcal{S}^\alpha(\mathbf{X}^\alpha, t_n) = \mathcal{S}_n^\alpha(\mathbf{X}^\alpha) \quad \text{and} \quad \mathcal{I}^\alpha(\mathbf{X}^\alpha, t_n) = \mathcal{I}_n^\alpha(\mathbf{X}^\alpha)\,. \tag{4.4}$$

We consider *boundary conditions* for the deformation and the temperature on $\Gamma^\alpha_\varphi \subset \partial\mathcal{B}^\alpha$ and $\Gamma^\alpha_\Theta \subset \partial\mathcal{B}^\alpha$

$$\begin{aligned} \varphi^\alpha(\mathbf{X}^\alpha,t) &= \bar{\varphi}^\alpha(\mathbf{X}^\alpha,t) \text{ on } \Gamma^\alpha_\varphi \\ \Theta^\alpha(\mathbf{X}^\alpha,t) &= \bar{\Theta}^\alpha(\mathbf{X}^\alpha,t) \text{ on } \Gamma^\alpha_\Theta \end{aligned} \tag{4.5}$$

as well as for the traction vector and the heat flux on $\Gamma^\alpha_t \subset \partial\mathcal{B}^\alpha$ and $\Gamma^\alpha_q \subset \partial\mathcal{B}^\alpha$

$$\begin{aligned} \mathbf{t}^\alpha\,(\,\varphi^\alpha_t(\mathbf{X}^\alpha),t\,) &= \bar{\mathbf{t}}^\alpha\,(\,\varphi^\alpha_t(\mathbf{X}^\alpha),t\,) \text{ on } \varphi^\alpha_t(\,\Gamma^\alpha_t\,) \\ q^\alpha\,(\,\varphi^\alpha_t(\mathbf{X}^\alpha),\Theta^\alpha_t(\mathbf{X}^\alpha),t\,) &= \bar{q}^\alpha\,(\,\varphi^\alpha_t(\mathbf{X}^\alpha),\Theta^\alpha_t(\mathbf{X}^\alpha),t\,) \text{ on } \varphi^\alpha_t(\,\Gamma^\alpha_q\,) \end{aligned} \tag{4.6}$$

The initial–boundary–value problem is completed by the thermomechanical *constitutive equations on the current slave contact surface* $\varphi^1_t(\Gamma^1_c)$

$$\begin{aligned} p_N &= \hat{p}_N\,[\,\mathcal{S}^{\alpha,\beta},\,\mathcal{I},\,\lambda\,] \\ t_T &= \hat{t}_T\,\,[\,\mathcal{S}^{\alpha,\beta},\,\mathcal{I},\,\lambda\,] \\ q_N &= \hat{q}_N\,[\,\mathcal{S}^{\alpha,\beta},\,\mathcal{I},\,\lambda\,] \\ \mathcal{D}^p &= \hat{\mathcal{D}}^p\,[\,\mathcal{S}^{\alpha,\beta},\,\mathcal{I},\,\lambda\,] \end{aligned} \tag{4.7}$$

for the contact pressure, the contact tangential stress, the contact heat flux and the frictional dissipation and the evolution equation for the plastic slip with structure

$$\frac{\partial}{\partial t}\mathcal{I} \;=\; \lambda\,\hat{\mathcal{E}}\,[\,\mathcal{S}^{\alpha,\beta},\,\mathcal{I}\,] \tag{4.8}$$

which is constraint by the Kuhn–Tucker loading–unloading conditions

$$\lambda \geq 0 \quad ; \quad \hat{f}_p\,[\,\mathcal{S}^{\alpha,\beta},\,\mathcal{I}\,] \leq 0 \quad ; \quad \lambda\,\hat{f}_p\,[\,\mathcal{S}^{\alpha,\beta},\,\mathcal{I}\,] = 0\,. \tag{4.9}$$

The superscript (α,β) indicates the dependency of the constitutive functions on both contacting surfaces. The equations (4.7)–(4.8) are within this paper of central interest and have already been specified for model problems in section 3, compare equation (3.1)–(3.7).

5 Solution Algorithms

The main solution strategies for coupled problems are *monolithic schemes* where the differential equations for the different variables are all solved together, or *staggered schemes* where the different variables are solved separately, see WOOD (1990) for an overview. Global solution algorithms for coupled thermomechanical analysis have been formulated by ARGYRIS & DOLTSINIS (1981), MIEHE (1988), DOLTSINIS (1990), SIMO & MIEHE (1991), among others. In what follows we adopt the staggered scheme proposed by SIMO & MIEHE (1991) based on an *operator split* of the global thermomechanical evolution operator discussed in section 4. This strategy yields an algorithmic decoupling of the thermomechnical equations within a time step on the basis of (in the frictionless case) symmetric subprobelems.

A further objective of this section is to formulate the local update algorithm for the contact tangential stress and the frictional dissipation. Here we extend the return mapping algorithm proposed e.g. by WRIGGERS (1987) for isothermal frictional contact to the nonisothermal case.

5.1 Global Thermomechanical Operator Split Algorithm. The central idea is a split of the evolution operator $\hat{\mathbf{A}}^{\alpha}$ in (4.2) into its natural mechanical and thermal parts as indicated in equation (4.1): $\hat{\mathbf{A}}^{\alpha}(\mathcal{S}^{\alpha}) = \hat{\mathbf{A}}^{\alpha}_{M}(\mathcal{S}^{\alpha}) + \hat{\mathbf{A}}^{\alpha}_{T}(\mathcal{S}^{\alpha})$. This split defines within the time step $\Delta t_{n+1} := t_{n+1} - t_n$ two subproblems

$$\text{(M):} \quad \frac{\partial}{\partial t}\bar{\mathcal{S}}^{\alpha} = \hat{\mathbf{A}}^{\alpha}_{M}(\bar{\mathcal{S}}^{\alpha}) \qquad \text{and} \qquad \text{(T):} \quad \frac{\partial}{\partial t}\mathcal{S}^{\alpha} = \hat{\mathbf{A}}^{\alpha}_{T}(\mathcal{S}^{\alpha}) \, , \tag{5.1}$$

a *purely mechanical subproblem* (M) *at frozen temperature* along with the mechanical initial and boundary conditions in (4.4)–(4.7) followed by a *purely thermal subproblem* (T) *at frozen configuration* along with the thermal initial and boundary conditions in (4.4)–(4.7). Both subproblems are constraint by the evolution of the plastic variables (4.8)–(4.9). We consider the *quasistatic problem* and integrate both phases (5.1) with the *backward Euler* algorithm. Then the algorithmic counterpart $ALGO_M$ for the mechanical subproblem (M) takes the form (integration of $(4.1)_{1,2}$)

$$\mathrm{DIV}\,\hat{\mathbf{P}}^{\alpha}(\varphi^{\alpha}_{n+1};\Theta^{\alpha}_{n}) = \mathbf{0} \qquad \text{and} \qquad \mathbf{V}^{\alpha}_{n+1} = \tfrac{1}{\Delta t_{n+1}}(\varphi^{\alpha}_{n+1} - \varphi^{\alpha}_{n}) \tag{5.2}$$

in $\mathcal{B}^{\alpha}$ at *frozen thermal primary variable* Θ^{α}_{n}. The thermal subproblem (T) obtains the algorithmic form $ALGO_T$ (integration of $(4.1)_3$)

$$\tfrac{c^{\alpha}}{\Delta t_{n+1}}(\Theta^{\alpha}_{n+1} - \Theta^{\alpha}_{n}) = -\mathrm{DIV}\,\hat{\mathbf{Q}}^{\alpha}(\Theta^{\alpha}_{n+1};\varphi^{\alpha}_{n+1}) + \hat{\mathrm{S}}^{\alpha}(\Theta^{\alpha}_{n+1};\varphi^{\alpha}_{n+1},\mathbf{V}^{\alpha}_{n+1}) \tag{5.3}$$

in $\mathcal{B}^{\alpha}$ at *frozen mechanical primary variables* $\varphi^{\alpha}_{n+1}, \mathbf{V}^{\alpha}_{n+1}$. Thus we solve within a time step Δt_{n+1} first the mechanical problem (5.2) for the actual configuration field φ^{α}_{n+1}. Next we compute the actual temperature field Θ^{α}_{n+1} by solving the thermal problem (5.3). The overall global thermomechanical solution algorithm has the product structure $ALGO_{TM} = ALGO_T \circ ALGO_M$. Thus the operator split algorithm results in an *algorithmic decoupling* of the thermomechanical problem within the time interval.

In view to the finite element treatment we construct the weak forms of the time discrete field equations $(5.2)_1$ and (5.3) by standard Galerkin procedures. The *weak form of the mechanical subproblem* $ALGO_M$ takes the form

$$\begin{aligned} \hat{G}_M(\varphi^{\alpha}_{n+1}, \overset{*}{\mathbf{u}}^{\alpha}; \Theta^{\alpha}_{n}) := & - \int_{\varphi^{1}_{n+1}(\Gamma^{1}_{c})} [\delta g_N(-p_N) + \delta g_T\, t_T]\, ds^1 \; + \\ & \sum_{\alpha=1}^{2} \Big\{ \int_{\mathcal{B}^{\alpha}} \nabla\overset{*}{\mathbf{u}}^{\alpha} \cdot \boldsymbol{\tau}^{\alpha}\, dV - \int_{\varphi^{\alpha}_{n+1}(\Gamma^{\alpha}_{t})} \overset{*}{\mathbf{u}}^{\alpha} \cdot \bar{\mathbf{t}}^{\alpha}\, da \Big\} = 0 \end{aligned} \tag{5.4}$$

where we have dropped for convenience the subscript $n+1$ on the right hand side. $\overset{*}{\mathbf{u}}^{\alpha}$ denotes a mechanical test function field defined on the current configuration (virtual displacements). The first term in (5.4) describes the contributions due to the contact with the relative (Lie–type) variations $-\delta g_N = [\,\overset{*}{\mathbf{u}}{}^1_t - \overset{*}{\mathbf{u}}{}^2_t(\bar{\xi})\,] \cdot \bar{\mathbf{e}}^2_N$ analogous to (2.2) and $\kappa\, \delta g_T = [\,\overset{*}{\mathbf{u}}{}^1 - \overset{*}{\mathbf{u}}{}^2(\bar{\xi})\,] \cdot \bar{\mathbf{e}}^2_T - \frac{g_N}{l}\, \overset{*}{\mathbf{u}}{}^2_{,\xi}(\bar{\xi}) \cdot \bar{\mathbf{e}}^2_N$ anologous to (2.4). The second term contains the standard domain contributions with the Kirchhoff stress tensor $\boldsymbol{\tau}^\alpha$. The *weak form of the thermal subproblem* $ALGO_T$ takes the form

$$\hat{G}_T(\Theta^\alpha_{n+1}, \overset{*}{\vartheta}{}^\alpha; \varphi^\alpha_{n+1}) := - \int\limits_{\varphi^1_{n+1}(\Gamma^1_c)} [\,\delta\vartheta_q\,(-q_N) + \delta\vartheta_{\mathcal{D}}\, \mathcal{D}^p\,]\, ds^1 \; +$$

$$\sum_{\alpha=1}^{2} \{ \int\limits_{\mathcal{B}^\alpha} [\,\nabla\overset{*}{\vartheta}{}^\alpha \cdot (-\mathbf{q}^\alpha) - \overset{*}{\vartheta}{}^\alpha\,(\,S^\alpha - \tfrac{c^\alpha}{\Delta t}(\Theta^\alpha - \Theta^\alpha_n))\,]\, dV - \int\limits_{\varphi^\alpha_{n+1}(\Gamma^\alpha_q)} \overset{*}{\vartheta}{}^\alpha\,(-\bar{q}^\alpha)\, da \,\} = 0 \tag{5.5}$$

Here $\overset{*}{\vartheta}{}^\alpha$ denotes a thermal test function field defined on the current configuration (virtual temperatures). The first term in (5.5) describes the contributions due to the thermal contact conductance and frictional heating with the variations $\delta\vartheta_q = \overset{*}{\vartheta}{}^1 - \overset{*}{\vartheta}{}^2$ and $\delta\vartheta_{\mathcal{D}} = \frac{1}{2}(\,\overset{*}{\vartheta}{}^1 + \overset{*}{\vartheta}{}^2\,)$ on $\varphi^1_t(\Gamma^1_c)$. The second term characterizes the standard domain contributions with Kirchhoff heat flux vector $\mathbf{q}^\alpha$. The nonlinear problems (5.4) and (5.5) have to be solved by an iterative solver, e.g. Newton's algorithm.

5.2 Local Update Algorithm for Contact Tangential Stress and Dissipation. The algorithmic update of the tangential stress $t_{T\,n+1}$ and dissipation $\mathcal{D}^p_{n+1}$ in (5.4) and (5.5) is performed by the return algorithm based on an *objective (backward Euler) integration* of the evolution equation (3.5) for the plastic slip, see e.g. WRIGGERS (1987), JU & TAYLOR (1988), GIANNIOPOULOS (1989), WRIGGERS et.al. (1990). The results can be summarized as follows: Integration of (2.3) gives the *increment of the total slip* within the time step Δt_{n+1}

$$\Delta g_{T\,n+1} := (\,\bar{\xi}_{n+1} - \bar{\xi}_n\,)\, l_{n+1} \tag{5.6}$$

which has to be decomposed into an elastic and an plastic part Thus in the case of contact, i.e. for $g_{N\,n+1} > 0$, we compute the elastic trial state

$$\begin{aligned} t^{trial}_{T\,n+1} &:= t_{T\,n} + c_T\, \Delta g_{T\,n+1} \\ f^{trial}_{p\,n+1} &:= |t^{trial}_{T\,n+1}| - \mu\, p_{N\,n+1} \end{aligned} \tag{5.7}$$

and perform the return algorithm

$$t_{T\,n+1} = \begin{cases} t^{trial}_{T\,n+1} & \text{for } f^{trial}_{p\,n+1} \le 0 \\ t^{trial}_{T\,n+1}\,(\mu\, p_{N\,n+1} / |t^{trial}_{T\,n+1}|) & \text{otherwise} \end{cases} \tag{5.8}$$

where we have chosen $t_{T\,n}$ as *history variable*. The dissipation due to the plastic slip is given by

$$\mathcal{D}^p_{n+1} = \begin{cases} 0 & \text{for } f^{trial}_{p\,n+1} \le 0 \\ t_{T\,n+1}\,(\,t^{trial}_{T\,n+1} - t_{T\,n+1}\,)/c_T\, \Delta t_{n+1} & \text{otherwise} \end{cases} \tag{5.9}$$

6 Finite–Element–Discretization

The discretization of the domain contributions in (5.4) and (5.5) is not objective of this work. Within this context we refer to the finite element implementations of coupled finite thermoelastic initial-boundary-value-problems documented in MIEHE (1988) and coupled finite strains thermoplastic problems in SIMO & MIEHE (1991), see also references therein.

In this paper we focus on the contribution due to the presence of the contact constraints, i.e. the discretization of the integrals in (5.4) and (5.5) over the moving master surface. The basis for our subsequent finite element treatment provides a two-dimensional contact element formulation which has been proposed by WRIGGERS & SIMO (1985) for the frictionless case and extended by WRIGGERS et. al. (1990) to finite deformations including frictional slip. In what follows we extend this element formulation to the thermomechanically coupled case under consideration involving contact heat exchange and frictional heating. Within the context of formulating the contact element geometry it is shown that both strategies, (i) starting directly from the discrete formulation and (ii) starting from the continuous formulation outlined in section 2, result in the same discretized formulation of the local contact geometry.

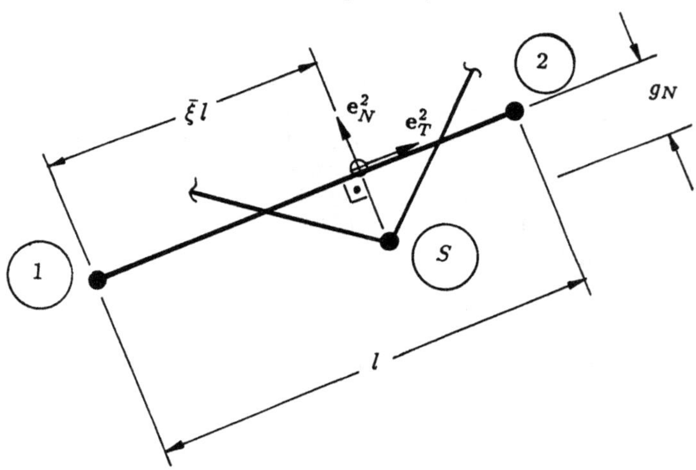

Figure 6.1: Approach in the contact area: 3–node contact element.

Consider the contact element in Figure 6.1 and assume the case of contact, i.e. the *discrete slave point* (s) penetrates into the *master segment* (1)–(2). $l := \| \mathbf{x}_2^2 - \mathbf{x}_1^2 \|$ denotes the current length of the master segment and $\{ \mathbf{e}_T^2 := (\mathbf{x}_2^2 - \mathbf{x}_1^2)/l \; ; \; \mathbf{e}_N^2 := \mathbf{E}_3 \times \mathbf{e}_T^2 \}$ a local orthonormal base attached to the master segment. $\bar{\xi}$ and g_N are *given by the solution of the minimal distance problem*, i.e. by the projection of the slave node (s) onto the master segment (1)–(2). We obtain in the case of penetration

$$\bar{\xi} = \frac{1}{l}(\mathbf{x}_s^1 - \mathbf{x}_1^2) \cdot \mathbf{e}_T^2 \qquad \text{and} \qquad g_N = \| \mathbf{x}_s^1 - (1 - \bar{\xi})\, \mathbf{x}_1^2 - \bar{\xi}\, \mathbf{x}_2^2 \| \; . \tag{6.1}$$

From these equations we compute directly

$$\begin{aligned}\dot{g}_N &= [\,\dot{\mathbf{x}}^1_s - (1-\bar{\xi})\,\dot{\mathbf{x}}^2_1 - \bar{\xi}\,\dot{\mathbf{x}}^2_2\,]\cdot\mathbf{e}^2_N \\ \dot{\xi}\,l &= [\,\dot{\mathbf{x}}^1_s - (1-\bar{\xi})\,\dot{\mathbf{x}}^2_1 - \bar{\xi}\,\dot{\mathbf{x}}^2_2\,]\cdot\mathbf{e}^2_T - \frac{g_N}{l}\,[\,\dot{\mathbf{x}}^2_2 - \dot{\mathbf{x}}^2_1\,]\cdot\mathbf{e}^2_N\end{aligned} \tag{6.2}$$

The equations (6.1) and (6.2) characterize the main kinematical relations of the contact element in Figure 6.1. We remark that the equations (6.2) can be derived *directly from the local formulations* (2.2) and (2.4) by assuming the *linear interpolations* $\hat{\mathbf{x}}^2_t(\xi) = \mathbf{x}^2_1 + \xi\,(\mathbf{x}^2_2 - \mathbf{x}^2_1)$ and $\hat{\mathbf{v}}^2_t(\xi) = \dot{\mathbf{x}}^2_1 + \xi\,(\dot{\mathbf{x}}^2_2 - \dot{\mathbf{x}}^2_1)$ on the straight master segment (1)–(2).

In what follows we formulate the constitutive relationships for the *normal force* P_N , *tangential force* T_T, *the discrete heat flux* Q_N and the discrete dissipation $\mathcal{D}^p$ at the *discrete contact point* (*s*) of the contact element under consideration. Assume that $\bar{\xi}_{n+1}$ and $g_{N\,n+1}$ are known from the exploitation of (6.1). Then we compute analogous to (3.1) the *normal force*

$$P_{N\,n+1} = C_N\,(\,g_{N\,n+1}\,)^m \tag{6.3}$$

The algorithmic update of the *tangential force* is performed by the return algorithm as described in section 5.2. We compute analogous to (5.6)–(5.9) the elastic trial state

$$\begin{aligned}T^{trial}_{T\,n+1} &:= T_{T\,n} + C_T\,\Delta g_{T\,n+1} \\ F^{trial}_{P\,n+1} &:= |T^{trial}_{T\,n+1}| - \mu\,P_{N\,n+1}\end{aligned}\;, \tag{6.4}$$

perform the return algorithm

$$T_{T\,n+1} = \begin{cases} T^{trial}_{T\,n+1} & \text{for } F^{trial}_{P\,n+1} \le 0 \\ T^{trial}_{T\,n+1}\,(\mu\,P_{N\,n+1}/\,|T^{trial}_{T\,n+1}|\,) & \text{otherwise}\end{cases} \tag{6.5}$$

and compute the frictional dissipation

$$\mathcal{D}^p_{n+1} = \begin{cases} 0 & \text{for } F^{trial}_{P\,n+1} \le 0 \\ T_{T\,n+1}\,(\,T^{trial}_{T\,n+1} - T_{T\,n+1}\,)/C_T\,\Delta t_{n+1} & \text{otherwise}\end{cases} \tag{6.6}$$

Finally the *discrete heat flux* takes the form analogous to (3.7)

$$Q_{N\,n+1} = H_{S_0}\left[\frac{P_{N\,n+1}/l_{n+1}}{H}\right]^{\epsilon}\,[\,\Theta^1_{s\,n+1} - (1-\bar{\xi}_{n+1})\,\Theta^2_{1\,n+1} - \bar{\xi}_{n+1}\,\Theta^2_{2\,n+1}\,] \tag{6.7}$$

The contributions of the thermomechanical contact in the mechanical and thermal weak forms (5.4) and (5.5) take the contact element the form

$$\begin{aligned}G^c_M &:= \delta g_{N\,n+1}\,P_{N\,n+1} - \delta g_{T\,n+1}\,T_{T\,n+1} \\ G^c_T &:= \delta g_{Q\,n+1}\,Q_{N\,n+1} - \delta g_{\mathcal{D}\,n+1}\,\mathcal{D}^p_{n+1}\end{aligned} \tag{6.8}$$

for the disctrete contact point (s) with the mechanical relative (Lie–type) variations analogous to (6.2)

$$\begin{aligned}\delta g_{N\,n+1} &= [\,\overset{*}{\mathbf{u}}{}^1_s - (1-\bar{\xi}_{n+1})\,\overset{*}{\mathbf{u}}{}^2_1 - \bar{\xi}_{n+1}\,\overset{*}{\mathbf{u}}{}^2_2\,]\cdot\mathbf{e}^2_{N\,n+1} \\ \delta g_{T\,n+1} &= [\,\overset{*}{\mathbf{u}}{}^1_s - (1-\bar{\xi}_{n+1})\,\overset{*}{\mathbf{u}}{}^2_1 - \bar{\xi}_{n+1}\,\overset{*}{\mathbf{u}}{}^2_2\,]\cdot\mathbf{e}^2_{T\,n+1} - \frac{g_{N\,n+1}}{l_{n+1}}\,[\,\overset{*}{\mathbf{u}}{}^2_2 - \overset{*}{\mathbf{u}}{}^2_1\,]\cdot\mathbf{e}^2_{N\,n+1}\end{aligned} \tag{6.9}$$

and thermal variations

$$
\begin{aligned}
\delta g_{q\,n+1} &= [\,\overset{*}{\vartheta}{}^1_s - (1-\bar{\xi}_{n+1})\,\overset{*}{\vartheta}{}^2_1 - \bar{\xi}_{n+1}\,\overset{*}{\vartheta}{}^2_2\,] \\
\delta g_{\mathcal{D}\,n+1} &= [\,\overset{*}{\vartheta}{}^1_s + (1-\bar{\xi}_{n+1})\,\overset{*}{\vartheta}{}^2_1 + \bar{\xi}_{n+1}\,\overset{*}{\vartheta}{}^2_2\,]/2
\end{aligned}
\tag{6.10}
$$

The matrix formulation of the mechanical part in (6.9) may be found in WRIGGERS et. al. (1990) which also contains all necessary linearizations. The matrix formulation of the thermal part in (6.9) is similar to the mechanical part. As a consequence of the global operatorsplit algorithm discussed in section 5 and the assumed simplyfied constitutive equation (6.8) is the thermal part linear.

7 Numerical Examples

7.1 Heat Transfer through the Contact Surface

The proposed model has been implemented in the finite element code FEAP and was tested by means of several examples. The first example is a test for the developed and implemented pressure dependent thermomechanical constitutive equation. The system and its finite element discretization is depicted in Figure 7.1. A temperature of $100\,K$ and a pressure p is applied to the upper surface of $\mathcal{B}^2$ whereas the bottom of $\mathcal{B}^1$ is fixed and has a temperature of $0\,K$, the vertical boundaries are thermally isolated. Body $\mathcal{B}^1$ is assumed to be rigid but can conduct heat. Body $\mathcal{B}^2$ is thermoelastic. The thermoelastic constitutive parameters for aluminium are given in Table 7.1.

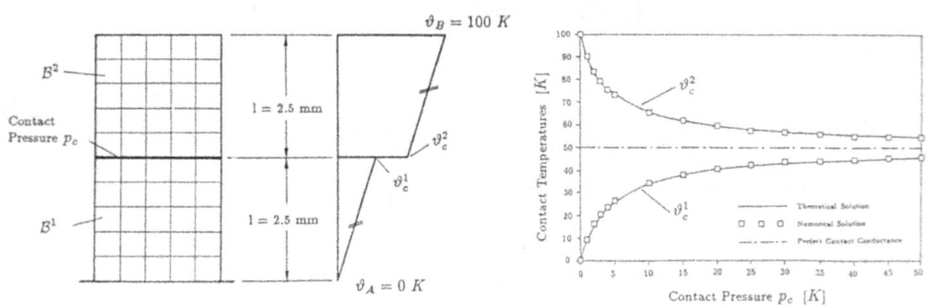

Figure 7.1: System and discretization. Contact pressure versus contact temperature.

Figure 7.1 shows the comparison with the analytical solution which is given for this one dimensional steady state example by $\vartheta^1_c = (1+\eta)\,\vartheta_A + \eta\,\vartheta_B\,/\,(1+2\,\eta)$ and $\vartheta^2_c = (1+\eta)\,\vartheta_B + \eta\,\vartheta_A\,/\,(1+2\,\eta)$ with $\eta = h_c\,l\,/k$. These results show clearly the pressure dependency of the constitutive law for the heat transfer in the contact area. It depicts also that the finite element approximation is in very good accordance with the analytical solution.

7.2 Frictional Heating of a Block on a Rigid Surface

The mechanism of heating due to friction is studied in this example. For this purpose we consider an elastic block $\mathcal{B}^2$ which is slid over a rigid block $\mathcal{B}^1$. Both bodies are heat conductors. The upper body moves within $3.75 \cdot 10^{-3}\, s$ from the left to the right end of the lower block, the displacements are prescribed at the top of $\mathcal{B}^2$. During this process a pressure of $p = 10\, N \,/\, mm^2$ is applied on block $\mathcal{B}^2$. The system and its finite element discretization are shown in Figure 7.2. In order to check the numerical result we consider idealized thermally isolated surfaces of both bodies. Again we assume that both bodies consists of aluminium with the material parameters given in Table 7.1. The frictional coefficient is assumed to be $\mu = 0.2$.

Bulk Modulus	κ	$58333\ N \,/\, mm^2$
Shear Modulus	G	$26926\ N \,/\, mm^2$
Expansion Coefficient	α	$23.86 \cdot 10^{-6}\ K^{-1}$
Conductivity	k	$150\ N \,/\, s\, K$
Capacity	c	$0.9 \cdot 10^{9}\ mm^2/s^2 K$
Resistance Coefficient	h_{So}	$150\ N \,/\, s\, K$
Vickers Hardness	H	$932\ N \,/\, mm^2$
Exponent	ϵ	0.95 –

Table 7.1: Material parameters.

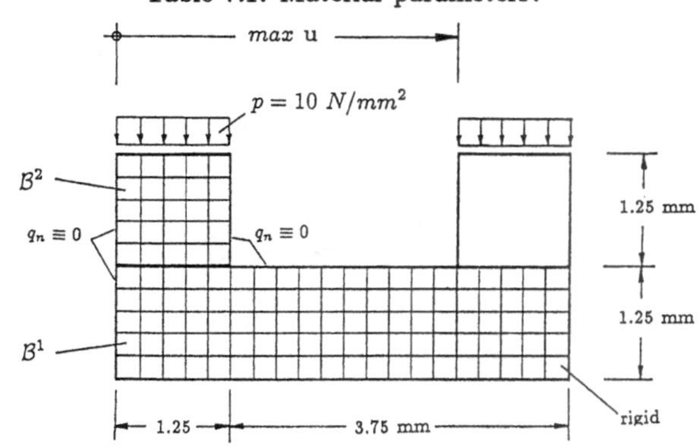

Figure 7.2: System of frictional heating problem.

100 time steps have been used to compute the evolution of the temperature during the process. The temperature distributions at time $t = 0.15 \cdot 10^{-3}\ s$ and $t = 3.75 \cdot 10^{-3}\ s$ are shown in Figure 7.3. After that time the movement of the upper body was stopped.

Another 50 time steps with increasing step length were applied to compute the temperature evolution up to the final homogeneous steady state at the process time of 10 s.

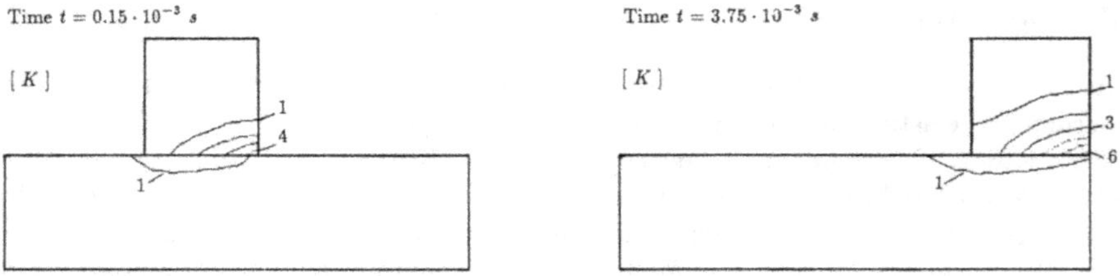

Figure 7.3: Temperature distribution at different time steps.

After that time the temperature is balanced and thus equal at all elements of the bodies $\mathcal{B}^{\alpha}$. Since we wanted to evaluate in this example only the frictional heating mechanism no conductance of the heat through the contact surface was allowed. This made it possible to compare the finite element result with a global energy check. We compute the dissipative mechanical work: $W_{diss} = max\, u\, F_T = max\, u\, [\mu\, A\, p] = 9.375\, N\, mm$. From this quantity we obtain the average temperature of the isolated system $\mathcal{B}^{\alpha}$ for an infinetly process time $\vartheta^{\alpha}_m = W_{diss} \,/\, (2\, \rho^{\alpha}\, c^{\alpha}\, V^{\alpha})$ which yields

–	Finite element result	Global Energy check
ϑ^1	$0.317\,K$	$0.309\,K$
ϑ^2	$1.267\,K$	$1.235\,K$

Again we observe the good agreement of the global energy check with the finite element solution based on the proposed model.

Literature

J.H. Argyris, J. St. Doltsinis (1981), On the Natural Formulation and Analysis of Large Deformation Coupled Thermomechanical Problems, Comp. Meth. Appl. Mech. Eng. **25**, 195–253.

A. B. Chaudhary, K. J. Bathe (1985), A Solution Method for Static and Dynamic Analysis of Three-dimensional Contact Problems with Friction, Computers & Structures, **24**, 855–873.

A. Curnier, R. L. Taylor (1982), A Thermomechanical Formulation and Solution of Lubricated Contacts between Deformable Bodies, J. of Lubrication Technology, **104**,

J. St. Doltsinis (1990), Aspects of Modelling and Computation in the Analysis of Metal Forming, Engineering Computations, **7**, 2–20.

A. E. Giannokopoulos (1989), The Return Mapping Method for the Integration of Friction Constitutive Relations, Computers & Structures, **32**, 157–168.

J. Hallquist (1979), NIKE2D: An Implicit, Finite Deformation, Finite- Element Code for Analysing the Static and Dynamic Response of Two–Dimensional Solids, Rept. UCRL–52678, UC–Lawrence Livermore National Laboratory.

T. R. J. Hughes, R. L. Taylor, J. L. Sackman, A. Curnier, W. Kanoknukulchai (1976) A Finite Element Method for a Class of Contact–Impact Problems Comp. Meth. Appl. Mech. Engng., **8**, 249–276.

L. Johannson, A. Klarbring, Thermoelastic Frictional Contact Problems: Modelling, Finite Element Approximation and Numerical Realization, preprint.

W. Ju, R. L. Taylor (1988), A Perturbed Lagrangian Formulation for the Finite Element Solution of Nonlinear Frictional Contact Problems, Journal of Theoretical and Applied Mechanics, **7**, 1–14.

I. V. Kragelsky, M. N., Dobychin, V. S., Kombalov (1982), Friction and Wear — Calculation Methods, (Translated from The Russian by N. Standen), Pergamon Press.

T. A. Laursen, J. C. Simo (1991), On the Formulation and Numerical Treatment of Finite Deformation Frictional Contact Problems, in Computational Methods in Nonlinear Mechanics, eds. P. Wriggers, W. Wagner, Springer, Berlin.

C. V. Madhusudana, L. S. and Fletcher (1986) Contact heat transfer - the last decade, A AIAA Journal, **24**, 510–523,.

R. Michalowski, Z. Mroz (1978), Associated and Non–associated Sliding Rules in Contact Friction Problems, Arch. of Mechanics, **30**, 259–276.

C. Miehe (1988), On the Numerical Treatment of Thermomechanical Processes (in German), Forschungs– und Seminarberichte aus dem Bereich der Mechanik der Universität Hannover, Report–No. F 88/6.

B. Nour–Omid, P. Wriggers (1986), A Two–Level Iteration Method for Solution of Contact Problems, Comp. Meth. Appl. Mech. Engng., **54**, 855–873.

J. T. Oden, J.A.C. Martins (1986), Models and Computational Methods for Dynamic Friction Phenomena, Comp. Meth. Appl. Mech. Engng. **52**, 527–634.

J. C. Simo, R. L. Taylor (1985), Consistent Tangent Operators for Rate-independant Elastoplasticity., Comp. Meth. Appl. Mech. Engng., **48**, 101–118.

J. C. Simo, C., Miehe (1991), Coupled Associative Thermoplasticity at Finite Strains: Formulation, Numerical Analysis and Implementation, Comp. Meth. Appl. Mech. Engng., in press.

S. Song, M. M. Yovanovich, M. M. (1987), Explicit relative contact pressure expression: dependence upon surface roughness parameters and Vickers microhardness coefficients, AIAA Paper 87-0152.

W. L. Wood (1990), Practical Time–Stepping Schemes, Oxford Applied Mathematics and Computing Science Series, Clarendon Press, Oxford.

P. Wriggers, J. C.Simo (1985), A Note on Tangent Stiffness for Fully Nonlinear Contact Problems, Communications in Applied Numerical Methods, **1** , 199–203.

P. Wriggers (1987), On Consistent Tangent Matrices for Frictional Contact Problems, Proc. NUMETA 87 Conf., Ed. J. Middleton, G.N. Pande.

P. Wriggers, T., Vu Van, E. and Stein (1990), Finite Element Formulation of Large Deformation Impact-Contact Problems with Friction, Computers & Structures, **37**, 319–331.

G. Zavarise, P. Wriggers , E. Stein, B. A. Schrefler (1992), A Numerical Model for Thermomechanical Contact based on Microscopic Interface Laws, accepted for publication in Mechanics Research Communications.

On Solving Problems of Advanced Plastic Flow

M.KLEIBER

Research Center for Advanced Science and Technology, University of Tokyo, Japan*

Summary

An algorithm applicable to incremental finite element analysis of a very general class of elastic-plastic problems is discussed. The approach is based on solving a rigid-plastic problem first followed by a modification due to the elastic effects. Some numerical results are discussed.

Introductory Remarks

The paper is concerned with some basic aspects of incremental solution algorithms for solving problems of advanced plastic flow in metals. As a matter of fact, looking for best possible ways to solve the rate boundary-value problems of finitely deformed inelastic solids has always been one of the central issues in the field of computational mechanics. In the earlier days the dominating approach used in the majority of finite element programs had been the so-called initial load method based on the incremental elastic solution (global predictor) supplemented with the iterative correction due to plastic effects translated entirely into the corresponding nodal forces. Later, the (consistent) tangent stiffness methodology supported by Newton-Raphson iterations was adopted as the basis algorithmic strategy; still it has been widely agreed to think of the incremental solution process for the compound elastic-plastic problem at the local (integration point) level as consisting of two phases - elastic solution (local predictor phase) followed by a plastic

---

* On leave of absence from the Institute of Fundamental Technological Research, Polish Academy of Sciences, Warsaw.

D. Besdo, E. Stein (Eds.)
Finite Inelastic Deformations -
Theory and Applications
IUTAM Symposium Hannover/Germany 1991

modification accounting for the response 'softening' (corrector phase). Such a solution philosophy has been generally employed even for problems in which the plastic strain exceeds the elastic one by, say, two orders of magnitude.

This paper attempts to explore in some numerical detail an approach which has at its core the basic observation that for advanced plastic flow problems the inelastic behaviour should be given an 'algorithmic priority' over the elastic behaviour as the former contributes to the overall behaviour of the material in a definitely more substantial manner. Consequently, rigid-plastic incremental solution will be treated here as a global predictor followed by the 'elastic' corrector phase rather than the other way around as considered in the past. The analogous alternative to the tangent stiffness approach with local plastic predictor/elastic corrector scheme remains to be explored.

Adopting the assumption of additivity of elastic and plastic strain contributions (the basic idea of the approach may be also applied to the so-called large strain elasto-plasticity case based on the multiplicative split of the deformation gradient) the main algorithmic idea advocated here is illustrated in Fig.1, in which the notation is believed to be

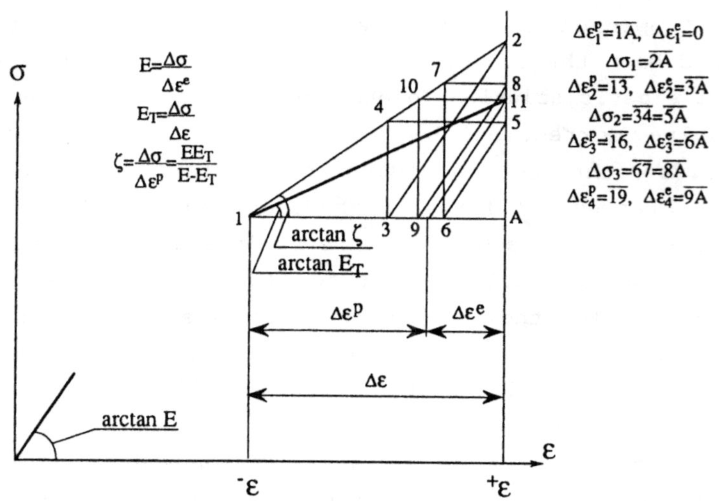

Fig.1 Plastic predictor/elastic corrector algorithm

self-explanatory. At a typical incremental step in the simple tension, piecewise-linearized, strain-driven problem the incremental plastic strain $\Delta\varepsilon_1^p$ is initially assumed to be equal to the total incremental strain $\Delta\varepsilon$, $\Delta\varepsilon_1^p=\overline{1A}$; consequently $\Delta\varepsilon_1^e=0$. The predictor-type rigid-plastic solution with the plastic modulus ζ gives the incremental stress $\Delta\sigma_1=\overline{2A}$. The second estimate for the incremental elastic strain $\Delta\varepsilon_2^e$ is obtained by moving from point 2 down to point 3, $\Delta\varepsilon_2^e=\overline{3A}$ which in turn generates $\Delta\varepsilon_2^p=\overline{13}$ and, using the rigid-plastic solution, the incremental stress $\Delta\sigma_2=\overline{34}=\overline{5A}$, etc. The process shown in Fig.1 is hypothetically assumed to converge in 4 iterations.

As we shall see below, this basic algorithmic idea can easily be extended to cover very sophisticated material idealizations often unavoidable in modelling advanced plastic flow. Description of the algorithm together with the results obtained for some non-trivial test examples are the main objectives of this note.

Formulation of the Problem

In accordance with the comments of the preceding section we shall now start by considering the predictor phase of the incremental solution process corresponding to the rigid-plastic material idealization. The term 'rigid-plastic' is understood here very broadly as a hardening or non-hardening, viscous or non-viscous response with or without damage and/or temperature effects (i.e. only elastic effects are excluded). Deformation processes in such rigid-plastic solids can be conveniently and uniformly analyzed by resorting to the flow-type formulation taking advantage of the formal analogy between the flow equations and those of linear elasticity; any other approach to rigid-plastic analysis can clearly be used as well. The idea of using the analogy is certainly not new - in the field of metal forming it has been successfully employed in the finite element context for almost two decades, [1-3]. It is an attempt to effectively include into this formulation the whole variaty of sophisticated material effects with the particular attention paid to the elastic 'corrector' term, and to confront the algorithm with the more conventional elastic-plastic analysis, which is the subject of this note.

The aforementioned analogy may be presented in the following 3D form related to a fixed cartesian coordinate frame, [4]:

Compressible 3D elasticity	Rigid-viscoplastic 3D flow of void-containing material (Gurson model)

$$\sigma_{ij,j} = 0$$

$$\varepsilon_{ij}=\frac{1}{2G}(\sigma_{ij}-\frac{3\nu}{1+\nu}p\delta_{ij}) \leftrightarrow d^p_{ij}=\frac{1}{2G^*}(\sigma_{ij}-\frac{3\nu^*}{1+\nu^*}p\delta_{ij})$$
$$\varepsilon_{ij}=\frac{1}{2}(u_{i,j}+u_{j,i}) \leftrightarrow d^p_{ij}=\frac{1}{2}(v_{i,j}+v_{j,i}) \qquad (1)$$
$$u_i \leftrightarrow v_i$$

with

$$G^*=\frac{\left(\frac{\bar{\sigma}^2}{\sigma_M^2}+\beta\sigma_{kk}\right)\sigma_0^{1/m}\left(1+\frac{\bar{\varepsilon}^p_M}{\varepsilon_0}\right)^{n/m}}{3(1-\xi)\sigma_M^{1/m}},$$
$$\nu^*=\frac{\sigma_{kk}-2\beta\sigma_M}{2(\sigma_{kk}-\beta\sigma_M)},\quad p=\frac{1}{3}\sigma_{kk},\quad \beta=\sqrt{\frac{2}{3}}\xi\frac{\sin h(\frac{\sigma_{kk}}{2\sigma_M})}{2} \qquad (2)$$

Most of the notation here is believed to be self-explanatory; v_i, d^p_{ij} are the spatial velocity vector and the rate of the deformation tensor typical of the flow formulation, the material 'coefficients' G^*, ν^* are in fact functions of the current state variables and thus they vary during the deformation process, the subscript M stands for the matrix material, n and m are strain-hardening and strain-rate-hardening exponents in a power law flow rule, [4], $\bar{\sigma}$ and $\bar{\varepsilon}^p_M$ are equivalent aggragate (average) stress and matrix plastic strain, respectively and ξ is the porosity (or the void-volume fraction) of the material. A comprehensive discussion of the above equations can be found in [4].

By neglecting the effects of porosity the above analogy reduces to the form known in the literature [1,2] with

$$G^*=\frac{\sigma_y}{3\dot{\bar{\varepsilon}}^p},\quad \nu^*=\frac{1}{2} \qquad \text{(rigid-plastic material)} \quad (3)$$

$$G^*=\frac{\sigma_y+\left(\frac{\dot{\bar{\varepsilon}}^p}{\gamma}\right)^{1/n}}{3\dot{\bar{\varepsilon}}^p},\quad \nu^*=\frac{1}{2}$$ (rigid-viscoplastic material)

γ, n being the parameters of the so-called over-stress viscous model defined by

$$d^p_{ij}=\bar{\gamma}\left(\frac{\bar{\sigma}-\sigma_y}{\sigma_y}\right)^n\frac{3}{2}\frac{\sigma^D_{ij}}{\bar{\sigma}},\quad \dot{\bar{\varepsilon}}^p=\gamma(\bar{\sigma}-\sigma_y)^n,$$
$$\gamma=\bar{\gamma}\frac{\sqrt{3}}{2\sigma_y^n},\quad \bar{\sigma}=\sigma_y+\left(\frac{\dot{\bar{\varepsilon}}^p}{\gamma}\right)^{1/n}. \tag{4}$$

The above analogies have a great computational potential since they allow to treat advanced plastic flow problems using computer software developed for linear elasticity. To do so one has to simply allow the elastic shear modulus G and the Poisson ratio ν to be given functions of the current state variables (and, in the non-viscous case, of their rates which then calls for an additional iterative treatment) and to supplement the algorithm with an appropriate accumulation procedure at the end of the incremental step.

It has to be kept in mind that the flow formulation is based on the Eulerian description of motion so that any analysis of boundary-value problems defined in a region moving in space requires appropriate care. In particular, for sufficiently small displacement increments the formulation can be approximately employed 'as-is' on the assumption that the so-called updated Lagrangian and Eulerian formulations become indistinguishable.

As far as the coupled thermo-mechanical effects are concerned it seems consistent with the assumed rigid-plastic behaviour to neglect the thermal strains and to account only for: (i) temperature induced changes in material characteristics in the equations describing the mechanical part of the coupled problem, and (ii) internal heat generation effects due to the dissipated plastic work in the heat transfer equation. Thus, at a typical step we have to solve the heat transfer equation coupled with the linear compressible elasticity equations in which the material constants are replaced by G^*, ν^* appropriately modified to account for the temperature influence upon the plastic flow rule, [4], i.e. $G^*=G^*(\vartheta)$, $\nu^*=\nu^*(\vartheta)$, ϑ being the temperature.

Without going into further details of the 'analogy'-based computations the reader is referred to [1-4] for the relevant discussions.

We note in passing that the apparently weak point of the analysis is the way in which to treat instanteneously rigid parts of the solid considered. Leaving them entirely out of the incremental solution process is clearly not advisable as it implies necessity of essential finite element remodelling of the problem. This problem will be commented upon later in this note.

Having solved the rigid-plastic incremental problem we are now to 'correct' the solution by supplementing it with the elastic effects. Their inclusion must take different forms depending on whether the plastic or instanteneously rigid region of the body is considered. Thus, in the most general thermo-viscoplastic case we should:

(a) supplement the viscoplastic analysis with the thermo-elastic contributions,

(b) account for the thermo-elastic deformation in the region which at the particular moment considered is not deforming plastically (i.e. it is rigid according to the rigid-plastic material idealization).

The correction relevant to (a) follows by noting that eq.(1) can be inverted to read

$$\sigma_{ij}=C^{*}_{ijkl}d^{p}_{kl} \tag{5}$$

while by the strain additivity assumption

$$d^{p}_{ij}=d_{ij}-d^{e}_{ij}-d^{\vartheta}_{ij} \tag{6}$$

where C^{*}_{ijkl} is the tensor of 'rigid-plastic' constitutive moduli (i.e. elasticity tensor with G, ν replaced by G^{*}, ν^{*}) and d^{e}_{ij} and d^{ϑ}_{ij} are the elastic and thermal strain rate contributions, respectively, assumed to be iteratively determined as

$$d^{e}_{ij}=C^{-1}_{ijkl}\overset{\circ}{\tau}_{kl}, \quad d^{\vartheta}_{ij}=\alpha\dot{\vartheta}\delta_{ij} \tag{7}$$

$\overset{\circ}{\tau}_{kl}$ and $\dot{\vartheta}$ being the most recently computed Jaumann derivative of the Kirchhoff stress and temperature rate, respectively; α is the thermal expansion coefficient. Eqs.(5),(6) yield

$$\sigma_{ij}=C^{*}_{ijkl}d_{kl}-C^{*}_{ijkl}(d^{e}_{kl}+d^{\vartheta}_{kl})=C^{*}_{ijkl}d_{kl}-\sigma^{\#}_{ij} \tag{8}$$

where $\sigma^{\#}_{ij}$ is the stress-type tensor due to the thermo-elastic effects. Its inclusion in the FEM formulation follows the standard 'initial stress' methodology which results in the spatially discretized FEM equations of the form

$$\mathbf{K}^{*}\dot{\mathbf{q}}=\mathbf{Q}^{ext}+\mathbf{Q}^{\#} \tag{9}$$

where $\mathbf{K}^{*}$ is the stiffness matrix involving G^{*}, ν^{*}; $\mathbf{q}$ is the generalized nodal displacement vector, $\mathbf{Q}^{ext}$ is the nodal external load vector and $\mathbf{Q}^{\#}$ is the 'initial' load vector due to the thermo-elastic effects.

Accounting for thermo-elastic effects in the non-plastic (i.e. rigid or elastic, depending on the way we look at it) region necessitates a different approach. Dividing the finite element mesh into the zone P (plastic) and the zone E (non-plastic) we may partition and modify the governing FEM equations so that they may be symbolically written as follows:

$$\begin{bmatrix} \mathbf{K} & & \\ & \ddots & \\ & & \mathbf{K}^{*} \end{bmatrix} \begin{bmatrix} \dot{\mathbf{q}}_{E} \\ \dot{\mathbf{q}}_{EP} \\ \dot{\mathbf{q}}_{P} \end{bmatrix} = \begin{bmatrix} \mathbf{Q}^{ext}-\sum_{el\in E} {}^{-}\mathbf{Q}^{(el)int} \\ \mathbf{Q}^{ext}+\sum_{el\in P} \mathbf{Q}^{(el)\#}-\sum_{el\in E} {}^{-}\mathbf{Q}^{(el)int} \\ \mathbf{Q}^{ext}+\sum_{el\in P} \mathbf{Q}^{(el)\#} \end{bmatrix} \tag{10}$$

where $\mathbf{q}_{E}$, $\mathbf{q}_{P}$, $\mathbf{q}_{EP}$ are the vectors collecting the nodal displacements in the zones E, P and at the interface between the zones E and P, respectively, $\mathbf{Q}^{(el)}$ stands for the el-th element contribution to the nodal vector $\mathbf{Q}$ and ${}^{-}\mathbf{Q}^{int}$ indicates the beginning of the step value of the internal nodal force vector. The violated compatibility conditions at the elastic-plastic interface are recovered during the iteration process.

For the reasons indicated earlier it is highly desirable to include the effects listed above as (b) in the global predictor phase; this makes it possible to consider this phase of the incremental solution as defined in the whole region regardless the size of the currently plastic zone. In other words, the algorithm suggested here is as follows:

(i) the global predictor phase consists in solving (using the analogy) the boundary-value problem assuming the rigid-plastic behaviour in the plastic zone and the thermo-elastic behaviour in the remaining region in which the local loading conditions are not met,

(ii) the corrector phase imposes on the so obtained solution the modification due to the thermo-elastic effects in the zone previously assumed rigid-plastic.

In the framework of this algorithm changes in the shape of the elastic and plastic zones can be handled easily.

Numerical illustration

The performance of the conventional algorithm (tangent stiffnes, modified Newton-Raphson, radial return) available to the author referred to as A1 below vs the algorithm advocated in this paper has been assessed by carrying out numerical simulation of the axisymmetric tension test with necking effects included. The latter approach has been employed in two versions: with and without the elastic corrector phase, referred to as the algorithms A2 and A3, respectively. The iterative algorithm is based on a direct iteration scheme which has proved in the past on many practical problems to converge in a reasonable number of iterations.

Generally, existence of free surfaces poses a difficult computational problem for flow-type formulations. As an assumed location of a free surface will likely not be correct, material may enter (or leave) the control volume and disrupt the iterative procedure. Here, the velocity was specified to be tangent to the assumed free surface and adjusted iteratively if the analysis indicated non-zero tractions along the boundary. The procedure is known to be generally unstable but using the free boundary shape taken from the corresponding solid-type solution proved to be an effective cure.

The geometry and material properties as well as the discretization mesh based on the linear displacement triangular elements are shown in Fig.2. To trigger the initiation of necking the specimen was assumed to have an initial sinusoidal imperfection with the maximum diameter reduction of 2% at the bar mid-cross-section.

Each of the basic algorithms tested in this study was used to solve a sequence of problems for the non-viscous von Mises inelastic material with isotropic power law hardening under the following additional assumptions:

I : elastic-plastic material,

II : thermo-elastic-plastic material with adiabatic decoupling,

III: thermo-elastic-plastic material - coupled problem.

We may note that the case III implies rate-sensitive behaviour even for the non-viscous material due to the time dependent heat flux generating a non-uniform temperature which in turn influences the distribution of thermal strains and changes in the material properties.

Each of the analyses was further differentiated by assuming: (a) material without voids, (b) material with void growth effects following the uniform initial porosity of ξ=0.01.

In each case the overall elongation was imposed in equal steps of prescribed specimen end displacement increments Δu=0.03 mm at the strain rate $\dot{\varepsilon}$=0.01 1/s, 284 steps had to be employed to finally reach the engineering strain of ε^{max}=0.213. To facilitate comparisons the global stiffness matrix was triangularized anew at each iteration in every analysis.

The results obtained by the methods A1 and A2 were very closed as measured in terms of the elongation vs force and elongation vs mid-cross-section reduction over the range of the deformation process up to some 11% of elongation (necking takes place at the elongation of some 8%, [4]), cf.Fig.3 for the process with full account for heat conduction effects. The iterative process A2 breaks down at about 13% elongation; no easy cure (smaller step size, underrelaxation factor) could be found. The lack of convergence at larger elongations (when the material rotations in the necked region become significant) puts limitations on the current effectiveness of the method. Some deviation from the equilibrium path has been observed in the analysis by the algorithm A3 (no elastic effects in the plastic region); the decrease in accuracy measured in terms of the specimen end reaction at a given elongation has never exceeded 5%, though. Further algorithmic details were discussed elsewhere,[4-6], and are not repeated here.

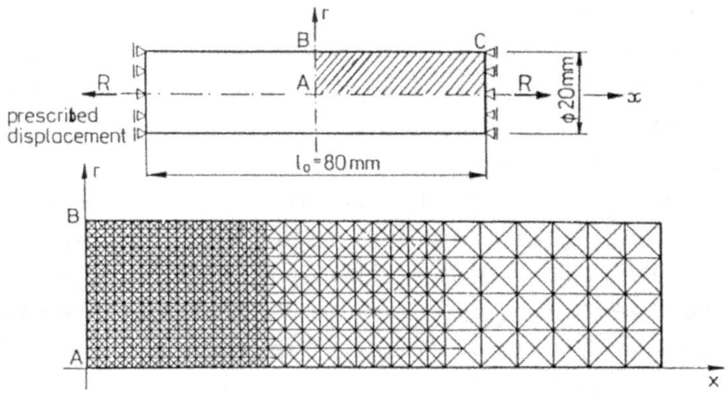

$$\sigma_y=1.1\sigma_M^0\left(\frac{8E\bar{\varepsilon}^p}{1.1\sigma_M^0}+\left(\frac{1.0}{1.1}\right)^8\right)^{1/8}(1-\beta\vartheta)$$

E=207 GPa, ν=0.3, σ_M^0=196 MPa, ξ_0=0.01, β=0.002

mass density=$7.8*10^{-6}$ kg/mm^3, thermal conductivity=0.0045 J/s K mm

specific heat=460 J/kg K, fraction of plastic work converted to heat=0.9

thermal expansion coeficient=$1.2*10^{-5}$

Fig.2 Axisymmetric necking:problem geometry, discretization mesh, material data

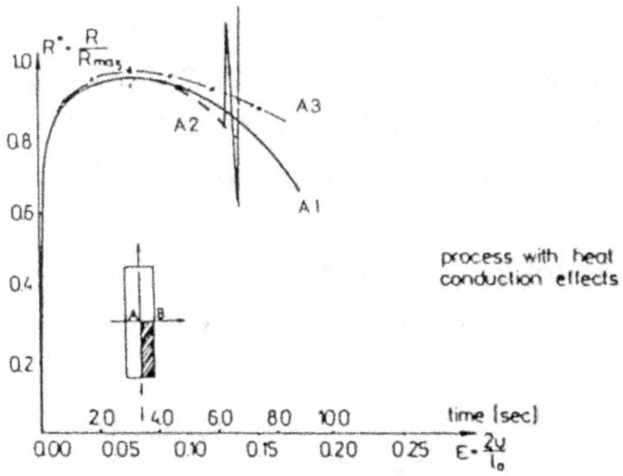

Fig.3 Elongation vs reaction for thermo-mechanical necking

No direct comparison of the CPU times was possible because of essential differences in the organization of programs used to perform computations. Also, the CPU times for problems III were much higher because of the need to solve the heat transfer equations in addition to the mechanical FEM equations of continuing equilibrium. On the other hand, it seems that a reasonable assessment of the algorithms performance can be achieved by comparing the total number of iterations needed in each of the analyses. These are given in Tabl.1 as a relative increase (decrease) of the number of iterations over the elongation range 0-12% taken with respect to the 'most standard' analysis, i.e. A1 (for which the avarage of 4.1

		Conventional algorithm	This paper's algorithm	
		A1	A2	A3
I	a	1.00	1.45	0.79
	b	1.17	1.61	0.85
II	a	1.34	1.75	1.07
	b	1.49	1.90	1.11
III	a	1.61	2.13	1.22
	b	1.80	2.48	1.30

Tabl.1 Algorithm performance

iteration per step was needed). Factor 2 in Tabl.1 means, for instance, that some 8 iterations at a step would be needed on the average in this specific case. The analysis cost for problems involving damage effects is only slightly higher than that for non-damaged materials which is due to rather small porosities considered; this may look quite differently for large porosities,cf.[5]. The substantial increase in cost for coupled problems results from the iterative treatment of temperature distributions involved in the solution process.

Conclusions

No matter how imperfect the assessment given in Tabl.1 may seem, the following conclusions are likely to be valid:

1. For advanced plastic flow solid mechanics problems solution process based on global rigid-plastic predictor approach is an alternative to the standard methodology. However, the convergence properties of the iterative scheme are not yet satisfactory and further work in this area in necessary.
2. Numerical experiences (limited in scope, though) indicate that the alternative approach advocated in this paper may be both: easier to implement and numerically more effective to employ. Since only a very primitive version of the 'new' algorithm has been programmed so far (direct iterations, ad hoc free boundary treatment), substantial improvements in its performance may be expected in the future.
3. Neglecting elastic effects in the plastic zone of the material results in significant CPU time saving without a significant decrease in the solution accuracy. This seems to be a confirmation of our initial guess that the properly adopted rigid-plastic idealization is all what is required in the analysis of many advanced plastic flow problems. However, improvements in both the formulation and algorithmic treatment of such problems are highly desirable before the approach can be used for general applications in a reliable manner.

Acknowledgment

The major part of this work was carried out while the author was a Visiting Professor of Systems Analysis of the Mitsubishi Heavy Industries Chair at the Research Center for Advanced Science and Technology, University of Tokyo, Japan. The support provided during this stay is gratefully acknowledged.

References

1. Zienkiewicz,O.C.,Godbole,P.N., Flow of plastic and visco-plastic solids with special reference to extrusion and forming processes, IJNME 8,3-16,1974.
2. Zienkiewicz,O.C., Jain,P.C.,Onate,E., Flow of solids during forming and extrusion, Int.J.Sol.Struct.14,1538,1978.
3. Onate,E., Kleiber,M., Agalet de Saracibar,C.,Plastic and visco-plastic flow of void-containing metals, IJNME25, 227-251 1978.
4. Kleiber,M., Computational coupled non-associative thermo-plasticity, Comp.Meth.Appl.Mech.Eng.(in print).
5. Kleiber,M., Numerical study on necking-type bifurcations in void-containing elastic-plastic material, Int.J.Sol.Struct.20, 191-210,1984.
6. Wriggers,P., Miehe,C., Kleiber,M., Simo,J.C., On the coupled thermo-mechanical treatment of necking problems via FEM, Proc.COMPLAS'89,Barcelona, 527-542.

Modelling and Computation of Finite Viscoplastic Strains

N. MÜLLER-HOEPPE

DBE,
Peine, FRG

E. STEIN

Institut für Baumechanik und Numerische Mechanik
Universität Hannover, Hannover, FRG

Summary

Recent developments in viscoplastic material behaviour at finite strains based on a multiplicative decomposition of the deformation gradient and employing hyperelastic stress–strain relations are summarized. The structure of an associative flow rule is uniquely defined as optimality condition emanting from the principle of maximum inelastic dissipation.

Computationally, the integration of the flow rule is performed by the Euler corrector method. The proposed algorithm is amenabble to exact linearization. This results in a closed form expression of the consistent tangent moduli.

1 Introduction

Recent developments in the formulation of viscoplastic material behaviour at finite strains based on a multiplicative decomposition of the deformation gradient, as proposed by Lee & Liu [4], are summarized. The approach proposed by Simo [10], [11] for the case of finite strain elastoplasticity is extended to finite strain elastoviscoplasticity, see Müller–Hoeppe [5], employing at the outset hyperelastic stress–strain relations and the maximum dissipation principle and leading at the outcome to an algorithm suitable for large scale computation. From a theoretical standpoint novel steps that differentiate the proposed approach from previous formulations are the following

- An outline of an associative framework of viscoplasticity with respect to the intermediate configuation is given. The derivation of the viscoplastic flow rule is based on exploition of the maximum dissipation principle in *consequent spatial description*, i.e. a formulation exclusively with respect to the current configuration.
- For example a finite strain version of the classical power law is developed and a numerical algorithm for integration of the viscoplastic flow rule is presented.

From a computational standpoint the proposed approach enjoys the following advantage

- As in the infinitesimal theory, for the Euler backward integration algorithm the stress–point integration algorithm can be linearized exactly leading to a closed form expression of the consistent tangent moduli.

D. Besdo, E. Stein (Eds.)
Finite Inelastic Deformations –
Theory and Applications
IUTAM Symposium Hannover/Germany 1991

2 Multiplicative splits

In this section necessary results are recalled concerning the kinematics of finite inelastic strains, see e.g. Simo [10].

Let $B_0 \in \mathbf{R}^3$ be the reference configuration of the body of interest. $\boldsymbol{\phi}: B_0 \longrightarrow B_t \in \mathbf{R}^3$ denotes the deformation map and $\mathbf{F} = \frac{\partial \boldsymbol{\phi}}{\partial \mathbf{X}}$ is the deformation gradient. The local multiplicative decomposition

$$\mathbf{F} = \mathbf{F}^e\, \mathbf{F}^i \tag{1}$$

is assumed. The indices e and i characterize the elastic and the inelastic part of the deformation, respectively.

With respect to an effective numerical treatment additionally the multiplicative decomposition of the deformation in volume–preserving and spherical parts is introduced,

$$\mathbf{F} = J^{\frac{1}{3}}\, \hat{\mathbf{F}}. \tag{2}$$

Since $J = det\, \mathbf{F}$, it is shown easily, that $\hat{\mathbf{F}}$ describes the isochoric part of the deformation. Under the assumption of inelastic incompressibility $det\, \mathbf{F}^i = J^i = 1$, the volume change is considered to be purely elastic, that is $J = J^e$. Thus the multiplicative decomposition takes the form

$$\mathbf{F} = J^{\frac{1}{3}}\, \hat{\mathbf{F}}^e\, \mathbf{F}^i. \tag{3}$$

If only theoretical aspects are recognized, it is not necassary to introduce decomposition (2).

3 Strain tensors, rate of deformation tensors

Relative to the current configuration B_t emanating from the decomposition (1) one has the left Cauchy–Green tensors

$$\mathbf{b} = \mathbf{F}\, \mathbf{F}^T\,, \qquad \mathbf{b}^e = \mathbf{F}^e\, \mathbf{F}^{eT} = J^{\frac{2}{3}}\, \hat{\mathbf{b}}^e \tag{4}$$

and the Almansi strain tensors

$$\mathbf{e} = \frac{1}{2}(\mathbf{g} - \mathbf{b}^{-1})\,, \qquad \mathbf{e}^e = \frac{1}{2}(\mathbf{g} - \mathbf{b}^{e-1}), \tag{5}$$

where $\mathbf{g}$ with components g_{ij} denotes the spatial metric tensor. By definition, see Simo [10], one can specify the inelastic Almansi strain tensor

$$\mathbf{e}^i := \mathbf{e} - \mathbf{e}^e = \frac{1}{2}(\mathbf{b}^{e-1} - \mathbf{b}^{-1}), \tag{6}$$

in (6) the $\mathbf{b}^{e-1}$ plays the role of the spatial inelastic metric tensor.

The spatial rate of deformation tensor is obtained by Lie–derivation of the total strain $\mathbf{e}$ or the spatial metric tensor $\mathbf{g}$

$$L_v\, \mathbf{e} = \mathbf{d}\,, \qquad L_v\, \mathbf{g} = 2\mathbf{d}, \tag{7}$$

with $L_v\, \mathbf{b} = \mathbf{0}$.

Finally the elastic and the inelastic strain rate tensors are defined as Lie–derivatives of $\mathbf{e}^e$ and $\mathbf{e}^i$, Simo & Ortiz [12]. That is

$$\mathbf{d}^e := L_v \ \mathbf{e}^e \ , \qquad \mathbf{d}^i := L_v \ \mathbf{e}^i = \frac{1}{2} L_v \mathbf{b}^{e-1}. \tag{8}$$

4 Constitutive equations

4.1 Elastic material law

We start assuming Helmholtz free energy ψ^e for an isotropic hyperelastic Neo-Hookean material on the intermediate configuration, Simo, Taylor & Pister [14]

$$\rho_0 \psi^e = \frac{1}{2}\mu(tr\hat{\mathbf{b}}^e - 3) + Kf(J), \tag{9}$$

K denotes the bulk modulus, μ the shear modulus and $\hat{\mathbf{b}}^e = \hat{\mathbf{F}}^e \hat{\mathbf{F}}^{eT}$.

The Doyle–Erickson formula [1] leads to the elastic constitutive law

$$\boldsymbol{\tau} = Jp^e \mathbf{g}^{-1} + \mu \ dev \ \hat{\mathbf{b}}^e \ , \tag{10}$$

where $p^e = K\frac{\partial f(J)}{\partial J} = K \ lnJ$ denotes the true pressure.

4.2 Constraint, evolution equation

The second ingredient of the theory is a constraint (yield condition) limiting the elastic domain

$$\phi\left(\mathbf{g}, \mathbf{b}^{e-1}, q; \mathbf{F}\right) \leq 0 \tag{11}$$

where q is a scalar valued suitable set of internal variables, whose evolution is given with the rate equation

$$L_v q = \dot{q} = \gamma h(\mathbf{g}, \mathbf{b}^{e-1}, q; \mathbf{F}), \tag{12}$$

$h(\mathbf{g}, \mathbf{b}^{e-1}, q; \mathbf{F})$ is a prescibed function, which describes inelastic hardening, typically the equivalent inelastic strain.

4.3 Maximum inelastic dissipation power

An associative flow rule for isothermal processes compatible with the multiplicative decomposition is derived making use of the principle of inelastic dissipation power. This principle, often credited to von Mises, e.g. Hill [3], combined with a penalty regularization yields an associative flow rule as an optimality condition. This approach entirely bypasses kinematic definitions of the inelastic rate of deformation. In the case of rate independent plasticity, which is adopted as a Lagrange regularization of the inelastic dissipation power, see Simo's [10], [11] recent work on elastoplasticity.

Uncoupled Helmholtz free energy in the internal state variables q is assumed, formulated in a spatial setting

$$\psi = \psi^e\left(\mathbf{g}, \mathbf{b}^{e-1}; \mathbf{F}\right) + \psi^i(q) \, . \tag{13}$$

At the state defined by $\{\mathbf{g}, \mathbf{b}^{e-1}, q\}$ the inelastic dissipation power is given by

$$\begin{aligned} \mathcal{D}^i\left(\mathbf{g}, \mathbf{b}^{e-1}, q, \mathbf{F}; L_v\mathbf{b}^{e-1}, L_v q\right) : &= -\mathbf{b}^{e-1}\frac{\partial \psi}{\partial \mathbf{b}^{e-1}} : L_v \mathbf{b}^{e-1} - \frac{\partial \psi}{\partial q} L_v q \\ &= \frac{1}{\rho_0}\boldsymbol{\tau} : \mathbf{d}^i - \frac{\partial \psi}{\partial q}\dot{q}. \end{aligned} \tag{14}$$

In local form, the principle of maximum inelastic dissipation power extended to finite strains, may be formulated as follows. *Given a state* $\{\mathbf{g}, \mathbf{b}^{e-1}, q\}$ *among all possible spatial metrics* $\tilde{\mathbf{g}}$*, satisfying the constraint, the actual spatial metric tensor* $\mathbf{g}$ *is the one for which the inelastic dissipation power attains its maximum.* For fixed inelastic variables $\{\mathbf{b}^{e-1}, q\}$ introduce

$$C := \left\{\tilde{\mathbf{g}} \in \mathbf{R}^6 | \phi\left(\tilde{\mathbf{g}}, \mathbf{b}^{e-1}, q; \mathbf{F}\right) \leq 0\right\}. \tag{15}$$

Maximum inelastic dissipation power yields for any $\tilde{\mathbf{g}} \in C$,

$$\mathcal{D}^i\left(\mathbf{g}, \mathbf{b}^{e-1}, q, \mathbf{F}; L_v\mathbf{b}^{e-1}, L_v q\right) - \mathcal{D}^i\left(\tilde{\mathbf{g}}, \mathbf{b}^{e-1}, q, \mathbf{F}; L_v\mathbf{b}^{e-1}, L_v q\right) \geq 0. \tag{16}$$

Consequently, the solution of the varational inequality is the flow rule

$$L_v\mathbf{b}^{e-1} = \arg\left[\max_{\tilde{\mathbf{g}} \in C}\left\{-\mathbf{b}^{e-1}\frac{\partial\psi^e(\tilde{\mathbf{g}}, \mathbf{b}^{e-1}, \mathbf{F})}{\partial\mathbf{b}^{e-1}} : L_v\mathbf{b}^{e-1}\right\}\right]. \tag{17}$$

Instead of the variational inequality (16) a constrained minimization problem is solved.

4.4 Elastoviscoplasticity

In the infinitesimal theory elastoviscoplasticity is often expressed mathematically as a penalty regularization of the principle of inelastic dissipation power. Next, also in the case of finite strains the constraint condition is removed by introducing a penalty functional

$$\mathcal{P}^i := -\rho_0\mathcal{D}^i + \frac{1}{2}\gamma(\phi^+)^2 \longrightarrow \operatorname*{stat}_{\mathbf{g}} \tag{18}$$

where

$$\phi^+ := \begin{cases} 0 & \text{if} \quad \phi \leq 0 \\ \phi & \text{if} \quad \phi > 0 \end{cases}$$

and $\gamma \geq 0$ is the penalty factor or viscosity.
Now the optimality conditions for the problem are formulated in terms of the penalty functional $\mathcal{P}^i$

$$\frac{\partial\mathcal{P}^i}{\partial\mathbf{g}} = 2\rho_0\mathbf{b}^{e-1}\frac{\partial^2\psi^e}{\partial\mathbf{b}^{e-1}\partial\mathbf{g}} : L_v\mathbf{b}^{e-1} + 2\gamma\phi^+\frac{\partial\phi^+}{\partial\mathbf{g}} = \mathbf{0}. \tag{19}$$

yields the viscoplastic flow rule

$$L_v\boldsymbol{\tau}^i = 2\gamma\phi^+\frac{\partial\phi^+}{\partial\mathbf{g}}. \tag{20}$$

with the definition

$$L_v\boldsymbol{\tau}^i := -2\rho_0\mathbf{b}^{e-1}\frac{\partial^2\psi^e}{\partial\mathbf{b}^{e-1}\partial\mathbf{g}} : L_v\mathbf{b}^{e-1}. \tag{21}$$

Application: ϕ is specified as a power law

$$\phi = \left(\frac{\sqrt{\frac{2}{3}}\kappa}{m}\right)^{\frac{1}{2}} F^m \tag{22}$$

including the dimensionless von Mises yield condition

$$F = \frac{|\mathbf{s}|}{\sqrt{\frac{2}{3}}\kappa} - 1, \tag{23}$$

with the yield stress κ and $\mathbf{s} = dev\,\boldsymbol{\tau}$. In this special case $\phi \geq 0$ holds whenever $F \geq 0$ is obtained. Thus, the constraint $\phi^+ := \begin{cases} 0 & \text{if} \quad \phi \leq 0 \\ \phi & \text{if} \quad \phi > 0 \end{cases}$ is replaced by the classical approach, that the v. Mises yield condition F has to be fullfilled. The factor $\left(\frac{\sqrt{\frac{2}{3}}\kappa}{m}\right)^{\frac{1}{2}}$ is introduced to obtain a viscoplastic flow rule, equivalent to that, Perzyna [9] suggested for small strains. The resulting flow rule is

$$\begin{aligned} L_v\boldsymbol{\tau}^i &= \frac{2}{3}\mu\gamma\, tr\hat{\mathbf{b}}^e \langle \left(\frac{|\mathbf{s}|}{\sqrt{\frac{2}{3}}\kappa} - 1 \right) \rangle^{2m-1} \left(\mathbf{n} + \frac{|\mathbf{s}|}{\frac{\mu}{3}\, tr\hat{\mathbf{b}}^e}\, dev\mathbf{n}^2 \right) , \qquad (24) \\ \mathbf{n} &= \frac{\mathbf{s}}{|\mathbf{s}|}, \\ \langle F \rangle^{2m-1} &= \begin{cases} 0 & \text{if} \quad F \leq 0 \\ F^{2m-1} & \text{if} \quad F > 0 \end{cases} . \end{aligned}$$

To simplify flow rule (24) the second term has been neglected, because $|\mathbf{s}| \ll \mu$ for most metals. So the expression

$$\begin{aligned} L_v\boldsymbol{\tau}^i = L_v\mathbf{s}^i &\approx \frac{2}{3}\mu\gamma\, tr\hat{\mathbf{b}}^e \langle \left(\frac{|\mathbf{s}|}{\sqrt{\frac{2}{3}}\kappa} - 1 \right) \rangle^{2m-1}\mathbf{n} \qquad (25) \\ &= \frac{2}{3}\mu\gamma\, tr\hat{\mathbf{b}}^e \langle F \rangle^k \mathbf{n} \end{aligned}$$

with $k = 2m - 1$ holds, establishing an nonassocited flow rule.

5 Numerical treatment of the proposed formulation

5.1 Integration of the flow rule

Let $[0, T]$ be the time intervall of interest. At time t_n the following quantities are assumed to be known: The current configuration $\boldsymbol{\phi}_n$, the placement of the intermediate configuration up to a rotation, i.e. $\mathbf{b}_n^e$. Computationally, the basic problem is to update the variables at time t_{n+1} for a given incremental displacement $\Delta\mathbf{u}$ applied during the finite time increment $\Delta t = t_{n+1} - t_n$. A backward Euler differencing scheme will be used to integrate the flow rule (25), see e.g. Wilkins [17]. This algorithm is summarized in box 1.

Box 1: Stress computation algorithm

1. Update of geometry

$$\boldsymbol{\phi}_{n+1} = \boldsymbol{\phi}_n + \Delta \mathbf{u}\,, \quad \Delta \mathbf{F} = \mathbf{I} + \nabla_n \Delta \mathbf{u}$$

2. Elastic predictor

$$\hat{\mathbf{b}}^e_n = J_n^{-\frac{2}{3}} \mathbf{b}^e_n\,, \quad \Delta\hat{\mathbf{F}} = (det \Delta \mathbf{F})^{-\frac{1}{3}}\, \Delta \mathbf{F}\,, \quad \hat{\mathbf{b}}^{etrial}_{n+1} = \Delta\hat{\mathbf{F}} \hat{\mathbf{b}}^e_n \Delta \hat{\mathbf{F}}^T\,,$$
$$\mathbf{s}^{trial}_{n+1} = \mu\, dev\, \hat{\mathbf{b}}^{etrial}_{n+1}\,.$$

3. Check for yielding

$$\text{IF} \quad F_{n+1} = \frac{|\mathbf{s}^{trial}_{n+1}|}{\sqrt{\frac{2}{3}}\kappa(\bar{e}^i_n)} - 1 \leq 0 \text{ THEN}$$

$$\textit{set: } \mathbf{s}_{n+1} = \mathbf{s}^{trial}_{n+1} \textit{ and exit} \qquad \text{ELSE}$$

4. Stress iteration

$$\bar{\mu} = \frac{\mu}{3} tr\,(\hat{\mathbf{b}}^e_{n+1})\,, \quad \mathbf{n}_{n+1} = \frac{\mathbf{s}^{trial}_{n+1}}{|\mathbf{s}^{trial}_{n+1}|}$$

$$\mathbf{s}_{n+1}\,, \quad \Delta \bar{e}^i_{n+1} \textit{ by local iteration of}$$

$$\mathbf{s}^{trial}_{n+1} - \mathbf{s}_{n+1} - 2\bar{\mu}\gamma\Delta t \left(\frac{|\mathbf{s}_{n+1}|}{\sqrt{\frac{2}{3}}\kappa(\bar{e}^i_n + \Delta \bar{e}^i)} - 1 \right)^k \mathbf{n}_{n+1} = \mathbf{0}$$

$$\Delta \bar{e}^i - \sqrt{\frac{2}{3}}\gamma\Delta t \left(\frac{|\mathbf{s}_{n+1}|}{\sqrt{\frac{2}{3}}\kappa(\bar{e}^i_n + \Delta \bar{e}^i)} - 1 \right)^k = 0$$

5. Update of state variables

$$\begin{aligned}
\mathbf{s}_{n+1} &= \mathbf{s}^{trial}_{n+1} - 2\bar{\mu}\gamma\Delta t \left(\frac{|\mathbf{s}_{n+1}|}{\sqrt{\frac{2}{3}}\kappa(\bar{e}^i_{n+1})} - 1 \right)^k \mathbf{n}_{n+1} \\
\bar{e}^i_{n+1} &= \bar{e}^i_k + \Delta \bar{e}^i_{n+1} \\
\hat{\mathbf{b}}^e_{n+1} &= \frac{\mathbf{s}_{n+1}}{\mu} + \frac{1}{3} tr\,(\hat{\mathbf{b}}^e_{n+1})\, \mathbf{I}
\end{aligned}$$

6. Kirchhoff stresses

$$\boldsymbol{\tau}_{n+1} = J_{n+1} p^e \mathbf{I} + \mathbf{s}_{n+1}$$

7. Compute consistent tangent moduli (if required)
see box 2

The algorithm summarized in box 1 is amenable to exact linearization leading to a closed form of the consistent elastoviscoplastic tangent moduli (box 2), e.g. Simo [11], Simo & Taylor [13], Gruttmann & Stein [2], Nagtegaal [6]. The use of the consistent moduli is essential to preserve the asymptotic rate of quadratic convergence in Newton methods.

Box 2: Consistent tangent elastoviscoplastic moduli

Consistent tangent modulus

$$J\,\mathbb{c} = J\,\mathbb{c}^S + J\,\mathbb{c}^D$$

Modulus associated with the hydrostatic stress state

$$J\,\mathbb{c}^S = J^2 p^{e\prime}\mathbf{g}^{-1}\otimes\mathbf{g}^{-1} + Jp^e(\mathbf{g}^{-1}\otimes\mathbf{g}^{-1} - 2\mathbb{1})$$

with the hydrostatic pressure $p^e = K\, ln\, J,\ \ p^{e\prime} = \frac{K}{J}$

Modulus associated with the deviatoric stress state

$$J\,\mathbb{c}^D = f_1\Big(\mathbb{1} - \frac{1}{3}\mathbf{g}^{-1}\otimes\,\mathbf{g}^{-1}\Big) - f_2\Big(\mathbf{n}\otimes\mathbf{g}^{-1} + \mathbf{g}^{-1}\otimes\mathbf{n}\Big)$$
$$-f_3\mathbf{n}\otimes\mathbf{n} - f_4\mathbf{n}\otimes\ dev\ \mathbf{n}^2 - f_5\mathbf{n}\otimes\mathbf{g}^{-1}$$

Scaling factors ($\kappa' = \frac{\partial\kappa}{\partial\bar{e}^i}$)

$$\begin{aligned}
\bar{\mu} &= \frac{1}{3}\mu J^{-\frac{2}{3}} tr\ \mathbf{b}^e \\
f_1 &= \frac{2\bar{\mu}}{1+\delta_1\delta_2}, \\
f_2 &= \frac{2}{3}\frac{|\mathbf{s}^{trial}|}{1+\delta_1\delta_2}, \\
f_3 &= \frac{4}{3}\frac{\gamma\Delta t\delta_2|\mathbf{s}^{trial}||\mathbf{s}|}{1+\delta_1\delta_2} + \frac{2\bar{\mu}}{1+\delta_1\delta_2}\frac{\beta}{1+\beta} - \frac{4}{3}\frac{\gamma\Delta t\delta_2|\mathbf{s}^{trial}||\mathbf{s}|}{1+\delta_1\delta_2}\frac{\beta}{1+\beta} \\
f_4 &= \frac{2\delta_1\delta_3|\mathbf{s}|^3}{1+\delta_1\delta_2} - \frac{2\delta_1\delta_3|\mathbf{s}|^3}{1+\delta_1\delta_2}\frac{\beta}{1+\beta} \\
f_5 &= -\frac{2}{3}\frac{|\mathbf{s}^{trial}|}{1+\delta_1\delta_2}\frac{\beta}{1+\beta} \\
\delta_1 &= 2\bar{\mu}\gamma\Delta t \\
\delta_2 &= \frac{\left(\frac{|\mathbf{s}|}{\sqrt{\frac{2}{3}}\kappa} - 1\right)^k}{|\mathbf{s}|} \\
\delta_3 &= \frac{1}{|\mathbf{s}|^2}\frac{1}{1+|\mathbf{s}|\frac{\gamma\Delta t\kappa'}{\kappa^2}F^{k-1}}\left(\frac{F^{k-1}}{\sqrt{\frac{2}{3}}\kappa} - \frac{F^k}{|\mathbf{s}|} - \frac{\gamma\Delta t\kappa'}{\kappa^2}F^{2k-1}\right) \\
\beta &= \frac{\delta_1\delta_3|\mathbf{s}|^2}{1+\delta_1\delta_2}
\end{aligned}$$

5.2 Remarks on finite element implementation

Since the fundamental work of Nagtegaal, Parks & Rice [7] it is wellknown that within the finite element method the isochoric nature of inelastic flow results in overconstrained displacement formulations that may lead to locking. The approach followed in this work is based on a special Hu–Washizu principle, as proposed in Simo, Taylor & Pister [14]. This Hu–Washizu functional is formulated in terms of three variables p, Θ and the configuration $\boldsymbol{\phi}$. p denotes the pressure conjugated to the volumetric deformation Θ. For the finite element implementation the classical isoparametric bi-linear quadrilateral for plane strain and axisymmetric problems with discontinous, i.e. constant, approaches for pressure p and Jacobian Θ was chosen. Details of this element are described in Simo [11]. The variables p and Θ are eliminated on the element level, so the tangential stiffness takes the standard form of the displacement method.
To integrate the accleration term, arising from the calculated example, standard Newmark's method, Newmark [8], is applied.

6 Numerical simulation

In this section we present numerical calculations of the impact of steel cylinders on a rigid boundary. The finite element formulation was implemented in the finite element program FEAP, developed by Taylor [16], see Zienkiewicz & Taylor [19].
The experimental data underlying the computations were published by Taylor [15]. Wilkins & Guinan [18] already calculated these examples using a rigid plastic model.

For computation two bars with the common radius of 3,81mm and the initial lengths of 46,94mm and 23,47mm are considered, which are supported at one end to model the rigid boundary. The finite element discretization consists of 150 and 75 elements. Due to symmetry only half of the specimens were discretized.
As the steel shows only little hardening for inelastic strains above 10%, a saturation type hardening law was chosen

$$\kappa = \kappa_0 + \xi \bar{e}^i + (\kappa_\infty - \kappa_0)(1 - e^{-\omega \bar{e}^i}) \tag{27}$$

The material data, which was used for the calculations, is given in Tab. 1. The viscosity parameter γ and the exponent k of the transition function (power law) were achieved by fitting heuristically the computed results to the experiments.

$\rho_0 = 7,85g/cm^3$	$K = 17200kN/cm^2$	$\mu = 8100kN/cm^2$
$\kappa_0 = 54kN/cm^2$	$\kappa_\infty = 72,5kN/cm^2$	$\xi = 0,1kN/cm^2$
$\omega = 17$	$\gamma = 0,0125s^{-1}$	$n = 7$

Table 1: Material data

The loading was modelled applying nodal velocities according to the experiments. The Newmark parameters were chosen $\alpha = 0,4$ and $\delta = 0,7$, i.e. numerical damping was included.

The comparison of the experiments and calculated results are given in fig. 1. In a wide range of velocities the numerical results agree with the experimental data, but it should be noted, that on the high speed level, the solutions become too stiff. Wilkins' and Guinan's calculations are characterized by the same effect, see fig. 1. Wilkins and Guinan [18] explain this effect on the high speed level with the influence of thermomechanical softening, which has not been taken into account.

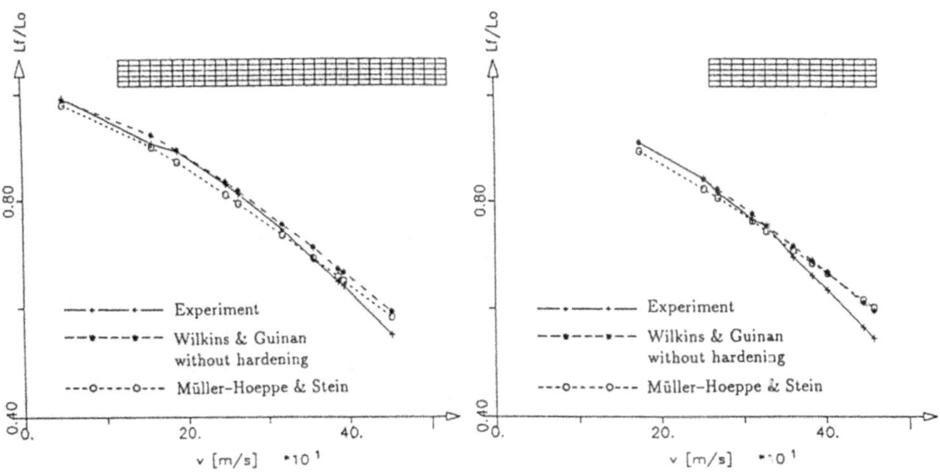

Fig. 1: Impact of steel cylinders. Ratio final lenght/initial lenght of the long cylinder (left) and the short cylinder (right) versus impact velocity

Finally, the final configurations of the specimens are given in fig. 2.

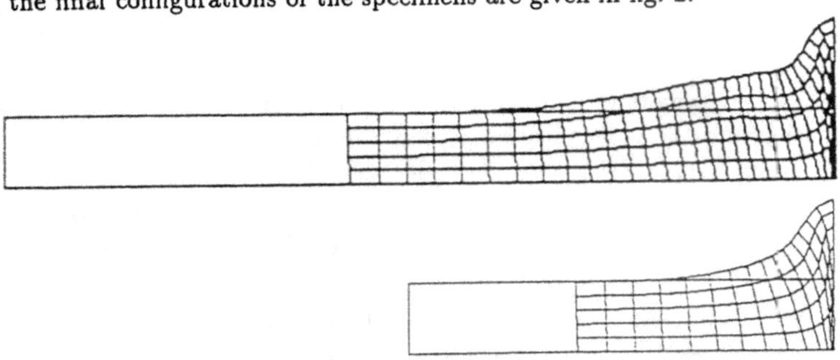

Fig. 2: Impact of steel cylinders. Final configurations of the long cylinder (impact velocity $452m/s$) and the short cylinders (impact velocity $458m/s$)

7 Conclusion

This paper contains a coherent and systematic mathematical modelling and numerical approach of finite strain elastoviscoplasticity based on an associative framework of viscoplasticity in the intermediate configuration theory. The derivation of the viscoplastic flow rule is performed on the basis of the maximum dissipation principle in a consequent spatial description. Application is made to the classical power law extended to finite strains. An algorithm for integration of the viscoplastic flow rule is given and some remarks on the finite element implementation are added. Finally, a numerical example demonstrates the good performance of the proposed algorithm.

References

1. Doyle, T.C.; Erickson, J.L.: Nonlinear elasticity. In: Dryden, H.L., Kármán, T. (eds.): Advances in Applied Mechanics IV, pp. 53–115. New York: Academic Press 1959.
2. Gruttmann, F.; Stein, E.: Tangentiale Steifigkeitsmatrizen bei Anwendung von Projektionsverfahren in der Elastizit"atstheorie. Ingenieur–Archiv 58 (1988) 15–24.
3. Hill, R.: The mathematical theory of plasticity. Oxford: Claredon Press 1956.
4. Lee, E.H.; Liu, D.T.: Finite strain elastic–plastic theory particularly for plane–wave analysis. J. Appl. Phys. 38 (1967) 19–27.
5. Müller–Hoeppe, N.: Beiträge zur Theorie und Numerik finiter inelastischer Deformationen, Bericht F90/4. Hannover: IBNM, Universität Hannover, 1990.
6. Nagtegaal, J.C.: On the implementation of inelastic constitutive equations with special reference to large deformation problems. Comp. Meth. Appl. Mech. Eng. 33 (1982) 469–484.
7. Nagtegaal, J.C.; Parks, D.M.; Rice, J.R.: On numerically accurate finite element solutions in the fully plastic range. Comp. Meth. Appl. Mech. Eng. 4 (1974) 153–177.
8. Newmark, N.M.: A method of computation for structural mechanics. In: Proc. ASCE 85, EM3 (1959) 67–94.
9. Perzyna, P.: Fundamental problems in viscoplasticity. Adv. Appl. Mech. 9 (1966) 243–377.
10. Simo, J.C.: A framework for finite strain elastoplasticity based on maximum plastic dissipation and the multiplicative decomposition: Part I: Continuum formulation. Comp. Meth. Appl. Mech. Eng. 66 (1988) 199–219.
11. Simo, J.C.: A framework for finite strain elastoplasticity based on maximum plastic dissipation and the multiplicative decomposition: Part II: Computational aspects. Comp. Meth. Appl. Mech. Eng. 68 (1988) 1–31.
12. Simo, J.C.; Ortiz, M.: A unified approach to finite deformation elastoplastic analysis based on the use of hyperelastic constitutive equations. Comp. Meth. Appl. Mech. Eng. 49 (1985) 221–245.
13. Simo, J.C.; Taylor, R.L.: Consistent tangent operators for rate–independent elastoplasticity. Comp. Meth. Appl. Mech. Eng. 46 (1985) 101–118.
14. Simo, J.C.; Taylor, R.L.; Pister, K.S.: Variational and projection methods for the volume constraint in finite deformation elasto–plasticity. Comp. Meth. Appl. Mech. Eng. 51 (1985) 177–208.
15. Taylor, G.I.; Proc. Roy. Soc. Ser. A 194 (1948) 289.
16. Taylor, R.L.: FEAP – A finite element analysis program. User manual. Dept. Civil Eng., University of California, Berkeley: 1984.
17. Wilkins, M.L.: Calculation of elastic–plastic flow. In: Alder, B. et al. (eds.): Methods of Computational Physics 3. New York: Academic Press 1964.
18. Wilkins, M.L.; Guinan, M.W.: Impact of cylinders on a rigid boundary. J. Appl. Phys. 44 (1973) 1200–1206.
19. Zienkiewicz, O.C.; Taylor, R.L.: The finite element method I. London: Mc Graw–Hill 1989.

Finite Inelastic Deformations of Three-Dimensional Shells with Applications to Sheet Metal Forming Processes

J.C. GELIN, P. BOISSE

Laboratoire de Mécanique Appliquée - UA CNRS 004, Université de Franche-Comté, La Bouloie, Route de Gray, 25030 BESANCON France

ABSTRACT

The new developments presented in the paper concern the formulation of a fully non-linear shell theory including both membrane and bending components in the field of large elasto-plastic deformations. The kinematic description is a fully non-linear one that includes large rotations and large strains. The plastic behaviour is based on Hill's theory for initial anisotropy and is formulated in using a general curvilinear system. The rotation of anisotropy axes is taken into account and an update procedure have been defined. Furthermore some aspects concerning contact and friction with non-linear rigid surfaces including large displacement are described. Classical non-linear shell benchmark problems have been treated that show the efficiency of the non-linear shell formulation. The procedures developed have been applied to solve sheet metal forming problems. Two numerical examples in that field are proposed. The results obtained are compared with results of the litterature or with experimental ones, in all the cases the agreement is very good.

1. INTRODUCTION

The accurate simulation of processes such as sheet metal forming needs efficient structural shell elements for the finite inelastic strain analysis. As an example it becomes now possible to simulate, with a good accuracy laboratory type processes mainly with spherical punch and the sheet firmly clamped on the boundary. Despite this success the simulation of real, arbitrary shaped components is still an opened challenging problem. Particularly many cases have been surfaced where the shell approach based on a membrane formulation has failed to produce a convergent solution as remarked by Wang and Tang [1].

The first subject that the paper focuss on is the formulation of fully non linear three node and four node shell elements including both membrane and bending components in the field of large deformations. The point of departure of the elements developed is the so-called degenerated solid approach combined with an assumed strain method. Using the covariant base vectors and an intermediate, non orthogonal frame constructed with reference to the element sides that joint a given node, the out of plane shear strain components are evaluated at each node taking the mean value of the components along the sides of the node under

D. Besdo, E. Stein (Eds.)
Finite Inelastic Deformations -
Theory and Applications
IUTAM Symposium Hannover/Germany 1991

consideration. Then the transverse shear strain are interpolated using linear and bilinear shape functions. The formulation does not refer to special assumption as Kirchhoff-Love and is applicable to thin and moderately thick shells to model arbitrary geometries. The procedure is applicable both in the linearized geometric case or in the fully non linear case. The three node element is a very simple one, it can be used with fully integration or with reduced integration techniques. Triangular elements as developed by Boisse et al. [2] are very attractive in practical cases because they allow automatic meshing for various geometries. The four node element developed by Boisse et al. [3] can be compared with the MICT4 element developed by Bathe and Dvorkin [4] [5] and give satisfactory results for a large variety of tests.

The second part of the paper concerns the large strain elasto-plastic formulation to takes into account material non-linearities. A modified version of the well-known radial return method is proposed that takes into account the stress constraint $\sigma_{33} = 0$. The isotropic elasto-plastic as well the anisotropic elasto-plastic behaviour are taken into account. In the anisotropic case the orientation of the anisotropy axes is updated taken into account the material rotation.

The third part of the paper concerns the formulation of contact and friction conditions. The paper describes penalty and Lagrangian formulations to incorporate the contact and friction constraints in the variational formulation of the problem.

The last parts of the paper presents as well numerical examples that illustrate the non-linear behaviour of the shell theory and shell elements developed as examples in the field of sheet metal forming.

2. KINEMATIC ASPECTS

In the classical degenerated shell theory, the position of a material point in the thickness of the shell is assumed to be a linear function of the displacement and rotation of the mean surface. An extension is proposed that directly takes into account large structural rotations arising when the shell is subjected to large deformations, as example in sheet metal forming. So the position of a material point is expressed in a cartesian coordinates system as:

$$\mathbf{x}(\xi,\eta,\zeta) = \bar{\mathbf{x}}(\xi,\eta) + \zeta \frac{h}{2} \mathbf{R}(\xi,\eta)\, \mathbf{X}_0(\xi,\eta) \tag{1}$$

where $\bar{\mathbf{x}}(\xi,\eta)$ is the position of the corresponding point on the mean surface and $\mathbf{X}_0$ is a unit vector normal to the shell in the undeformed configuration. $\mathbf{R}$ is an orthogonal tensor obtained from the solution of the first order differential equation:

$$\dot{\mathbf{R}}\,\mathbf{R}^T = \boldsymbol{\Theta} \tag{2}$$

where $\boldsymbol{\Theta}$ is the matrix representation of the incremental rotation vector (two independent components in the case of thin shells), see Simo and Fox [6]

Equation (1) is interpolated using the classical isoparametric approach and is expressed as

$$\mathbf{x}(M) = \sum_{i=1}^{n} N^i(\xi,\eta)\, \overline{x^i} + \zeta \sum_{i=1}^{n} N^i(\xi,\eta) \frac{h^i}{2} X^i \tag{3}$$

where h^i is the thickness at node i. The shape functions are the classical linear (3 nodes) and bilinear (4 nodes) shape functions.

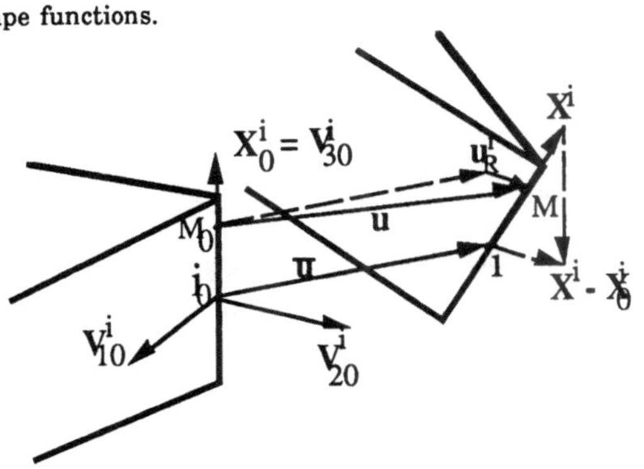

Fig. 1 Schematic representation of the deformation of the shell

The displacement of a point M of the shell (figure 1) is the sum of the displacement of the related point H on the mid-surface and of the displacement given by the pseudonormal rotation

$$\mathbf{u}(M) = \bar{\mathbf{u}}(H) + \mathbf{u}_R(M) \tag{4}$$

and is interpolated by :

$$\bar{\mathbf{u}}(H) = \sum_{i=1}^{n} N^i(\xi,\eta)\, \bar{\mathbf{u}}^i \tag{5}$$

$$\mathbf{u}_R(M) = \sum_{i=1}^{n} N^i(\xi,\eta)\, \mathbf{u}_R^i(\zeta) \tag{6}$$

where $\mathbf{u}_R^i$ is the displacement given by the rotation of the point M_0^i initially on the pseudonormal $\mathbf{X}_0^i$ at node i and at the same position ζ as M, $\mathbf{u}_R^i$ is expressed as:

$$\mathbf{u}_R^i = \frac{h^i}{2}\,\zeta\,(\mathbf{X}^i - \mathbf{X}_0^i) = \frac{h^i}{2}\,\zeta\,(\mathbf{R}^i - \mathbf{I})\mathbf{X}_0^i \tag{7}$$

$\mathbf{R}^i$ is the rotation tensor that transforms $\mathbf{X}_0^i$ to $\mathbf{X}^i$. If the rotation components are small enough in a loading step it is possible to linearize equation (7) as:

$$(\mathbf{R}^i - \mathbf{I})\mathbf{X}_0^i = \theta^i \wedge \mathbf{X}_0^i \tag{8}$$

where θ^i is the rotation vector at node i. To define θ^i an orthonormal frame $(\mathbf{V}_{10}^i, \mathbf{V}_{20}^i, \mathbf{V}_{30}^i)$ is constructed at each node with $\mathbf{X}_{30}^i = \mathbf{V}_{30}^i$ so that $\theta^i = \theta_1^i \mathbf{V}_{10}^i + \theta_2^i \mathbf{V}_{20}^i$.

From the kinematics above described the in-plane components of the Green-Lagrange strain tensor in the material covariant base related to the membrane and bending strains are given by:

$$E_{\alpha\beta} = \frac{1}{2}\left(\frac{\partial \mathbf{x}}{\partial a_\alpha}\frac{\partial \mathbf{x}}{\partial a_\beta} - \frac{\partial \mathbf{x}^0}{\partial a_\alpha}\frac{\partial \mathbf{x}^0}{\partial a_\beta}\right) \qquad \alpha = 1,2\ ;\ \beta = 1,2 \text{ and } a = [\xi, \eta, \zeta] \tag{9}$$

The out of plane strain components or shear strains components are given by:

$$E_{\alpha\zeta} = \frac{1}{2}\left(\frac{\partial \mathbf{x}}{\partial a_\alpha}\frac{\partial \mathbf{x}}{\partial a_\zeta} - \frac{\partial \mathbf{x}^0}{\partial a_\alpha}\frac{\partial \mathbf{x}^0}{\partial a_\zeta}\right) \qquad \alpha = 1,2 \text{ ; and } a = [\xi, \eta, \zeta] \tag{10}$$

Components $E_{\alpha\beta}$ or $E_{\alpha\zeta}$ are easily defined as functions of the contravariant vectors $g_i{}^0$ and g_i corresponding to:

$$g_i{}^0 = \frac{\partial \mathbf{x}^0}{\partial a_i} \quad \text{and } g_i = \frac{\partial \mathbf{x}}{\partial a_i} \quad \text{where } a = [\xi, \eta, \zeta] \tag{11a,b}$$

The model above described consider no extension along the thickness of the shell and can be linearized to obtain the classical shell theory corresponding to the degenerated three dimensional approach.

3. TRIANGULAR AND QUADRANGULAR SHELL ELEMENTS

The point of departure of the developed elements is the so-called degenerated solid approach, see Ahmad , Irons and Zienkiewicz [7]. Many authors have developed elements based on such an approach which have been proved to be among the most effective ones. The elements presented here are a triangular one and a quadrangular one, figure 2.

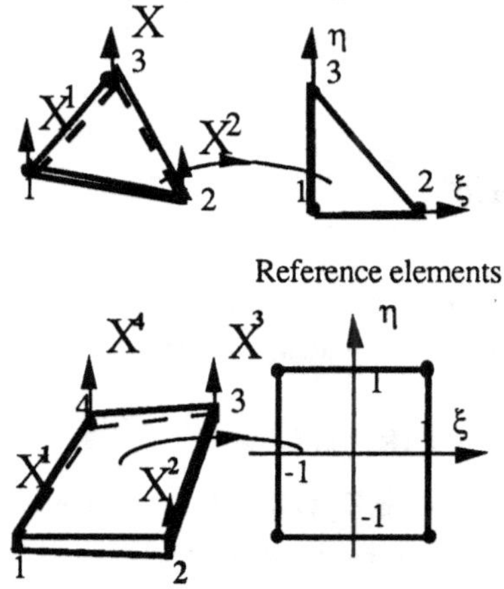

Fig. 2 Triangular and quadrangular shell finite elements

The kinematics of the elements is described with five degrees of freedom per mid-surface node. Three displacements corresponding to global co-ordinate directions and two rotations along the axes V^i_{10} and V^i_{20} axes. The elements built on such Mindlin kinematics are known for their propensity to lock, i.e to give a poor solution when the thickness of the element is small compared to other dimensions. Kirchhoff solution expected in such cases imposes too many constraints per element in regard to its number of degrees of freedom. In the case of low order elements the solution is very inaccurate even for medium thickness and the reduced-selective integration is proved to be ineffective. A mixed interpolation of shear strain components is used [4] [5] [7] [8] [9]. Transverse shear strain components are not classically calculated from the nodal displacements but are directly interpolated from their own nodal values.

The in plane strain components defined from equation (9) are classically obtained from (g_1, g_2, g_3) and nodal displacements as:

$$E_{\alpha\beta} = \frac{1}{2}\left(\frac{\partial u}{\partial \xi_\alpha} g_{\beta 0} + \frac{\partial u}{\partial \xi_\beta} g_{\alpha 0} + \frac{\partial u}{\partial \xi_\alpha}\frac{\partial u}{\partial \xi_\beta}\right) \tag{12}$$

Displacement interpolation (12) leads to the in plane linear and non-linear interpolation matrix:

$$D_u[E_{\alpha\beta}]\,\eta = {}^{L}\mathbf{B}_{mb}\,\eta_n^e + {}^{NL}\mathbf{B}_{mb}(\mathbf{u}_n^e)\,\eta_n^e \tag{13}$$

As above mentionned a mixed interpolation of the shear strain components is made from the value of those components at nodes of the element

$$E_{\alpha 3} = \sum_{i=1}^{n} N^i(\xi,\eta)\, E^i_{\alpha 3} \qquad \alpha = 1 \text{ or } 2 \tag{14}$$

where N^i are the shape functions of the element (linear or bilinear).

The main problem is now to get the values of the nodal shear strains. To do this the following condition is imposed : the out of plane shear strain along the element side is assumed to be constant. At each node the following side vectors are defined, see figure 3:

$$f_1^i = \frac{\partial x}{\partial r_1^i} \qquad f_2^i = \frac{\partial x}{\partial r_2^i} \qquad f_3^i = g_3^i \tag{15}$$

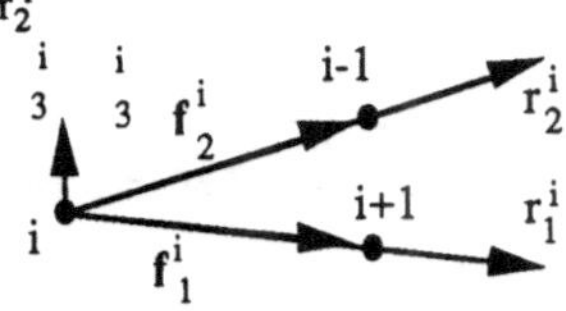

Fig. 3 Definition of the side vectors f_α^i

where r_1^i is the convected co-ordinate along the side (i, i+1) with:

$$r_1^i = 0 \text{ at node } i \quad ; \quad r_1^i = 1 \text{ at node } i{+}1 \tag{16}$$

and where r_2^i is the convected co-ordinate along the side (i, i-1) with:

$$r_2^i = 0 \text{ at node } i \quad ; \quad r_2^i = 1 \text{ at node } i{-}1 \tag{17}$$

Using the covariant frame $(\mathbf{f}_1, \mathbf{f}_2, \mathbf{f}_3)$ the strain tensor can be expressed :

$$E = E^*_{kl}\, \mathbf{f}^k \otimes \mathbf{f}^l \tag{18}$$

$$E^*_{13} = \frac{1}{2}(\mathbf{f}_1 \mathbf{g}_3 - \mathbf{f}_{10}\mathbf{g}_{30}) \qquad E^*_{23} = \frac{1}{2}(\mathbf{f}_2 \mathbf{g}_3 - \mathbf{f}_{20}\mathbf{g}_{30}) \tag{19a,b}$$

where $(\mathbf{f}^1, \mathbf{f}^2, \mathbf{f}^3)$ are the contravariant vectors related to $(\mathbf{f}_1, \mathbf{f}_2, \mathbf{f}_3)$.

Along each side these shear strains are assumed to be constant and equal to the average value at the midpoint of the side, that is:

$$E^{*i}_{\alpha 3}=\frac{1}{2}\left[(\frac{\partial \mathbf{u}}{\partial \zeta})_m(\mathbf{f}^i_{\alpha 0})_m+(\frac{\partial \mathbf{u}}{\partial r^i_\alpha})_m(\mathbf{g}_{30})_m+(\frac{\partial \mathbf{u}}{\partial \zeta})_m(\frac{\partial \mathbf{u}}{\partial r^i_\alpha})_m\right] \tag{20}$$

$(.)_m$ means medium value on the side of the element.

Any part of this relation can be related to nodal displacements which leads to:

$$D_{\mathbf{u}}\left[E^{*i}_{\alpha 3}\right]\eta = {}^{L}\mathbf{C}^i\,\eta^e_n + {}^{NL}\mathbf{C}^i(\mathbf{u}^e_n)\,\eta^e_n \tag{21}$$

A convected base transformation is used that relates the transverse shear in $(\mathbf{g}_1,\mathbf{g}_2,\mathbf{g}_3)$ to nodal displacements:

$$\mathbf{E} = E^{*}_{kl}\,\mathbf{f}^k \otimes \mathbf{f}^l = E_{mn}\,\mathbf{g}^k \otimes \mathbf{g}^l \tag{22}$$

and gives $\mathbf{D}^i$ as:

$$\left[E^i_{\alpha 3}\right] = \mathbf{D}^i\left[E^{*i}_{\beta 3}\right] \tag{23}$$

The interpolation (14) and the equations (21)(23) lead to the expression of modified interpolation matrix of transverse shear strain:

$$D_{\mathbf{u}}[E_{\alpha 3}]\,\eta = \sum_{i=1}^{n} N^i\,\mathbf{D}^i\,{}^{L}\mathbf{C}^i\,\eta^e_n + \sum_{i=1}^{n} N^i\,\mathbf{D}^i\,{}^{NL}\mathbf{C}^i(\mathbf{u}^e_n)\,\eta^e_n \tag{24}$$

$$D_{\mathbf{u}}[E_{\alpha 3}]\,\eta = {}^{L}\overline{\mathbf{B}}_s\,\eta^e_n + {}^{NL}\overline{\mathbf{B}}_s(\mathbf{u}^e_n)\,\eta^e_n \tag{25}$$

4. UPDATED LAGRANGIAN FORMULATION

The principle of virtual work is expressed at the end of each loading step as:

$$\int_{V_0}\left(D_{\mathbf{u}}(E)\eta\right)^T S\,dV - \int_{V_0}\eta^T \mathbf{f}\,dV - \int_{\Gamma_0}\eta^T \bar{\mathbf{t}}\,dS = G(\mathbf{u},\eta) = 0 \tag{26}$$

where η is the homogeneous displacement field on the part of the boundary with prescribed displacements, $\mathbf{f}$ is the body loads vector and $\bar{\mathbf{t}}$ is the surface loads vector.

A Newton method is used in order to linearize the equation (26), see Gelin and Picart [10].The solution $\mathbf{u}^{j+1} = \mathbf{u}^j + \Delta\mathbf{u}^j$ at iteration j+1 is calculated as a solution of:

$$D_{\mathbf{u}}\left(G(\mathbf{u},\eta)^j\right)\Delta\mathbf{u}^j = -G(\mathbf{u},\eta)^j \tag{27}$$

The uncoupled membrane and bending mechanical behavior allows to separate the virtual work of internal membrane-bending in one hand and the virtual work of internal shear forces in the other hand. Principle of virtual work (27) can be written:

$$T^*_{mb} + T^*_s - T^*_{ext} = 0 \tag{28}$$

with:

$$T^*_{mb} = \int_{V_0} \left(D_u[E_{\alpha\beta}]\eta\right)^T C_{mb}\left[E_{\alpha\beta}\right] dV \qquad T^*_s = \int_{V_0} \left(D_u[E_{\alpha 3}]\eta\right)^T C_s\left[E_{\alpha 3}\right] dV \quad (29a,29b)$$

Equation (27) becomes:

$$D_u(G(\mathbf{u},\eta))\Delta\mathbf{u} = D_u\left(T^*_{mb}\right)\Delta\mathbf{u} + D_u\left(T^*_s\right)\Delta\mathbf{u} \qquad (30)$$

Linear and non linear strain interpolation matrix relates the linearized derivatives of strains to the nodal values of the homogeneous displacement field:

$$D_u\left[E_{\alpha\beta}\right]\eta = {}^{L}\mathbf{B}_{mb}\,\eta^e_n + {}^{NL}\mathbf{B}_{mb}(\mathbf{u}^e_n)\,\eta^e_n \qquad (31)$$

that leads to the discretized expression of (26):

$$D_u\left(T^*_{mb}\right)\Delta\mathbf{u} = \eta^T_n\left[\mathbf{K}_{mbL} + \mathbf{K}_{mbNL}\right]\Delta\mathbf{u}_n \qquad (32)$$

where $\mathbf{K}_{mbL}$ and $\mathbf{K}_{mbNL}$ are the linear and non-linear membrane-bending stiffness matrix calculated from ${}^{L}\mathbf{B}_{mb}$ and ${}^{NL}\mathbf{B}_{mb}$ .As previously mentionned a direct interpolation of the shear strain components is made from values at nodes which gives a modified shear strain interpolation matrix:

$$D_u[E_{\alpha 3}]\,\eta = {}^{L}\overline{\mathbf{B}}_s\,\eta^e_n + {}^{NL}\overline{\mathbf{B}}_s(\mathbf{u}^e_n)\,\eta^e_n \qquad (33)$$

Achievement of these out of plain shear strain interpolation matrix is detailed in next section. Modified shear stiffness matrix are calculated from ${}^{L}\overline{\mathbf{B}}_s$ and ${}^{NL}\overline{\mathbf{B}}_s$:

$$D_u\left(T^*_{sb}\right)\Delta\mathbf{u} = \eta^T_n\left[\mathbf{K}_{sL} + \mathbf{K}_{sNL}\right]\Delta\mathbf{u}_n \qquad (34)$$

The displacement field $\mathbf{u}^{j+1}$ calculated at iteration j+1 of the Newton method is solution of:

$$\mathbf{F}_{ext} - \sum_{elt} \int_{V^e_0} \left({}^{L}\mathbf{B}^T_{mb} + {}^{NL}\mathbf{B}^T_{mb}(\mathbf{u}^e_n)^j\right)\left[S_{\alpha\beta}\right]^j + \left({}^{L}\overline{\mathbf{B}}^T_s + {}^{NL}\overline{\mathbf{B}}^T_s(\mathbf{u}^e_n)^j\right)\left[S_{\alpha 3}\right]^j dV$$

5. ANISOTROPIC ELASTIC-PLASTIC CONSTITUTIVE EQUATIONS

The classical multiplicative decomposition is adopted to describe the elastic-plastic behaviour, see Gelin [11]:

$$\mathbf{F} = \mathbf{F}^e\mathbf{F}^p \qquad (35)$$

where $\mathbf{F}^e$ is the elastic deformation gradient and where $\mathbf{F}^P$ is the plastic deformation gradient. The elastic behaviour is considered to be an isotropic one and it is defined with the elastic logarithmic strain measure associated with the elastic deformation gradient, and the Cauchy stress tensor is defined by:

$$\sigma = 2\mu\, \mathrm{Ln}\mathbf{V}^e + \lambda\, \mathrm{tr}\,(\mathrm{Ln}\mathbf{V}^e)\,\mathbf{1} \qquad (36)$$

where λ and μ are the Lamé coefficients.

The plastic behaviour is defined with the Hill yield condition expressed as:

$$\Phi(\sigma) = \frac{1}{2}\sigma : \mathbf{H} : \sigma - \frac{1}{3}\,\sigma_0(\bar{\varepsilon}^p)^2 = 0 \tag{37}$$

where the anisotropy tensor **H** takes into account of initial anisotropy of the material, σ_0 is the flow strength of the material assumed to be a function of the cumulated effective plastic strain rate, $\bar{\varepsilon}^p$.

In using the projection matrix **P** that takes into account of the constraint on the normal stress component, sse Simo and Taylor [12], relation (39) can be written in a matrix form as:

$$\Phi(\sigma) = \frac{1}{2}\sigma^T \mathbf{P}^T \mathbf{H} \mathbf{P} \sigma - \frac{1}{3}[\sigma_0(\bar{\varepsilon}^p)]^2 \tag{38}$$

where σ is the reduced stress vector, H is the matrix form of **H**.

In supposing that the material is a standard one and obeys to the normality rule, it results immediately from relation (38) that the plastic strain rate is given by:

$$\mathbf{D}^p = \lambda \mathbf{P}^T \mathbf{H} \mathbf{P} \sigma \tag{39}$$

where λ is the time independent plastic multiplier.The elastic-plastic loading/unloading condition is the following: $\Phi \leq 0$ and $\lambda \geq 0$ or in a combined form $\Phi\lambda = 0$.

In order to calculate σ and $\bar{\varepsilon}^p$ at the end of the loading increment a two step procedure is adopted. First an elastic stress prediction is realized considering that $\mathbf{F}^p$ is constant and equal to the value obtained at the end of the previous increment. It results that a trial value $(\sigma_{n+1})^{trial}$ defined as :

$$(\sigma_{n+1})^{trial} = \mathbf{C}^e (\mathbf{e}_{n+1})^0 \text{ with } (\mathbf{e}_{n+1})^0 = \mathrm{Ln}(\mathbf{V}^e)^0{}_{n+1} \tag{40}$$

The second step consists to introduce a correction of the stress tensor defined as:

$$\sigma_{n+1} = (\sigma_{n+1})^{trial} - \mathbf{C}^e \Delta\lambda \mathbf{P}^T \mathbf{H} \mathbf{P} \sigma_{n+1} \tag{41}$$

Taking into account of yield condition, it results that equation (41) can be used to determine the consistent tangent plastic modulus:

$$C^{ep} = \frac{\partial \sigma_{n+1}}{\partial e_{n+1}} \tag{42}$$

The rotation of the axes of anisotropy during deformation is taken into account and the update procedure is based on the orthogonal tensor **Q** obtained from the solution of the first order differential equation $\dot{\mathbf{Q}}\mathbf{Q}^T = \omega$ where ω is the spin tensor.

6. CONTACT AND FRICTION DESCRIPTION

When the punch becomes in contact with the sheet metal a contact zone develops around the first contact point, whereas the contact surface between the sheet and the die generally increases.

Due to the fact that the contact surfaces are not known in advance, it is necessary to used an automatic scheme to treat the moving contact conditions.

A point M of the sheet that is in contact with the punch or with the die is subjected to a displacement of normal component u_n and tangential component u_t such that:

$$u_n = (G_n)^T \mathbf{u} \quad ; \qquad u_t = (G_t)^T \mathbf{u}. \tag{43a,b}$$

where G_n and G_t are respectively the normal and tangential projection operators, G_n and G_t are expressed directly as a function of the normal of the point under consideration.

The non-penetration condition and the sticking friction condition are expressed as:

$$(G_n)^T(\mathbf{u} - \bar{\mathbf{u}})=0 \quad ; \qquad (G_t)^T\mathbf{u} = 0 \tag{44a,b}$$

where $\bar{u}$ is the displacement imposed to the rigid die.

The Coulomb contact condition is expressed as:

$$|f_t| < \mu\, |f_n| \Rightarrow u_t = 0 \text{ or } G_t^T u = 0 \tag{45a}$$

$$|f_t| = \mu\, |f_n| \Rightarrow \exists\, \lambda \geq 0 \text{ with } \Delta u_t = -\lambda f_t \tag{45b}$$

where μ is the Coulomb friction coefficient.

The above conditions are taken into account with the penalty approach that consists to modify somewhat the potential energy expression that can be written as (for the frictionless contact case):

$$V^*(u) = V(u) + \frac{1}{2}\alpha_n u^T G_n (G_n)^T u + \frac{1}{2}\alpha_t u^T G_t (G_t)^T u \tag{46}$$

where u is the nodal unknowns vector and where α_n and α_t are two penalty parameters. In the case of frictional contact a perturbed Lagrangian procedure is used that consists to work with the following functional:

$$V^*(u) = V(u) + \Lambda_n^T (G_n)^T u - \frac{1}{2\alpha_n}\Lambda_n^T \Lambda_n + \Lambda_t^T (G_t)^T u - \frac{1}{2\alpha_t}\Lambda_t^T \Lambda_t \tag{48}$$

where u is the nodal unknown vector , Λ_n and Λ_t are the vectors of normal and tangential contact forces, α_n and α_t are the penalty terms. The Euler equations associated to functional (48) lead to the equilibrium equations and to

$$\Lambda_n = \alpha_n G_n^T u \qquad \text{and} \qquad \Lambda_t = \alpha_t G_t^T u \tag{49}$$

For the case of frictional contact, use of the Coulomb law impose relation $\Lambda_t = -\mu\Lambda_n$.

7. NON-LINEAR SHELL BENCHMARK PROBLEMS

7.1 Non linear analysis of a shallow circular arch

The shell is considered to be infinite extent (figure 4) and the out of plane degrees of freedom were set to zero. Only geometrical non linearities are taken into account. This non linear problem has been solved previously by Bathe [13] and Hughes[14]. Meshes of sixteen four node and 16x2 triangles are used for half of the shell. It is interesting to note that the solutions reached using the 3 and the 4 node elements are exactly the same but the CPU time for the three node element was about half that for the four node element. For small problems of this type, all other things being equal the CPU time is dominated by the number of numerical integration points and the triangle uses one set of gauss points while 4 node element uses 4 sets. The results are given for different load intensities but it must be noted that the solution can be reached for any load level using only one increment(figure4).

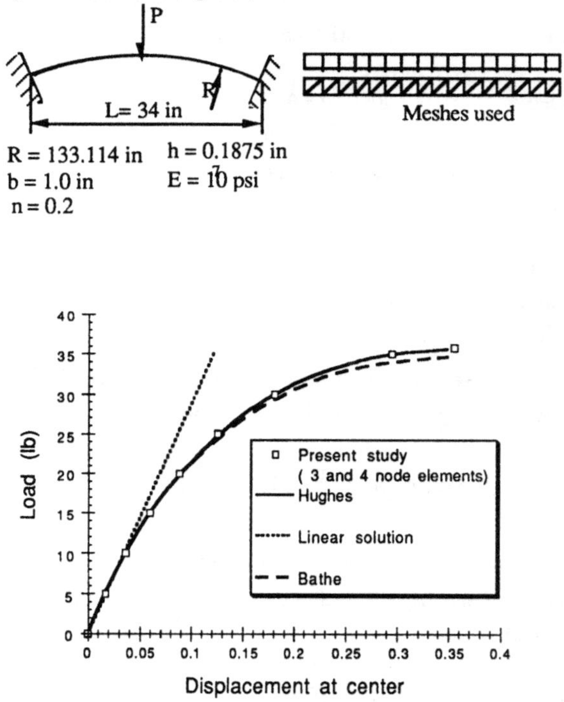

Fig. 4 A shallow circular arch problem

7.2 Non linear hinged cylindrical shell

A 4x4 quadrilateral mesh and the corresponding cross diagonal mesh are used, figure 5. Due to symmetry, only one quarter of the shell had to be modelled. Straight edges are hinged and immovable, curved edges are free. The results are compared with those obtained by Hughes [14] and Crisfield [15], figure 5. The maximum load before the snap through is 2.2 kN. There too the problem can be computed using a single increment for any load intensity.

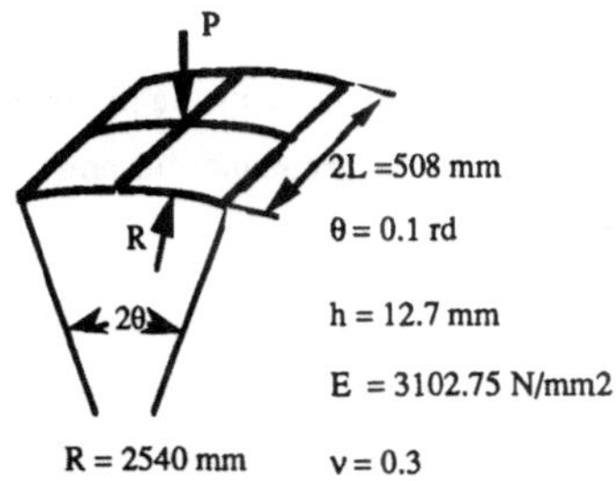

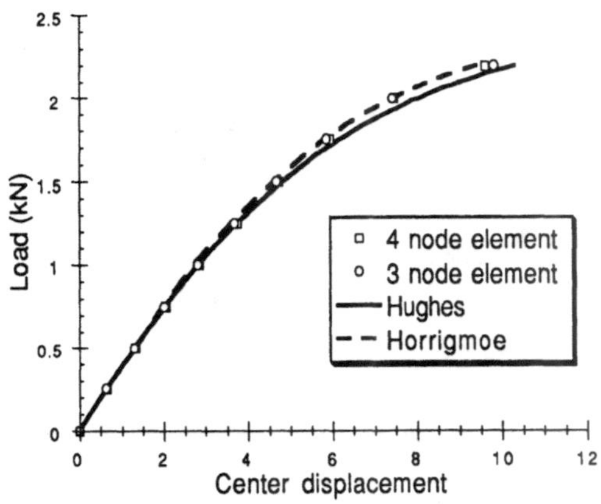

Fig 5 A hinged cylindrical shell problem

8. NUMERICAL SIMULATION OF SHEET METAL FORMING

In using the methods previously described different sheet metal forming cases have been simulated. First example concerns a deep drawing of an axisymmetric cup of a low carbon steel. Second example concerns the square cup drawing of a square blank where numerical results have been compared with experimental ones.

8.1 A deep drawing problem with a cylindrical punch

Figure 6 shows the initial geometry of the deep drawing problem with a cylindrical punch, the sheet have the possibility to slide under the blank holder. The material properties are the following: E=210000 Mpa, ν =0.3, σ_y= 220Mpa, and the hardening law of the material is given by $\sigma_0 = K(\bar{\varepsilon}^p)^n$ with K= 547 Mpa and n = 0.18. The normal anisotropy parameter is R=1.65. During the numerical simulation, the friction coefficient used is m =0.22. The problem is solved using quadrangular shell elements. Fig. 7 shows the punch load versus the punch displacement. We have compared our finite element results with experimental results reported in [16] and with FEM results obtained with a membrane approach also reported in [16]. There is a good agreement between our results and experimental ones. Fig. 8 shows the distribution of radial and circumferential strains .

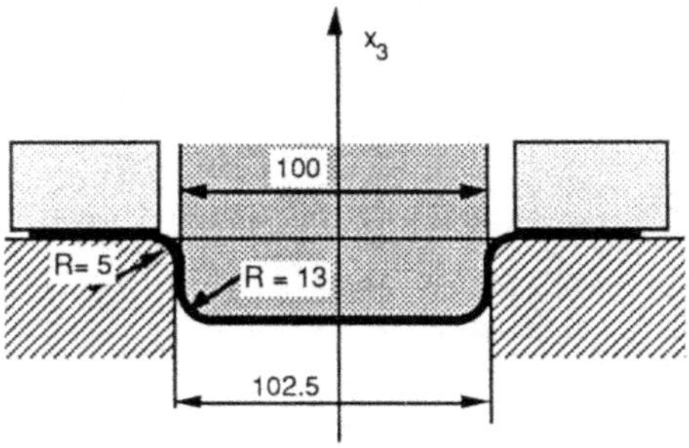

Fig. 6 Initial geometry for the deep drawing problem with a cylindrical punch

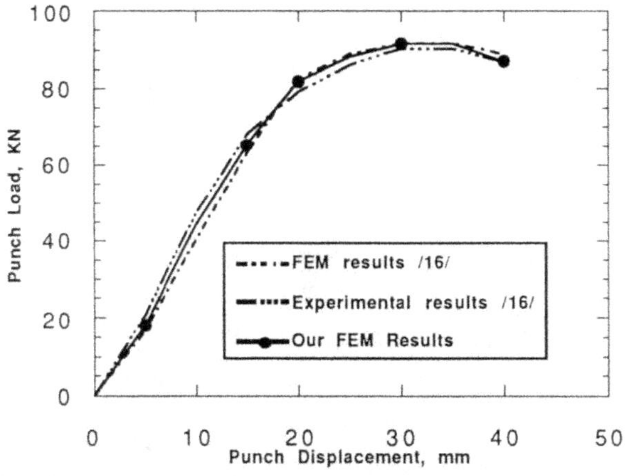

Fig. 7 A comparison between numerical and experimental results for the punch load in the deep drawing of a cylindrical cup.

The effects of the friction coefficient as well as the blank holder restraining load are very important and well revealed by the computations. The increasing of the friction coefficient or the increasing of the blank holder restraining load leads to an increasing of the thickness strain and to a displacement of the pic along the initial radius.

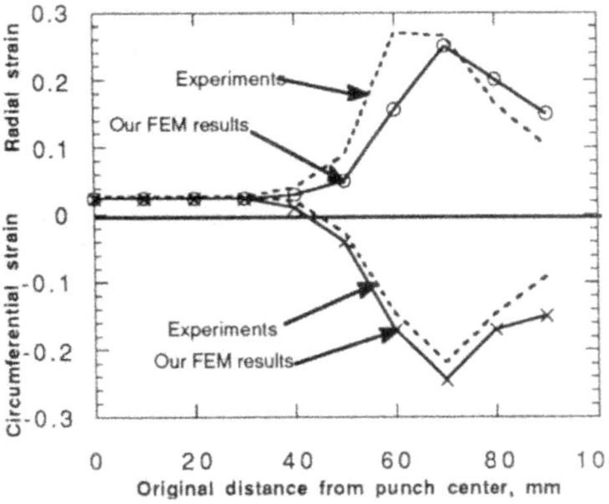

Fig. 8 A comparison between numerically and experimentally determined radial and circumferential strains along the initial radius

8.2 A square cup drawing with a square blank

Second example concerns the square cup drawing of a square blank. The geometry of the problem is represented in figure 9. The material is a steel and is considered as an isotropic one. The material properties are the following: E = 203000 MPa, ν =0.3 and σ_0 = $739(\bar{e}^p)^{0.228}$ Mpa. The punch size is 40mm x 40mm and the die size is 42.5 mm x 42.5mm. The punch radius is 5mm, the die radius is 5 mm and the corner radius 3.2 mm. The thickness of the sheet metal is 0.86 mm. The following friction coefficients have been used : m =0.2 for the punch and m = 0.04 for the die. the blank restraining force is 500DaN.

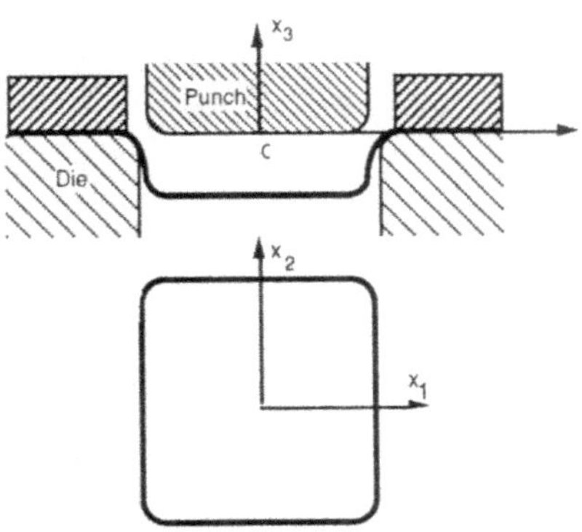

Fig. 9 A square cup drawing problem with a square blank

A quarter of the problem is discretized and figure 10a shows the deformed mesh corresponding to a punch displacement equal 15mm whereas figure 10b represents the deformed mesh for a punch displacement equal 30mm. Figure 11 shows the thickness strain distribution in the transverse direction of the cup drawn.We have compared our results with numerical results found in [17] and experimental results reported in [18]. The trends of the strain distribution predicted by our finite element program are consistent with experimental ones and with other numerical results.

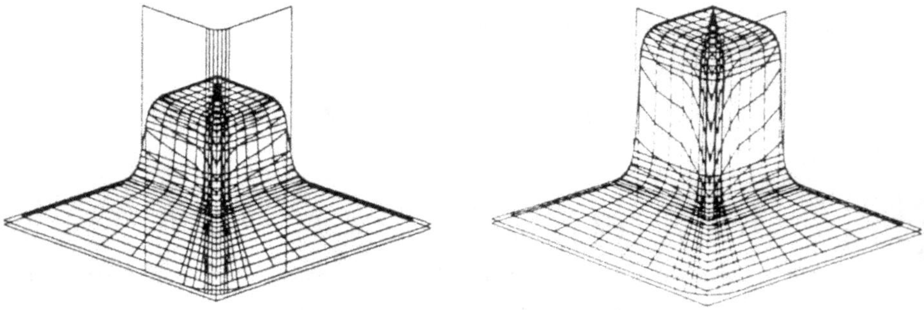

Fig. 10 (a) Deformed mesh for a punch displacement equal 15mm and (b) deformed mesh for a punch displacement equal 30mm

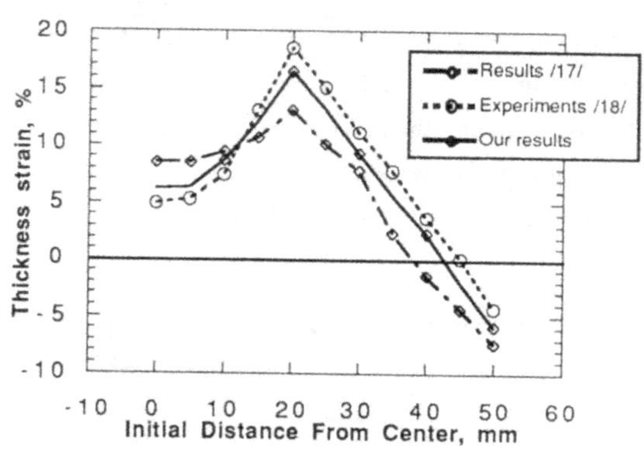

Fig. 11 Comparison of the thickness strain distributions in the transverse direction of the cup drawn

9. CONCLUSIONS

There are three significant features in this paper. The first one is the description of a fully non-linear shell kinematics that takes into account large structural rotations. The second one is the description of a new shell element very reliable for extensive non-linear shell analysis. And the third one is the accurate determination of stress state in the finite strain elastic-plastic range, taking into account stress constraints and combined with an exact update of orthotropy directions to take into account anisotropic elastic-plastic behavior.The results obtained in the classical non-linear shell benchmark problems are very attractive because they show that the triangular element is sufficiently accurate to give good results. The results obtained in the simulation of sheet metal forming problems show that the formulations used are suitable to solve extensively non-linear shell problems.

Aknowlegdments

This work is supported by RENAULT (R.N.U.R.) and PEUGEOT (P.S.A.) motor companies. We thank Mrs Detraux, Horkay, Jameux and Ronde-Ousteau for their collaboration and support.

REFERENCES

[1] Wang,N.M.; Tang,S.C. , Analysis of bending effects in sheet forming. Int J.Numer.Meth. Engrg., 25(1988) 255-267.

[2] Boisse,P.; Daniel,J.L.; Gelin,J.C. , A simple isoparametric three node element . Submitted to Computers and Structures (1991)

[3] Boisse P., Daniel, J.L.; Gelin JC., Sheet metal forming using three and four node shell elements. Int Conf "FE-Simulation of 3D sheet metal forming process in automotive industries" , VDI Report 894 (1991)343-355.

[4] Dvorkin,N.; Bathe,KJ.,A continuum mechanics based four node shell elements for general non linear analysis. Eng. Comput. 1(1984) 77-78.

[5] Dvorkin,N., Bathe KJ., A formulation of general shell elements - the use of mixed interpolation of tensorial components, Int.J.Numer.Meth.Engrg., 22(1986) 697-722.

[6] Simo, J.C.; Fox, D.D., On a stress resultant geometrically exact shell model, Part I: Formulation and optimal parametrization, Comp. Meths. in Appl. Mech. Engrg., 72(1989) 267-304.

[7] Ahmad,S.;Irons,B.M.;Zienkiewiecz;O.C., Analysis of thick and thin shell structures by curved finite elements. Int J.Numer.Meth. Engrg., 2(1971) 419-451.

[8] Stanley,G.,Continuum based shell elements, PhD Thesis, Stanford University(1985).

[9] Hughes,T.J.R.;Tezduyar,T.E., Finite Element Based upon Midlin Plate Theory with Particular Reference to the four node bilinear Isoparametric Element . J. of Applied Mech., (1981) 587-596.

[10] Gelin, J.C., Picart, P., Use of Newton and Quasi-Newton methods for large strain elastic-plastic finite elements computations, Comm. in Appl. Numer. Meth, 4 (1988), 457-486.

[11] Gelin JC., Recent advances in finite elastic-plastic computationnal algorithms. Proc of Strucome Int.Conf. Hermes, Paris (1988)

[12] Simo, J.C.; Taylor, R.L., A return mapping algorithm for plane stress elasto-plasticity, Int J.Numer.Meth. Engrg., 22 (1986) 649-670.

[13] Bathe, KJ.; Ramm. E.; Wilson E. L., Finite element formulations for large displacements and large strain analysis, Report UC-SESM 73-14 , Departement of Civil Engineering, University of California, Berkeley (1973).

[14] Hughes,T.J.R.;Liu,W.K., Non linear finite element analysis of shells. Part 1. Three dimensional shells. Com. Meth App. Mech. and Eng. , 26(1981)331-362.

[15] Crisfield, M.A., Non-linear Finite Element Analysis of Solids and Structures, Volume 1, Wiley Publishers, (1991).

[16] Saran,M.J. et al. , Numerical and experimental investigations of deep drawing of sheet metal, J; Eng. for Industry, Trans. ASME, 112(1990) 272-277.

[17] Toh, C.H., Kobayashi, S., deformation analysis and blank design in square cup drawing, Int. J. Mach. Tool Des. Res., 25-1(1985) 15-32.

[18] THOMSEN, T.R., Experimental results in square cup drawing, J. Australien Inst. Metals, 20 (1975), 106-123.

3D Finite Element Analysis of Thermomechanical Processes. Application to Forging and Casting

M.BELLET, N.SOYRIS, J.L.CHENOT
Centre de Mise en Forme des Matériaux. Ecole des Mines de Paris. Sophia Antipolis, France.

1. Introduction

In hot forming processes, the prominent role of the thermomechanical coupling has often been neglected in numerical simulations until recently, except in some pioneer work [1]. The purpose of this paper is to take the example of two very different forming processes, the hot forging process and the casting process, and to illustrate the role of the thermomechanical coupling and the way it can be treated numerically.

Basically, in hot forming processes, the coupling between heat transfer and mechanics stands in several points :
1. Metals in the solid state and at high temperature can be considered as viscoplastic materials, the behaviour of which depends highly on the temperature according an activation energy law. This is all the more true when the solidus temperature is approached, and a fortiori, in the solidification interval. This influence is also effective for the friction behaviour.
2. the thermal exchanges with the tools, or the moulds, proceed mainly through contact interfaces which are determined by the mechanical deformation of the workpiece, and which evolve with it.
3. the heat transfer equation includes the heat dissipated by the viscoplastic work as a source term which cannot be neglected as soon as the strains are finite and the strain rates are high. Moreover, friction dissipation at contact interfaces must be taken into account.

If we consider first the hot forging process, these three points can be shown to be very important [2,3]. In addition, after the forming stage, during the cooling stage, the temperature history will determine the residual stresses in the part through a thermo-elastoviscoplastic behaviour.

As regards the die-casting process, and in particular, the cooling stage, point 3 is negligible as the deformations are very low. Point 2 has long been dramatically simplified by considering constant conduction coefficients at interfaces. Consequently, the temperature has been considered so far and in first approximation as the governing variable. Starting from the temperature field, issued from pure thermal analysis [4-6], empirical shrinkage criteria can be computed in order to evaluate the soundness of the part and, concerning the residual stresses, they result from *a posteriori* mechanical computations where the temperature history is an input data [7]. The limitation of such an approach is clear : first, the thermal exchanges between the solidifying part and the mould are governed by local conditions (if contact, the normal stress ; if not, the air gap width) that can be deduced only from a mechanical computation. Second, the computation of residual stresses needs a mechanical approach again. Consequently, there is no possible predictive model for the casting process without taking into account the thermomechanical coupling.

In this paper, we will present our approach concerning the heat transfer computation, the mechanical computation and the coupling between them with examples of application to the hot forging process and the casting process.

2. Heat transfer analysis

2.1 Heat transfer equations

The Fourier classical conduction law is considered. On a computational domain Ω we have :

$$\rho c \frac{dT}{dt} = \text{div}\,(k\ \mathbf{grad}\ T) + \delta w/\delta t \qquad (1)$$

D. Besdo, E. Stein (Eds.)
Finite Inelastic Deformations -
Theory and Applications
IUTAM Symposium Hannover/Germany 1991

where k is the thermal conductivity, c the specific heat, ρ the density and $\delta w/\delta t$ an internal power source. The parameters k, c, and ρ may depend on T.
Different boundary conditions can be handled on the external faces of the domain Ω :

- prescribed heat flux : $\Phi_{imp} = -\,k\ \mathbf{grad}\ T\,.\,\mathbf{n}$ (2)
  where **n** is the outward normal unit vector
- convection : $-k\ \mathbf{grad}\ T\,.\,\mathbf{n} = \alpha\,(T - T_{ext})$ (3)
  where T_{ext} is the external temperature
- radiation : $-k\ \mathbf{grad}\ T\,.\,\mathbf{n} = \varepsilon\sigma\,(T^4 - T_{ext}^4)$ (4)
  where ε is the material emissivity, σ the Stefan constant. It is to be noticed that (4) is equivalent to (3) after defining, by linearization,
  $\alpha_r = \varepsilon\sigma\,(T^2 + T_{ext}^2)\,(T + T_{ext})$
- prescribed temperature : $T = T_{imp}$ (5)
- conduction exchange with another domain Ω_j : the conductive flux is then prescribed as :
  $-\,k\ \mathbf{grad}\ T\,.\,\mathbf{n} = 1\,/\,R_{ij}\,(T_i - T_j)$ (6)

where R_{ij} is the thermal resistance of the interface between Ω_i and Ω_j.

2.2 Numerical resolution scheme

The space discretization of equation (1) is done by the finite element Galerkin method, leading to the classical following set of non-linear differential equations in which **T** is the vector of nodal temperatures.

$$\mathbf{C}\,\frac{d\mathbf{T}}{dt} + \mathbf{K}\,\mathbf{T} + \mathbf{Q} = 0 \tag{7}$$

The solution is performed by three level (two step) schemes which are consistent to the second order [8]. The temperature **T** and its derivative dT/dt are expressed as :

$$\mathbf{T} = a\,\mathbf{T}_{t\text{-}\Delta t} + (\frac{3}{2} - 2a - g)\,\mathbf{T}_t + (a - \frac{1}{2} + g)\,\mathbf{T}_{t+\Delta t} \tag{8}$$

$$\frac{d\mathbf{T}}{dt} = (1\text{-}g)\,\frac{\mathbf{T}_t - \mathbf{T}_{t\text{-}\Delta t}}{\Delta t} + g\,\frac{\mathbf{T}_{t+\Delta t} - \mathbf{T}_t}{\Delta t} \tag{9}$$

As regards **C**, **K** and **Q**, equation (8) is not used because it would result in an iterative resolution of (7). This is avoided by using a linearization technique [9], each non-linear parameter **A** (**C**, **K** or **Q**) being written as

$$\mathbf{A} = (\frac{1}{2} - g)\,\mathbf{A}_{t\text{-}\Delta t} + (\frac{1}{2} + g)\,\mathbf{A}_t \tag{10}$$

In practice, the implicit scheme (a=0, g=3/2) is the more frequently used for stability reasons.

2.3 Phase change analysis

This problem is specific of the casting process. Casting alloys generally exhibit sharp variations of heat capacity in the solidification temperature interval. In order to take into account the possible presence of the solidification front inside finite elements, ρc is computed as follows, after Lemmon et al. [10]. The enthalpy function is given, point by point as material data.

$$H(T) = \int_{T_0}^{T} \rho c\, dT \tag{11}$$

At integration points ρc is computed as $\rho c = \frac{dH}{dT} = \frac{\|\,\mathbf{grad}\ H\,\|}{\|\,\mathbf{grad}\ T\,\|}$ (12)

The matrix **C** can then be computed. This technique prevents the underestimation of the capacity matrix if the solidification interval is found inside one element with no integration point in the mushy zone, where ρc is dramatically increased.

<u>2.4 Iterative thermal resolution</u>
In the case of the forging process, there is only one computation domain which is the workpiece. The volumic heat source term is given by the dissipated viscoplastic power during deformation. At the interface with the tools, a heat flux due to friction is prescribed (see section 4).
In die casting, there are several computation domains : the part itself in which the transition liquid-solid occurs, and the different regions (made of different materials) of the mould. At each time increment, the thermal evolution of the global system is determined by successive resolutions of the thermal equation (7) on each domain Ω_i. When computing the solution on Ω_i, the coupling with other domains Ω_j, through interface regions Σ_{ij} is taken into account by prescribing the thermal flux according to (6), in which T_j is the more recent surface temperature of the domain Ω_j and R_{ij} is the thermal resistance between the two domains i and j. If one of them is the part, the thermal resistance can be updated locally in function of the local air gap width according to a predefined law. The local air gap width is computed by the mechanical analysis (see further). The temperature evolution is then computed iteratively on all the domains until convergence is reached. As already mentioned, this method accounts for non-coincident meshes at the interfaces between the different domains. At the beginning of the computation, a procedure is run to define, for each integration point of interface regions Σ_{ij} of each domain Ω_i, which nodes of Σ_{ij} in Ω_j will serve as interpolation nodes to determine the external temperature T_j in (6). An other advantage that counter-balances the fact that the solution is iterative, is that thermal resolutions are performed on domains Ω_i that are smaller than a global finite element mesh. It should be also noted that the capacity and conductivity matrices are computed only once at the beginning of the increment, before starting the iterations.

3. Mechanical analysis

<u>3.1 General constitutive equations of metal alloys at high temperature : elasto-viscoplasticity</u>
The Perzyna's model is used to model the elastic-viscoplastic behaviour [11]. The equations of the model are as follows :

$$\dot{\varepsilon} = \dot{\varepsilon}^{el} + \dot{\varepsilon}^{vp} + \dot{\varepsilon}^{th} + \dot{\varepsilon}^{tr} \tag{13}$$

$$\dot{\varepsilon}^{th} = \alpha \, \dot{T} \, \mathbf{I} \tag{14}$$

$$\dot{\varepsilon}^{tr} = \frac{1}{3} \dot{f}_s \, \Delta\varepsilon_{tr} \, \mathbf{I} \tag{15}$$

$$\dot{\sigma} = \mathbf{E} \, \dot{\varepsilon}^{el} = 2\mu \, \dot{\varepsilon}^{el} + \lambda \, \mathrm{Trace}\,(\dot{\varepsilon}^{el}) \, \mathbf{I} \tag{16}$$

$$\dot{\varepsilon}^{vp} = \gamma \left\langle \frac{\sigma_{eq}}{\sigma_0} - 1 \right\rangle^{1/m} \frac{3}{2\sigma_{eq}} \, \mathbf{s} \tag{17}$$

$$\mathbf{s} = \mathbf{s} + p \, \mathbf{I} = \mathbf{s} - 1/3 \, \mathrm{Trace}\,(\mathbf{s}) \, \mathbf{I} \tag{18}$$

$$\sigma_{eq} = \sqrt{\frac{3}{2} s_{ij} s_{ij}} \tag{19}$$

Equation (15) gives the expression of the additional strain rate which is specific of the solidification of most of metal alloys. $\dot{f}_s$ is the rate of the volumic solid fraction and $\Delta\varepsilon^{tr}$ is the relative volume change due to the total liquid-solid transformation. Equation (17) gives the relation between the viscoplastic strain rate and the deviatoric stress tensors, in which σ_{eq} is the

Von Mises stress and σ_0 the static yield stress below which no viscoplastic deformation occurs. It can be shown easily that this leads to :

$$\sigma_{eq} = \sigma_0 + k\,\dot{\varepsilon}_{eq}{}^m \qquad \text{with} \qquad k = \sigma_0 \left(\frac{1}{\gamma}\right)^m \tag{20}$$

where $\dot{\varepsilon}_{eq}$ is the equivalent strain rate defined by :

$$\dot{\varepsilon}_{eq} = \sqrt{\frac{2}{3}\,\dot{\varepsilon}_{ij}\dot{\varepsilon}_{ij}} \tag{21}$$

The different material coefficients are temperature dependant. The resolution of the non-linear set of equations (13 to 19) is performed by a decoupling method between spherical and deviatoric terms. It can be demonstrated (details will be given in a next paper) that this system can be reduced to a scalar non-linear equation that it solved with the Newton method. Hence, the strain rate $\dot{\varepsilon}'$ being given, it is possible to compute easily σ' as well as the exact tangential modulus $d\sigma'/d\varepsilon'$.

3.2 Adapted constitutive equations for hot forging : pure viscoplasticity

In this case, the elastic part of the strain rate can be neglected. The stress deviator is directly related to the viscoplastic strain rate by the so called Norton-Hoff law :

$$\dot{\varepsilon} = \dot{\varepsilon}^{vp} + \dot{\varepsilon}^{th} \tag{22}$$

$$\dot{\varepsilon}^{th} = \alpha\,\dot{T}\,\mathbf{I} \tag{23}$$

$$\text{Trace}\,(\dot{\varepsilon}^{vp}) = 0 \tag{24}$$

$$s = 2\,K \left(\sqrt{3}\,\dot{\bar{\varepsilon}}\right)^{m-1} \dot{\varepsilon}^{vp} \tag{25}$$

In the above relations, the consistancy K depends on temperature and on the plastic deformation according to :

$$K = K_0\,(\bar{\varepsilon} + \bar{\varepsilon}_0)\,\exp(\beta/T) \tag{26}$$

A consistent viscoplastic friction law is used :

$$\mathbf{T} = -\,\alpha\,K\,\|\Delta\mathbf{v}_t\|^{q-1}\,\Delta\mathbf{v}_t \tag{27}$$

in which $\Delta\mathbf{v}_t$ is the relative velocity of the material with respect to the tool.

3.3 Expression of the mechanical equilibrium

The equilibrium is given by the virtual work principle :

$$\forall\,\mathbf{v}^* \quad \int_\Omega \sigma : \dot{\varepsilon}^*\,dV \;-\; \int_\Omega \rho\mathbf{g}\,.\,\mathbf{v}^*\,dV \;-\; \int_{\partial\Omega} \tau\,.\,\mathbf{v}^*\,dS \;=\; 0 \tag{28}$$

where $\mathbf{g}$ is the gravity vector. It can be neglected in forging but not in casting as we are interested in air gap formation.

In the casting case, equation (28) is written at the end of the increment, on the new configuration Ω' that is defined by the following integration scheme of the velocity field :

$$\mathbf{X}' = \mathbf{X} + [\,(1-\eta)\,\mathbf{v} + \eta\,\mathbf{v}'\,]\,\Delta t \tag{29}$$

in which η is an arbitrary parameter. It appears then that equation (28), when written at the end of the increment, has only one unknown which is the new velocity field v' at time t+Δt. Hence, we have to solve an equation of the form :

$$\mathbf{R}(\mathbf{v}') = \mathbf{0} \tag{30}$$

Concerning now the forging process, it can be demonstrated [12] that the mechanical equilibrium (28) is reached for the velocity field **v** which minimizes the following functional (neglecting body forces, this time) :

$$\Phi(\mathbf{v}) = \int_{\Omega} \frac{K}{m+1} (\sqrt{3}\,\dot{\bar{\varepsilon}})^{m+1} dV + \int_{\partial\Omega_c} \frac{\alpha K}{q+1} \|\Delta \mathbf{v}_t\|^{q+1} dS + \frac{\rho}{2} \int_{\Omega} K(\mathrm{div}\,\mathbf{v} - 3\alpha \dot{T})^2 dV \tag{31}$$

in which the third term is an additional penalty term to enforce the incompressibility condition, ρ being a large positive constant. The problem is then to solve for the velocity field **v** the equation :

$$\mathbf{R}(\mathbf{v}) = \frac{\partial \Phi}{\partial \mathbf{v}} = 0 \tag{32}$$

The velocity field solution is in this case integrated with an explicit scheme :

$$\mathbf{X}' = \mathbf{X} + \mathbf{v}\,\Delta t \tag{33}$$

3.4 Finite element resolution of the mechanical problem

In both cases (elastic-viscoplastic or viscoplastic behaviour), the equilibrium equation (30 or 32) is a non-linear equation for the velocity field which is solved, after discretization into finite elements, by the Newton-Raphson method. The meshes are the same as those used for the thermal analysis.

3.5 Contact

Contact algorithm : algorithms for evolving contact in forging have been analysed in [13,14]. The algorithm used consist in a discrete nodal contact approach, the surface of the forging tools being discretized in triangles. The contact between the workpiece and the tool is enforced until the normal stress is no more compressive. The algorithm used for casting is about the same : at the beginning, all surface nodes of the liquid part are considered in contact with the mould. The conjugate effects of the alloy shrinkage due to solidification and of the gravity can result in non-compressive normal stresses at the interface, and then in the air gap formation.

4. The coupled problem

The mechanical problem and the thermal one are solved independantly at each time step. At each time increment, a mechanical computation is done first in the workpiece (forging) or in the solidifying part (casting). This will modify the parameters and the boundary conditions of the thermal computation. First, the dissipated plastic work, which is given by

$$\sigma : \dot{\varepsilon} = \sigma_{eq}\,\dot{\varepsilon}_{eq} = K\left(\sqrt{3}\,\dot{\bar{\varepsilon}}\right)^{m+1} \tag{34}$$

will appear as the heat source term $\delta w/\delta t$ in (1) for forging (it is negligible in casting). Second, in forging analysis also, the dissipated work due to friction and which is given by

$$\frac{b}{b+b_{die}}\,\alpha K \,\|\,\Delta \mathbf{v}_t\,\|^{q+1} \tag{35}$$

will appear as an additional external heat flux (with negative sign) to the second term of (6). Finally the evolution of the contact areas is known, in particular the air gap width between the part and the mould in casting, which allows to update the local heat exchange coefficients R_{ij} in (6). The thermal computation is then run on the previous basis : only in the workpiece in forging, in the part and all the subdomains of the mould in casting. After the thermal problem has been solved, the local rheology can be updated for the next mechanical computation at next increment.

5. Application to the hot forging process

5.1 Forging of a rocker arm

The workpiece made of aluminium and the forming dies are shown on figure 1 in position before forging. A thermomechanical computation of the forging has been carried out until degenerescence of the finite element mesh (50% height reduction).

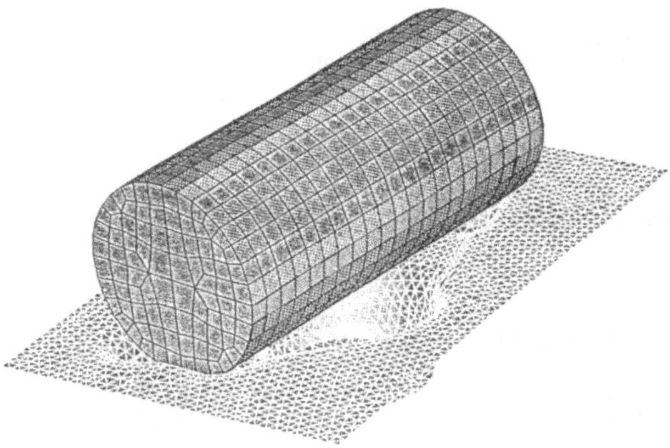

Figure 1 : initial workpiece on the lower die. The upper die is similar to the lower.

Figure 2 gives the temperature profile in the part at this stage, showing that the free surfaces are the coldest while the surfaces where sliding occurs are the warmest, due to the frictional heat dissipation. Detailed results can be found in [2].

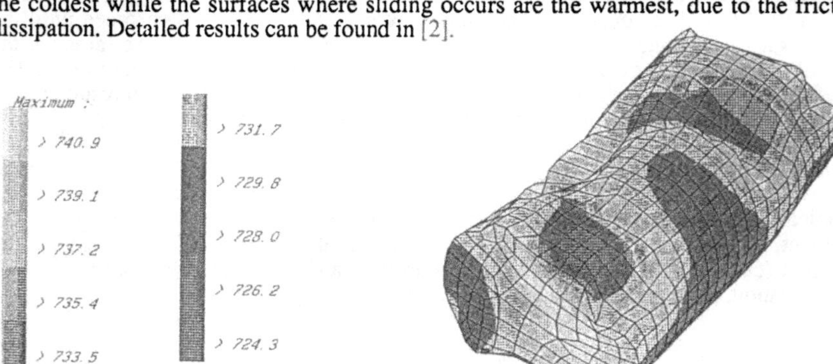

Figure 2 : temperature distribution (K) at 50% height reduction.

5.2 Simulation with remeshing

To avoid the major drawback of mesh degenerescence, a 3D remeshing procedure has been developed [15,16]. Ten node quadratic tetrahedra are used. The procedure is applicable to mesh the initial part as well as to remesh it when necessary. The remeshing procedure can be splitted into four major tasks :

1) discretization of the surface of the part in triangles. This information is deduced from the current surface of the deformed part : the triangles are the emerging faces of the tetrahedra. All the triangles are then splitted into several smaller triangles to get an over-discretization of the surface. The position of the different nodes with respect to the dies is then checked : corrections are done if necessary to avoid penetration.

2) creation of a linear surface mesh. Starting from the over-discretized surface mesh, a better mesh is computed with an appropriate (and reduced) number of nodes to ensure a good precision of the geometrical description.

3) creation of a quadratic surface mesh. The previous triangles elements are changed to 6 node quadratic ones. The position of nodes is determined in order that the surface fits as precisely as possible the initial over-discretized surface.

4) creation of a quadratic volumic mesh. Starting from the linear triangles of the surface mesh, linear tetrahedra are created in the internal volume. The addition of intermediate nodes yields a 10 node quadratic tetrahedral mesh.

At the moment, these procedures allow to remesh the part during the forging with the control of the user. We give hereunder the example of an automotive transmission part calculated in isothermal conditions [17]. Figure 3 presents the workpiece and the tools in initial position.

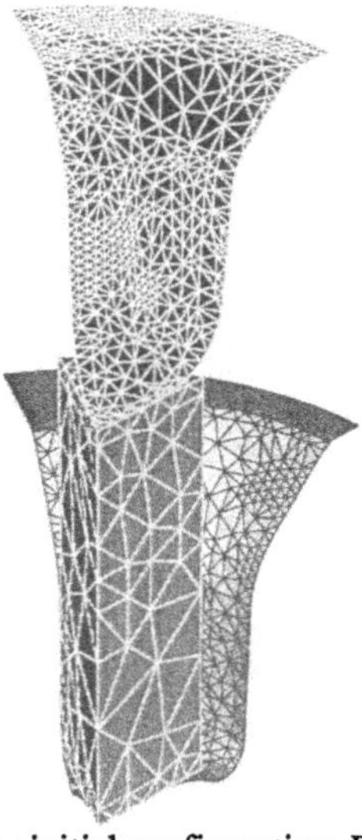

Figure 3 : initial configuration. Punch (surface), workpiece, lower die (surface).

The part is made of steel and has symmetry planes oriented at 0 and $\pi/3$, so that one sixth of the part is computed. The initial height is 84 mm and the diameter 55.7 mm. 9 remeshings were necessary to achieve the forging with a total of about 400 increments (figure 4).

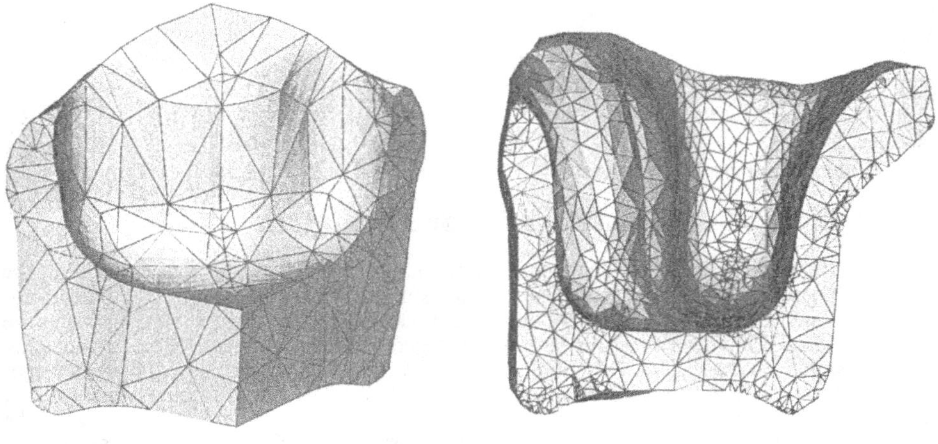

Figure 4 : one third of the workpiece (left). Half of the part at the end of forging (right).

6. Application to the casting process

6.1 Casting of a suspension cylinder

This three-dimensional sand casting involves six different subdomains including the part itself which is made of aluminium alloy. A pure heat transfer analysis has been performed. Figure 5 shows the global enmeshment of the system, made of eight node hexaedra. Figure 6 shows the temperature profile in the part and in the sand upper core after 220 s. The part is almost completely solidified (solidus temperature at 577°C), except in the upper region of the riser.

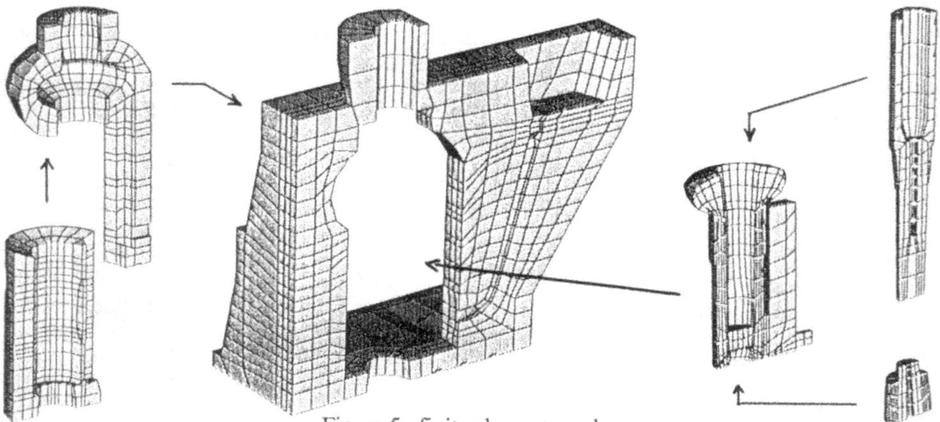

Figure 5 : finite element meshes

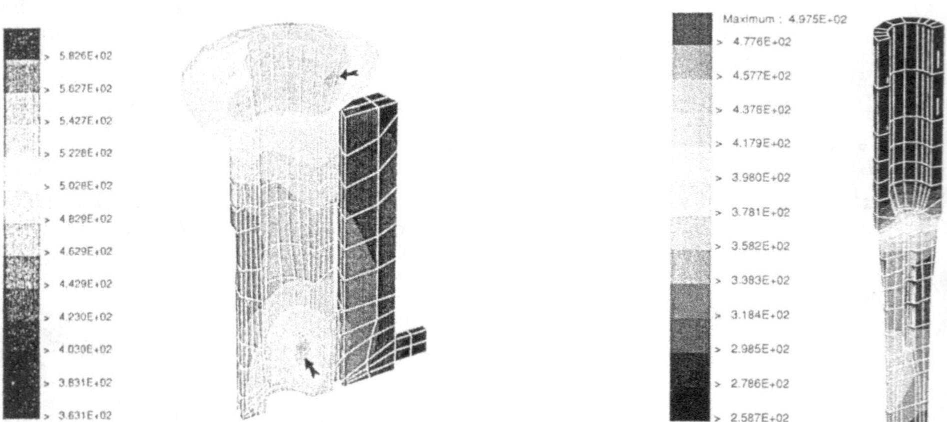

Figure 6 : temperature distribution (°C) in the part (left) and in the upper core (right).

6.2 Two examples of coupled thermomechanical computation : air gap formation

These examples are used to test the coupled formulation and to analyse the sensitivity of the model to the material data. The geometries that are studied are simple ones, in two dimensions. The enmeshments are 3D but with only one layer of elements in the direction orthogonal to the symmetry plane. The first example is illustrated in figure 7. The part is made of aluminium alloy and the mould is made of steel. The specimen is 100 mm high. The figure shows the evolution of the air gap occurring after solidification has begun in the part. The model accounts quite well for the competition between gravity forces, that tends to maintain contact, and the shrinkage of the material which genarates the gap. The material behaviour in the solidification interval and at high temperature in the solid state (i.e. the ability to adapt to shrinkage by creep) is found determinant in the air gap evolution.

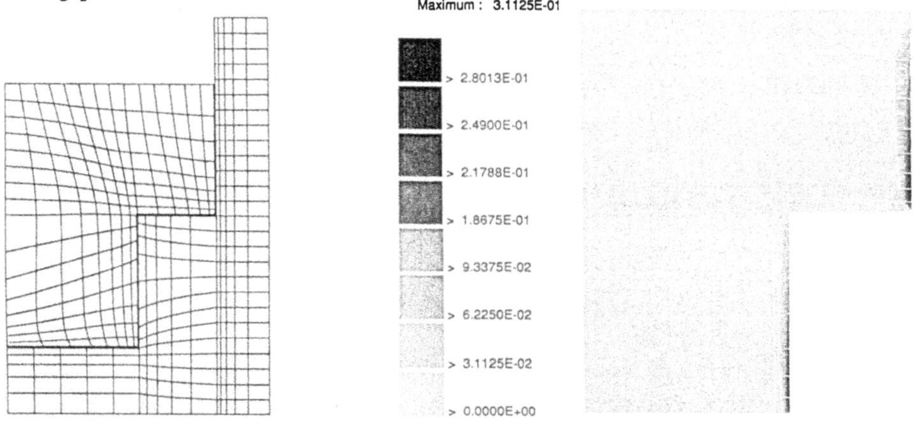

Figure 7 : finite element enmeshment (left); iso-air gap widths (in mm) at the boundary of the part: the air gap appears on the lateral sides first (right).

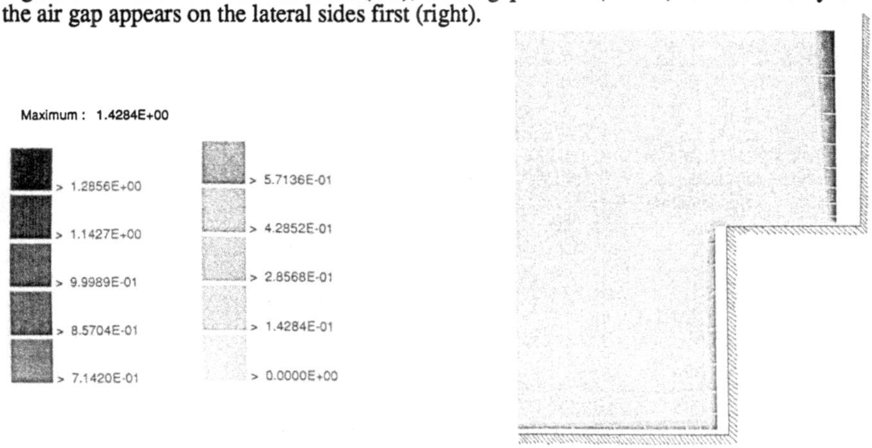

Figure 8 : final air gap distribution. Air gap also appears at the bottom of the specimen. The air gap has been post-designed and magnified by a ratio of 5.

The second example deals with another sand casting (figure 9). The part is 40 mm high. Figure 10 shows the air gap evolution, first in the central region and then in the lateral ends. In the central part, tensile stresses appear since the ends are blocked by the mould.

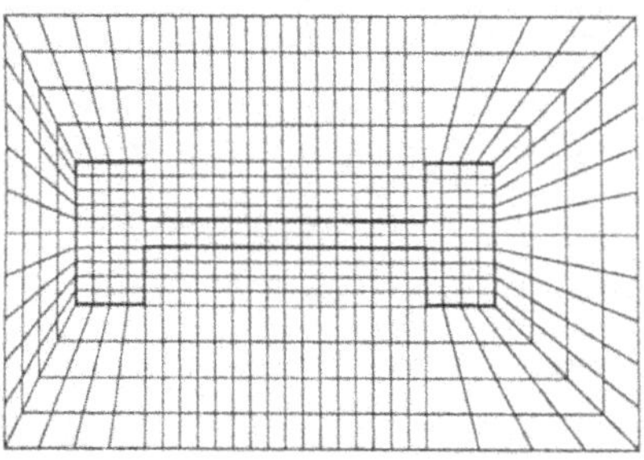

Figure 9 : global mesh (2 subdomains).

Acknowledgements

The authors would like to thank the French Ministère de la Recherche et de la Technologie, the Centre Technique des Industries de la Fonderie, and the companies Citroën, Creusot Loire Industrie, Forges de Courcelles, Montupet, Pechiney, Peugeot, Teksid for their financial support.

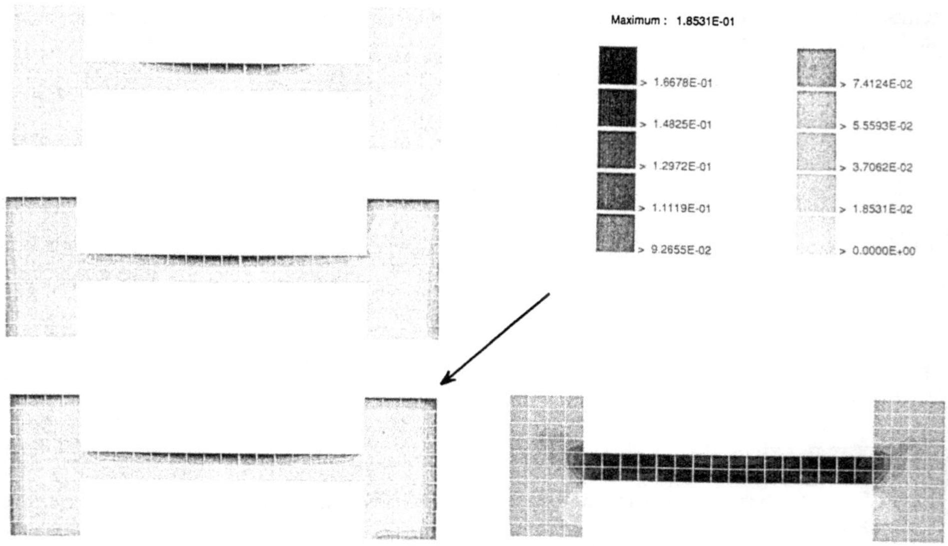

Figure 10 : a) b) c) iso-air gap width distribution at the boundary, d) σ_{xx} distribution in the part.

References

1. Rebelo N., Kobayashi S. A coupled analysis of viscoplastic deformation and heat transfer. Theoretical considerations, applications. Int. J. Mech. Sci., 22, p.699 (1980)
2. Soyris N., Brioist J.J., Brachotte G., Chenot J.L. Three dimensional finite element calculation of the whole forging process of an automotive part. Proc. ICTP 1990, 1, p.165 (1990)
3. Soyris N. Modélisation tridimensionnelle du couplage thermique en forgeage à chaud (3D modelling of the thermal coupling in hot forging), PhD dissertation (in french). Ecole des Mines de Paris (1990)
4. Giamei A.F., Abbaschian R. (eds.). Modelling of casting and welding processes. Warrendale, PA, Metallurgical Society of AIME (1988)
5. Lewis R.W., Roberts P.M. Finite element simulation of solidification problems. Applied Scientific Research 44, p.61 (1987)
6. Rappaz M. Modelling of microstructure formation in solidification processes. Int. Material Reviews, 34, n°3, p.93 (1989)
7. Chandra U. Benchmark problems and testing of a finite element code for solidification in investment castings. Int. J. Num. Meth. Eng., 30, p.1301 (1990)
8. Hogge M.A. A comparison of two and three level integration schemes for non-linear heat conduction. Proc. Numerical Methods in Heat Transfer (Wiley), p.75 (1981)
9. Zlamal M. Finite element methods for non-linear parabolic equations. RAIRO Num. Anal. 11 p.93 (1977)
10. Lemmon E.C. Multidimensional integral phase change approximations for finite conduction codes. Proc. Numerical Methods in Heat Transfer (Wiley), p.201 (1981)
11. Perzyna P. The constitutive equations for rate sensitive plastic materials. Quat. Appl. Math., 20, p.321 (1963)
12. Bohatier C., Chenot J.L. Finite element formulations for non steady state large viscoplastic deformation. Int. J. Num. Meth. Eng., 21, p.1697 (1985)
13. Chenot J.L. Finite element calculation of unilateral contact with friction in non steady state problems. Proc. 2nd Int. Conf. on Advances in Numerical Methods in Engineering, Pande & Middleton (eds), Martinus Nijhoff Publishers (1987)
14. Montmitonnet P., Germain Y., Chenot J.L. Application of the finite element method to contact friction and lubrication in metal forming. JMTA suppl. n°1, 7, p.193 (1988)
15. Coupez T. Grandes transformations et remaillage automatique (Finite transformations and automatic remeshing), PhD dissertation (in french). Ecole des Mines de Paris (1991)
16. Coupez T., Soyris N., Chenot J.L. 3-D finite element modelling of the forging process with automatic remeshing. J. Mat. Proc. Tech., 27, p.119 (1991)
17. Gay C., Modélisation du forgeage 3D (Numerical modelling of 3D forging). Internal CEMEF report, in french (1991)

A Discussion of the Mechanics of Superplastic Forming

I.St. Doltsinis
Institute for Computer Applications, University of Stuttgart, F.R.G.

Summary

Several tasks essential to the process design for superplastic forming are treated in conjunction with a computer simulation methodology. In this context, the optimisation of the history of the forming pressure is considered, and the consequences of variations of the material properties. Furthermore, a numerical procedure is presented for the determination of the initial thickness profile of sheets, required for the production of a prescribed distribution of thickness in the formed component. The theory is illustrated by an example.

Superplastic deformation

The process of superplastic deformation is characterised by the production of high permanent elongations in metallic materials subject to tension. It can be induced under specific thermal and mechanical conditions in materials possessing a stable, ultrafine microstructure [1]. Usually, superplastic deformation is a slow process performed at a constant temperature $T \geq 0.4T_M$, T_M denoting the absolute melting temperature of the material. Essential to superplasticity is the dependence of the uniaxial flow stress σ on the rate of deformation δ, to a higher degree than in conventional hot forming. The functional form

$$\sigma = f(\delta, d) \tag{1}$$

for the flow stress accounts also for the influence of the grain size d which develops with time t in accordance with the grain growth kinetics of the material

$$\dot{d} = g(t, \delta) \tag{2}$$

Figure 1 depicts the logarithmic σ, δ-diagram for the superplastic Ti-6Al-4V alloy frequently used in aerospace engineering. Detailed experimental data for this material may be found in [2], an analytical interpretation was given in [3].

The material behaviour under multiaxial conditions may be described by a nonlinear viscous law relating the Cauchy stress $\boldsymbol{\sigma}$ (6×1 vector array) to the corresponding rate of deformation $\boldsymbol{\delta}$ defined by the symmetric part of the velocity gradient in the solid. The deformation is subject to the isochoric condition so that $\boldsymbol{\delta}$ is purely deviatoric. Isochoric response is characterised by the viscosity coefficient μ which gives rise to the deviatoric stress

$$\boldsymbol{\sigma}_D = 2\mu\boldsymbol{\delta} \tag{3}$$

D. Besdo, E. Stein (Eds.)
Finite Inelastic Deformations –
Theory and Applications
IUTAM Symposium Hannover/Germany 1991

If the stress state is two-dimensional as in superplastic sheet forming, the rate of deformation specifies the hydrostatic stress as well, under strict observance of the isochoric condition, cf. [4]. An analogous complement to (3) for three-dimensional situations requires the introduction of a penalty approach implying a slight relaxation of the isochoric condition, cf. [3].

For a specification of the deviatoric viscosity coefficient the formal relation between the von Mises equivalent stress $\bar{\sigma}$ and the rate of deformation $\bar{\delta}$ as deduced from (3), is compared with the uniaxial characteristic (1). One obtains

$$\bar{\sigma} = 3\mu\bar{\delta} = f(\bar{\delta}, d) \tag{4}$$

and $\mu(\bar{\delta}, d)$ may thus be expressed in terms of uniaxial material data.

The advantage of superplastic deformation is utilized in the manufacturing of structural components requiring high permanent deformations in a single fabrication cycle and appears particularly suitable for specific applications in aerospace engineering. In the context of a numerical computer simulation of the forming process, we consider a finite element representation of the work-piece material. The geometry is then specified by the vector array $\boldsymbol{X}$ which comprises the coordinates of the nodal points of the finite element mesh; the velocities are accordingly collected in the vector $\boldsymbol{V}$. The applied forces constitute the vector $\boldsymbol{R}$ and the forces resulting at the nodal points from stresses in the elements, the vector $\boldsymbol{S}$. The process of forming can be considered quasistatic and isothermal; it is governed by the equation,

$$\boldsymbol{S}(\boldsymbol{V}, \boldsymbol{X}) = \boldsymbol{D}(\boldsymbol{V}, \boldsymbol{X})\boldsymbol{V} = \boldsymbol{R}(t, \boldsymbol{X}) \tag{5}$$

which expresses the equilibrium condition in the finite element representation of a viscous solid [3]. As a consequence of the nonlinear constitution of the material, cf. (3) and (4), the viscosity matrix $\boldsymbol{D}$ of the system depends on the velocity and may require knowledge of the actual grain size; the geometry is associated with the deformed configuration of the solid.

The motion of the material during the course of the process is linked to the velocity by the time integration. For the time increment $\tau = {}^{b}t - {}^{a}t$ an approximate scheme yields

$${}^{b}\boldsymbol{X} = {}^{a}\boldsymbol{X} + (1 - \zeta)\tau\,{}^{a}\boldsymbol{V} + \zeta\tau\,{}^{b}\boldsymbol{V} \tag{6}$$

with the parameter $0 \leq \zeta \leq 1$. Explicit integration ($\zeta = 0$) merely requires the velocity $\boldsymbol{V} = {}^{a}\boldsymbol{V}$ at the beginning of the increment ($t = {}^{a}t$) as governed by (5) with $\boldsymbol{X} = {}^{a}\boldsymbol{X}$ known. The implicit case ($\zeta > 0$) starts from a known state and requires a solution of (5) at the end of the increment ($t = {}^{b}t$) for $\boldsymbol{V} = {}^{b}\boldsymbol{V}$ with $\boldsymbol{X} = \boldsymbol{X}({}^{b}\boldsymbol{V})$ in accordance to (6).

In either case (5) may be solved by the iteration procedure

$$\boldsymbol{V}_{i+1} = \boldsymbol{V}_i + \boldsymbol{D}_i^{-1}\left[\boldsymbol{R}_i - \boldsymbol{S}_i\right] \tag{7}$$

as applied to the residual form of the equilibrium condition in conjunction with the secant viscosity matrix of the system, which usually proves satisfactory.

When the material is formed onto a die, the latter may be assumed rigid and is defined as a geometric surface limiting the deformation. In the computation algorithm, nodal points contacting the die experience a contact force

$$\boldsymbol{F}_n = -k_n \boldsymbol{v}_n \quad , \quad k_n \rightarrow \infty \tag{8}$$

opposed to the normal velocity $\boldsymbol{v}_n$ towards the die (penalty approach), and a friction force

$$\boldsymbol{F}_t = -k_t \boldsymbol{v}_t \tag{9}$$

opposed to the sliding velocity $\boldsymbol{v}_t$ tangential to the die. The determination of the factor

$$k_t = |\boldsymbol{F}_t| / |\boldsymbol{v}_t| \leq k_{max} \tag{10}$$

implies the specification of the intensity $|\boldsymbol{F}_t|$ of the friction force by a friction law (e.g. Coulomb friction) and thus allows a general description of friction by (9); its limitation by k_{max} corresponds to the penalty approach to the condition of sticking.

The above procedure was introduced in [4] in order to treat the contact developing between work-piece material and die during the process of forming in complete conformity with the viscous computation. In fact, the contact forces expressed by (8) and (9) as functions of the velocities are added to the stress resultants $\boldsymbol{S}$ in (5) and simply modify the viscosity matrix $\boldsymbol{D}$ by the factors k_n, k_t while the structure of the system matrix remains unaffected throughout the process. The above remarks transfer to the respective entities in the iterative solution procedure (7).

Optimisation of the forming pressure

Typical superplastic forming processes produce shell components from initially flat sheets by the application of gas pressure. The time history of the forming pressure controls the process. At each time instant, the rate of mechanical work performed for the deformation of the work-piece material satisfies the equation

$$p\dot{G} = \int_V \boldsymbol{\sigma}_D^t \boldsymbol{\delta} \, \mathrm{d}V + \int_S \tau v_t \, \mathrm{d}S \tag{11}$$

In (11), $\dot{G}$ denotes the rate of space volume displaced by the gas pressure p, v_t is the sliding velocity at the parts of the surface contacting the die and may be opposed by τ, the friction force per unit surface area. As a consequence of the isochoric deformation, only the deviatoric stress contributes to the work expressed by the volume integral in (11).

Optimised processes are characterised by a minimum process time under observance of the superplastic conditions. The latter ensure that the material will not fail by the appearence of localised necks and for this purpose it is required that the maximum equivalent rate of deformation in the work-piece material must not exceed the value δ_o corresponding to the maximum rate sensitivity of the flow stress

$$m = \frac{\mathrm{d} \ln \sigma}{\mathrm{d} \ln \delta} \tag{12}$$

The differential quotient in (12) is defined for a constant grain size, and both its maximum value and the associated optimum rate of deformation decrease as the grains grow during the course of the forming, cf. Fig.1. In the work-piece material, the rate of grain growth as given by (2) is highest where the maximum rate of deformation appears. Therefore, the optimum value δ_o appertaining to the actual grain size at this place may be considered as the critical one rather than the higher value associated with the initial uniform grain size in the work-piece material.

For an assessment of the optimum pressure at a certain stage of the forming process, let the loading with a trial pressure p_t result in the distribution $\boldsymbol{\sigma}_t$ and $\boldsymbol{\delta}_t$ of the stress in the work-piece material and of the rate of deformation respectively. The maximum equivalent rate of deformation in the material is $\bar{\delta}_{max}$ whilst for the optimum pressure it must attain the prescribed value δ_o. A kinematically admissible rate of deformation field satisfying this requirement reads

$$\boldsymbol{\delta}_o = \frac{\delta_o}{\bar{\delta}_{max}} \boldsymbol{\delta}_t \tag{13}$$

The deviatoric stress associated with (13) follows from (3) and (4) as

$$\boldsymbol{\sigma}_{Do} = \frac{f(\bar{\delta}_o, d)}{f(\bar{\delta}_t, d)} \boldsymbol{\sigma}_{Dt} \tag{14}$$

where the indices t and o refer again to the trial pressure and the envisaged optimum value respectively. It may be seen that if (13) is combined with a uniaxial material law of the form,

$$\sigma = f(\delta, d) = k(d)\delta^m \tag{15}$$

the proportionality factor for the stresses in (14)

$$\frac{f(\bar{\delta}_o, d)}{f(\bar{\delta}_t, d)} = \left(\frac{\bar{\delta}_o}{\bar{\delta}_t}\right)^m = \left(\frac{\delta_o}{\bar{\delta}_{max}}\right)^m \tag{16}$$

is constant within the solid at the instant under consideration.

Since $\boldsymbol{\sigma}_t$ is an actual stress state, a stress system $\boldsymbol{\sigma}_o$ proportional to $\boldsymbol{\sigma}_t$ with a constant factor satisfies the condition of internal equilibrium but is associated with a different intensity of the applied pressure, which may be obtained by an application of the virtual work principle. To this end we consider first frictionless situations, and utilizing the trial state for the virtual kinematics, we deduce the new pressure as

$$p_o = \frac{1}{\dot{G}_t} \int_V \boldsymbol{\sigma}^t_{Do} \boldsymbol{\delta}_t \mathrm{d}V = \left(\frac{\delta_o}{\bar{\delta}_{max}}\right)^m p_t \tag{17}$$

The final expression in (17) is obtained by substituting (14) for the stress in conjunction with (16) and using the work equivalence (11) as applied to the trial state. While contact friction was excluded so far, its consideration does not change the ultimate result for the pressure in (17) as long as the intensity of the surface friction τ is related to the contact pressure by a constant friction coefficient.

By (17), the forming pressure may be adapted to a prescribed value δ_o of the maximum equivalent rate of deformation in the work-piece material. This, however, presumes a material obeying the power constitutive law of (15) which is based on a rate sensitivity m, cf. (12), not varying with the rate of deformation. Despite the above essential deviation from our initial motivation, the result reflected by (17) is useful. It may be concluded, that for the particular material described by (15) a variation of the pressure from p_t to p_o changes by (13) the rate of deformation to

$$\boldsymbol{\delta}_o = \left(\frac{p_o}{p_t}\right)^{1/m} \boldsymbol{\delta}_t \tag{18}$$

and following (14), the stress to

$$\boldsymbol{\sigma}_o = \frac{p_o}{p_t}\boldsymbol{\sigma}_t \tag{19}$$

As a matter of fact, relation (18) applies also to the velocity field.

It may be noticed, that for a statically determined system, relation (19) between stresses applies regardless the form of the material constitutive law and is not restricted to the particular case of (15). Utilizing (19) for the von Mises equivalent stresses we obtain with reference to (4)

$$\frac{p_o}{p_t} = \frac{\bar{\sigma}_o}{\bar{\sigma}_t} = \frac{f(\bar{\delta}_o, d)}{f(\bar{\delta}_t, d)} \tag{20}$$

Consequently, the forming pressure may be adjusted to a prescribed optimum value δ_o for the maximum equivalent rate of deformation $\bar{\delta}_{max}$ as by

$$p_o = \frac{f(\delta_o, d)}{f(\bar{\delta}_{max}, d)} p_t \tag{21}$$

Relation (21) for the optimum pressure does not imply any restriction with respect to the material but is applicable only to statically determined situations. It was first presented in [4] and replaces (17) which is not restricted to any particular system but implies the specific material type of (15). In the latter case, both approaches of course coincide and relation (18) commonly applies to the instantaneous kinematics.

The above discussion may be utilized under more general conditions for the adjustment of the forming pressure to an optimum rate of deformation during the course of the computation. For loading by a gas pressure p the force vector in (5) reads

$$\boldsymbol{R}(t, \boldsymbol{X}) = p(t)\boldsymbol{R}_p(\boldsymbol{X}) \tag{22}$$

where $\boldsymbol{R}_p$ denotes the force vector associated with the unit pressure acting on the actual geometry of the work-piece material. In the present context the pressure is considered unknown and may be obtained at each time instant together with the velocities by the recursive scheme

$$\boldsymbol{V}_{j+1} = \boldsymbol{V}_j + \boldsymbol{D}_j^{-1}\left[p_i \boldsymbol{R}_p - \boldsymbol{S}_j\right] \tag{23a}$$

$$p_{i+1} = \frac{f(\delta_o, d)}{f(\bar{\delta}_{max\, i+1}, d)} p_i \tag{23b}$$

The first of the operations in (23) is performed for a constant pressure p_i and corresponds to (7). The convergent result of this process, $\boldsymbol{V}_{i+1}$, is used for the adjustment of the pressure by the auxiliary second formula in (23), which is suggested by (17) and (21). If (23) is operated in conjunction with an implicit integration scheme for the deformed geometry ($\zeta > 0$ in (6)), the quantities $\boldsymbol{R}_p$ and d are taken at the end of the current time increment and change accordingly during iteration. This may extend also to the optimum rate of deformation δ_o. Satisfaction of the condition of equilibrium at the time instant considered under observance of the constraint $\bar{\delta}_{max} = \delta_o$, requires a convergent solution for the velocities and the pressure. It often proves, however, efficient to proceed to the next time increment after a single correction of the pressure [4].

The adjustment of the forming pressure was attempted so far via a trial and error approach supplementing the standard simulation procedure based on a prescribed time history of the applied load. The actual set of equations governing the problem at each time instant during the course of the deformation process, consists of the condition of equilibrium (5) under the action of the gas pressure, cf. (22), and the constraint requirement for the maximum equivalent rate of deformation; it reads

$$\boldsymbol{D}(\boldsymbol{V})\boldsymbol{V} = p\boldsymbol{R}_p \tag{24a}$$

$$\bar{\delta}_{max} = \delta_o \tag{24b}$$

The type of the equation set (24) was encountered in [5] in connection with the control of the loading of nonlinear systems towards a numerical response following the equilibrium path of quasistatic deformation. The proposed treatment may be transferred to the present situation on the basis of a linearisation of (24a) with respect to the velocities as by (7) in conjunction with any iteration matrix in place of $\boldsymbol{D}$ as for instance the gradient of the system.

The customary case of the secant matrix $\boldsymbol{D}$ in (7) is equivalent to the successive approximation

$$\boldsymbol{V}_{i+1} = p_{i+1}\left[\boldsymbol{D}_i^{-1}\boldsymbol{R}_{pi}\right] = p_{i+1}\boldsymbol{V}_{pi} \tag{25a}$$

to the solution vector, when the pressure is prescribed. In the present case the vector,

$$\boldsymbol{V}_{pi} = \boldsymbol{D}_i^{-1}\boldsymbol{R}_{pi} \tag{25b}$$

which arises from the unit pressure is computed first and leads to a distribution of the equivalent rate of deformation $\bar{\delta}_{pi}$. With reference to (25a), the constraint (24b) must be satisfied for $\bar{\delta}_{i+1} = p_{i+1}\bar{\delta}_{pi}$ and yields the associated intensity of the pressure,

$$p_{i+1} = \frac{\delta_o}{\bar{\delta}_{pmaxi}} \tag{25c}$$

which completes the determination of the velocity $\boldsymbol{V}_{i+1}$ as a result of the current iteration cycle.

Variation of material properties

For an assessment of the significance of the material properties for the superplastic forming process, we discuss in the following consequences of their variation on the basis of the power law (15) for the uniaxial flow stress. In terms of equivalent quantities this law reads

$$\bar{\sigma} = k\bar{\delta}^m \tag{26}$$

and may be subject to variations of the parameters k and m.

Let at a certain stage of the process the forming pressure induce in the work-piece material the stress distribution $\boldsymbol{\sigma}$, and the rate of deformation $\boldsymbol{\delta}$, which satisfy (26) with the parameters k_1, m_1. Quantities associated with the modified material are marked with the index 2. Under the assumption of a statically determined system at constant pressure, the variation of the material at the given deformed state leaves the stress unaffected, i.e. $\boldsymbol{\sigma}_2 = \boldsymbol{\sigma}_1$. For the rate of deformation we deduce then from (3) and the first equation (4) the proportionality,

$$\boldsymbol{\delta}_2 = \frac{\bar{\delta}_2}{\bar{\delta}_1}\boldsymbol{\delta}_1 \tag{27}$$

between the kinematic quantities $\boldsymbol{\delta}_2$ and $\boldsymbol{\delta}_1$.

We consider next a variation of the rate sensitivity of the material as from m_1 to m_2, whilst $k_2 = k_1 = k$. In this case, the proportionality factor in (27) is obtained from (26) as

$$\left(\frac{\bar{\delta}_2}{\bar{\delta}_1}\right)_{k=const.} = \bar{\delta}_1^{(m_1/m_2-1)} \tag{28}$$

It may be seen that for an increase of the rate sensitivity, i.e. $m_2 > m_1$, the exponent on the right-hand side of (28) becomes negative and since $\bar{\delta} < 1$ in superplasticity, the rate of deformation increases to $\bar{\delta}_2 > \bar{\delta}_1$. At the same time, the quotient $\bar{\delta}_2/\bar{\delta}_1$ becomes lower where $\bar{\delta}_1$ is higher and thus an increase of the rate sensitivity m tends to diminish inhomogeneities in the distribution of the rate of deformation. For a variation $m_2 < m_1$, on the other hand, the rate of deformation decreases everywhere in the statically determined system whilst inhomogeneities in the distribution of the rate of deformation are enhanced.

Alternatively, the value of the parameter k in (26) is changed from k_1 to k_2, whilst the rate sensitivity m remains unaffected. The proportionality factor between the respective rates of deformation in (27), referring to a common stress, then reads

$$\left(\frac{\bar{\delta}_2}{\bar{\delta}_1}\right)_{m=const.} = \left(\frac{k_1}{k_2}\right)^{1/m} \tag{29}$$

and is considered unique in the entire work-piece material. In this case the kinematic compatibility of $\boldsymbol{\delta}_1$ ensures also that of $\boldsymbol{\delta}_2$. Since, in addition, the rate of deformation $\boldsymbol{\delta}_2$ was related via the modified material law to the original stress field $\boldsymbol{\sigma}_1 = \boldsymbol{\sigma}_2$, the set $\boldsymbol{\sigma}_1$, $\boldsymbol{\delta}_2$ is seen to constitute here the new solution independently of the assumption of a statically determined system.

It may be concluded, that at each fixed stage of deformation and loading, a variation of the parameter k at constant m, does not affect the stress and the distribution at the rate of deformation in the work-piece material. The rate of deformation is modified by the proportionality factor of (29), and may be derived from a velocity field

$$\boldsymbol{v}_2 = \left(\frac{k_1}{k_2}\right)^{1/m} \boldsymbol{v}_1 \tag{30}$$

As a consequence of (30) a reproduction of the original deformation process in conjunction with the new value of the material parameter k, requires a modification of the time scale of the process as by

$$t_2 = \left(\frac{k_2}{k_1}\right)^{1/m} t_1 \tag{31}$$

Determination of the initial geometry

It is often requested that the superplastic process directly produces the detailed shape of the ultimate component. This raises the question of finding the associated initial geometry of the work-piece material. In order to be specific, we consider in the following the forming of shells with a prescribed thickness distribution, by the superplastic deformation of sheets. It is assumed that the component is formed onto a die, which shapes the surface of the shell, whilst the initial thickness profile in the sheet is decisive for the achievement of the prescribed distribution of thickness in the product.

Let the required surface of the shell component be defined by an equation of the form $\phi(\boldsymbol{x}) = 0$, which at the same time specifies the geometry of the surface limiting the deformation. The dependence of the thickness on the position in the shell follows a prescribed function $s(\boldsymbol{x})$. At the beginning of the process, the surface of the sheet - not necessarily flat- is given by ${}^0\phi(\boldsymbol{x}) = 0$, whilst the thickness profile ${}^0s(\boldsymbol{x})$ is to be determined. At that stage, the properties of the material are considered known if the grain size is specified.

Material particles on the surface may be conveniently marked by embedded Gaussian coordinates $\boldsymbol{u}$ with Cartesian positions $\boldsymbol{x}$ at the deformed configuration and ${}^0\boldsymbol{x}$ at the initial configuration. Since material elements conserve their volume during the course of superplastic deformation, one may verify the relation

$$s = \frac{\mathrm{d}\,{}^0A}{\mathrm{d}A}\,{}^0s = \alpha^{-1}\,{}^0s \tag{32}$$

for the thickness in the deformed state. In (32), $\mathrm{d}A, \mathrm{d}\,{}^0A$ denote the area of material elements of the surface at the deformed and the initial state respectively. The quotient

$$\alpha = \frac{\mathrm{d}A}{\mathrm{d}\,{}^0A} \tag{33}$$

may be determined on the basis of the deformation mapping $\boldsymbol{x}({}^0\boldsymbol{x})$ requiring knowledge of the coordinate functions $\boldsymbol{x}(\boldsymbol{u})$, ${}^0\boldsymbol{x}(\boldsymbol{u})$.

Here we deal with the case of a completely specified ultimate geometry of the shell where the thickness

$$s = \alpha^{-1}\,{}^{0}s = {}^{u}s \tag{34}$$

is given as a function of the position $\boldsymbol{x}$. The associated initial thickness of the component may then be calculated from (32) once the mapping ${}^{0}\boldsymbol{x}(\boldsymbol{x})$ is known. The definition of the problem might suggest to obtain the required mapping by following the deformation process back to the initial geometrical state of the surface of the shell, which is considered to be given as well. In addition to numerical instabilities, however, the above back calculation is in general not feasible because *i*) the final material state is not known in advance and *ii*) contact and friction are unidirectional phenomena.

For the above reasons an iteration procedure is proposed for the determination of the initial thickness profile ${}^{0}s\,({}^{0}\boldsymbol{x})$ of the sheet, as follows. The given surface ${}^{0}\boldsymbol{x}(\boldsymbol{u})$ is supplemented by an estimate ${}^{0}s_i(\boldsymbol{u})$ for the thickness distribution which completely defines the initial geometry. On this basis, a computation of the forming process can be performed by a finite element model in accordance with (7) and (6), once the material properties at the initial state are specified. As a result the deformed geometry of the model is obtained and determines the thickness $s_i(\boldsymbol{u})$ at each representative material position $\boldsymbol{x}_i(\boldsymbol{u})$ on the surface where the ultimately prescribed thickness is ${}^{u}s(\boldsymbol{x}_i)$. A new estimate for the initial thickness is then supplied by the recursive formula

$${}^{0}s_{i+1} = {}^{0}s_i + h_i\,[{}^{u}s(\boldsymbol{x}_i) - s_i] \tag{35}$$

and the next iteration can be started. For a suitable choice of the relaxation factor h in (35), it is helpful to assess the derivative of the residual variant of (34) with respect to the initial thickness. The expression for the above tangent is required for the construction of a Newton scheme from (35) and coincides with the secant of (34) if the final position of surface particles does not depend on the initial thickness profile, cf. (32).

Operation of (35) with the secant is effected by the choice $h = \alpha$ and is equivalent to the successive approximation,

$${}^{0}s_{i+1} = \alpha_i\,{}^{u}s(\boldsymbol{x}_i) = \frac{{}^{0}s_i}{s_i}\,{}^{u}s(\boldsymbol{x}_i) \tag{36}$$

which follows immediately from (34) whilst the second expression may be verified by (32).

Superplastic forming of a hemispherical shell

In order to illustrate the theory we deal here with an example of a rather simple geometrical shape. For geometrically more complex applications the reader may be referred to [4,6]. Our investigations start with the determination of the initial thickness profile of a flat circular sheet formed onto a hemispherical die to a shell with a prescribed thickness distribution. The sheet is from Ti-6Al-4V alloy with an initial grain size of 15μm; the process temperature is 1200K. The uniaxial material characteristic is depicted in Fig.1 and provides together with the grain growth kinetics of the material the basis for a description as by (1) and (2), cf. [3]. The computer simulation is controlled by a prescribed value $\delta_o = 3 \times 10^{-4}\mathrm{s}^{-1}$ for the maximum equivalent rate of deformation in the work-piece material; the time history of the forming pressure is accordingly adjusted. In the process, the contact between sheet material and die may be considered frictionless.

The requested initial profile of the sheet is obtained by an application of the iteration procedure outlined before. Thereby, the process of superplastic forming is simulated for an assumed distribution of the initial thickness. The final thickness resulting from the computation is compared with the prescribed one, the difference defines a new estimate of the thickness profile at the outset and the process is repeated until convergence is achieved.

Figure 2 shows a sequence of deformed states of the axisymmetric component at approximately equidistant times. This result refers to the initial thickness profile in Fig.3 obtained as the convergent solution of the iteration algorithm; the procedure was started with the assumption of a constant thickness of the undeformed sheet. After completion of the superplastic forming with the determined initial thickness, the computation reproduces the prescribed thickness distribution in the component shown in Fig.3 as a function of the distance from the pole. The transfer of the discrete initial geometry obtained by the finite element model to the sheet actually to be formed may require some caution, in particular at sharp corners.

In a second stage of the study, a slightly smoothed variant of the determined initial thickness profile was used in conjunction with the power material law of (15) aiming at a verification of the respective theoretical results. The material parameters were chosen as $k = 36.3, m = 0.38$, and a first computation examines the adjustment by eq. (17) of the forming pressure to the prescribed rate of deformation $\delta_o = 2.85 \times 10^{-4}s^{-1}$. Whilst the calculated time history of the forming pressure is depicted in Fig.5, Fig.4 confirms that single correction at each time instant of the trial pressure intensity by (17) ultimately leads to $\bar{\delta}_{max} = \delta_o$, where allowed by the numerical behaviour of the solution. Early discrepancies are due to the initiation of the deformation, whilst the contact with the die disturbes at the end of the process.

The subsequent investigations deal with the effect of variations of the material parameters. In the material law (15), first the factor is changed to $0.886k$ whilst the exponent m is kept constant. The theoretical framework requires a modification of δ_o as by (29), and the numerical analysis of the process for the modified conditions practically confirms the prediction that the thickness profile of the product and the intensity of forming pressure are not affected by a variation of k. As demonstrated in Fig.5, the time scale of the process changes by 0.5 in accordance with (31). The consequence of a variation of the exponent to $2m$ at constant factor k on the resulting deformation is shown in Fig.6. In conformity with the discussion of expression (28), this increase of the rate sensitivity tends to reduce differences in the thickness of the final component.

References

[1] Padmanabhan, K.A. and Davies, G.J.: Superplasticity. Berlin: Springer 1980.

[2] Ghosh, A.K. and Hamilton, C.H.: Mechanical behaviour and hardening characteristics of a superplastic Ti-6Al-4V alloy. Metallurgical Trans. A. 10A (1979) 699-706.

[3] Argyris, J. and Doltsinis, I.St.: A primer on superplasticity in natural formulation. Comput. Meths. Appl. Mech. Engrg. 46 (1984) 83-131.

[4] Doltsinis, I.St., Luginsland, J. and Nölting, S.: Some developments in the numerical simulation of metal forming processes. Engrg. Comput. 4 (1987) 266-280.

[5] Doltsinis, I.St.: Nonlinear concepts in the analysis of solids and structures. In: Advances in Computational Nonlinear Mechanics. Wien, New-York: Springer 1989 1-80.

[6] Doltsinis, I.St.: Aspects of modelling and computation in the analysis of metal forming, Engrg. Comput. 7 (1990) 2-20.

Figures

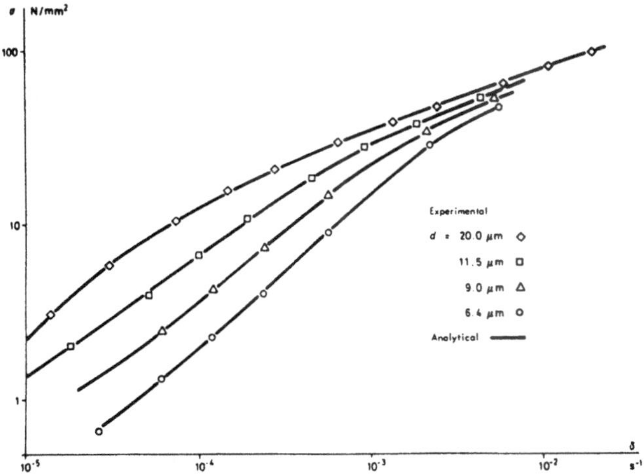

Fig.1 Dependence of flow stress σ on rate of deformation δ (Ti-6Al-4V, T=1200K)

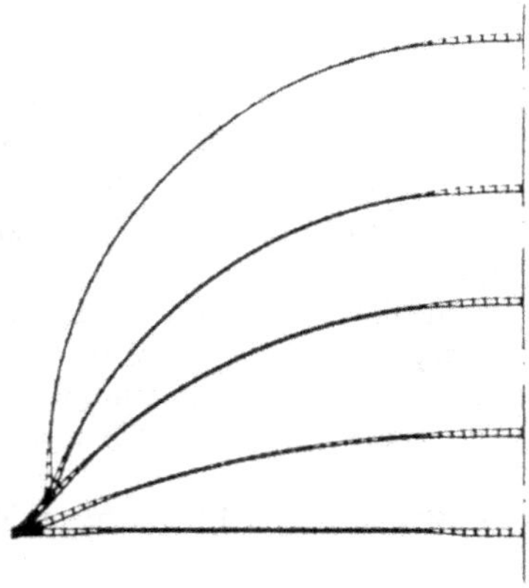

Fig.2 Superplastic forming of a shell (sequence of deformed states)

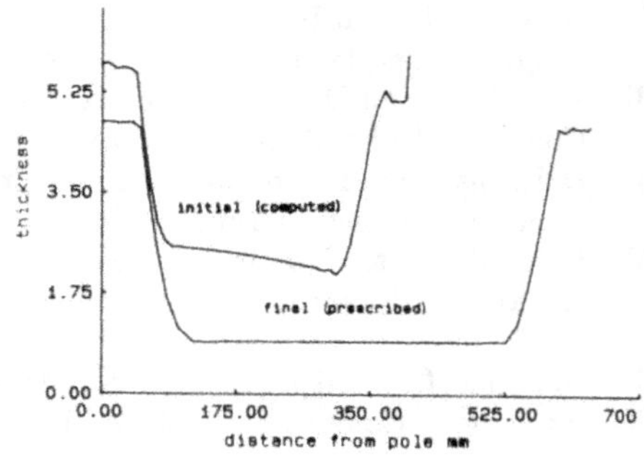

Fig.3 Determined distribution of initial thickness for prescribed distribution of final thickness

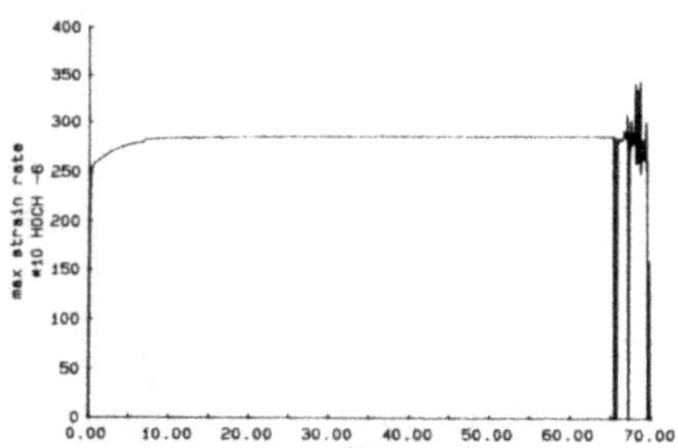

Fig.4 Control of $\bar{\delta}_{max}$ during forming

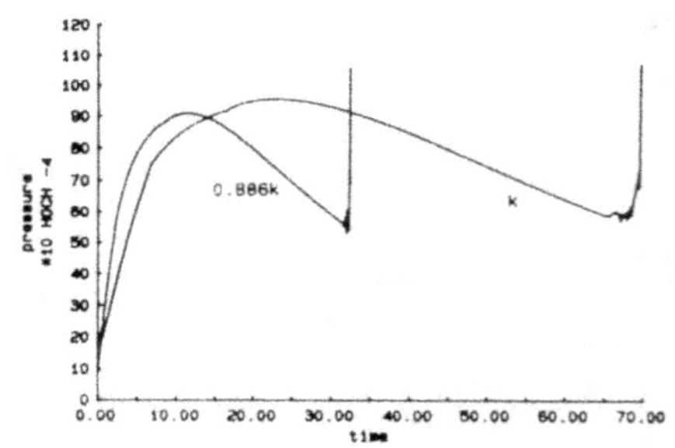

Fig.5 Effect of k on time history of forming pressure

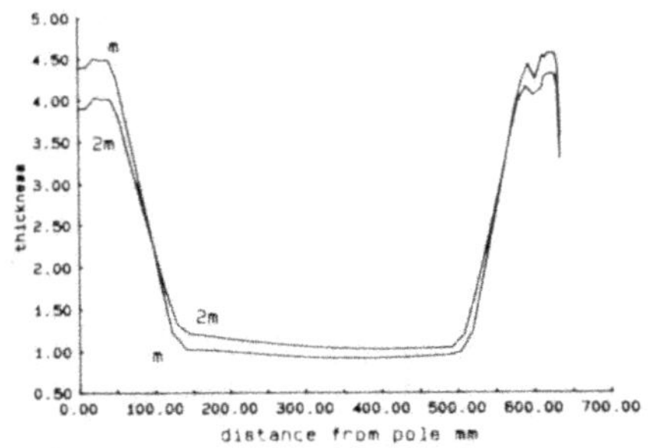

Fig.6 Effect of m on final thickness profile

A Quasi-Dynamic Approach to the Analysis of Sheet Metal Forming

L. BERNSPÅNG, K. MATTIASSON, A. SAMUELSSON

Department of Structural Mechanics,
Chalmers University of Technology,
Göteborg, Sweden

Summary

Sheet metal forming by punching is a quasistatic process with large elastoplastic deformations. Complicating factors are moving contact boundaries with friction, instabilities in compressed regions and springback after the punching process. Following an idea presented in 1989 by Honecker and Mattiasson [7] this paper presents a quasi–dynamic approach to this problem. The calculations with this approach are shown to be fast and more stable than with a quasi–static approach.

Introduction

In automotive industry as in many other industries sheet metal parts are formed by a cold punching process. As it shall produce the right form of the sheet with no thin parts and no wrinkles the design of the tool surfaces is a difficult task . The geometry of these surfaces are in most factories obtained in a trial and error way by highly experienced workers and engineers. The design is then very time–consuming.

In a typical punching process a thin plane metal sheet is placed on dies and kept there with so–called blankholders, Fig. 1. The blankholders are pressed by forces which can be strong enough to prevent all movement or less strong so that sheet material is allowed to flow in under friction during the punching process. The first type of punching is called stretch forming as the sheet gets only positive membrane strain components. The other method is called deep–drawing during which process some parts of the sheet get compressive strains.

The punching process is a relatively slow process with a speed of 0.1–1 m/s. It should give the desired form and thickness within strict tolerances.

D. Besdo, E. Stein (Eds.)
Finite Inelastic Deformations -
Theory and Applications
IUTAM Symposium Hannover/Germany 1991

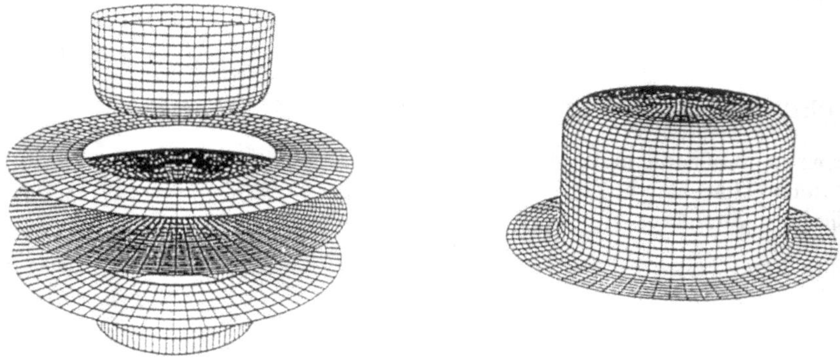

Fig. 1 Sheet and tools in sheet metal forming process

During the 1980th research has been taken up in USA, Japan and Europe with the aim of finding out if large strain continuum mechanics could be used for predicting the punching effect on thin metal sheets. One of the first important investigations was reported by Wang and Budiansky [1]. They used a quasistatic elastoplastic large strain theory proposed by Hutchinson with a cross anisotropy assumption according to Hill. They solved the associated membrane shell problem by a finite element total Lagrange approach. By advancing with very small punch displacement increments they could follow the deformation of the sheet including the development of the contact surface between the punch and the sheet with the assumption of Coulomb friction.

This approach has been improved and made efficient by many research groups. Among these are the groups led by Wang [2] and Tang [3]. Much of this work has been reported in the proceedings of the NUMIFORM conferences in Swansea, Wales, 1982, Göteborg, Sweden, 1986 and Fort Collins, Colorado, 1989.

In our department at Chalmers University of Technology in Göteborg we have during the 1980th worked on design and improvement of a large deformation membrane code for stretching and deep–drawing. The work has been performed in close cooperation with the Swedish Institute for Metals Research in Stockholm where they have made experiments and measurements on punching of metal sheets of many materials and different forms [4]. Our code predicts for cylindrical cups with sufficient accuracy relevant information about the effect of both punching at stretching [5] and at deep–drawing [6]. At the third NUMIFORM in 1989 Tang reported that his program based on a similar strategy and theory as ours now was in full use at the Ford Company.

However, at the same conference Honecker and Mattiasson [7] presented an interesting comparison between sheet forming calculations made with quasistatic and dynamic approaches. They showed that a dynamic calculation could be performed in a way that makes it more effective than a quasistatic. During 1990 a number of research projects based on this idea have been started. One of these is a EC–project with car factories, university departments and a software company involved. Another project has been developed by ABAQUS software company. Our department takes part in a ECSC project together with some European institutes for steel research and some manufacturing companies.

Basic equations

The Cauchy law of motion of a 3–dimensional deformed continuous medium is

$$\sigma^{ij}_{;j} + \varrho F^{j} = \varrho a^{j} \qquad (1)$$

where σ^{ij} are the components of the Cauchy stress tensor in convected coordinates. The density ϱ and the body force per unit mass F refer to the deformed medium.

Sheet punching is a quasistatic operation so the inertia term ϱa^j can be disregarded. For numerical reasons, though, it is often kept, as discussed below.

We prefer to have (1) expressed in quantities that are referred to the initial state, that is a flat, thin plate. Thus, we replace the Cauchy stress by the Kirchhoff stress

$$\tau^{ij} = (\varrho_0/\varrho)\sigma^{ij} \qquad (2)$$

where ϱ_0 is the density in the initial state. This stress is measured per unit area of the undeformed body but refer to the convected coordinates. Eq. (1) becomes after transformations

$$(\tau^{ij}x_{m,j})_{,j} + U_m = 0 \qquad (3)$$

referred to a Cartesian system in the undeformed sheet. Here U_m is a Cartesian component of the body load per unit original volume.

The plane stress case of (3) can in matrix form be written

$$(\bar{\nabla}_a + \bar{\bar{\nabla}}_b)^T N + U = 0 \qquad (4)$$

where $N = [N^{x}, N^{y}, N^{xy}]^T$ and $N^{\alpha\beta} = h_0\tau^{\alpha\beta}$. Here $\bar{\nabla}_a$ is the linear operator and

$$\bar{\bar{\bar{\nabla}}}_b = \begin{bmatrix} \frac{\partial}{\partial x}\frac{\partial u}{dx} & \frac{\partial}{\partial x}\frac{\partial v}{\partial x} & \frac{\partial}{\partial x}\frac{\partial w}{\partial x} \\ \frac{\partial}{\partial y}\frac{\partial u}{\partial y} & \frac{\partial}{\partial y}\frac{\partial v}{\partial y} & \frac{\partial}{\partial y}\frac{\partial w}{\partial y} \\ \frac{\partial}{\partial x}\frac{\partial u}{\partial y} + \frac{\partial}{\partial y}\frac{\partial u}{dx} & \frac{\partial}{\partial x}\frac{\partial v}{\partial y} + \frac{\partial}{\partial y}\frac{\partial v}{\partial x} & \frac{\partial}{\partial x}\frac{\partial w}{\partial y} + \frac{\partial}{\partial y}\frac{\partial w}{\partial x} \end{bmatrix} \tag{5}$$

where u, v, w are the displacement components in the Cartesian system. The rate of Lagrangian strain in matrix form looks very much the same

$$\dot{\eta} = (\bar{\nabla}_a + \bar{\nabla}_b)\dot{u} \tag{6}$$

where

$$\bar{\nabla}_b = \begin{bmatrix} \frac{\partial u}{\partial x}\frac{\partial}{\partial x} & \frac{\partial v}{\partial x}\frac{\partial}{\partial y} \\ & \text{etc} \end{bmatrix} \tag{7}$$

The constitutive equation is chosen according to the J_2-flow theory with the "old" Hill anisotropy formula applied to the thickness direction. With a parameter R defined as the ratio of the plastic width strain rate to the plastic thickness strain rate in a uniaxial test the yield function becomes

$$F = [\sigma_{11}^2 + \sigma_{22}^2 - \frac{2R}{R+1}\sigma_{11}\sigma_{22} + 2\frac{2R+1}{R+1}\sigma_{12}^2]^{1/2} \tag{8}$$

For sheet forming the elastic strains are in general small compared to the plastic ones so no big error is obtained if the material is assumed to be elastically incompressible and if the elastic width and thickness strains are assumed to be related as the plastic strain rates.

The constitutive relation is expressed in the corotational Jaumann rate of Kirchhoff stress

$$\overset{*}{\tau}^{\alpha\beta} = \overset{*}{C}^{\alpha\beta\gamma\varrho}\dot{\eta}_{\gamma\varrho} = \dot{\tau}^{\alpha\beta} + G^{\alpha\xi}\tau^{\beta\zeta}\dot{\eta}_{\xi\zeta} + G^{\beta\xi}\tau^{\alpha\zeta}\dot{\eta}_{\xi\zeta} \tag{9}$$

The yield function can be written

$$F = \left\{ \frac{1+2R}{1+R} [G_{\alpha\gamma} G_{\beta\varrho} \tau^{\alpha\beta} \tau^{\gamma\varrho} - \frac{R}{1+R} G_{\alpha\beta} G_{\gamma\varrho} t^{\alpha\beta} \tau^{\gamma\varrho}] \right\}^{1/2} \tag{10}$$

and the constitutive equation in tensor form

$$\dot{N}^{\alpha\beta} = h_0 C^{\alpha\beta\gamma\varrho} \dot{\eta}_{\gamma\varrho} \tag{11}$$

Here the Kirchhoff stress is identified with the Cauchy stress which is a good approximation for the metals here considered. An explicit formula for $C^{\alpha\beta\gamma\varrho}$ is given by Wang and Budiansky [1]. We do not use it directly in this form in our quasistatic code since the satisfaction of the yield condition is not quite clear. Instead we use an iterative procedure including an elastic predictor and a corrector to the yield surface in each stage, see [8].

Preparing for a finite element discretization we time differentiate the equilibrium equation (4), multiply with a row matrix ϕ^T of weighting functions and integrate over the domain, obtaining

$$\int_A \phi^T (\bar{\nabla}_a + \bar{\bar{\nabla}}_b^T) \dot{N} dA + \int_A \phi^T \dot{\bar{\bar{\nabla}}}_b^T N dA = 0 \tag{12}$$

This can by Green's formula be transformed to

$$\int_A [(\bar{\nabla}_a + \bar{\nabla}_b)\phi]^T \dot{N} dA + \int_A (\dot{\bar{\nabla}}_b \phi)^T N dA \tag{13}$$
$$+ \textit{ boundary terms}$$

Observe that the kinetic nonlinear operator becomes replaced by the kinematic nonlinear operator.

With a Galerkin finite element approximation

$$\dot{u} = \phi \dot{\bar{u}} \tag{14}$$

where $\dot{\bar{u}}$ is a column matrix of node values, (13) together with the kinematical and constitutive equations yields a set of nonlinear equations

$$(K_1 + K_2 + K_3)\dot{\bar{u}} = \textit{boundary terms} \tag{15}$$

where K_1 is an integral with constant geometrical operators, K_2 has both constant and displacement dependent geometrical operators and K_3 has constant geometrical operators but a kernel of the current stresses.

The punching process is steered by a vertically moving, rigid punch with a smooth surface. At the interface between sheet and punch three situations are possible: no contact, contact with slip, or contact without slip (stick).

The friction may be assumed to be of Coulomb type with the normal force as an additional variable. The additional equations expresses that the slip is tangential to the punch surface.

Numerical solution

In a quasistatic analysis small steps in the vertical displacement of the punch are taken. In each step the finite element version of the global governing equations are solved with a Newton–Raphson iteration combined with local projection on the yield surface thus satisfying equilibrium conditions as well as contact, friction and plasticity conditions. The size of the punch displacement increments is chosen by experience. Not only accuracy considerations must be taken into account but also instabilities and failure of convergence occur frequently. There are pure numerical instabilities due to a large number of contact changes and physical instabilities due to wrinkling in pressure zones. However, for sheet stretching and relatively smooth geometries the quasistatic approach may work well. The instabilities may be partly overcome by adding bending to the membrane action.

We will, however, show here by experiments that a quasidynamic procedure may be more efficient. In the dynamic approach the explicit central difference integration rule is applied to a time propagation problem. The governing equation is applied at the start of a time increment Δt and only the inertia term is kept on the left hand side

$$M\ddot{u}^{(i)} = load^{(i)} - inner\ forces^{(i)} \tag{16}$$

Simple interpolation gives

$$\dot{u}^{(i+\frac{1}{2})} = \dot{u}^{(i-\frac{1}{2})} + \Delta t\ddot{u}^{(i)} \tag{17}$$

$$u^{(i+1)} = u^{(i)} + \Delta t\dot{u}^{(i+\frac{1}{2})}$$

It is well known that the mass matrix can without important loss of accuracy be diagonalized so (1) becomes a set of separated equations which can be solved very efficiently with vector or parallel computers.

What is the cost for this? Numerical stability requires small time–steps. Approximately for linear equations, the time–step should be less than $h^{\min}/c$ where $h^{\min}$ is the smallest characteristic length of an element in the finite element mesh and c is the dilatational velocity of sound for the material. For non–linear equations the time–step should be somewhat reduced. Thus, if the speed of the punch is 1 m/s, the punch depth is 0.2 m, $c = 500$ m/s, and $h^{\min} = 0.01$m, the required number of time–steps is 10^5. It is, however, possible to reduce that number substantially. The real punching process is so slow that inertia effects can be disregarded and this is so even for higher punch velocities. If the speed is increased with a factor 20 a high oscillation in the punch force is observed but the mean value is almost correct.

Due to the small time–steps taken the changes in each step of the geometry is very small so simplifications can be made in the governing equations. Particularly, the friction – contact conditions can be treated in a simple way, see [9].

It is important for the calculation that the movement of the punch is prescribed with a smooth start and stop to avoid exaggerated oscillations.

Numerical experiments

We have performed numerical experiments with two finite element codes, one for static conditions specially developed at our department for sheet metal forming and one general code for dynamic structural analysis (DYNA3D) developed at Lawrence–Livermore. In the static code three–node triangular membrane elements and in the dynamic code four–node Reissner–Mindlin elements given by Belytschko and Tsay are used.

Forming of cylindrical cups of many different materials have been simulated by both methods and compared to experimental results. Forming of rectangular boxes has been simulated only by the dynamic approach also for different materials and compared to experimental results. Both stretch forming and deep drawing have been simulated. The friction coefficients at the punch and at the flanges have been adjusted so calculations fit to experimental data at the flat part under the punch.

Here results from simulations of deep–drawing of one cup from a carbon steel with an R-value of 1.65 and one box, fig 4 from a hot–dip galvanized steel with an R–value of 1.80.

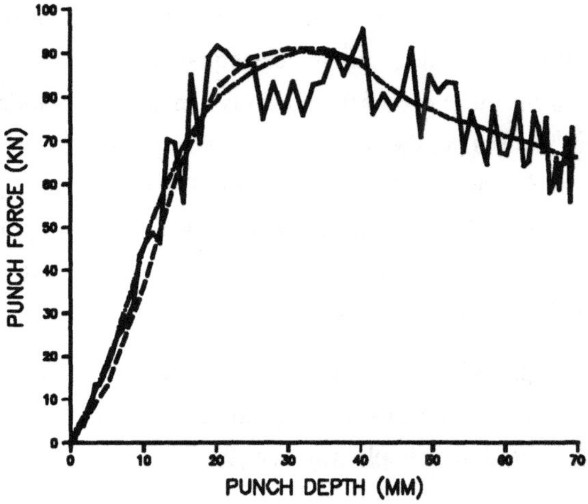

Fig. 2 Punch force. Solid line – explicit calculation
Dashed line – implicit calculation
Dash dotted line – experiment

Fig. 2 shows strains and Fig. 3 punch forces in the cup obtained by simulation and from experiments. At the dynamic simulation the punching time is 4.0 ms. The dynamic effects cause perturbations in the punch force but its mean value follows the experimental values very well. We also observe that the dynamic simulation can follow the process much longer than the static one.

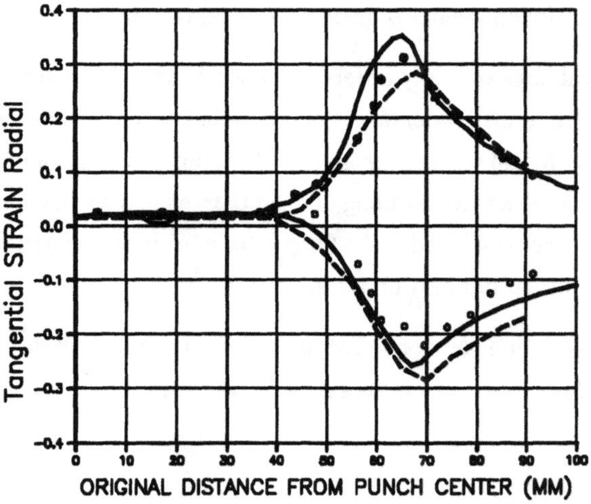

Fig. 3 Tangential and radial strains. Solid line – explicit calculation
Dashed line – implicit calculation

The box forming simulation is illustrated in Fig. 4 with the deformation of the original flat sheet at a punch depth of 59.7 mm.

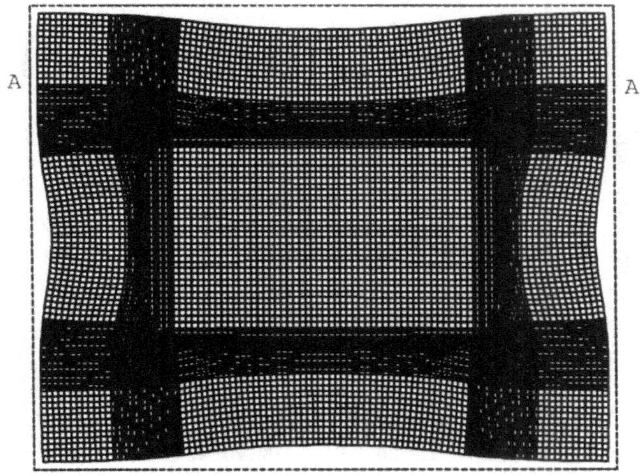

Fig. 4 Mesh in calculation of box (One quarter is used due to symmetry)

Corresponding calculated and measured principal strains in the plane of the sheet (EPS1 and EPS2) along the line AA, Fig. 4, are displayed in Fig. 5 together with the thickness strain (EPS3). The discrepancies demonstrate the difficulties met both in the experiments and the computer simulations in more complicated forming.

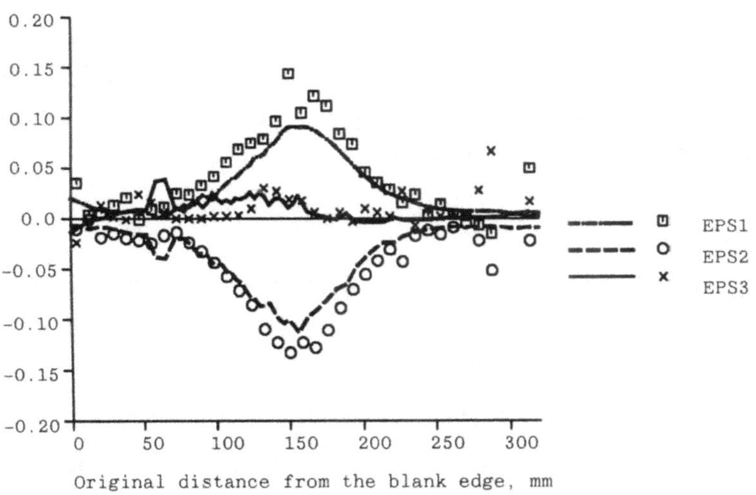

Fig. 5 Strains along line AA

The calculated punch force as a function of the punch depth is in Fig. 6 compared with the measured force. Again a good correlation is obtained up to a stage close to rupture.

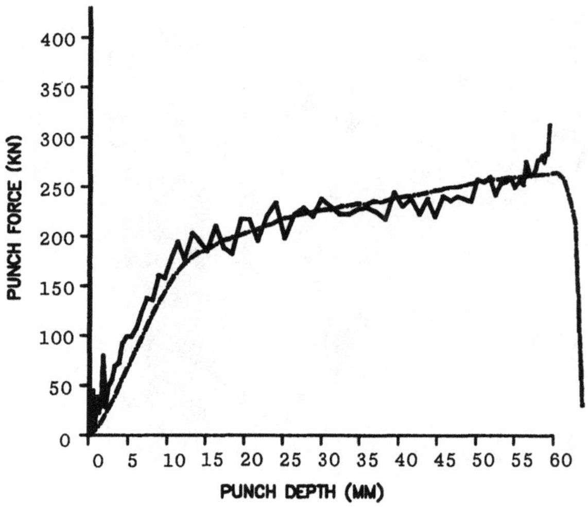

Fig. 6 Punch force. Solid line – explicit calculation
Dash dotted line – experiment

Final remarks

The quasidynamic approach seems to be useful for the type of sheet metal forming studied. The computer time is considerably reduced compared to simulations under static conditions. The modelling of friction, anisotropy and rupture has to be looked into more deeply. The unloading phase also needs much consideration.

References

1. Wang. N.M.; Budiansky, B.: Analysis of sheet metal stamping by a Finite Element Method. J. of Appl. Mechanics 45 (1978) 73–82.

2. Wang, M.–N.; Tang, S.C.: Analysis of bending effects in sheet forming operations. NUMIFORM 86, ed. Mattiasson et al, Balkema (1986) 71–76.

3. Tang, S.C.; Chappuis, L.B.: Thin shell element simulation of sheet metal forming processes. Advanced Technology of Plasticity, Proc. Third International Conference on Technology of Plasticity, Kyoto, 4, 1757–1762

4. Schedin, E.; Melander, A.: The evaluation of large strains from industrial sheet metal stamping with a square grid. J. Appl. Metalworking 4 (1986) 143–156

5. Saran, M.; Schedin, E.; Samuelsson, A.; Melander, A.; Gustafsson, C.: Computational and experimental modelling of stretch forming operations. J. mater. shaping techn. 8 (1990) 31–38.

6. Saran, M.; Schedin; E.; Samuelsson; A.; Melander, A.; Gustafsson, C.: Numerical and experimental investigations of deep drawing of metal sheets. J. eng. indust. ASME (1990) 272–277.

7. Honecker, A.; Mattiasson, K.: Finite element procedures for 3D sheet forming simulation. Proc. NUMIFORM 89, Balkema 1989.

8. Saran, M.; Samuelsson, A.: Elastic–viscoplastic implicit formulation for finite element simulation of complex sheet forming processes. Int. j. numer. methods eng. 30 (1990) 1675–1697.

9. Nagtegaal, J.C.; Taylor, L.M.: Comparison of implicit and explicit Finite Element Methods for analysis of sheet forming problems. VDI Berichte nr 894 (1991) 705–725.

Shell Finite Element for Deep Drawing Problems: Computational Aspects and Results

J.M. ROELANDT and J.L. BATOZ

Laboratoire de Génie Mécanique pour les Matériaux et les Structures
UTC - BP 649 - 60206 Compiègne Cédex - France.

Abstract

We present an implicit finite element approach for the simulation of the deep-drawing of arbitrary shaped workpieces based on an updated Lagrangian formulation. Two triangular shell elements have been created (DKT18 and DKT12) taking into account bending effects, large displacements, large rotations and large strains, anisotropic elasto-plastic behaviour. Penalization methods are used to simulate contact conditions with Coulomb friction. Various numerical results are presented and compared with experimental ones.

Introduction

In this paper we present an implicit finite element approach for sheet metal forming associated to a model with the following characteristics :

- updated Lagrangian formulation
- Newton-Raphson iterations with load, displacement or arc-length control (and approximate tangent stiffness matrix).
- two triangular shell elements constructed by superposition of CST membrane element and DKT bending elements (DKT18 and DKT12),
- the computation of strain increments is based on the assumptions of Rolph and Bathe [1] and Braudel [2],
- the anisotropic elastic-plastic laws are computed using a Simo-Taylor scheme [3],
- contact conditions with Coulomb friction are taken into account through penalization techniques.

Numerical results are presented with comparison with experimental results concerning the bulging of a clamped sheet with ellipsoïdal boundaries and the deep drawing of a sheet subjected to an hemispherical punch with draw-in under the blankholder.

This study is an extension of our previous works [4], [5], [6].

D. Besdo, E. Stein (Eds.)
Finite Inelastic Deformations -
Theory and Applications
IUTAM Symposium Hannover/Germany 1991

2. Theoretical aspects

2.1. Principle of virtual work

The shell structure is discretized with triangular elements. A local orthogonal coordinate system(x_t, y_t, z_t) is associated to each element at each step (fig. 1) defined at time t by the rotation matrix $[Q]_t$:

$$[Q]_t = [\, \vec{e}_{xt}, \vec{e}_{yt}, \vec{e}_{zt} \,]$$

where $\vec{e}_{xt}$, $\vec{e}_{yt}$ are unit vectors in the element plane
$\vec{e}_{zt}$ is a unit vector normal to the element plane

fig.1

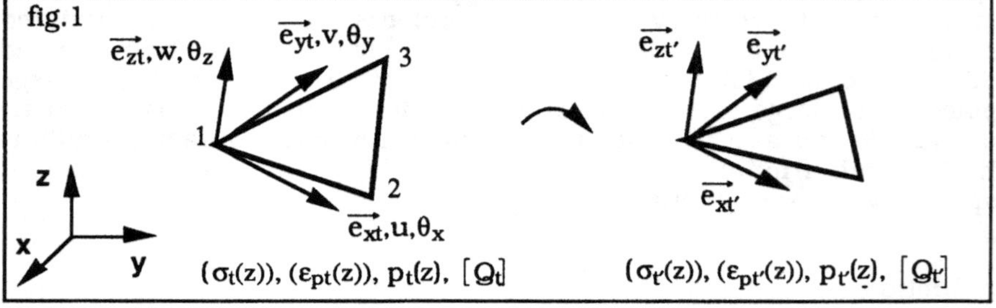

The known configuration of each element at time t is defined by $[Q]_t$, $(\sigma_t(z))$, $(\varepsilon_{pt}(z))$, $p_t(z)$. $(\sigma_t(z))$, $(\varepsilon_{pt}(z))$, $p_t(z)$ are respectively the distributions of stresses, plastic strains, cumulated plastic strain in the element through the thickness.

We use a Newton-Raphson procedure to calculate the equilibrium of the configuration at time t' ($t' = t + \Delta t$) defined by $[Q]_{t'}$, $(\sigma_{t'}(z))$, $(\varepsilon_{pt'}(z))$, $p_{t'}(z)$. Neglecting transverse shear strains the principe of virtual work on the discretized structure is :

$$\sum_{elts} \int_{V^e_{t'}} \langle D^* \rangle \, (\sigma_{t'}(z))\, dV - \int_{S^e_{t'}} \langle u^* \rangle \, (F_{t'})\, dS = 0$$

$V^e_{t'}$, $S^e_{t'}$ volume and mid plane surface of an element at time t'
$(F_{t'})$ surface tractions at time t'

$\langle D^* \rangle = \langle D_m^* \rangle + z \langle \chi^* \rangle$ (virtual strains)

$\langle D_m^* \rangle = \langle u^*_{,x_{t'}}, v^*_{,y_{t'}}, u^*_{,y_{t'}} + v^*_{,x_{t'}} \rangle$ (virtual membrane strains)

$\langle \chi^* \rangle = \langle \beta^*_{x,x_{t'}}, \beta^*_{y,y_{t'}}, \beta^*_{x,y_{t'}} + \beta^*_{y,x_{t'}} \rangle$ (virtual curvatures)

u^*, v^* : virtual displacement field components in the element plane

$\beta^*_x = \theta^*_y$, $\beta^*_y = -\theta^*_x$: virtual rotation components

$-h_{t'}/2 \le z \le h_{t'}/2$ $\quad h_{t'}$: thickness at time t'

2.2. Mid-plane computation

For the ith Newton's iteration the following computations are performed using the CST relations :

- increment of mid-plane displacement field on the step (between t and $t + \Delta t$) :

$$(\Delta u) = \sum_{j=1}^{i} (\Delta u_j) \qquad (\Delta u_i : \text{iteration increment of displacement})$$

- gradient deformation matrix

$$[\Delta F] = \left[\frac{\partial (x_t + \Delta u)}{\partial x_t} \right]$$

- decomposition of incremental Cauchy-Green deformations

$$[\Delta C] = [\Delta F]^T [\Delta F] = [M] [\Delta\lambda^2] [M]^T$$

$[\Delta\lambda]$: diagonal matrix of principal stretch increments
$[M]$: eigenvectors

$$[\Delta U] = [M] [\Delta\lambda] [M]^T \qquad \text{(pure strech matrix)}$$

$$[\Delta R] = [\Delta F] [\Delta U]^{-1} \qquad \text{(pure rotation matrix)}$$

$$[\Delta \varepsilon_l] = [M] [l_n \, \Delta\lambda] [M]^T \qquad \text{(increments of logarithmic strains)}$$

The local coordinate system and the thickness are updated as :

$$[Q]_{t'} = [Q]_t [\Delta R] \quad , \quad h_{t'} = h_t \,/\, \Delta\lambda_1 \, \Delta\lambda_2 \qquad \text{(incompressibility)}$$

2.3. Bending effects computations

Along the thickness the strains are simply defined by the relation :

$$(\Delta\varepsilon (z)) = (\Delta\varepsilon_l) + z (\Delta\chi) \qquad \text{where} \qquad (\Delta\chi) = [B_f] (\Delta u)_e$$

$[B_f]$ is the matrix of derivation of the bending element.

Two triangular shell elements are built by superposition of CST element (6 d.o.f.) and DKT9 (9 d.o.f.) [7] or DKT6 (6 d.o.f.) [8] (fig. 2).

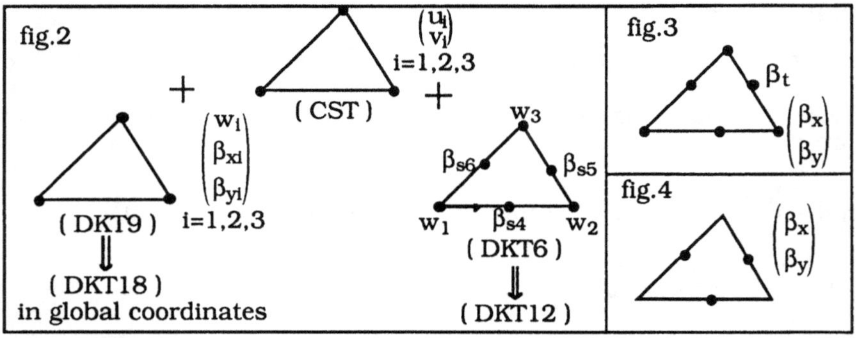

DKT9 and DKT6 are built assuming the following assumptions :

- β_x and β_y are compatible incomplete quadratic functions for DKT9 (fig. 3) and incompatible linear functions for DKT6 (fig. 4).

- The Kirchhoff-Love assumptions are satisfied in the discrete form :

$$\int \left(w_{,s} + \beta_s\right) ds = 0 \qquad \text{on each side of the element}$$

DKT12 and Morley's element [9] give the same linear stiffnesss matrix.

2.4. Stresses and element internal force vector

The anisotropic elasto-plastic behaviour is represented through an Hill yeld function and a normal flow rule :

$$2\phi\left((\sigma), p\right) = \langle\sigma\rangle \left[P(p)\right](\sigma) - 1 \leq 0 \qquad \text{(yield function)} \qquad (1)$$

$$\left(\dot{\varepsilon}_p\right) = \dot{\lambda}\left(\frac{\partial\phi}{\partial\sigma}\right) \qquad \text{(normal flow rule)} \qquad (2)$$

$$\dot{p} = \sqrt{2/3 \langle\dot{\varepsilon}^P\rangle\left[P(p)\right]^{-1}\left(\dot{\varepsilon}^P\right)(F+G+H)} \qquad (3)$$

$$\text{where : } \left[P(p)\right] = \begin{bmatrix} G+H & -H & 0 \\ -H & F+H & 0 \\ 0 & 0 & 2N \end{bmatrix}$$

F, G, H, N : Hill's coefficients (functions of p)

$$(\dot{\sigma}) = \left[C_e\right]\left((\dot{\varepsilon}) - (\dot{\varepsilon}^P)\right), \left[C_e\right] : \text{elastic matrix} \qquad (4)$$

At each integration point in the thickness (at z) the stresses ($\sigma_{t'}$) are computed using a return mapping algorithm proposed by Simo and Taylor [3] the computational steps are :

- total strain increment : $(\Delta\varepsilon) = (\Delta\varepsilon_l) + z\,(\Delta\chi)$
- elastic estimation : $(\sigma_E) = (\sigma_t) + [C_e]\,(\Delta\varepsilon)$
- using (1), the relations (2), (3) become :

$$(\Delta\varepsilon_p) = \lambda\,[P]\,(\sigma_{t'}),\ \Delta p = \lambda\sqrt{\frac{2}{3}<\sigma_{t'}>[P]\,(\sigma_{t'})\,(F+G+H)} \qquad (5)$$

- stress at time t' (using (2), (5)) :

$$(\sigma_{t'}) = [A]\,(\sigma_E) \qquad [A] = \left([I] + \lambda\,[C_e]\,[P\,(p_{t'})]\,\right)^{-1} \qquad (6)$$

- relation (1) becomes (using (6)) :

$$2\phi(\lambda) = a(\lambda) - 1 = 0 \qquad \text{with :} \qquad a(\lambda) = <\sigma_E>[A]^T[\,P\,(p_{t'})\,]\,[A]\,(\sigma_E) \qquad (7)$$

where : $p_{t'} = p_t + \lambda\sqrt{\frac{2}{3}a(\lambda)\,(F+G+H)}$

- λ is obtained by Newton iteration in order to satisfy (7),
- $(\sigma_{t'})$, $(\varepsilon_{pt'})$, $p_{t'}$ are updated using (6), (5).

The element internal force vector in the global coordinate system is :

$$(f_{int}) = [T]_{t'}\int_{V_{t'}} ([B_m] + z\,[B_f]\,)^T\,(\sigma_{t'})\,dV$$

where $[B_m]$(from CST), $[B_f]$ are evaluated in the updated configuration.

$$[T'] = \begin{bmatrix} [Q]_{t'} & & & \\ & [Q]_{t'} & & \\ & & \ddots & \\ & & & [Q]_{t'} \end{bmatrix} \qquad \text{for DKT18 element}$$

$$[T'] = \begin{bmatrix} [Q]_{t'} & & & & & \\ & 1 & & & & \\ & & [Q]_{t'} & & & \\ & & & 1 & & \\ & & & & [Q]_{t'} & \\ & & & & & 1 \end{bmatrix} \qquad \text{for DKT12 element}$$

The stiffness matrix $[K_T]$ is defined by finite differences from (f_{int}). Finally the iterative displacement increment (Δu_i) is the result of :

$$[K_T]\,(\Delta u_i) = (R) \qquad \text{where :} \qquad (R) = (F_{t'}) - \sum_e (f_{int})$$

Penalizations procedures are for the treatment of contact conditions. Tools are defined analytically or by finite isoparametric elements.

3. Numerical results

3.1. Bulging of an ellipsoïdal sheet

The sheet (fig. 5) is made of an anisotropic steel. The constitutive equation is defined by :

$E = 208000$ MPa (Young modulus)
$\nu = 0.3$ (Poisson ratio)
$\sigma = c\,(p + p_0)^{\alpha}$ (hardening law in x - axis)
where : $\sigma_0 = 150$ MPa, $c = 544$ MPa, $\alpha = 0.2314$, $p_0 = (\sigma_0/c)^{1/\alpha}$

The Landkford's coefficients are :

$R_0 = 1.88$, $R_{45} = 2.18$, $R_{90} = 1.37$

$$\left(R_0 = \frac{H}{G},\ R_{90} = \frac{H}{F},\ R_{45} = \frac{N}{F+G} - \frac{1}{2}\right)$$

A quarter of the blank is discretized using 45 nodes and 66 elements.

In figure 5, we present the distribution of the thickness strain along the short axis for different levels of pressure.

The same results are obtained using DKT18 and DKT12 elements. They are compared with experimental results.

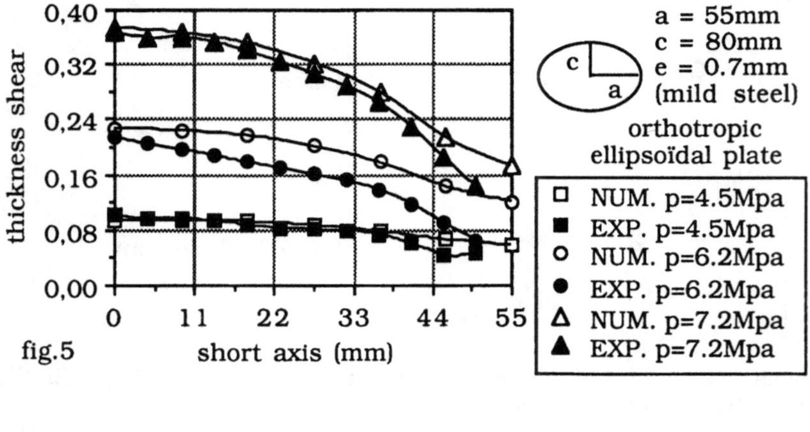

fig.5

3.2. Deep drawing of a sheet subjected to an hemispherical punch with draw-in under the blankholder

The geometry and the load conditions are described in figure 6.

The elasto-plastic material is assumed isotropic through the thickness with Lankford's coefficient equal 1.8.The other material caracteristics are:

$E = 200\,000$ MPa $\qquad \nu = 0.3$

$\sigma = c\,(p + p_0)^{\alpha} \qquad \sigma_0 = 400$ MPa, $C = 845$ MPa, $\alpha = 0.2177$

The friction coefficient is 0.25. Blankholder pressures are updated at the end of each step.
A quarter of the blank is discretized using 180 nodes and 312 elements.
In figure 7 we present the repartition of the thickness strain along radius for a maximal deepth of 34.7 mm.
For the same level we have reported on figure 8 for each node along a radius the points corresponding to the values of circumferential strain versus radial strain.
The numerical results using DKT12 elements and CST elements are compared with experimental results.

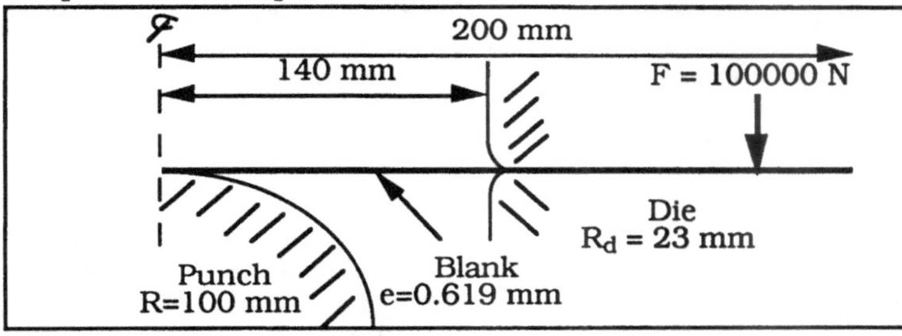

fig.6

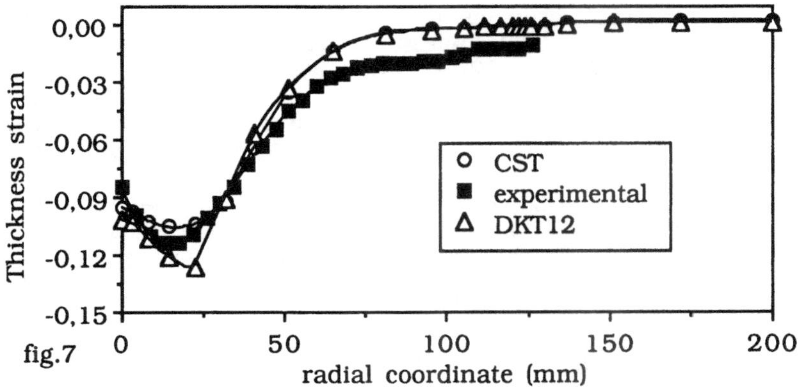

fig.7

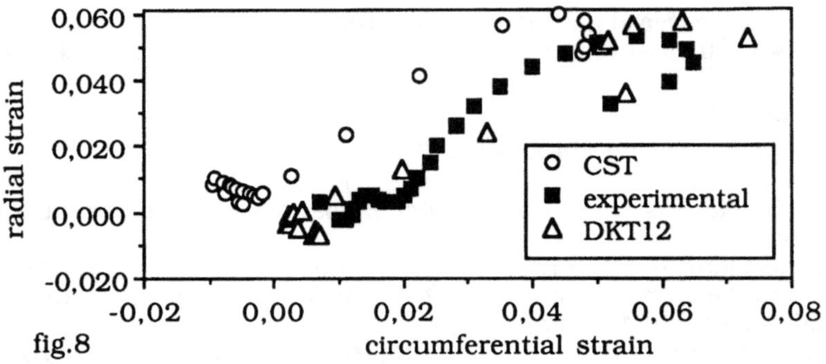

fig.8

4. Conclusion

Two triangular shell elements have been formulated for the analysis of large strain problems encountered in deep drawing of anisotropic metallic sheets. Good agreement is obtained for the cases presented. The present elements and formulation will be used for the simulation of complex industrial problems.

5. References

1. Rolph,W.D., Bathe,K.J. 1984. On a large strain finite element formulation for elasto-plastic analysis. Proceedings ASME-WAM Symposium on Constitutive Equations. 17 pages.
2. Braudel, H.J. 1986. Modélisation des grandes déformations élasto-plastiques d'un solide isotrope par la méthode des éléments finis - Application à l'extrusion à froid des métaux. D.Sc. Thesis. Lyon. France.
3. Simo, J.C., Taylor, R. 1986. A return mapping algorithm for plane stress elasto-plasticity. IJNME. 22 : 649-670.
4. Batoz, J.L., Roelandt, J.M., Pol, P., Duroux, P., 26-30 June 1989. A membrane bending finite model for sheet forming., Proceedings 3rd Int. Conf. on NUMIFORM 89, Fort Collins, Col., USA, Balkema Editeur, pp. 389-394.
5. Roelandt, J.M., Batoz. J.L., Duroux, P., Detraux, J.M., 11-15 September 1989. Deep drawing simulations using triangular shell element. 5th Int. Sym. on Numerical Methods in Engineering, Lausanne, Switzerland.
6. Roelandt, J.M., Liu, X.J., Batoz, J.L., Jameux, J.P., 7-9 August 1991. Axisymmetric and general shell elements for large transformations (formulation and applications).Int. sem. on large plastic deformations MECAMAT 91, Balkema Editeur, Fontainebleau , France.
7. Batoz, J.L., Bathe, K.J., Ho, L., 1980. Three node triangular plate bending elements. Int. J. for Num. Meth. in Eng. 15 : 1771-1812
8. Batoz, J.L., in preparation. A triangular discrete Kirchhoff with six dof. Division MNM, UTC.
9. Morley, L.S.D., 1968. The triangular equilibrium element in the solution of plate bending problems. Aero. Quart. 19, 149-169.

Acknowledgments : this research is undertaken with financial support from RENAULT and PSA.

Stress Localization and Deformations of Surface Asperities in Metal Forming

M. RAOUS, P. CHABRAND

Laboratoire de Mécanique et d'Aoustique-C.N.R.S.
13402 Marseille Cedex 9 - FRANCE

Summary

A finite strain model is used to analyse the behavior of surface asperities such as plateaus in a Lasertex sheet metal. A unilateral contact with friction is considered for the contact between the sheet and the tool. Formulation and numerical methods are given. Mechanical analysis of the phenomenon is carried out on the basis of experimental results from the literature.

Introduction

In metal forming, friction plays an essential role both in the contact between the sheet and the blankholder and in the contact between the sheet and the tool. It provides the necessary retaining forces on the boundary of the sheet but it can also produce debris or galls. In previous studies, the role of the blankholder was investigated using a model for unilateral contact with friction in the context of finite elastoplastic strains (1)

The present contribution deals with the understanding of friction as plastic deformation of surface asperities. The large stress concentration on these asperities makes it necessary to use a finite elastoplastic model.

A numerical simulation of the microgeometrical deformation of the surface is presented for a Lasertex sheet metal. In this case, the geometry is composed of plateaux and valleys. We focus on the deformation of an isolated plateau.

Several loading cases are taken into consideration : they correspond to various hypotheses as to the size of the effective surface acting during the contact. The highest zones are subjected to large overloadings. Deformations of the plateau are analysed in these various cases and a conclusion as to their role in the debris formation is proposed.

The mechanical model

Let Ω be the reference configuration of an isotropic elastoplastic body. We assume that Ω is open and bounded with smooth boundary $\partial\Omega$. $\partial\Omega$ contains three open subsets $\partial_D\Omega$, $\partial_L\Omega$, $\partial_C\Omega$ such that :

D. Besdo, E. Stein (Eds.)
Finite Inelastic Deformations -
Theory and Applications
IUTAM Symposium Hannover/Germany 1991

$$\partial\Omega = \partial_D\Omega \cup \partial_L\Omega \cup \partial_C\Omega \qquad (1)$$

$\partial_L\Omega$ or $\partial_D\Omega$ is the part of $\partial\Omega$ on which the loads $\bar{t}$ or the displacements $\bar{u}$ are prescribed respectively. $\partial_C\Omega$ is the candidate contact part of the boundary.

The assumption of plane strain is considered.

To characterize the large strain elasto-plastic deformation of the body we use the multiplicative decomposition of the deformation gradient F introduced by Lee (2) :

$$F = F^e.F^p \qquad (2)$$

where F^e and F^p represents the elastic and plastic parts of the deformation,.respectively.

Since the elastic strain in the body is small in comparison with the plastic strain, we use an additive decomposition of the strain rate tensor D :

$$D = D^e + D^p \qquad (3)$$

The constitutive equations considered in this paper are the generalized Prandtl-Reuss equations for large deformations proposed by Hill (3),(4) associated with a Von-Mises yield condition.

The weak form of the equilibrium equations is written using an incremental updated Lagrangian formulation. The resulting variational problem is then solved within the context of the finite element analysis using a modified Newton-Raphson method.

Integration of the constitutive equations

Taking the elastic logarithmic strain tensor :

$$e = \frac{1}{2}\log(F^e.F^{eT}) \qquad (4)$$

involving the elastic deformation gradient, the stress-strain law is written using the Kirchhoff stress tensor as :

$$\tau = 2\mu.e + \lambda tr(e).\mathbb{1} \qquad (5)$$

where μ and λ are the Lame coefficient, $\mathbb{1}$ is the second unit tensor and tr(.) indicates the trace operator.

At each increment , the computation of the incremental displacement Δu, is computed to determine the incremental deformation gradient $F_{\Delta u}$:

$$F_{\Delta u} = \frac{\partial \Delta u}{\partial x_k} + \mathbb{1} \qquad (6)$$

And the total deformation gradient at the end of the increment is then expressed with the deformation gradient F_k at the start of the increment by :

$$F_{k+1} = F_{\Delta u}F_k \qquad (7)$$

The integration of the constitutive equations is performed by means of a direct stress update algorithm (5), (6), (7). Since F_k, F^p_k, e_k, τ_k are known on the configuration Ω_k at the start of the increment, the updated values F_{k+1}, F^p_{k+1}, e_{k+1}, τ_{k+1} in the configuration Ω_{k+1} at the end of the increment are calculated as shown below.
First an elastic predictor is calculated :

$$F^{p(0)}_{k+1} = . F^p_k \tag{8}$$

$$F^e_{k+1} = F_{k+1} . F^{p(0)-1}_{k+1} \tag{9}$$

τ^e_{k+1} and e_{k+1} are then obtained .

In the second stage, if τ^e_{k+1} is outside the elastic region enclosed by the yield surface, a radial return algorithm is then used. The stress tensor is projected onto the yield surface. And since we are dealing with an isotopic case, we can set :

$$F^e_{k+1} = V^e = \exp(e_{k+1}) = \exp(\frac{1}{2}(\tau_{k+1} - \frac{\lambda}{\lambda+2\mu}\mathrm{tr}(\tau_{k+1}).1) \tag{10}$$

and

$$F^p_{k+1} = F^{e\ -1}_{k+1} . F_{k+1} \tag{11}$$

in order to calculate the elastic and plastic deformation

Description of the contact and friction

We consider in this part a contact problem between a deformable workpiece and an undeformable tool. In the contact area the friction forces and the displacement situations (contact, no-contact, sliding, sticking) are unknown a priori. This contact area is characterized by unilateral conditions (no-penetration conditions) and the Coulomb inequalities are used for the friction.

The incremental form of the unilateral condition is written on $\partial_C\Omega$ as :

$$u^{k-1}_N + \Delta u^k_N \leq d \quad R^k_N \leq 0 \quad (u^{k-1}_N + \Delta u^k_N).R^k_N = 0 \tag{12}$$

where Δu^k_N is the displacement increment between configuration Ω_{k-1} to configuration Ω_k and superscript k indicates increment k.

On $\partial_C\Omega$ the displacements u and the contact forces R are written in a local reference coordinate system (n,t) in which n is the exterior normal unit vector to $\partial_C\Omega$. Subscript N indicates the normal component of the variable.

Coulomb's friction law can be written on $\partial_C\Omega$, using a forward finite difference approximation of the displacemnt velocity $\dot{u}$, as follows :

$$|R^k_T| \leq \mu \ |R^k_T| \ \text{if} \begin{cases} |R^k_T| < \mu \ |R^k_T| \Rightarrow \Delta u^k_T = 0 \\ |R^k_T| = \mu \ |R^k_T| \Rightarrow \Delta u^k_T = -\lambda.R^k_T \quad \lambda > 0 \end{cases} \tag{13}$$

We have studied two kind of methods to solve the contact with friction problem :

- A successive over-relaxation method with projections, used to solve the variational form of the problem (8)
- A Lemke's method which is a mathematical programming method applied to a complementary problem (9).

Successive over relaxation method with projections

The contact and fiction contributions are written, in the variational formulation, as has been done in the case of the elastic small deformations hypothesis. Using a fixed point method on the sliding limit the problem is then transformed at each increment, into a sequence of problems each corresponding to a Tresca's formulation. These problems are then solved by projection techniques along with an over relaxed Gauss-Seidel method.

To improve the efficiency of these methods convergence acceleration procedures are used :

- Partial resolution using the Gauss-Seidel method for the first iterations of the fixed point process,
- Condensing the problem to the contact variables only. The relaxation method is then used to solve a system which is greatly reduced,
- Use of an Aitken procedure to avoid the need to determine of an optimum relaxation coefficient.

Lemke's method

To be able to use Lemke's method, Coulomb's friction law has to be written in a complementary form similar to the expression for the unilateral conditions :

Fist we define (λ_1,λ_2) and (T_1,T_2) as follows :

$$\lambda_1^k - \lambda_2^k = \Delta u_T^k \tag{14}$$

$$T_{T1}^k = -R_T^k + \mu.R_N^k \tag{15}$$

$$T_{T2}^k = R_T^k + \mu.R_N^k \tag{16}$$

$$T_N^k = R_N^k \tag{17}$$

With these notations the Coulomb friction law may then be written :

$$\lambda_1^k \leq 0 \quad \lambda_2^k \leq 0 \tag{18}$$

$$T_{T1}^k \leq 0 \quad T_{T2}^k \leq 0 \tag{19}$$

$$\lambda_1^k.T_{T1}^k = \lambda_2^k.T_2^k = 0 \tag{20}$$

At each increment a modified Newton-Raphson method is used to solve the following system at iteration (i) :

$$K(u^k)^i.(du^k)^{i+1} = \bar{t}^k - (\bar{S}^k)^i + (\bar{T}^k)^{i+1} \quad \text{with :} \tag{21}$$

$$(\Delta u^k)^{i+1} = (\Delta u^k)^i + (du^k)^{i+1} \quad \text{and} \tag{22}$$

$$u_N^{k-1} + (\Delta u_N^k)_j^{i+1} \leq d^i \; ; \; (T_N^k)_j^{i+1} \leq 0 \tag{23}$$

$$(u_N^{k-1} + (\Delta u_N^k)_j^{i+1}).\,(T_N^k)_j^{i+1} \leq = 0 \quad \forall\, j = 1..p \tag{24}$$

$$(\lambda_{1}^{k})_{j}^{i+1} \leq 0;\ (\lambda_{2}^{k})_{j}^{i+1} \leq 0;\ (T_{T1}^{k})_{j}^{i+1} \leq 0;\ (T_{T2}^{k})_{j}^{i+1} \leq 0;\ \forall j = 1..p \quad (25)$$

$$(\lambda_{1}^{k})_{j}^{i+1}.(T_{T1}^{k})_{j}^{i+1} = (\lambda_{2}^{k})_{j}^{i+1}.(T_{T2}^{k})_{j}^{i+1} = 0 \ \forall j = 1..p \quad (26)$$

where p is the number of contact nodes. As the Lemke's method operates on systems in which all the components of the unknown vectors are subjected to constraints (p = total number of nodes) the same condensation procedure as that used in the overrelaxation method has to be applied.

In the applications presented below, we have used the Lemke's method which is more performant applied on a elastoplastic finite strain problem than the over relaxed Gauss-Seidel method.

Deformations of a plateau in a Lasertex sheet.

In a Lasertex sheet, the surface is composed of circular plateaux with a diameter of about 200 micrometers and valleys of a fairly similar size. The analysis consists of investigating the surface phenomena :

-Does a total collapse of the plateau occur during stamping ?
-Is the metal flow caused by the plateau deformation or by the asperity deformations?

The analysis is carried out under the approximation of plane strain hypothesis to avoid the need for a 3-D treatment in the present qualitative investigation. The geometry is given in figure 1.

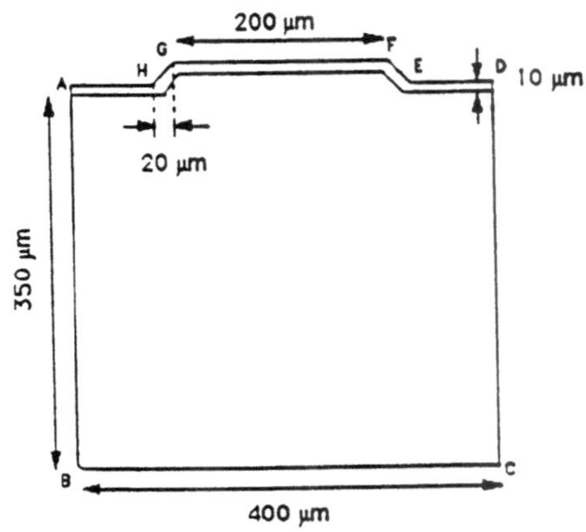

Fig. 1. The geometry

The sheet metal is 0.7 mm thick. Computations are performed using bilinear Q1-elements. The thin layer on the top of the plateau has been used in other calculations to take into account either the zinc in the case of covered sheet or an equivalent material in the case of the surface asperities.

Two kinds of computations have been carried out. First, by considering a unilateral frictional contact at the top, we analyse the behavior of the plateau in various loading cases. Since the experimental morphology ((9),(10)) is not obtained, displacements of the top of the plateau are prescribed in order to estimate the tangential forces that would be necessary to obtain the right shape.

The characteristics of the material are E = 210000 MPa (Young's modulus), $\nu = 0.3$ (Poisson coefficient). The equivalent stress is given by : $\bar{\sigma} = 200 + 500\ \bar{\varepsilon}^p$ ($\bar{\varepsilon}^p$ is the equivalent plastic strain). Computations have been carried out with a friction coefficient equal to 0.15 or 0.3. The following results were obtained taking the value 0.3.

In the first part, we use various hypotheses as to the magnitude of the overloadings which can occur on localized parts of the surface. The overloadings are due to the differences in height on the surface. Consequently, the magnitude of the overloadings is directly connected to various estimates of the effective contact area between the sheet and the tool.

The average of the normal pressure is taken to be about 30 MPa under the blankholder during stamping. The following normal pressures have been used for the computations. (φ denotes the normal pressure).

a- φ = 120 MPa : we assume that all the plateaus partipate in the contact. The effective contact surface, i.e. the surface of all the plateaus, is about 25% of the total surface. In the following cases, a tangential displacement of the obstacle (i.e. the tool) is prescribed until the total sliding of the surface of the plateau is obtained.

In this case, the deformations are only elastic even in the case of the local stress concentration caused by the frictional contact. It is clear that the plastic phenomenon observed during the experiments ((9),(10)) is caused by the height of the plateaus.

b- φ = 240 MPa : we assume in this case that one plateau out of two participates in the contact. The effective contact area is 12.5% of the total surface.
The situation is still far from fitting the experimental data.

c- φ = 960 MPa : we consider that only one plateau out of eight paticipates in the contact. The effective contact area is 3.125% of the total surface.
In this case, the plastic deformations reach 81% on the left side of the plateau. The displacements are shown in figure 2. The morphology is not really what was expected.

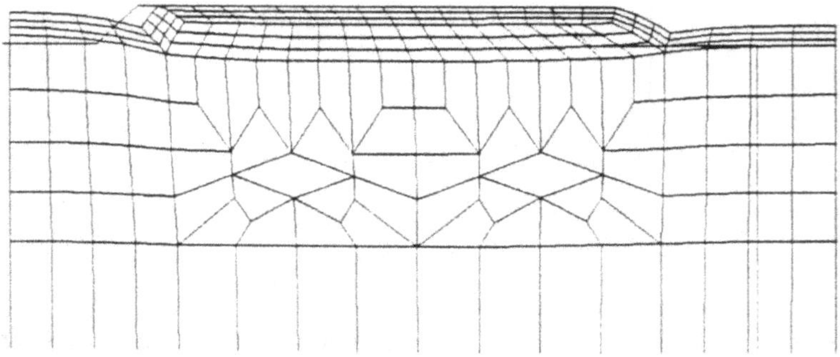

Fig. 2. : Plateau deformations for φ = 960 MPa

d- φ = 2000 MPa : This is the case of large overloading occuring on a very high zone. The effective contact is only 1.5% of the total surface.

The plastic deformation increases up to 142% on the left top side. A collapse of the plateau occurs but the shape does not look like a metal wave as observed experimentally (see figure 3).

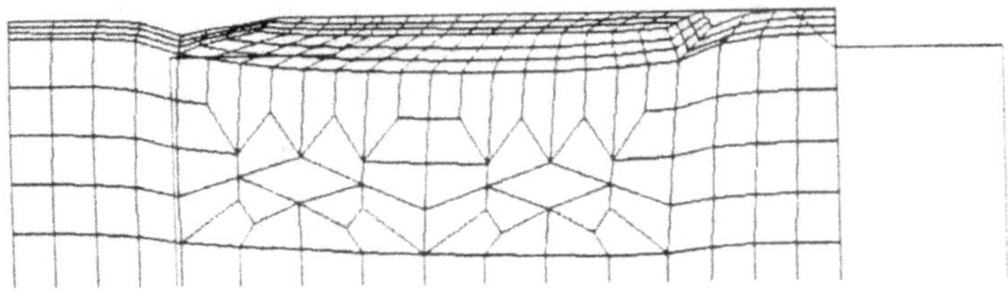

Fig. 3. : Plateau deformations for φ = 2000 MPa

Since the experimental morphology was not obtained in the previous cases, we will now consider a second kind of loading.

A tangential displacement is prescribed on the top of the plateau in order to obtain an estimation of the shear force necessary to generate the metal wave. The results in the case of a tangential displacement equal to 0.15mm are presented on figure 4.

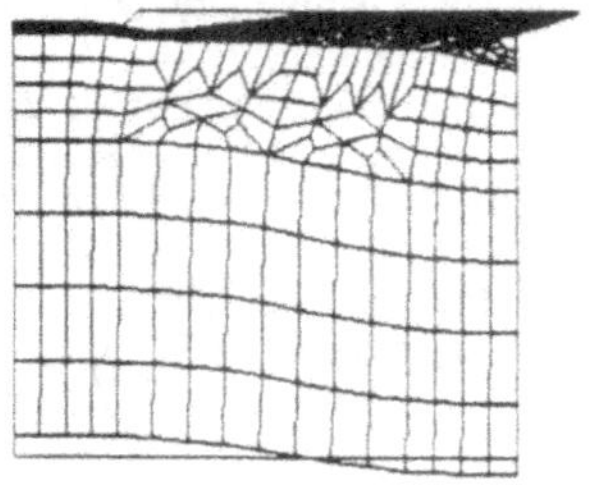

Fig. 4 : Plateau deformations φ = 240 MPa; Prescribed displacement α = 0.15mm

The wave is obtained but the tangential force would then correspond to a friction coefficient of about four : the prescribed normal pressure is 240 MPa, the prescribed tangential displacement is 0.15mm, and the resulting tangential force on the top of the plateau is computed to be 924 MPa. This is not a realistic possibility for the friction.

It can be concluded that the morphology observed experimentally at the surface of the metal sheet is not caused by the collapse of the plateau itself but by plastic effects on the smaller asperities located at the top of the plateau. A study on the deformation of these asperities is at present under way.

Acknowledgments : this study was supported by the MRT contrat "contact métal outil pour l'emboutissage" n° 87 A 0024 du 25/08/87

References

1. Chabrand, P.; Pinto, Y. : Numerical model for blankholders with drawbeads. Proceedings of MECAMAT'91 International Seminar on Large Plastic-deformations Fundamental Aspects and Applications to Metal Forming. Fontainebleau, France, 7-9 August 1991.

2. Lee,E. H.: elastic-plastic deformation at finite strains.j.Appl. Mech., 36, (1969) 1-6.

3. Hill, R. : The mathematical theory of plasticity. Oxford, Clarenton, (1967).

4. Sidoroff, F.: Incremental constitutive equations for large strain elastoplasticity. Int; j. Engng. Sci., Vol 20 (1982) 19-26.

5. Gelin, J.C.: Modèles numériques et expérimentaux en grandes déformations plastiques et endommagement de rupture ductile. Thesis, University P. and M. Curie Paris VI (1985).

6. Gelin, J.C.: Application of an implicit method for the analysis of damage with temperature effects in large strain plasticity problems. Numerical methods for nonlinear problems. Edts C Taylor and al., Pineridge press (1986) 494-508.

7. Simo J.C. and Ortiz M. : A unified approach to finite deformation elasto-plastic analysis based on the use of hyperelastic constitutive equations. Comp. meth. appl. mech. engng. vol 49 (1985) 221-245.

8. Raous, M.; Chabrand, P. and Lebon, F. : Numerical methods for frictional contact problems and applications. J. Theo. Appl. Mech. special issue, supplement n°1 to vol 7.(1988), 111-128.

9. Pinto, Y. 1990. Contact et frottement en grandes déformations plastiques. Application au serre-flan en emboutissage. Thesis, Provence University, Marseille.

10. Berthier,Y.; Vincent , L.; Godet, M.; Frottement et usure: approche multiéchelle. Proceedings of "Rencontres Scientifiques au LMA : Mécanique, Méthodes numériques et dynamique des matériaux", 8-10 Avril 1991, Publication LMA n°124, 1991.

11. Rapport de fin de contrat du Programme MRES : Contact métal outil lors de l'emboutissage. Contrat MRES n° 87 A 0024 du 25/08/87 , Octobre 1991.

Nonlinear Boundary Element Analysis for Elasto-Plastic Problem

Junji Kihara
Department of Metallurgy, University of Tokyo
7-3-1 Hongo, Bunkyo-ku, Tokyo 113

Abstract

This article concerns with the feasibility of the boundary integrtal equation or element technique to analyse finite nonliner elasto-plastic problems. The initial stress scheme is adopted to evaluate both the materials and geometrical nonlinearities, and the formulation and program implementation are performed. The numerical result on the cantilever bending considering the spin effect of previous stress state is compared with that without such consideration and it is found that the discrepancy between them is rather small even for the large deflection of the beam. It is concluded that the boundary element technique is a tough numerical method against to the effect of geometrical nonlinearlity in its nature, only if the shape of the boundary and the boundary condition are suitably updated.

1 INTRODUCTION

The boundary integral equation method or the boundary element method has been applied not only the linear elastic problems but also to such nonlinear problems as elasto-creep, elasto-plastic or visco-plastic analyses [1,2]. Since the boundary element approaches are formulated on the basis of the fundamental solution in linear elasticity, the nonlinear contributions of the whole strains besides the elastic strains are taken for the initial or the eigen strains in the incremental calculation. Hence, this type of evaluation is not always applicable to all nonlinear problems including the deeply plastic deforming solids with large rotation. Most of practical elasto-plastic problems can be described and understood by tracing the nonlinear deformations upto the moderately finite strains in computation. In these fields of nonlinearity, the above class of initial stress method becomes useful in prediction of elasto-plastic or elasto-viscoplastic behaviors [3,4,5].

In the present study, the author will make a brief introduction to the boundary element formulation for the elasto-plastic and the viscoplastic problems with some comments on the initial stress approach and the consideration of geometrical nonlinearities. The boundary element equations will be proposed together with the computational aspects for incremental procedure and iterative calculation. Three dimensional modeling

D. Besdo, E. Stein (Eds.)
Finite Inelastic Deformations –
Theory and Applications
IUTAM Symposium Hannover/Germany 1991

for beam structure is taken for example to consider the effect of geometrical nonlinearity. Furthermore, the mechanical behaviors under tension and bending in two dimensional situation will be discussed over the applicability of the present methods both in materials and geometrical nonlinearity.

2 FUNDAMENTAL FORMULATION FOR NONLINEAR ANALYSIS

Consider a two or three dimensional domain V and its boundary S. Then, the elasto-plastic problem can be formulated by the boundary integral approach in the rate form:

$$C_{ij}\dot{u}_j + \int_S P^*_{ij}\dot{u}_j dS = \int_S U^*_{ij}\dot{p}_j dS + \int_V \epsilon^*_{jki}\dot{\sigma}^p_{jk} dV, \tag{1}$$

where C_{ij} is a constant factor dependent on the topology of a point ($C_{ij} = 0$ when the point is outside V, $C_{ij} = 1/2$ when S is smooth there, or $C_{ij} = 1$ when it becomes an internal point). $\dot{u}_i$ denotes a velocity vector, $\dot{p}_i$ a time derivative of the traction vector, $\dot{\sigma}^p_{ij}$ a internal stress rate tensor due to plastic deformation, and U^*_{ij}, P^*ij and ϵ^*_{ijk} the fundamental solutions in elasticity. In the absence of the second term in the right-hand side, eq. (1) is equivalent to an integral equation appearing in the boundary integral formulation for elastic analysis; this term corresponds to the virtual work by the initial stress or the plastic contribution of stress rate tensor. The integration of this term is performed for the plastic region V_p in the incremental computation.

The stress rate tensor $\dot{\sigma}_{ij}$ at any internal point is explicitly expressed by the spatial differentiation of eq. (1) and the conventional BEM procedure. Then, the stress tensor σ_{ij} in the current state is obtained by the time integration $\int \dot{\sigma}_{ij} dt$ or the incremental updating $\sigma_{ij} = \sigma^{(0)}_{ij} + \dot{\sigma}_{ij}\Delta t$ for the previous stress state $\sigma^{(0)}_{ij}$. In the same manner, both the displacement u_i and the traction p_i vectors on the boundary S can be updated from the previous values $u^{(0)}_i$ and $p^{(0)}_i$ by $u_i = u^{(0)}_i + \dot{u}_i\Delta t$ and $p_i = p^{(0)}_i + \dot{p}_i\Delta t$, respectively. To be noted, the above explicit integration with use of constant time increment Δt works well for the infinitismal elasto-plastic analysis.

To cope with the visco-elastic problems, the viscous contribution of stresses is expressed by the initial stress term in the boundary integral eq. (1); in the frame of boundary element approach, the governing operator in the 2D/3D field is expressed by the elastic fundamental equation and the induced deviator or perturbation by the materials nonlinearities is approximated by the inhomogeneous data in the form of the initial stress.

With increase of plastic deformation, the change of geometrical configuration and physical quantities exceed the limit of tolerance where the elasto-plastic behaviors of solids can be described without consideration of the geometrical nonlinearities. The first approach in the boundary element method to deal with the finite strain - finite deformation behaviors is as follows:

1. The whole nonlinear terms including both materials and geometric nonlinearities are evaluated in the form of the initial stress.

2. The integral rate form in the boundary element equation is assumed to be invariant to the finite deformation; both the boundary profile and the inner point/cell are updated in the Lagrangian approach, and the same fundamental solution is utilized in the rate form or in the time integration during Δt like the Eulerian approach.

3. The stress rate tensor is evaluated with consideration of objectivity against finite deformation and rotation of solids; within the moderate range of strains and rotation, the stress rate tensor can be measured by the conventional Jaumann stress rate tensor, which is defined by

$$\check{\sigma}_{ij} = \dot{\sigma}_{ij} + \Omega_{ik}\sigma_{kj} + \sigma_{ik}\Omega_{kj} \tag{2}$$

where Ω_{ij} is the spin tensor in the rotating coordinate system. This approach enables us to consider the effect of geometrical nonlinearity in the elasto-plastic analysis by employing as the initial stress $\check{\sigma}^p_{ij}$ instead of $\dot{\sigma}^p_{ij}$ in eq. (1).

That is

$$C_{ij}\dot{u}_j + \int_S P^*_{ij}\dot{u}_j dS = \int_S U^*_{ij}\dot{p}_j dS + \int_V \epsilon^*_{jki}\check{\sigma}^p_{jk} dV. \tag{3}$$

The stress rate tensors at any inner points are also rewritten by using the Jaumann rate.

With respect to the validity of the above approach, two questionnaires are rising as follows:

1. How is the limit of tolerance for the present approach in the practical elasto-plastic analysis?

2. How much is the increase of computational cost and labors?

In the former problem, since the actual objectivity is invaded by the infiltration of rigid motion into thus evaluated stress/strain state, the tolerance of practical use for the present approach must be demonstrated by some numerical tests. In particular, the comparison of solutions with or without consideration of geometrical nonlinearities provides an idea to discuss the applicability or validity of the present approach. In the latter, the additional labors are relatively small for the elasto-plastic analysis without consideration of geometrical nonlinearities. However, attentions must be paid to the time integration in each time increment; time increment should be controlled to realize the moderate change of geometrical configuration, and time integration scheme must be also optimized to reduce the truncated errors in the time direction.

In other word, the updating of previous Lagrangian system is usually necessary to control the magnitude of rigid rotation increment and keep the deviation of stress invariant in the previous stage, say the Mises stress, within some tolerable level. This condition of the rotation increment should restrict the magnitude of time step.

Recently, a boundary element approach to three dimensional bending problem was succesfully performed based on the above mentioned concept [6]. The main numerical results is shown in Fig. 1, where the finite elastic calculations agree well with the reference solution. The boundary element approach taking account of geometrical nonlinearity must be at least valid for the finite elastic problems.

3 COMPUTATIONAL IMPLEMENTATION AND EVALUATION

At first, let us make brief introduction of the boundary element approximation in the elasto-plastic analysis. Discretizing the boundary S and the volume V into N boundary elements and M cell elements, we have the discrete from of eq. (3):

$$[H]\{\dot{U}\} = [G]\{\dot{P}\} + [QB]\{\check{\sigma}^p\}, \tag{4}$$

or arranging the known and unknown components of $\{\dot{U}\}$ and $\{\dot{P}\}$,

$$[A]\{\dot{X}\} = \{\dot{f}\} + [QB]\{\check{\sigma}^p\}. \tag{5}$$

In the same manner, the stress rate tensor can be expresses by

$$\{\check{\sigma}\} = [D]\{\dot{P}\} - [S]\{\dot{U}\} + [QQ]\{\check{\sigma}^p\}. \tag{6}$$

In the absence of matrices $[QB]$ and $[QQ]$, the above matrix equations are equivalent to the conventional boundary element equations. Assuming that the geometrical configuration, the plastic zones and the initial stress rate tensor are prescribed in the previous step, the essential part of incremental procedure in the present approach is listed as follows:

1. Calculate the unknowns in $\{\dot{U}\}$ and $\{\dot{P}\}$ by solving the simultaneous eq. (5).
2. Obtain the true stress rate tensor by eq. (6) for each inner cell element.
3. Examine whether the equivalent stress $\bar{\sigma} = \bar{\sigma}(\sigma_{ij})$ for $\sigma_{ij} = \sigma_{ij}^{(0)} + \dot{\sigma}_{ij}\Delta t$ exceeds the yield stress σ_Y or not. When $\bar{\sigma} < \sigma_Y$, the current stress state is updated by σ_{ij}. On the other hand, if $\bar{\sigma} > \sigma_Y$, calculate new initial stress rate to satisfy the yield condition and the related flow rules.

The above computation becomes iterative to determine the initial or plastic contributions of stress rate finally and to settle down the developing plastic regions in the current increment. In the practical computations, careful attentions must be paid to the following items:

1. A center of element is chosen an evaluation point in the current cell element, so that a cell becomes plastic if the equivalent stress at the center point exceeds the yield stress. Then, the plastic region grows cell element by element.

2. The displacement gradient tensor at the point is explicitly evaluated in computation. To be noted is the accurate evaluation of the singular kernels appearing in the integral equations.

3. Convergence of the iterative calculation is attained if the change of absolute values of $\{\dot{U}\}$ and $\{\dot{P}\}$ or the ratio of $|\Delta\dot{X}|/|\dot{X}|$ for the unknown vector $\dot{X}$ becomes less than the prescribed tolerance.

4. In the initial stage of incremental calculation, the coordinates of boundary nodes and cell nodes are updated and the boundary element coefficient matrices or $[H]$ and $[G]$ are recalculated.

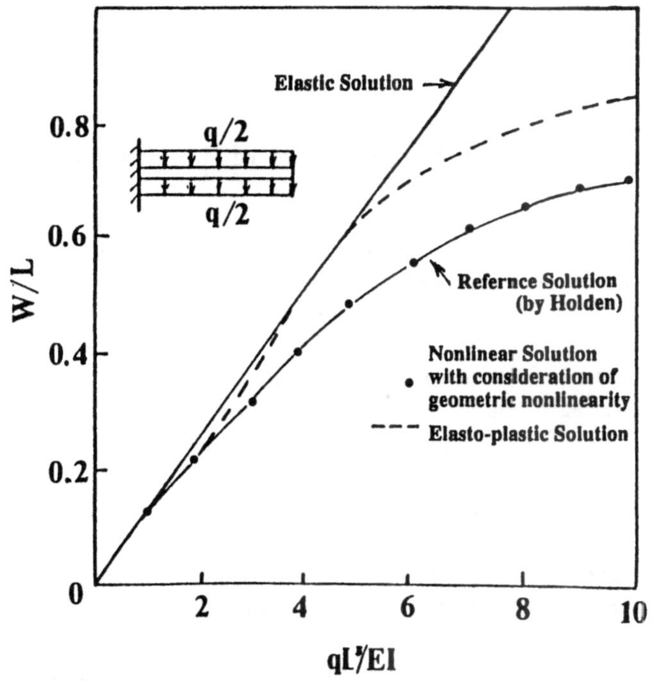

Fig.1 Non dimensional deflection of the center of free end depending of no dimensional loading parameter, W: deflection, L: length of beam, q: load density, E: Youngs Modulus and I: moment

4 NUMERICAL EXAMPLES

Now, two cases of two dimensional elasto-plastic problem, plane-strain tensile deformation and plane-strain bending of cantilever by end point loading are chosen for numerical examples. The material properties are as follows: Young's modulus is 210 GPa, Poison's ratio is 0.3, and the stress-strain curve is represented by the equation, $\sigma_Y = 403(1+\epsilon/\epsilon_Y)^{0.06}$ MPa. Fig. 2 shows the element model and the schematical boundary condition of the tesile problem. Fig. 3 shows how the plastic zone exapands with increase of displacement. The region resembles with that predicted by slip line theorem for rigid plastic body.

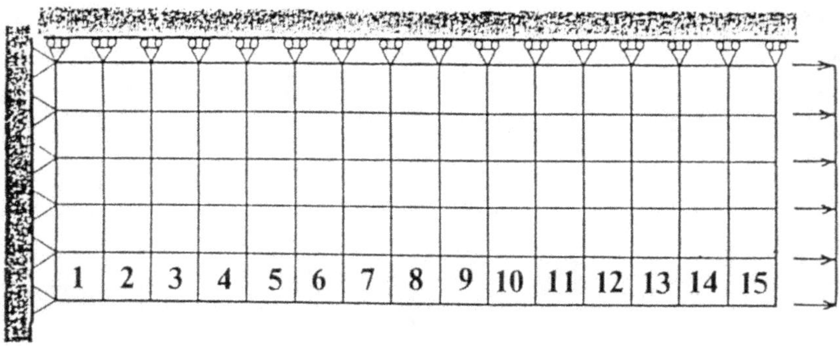

Fig.2 The boundary and cell element model and the schematical boundary condition in the plane strain tensile deformation.

Fig. 4 shows the element model and schematical booundary condition of the bending problem. The dimension of the cantilever is 300 mm in length and 100 mm in height. Fig. 5 shows the yielding region after 60 mm deflection at the end point. Fig. 6 shows the calculated stresses calculated by the two methods, one of which consider the spin effect, and the other does not. The point where the stress is calculated is one where the declination is considerably large. It is noted that the numerical results are not so different from each other, and this result represents the important characteristics of the boundary element approach. The reason why the difference is small is thought of as the fact that total elastic stress field is precisely calculated only if the shape of the boundary and the boundary condition are updated. In the case of elasto-plastic problem, the results will be also not so different from each other if the von Mise steress is controlled precisely, and this characteristic resembles with that of the rigid plastic FEM which is a tough numerical method agaist to geometrical nonlinearity.

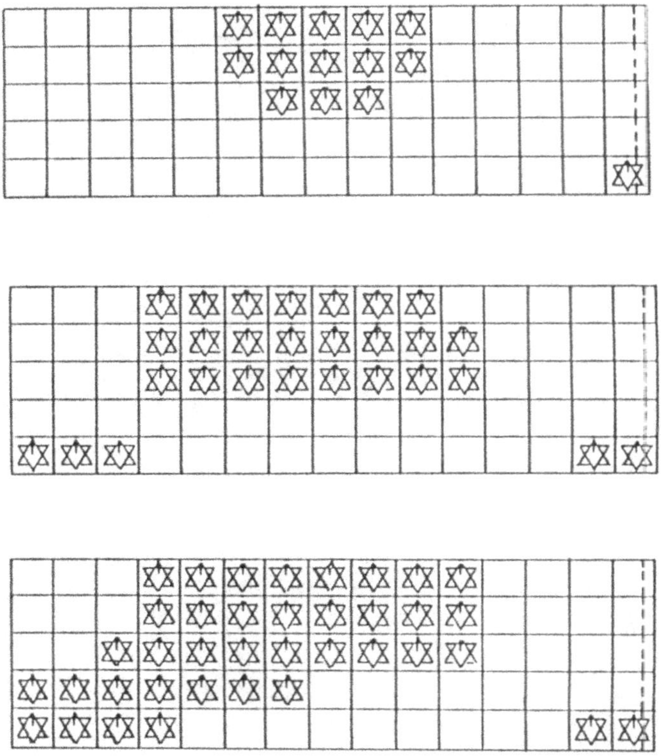

Fig.3 The expansion of yielding region with increase of tensile displacement

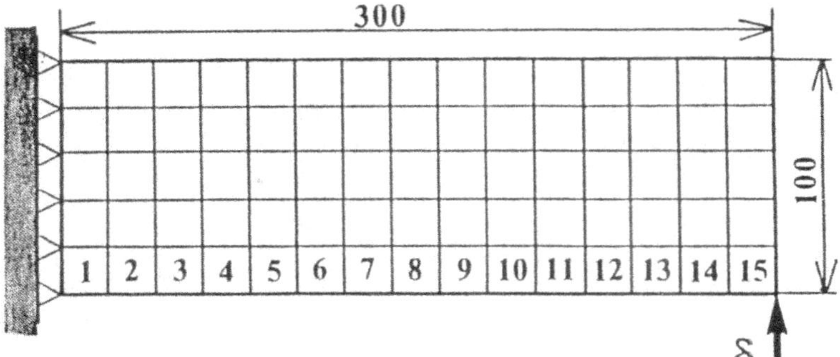

Fig.4 The beundary and cell element model and the boundary condition in the bending of rectangular cantilever

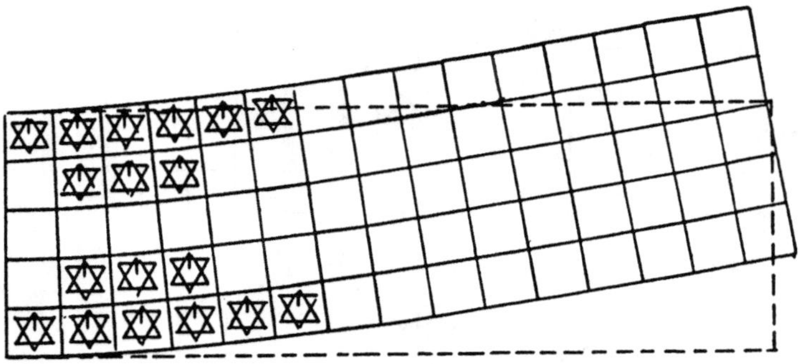

Fig.5 The yielding region and the deformed state at $\delta = 60$ mm calculated with consideration of spin effect

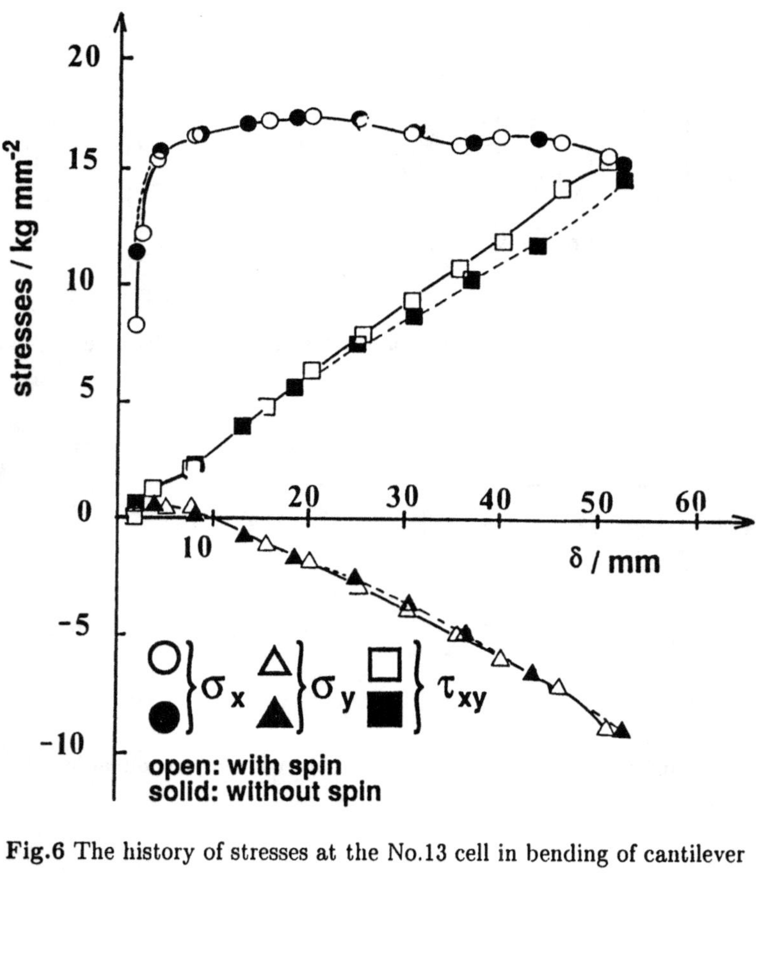

Fig.6 The history of stresses at the No.13 cell in bending of cantilever

5 CONCLUSION

In the frame of the nonlinear boundary element analysis on the basis of the initial stress method, relatively high accurate stress information can be obtained as far as we are concerned with moderately finite straining. Excluding the necking behaviors, uniform plastic deformations in tension/compression or bending state can be dealt with even in practical situation. It is noted that, only if the shape of boundary and the boundary condition are well updated and the increment of load including displacement is well controlled, even the result without consideration of the effect of geometrical nonlinearity represents a realistic solution. Of course, if the spin effect can be well modelled as one of various nonlinear eigen strains and can be taken into account of the original boundary element equations in the form of the initial stress, the more reliable results will be obtained.

6 REFERENCES

[1] D. Pape and P.K. Banerjee, J. Applied Mechanics(1987)313-326

[2] N. Atruli, Boundary Element Method in Applied Mechanics(1988)

[3] J. Kihara and K. Ishii, J. Japanese Soc. Tech. Plast.vol.26 (1985) 1046-1050

[4] T. Aizawa and J. Kihara, Theory and Applications of Bounary Element Methods (1987) 281-294

[5] T. Aizawa and J. Kihara, Boundary Element Methods in Applied Mechanics (1988) 475-482

[6] H. Xiao, Doctorate Thesis in Northeast Institute of Heavy Machinery, China (1991) 38

Rigid-Plastic Finite Element Simulator with Adaptive Element Control for Practical Forging Processes

T. Aizawa and J. Kihara

Department of Metallurgy, University of Tokyo
7-3-1 Hongo, Bunkyo-ku, Tokyo 113, JAPAN

Summary

New finite element control has been proposed and developed to cope with large plastic flow of materials in practical forging situation; since composite triangular or tetrahedral elements are employed in the present approach, completely automatic mesh control can be available to trace the history of deformation and stress/strain states with reduction. Numerical examples are taken to demonstrate the validity and effectiveness of our developing system.

Introduction

Both two and three dimensional rigid-plastic finite elements have been used to describe metal flow of work materials in rolling or forging and to predict accumulation of plastic strains. Since severe geometric deterioration of elements is expected to occur in practical computation, the finite element control together with use of robust element scheme should be indispensable to provide the reliable and accurate rigid-plastic finite element simulator [1,2]. Authors [3,4,5] are concerned with the finite element control on the basis of the computational geometry to realize completely automatic mesh generation and refinement/de-refinement and to accommodate ALE (Arbitrary Lagrangian Eulerian) control. In this paper, precise discussions are made over our developing rigid-plastic finite element simulator and methodology of adaptive element control by the computational geometry. Adaptivity of finite elements to rigid-plastic analysis is first discussed through benchmark test, and the frame of element control by the direct tessellation method is described with comments on its features.

Rigid-plastic finite element simulator

Most of the conventional rigid-plastic finite element simulators have common features

D. Besdo, E. Stein (Eds.)
Finite Inelastic Deformations -
Theory and Applications
IUTAM Symposium Hannover/Germany 1991

to each other; 1) Finite element scheme: Quadrilateral or brick-type element with constant pressure (although the pressure term is sometimes not explicitly evaluated, this pair of interpolations for both velocity and pressure must be implicitly considered in the form of the penalty function), 2) Control of coordinates: updated Lagrangian coordinate system (general coordinates for each node should be updated in every incremental computation), 3) Element control: None or less mesh control (element models are controlled only manually or far from automatic) and 4) Nonlinear solution: Incremental approach + Iteration. In practical forging analyses, a promising simulator must overcome the following items: [I1] Local plastic flow concentrations with shears and rotations, [I2] Severely large distortion of elements especially in three dimensional situation, or, [I3] Various types of singular points, edges or faces often occurring between work materials and tools. Hence, the conventional approaches often fail to provide adequate solutions in practical situation.

With respect to the item [I1] and [I2], robust and stable finite elements other than quadrilateral or hexahedral elements should be designed. The first step in this process is proper introduction of LBB condition into the mixed interpolations for both velocity and pressure or the penalty-type elements. As discussed in [6], a certain benchmark tests are necessary to compare the potentials of these kinds of candidate mixed finite elements since mathematical theory of the finite element method provides us only necessary condition in using the adequate pair of velocity and pressure elements. Fig. 1 (a) and (b) list most of the candidate mixed finite elements with piecewise continuous and discontinuous modes, respectively. Use of the elements with discontinuous pressure mode enables us to eliminate the degrees of freedom for pressure in each element and to solve each element finite element equation in nodal velocity vector in the sense of the Hybrid method or the penalty function method [5]. Roughly speaking, when the number of degrees of freedom for velocity is nearly equal to or not so larger than number of constraints, locking phenomena should take place [7]; consider a triangular element with piecewise linear interpolation for velocity and constant for pressure to be used in the two edge fixed problem. Due to the meshing locking, completely zero velocity is obtained even by using fine element meshing. On the other hand, elements with too many number of degrees of freedom become often sensitive to contamination of zero-energy modes. Let me compare such candidate finite elements as listed in Fig. 1 (b) by the ratio r of number of degrees of freedom to number of constraints in case of the two edge fixed problem in Table 1. In the present study, four elements are employed among the candidates since $2 < r < 4$: Fig. 2 shows a list of elements to be verified through benchmark test. Let us make some comments on the type [B5] of elements or the triangular composite element in Fig. 2. In other elements than [B5], the velocity distribution is assumed to be quadratic on the edge with the middle node. In the composite element, as shown in Fig. 3, an element is further subdivided into four subregions in which the velocity is approximated by the piecewise linear function; the element is allowed to be deformed in star-type hexagonal shape. While the pressure is assumed to be constant on the element; the same pressure is shaped with four subregions. For a benchmark test is taken a back-extrusion problem where rigid-plastic materials is reduced with constant velocity. As shown in Fig. 4, a tool is fixed to work materials, and on both the right and left side walls, the lateral velocity is only fixed. The mesh subdivision is also normalized: total number of nodes is prescribed by nearly constant, and the initial mesh patterns are arranged for comparison. The deformed shapes of materials in 50% reduction are shown for each type of element in Fig. 5. For comparison is added in figures the computational results by 4-node quadrilateral elements; in using that conventional element, severe locking inevitably occurs at the vicinity of the singular point of tool. In case of both triangular and quadrilateral quadratic/linear elements or so-called bubble elements, the free surface becomes a little irregular or wavy. The pair of quadratic/constant interpolations indicates more stiff response. Then, the triangular composite element is determined to be most suitable and robust in our developing simulator.

Adaptive element control by computational geometry

To cope with the complex configuration changes taking place in actual forging processes, the mesh modeling must be updated by controlling number of nodes and elements and their locations without loss of accuracy. Most of all approaches for the automatic adaptive control are based on the heuristic algorithms: r-/h-/p-methods are typical examples of those algorithms in realizing the mesh adaptivity. As strongly stressed in [1] and [2], the promising element control for nonlinear analysis or practical simulation should have the following three key-items: [K1] Local and global mesh refinement to reduce and equalize the error distribution, [K2] Regular or semi-regular mesh generation independently of number of nodes and elements, and [K3] Self-consistent element control algorithms to be available both in two and three dimensional situations. The above r-/h-/p-methods or the conventional remapping procedure by the least square method are never true to satisfaction of either item in the above. Our developing element control method or the direct tessellation method has the following five characteristic features: [F1] Contiguity (every node, edge or face is shared with corresponding element boundaries), [F2] Completeness (the whole region is automatically divided into elements without gaps or overlaps between elements), [F3] Self-Consistency (algorithm is controlled by searching for connecting edges or faces with automatic termination), [F4] Efficiency (nearly the same algorithms or functions can be applied to both 2D/3D situations with the polynomial time), and [F5] Reliability (remeshed elements are free from distortion). Due to [F3] and [F4], a large amount of ambiguity can be eliminated in construction of feasible and neutral element control algorithms and functions. The contiguity constraint [F1] provides us elements with semi-regular distribution of aspect ratio: little or no distorted elements can be generated. [F2] and [F5] enable us to control two types of tessellations or Delaunay and Voronoi tessellations when updating element subdivision in global or local. Let us summarize the standard procedures for both mesh generation and element control in what follows; for simplicity, two dimensional procedure is listed in Figs. 6 and 7. Through search for true neighboring nodes for the current node or generation of its Voronoi tessellation, the current two dimensional subregion is transformed into a Delaunay tessellation with each node as an origin. Once both two tessellations are created for every node, such physical quantities as accumulated strains cab be transferred between old and new tessellations with ease. [F3] assures that nearly the same algorithms and programs are available to 3D element control.

Discussions

Two and three dimensional examples are taken for examples to evaluate our developing element control in actual situation. In the closed forging, the preform is directly processed into near-net shaped product; in the final stage, the dimensional accuracy must be considered into die design. Fig. 8 depicts the mesh refinement to deal with accurate prediction of work materials deformation in the final reduction.

Through completely automatic element model handling, both nodes and elements are generated with semi-regular alignment. In three dimensional situation, the geometric modeler is necessary to provide the surface data to the present element control. Fig. 9 shows the geometric surface model, the generated nodes and the refined element subdivision. Due to [F1] and [F2], element adaptivity is strictly preserved even in three dimensional situation.

Conclusion

Our developing methodology of finite element control for rigid-plastic analysis has been proposed and discussed: composite triangular elements are generated and controlled by using the node density with completely automatic way. Estimate of error distribution leads to the node density distribution; owing to thus obtained density, nodes are fully updated or generated. After remapping the previously calculated physical quantities, composite triangular elements are automatically generated for further computation. At the present stage, our system is on the way to deal with the three dimensional finite element control [8,9]. New methodology for generation and control of tetrahedral elements will be reported in near future [10,11].

[References]

[1] N. Yukawa, N. Kikuchi and A. Tezuka, Advanced Technology of Plasticity (1990) 1719-1728.

[2] O.C. Zienkiewicz, G.C. Huang and Y.C. Liu, International Journal for Numerical Methods in Engineering, 30 (1990) 1527-1553.

[3] T. Aizawa and J. Kihara, Proc. European Conference on New Advances in Comp. Struc. Mech. (1991) 121-128.

[4] T. Aizawa, Proc. 5th International Conference on CAD/CAM Robotics & Factories of Future (1990) 821-826.

[5] T. Aizawa and J. Kihara, Journal of Materials Processing Technology, 27 (1991) 55-71.

[6] O.C. Zienkiewicz, S. Qu, R.L.Taylor and S. Nakazawa, International Journal for Numerical Methods in Engineering, 23 (1986) 1873-1883.

[7] T. J.R. Hughes, The Finite Element Method (1988) Prentice-Hall.

[8] T. Aizawa, Proc. ICES 88; Post Conference Seminar on Computer-Aided Engineering and Educational Aspects of Computational Mechanics (1988) 29-48.

[9] T. Aizawa, Computer Aided Innovation of New Materials. North-Holland (1991) 603-608.

[10] T. Aizawa, to be published in the International Conference on the numerical modeling for welding and casting (1991, Soule).

[11] T. Aizawa, to be submitted in Comp. & Struct. (1992).

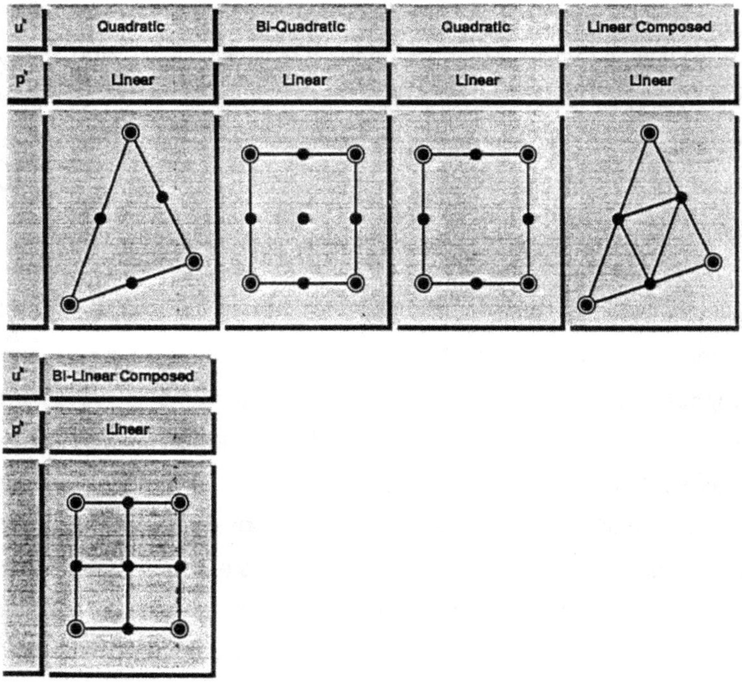

Fig. 1(a) List of mixed type elements with continuous pressure mode.

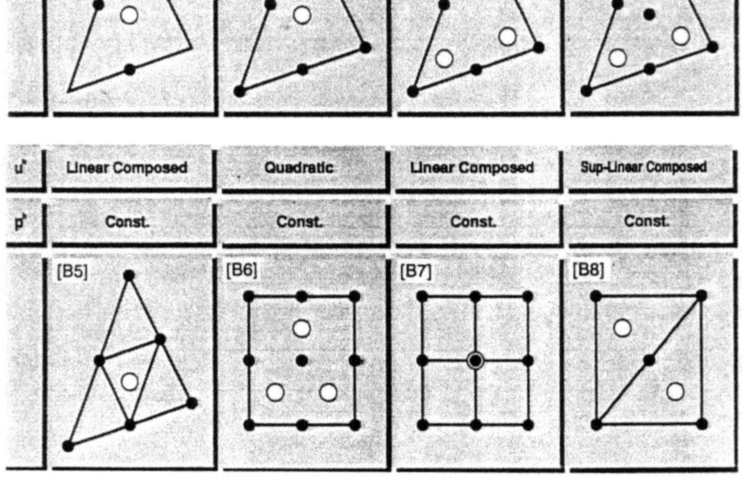

Fig. 1(b) List of mixed type elements with discontinuous pressure mode.

Table 1 Ratio of degrees of freedom for velocity and pressure interpolations.

Type of elements	Ratio of DOF for v_i and p
[B1]	3
[B2]	4
[B3]	4/3
[B4]	2
[B5]	4
[B6]	8/3
[B7]	8
[B8]	2

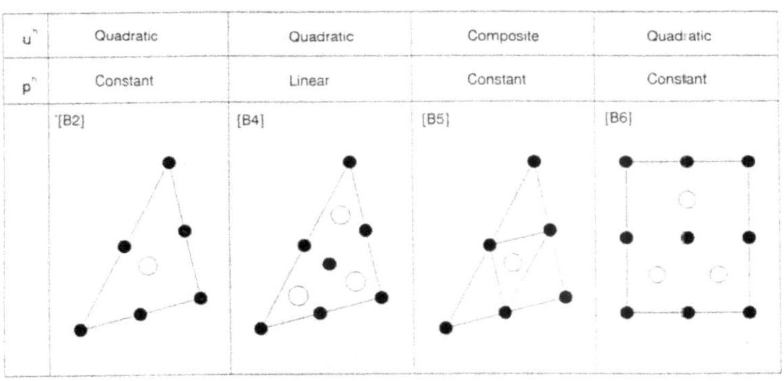

Fig. 2 List of feasible elements to be employed in the present benchmark test.

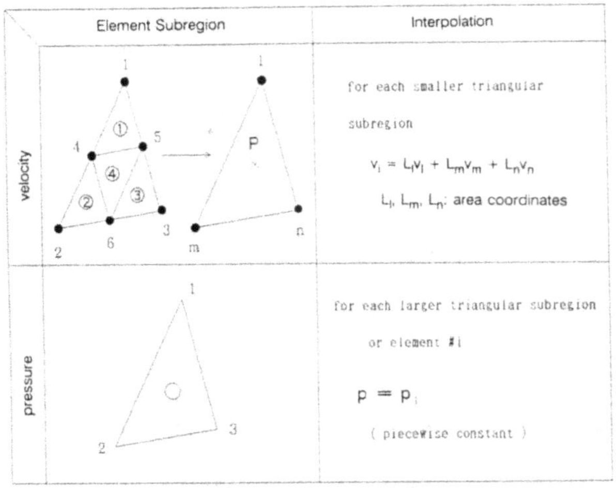

Fig. 3 An element design frame of composite triangular element.

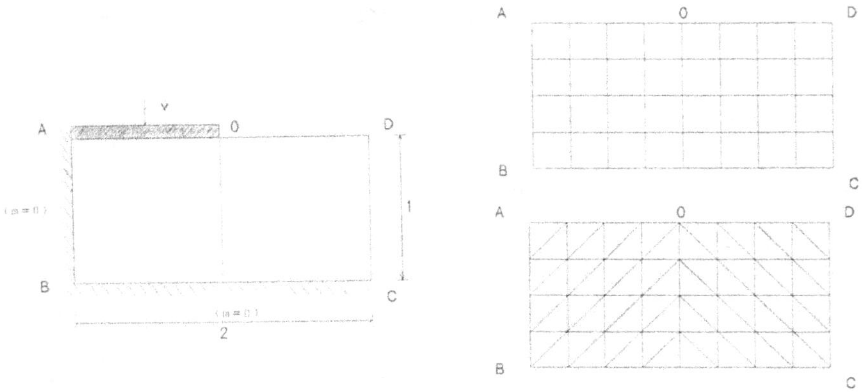

Fig. 4 A benchmark test: uniaxial compression test.

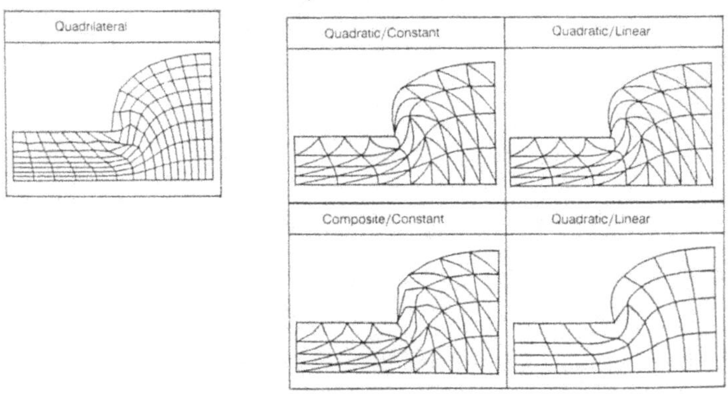

Fig. 5 Comparison of numerical behaviors in the benchmark test.

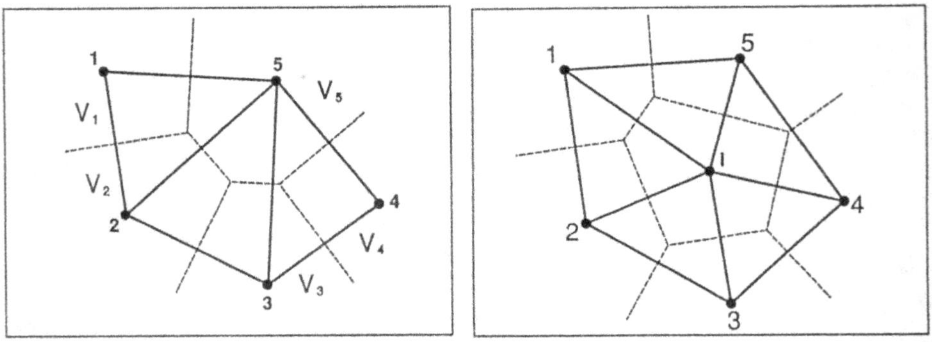

Old DT-Tessellation and Voronoi Polygons **New DT-Tessellation and Voronoi Polygons**

Fig. 6 Complete element refinement through mapping/remapping between old and new Delaunay/Voronoi tessellations.

[1] Generation of geometric data with initial mesh density
- Wire frame modeling
- Standard element length h

[2] Generation of boundary and inner nodes
- Vertex on the boundary --> a node
- Edge on the boundary --> subdivided by h
- Domain --> orthogonal meshing with h

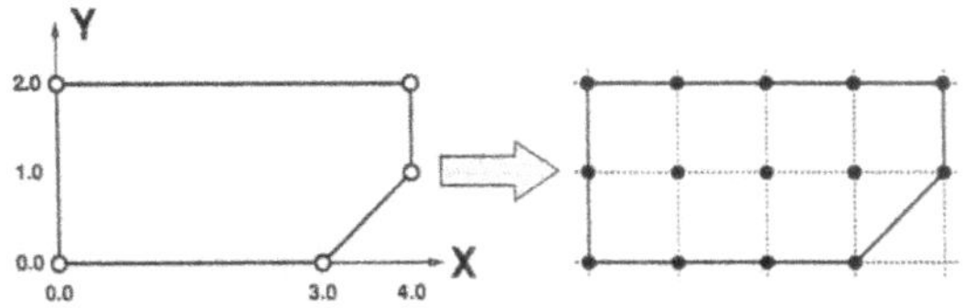

[3] Assembly of approximately neighboring node set SAN(i)
- Open circular overlap is generated to include nodes for each i node.
- Radius of overlapping circle is controlled in order that SAN(i) should have nearly constant density irrespectively of node density and node location in global.

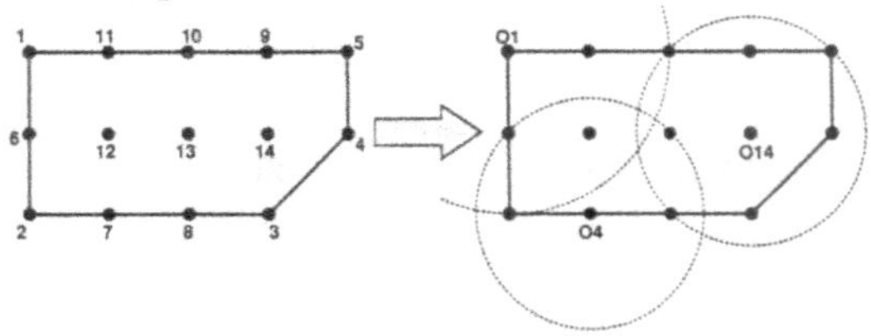

[4] Construction of elements
(Generation of both Delaunay and Voronoi Tessellations)
(1) Construct the initial Delaunay Tessellation for each i by two minimization theorems
- Minimize the distance between i and j for j STN(i)
- Minimize the radius of circles running through three nodes i, j and j1

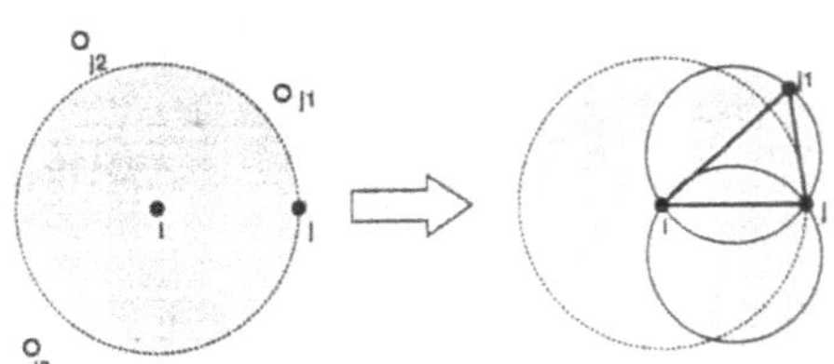

(2) Construct both a series of elements and a set of truly neighboring nodes STN(i): every edge starting from i is shared with neighboring elements and boundary.

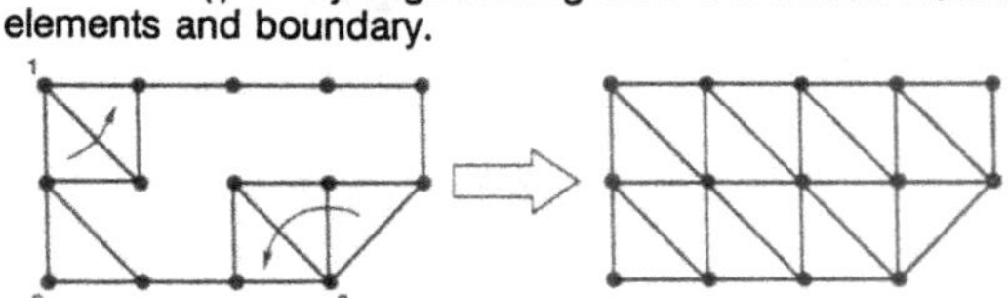

Fig. 7 Standard data handling in the generation phase of elements.

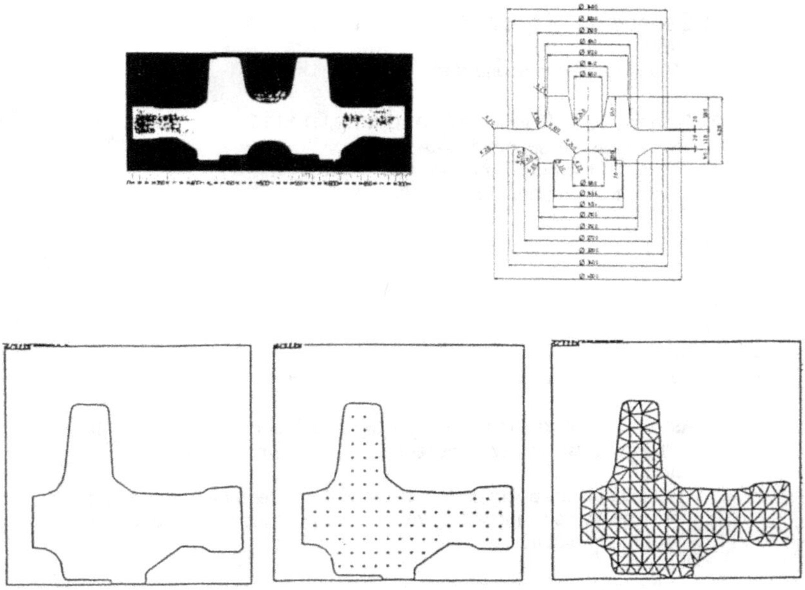

(1) Geometric model (2) Generation of nodes (3) Mesh generation

Fig. 8 Two dimensional element control in case of rotor forging.

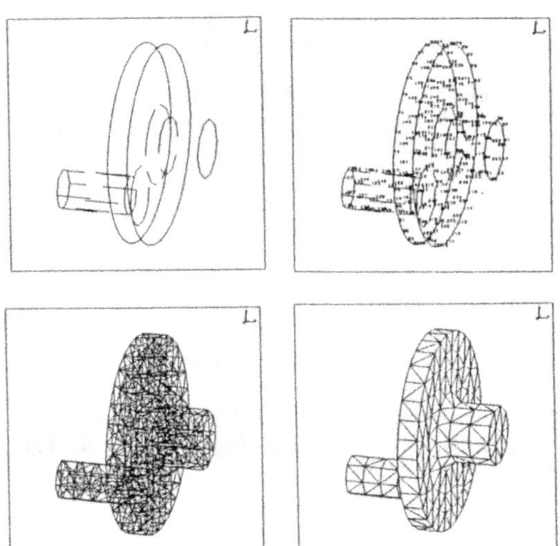

Fig. 9 Three dimensional element control.

A Boundary/Interior Element Discretization Method for the Analysis of Two- and Three-Dimensional Elastic-Plastic Structures

H. Andrä[1], C. Polizzotto[2], E. Schnack[1]

Summary

A coupled boundary/interior element method is presented for the analysis of elastic-plastic structures with material models endowed of dual internal variables. The domain field modelling is limited to the only plastic strains and strain-like internal variables, represented by their node values at a set of strain points in each interior element. The formulation, based on a Galerkin-type approach, is variationally consistent and leads to a fully symmetric-definite equation system. The backward difference method is adopted for the step-by-step integration procedure, and each step is addressed by an iterative predictor/corrector solution scheme. The analysis method is expected to be most appropriate and effective for a large class of engineering problems for which an accurate description of the stress in all, or part of, the domain is paramount.

Introduction

The boundary element method (BEM) has been widely recognized as an effective tool for analysing homogeneous elastic structures, competitive with respect to the finite element method (FEM) due to the reduction by one of the problem's dimensions, and to consequent advantages in the mesh generation. BEM applications to elastic-plastic problems have been successfully achieved (see e. g. [2, 29]) and are of continuing increasing interest (see e. g. [9]), showing the BEM capability of being competitive with FEM also for elastic-plastic analysis. To this concern, it is worth remarking that — contrary to the usual collocation BEM that destroys the continuum's self-adjointness properties — the more recent symmetric-definite Galerkin-type BEM [4, 12, 18] possesses full variational consistency [18 – 22]. This fact, besides making easier convergence and error analysis [31], entitles the BEM as a potentially effective alternative to the FEM for elastic-plastic analysis, in particular when plastic strains develop in a small region of the whole domain.

[1]Institute of Solid Mechanics, University of Karlsruhe, Karlsruhe, Germany.

[2]On leave from the Departimento di Ingegneria Strutturale & Geotecnica, Università di Palermo, Palermo, Italy.

D. Besdo, E. Stein (Eds.)
Finite Inelastic Deformations –
Theory and Applications
IUTAM Symposium Hannover/Germany 1991

The main purpose of the present paper is to address elastic-plastic analysis problems in small displacements, and to this aim a discretization method will be presented, with the use of boundary elements (BE) to model the elastic structural behaviour, and interior elements (IE) (elsewhere [19 - 22] called "cell elements") to approximate the plastic strain field. The resulting analysis method, entirely derivable through energy methods and endowed with symmetry and sign definiteness properties, constitutes a suitable means to solve a variety of problems in structural plasticity, e. g. limit and shakedown analysis, bounding techniques and the like, besides evolutive analysis.

The proposed method may also be applied to a class of problems for which the FEM/BEM domain decomposition has been recommended [5, 8, 24, 25], for instance crack tip problems combined with plastic yielding. It will emerge that the use of IEs will be sufficient to the coupling purposes, that is the modelling of only the plastic strain field, thus without the necessity of modelling the displacement field by the use of FEs — if the latter are not dictated by other considerations, e. g. inhomogeneities, or anisotropy.

Elastic-plastic standard material models with dual internal variables and quadratic thermodynamic potential will be considered [7, 11]. As in [13, 18], a Galerkin-type formulation will be adopted for both BE and IE discretizations, but the resulting formulation is fully equivalent to a variational formulation [18 - 22].

The governing equations

We consider a solid body or structure occupying the (finite, simply connected, open) domain $\Omega \subset R^\alpha$ ($\alpha = 2, 3$), with Lipschitzian boundary surface $\Gamma = \Gamma_1 \cup \Gamma_2$, $\Gamma_1 \cap \Gamma_2 = \emptyset$, mes $\Gamma_1 > 0$, with given Dirichlet data on Γ_1, and Neumann data on Γ_2. The body is referred to a Cartesian orthogonal co-ordinate system $\boldsymbol{x} = (x_1, ..., x_\alpha)$, $\alpha = 2, 3$, and it is acted upon by body forces $\bar{\boldsymbol{b}}$ in Ω, tractions $\bar{\boldsymbol{t}}$ on Γ_2, initial thermal-like strains $\bar{\boldsymbol{\varepsilon}}^\theta$ in Ω, and imposed displacements $\bar{\boldsymbol{u}}$ on Γ_1, all actions being quasi-statically variable with a pseudo-time $t \geq 0$. The material is a standard material [7, 11, 22], with the constitutive equations:

$$\boldsymbol{\sigma} = \boldsymbol{E}\boldsymbol{\varepsilon}^e, \qquad \boldsymbol{\chi} = \boldsymbol{H}\boldsymbol{\eta}, \tag{1}$$

$$\dot{\boldsymbol{\varepsilon}}^p = \dot{\lambda}\frac{\partial \phi}{\partial \boldsymbol{\sigma}}, \qquad -\dot{\boldsymbol{\eta}} = \dot{\lambda}\frac{\partial \phi}{\partial \boldsymbol{\chi}}, \tag{2}$$

$$\phi(\boldsymbol{\sigma}, \boldsymbol{\chi}) \leq 0, \qquad \dot{\lambda} \geq 0, \qquad \dot{\lambda}\phi(\boldsymbol{\sigma}, \boldsymbol{\chi}) = 0, \tag{3}$$

$$\boldsymbol{\varepsilon} = \boldsymbol{\varepsilon}^e + \boldsymbol{\varepsilon}^p + \bar{\boldsymbol{\varepsilon}}^\theta, \tag{4}$$

where $\boldsymbol{\sigma}$ is the Cauchy stress tensor, $\boldsymbol{\varepsilon}$ the (small) total strain tensor, $\boldsymbol{\varepsilon}^p$ the plastic strain tensor, $\boldsymbol{\chi}$ and $\boldsymbol{\eta}$ dual internal variable vectors (or tensors), $\dot{\lambda}$ the plastic coefficient, whereas $\boldsymbol{E}$ and $\boldsymbol{H}$ (both symmetric and positive definite) are the elastic stiffness fourth order tensor and, respectively, the work hardening coefficient matrix.

At time t of the loading process some plastic strains $\boldsymbol{\varepsilon}^p$ have been accumulated within Ω, together with some strain-like internal variable value $\boldsymbol{\eta}$. The response of the body to the loads acting at this time, plus the latter strains $\boldsymbol{\varepsilon}^p$ considered as initial strains, can be described by means of the Somigliana formulae [1, 3, 12, 18], i. e.

$$\boldsymbol{u}(\boldsymbol{x},t) = \mathcal{R}_u[\boldsymbol{t}_{\Gamma_1}, \boldsymbol{u}_{\Gamma_2}, \boldsymbol{\varepsilon}^p; \boldsymbol{x}] + \hat{\boldsymbol{u}}(\boldsymbol{x},t), \qquad \boldsymbol{x} \in \Omega, \quad t \in (0,T), \tag{5}$$

$$\boldsymbol{t}_{(n)}(\boldsymbol{x},t) = \mathcal{R}_t[\boldsymbol{t}_{\Gamma_1}, \boldsymbol{u}_{\Gamma_2}, \boldsymbol{\varepsilon}^p; \boldsymbol{x}, \boldsymbol{n}] + \hat{\boldsymbol{t}}_{(n)}(\boldsymbol{x},t), \qquad \boldsymbol{x} \in \Omega, \quad t \in (0,T), \tag{6}$$

$$\boldsymbol{\sigma}(\boldsymbol{x},t) = \mathcal{R}_\sigma(\boldsymbol{t}_{\Gamma_1}, \boldsymbol{u}_{\Gamma_2}, \boldsymbol{\varepsilon}^p; \boldsymbol{x}) + \hat{\boldsymbol{\sigma}}(\boldsymbol{x},t), \qquad \boldsymbol{x} \in \Omega, \quad t \in (0,T), \tag{7}$$

where $\boldsymbol{t}_{(n)}$ denotes the traction vector relative to the surface element of unit normal $\boldsymbol{n}$, and furthermore

$$\mathcal{R}_r[\boldsymbol{t}, \boldsymbol{u}, \boldsymbol{\varepsilon}^p; \boldsymbol{x}] := \int_{\Gamma_1} \boldsymbol{G}_{ru}(\boldsymbol{y},\boldsymbol{x})\boldsymbol{t}(\boldsymbol{y},t)d\Gamma_1(\boldsymbol{y}) - \int_{\Gamma_2} \boldsymbol{G}_{rt}(\boldsymbol{y},\boldsymbol{x})\boldsymbol{u}(\boldsymbol{y},t)d\Gamma_2(\boldsymbol{y})$$

$$+ \int_{\Omega} \boldsymbol{G}_{r\sigma}(\boldsymbol{y},\boldsymbol{x})\boldsymbol{\varepsilon}^p(\boldsymbol{y},t)d\Omega, \qquad (r = u, t, \sigma). \tag{8}$$

Here, the simpler notation $\boldsymbol{t}$ indicates the traction vector upon the boundary surface, with the unit normal $\boldsymbol{n}$ coincident with the unit external normal to the surface. The two-point tensor valued functions $\boldsymbol{G}_{rs}(\boldsymbol{y},\boldsymbol{x}) = \boldsymbol{G}^T_{sr}(\boldsymbol{x},\boldsymbol{y})$ collect different types of fundamental solutions. For 3D infinite domain, the above $\boldsymbol{G}_{rs}$ read (Kelvin solution [1, 3]):

$$\boldsymbol{G}_{uu}(\boldsymbol{y},\boldsymbol{x}) = \frac{\lambda + 3\mu}{8\pi\mu(\lambda + 2\mu} \left(\frac{1}{|\boldsymbol{y} - \boldsymbol{x}|} \boldsymbol{I}_{3\times 3} + \frac{\lambda + \mu}{\lambda + 3\mu} \frac{(\boldsymbol{y}-\boldsymbol{x})(\boldsymbol{y}-\boldsymbol{x})^T}{|\boldsymbol{y}-\boldsymbol{x}|^3} \right),$$

$$\boldsymbol{G}_{ut}(\boldsymbol{y},\boldsymbol{x}) = (T_y \boldsymbol{G}_{uu}(\boldsymbol{x},\boldsymbol{y}))^T, \qquad \boldsymbol{G}_{tu}(\boldsymbol{y},\boldsymbol{x}) = T_x \boldsymbol{G}_{uu}(\boldsymbol{y},\boldsymbol{x}),$$

$$\boldsymbol{G}_{u\sigma}(\boldsymbol{y},\boldsymbol{x}) = (\boldsymbol{E} \;\; def_x \; \boldsymbol{G}_{uu}(\boldsymbol{y},\boldsymbol{x}))^T, \qquad \boldsymbol{G}_{\sigma u}(\boldsymbol{y},\boldsymbol{x}) = \boldsymbol{E} \;\; def_y \; \boldsymbol{G}_{uu}(\boldsymbol{x},\boldsymbol{y}),$$

$$\boldsymbol{G}_{t\sigma}(\boldsymbol{y},\boldsymbol{x}) = (\boldsymbol{E} \;\; def_x \; \boldsymbol{G}_{tu}(\boldsymbol{y},\boldsymbol{x}))^T, \qquad \boldsymbol{G}_{\sigma t}(\boldsymbol{y},\boldsymbol{x}) = \boldsymbol{E} \;\; def_y \; \boldsymbol{G}_{ut}(\boldsymbol{x},\boldsymbol{y}),$$

$$\boldsymbol{G}_{tt}(\boldsymbol{y},\boldsymbol{x}) = T_x(T_y \boldsymbol{G}_{uu}(\boldsymbol{x},\boldsymbol{y}))^T, \qquad \boldsymbol{G}_{\sigma\sigma}(\boldsymbol{y},\boldsymbol{x}) = \boldsymbol{E} \;\; def_y \; \boldsymbol{G}_{u\sigma}(\boldsymbol{x},\boldsymbol{y})$$

with the trace operator

$$(T_z \boldsymbol{v}(\boldsymbol{z}))_i := \lambda \boldsymbol{n}_i \;\; div \; \boldsymbol{v} + \mu \frac{\partial}{\partial \boldsymbol{n}} \boldsymbol{v}_i + \mu \left\langle \frac{\partial \boldsymbol{v}}{\partial \boldsymbol{z}_i}, \boldsymbol{n} \right\rangle, \qquad 1 \le i \le 3,$$

and

$$def_z\ \boldsymbol{v}(\boldsymbol{z}) := \frac{1}{2}(\partial_i \boldsymbol{v}_j + \partial_j \boldsymbol{v}_i), \qquad i,j = 1,2,3.$$

Remark:

The subscript notation in (8) indicates the effect at $\boldsymbol{y}$ by the first subscript (i. e. displacement for $r = u$, traction for $r = t$, stress for $r = \sigma$) and the singularity at $\boldsymbol{x}$ producing that effect by the second subscript through a duality rule (i. e. unit force for $s = u$, unit concentrated relative displacement for $s = t$, unit concentrated distortion for $s = \sigma$) [12, 18].

The terms $\hat{\boldsymbol{u}}, \hat{\boldsymbol{\sigma}}, \hat{\boldsymbol{t}}_{(n)} = \hat{\boldsymbol{\sigma}} \cdot \boldsymbol{n}$ on the right hand side of (5), (6), (7) represent the displacement, the stress and traction fields produced by the loads in the infinite elastic domain Ω_∞ at time t. The functions $\hat{\boldsymbol{u}}$ and $\hat{\boldsymbol{t}}_{(n)}$ are discontinuous across Γ_1 and Γ_2, respectively, that is:

$$\hat{\boldsymbol{u}}^+ - \hat{\boldsymbol{u}}^- = -\bar{\boldsymbol{u}} \text{ on } \Gamma_1, \qquad \hat{\boldsymbol{t}}^+|_{\Gamma_2} - \hat{\boldsymbol{t}}^-|_{\Gamma_2} = -\bar{\boldsymbol{t}} \text{ on } \Gamma_2, \qquad \forall\, t \geq 0, \tag{9}$$

where $(.)^+$ and $(.)^-$ are values at points infinitly close to Γ from outside and inside, respectively.

Equations (5), (6), (7) satisfy equilibrium, compatibility and Hooke's law everywhere within Ω. In order to enforce the boundary conditions too, the following equations must be satisfied [17]:

$$\int_{\Gamma_1} \boldsymbol{v}^T(\boldsymbol{x}) \mathcal{R}_u[\boldsymbol{t}, \boldsymbol{u}, \boldsymbol{\varepsilon}^p; \boldsymbol{x}]\, d\Gamma_1(\boldsymbol{x}) = \int_{\Gamma_1} \boldsymbol{v}^T(\boldsymbol{x})\big(\, \frac{1}{2}\, \bar{\boldsymbol{u}}(\boldsymbol{x}) - \hat{\boldsymbol{u}}(\boldsymbol{x})\big)\, d\Gamma_1(\boldsymbol{x}), \tag{10}$$

$$\int_{\Gamma_2} \boldsymbol{w}^T(\boldsymbol{x}) \mathcal{R}_t[\boldsymbol{t}, \boldsymbol{u}, \boldsymbol{\varepsilon}^p; \boldsymbol{x}, \boldsymbol{n}(\boldsymbol{x})|_{\Gamma_2}]\, d\Gamma_2(\boldsymbol{x}) = \int_{\Gamma_2} \boldsymbol{w}^T(\boldsymbol{x})\big(\, \frac{1}{2}\, \bar{\boldsymbol{t}}(\boldsymbol{x}) - \hat{\boldsymbol{t}}(\boldsymbol{x})\big)\, d\Gamma_2(\boldsymbol{x}) \tag{11}$$

for all functions $\boldsymbol{v}, \boldsymbol{w}$. This system of boundary integral equations provides, for any $\boldsymbol{\varepsilon}^p$ specified in Ω, the unknown boundary values $\boldsymbol{t}|_{\Gamma_1}$ and $\boldsymbol{u}|_{\Gamma_2}$ pertaining to the elastic response of the body to the loads and to the (unknown) plastic strains $\boldsymbol{\varepsilon}^p$ at time t.

Space discretization by BEs and IEs

A Galerkin approach is here employed to achieve the space discretization of the boundary surface by BEs e_m^Γ $(m = 1, ..., M_\Gamma)$, and of the domain by IEs e_m^Ω $(m = 1, ..., M)$. To this purpose let the boundary Γ be discretized with a general mesh parameter h and the region Ω with an independent volume mesh parameter H; furthermore, let $\boldsymbol{t}|_{\Gamma_1}$ and $\boldsymbol{u}|_{\Gamma_2}$ be approximated by

$$\boldsymbol{t}_h(\boldsymbol{x},t) := \boldsymbol{N}_1(\boldsymbol{x})\boldsymbol{X}_1(t) \qquad \text{on } \Gamma_1, \tag{12}$$

$$-\boldsymbol{u}_h(\boldsymbol{x},t) := \boldsymbol{N}_2(\boldsymbol{x})\boldsymbol{X}_2(t) \qquad \text{on } \Gamma_2, \tag{13}$$

where N_1 and N_2 are matrices of (piecewise $C^1(\Gamma)$) shape functions. Also, let $\boldsymbol{\varepsilon}^p$ and the associated strain internal variable $\boldsymbol{\eta}$ be approximated by

$$\boldsymbol{\varepsilon}^p_H(\boldsymbol{x},t) := \sum_{m=1}^{M} \delta(\boldsymbol{x}-\boldsymbol{x}_m)\boldsymbol{P}_m(t), \qquad \text{in } \Omega, \tag{14}$$

$$\boldsymbol{\eta}_H(\boldsymbol{x},t) := \sum_{m=1}^{M} \delta(\boldsymbol{x}-\boldsymbol{x}_m)\boldsymbol{Q}_m(t), \qquad \text{in } \Omega, \tag{15}$$

with the Dirac's delta $\delta(.)$ and the strain points $\boldsymbol{x}_m \in e^{\Omega}_m$. The use of trial functions (14), (15) makes it possible deriving the common strain point scheme [16, 32] through a formal Galerkin procedure.

Thus, applying upon equations (10), (11) and (7) such a procedure with

$$\boldsymbol{v}_h := \boldsymbol{N}_1\boldsymbol{Y}_1, \qquad \boldsymbol{w}_h = \boldsymbol{N}_2\boldsymbol{Y}_2,$$

gives the equations

$$\boldsymbol{K}_{11}\boldsymbol{X}_1 + \boldsymbol{K}_{12}\boldsymbol{X}_2 + \sum_{m=1}^{M} \boldsymbol{B}_{1m}\boldsymbol{P}_m = \boldsymbol{R}_1, \tag{16}$$

$$\boldsymbol{K}_{21}\boldsymbol{X}_1 + \boldsymbol{K}_{22}\boldsymbol{X}_2 + \sum_{m=1}^{M} \boldsymbol{B}_{2m}\boldsymbol{P}_m = \boldsymbol{R}_2, \tag{17}$$

$$\boldsymbol{\sigma}_{H,i} = \boldsymbol{B}^T_{1i}\boldsymbol{X}_1 + \boldsymbol{B}^T_{2i}\boldsymbol{X}_2 + \sum_{m=1}^{M}(\boldsymbol{G}_{\sigma\sigma})_{im}\boldsymbol{P}_m + \hat{\boldsymbol{\sigma}}_i, \qquad (i=1,...,M), \tag{18}$$

where $\boldsymbol{\sigma}_{H,i} = \boldsymbol{\sigma}_{H,i}(t) := \boldsymbol{\sigma}(\boldsymbol{x}_i,t)$, $\hat{\boldsymbol{\sigma}}_i = \hat{\boldsymbol{\sigma}}_i(t) := \hat{\boldsymbol{\sigma}}(\boldsymbol{x}_i,t)$, $(\boldsymbol{G}_{\sigma\sigma})_{im} = \boldsymbol{G}_{\sigma\sigma}(\boldsymbol{x}_i,\boldsymbol{y}_m)$,

$$\boldsymbol{K}_{11} := \int_{\Gamma_1}\int_{\Gamma_1} \boldsymbol{N}^T_1\boldsymbol{G}_{uu}\boldsymbol{N}_1 \, d\Gamma_1 \, d\Gamma_1, \qquad \boldsymbol{K}_{22} := \int_{\Gamma_2}\int_{\Gamma_2} \boldsymbol{N}^T_2\boldsymbol{G}_{tt}\boldsymbol{N}_2 \, d\Gamma_2 \, d\Gamma_2, \tag{19}$$

$$\boldsymbol{K}_{12} := \int_{\Gamma_1}\int_{\Gamma_2} \boldsymbol{N}^T_1\boldsymbol{G}_{ut}\boldsymbol{N}_2 \, d\Gamma_1 \, d\Gamma_2 = \boldsymbol{K}^T_{21}, \tag{20}$$

$$\boldsymbol{B}_{1i} := \int_{\Gamma_1} \boldsymbol{N}_1\boldsymbol{G}_{u\sigma}(\boldsymbol{x},\boldsymbol{y}_i) \, d\Gamma_1(\boldsymbol{x}), \quad \boldsymbol{B}_{2i} := \int_{\Gamma_2} \boldsymbol{N}_2\boldsymbol{G}_{t\sigma}(\boldsymbol{x},\boldsymbol{y}_i) \, d\Gamma_2(\boldsymbol{x}), \quad (i=1,...,M), \tag{21}$$

$$\boldsymbol{R}_1 := \int_{\Gamma_1} \boldsymbol{N}_1^T (\frac{1}{2}\,\bar{\boldsymbol{u}} - \hat{\boldsymbol{u}})\; d\Gamma_1, \qquad \boldsymbol{R}_2 := \int_{\Gamma_2} \boldsymbol{N}_1^T (\frac{1}{2}\,\bar{\boldsymbol{t}} - \hat{\boldsymbol{t}})\; d\Gamma_2, \tag{22}$$

We note that $\boldsymbol{K}_{11}$ and $\boldsymbol{K}_{22}$ are both symmetric, $\boldsymbol{K}_{11}$ positive definite and $\boldsymbol{K}_{22}$ negative definite. Also, the matrices $(\boldsymbol{G}_{\sigma\sigma})_{im}$ are symmetric and negative semidefinite for all $i, m = 1, 2, ..., M$ [18].

Let us rewrite equations (14), (15) in rate form, i. e.

$$\dot{\boldsymbol{\varepsilon}}^p_H(\boldsymbol{x}, t) := \sum_{m=1}^{M} \delta(\boldsymbol{x} - \boldsymbol{x}_m)\dot{\boldsymbol{P}}_m(t), \qquad \dot{\boldsymbol{\eta}}_H(\boldsymbol{x}, t) := \sum_{m=1}^{M} \delta(\boldsymbol{x} - \boldsymbol{x}_m)\dot{\boldsymbol{Q}}_m(t), \qquad \text{in } \Omega, \tag{23}$$

and let us apply the Galerkin procedure to equations (2) and (3) with the aid of the following approximation for $\dot{\lambda}$, consistent with (23):

$$\dot{\lambda}_H(\boldsymbol{x}, t) = \sum_{m=1}^{M} \delta(\boldsymbol{x} - \boldsymbol{x}_m)\dot{\Lambda}_m(t). \tag{24}$$

Thus, substitution of (23) and (24) into equations (2) and then integration over Ω gives:

$$\dot{\boldsymbol{P}}_i = \dot{\Lambda}_i \frac{\partial \phi}{\partial \boldsymbol{\sigma}}|_{\boldsymbol{\sigma}_{H,i}, \boldsymbol{\chi}_{H,i}}, \qquad -\dot{\boldsymbol{Q}}_i = \dot{\Lambda}_i \frac{\partial \phi}{\partial \boldsymbol{\chi}}|_{\boldsymbol{\sigma}_{H,i}, \boldsymbol{\chi}_{H,i}} \qquad (i = 1, 2, ..., M). \tag{25}$$

Furthermore, integration over Ω of the complementarity condition in (3) implies that

$$\sum_{m=1}^{M} \phi(\boldsymbol{\sigma}_m, \boldsymbol{\chi}_m)\dot{\Lambda}_m = 0,$$

such that equation (3) can be replaced by

$$\phi(\boldsymbol{\sigma}_{H,i}, \boldsymbol{\chi}_{H,i}) \le 0, \quad \dot{\Lambda}_i \ge 0, \quad \dot{\Lambda}\phi(\boldsymbol{\sigma}_{H,i}, \boldsymbol{\chi}_{H,i}) = 0 \;\; (i = 1, 2, ..., M). \tag{26}$$

Therefore, with the Galerkin procedure, equations (2) and (3) are approximated into the equations (25) and (26), respectively, which simply enforce the material equations at the selected strain points.

The above procedure, often employed in the literature — likely in different but equivalent manners [12, 13, 16] — may appear to be rather heuristic. This is not the case, as in fact:

1. equations (16), (17) can equally be derived through a variational approach (for instance using the boundary min-max principle [18 - 20, 22],
2. equations (25) and (26), together with equation (18) and corresponding terms in equations (16), (17), are specialized forms of more general IEs global variationally consistent yield laws [19, 20, 22], obtained on choosing the relevant shape function matrix as in (14), (15).

Backward-difference equation system and solution method

Equations (16) - (18), (25), (26), together with the second of (1), govern the BE/IE discretized problem, whose solution requires a time discretization. To this purpose, here the backward difference method, known for its variational consistency and convergence properties [13, 23], will be employed here. This implies subdividing the entire time interval $(0,T)$ into steps (for the sake of simplicity of equal length τ), enforcing all equations at the step ends, and replacing the rate quantities $\dot{\boldsymbol{P}}_i$ and $\dot{\boldsymbol{Q}}_i$ with the step increments $\Delta\boldsymbol{P}_i$, $\Delta\boldsymbol{Q}_i$.

For later use, let the stress $\boldsymbol{\sigma}_{H,i}$ be split — according to the first equation (1) and (2) — into two parts, namely $\boldsymbol{\sigma}_{H,i} = \tilde{\boldsymbol{\sigma}}_{H,i} + \boldsymbol{\sigma}^0_{H,i}$, where $\tilde{\boldsymbol{\sigma}}_{H,i} := \boldsymbol{E}(\boldsymbol{\varepsilon}_{H,i} - \boldsymbol{\varepsilon}^\theta_i)$ is the "elastic" part (i. e. not related to $\boldsymbol{\varepsilon}^p_{H,i}$) and $\boldsymbol{\sigma}^0_{H,i} := -\boldsymbol{E}\boldsymbol{\varepsilon}^p_{H,i}$ is the "initial" stress. Correspondingly, the matrices $(\boldsymbol{G}_{\sigma\sigma})_{im}$ can also be split into two parts, i. e.

$$(\boldsymbol{G}_{\sigma\sigma})_{im} = (\tilde{\boldsymbol{G}}_{\sigma\sigma})_{im} - \boldsymbol{E}\delta_{im} \qquad (i,m = 1,2,...,M), \tag{27}$$

where δ_{im} is the Kronecker symbol. Accounting for (27), equations (16), (17), (18), the second equation (1), (25), (26) are rewritten here in their incremental form for the n-th step and in a more compact form as follows:

$$\boldsymbol{K}\Delta\boldsymbol{X} + \sum_{m=1}^{M} \boldsymbol{B}_m \Delta\boldsymbol{P}_m = \Delta\boldsymbol{R}, \tag{28}$$

$$\Delta\tilde{\boldsymbol{\sigma}}_{H,i} = \boldsymbol{B}_i^T \Delta\boldsymbol{X} + \sum_{m=1}^{M} (\tilde{\boldsymbol{G}}_{\sigma\sigma})_{im} \Delta\boldsymbol{P}_m + \Delta\hat{\boldsymbol{\sigma}}_i \tag{29}$$

$$\Delta\boldsymbol{\sigma}^0_{H,i} = -\boldsymbol{E}\Delta\boldsymbol{P}_i, \qquad \Delta\boldsymbol{\chi}_{H,i} = \boldsymbol{H}\Delta\boldsymbol{Q}_i, \tag{30}$$

$$\boldsymbol{\sigma}_{H,i} = \boldsymbol{\sigma}_{H,i,(n-1)} + \Delta\tilde{\boldsymbol{\sigma}}_{H,i} + \Delta\boldsymbol{\sigma}^0_{H,i}, \qquad \boldsymbol{\chi}_{H,i} = \boldsymbol{\chi}_{H,i,(n-1)} + \Delta\boldsymbol{\chi}_{H,i}, \tag{31}$$

$$\Delta\boldsymbol{P}_i = \Delta\Lambda_i \frac{\partial\phi}{\partial\boldsymbol{\sigma}}\Big|_{\boldsymbol{\sigma}_{H,i},\boldsymbol{\chi}_{H,i}}, \qquad -\Delta\boldsymbol{Q}_i = \Delta\Lambda_i \frac{\partial\phi}{\partial\boldsymbol{\chi}}\Big|_{\boldsymbol{\sigma}_{H,i},\boldsymbol{\chi}_{H,i}}, \tag{32}$$

$$\phi(\boldsymbol{\sigma}_{H,i},\boldsymbol{\chi}_{H,i}) \leq 0, \quad \Delta\Lambda_i \geq 0, \quad \Delta\Lambda\phi(\boldsymbol{\sigma}_{H,i},\boldsymbol{\chi}_{H,i}) = 0, \tag{33}$$

where no index n has been appended for simplicity (except for the terms denoting values related to the previous $(n-1)$-th step).

Equations (28)-(33) govern the typical step problem. The latter can be solved using an iterative predictor/corrector scheme. Let $\Delta\boldsymbol{P}_i^{(k-1)}$ and $\Delta\boldsymbol{Q}_i^{(k-1)}$ be an estimate of $\Delta\boldsymbol{P}_i$ and $\Delta\boldsymbol{Q}_i$. Then, from equation (28) we derive $\Delta\boldsymbol{X}^{(k)}$, and from equation (29) $\Delta\tilde{\boldsymbol{\sigma}}^{(k)}_{H,i}$. The stresses $\boldsymbol{\sigma}_{H,i} = \boldsymbol{\sigma}_{H,i,(n-1)} + \Delta\tilde{\boldsymbol{\sigma}}^{(k)}_{H,i} - \boldsymbol{E}\Delta\boldsymbol{P}_i^{(k-1)}$ together with the estimated vectors

$\Delta \boldsymbol{P}_i^{(k-1)}$, $\Delta \boldsymbol{Q}_i^{(k-1)}$ and some $\Delta\Lambda_i$ likely do not comply with equations (30)–(33) at least somewhere, thus we proceed with a correction. This consists in finding, for every strain point, the set $(\Delta\boldsymbol{\sigma}_{H,i}^{0(k)}, \Delta\boldsymbol{P}_i^{(k)}, \Delta\boldsymbol{\chi}_{H,i}^{(k)}, \Delta\boldsymbol{Q}_i^{(k)}, \Delta\Lambda_i^{(k)})$ satisfying equations (30)–(33). This set of nonlinear equations can be solved in a number of ways [13, 16, 32], for instance by solving, for every strain point, the following equivalent (small size, convex) mathematical programming (MP) problem in terms of only $\Delta\boldsymbol{\sigma}_{H,i}^0$ and $\Delta\boldsymbol{\chi}_{H,i}$ [22, 23]:

$$\Pi(\Delta\boldsymbol{\sigma}_{H,i}^0, \Delta\boldsymbol{\chi}_{H,i}) := \frac{1}{2}(\Delta\boldsymbol{\sigma}_{H,i}^0)^T \boldsymbol{E}^{-1}\Delta\boldsymbol{\sigma}_{H,i}^0 + \frac{1}{2}(\Delta\boldsymbol{\chi}_{H,i})^T \boldsymbol{H}^{-1}\Delta\boldsymbol{\chi}_{H,i} = \min ! \qquad (34)$$

$$\text{subjects to} \quad \phi(\boldsymbol{\sigma}_{H,i,(n-1)} + \Delta\tilde{\boldsymbol{\sigma}}_{H,i}^{(k)} + \Delta\boldsymbol{\sigma}_{H,i}^0,\ \boldsymbol{\chi}_{H,i,(n-1)} + \Delta\boldsymbol{\chi}_{H,i}) \leq 0. \qquad (35)$$

The (unique) solution of (35) is the basis of a new iteration, if necessary in relation to an assumed tolerance. A MP formulation in terms of only $\Delta\boldsymbol{P}_i, \Delta\boldsymbol{Q}_i$ is also possible [22], but here we miss this point for lack of space.

Conclusions

We have presented a discretization method for elastic plastic analysis problems with standard material models, i. e. endowed of dual internal variables and quadratic thermodynamic potential (linear hardening laws), in the case of quasi-statically variable external actions (body forces, tractions, thermal strains, imposed displacements) and small displacements.

A Galerkin-type approach has been utilized for the space discretization by coupled boundary elements (BE) and interior elements (IE), the latter in order to modell the unknown plastic strains and the strain-like internal variables, with the material plastic flow laws enforced at a set of strain points within every IE. Time discretization has been given the shape of a standard backward difference method. The resulting finite step-BE-IE formulation enjoys of symmetry and sign definiteness properties, and is thus most suitable to solve a variety of (3D or 2D) plasticity problems.

The proposed method is in particular expected to be most appropriate for a class of elastic-plastic problems where an accurate evaluation of the stress field (for instance, around a singularity as a crack tip) and a realistic assessment of the material hardening laws may be of paramount importance. The method requires double integrations and the use of a wider set of fundamental solutions, including hypersingular ones, with some computational problems. Numerical applications for the assessment of the computational performance of the method are on the way and will be reported elsewhere.

References

1. Banerjee, P. K.; Butterfield R.: *Boundary Element Methods in Engineering Science.* London: McGraw-Hill 1981.

2. Banerjee, P. K.; Cathie, D. N.: *A direct formulation and numerical implementation of the boundary element method for two-dimensional problems in elastoplasticity.* Int. J. Mech. Sci. 22 (1988) 233-245.

3. Brebbia, C. A.; Telles, J. C. F.; Wrobel, L. C.: *Boundary Element Techniques.* Berlin, Heidelberg: Springer-Verlag 1984.

4. Costabel, M.: *Principles of Boundary Element Methods.* Lecture notes at the summer course "Finite Elements in Physics", Lausanne 1986.

5. Costabel, M.; Stephan, E. P.: *Coupling of finite element and boundary element methods for an elasto-plastic interface problem.* Preprint, Georgia Institute of Technologie 1988.

6. Guiggiani, M.; Krishnamy, G.; Rudolphi, T. J.; Rizzo, F. J.: *A General Algorithm for Numerical Solution of Hypersingular Boundary Integral Equations.* ASME J. Appl. Mech., to appear.

7. Halphen, B.; Nguyen, Q. S.: *Sur les matériaux standard génerallisés.* J. de Mécanique 14 (1975) 39-63.

8. Hsiao, G. C.: *The coupling of boundary element and finite element methods.* ZAMM 70 (1990) 493-503.

9. Kane, J. H.; Maier, G.; Tosaka, N.; Atluri, S. N. (eds.): *Advances in Boundary Element Techniques.* New York, Berlin, Heidelberg: Springer-Verlag (to appear).

10. Korneev, V. G., Langer, U.: *Approximate Solution of Plastic Flow Problems.* TEUBNER-TEXTE zur Mathematik, Bd. 69. Leipzig: BSB B. G. Teubner Verlagsgesellschaft, 1984.

11. Lemaitre, J.; Chaboche, J. L.: *Mécanique des Matériaux solides.* Paris: Dunod 1985.

12. Maier, G.; Polizzotto, C.: *A Galerkin approach to boundary element elasto-plastic analysis.* Comput. Meths. Appl. Mech. Engng. 60 (1987) 175-194.

13. Maier, G.; Novati, G.; Perego, U.: *Plastic analysis by boundary elements.* In: Stein, E.; Wendland, W. L.: Finite Element and Boundary Element Techniques from Mathematical and Engineering Point of View. Wien, New York: Springer-Verlag, 1988, pp. 213 - 272.

14. Hlaváček, I.; Haslinger, J.; Nečas, J.; Lovišek, J.: *Solution of Variational Inequalities in Mechanics.* Appl. Math. Sciences 66. New York, Berlin, Heidelberg, London, Paris, Tokyo: Springer-Verlag, 1988.

15. Ortiz, M.; Popov, E. P.: *Accuracy and stability of integration algorithms for elasto plastic constitutive equations.* Int. J. Num. Meth. Engng. 21 (1985) 1561 - 1675.

16. Owen, D. R. J.; Hinton, E.: *Finite Elements in Plasticity, Theory and Applications.* Swansea, UK: Pineridge Press 1980.

17. von Petersdorff, T.: *Randwertprobleme der Elastizitaetstheorie fuer Polyeder - Singularitaeten und Approximation mit Randelementmethoden.* Doctoral Thesis, Darmstadt University, 1989.

18. Polizzotto, C.: *An energy approach to the boundary element method. Part I: elastic solids.* Comput. Meths. Appl. Mech. Engng. 69 (1988) 167-184.

19. Polizzotto, C.: *An energy approach to the boundary element method. Part II: elastic-plastic solids.* Comput. Meths. Appl. Mech. Engng. 69 (1988) 263-276.

20. Polizzotto, C.: *A consistent formulation of the BEM within elasoplasticity.* In: Cruse, T. A. (ed.), Advanced Boundary Element Methods, 315-324, Berlin, Heidelberg: Springer-Verlag 1988.

21. Polizzotto, C.: *A symmetric-definite BEM formulation for the elastoplastic rate problem.* In: Brebbia, C. A.; Wendland, W. L.; Kuhn, G. (eds.): Boundary Elements IX; Vol. 2: Stress Analysis, 315-334. Southampton, Boston: Comput. Mech. Pubs. 1987.

22. Polizzotto, C.; Panzeca, T.; Zito, M.: *A consistent boundary/interior element method for evolutive elastic plastic structural analysis.* In: Kane, J. H. et al (eds.): Advances in Boundary Element Techniques. New York, Berlin, Heidelberg: Springer-Verlag (to appear).

23. Recontré, L. J., Bird, W. W.; Martin J. B.: *Internal variable formulation of a backward difference corrector algorithm for piecewise linear yield surfaces.* Meccanica (to appear).

24. Schnack, E.: *Coupling Strategies for BEM and FEM.* Proceedings of the Second World Congress on Comput. Mech. 1990, 27 - 31 August 1990 held in Stuttgart/Germany. In: Second World Congress on Comput. Mech. Int. Association of Comput. Mech., Stuttgart 1990, pp. 312 - 315.

25. Schnack, E.: *Macro-Elements for 2D- and 3D-Elasticity with BEM.* Proceedings of BEM 12, 24 - 27 September 1990, held in Sapporo/Japan. In: Tanaka, M.; Brebbia, C: A.; Honma, T. (eds.): Boundary Elements XII, Vol. 1, Comput. Mech. Pubs., Berlin, Heidelberg: Springer-Verlag 1990, pp. 21 - 31.

26. Schnack, E.: *Mixed Methods in BEM for Elasto-Plastic Problems.* Proceedings of the Boundary Elements in Engineering Conference in Tokyo, 22 - 25 September 1986. In: Tanaka, M.; Brebbia, C. A. (eds.): Boundary Elements VIII, Vol. II. A Computational Mechanics Publication. Berlin, Heidelberg, New York, Tokyo: Springer-Verlag, 1986, pp. 641-651.

27. Schwab, C.; Wendland, W. L.: *On the Numerical Integration of Singular Surface Integrals in the BEM.* Proceedings of the Seventh GAMM-Seminar on Numerical Techniques for Boundary Element Methods, 25 - 27 January 1991, held in Kiel/Germany. To appear in: Hirschel, E. H. et al (eds.): Notes on Numerical Fluid Mechanics. Vieweg-Verlag, Braunschweig.

28. Simo, J. C.; Taylor, R. L.: *A return mapping algorithm for plane stress elastoplasticity.* Int. J. Num. Meth. Engng. 22 (1986) 647 - 670.

29. Tanaka, M.; Tanaka, K.: *On a new boundary element solution scheme for elastoplasticity.* Ing. Arch. 50 (1981) 289-295.

30. Telles, J. C. F.; Brebbia, C. A.: *Boundary Elements: New Developments in Elastoplastic Analysis.* Appl. Math. Modelling 5 (1981) 376-382.

31. Wendland, W. L.: *Strongly elliptic boundary integral equations.* In: Iserles, A.; Powel, M. (eds.). The State of the Art in Numerical Analysis. Oxford: Oxford University Press 1987.

32. Zienkiewicz, O. C.: *The Finite Element Method.* London: McGraw-Hill 1977.

Experiments and Applications

On the Application of Plasticity Theory for Developing Metal Forming and Testing Processes

O. Pawelski

Max-Planck-Institut für Eisenforschung, Düsseldorf, Germany

Summary

Plasticity theory has proved to be an effective aid in developing metal forming and testing processes. The intention of this report is not to explain the various analytical and numerical methods, which today the engineer has at his disposal. Selected examples are rather discussed to demonstrate the practical use of the mathematical theory of plasticity in a special technical field, that is in the field of hot and cold rolling of steel products, backed up by laboratory experiments.

Introduction

Most of the conventional metal forming processes were developed empirically by trial and error and by the originality of our ancestors. The rapid progress of theory in our century enabled a deeper understanding of the questions how and why they function. At present we dispose of such powerful methods of computer simulation that the further development of the traditional forming processes and the creation of new ones may be brought about so to speak on the keyboard. But for an optimum effect it is now as before necessary to combine theoretical calculations with appropriate experiments in the laboratory and finally in the production shop, or to confirm this view with a word of Max Planck: A theory without experimental observations is empty, and an experiment without theory is blind.

The report gives some examples of the use of the mathematical theory of plasticity in the field of hot and cold rolling of

D. Besdo, E. Stein (Eds.)
Finite Inelastic Deformations –
Theory and Applications
IUTAM Symposium Hannover/Germany 1991

steel products, backed up by laboratory experiments. The well-known methods mainly used were the elementary theory of plastomechanics, the method of visioplasticity, the theory of upper bound of deformation energy, the theory of slipline field, and the finite element method. A rolling process, just as each other forming process, exerts an influence on the machine and on the product. The machine, i.e. the rolling mill, is loaded by forces and torques, sometimes in a certain vibration mode, and has to sustain high amount of heat. The quality of the product is characterized by accuracy of shape and dimensions, definite pattern of surface roughness, and controlled microstructure with associated mechanical properties, i.e. mainly strength and ductility.

Rolling within the continuous casting process

In the process of continuous casting of steel slabs it is of advantage to roll the slab already when the cross section is not yet fully solidified. As the effect of ferrostatic pressure can be neglected in a first approximation, this new process was investigated by use of a 3D-FEM-simulation of rolling hollow slabs with varied wall thickness. Fig. 1 shows the calculated v. Mises equivalent stress in the originally 60x150 mm^2 cross section of a hollow slab just entering the roll gap. For both

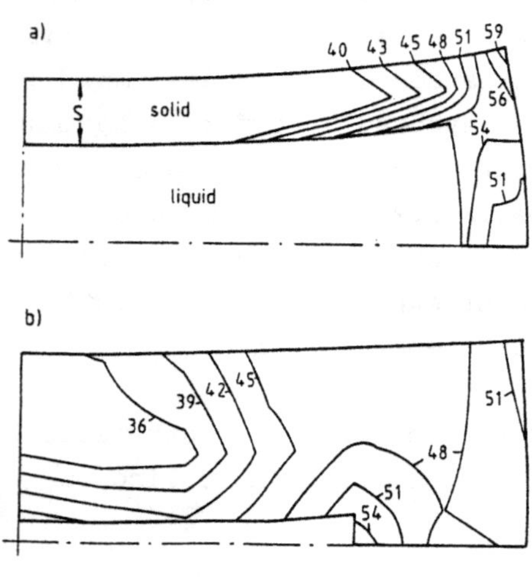

Fig.1. Equivalent stress (N/mm^2) in rolling of partially solidified slabs, one quarter of ingoing cross section; wall thickness s: a) 10 mm, b) 25 mm; reduction of height ε_h: a) 20%, b) 12%; outer dimensions before deformation: 60x150 mm^2

wall thicknesses of 10 and 25 mm a vertical inside bending and a lateral outside bending is to be noticed, so that only the edges of the profile remain in contact with the rolls. The same kind of deformation was found experimentally. Tests with pressurized water inside the specimen showed that a relatively high pressure is necessary to counterbalance this buckling phenomenon.

The described process is not free of tension stresses, which raise the danger of crack formation in the just solidified material. Fig. 2 presents a newly developed test method for the hot ductility of steels, the rolling-tension test /1/. Rolling compression and elongation of the side parts cause a free tension of the middle part until fracture. The reduction of area of the tension part gives a measure of ductility dependent on temperature and strain rate.

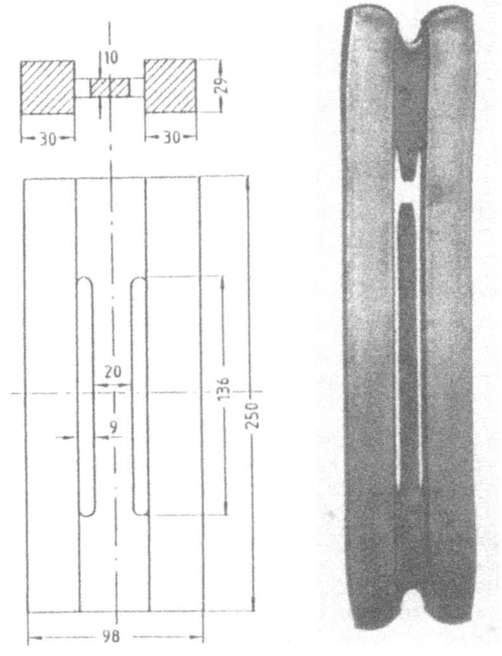

Fig.2. Rolling-tension test specimen before and after deformation

Strategy of rolling of wide hot strip

Rolling of wide hot strip on a tandem mill must be finished with exact data of thickness and temperature, in order to achieve the optimum microstructure and mechanical properties. Furthermore, the final product must be free of shape faults like edge or centre buckles. These demands having in mind, an optimising strategy for the pass schedule was developed /2/. Fig. 3 gives an example of roll force versus strip thickness, as calculated by elementary rolling theory. The first two passes are breakdown passes with the aim of maximum thickness reduction, resulting in high forces near the force limit of the plant. In the so-called form passes 3 to 7 the roll force is reduced proportional to the thickness. This ensures an adequate kind of elastic roll bending which does not affect the relative thickness profile across the strip width and thereby avoids the generation of strip buckles. The optimised changes of roll force, strip thickness, relative profile, and temperature from pass to pass for this special rolling example are given in detail in Fig. 4.

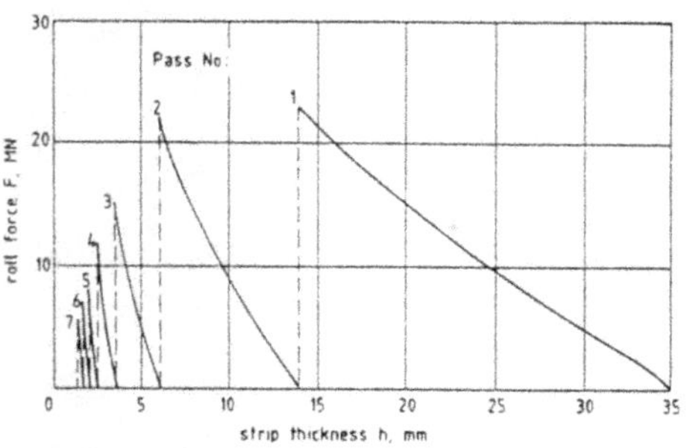

Fig.3. Change of roll force with strip thickness in wide hot strip rolling

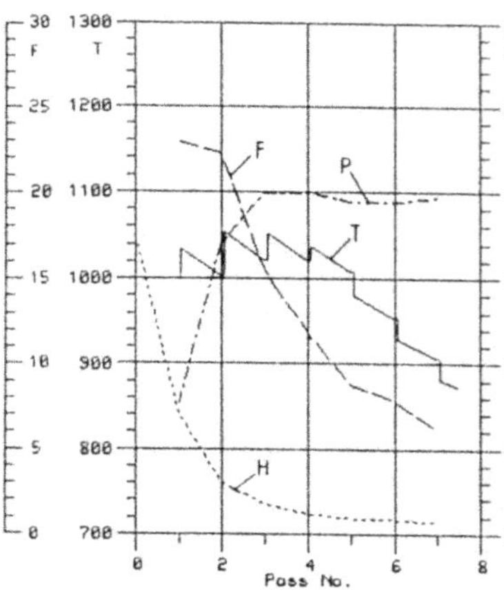

Fig.4. Optimised pass schedule for wide hot strip rolling: H thickness, mm; B width, mm; T temperature, ^{o}C; v roll speed, m/s; P profile number,%

	H	B	T	v
initial	35.00	1000	1000	0.3
final	1.50	1000	873	7.5

Simulation of hot deformation influences on material properties

A very wide field of research is the interrelationship between the parameters of the hot rolling process and the material properties. Since 15 years in the MPI für Eisenforschung the hot flat compression test has been used as a very effective method for the physical simulation of hot rolling in the "Warmumformsimulator WUMSI" /3/. Fig. 5 shows schematically that each rolling pass is replaced by a flat compression deformation, controlled to the same values of strain, strain rate, temperature, and time interval as in the technical rolling process.

Plasticity theory was again very helpful for analysing the stresses, strains, and temperatures in the specimen, and for deriving flow curves from the measured data. Fig. 6 gives an example of the ratio of local strain and nominal strain in a deformed WUMSI-specimen, as gained by the method of visioplasticity /4/. The shadowed areas with the ratio 1 are preferred regions for sampling of secondary specimens enabling a complete testing of microstructure, strength, and ductility after various processes of thermomechanical treatment.

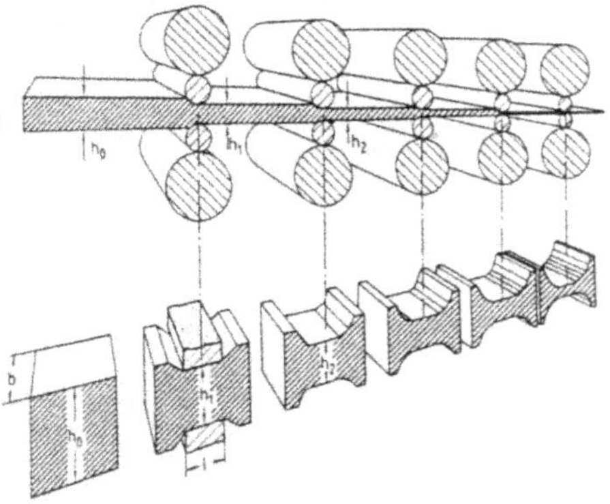

Fig.5. Simulation of hot rolling by plane strain compression test, principle of WUMSI

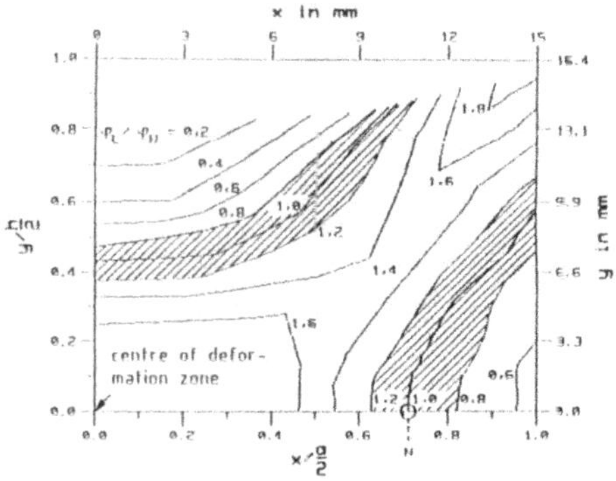

Fig.6. Ratio of local and nominal strain after flat compression, one quarter of deformation zone

Many comparison measurements have shown that a reliable transfer of the deformation mechanics and the material behaviour from flat compression to rolling is possible.

Hot deformation simulation of the described type is a good example for the urgently needed bridge between continuum mechanics and metal physics. In the last decade great progress has been made in better understanding the influences of metal working parameters on grain growth and grain refinement, hardening by deformation and softening by recovery and recrystallization, solution and precipitation of particles, and finally on the transformation of microstructure /5/.

Mathematical model of cold rolling processes

In the field of cold rolling plasticity theory has also proved to be a very effectful instrument for solving technical problems. Shape and thickness profile of cold-rolled strip depend

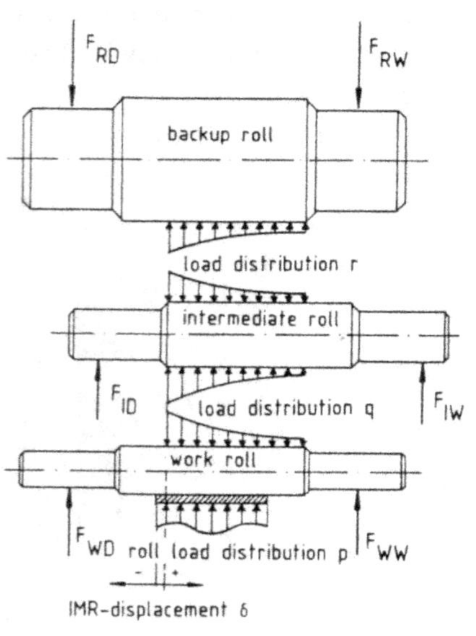

Fig.7. Distribution of forces acting in a six-high rolling mill, upper half

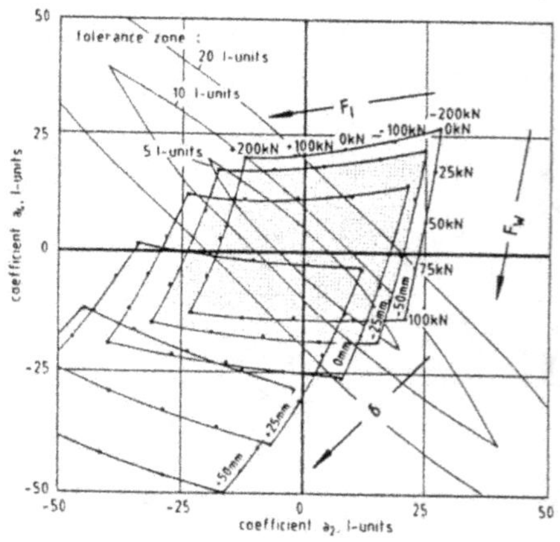

Fig.8. Performance graph for shape control of the six-high rolling mill; a_2, a_4 coefficients of $\Delta l/l$-distribution, F_W work roll bending force, F_I intermediate roll bending force, δ intermediate roll shift

on plastic deformations in the roll gap and elastic bending and flattening of work and backup rolls, Fig. 7. Complete mathematical models for 2-high-, 4-high-, and 6-high-rolling mills were developed /6/7/. In dependence on all relevant rolling parameters the pressure and the deformation across the strip width can be calculated. To minimize the computing time, an analytical approach in the rolling direction and a numerical one in the width direction was used.

Fig. 8 as an example shows the calculated possibilities of shape control of a 6-high-rolling mill. Shape faults are described by the different plastic elongation of the strip fibres L across the width B:

$$\Delta L/L = a_2\xi^2 + a_4\xi^4 \quad , \quad \xi = 2x/B \tag{1}$$

with the origin of the coordinate x in the strip centre. Ideal shape is defined by $a_2 = a_4 = 0$. Shape faults, expressed by the pair of coefficients a_2 and a_4, can be minimized by controlling work roll bending force, intermediate roll bending force, and intermediate roll shift.

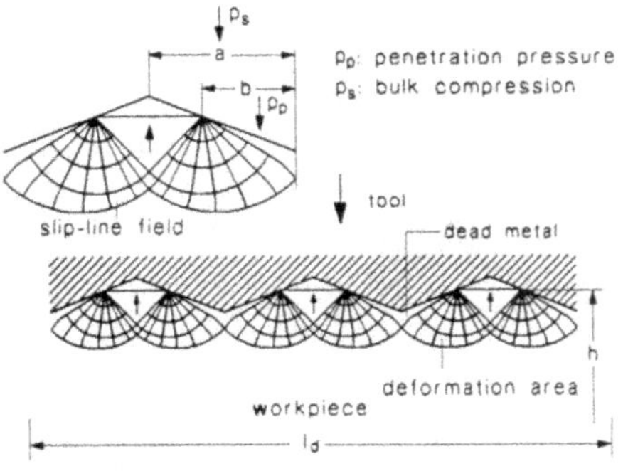

Fig.9. Slip line field model of surface roughness transfer from roll to strip

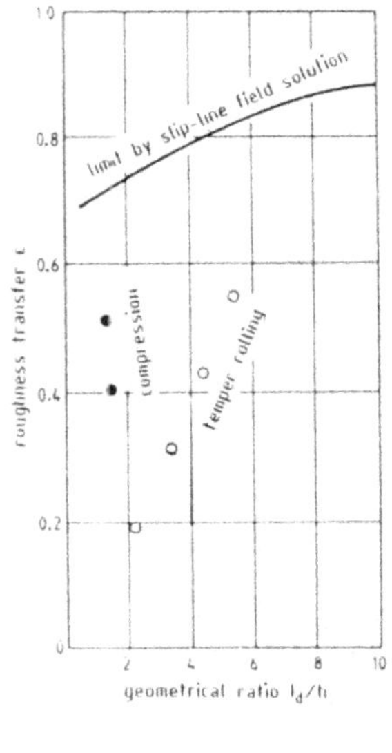

Fig.10. Comparison of experimental and theoretical roughness transfer depending on the deformation geometry

Surface roughness is a very important characteristic of cold rolled strip, especially for use in the automotive industry. By use of slip line field theory the influences on roughness transfer from the work roll to the strip could be calculated qualitatively, Figrs. 9 and 10, suggesting a competition between surface stamping and bulk plastic deformation /8/.

Vibrations in cold rolling tandem mills

A great danger in running a cold rolling tandem mill are roll stand vibrations with a frequency of around 150 Hz. To describe their effect on roll force and strip thickness, the conventional rolling theory was modified by including vertical movements of the roll gap /9/. The used mass-spring model of the roll stand and the rolls is shown in Fig. 11. With the periodic change of the roll gap exit also the length of the roll gap and the position of the neutral point is altered. Fig. 12 illustrates the change of the well-known "friction hill" of the pressure distribution in the roll gap due to the vibration effect.

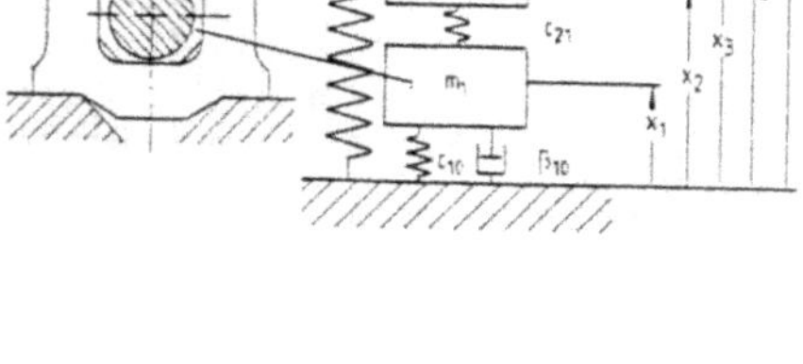

Fig.11. Mass-spring model for vibration of a four-high rolling mill

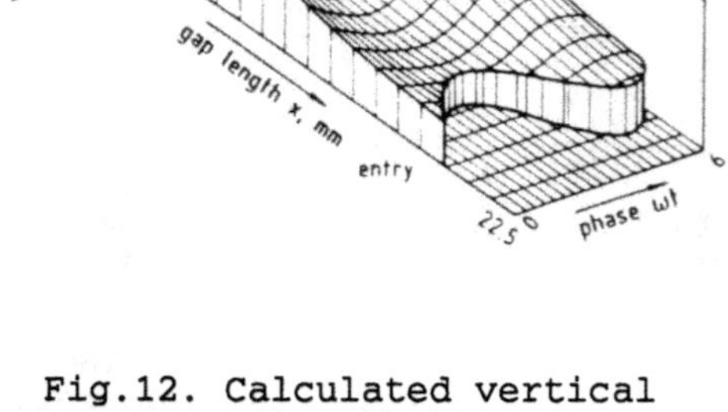

Fig.12. Calculated vertical pressure distribution in case of roll stand vibration

Tribological instability in high-speed cold rolling

One of the causes for the described vibrations may be a tribological instability in the roll gap. The lubrication film thickness increases with roll speed, so that finally the metallic contact between roll and strip is lost. To calculate this limiting condition, the equations of plasticity governing the roll gap must be combined with the Navier-Stokes differential equations for the hydrodynamic behaviour of the lubricant at the roll gap entry. The roll pressure is enlarged by a hydrodynamical part, Fig. 13. A simplified model gives the velocity distribution in the lubricating wedge, Fig. 14. The danger of roll slip is comparable to the well-known phenomenon of aquaplaning in car driving on a wet road /10/.

Stress-strain curves in the elastic-plastic transition region

Some processes of metal working technology take place in the transition region between elastic and plastic deformation. Restricting to rolling technology, mainly in temper rolling and in straightening by stretching and bending the plastic deformations are in the same order of magnitude as the elastic ones.

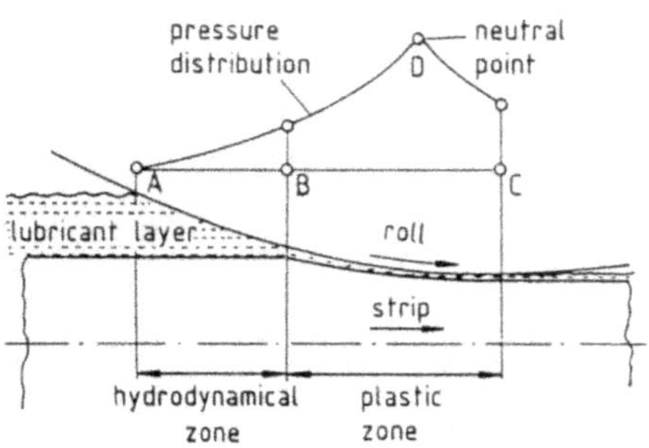

Fig.13. Pressure distribution in the hydrodynamical and plastic zone in strip rolling (schematically)

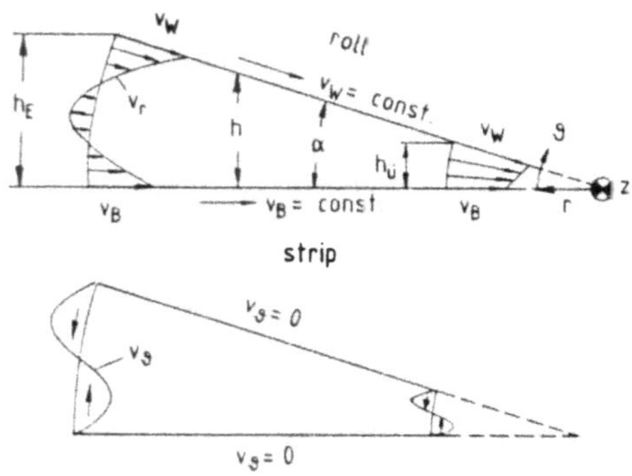

Fig.14. Velocity distribution in the lubrication wedge at the entry of roll gap; v_r radial velocity, v_ϑ tangential velocity

For a theoretical analysis of these processes it is useful to dispose of an adequate mathematical description of the elastic-plastic stress-strain curve. A very helpful method is to link the elastic straight line with the plastic part of the curve by use of a cut-off function. Using an arc tan-function for this purpose, the author has derived the following formula:

$$\frac{\sigma}{\sigma_F} = \frac{V}{E}\frac{\varepsilon}{\varepsilon_F} + \frac{1}{\pi}\left(1 - \frac{V}{E}\right) \cdot$$

$$\cdot \left\{\left(1 + \frac{\varepsilon}{\varepsilon_F}\right) \arctan\left[C\left(1 + \frac{\varepsilon}{\varepsilon_F}\right)\right] - \left(1 - \frac{\varepsilon}{\varepsilon_F}\right) \arctan\left[C\left(1 - \frac{\varepsilon}{\varepsilon_F}\right)\right]\right\} \quad (2)$$

Fig. 15 shows that the parameter C controls the degree of transition from elastic to plastic deformation. $C \rightarrow \infty$ gives the ideal case of elastic-linear strain hardening material, with the yield point (ε_F, σ_F). E and V are the elasticity modulus resp. strain hardening modulus only in case of $C \rightarrow \infty$. For finite values of C, mainly for $C < 2$, the slope of the curve in the origin has to be equated to the elasticity modulus. The

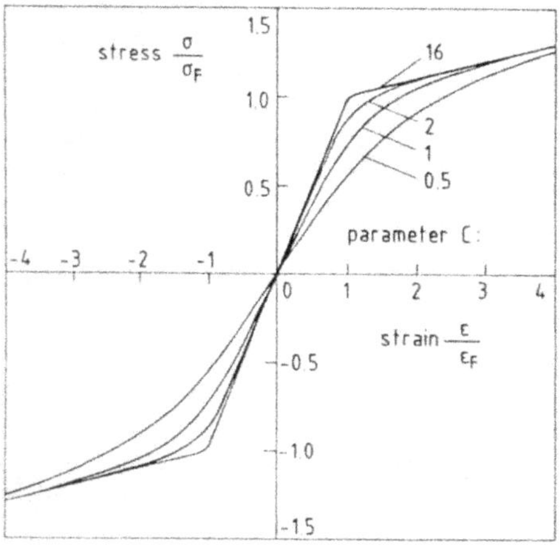

Fig.15. Mathematical description of stress-strain curves in the elastic-plastic transition region

proposed formula has the great advantage that by varying the parameter C a very accurate adaption to measured stress-strain curves is possible, and that the selected function is continuously differentiable, and integrable by use of elementary functions. Thus, e.g. in solving elementary problems of elastic-plastic bending, no discrimination between elastic and plastic zones is necessary.

Development of new methods for materials testing

Finally, plasticity theory does not only help to optimise metal working processes but also to develop new materials. An example is given by intermetallic compounds for high temperature applications, being of great industrial interest at time. Some new testing methods, the small-cup drawing test, the mantle compression test, and the ball compression test were developed for measuring yield stress and ductility of mainly brittle materials /11/. They were founded on the theory of upper bound of deformation energy and FEM calculations. The examples in Figrs. 16 and 17 show that the inside bulging of the wall in the small-cup drawing test was very well predicted by an FEM analysis.

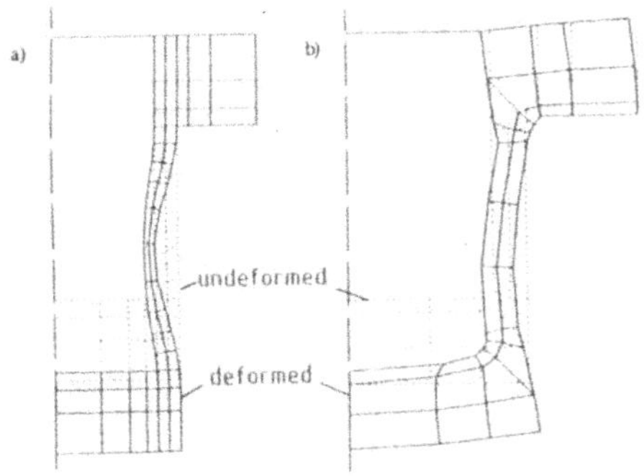

Fig.16. Calculated deformation mode of small-cup drawing test; FEM: a) FARM, b) Macroplast

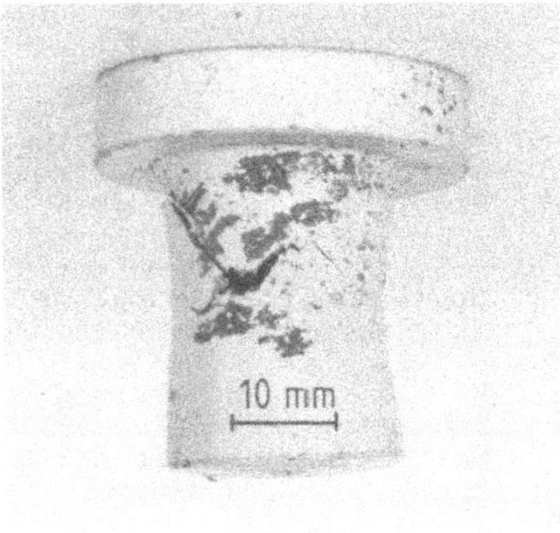

Fig. 17. TiAl-specimen, deformed at 990 °C by small-cup drawing test

References

1. Wihler, V.: Beurteilung des Umformvermögens unterschiedlicher Werkstoffe durch Walzversuche. Diplomarbeit, Fachhochschule Bochum, Abt. Gelsenkirchen, 1990.

2. Rasp, W.: Profile and shape control in hot strip rolling. Proc. 5th Intern. Rolling Conf., London, 1990, 349-356.

3. Pawelski, O.; Rüdiger, U.; Kaspar, R.: Der Warmumformsimulator des Max-Planck-Instituts für Eisenforschung - ein neues Konzept zur Erforschung schneller Warmumformvorgänge. Stahl u. Eisen 98 (1978) 181-189.

4. Distl, J. S.: Untersuchung der grundlegenden Einflüsse der Verfahrensparameter bei der thermomechanischen Behandlung auf Gefüge und mechanische Eigenschaften mikrolegierter Grobbleche. Umformtechn. Schriften, Bd. 8, Düsseldorf, Stahleisen, 1987.

5. Koschlig, M.: Einfluß einer thermomechanischen Behandlung auf die Werkstoffeigenschaften von Warmband aus Stählen mit mittleren Kohlenstoffgehalten und einer zusätzlichen Mikrolegierung. Umformtechn. Schriften, Bd. 23, Düsseldorf, Stahleisen, 1990.

6. Teutsch, H.: Ein mathematisches Modell mit numerischem und analytischem Ansatz zur Berechnung der Profilform beim Kaltwalzen von Band. Dr. rer. nat.-Diss., Rheinisch-Westf. Techn. Hochschule Aachen, 1984.

7. Rieckmann, J.: Berechnung von Bandprofil und Planheit beim Kaltwalzen auf Sechs-Walzen-Walzwerken. Umformtechn. Schriften, Bd. 21, Düsseldorf, Stahleisen, 1989.

8. Löffler, L.: Untersuchungen zum Aufrauhen von Bandstahl. Umformtechn. Schriften, Bd. 17, Düsseldorf, Stahleisen, 1988.

9. Friedewald, K.: Ein mathematisches Modell zur Berechnung des Schwingverhaltens von Tandem-Walzstraßen. Umformtechn. Schriften, Bd. 14, Düsseldorf, Stahleisen, 1988.

10. Pawelski, O.: Aquaplaning-Analogon für die Vorgänge im Schmierspalt beim Kaltwalzen. In: Berichtsband "Tribologie in Theorie und Praxis". Essen: Ges. für Tribologie 1988.

11. Hop, R. M.: Entwicklung von Prüfverfahren zur Bestimmung von Umformbarkeit und Fließspannung von Werkstoffen auf der Basis intermetallischer Verbindungen. Umformtechn. Schriften, Bd. 27, Düsseldorf, Stahleisen, 1991.

The Thick-Walled Tube Under Internal Pressure. An Exact Solution for Plain Strain

R. Bonn, P. Haupt

University of Kassel, Institute of Mechanics
Mönchebergstr. 7, D - 3500 Kassel

Summary

A theory of rate-independent plasticity based on the concept of dual variables and derivatives is utilized to describe finite elastic-plastic deformations including kinematic hardening effects. Application of this theory to the problem of the thick-walled tube under internal pressure leads to a system of partial differential equations for which an exact solution can be obtained. The influence of kinematic hardening is studied by calculating the stresses and strains for different values of the tube thickness. The results are compared with the rigid plastic solution.

1. General theory

In [1], HAUPT and TSAKMAKIS have proposed a concept of dual variables and derivatives which implies a rational rule to relate tensor valued kinematic and dynamic variables (strain and stress tensors, or strain and stress rates). Strain tensors $\mathbf{\Pi}$ and their derivatives $\overset{\Delta}{\mathbf{\Pi}}$ are derived from the GREEN strain tensor $\mathbf{E}$ by certain transformations $\mathbf{\Psi}$ which map material line elements $d\mathbf{X}$ from the reference configuration into physically meaningful intermediate states $d\hat{\mathbf{x}} = \mathbf{\Psi}\, d\mathbf{X}$:

$$\mathbf{\Pi} = \mathbf{\Psi}^{T-1}\, \mathbf{E}\, \mathbf{\Psi}^{-1} , \qquad \overset{\Delta}{\mathbf{\Pi}} = \mathbf{\Psi}^{T-1}\, \dot{\mathbf{E}}\, \mathbf{\Psi}^{-1} = \dot{\mathbf{\Pi}} + (\dot{\mathbf{\Psi}}\mathbf{\Psi}^{-1})^T\, \mathbf{\Pi} + \mathbf{\Pi}\, (\dot{\mathbf{\Psi}}\mathbf{\Psi}^{-1}) \qquad (1.1)$$

In the context of the invariance of the stress power and the complementary stress power

$$\rho_R\, w = \tilde{\mathbf{T}} \bullet \dot{\mathbf{E}} = \mathbf{\Sigma} \bullet \overset{\Delta}{\mathbf{\Pi}} \qquad \text{and} \qquad \dot{\tilde{\mathbf{T}}} \bullet \mathbf{E} = \overset{\nabla}{\mathbf{\Sigma}} \bullet \mathbf{\Pi} \quad , \qquad (1.2)$$

one is lead in a natural way to the definition of dual stress tensors $\mathbf{\Sigma}$ and dual stress derivatives $\overset{\nabla}{\mathbf{\Sigma}}$ on the basis of the 2nd PIOLA-KIRCHHOFF stress tensor $\tilde{\mathbf{T}}$:

$$\mathbf{\Sigma} = \mathbf{\Psi}\, \tilde{\mathbf{T}}\, \mathbf{\Psi}^T , \qquad \overset{\nabla}{\mathbf{\Sigma}} = \mathbf{\Psi}\, \dot{\tilde{\mathbf{T}}}\, \mathbf{\Psi}^T = \dot{\mathbf{\Sigma}} - (\dot{\mathbf{\Psi}}\mathbf{\Psi}^{-1})\, \mathbf{\Sigma} - \mathbf{\Sigma}\, (\dot{\mathbf{\Psi}}\mathbf{\Psi}^{-1})^T \qquad (1.3)$$

D. Besdo, E. Stein (Eds.)
Finite Inelastic Deformations - Theory and Applications
IUTAM Symposium Hannover/Germany 1991

Applying this concept to the multiplicative decomposition of the total deformation gradient into elastic and plastic parts,

$$\mathbf{F} = \hat{\mathbf{F}}_e \mathbf{F}_p \,, \tag{1.4}$$

additive decompositions of the total strain and strain rate result. With $\mathbf{\Psi} = \mathbf{F}_p$, the tensors

$$\hat{\Gamma} = \mathbf{F}_p^{T-1} \mathbf{E} \, \mathbf{F}_p^{-1} = \frac{1}{2} (\hat{\mathbf{F}}_e^T \hat{\mathbf{F}}_e - 1) + \frac{1}{2} (1 - \mathbf{F}_p^{T-1} \mathbf{F}_p^{-1}) = \hat{\Gamma}_e + \hat{\Gamma}_p \tag{1.5}$$

$$\overset{\Delta}{\hat{\Gamma}} = \dot{\hat{\Gamma}} + \hat{\mathbf{L}}_p^T \hat{\Gamma} + \hat{\Gamma} \hat{\mathbf{L}}_p = \overset{\Delta}{\hat{\Gamma}}_e + \overset{\Delta}{\hat{\Gamma}}_p \,, \qquad \hat{\mathbf{D}}_p \overset{\text{def}}{=} \overset{\Delta}{\hat{\Gamma}}_p = \frac{1}{2} (\hat{\mathbf{L}}_p + \hat{\mathbf{L}}_p^T) \tag{1.6}$$

refer to the intermediate state, where $\hat{\mathbf{L}}_p = \dot{\mathbf{F}}_p \mathbf{F}_p^{-1}$ is the plastic velocity "gradient". The dual stress of the intermediate state $\hat{\mathbf{S}}$ and its dual time derivative follow directly from (1.3) :

$$\hat{\mathbf{S}} = \mathbf{F}_p \tilde{\mathbf{T}} \mathbf{F}_p^T \quad \text{and} \quad \overset{\nabla}{\hat{\mathbf{S}}} = \dot{\hat{\mathbf{S}}} - \hat{\mathbf{L}}_p \hat{\mathbf{S}} - \hat{\mathbf{S}} \hat{\mathbf{L}}_p^T \tag{1.7}$$

The following formulation of rate-independent plasticity is based on the decomposition of the deformation (1.4), and the concept of dual variables and derivatives [1]. In view of the principal idea that the intermediate state may be approached by a local unloading process, it appears evident to formulate all constitutive equations with respect to the plastic intermediate configuration:

plastic incompressibility :
$$\operatorname{tr} (\overset{\Delta}{\hat{\Gamma}}_p) = 0 \iff \det \mathbf{F}_p = 1 \tag{1.8}$$

hyperelasticity:
$$\hat{\mathbf{S}} = 2\mu \left(\hat{\mathbf{E}}_e + \frac{\nu}{1 - 2\nu} (\mathrm{Sp}\hat{\mathbf{E}}_e) \, \mathbf{1} \right) \tag{1.9}$$

yield criterion :
$$(\hat{\mathbf{S}}^D - \hat{\mathbf{X}}^D) \bullet (\hat{\mathbf{S}}^D - \hat{\mathbf{X}}^D) = \frac{2}{3} \, g_0^2 \tag{1.10}$$

associated flow rule :
$$\hat{\mathbf{D}}_p = \lambda (\hat{\mathbf{S}}^D - \hat{\mathbf{X}}^D) \tag{1.11}$$

plastic arc length :
$$\dot{s} = \sqrt{\frac{2}{3} \hat{\mathbf{D}}_p \bullet \hat{\mathbf{D}}_p} \tag{1.12}$$

kinematic hardening :
$$\overset{\nabla}{\hat{\mathbf{X}}} = c \, \hat{\mathbf{D}}_p - b \, \dot{s} \, \hat{\mathbf{X}} \tag{1.13}$$

The proportionality factor λ in (1.11) is evaluated from the consistency condition. We get

$$\lambda = \frac{2\mu\,(\hat{\mathbf{S}}^D - \hat{\mathbf{X}}^D)\bullet\overset{\Delta}{\hat{\mathbf{\Gamma}}}}{\hat{N}} \qquad \text{with} \tag{1.14}$$

$$\begin{aligned}\hat{N} = {} & \frac{2}{3}\,g_0^2\,(2\mu + c) + \frac{8}{3}\,g_0^2\,\frac{\nu}{2(1+\nu)}\,\mathrm{tr}\,(\hat{\mathbf{S}}) - \frac{4}{9}\,g_0^2\,\mathrm{tr}\,(\hat{\mathbf{S}} - \hat{\mathbf{X}}) \\ & - \frac{2}{3}\,g_0\,b\,(\hat{\mathbf{S}}^D - \hat{\mathbf{X}}^D)\bullet\hat{\mathbf{X}} + 4\,(\hat{\mathbf{S}}^D - \hat{\mathbf{X}}^D)^2\bullet(\hat{\mathbf{S}} - \frac{\nu}{1+\nu}\,\mathrm{tr}\,(\hat{\mathbf{S}})\,\mathbf{1}) \\ & - 2\,(\hat{\mathbf{S}}^D - \hat{\mathbf{X}}^D)\bullet[(\hat{\mathbf{S}}^D - \hat{\mathbf{X}}^D)(\hat{\mathbf{S}}^D - \hat{\mathbf{X}}^D)]\end{aligned} \tag{1.15}$$

The superscipt $()^D$ indicates the deviatoric part of the tensor ().

2. Thick-walled tube under internal pressure

In the following we consider the problem of a thick-walled tube subject to internal pressure under plane strain conditions. Taking into account the cylindrical symmetry, convective coordinates are used:

$$\mathbf{g}_i = \mathbf{F}\,\mathbf{G}_i \tag{2.1}$$

($\mathbf{g}_i$, $\mathbf{G}_i$ are the covariant basis vectors referred to the actual state and reference state, respectively). All vectors and tensors are formulated with respect to the basis vectors $\mathbf{G}_i$ of the reference state. Thus, for the displacement vector, we obtain the reprensentation

$$\mathbf{u} = u(R,t)\,\mathbf{G}_1 \quad , \tag{2.2}$$

with R being the radius in the reference state, and t time, or an equivalent loading parameter. Eq. (2.2) is based on the assumption, that only a radial expansion of the tube will be considered. From the displacement vector **u** we calculate the following matrix representation of the total deformation gradient **F** and its plastic part $\mathbf{F}_p$:

$$\mathbf{F} \mathrel{\hat{=}} \begin{pmatrix} \alpha & 0 & 0 \\ 0 & \beta & 0 \\ 0 & 0 & 1 \end{pmatrix} , \quad \mathbf{F}_p \mathrel{\hat{=}} \begin{pmatrix} \alpha_p & 0 & 0 \\ 0 & \beta_p & 0 \\ 0 & 0 & (\alpha_p\beta_p)^{-1} \end{pmatrix} \tag{2.3}$$

$$\alpha \overset{\text{def}}{=} (1 + \frac{\partial u}{\partial R}) = \frac{\partial r}{\partial R} \qquad \text{and} \qquad \beta \overset{\text{def}}{=} (1 + \frac{u}{R}) = \frac{r}{R} \tag{2.4}$$

Here, r represents the position of a material point in the actual configuration. We note that the diagonal representation (2.3b) of the plastic deformation gradient $\mathbf{F}_p$ is the simplest one not leading to inconsistencies in the constitutive equations (1.8) - (1.14). Equations (2.3) satisfy the assumption of plastic incompressibility (1.8) and imply the following special form of the constitutive equations:

plastic arc length

$$\dot{s} = \frac{2}{3} g_0 \frac{2\mu}{\hat{N}} \left(\frac{\alpha \dot{\alpha}}{\alpha_p^2} (\hat{S}^D - \hat{X}^D)^1{}_1 + \frac{\beta \dot{\beta}}{\beta_p^2} (\hat{S}^D - \hat{X}^D)^2{}_2 \right) \tag{2.5}$$

$\hat{N}$ is defined by (1.15).

flow rule

$$\frac{\dot{\alpha}_p}{\alpha_p} = \frac{3}{2} \frac{1}{g_0} (\hat{S}^D - \hat{X}^D)^1{}_1 \, \dot{s} , \quad \frac{\dot{\beta}_p}{\beta_p} = \frac{3}{2} \frac{1}{g_0} (\hat{S}^D - \hat{X}^D)^2{}_2 \, \dot{s} , \tag{2.6}$$

kinematic hardening

$$\dot{\hat{X}}^1{}_1 = \left[\frac{3}{2} \frac{1}{g_0} (\hat{S}^D - \hat{X}^D)^1{}_1 \; (c + 2 \hat{X}^1{}_1) - b \hat{X}^1{}_1 \right] \dot{s}$$

$$\dot{\hat{X}}^2{}_2 = \left[\frac{3}{2} \frac{1}{g_0} (\hat{S}^D - \hat{X}^D)^2{}_2 \; (c + 2 \hat{X}^2{}_2) - b \hat{X}^2{}_2 \right] \dot{s}$$

$$\dot{\hat{X}}^3{}_3 = \left[\frac{3}{2} \frac{1}{g_0} (\hat{S}^D - \hat{X}^D)^3{}_3 \; (c + 2 \hat{X}^3{}_3) - b \hat{X}^3{}_3 \right] \dot{s} , \tag{2.7}$$

elasticity

$$\hat{S}^1{}_1 = \frac{\mu}{1 - 2\nu} \left((1 - \nu)(\frac{\alpha^2}{\alpha_p^2} - 1) + \nu (\frac{\beta^2}{\beta_p^2} - 1) + \nu \, (\alpha_p^2 \beta_p^2 - 1) \right)$$

$$\hat{S}^2{}_2 = \frac{\mu}{1 - 2\nu} \left(\nu (\frac{\alpha^2}{\alpha_p^2} - 1) + (1 - \nu)(\frac{\beta^2}{\beta_p^2} - 1) + \nu \, (\alpha_p^2 \beta_p^2 - 1) \right)$$

$$\hat{S}^3{}_3 = \frac{\mu}{1-2\nu}\left(\nu\,(\frac{\alpha^2}{\alpha_p{}^2} - 1) + \nu\,(\frac{\beta^2}{\beta_p{}^2} - 1) + (1-\nu)\,(\alpha_p{}^2\beta_p{}^2 - 1) \right) \tag{2.8}$$

Here, μ is the shear modulus, ν POISSON's number, and g_0 the yield stress. Equations (2.5) - (2.8) lead to a system of five differential equations in the seven unknown variables α, α_p, β, β_p, $\hat{X}^1{}_1$, $\hat{X}^2{}_2$, $\hat{X}^3{}_3$. The material time derivative of the unknowns are evaluated with respect to the reference state, and therefore reduce to partial time derivatives. The only equilibrium condition to consider, formulated in the spatial description ($T^k{}_k$ components of CAUCHY stress tensor), is

$$\frac{\partial T^1{}_1}{\partial r} + \frac{1}{r}(T^1{}_1 - T^2{}_2) = 0 \ . \tag{2.9}$$

Transforming (2.9) to the intermediate state we obtain the more complex form

$$\left[2\mu\frac{1-\nu}{1-2\nu}\frac{\alpha}{\alpha_p{}^2} + \frac{\hat{S}^1{}_1}{\alpha} \right]\frac{\partial\alpha}{\partial R} + 2\mu\frac{\nu}{1-2\nu}\left[\alpha_p{}^2\beta_p - \frac{\beta^2}{\beta_p{}^3} \right]\frac{\partial\beta_p}{\partial R} +$$
$$\left[\frac{2\mu}{1-2\nu}\left\{ \nu\,\alpha_p\,\beta_p{}^2 - (1-\nu)\frac{\alpha^2}{\alpha_p{}^3} \right\} - 2\frac{\hat{S}^1{}_1}{\alpha_p} \right]\frac{\partial\alpha_p}{\partial R} + \frac{2\mu\,\nu}{1-2\nu}\frac{\beta}{\beta_p{}^2}\frac{\partial\beta}{\partial R}$$
$$= -\frac{\hat{S}^1{}_1 - \dfrac{\beta}{\alpha}\dfrac{\alpha_p^2}{\beta_p^2}\hat{S}^2{}_2}{R} \ . \tag{2.10}$$

This is a partial differential equation with respect to the material coordinate R. Finally, the following compatibility relation between the components of the total deformation gradient must be incorporated:

$$\frac{\partial\beta}{\partial R} = \frac{\alpha-\beta}{R} \tag{2.11}$$

Summarizing, we get a system of partial differential equations (PDE's) in two independent and seven dependent variables which has to be completed by appropriate boundary conditions. At the external surface of the tube, the radial stress (external pressure) is presrcribed throughout the complete loading process, e.g. $\hat{S}^1{}_1\,(R_e) \equiv 0$. The remaining boundary conditions are given in the elastic case by the loading parameter (internal pressure or displacement), and in the case of the partly plastified tube by the calculated elastic solution. If the tube is completely plastified we obtain the boundary conditions at the external surface ($R = R_e$) by integrating the constitutive equations (2.6)-(2.7) with respect to the plastic arc length s_e at the external radius.

3. Numerical integration

Mathematically, the governing equations of the problem under consideration are a special case of the following quasilinear system:

$$\sum_{\mu=1}^{N} A_{\nu\mu} \frac{\partial z_\mu}{\partial t} = g_\nu \qquad \nu = 1,\dots,n$$

$$\sum_{\mu=1}^{N} A_{\nu\mu} \frac{\partial z_\mu}{\partial R} = g_\nu \qquad \nu = n+1, \dots, N \tag{3.1}$$

This system has some remarkable general properties :

(1) For $\det A_{\nu\mu} \neq 0$ the system (3.1) is hyperbolic in each point of (R,t)- space. This condition can be guaranteed for the special system (2.5)-(2.11).

(2) Independent of the number of unknown variables z_μ, only two distinct characteristic directions exist.

(3) For each point in (R,t)-space these two characteristics coincide with the coordinate lines R = const and t = const, respectively (see [2]).

(4) The existence and uniqueness of the solution of (3.1) can be proven [3].

An approximation to the exact solution of (3.1) can be calculated with high accuracy utilizing the *method of extrapolation to the limit* (RICHARDSON-Extrapolation). This method has been introduced by STOER&BULIRSCH [6] and GRAGG [7] to solve ordinary differential equations. The mathematical theory states, that extrapolation can only be applied if the following two criteria hold

(1) the solution of the dicretisized problem $\mathbf{z}_{dis}$ converges to the exact solution $\mathbf{z}_{exact}$

(2) the numerical solution possesses an asymptotic expansion of the form

$$\mathbf{z}_{dis}(R,t;h) = \mathbf{z}_{exact}(R,t) + \mathbf{c}_1(R,t)\, h^1 + \dots + \mathbf{c}_N(R,t)\, h^N + O(h^{N+1}) \tag{3.2}$$

where the coefficients $\mathbf{c}_1$ to $\mathbf{c}_N$ are independent of the stepsize h, and the largest expansion coefficient N can be chosen arbitrarily.

Calculating the solution $\mathbf{z}_{dis}(R,t;h_i)$ for k different stepsizes h_i ($h_0=h$, $h_1=\frac{h}{2}$, ...), extrapolation leads to the following approximate solution

$$\mathbf{z}_{approxi}(R,t) = \mathbf{z}_{exact}(R,t) + O(h_0^k) \ . \tag{3.3}$$

The order k of the truncation error depends on how often the extrapolation has been applied. HACKBUSCH [4] and SMITH [5] proved that for quasilinear hyperbolic systems in two independent variables, an asymptotic expansion of the form (3.2) exists. In particular, SMITH [5] investigated the system (3.1) in combination with the method of characteristics as an integration scheme.

In summary, we arrive at an *exact solution* in the following sense: The existence and uniqueness of the solution of the system (3.1) has been proven in the mathematical literature, as well as the convergence of numerical integration schemes. Moreover, the dependence of the numerical approximation on the stepsize h has special analytical properties which justify the application of an extrapolation technique to accelerate significantly (3.3) the rate of convergence.

4. Numerical results

The influence of kinematic hardening is studied for different values of the hardening parameters b and c, and for different values of the wall thickness $Q = \frac{R_e}{R_i}$. The internal pressure as function of the displacement u_e at the external radius is shown in Fig. 1. With increasing hardening parameter c the pressure increases, and with increasing saturation parameter b it decreases. If the value of b is chosen too large the tensor of kinematic hardening immediately approaches its saturation value, leading qualitatively to the same pressure-displacement dependency as in the case of ideal plasticity. As for the pressure, the same is true for the circumferential stress (Fig. 2). Figs. 3 and 4 illustrate the stress distribution of the circumferential stress across the wall thickness. The total strain at the inner radius for these examples is 40% and 80%, respectively. With increasing loading, the elastic-plastic solution approaches the rigid-plastic one. For a tube which is still elastic at the extrenal radius, this fact is illustrated in Fig.5. If hardening is not taken into account, rigid-plastic behaviour implies the following relation between the stress components: $T^3{}_3 = \frac{1}{2}\,(T^1{}_1 + T^2{}_2)$

For different values of the hardening parameters, the evolution of the circumferential stress during loading is plotted in Fig.6. All results, were calculated with values of the stepsize $h = 5 \cdot 10^{-3}$ to $h = 1 \cdot 10^{-3}$ and then extrapolated to the limit $h \to 0$. In all figures the stresses $T^1{}_1$, $T^2{}_2$, $T^3{}_3$ and g_0 are non-dimensionalized by the factor 2μ.

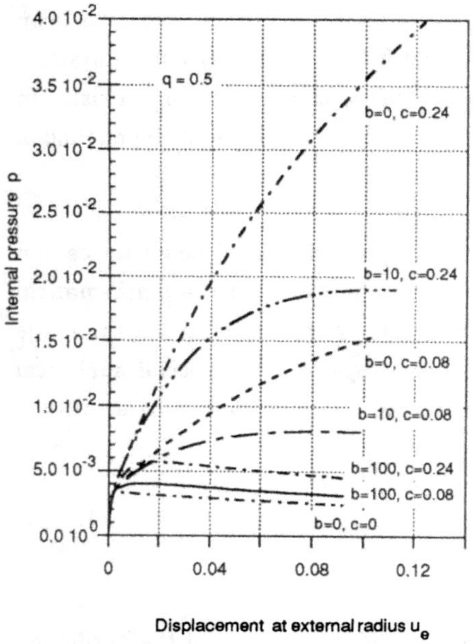

Figure 1

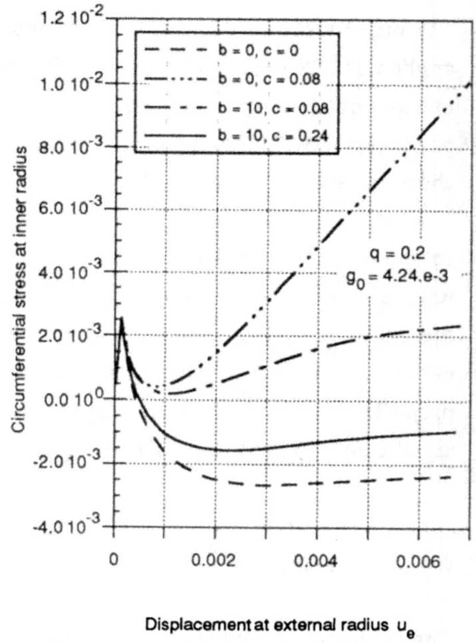

Figure 2

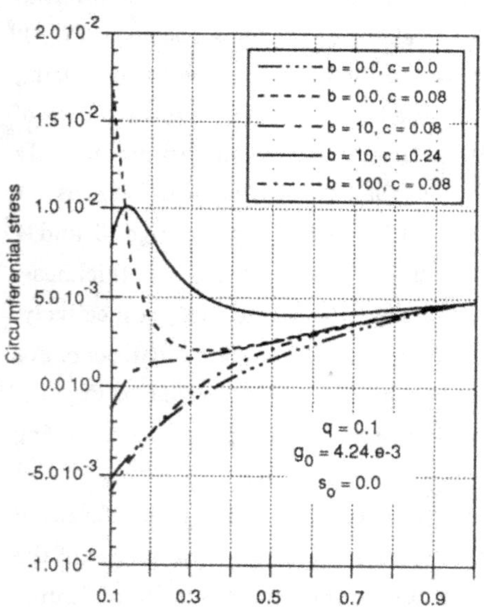

Figure 3

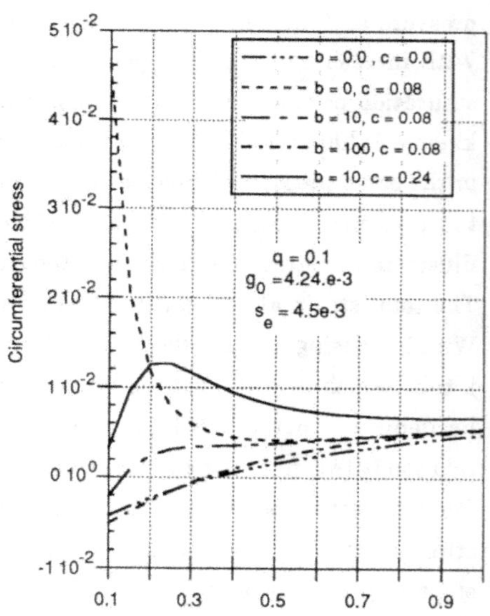

Figure 4

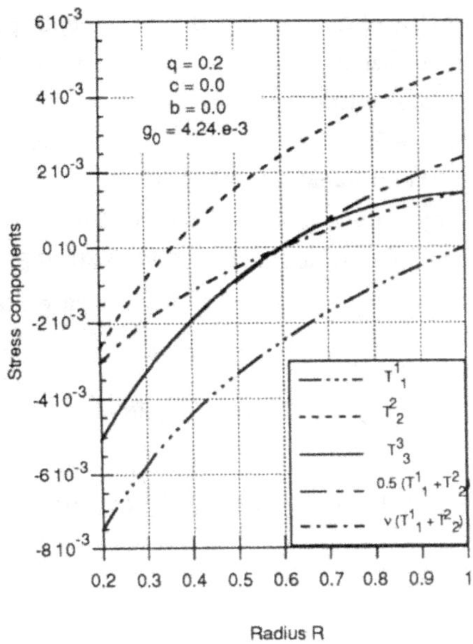

Figure 5

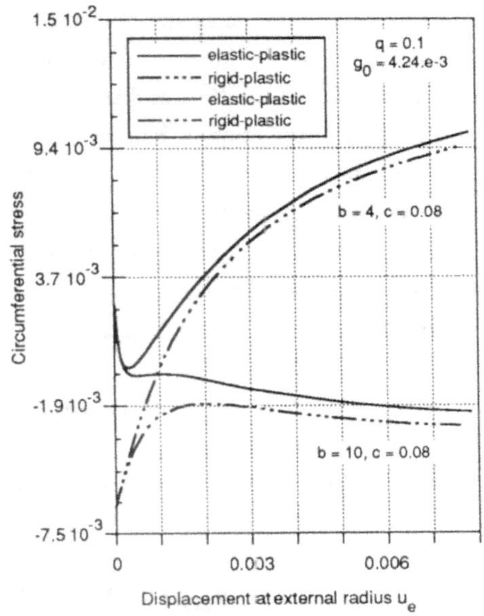

Figure 6

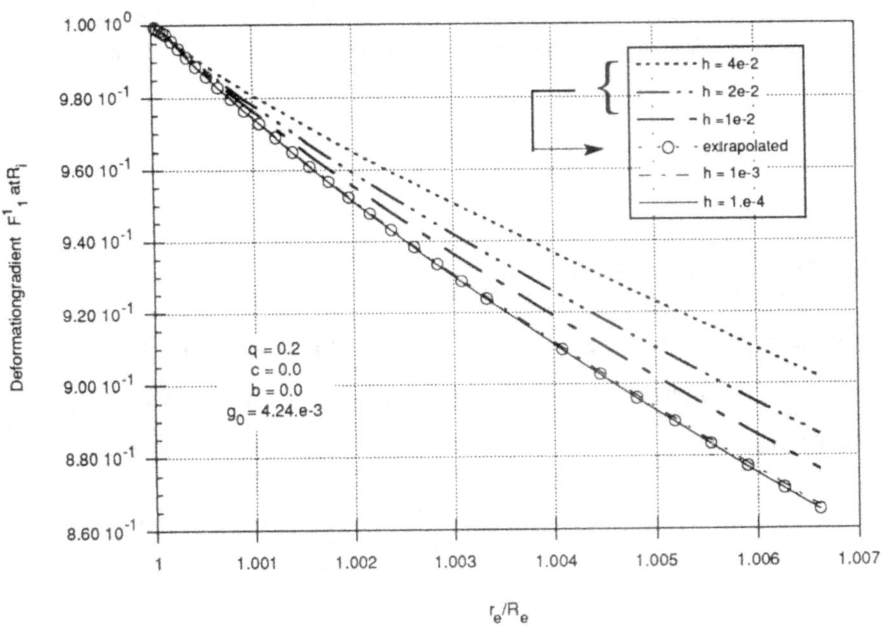

Figure 7

The method of characteristics as a basis for the extrapolation implies that the computing time t_c increases quadratically with decreasing stepsize h: $t_c \sim h^{-2}$. In the case of Fig. 7, the extrapolated solution, and the solution calculated with stepsize $h=1\cdot10^{-4}$ are almost identical; extrapolation is, however about 7000 times faster. Besides the high accuracy of the extrapolation method, it is possible to keep the influence of round-off errors small, because calculations can be done with rather large values of the stepsize h. The rate of convergence can even be improved if a modified midpoint-rule (s. GRAGG [7], SMITH [4]) is used as integration scheme instead of the method of characteristics. In this case the asymptotic expansion (3.2) has the form

$$\mathbf{z}_{dis}(R,t;h) = \mathbf{z}_{exact}(R,t) + \mathbf{c}_1(R,t)\,h^2 + \ldots\ldots + \mathbf{c}_N(R,t)\,h^{2N} + O(h^{2N+2})$$

and applying k times the extrapolation technique leads to an approximative solution (s. (3.3))

$$\mathbf{z}_{approxi}(R,t) = \mathbf{z}_{exact}(R,t) + O(h_0^{2k}) \ .$$

The applied method is restricted to quasilinear hyperbolic systems in two independent variables but not to the special form (3.1). More general kinematic hardening laws as well as isotropic hardening behaviour can be included in exactly the same way as the special hardening rule (1.13).

References

[1] HAUPT, P , TSAKMAKIS, CH.: On the Application of Dual Variables in Continuum Mechanics. Continuum Mechanics and Thermodynamics 1 (1989), 165-196

[2] COURANT, R. ; HILBERT, D. : Methoden der mathematischen Physik II, Springer 1968

[3] HARTMAN, P., WINTNER, A.: On hyperbolic partial differential equations. An. J. Math., Vol. 74, 1952, 834-864

[4] SMITH, R. R.: Extrapolation Applied to the Numerical Solution of Hyperbolic Partial Differential Equations. Dissertation University of California, San Diego 1970

[5] HACKBUSCH, W. : Extrapolation to the limit for numerical solutions of hyperbolic systems, Numerische Mathematik, Vol 28, 1977

[6] BULIRSCH,R.; STOER, J: Numerical Treatment of Ordinary Differential Equations by Extrapolation Methods, Numer. Mathematik 8, 1-13 (1966)

[7] GRAGG, W.: On Extrapolation Algorithms for Ordinary Initial Value Problems, J. SIAM Numer. Anal., Ser B, Vol 2, No. 3, 1965

Optimization Techniques for Numerical Identification of Forging Parameters

E. Doege; N. Werner; P. Groche

Institut für Umformtechnik und Umformmaschinen
University of Hannover, Germany

Summary
For shortening and optimizing the design of forging processes, the Finite-Element-Method (FEM) gains more and more importance. Since non-measurable and insufficiently known parameters influence the FEM-results substantially, the method of Numerical Identification was developed for their determination. This paper presents the definition and the theoretical as well as the practical proof of the reliability of the method.

Numerical Identification in the Manufacturing Process

The importance of the forging industry can clearly be shown by the wide variety of forged pieces. Nevertheless, the periods of development have to be shortened and the accuracy of the processes have to be improved to compete furtheron. Since the forging process is furthermore influenced by a couple of critical factors, like material flow, heat transfer, shrinking, and elastic die deformation, the design process (Fig.1) is of high importance.

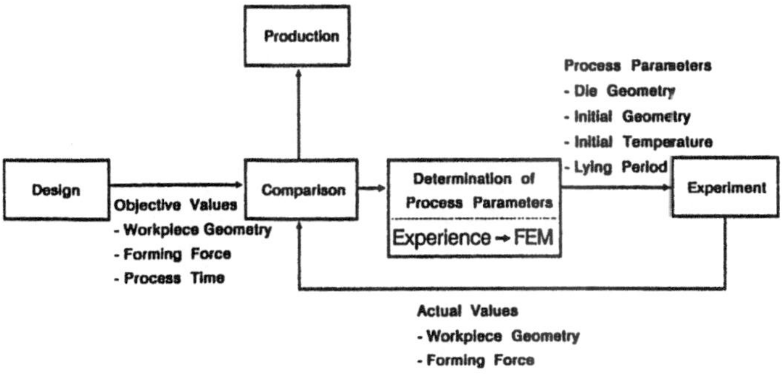

Fig.1. Design of Forging Processes

D. Besdo, E. Stein (Eds.)
Finite Inelastic Deformations –
Theory and Applications
IUTAM Symposium Hannover/Germany 1991

Hereby the department of design first appoints the objective values, for example workpiece geometry and forming force. Subsequently, the process parameters, like die geometry or initial temperature, have to be determined. Up to the present time, this is mainly based on the experience of the forging engineer. During and after the first experimental test, the actual values are measured for comparison with the objective values. Since a non-desirable difference occurs normally, the above shown iteration cycle has to be repeated until accordance is gained. Recently, the use of the Finite-Element-Method is obtaining greater importance in regard to minimize this time-consuming and consequently expensive iteration process.

Nevertheless the FEM-supported design can only yield to better results, if the FEM-model corresponds with the forging process. Since the results of FEM-calculations are influenced by FEM-specific parameters, such as element type or structure discretization, and by the physical models themselves, the assumption of correspondence can not be regarded as fulfilled up from the beginning of the simulation. By this, particulary the physical models are strongly influenced by non-measurable, insufficiently known process parameters, for instance the coefficients of friction and heat transfer or the yield function. Consequently, the conclusion of the above explained is to transfer an initially unconfirmed calculated process model to a verified one.

Definition

The method of Numerical Identification (NI) [1][2] now offers a suitable method to gain a verified FEM-model (Fig.2). The objective is to identify non-measurable process parameters by matching the calculated process results to the measured results. For this, forging process and FEM-calculation have to use the same measurable input data. For gaining correspondence after a lowest amount of iteration cycles, the variation of the unknown parameters has to be supported by suitable optimization methods [3],[4],[5].

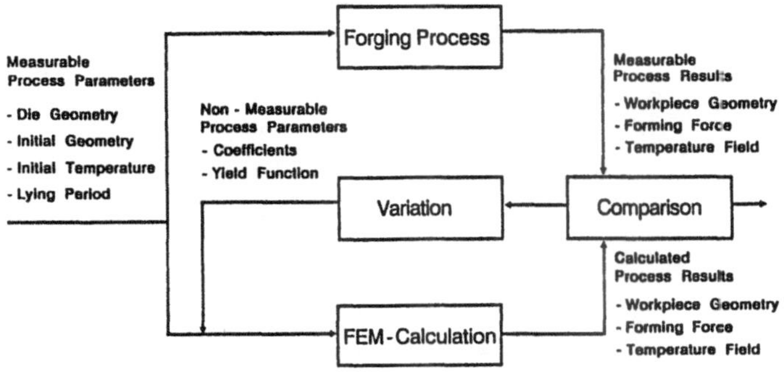

Fig.2. Control circuit of Numerical Identification

Using NI requires the following assumptions in addition:
- correct measuring results
- suitable physical models for simulating the process characteristics
- invariable FEM-model for the identification- and optimization-process

Theoretical Validation

Since the determination of a parameter set for a particular metal forming process is only a means to an end to design and optimize other forming processes faster and in a more secure way, the following questions have to be considered.
- Can the identified parameter set claim mathematical unambigiousness and physical exactness?
- Can the identified parameter set be transfered to other forming processes?

Since the considered target presents a minimum problem (Fig.3), the identified parameter set cannot claim mathematical unambigousness. It is however satisfying for the central problem, the process optimization, when the result fulfills the convergence criteria. In addition, physical sensible parameter sets will be gained by using mathematical restrictions (Fig.3).

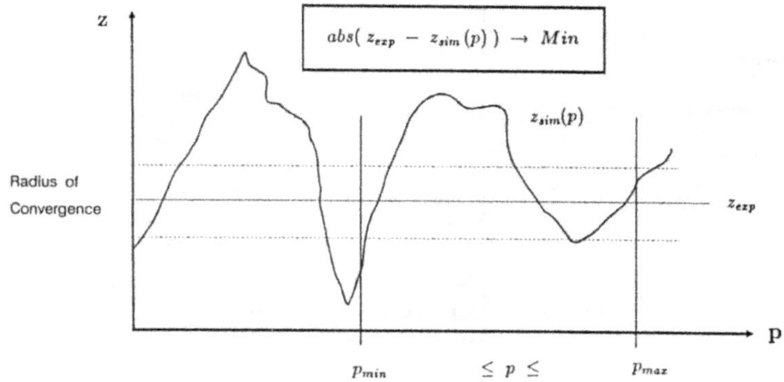

Fig.3. Mathematical unambigousness and physical exactness of the identified parameter set

However NI can not support a faster and less expensive design of forging proceses, if it is not possible to transfer an identified parameter set to other forming processes. Studies showed that transferring an identified parameter set to geometrical similar processes results in suitable mappings of the forming process. Geometrical similarity to the reference process, here an upsetting process, is clearly shown in Fig.4.

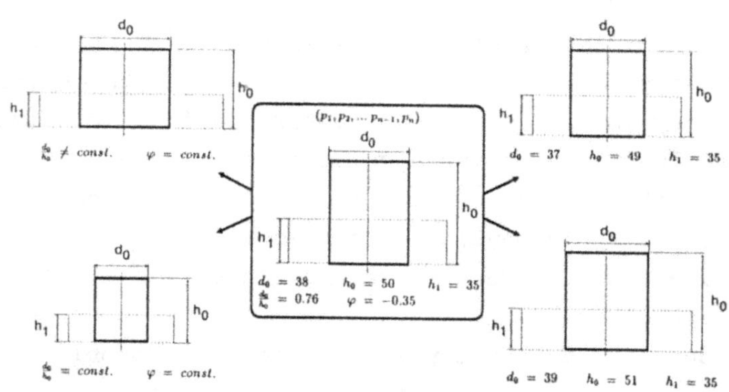

Fig.4. Geometrical similar upsetting processes

In the following, transferability is proved theoretically first, and by practical application afterwards.

An elastically stretched tension bar of two different materials was chosen for the theoretical proof (Fig.5). This example could be exactly described mathematical and claims low calculation costs in addition. Because of the cooperation of NI in the entire design process, four different groups of parameters can be defined (Fig.5). Depending on the forging engineer, the measurable process parameters can be fixed or be variable for the process optimization. Transferring this on the here considered tension bar, the cross section A, the displacement u, and the total length l were kept constant, while the single lengths l_A and l_B were varied for optimizing the process.

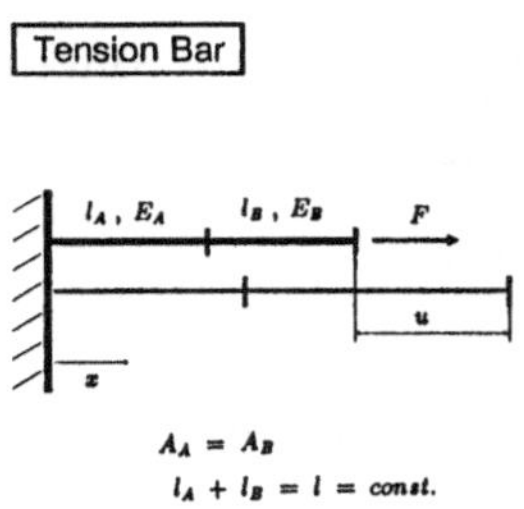

Analogy

	Tension Bar	Forging Process
Measurable Process Results	F	Workpiece Geometry Temperature Field Forming Force
Unknown Process Parameters	E_A, E_B	Coefficients μ, α, λ Yield Function $k_f = k_f(\varphi)$
Measurable, Variable Process Parameters	l_A, l_B	Die Geometry Initial Geometry
Measurable, Fixed Process Parameters	A, u, l	Initial Temperature Lying Period

Fig.5. Validation of transferability by tension bar analogy

Assuming for the moment constant l_{A_0} and l_{B_0} as well as a particular displacement u, while the force F_{real} was measured, it was the first aim to identify the Young's moduli $E_{A_{ident}}$ and $E_{B_{ident}}$. When $E_{A_{ident}}$ was given, $E_{B_{ident}}$ could easily be calculated (1). F_{real} was measured by a "calculated experiment" (2) by using the actually known $E_{A_{real}}$ and $E_{B_{real}}$.

$$E_{B_{ident}} = \frac{l_{B_0} \cdot E_{A_{ident}}}{\frac{A \cdot E_{A_{ident}} \cdot u}{F_{real}} \cdot \frac{F_{real}}{F_{ident}} - l_{A_0}} \qquad (1)$$

$$F_{real} = \frac{A \cdot E_{A_{real}} \cdot u}{\frac{E_{A_{real}}}{E_{B_{real}}} \cdot l_{B_0} + l_{A_0}} \tag{2}$$

Depending on the chosen convergence criteria $\frac{F_{ident}}{F_{real}}$, a wide range of parameter sets $E_{A_{ident}}$, $E_{B_{ident}}$ was identified (Fig.6).

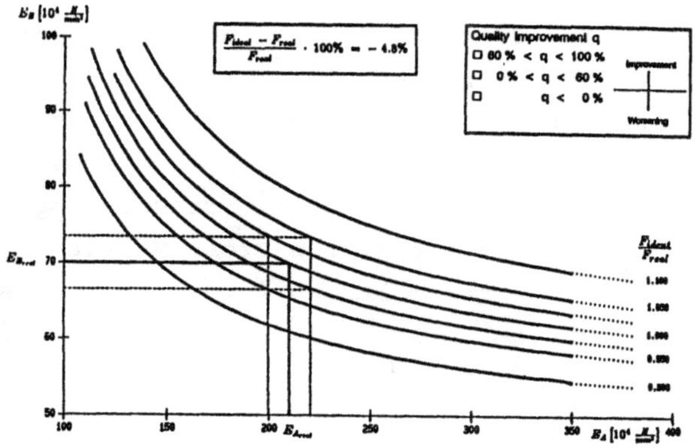

Fig.6. Identified parameter sets in dependence on convergence radius and the quality improvement of the process in dependence on the identified parameter sets used for optimization

The objective of the second step was to optimize the process according to a force F_{ideal} by varying $L_{B_{opti}}$, $L_{A_{opti}}$ respectively (3).

$$l_{B_{opti}} = \frac{\frac{A \cdot E_{A_{ident}} \cdot E_{B_{ident}} \cdot u}{F_{ideal}} - E_{B_{ident}} \cdot l}{E_{A_{ident}} - E_{B_{ident}}} \tag{3}$$

For checking the quality of the hereby gained model, first an "experimental check" (4) was done again. The so yielded force F_{opti} was used in the below defined quality improvement factor q (5).

$$F_{opti} = \frac{A \cdot E_{A_{real}} \cdot u}{\frac{E_{A_{real}}}{E_{B_{real}}} \cdot l_{B_{opti}} + l_{A_{opti}}} \tag{4}$$

$$q = \left(1 - abs\left(\frac{F_{opti} - F_{ideal}}{F_{real} - F_{ideal}}\right)\right) \cdot 100\% \tag{5}$$

Hereby 100% expresses total correspondence, 0% no improvement and negative values the worsening of the initial process. Choosing suitable restrictions for E_A (steel) and E_B (aluminium), Fig.6 clearly shows that about 70% of the identified parameter sets yield to quality improvement factors of 60% and more.

Practical Application

Since it was the objective to avoid high calculation times, especially in the testing phase cylindrical AlZn 4.5 Mg1 samples were formed by a cold and non-greased upsetting process on a screw press with direct drive. For proving the transferability to geometrical similar processes, a parameter set for the reference process was identified and transfered to the FEM-models of geometrical similar processes (Fig.4). The hereby gained results were finally compared with the experimental results. In contrast to the theoretical proof, the process tolerances and the measured inaccuracies had to be considered. While the accuracy of the measurements and the quality of the initial sample geometry was very high, the tolerance of the final sample diameter lies within a range of IT 12. Therefor, it could not be expected to identify a single parameter set but a range of parameter sets or an average mean.

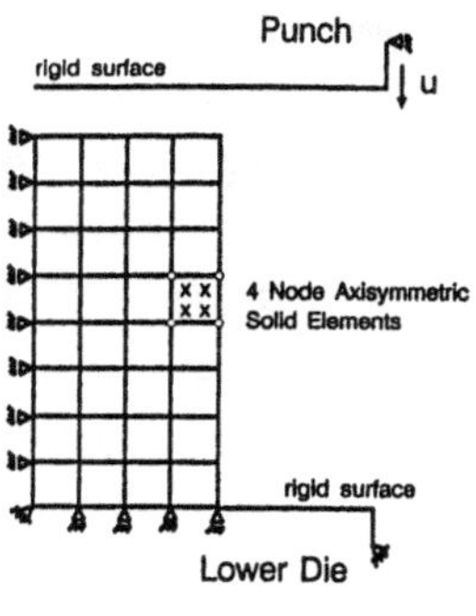

Fig.7. Model for FEM-simulation

The model shown in Fig.7 was chosen for the FEM-simulation after testing influencing variables, for example number of elements, element type or time discretization. Hereby the commercial FEM-package ABAQUS 4.8 was used.
First calculations showed that the used yield function, independent of the friction parameter, produces good correspondence between calculation and experiment. Moreover, horizontal symmetry of the sample could be proved by geometrical measurement. Consequently only one global friction coefficient μ and one global stiffness in stick factor sis had to be condsidered in the used classic COULOMB friction model.
Before presenting the results gained by NI for this application the most important hereby used algorithms shall be explained.
For comparison of the measured and calculated sample profile, linear interpolation betweeen the points of support was used. Thus 1000 equidistant pairs of variates r_{exp} and r_{sim} were calculated to compute the comparison factor c (6).

$$c = \frac{\sum_{i=1}^{n} abs\,(r_{exp} - r_{sim})}{n \cdot Volume} \cdot 10^5 \qquad (6)$$

Two different optimization algorithms were tested: first, a standard hill-climbing algorithm with constant step size and secondly, a knowledge based hill-climbing with decreasing step size, when no improvement was gained. "Knowledge based" stands for a optimized sequence of searching based on experience. Defining suitable interval limits for μ and sis as well as the required values for the optimization algorithms, both algorithms were tested by means of three different starting points. It can be recognized that the knowledge based hill-climbing algorithm yields always to better comparison values while the number of iterations do not increase substantially (Fig.8).

Interval Limits

μ: 0.02 — 0.5

sis: 10.0 — 5000.0

Step Sizes

- Standard Hill Climbing
  $\Delta\mu = 0.02 \quad \Delta\, sis = 20.0$
- Knowledge Based Hill Climbing
  $\Delta\mu_0 = 0.02 \quad \Delta\,\mu_{min} = 0.0025$
  $\Delta sis_0 = 20.0 \quad \Delta\, sis_{min} = 2.5$

Starting Point			Standard Hill Climbing				Knowledge Based Hill Climbing			
Comparison	μ	sis	Comparison	Iterations	μ	sis	Comparison	Iterations	μ	sis
1.868	0.050	4340.	1.868	5	0.050	4340.	1.105	32	0.0575	4340.
1.882	0.050	920.	0.937	182	0.070	40.	0.693	238	0.150	30.
18.534	0.440	920.	1.162	222	0.080	40.	0.693	248	0.150	30.

Experimenal Results : Arimethic Mean

Fig.8. Identification results in dependence of optimization method and starting point

Considering the above shown table as well as the process tolerance, it can be stated as expected that not a single parameter set can be identified but an area of parameter sets has to be defined (Fig.9). The existence of two totally different areas can be seen clearly:

1. $\mu = 0.053 \ldots 0.062$ $\quad sis > 1000.0$
2. $\mu > 0.09$ $\quad sis = 27.0 \ldots 31.0$

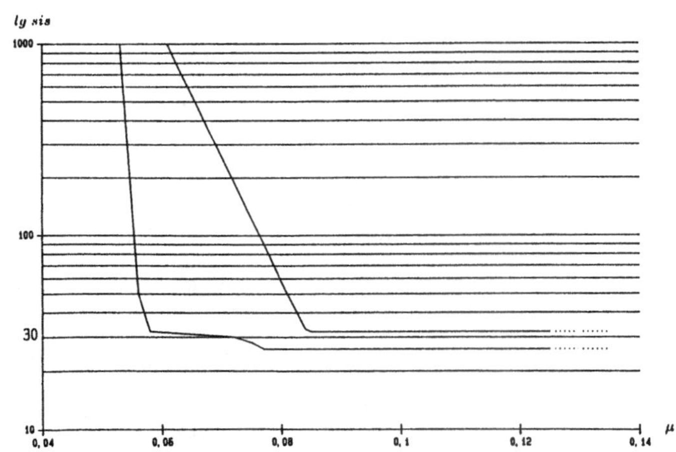

Fig.9. Identified area of parameter sets

For proving the transferability of both parameter sets to geometrical similar processes, the comparison factors c were computed for the reference process as well as for the geometrical similar processes. Particularly the processes where the initial sample geometry differs only a little bit, the comparison factors show high accordance relating to the reference process. Even modifying the diameter-heigh-ratio substantially yields more than satisfying results (Fig.10).

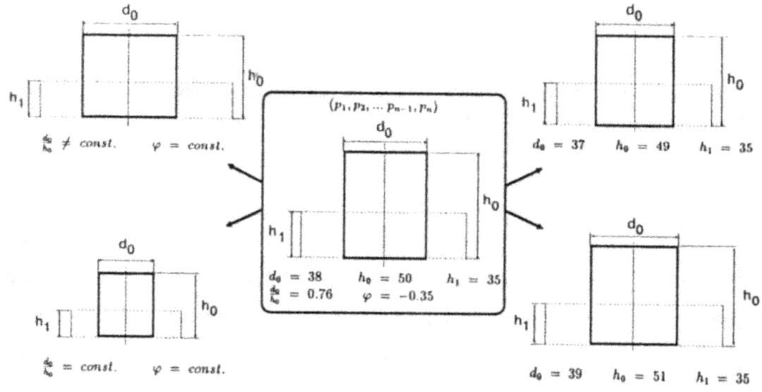

Fig.10. Comparison factors gained by transfering identified paramters sets to geometrical similar processes

References

1. Groche, P.; Weiss, U.: Numerical Identification of Forging Parameters. Chenot, J.L.; Onate, E. (ed.), Modelling of Metal Forming Processes. Kluwer Academic Publishers 1988, 237-244.

2. Weiss, U.: Numerische Simulation von Präzisionsschmiedeprozessen mit der Finite-Elemente-Methode. Fortschritt-Berichte VDI, Reihe 2 Nr. 146. Düsseldorf: VDI Verlag 1988.

3. Becker, M.: Anwendung von Höheren Optimierungsmethoden in der Umformtechnik. Diss. Aachen 1991.

4. Schwefel, H.-P.: Numerische Optimierung von Computer-Modellen mittels der Evolutionsstrategie. Mit einer vergleichenden Einführung in die Hill-Climbing- und Zufallsstrategie. Basel, Stuttgart: Birkhäuser Verlag 1977.

5. Vanderplaats, G. N.: Numerical Optimization Techniques For Engineering Design. New York u.a.: McGraw-Hill Book Company 1984.

Development of an Element for the Three Dimensional Finite Element Simulation of Sheet Metal Forming Processes

Prof. Dr.-Ing. K. Siegert, Dr.-Ing. T. Chatzikonstantinou,
Dipl.-Ing. J. Dehgan-Manshadi

Institut für Umformtechnik (IFU), Universität Stuttgart

Abstract

A simple flat triangular finite element for exact three dimensional finite element simulation of sheet metal forming processes has been developed by the Institut für Umformtechnik, Universität Stuttgart. It has bending characteristics like a shell element but it also has stress components in thickness direction and can discribe the continuum problem in the sheet accurate enough with a minimum of degrees of freedom. The computational efficiency of the element will help to simulate big problems of industrial interest at a minimum of computational costs.

1. Introduction

The simulation of sheet metal forming processes with the finite element method is possible today, but very high computation times or difficulties in the geometrical and physical description of the problem hinder the practical usage in industrial applications. In addition to efficient material laws and correct contact algorithms the formulation of the finite element for the discretisation of the sheet has a large significance for fast and correct solution of the whole problem.

The modelling of the sheet with brick elements (Fig. 1) is practically impossible, because the computation time is very high even with supercomputers. The reason for this is the necesserity to modell with more than one sheets of brick elements the sheet for accurate bending characteristics of the structure. This has a big significance for the planar discretisation of the sheet because the restriction of a good aspect ratio of the elements must be sutisfied. The result is a model of many thousands of elements even for small parts of industrial interest. Brick elements with better bending characteristics like the so named "assumed strain formulation" elements cannot eliminate this disadvendage.

The usage of shell elements (Fig .1) eliminates the problem of accurate bending characteristics of the sheet, because by use of shell elements only one sheet of elements is neccesery for good results. The problem here is, that shell elements do not have a stress in thickness direction of the sheet and so an interaction with the dies in this direction is impossible. The pressure on both sides of the sheet cannot be calculated and for some kinds of contact algorithms the calculation of Coulomb friction on the sheet is impossible. With this disadvantage remeshing strategies for faster calculation occur errors, because after transformation of the stress tensor stress components in thickness direktion cannot be interpreted.

D. Besdo, E. Stein (Eds.)
Finite Inelastic Deformations –
Theory and Applications
IUTAM Symposium Hannover/Germany 1991

Modelling the sheet with Membrane elements (Fig. 1) is indiscutable, because there is no bending and the other kinematic problems are the same as for schell elements.

The finite element developed by the Institut of Umformtechnik, a flat triangle element with the minimum of degrees of freedom possible for the class of the continuum problem in the sheet, works like a shell element but it also has stress components in thickness direktion. So only one sheet of elements is needed for modelling the sheet. The usage of this element in the simulation of sheet metal forming processes has no limits in the cinematics of the whole problem. It works with Coulomb friction in every kind of contact algorithm used today and there are no problems in remeshing strategies. The total Lagrangian formulation of the element brings advantages in the effectiveness of the computation and sometimes in the implementation of material laws.

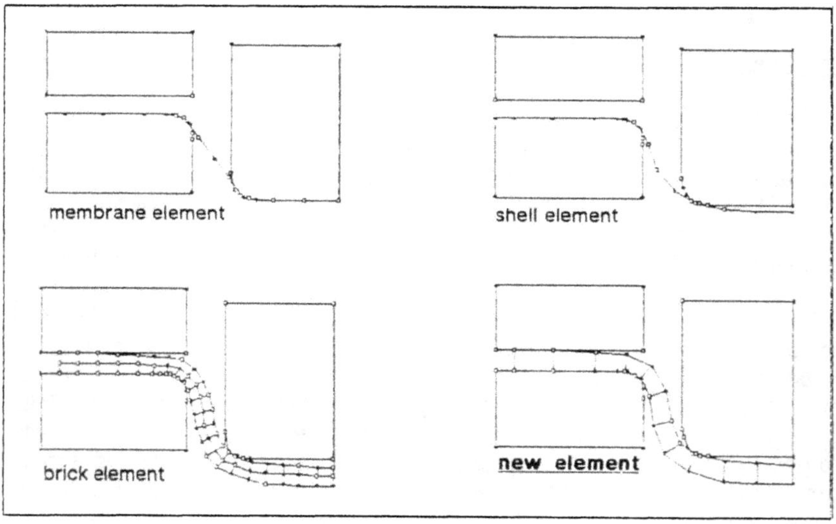

Fig. 1. New element in comparison with traditional elements in the simulation of sheet metal forming processes

2. Kinematics of the element

In sheet metal forming processes the undeformed configuration of the sheet is a plane one and has everywhere a constant thickness. So it is advantageous to use an element with a two dimensional reference configuration with originally constant thickness "t" over the element area "A" and a Total Lagrangian description of the problem (Fig. 2).

To keep the computational costs low, we use a middle surface formulation and formulate a simple triangular element with 6 nodes and one position vector per node in the Cartesian space (Fig. 2). With these 18 degrees of freedom we have to define at least 12 discrete element strains. This is enough to describe the most important of the real situations in the sheet during the forming process, such as streching, bending and changes of the thickness.

The first three strains are the middle surface strains. They are equal to the well known constant strains of a triangle membrane element (middle surface) with linear interpolation of its position vektors.

$$\varepsilon^{M}_{\alpha\beta} = 1/2 \; (F_{\alpha i} F_{\beta i} - \delta_{\alpha\beta}) \qquad (1)$$

$F_{\alpha n}$ is the deformation gradient of the middle surface.

$$F_{\alpha i} = 1/2 \; a_{\alpha n} \; (x^{u}_{in} + x^{l}_{in}) \qquad (2)$$

$a_{\alpha n}$ contains only reference values and is constant for each element.

$$a_{\alpha n} = 1/2A \begin{pmatrix} X_{22} - X_{23} & X_{23} - X_{21} & X_{21} - X_{22} \\ X_{13} - X_{12} & X_{11} - X_{13} & X_{12} - X_{11} \end{pmatrix} \qquad (3)$$

To describe element bending we follow Schoop /1/ and define six transverse strains at the corners of the element.

$$\theta^{B}_{\alpha n} = F_{\alpha i} \; (x^{u}_{in} - x^{l}_{in})/t \qquad (4)$$

$\theta^{B}_{\alpha n}$ is, for small element internal rotations and under Kirchhoff conditions, the angle between the middle surface of the element and the tangential plane at the corner n.

The next three strains are the transverse strains at the three corners of the element.

$$\varepsilon^{T}_{33n} = 1/2 \; (b_{in} b_{in} - 1) \qquad (5)$$

$b_{(i)n}$ are dot products of the three element corner edges, divided by the initial thickness t, and the middle surface unit vector a_i/a in the actual cinfiguration.

$$b_{in} = (x^{u}_{in} - x^{l}_{in}) \cdot a_i \; a_i/ta^2 \qquad (6)$$

Note, that all these strains vanish for arbitrary rigid body movements.

The emphasis in this paper lies specially in the definition of the element transverse strain. The definition of this strain guarantees, that the element activates only its transverse stifness in the case of double side contact. The definition of the transverse strain in /1/ makes a correct element behaviour in this point impossible, because double side contact would activate here, mistakenly, in most cases the bending stiffness of the element.

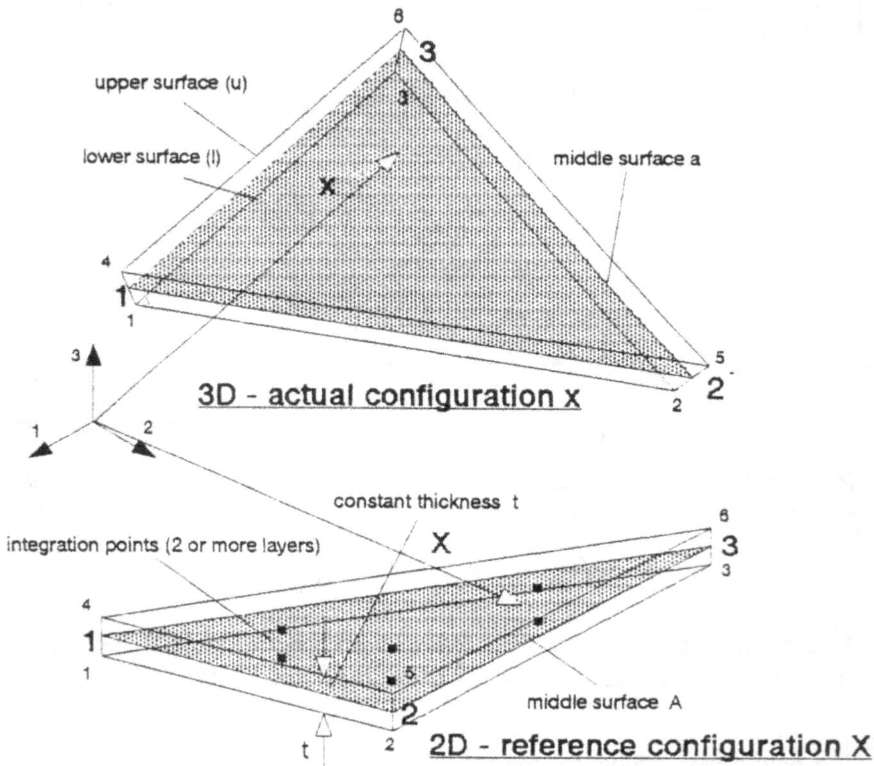

Fig. 2. 6 node (18 dof) element for three dimensional simulation of sheet metal forming processes.

---

1) We can very simple transform the element of Schoop /1/ to another one with its original strains but with our degrees of freedom. This element would have an upper and a lower surface and double side contact could be theoretically possible.

3. Element internal force vector and stiffness matrix

For calculation of internal force vector and stiffness matrix we use a classical numerical integration scheme with layers and three integration points per layer. Hereby we need the strains and their derivatives at the integration point "m" and layer "g". Here only the strains are given.

$$\varepsilon_{\alpha\beta mg} = \varepsilon^{M}{}_{\alpha\beta} + z_g \kappa^{B}{}_{\alpha\beta m} \tag{7}$$

$$\varepsilon_{33mg} = \varepsilon^{T}{}_{33m} \tag{8}$$

$$\varepsilon_{\alpha 3mg} = \varepsilon_{3\alpha mg} = 0 \tag{9}$$

$\varepsilon_{\alpha\beta mg}$ is the Green membrane-bending strain at integration point "m" / layer "g".

ε_{33mg} is the Green transverse strain at integration point "m" / layer "g".

$\varepsilon_{\alpha 3mg}$ is the Green transverse shear strain.

To calculate $\kappa_{\alpha\beta m}$ we use a quadratic trial function for the angle $\theta^{B}{}_{\alpha}$.

$$\theta^{B}{}_{\alpha} = L_n \theta^{B}{}_{\alpha n} + L_{n+1} L_{n-1} C_{\alpha n}$$

$$= L_n a_{\alpha k} (x^{u}{}_{ik} + x^{l}{}_{ik})(x^{u}{}_{in} - x^{l}{}_{in})/2t + L_{n+1} L_{n-1} C_{\alpha n} \tag{10}$$

The 6 constants $C_{\alpha n}$ are calculated from the well known DKT hypothesis in /2/: At the midpoint of side "s" of the middle surface the components of $\theta^{B}{}_{\alpha}$ along the side "$\theta^{B}{}_{\underline{a}s}$" and normal to the side "$\theta^{B}{}_{ns}$" are given as

$$\theta^{B}{}_{\underline{a}s} = -1/4\,(\theta^{B}{}_{\underline{a}(s-1)} + \theta^{B}{}_{\underline{a}(s+1)}) \tag{11}$$

$$\theta^{B}{}_{\underline{n}s} = 1/2\,(\theta^{B}{}_{\underline{n}(s-1)} + \theta^{B}{}_{\underline{n}(s+1)})\,. \tag{12}$$

With equations (10), (11) and (12) the angle $\theta^{B}{}_{\alpha}$ and the curvature $\kappa^{B}{}_{\alpha\beta}$

$$\kappa^{B}{}_{\alpha\beta} = \theta^{B}{}_{\alpha,\beta} \tag{13}$$

are known.

For $\kappa^B_{\alpha\beta}$ at the integration point "m" follows (after some transformations)

$$\kappa^B_{\alpha\beta m} = v_{\alpha\beta mnk}\,(x^u_{in}+x^l_{in})\,(x^u_{ik}-x^l_{ik})/2t \ . \tag{14}$$

Note, that t, **A**, $\mathbf{a}_{\alpha n}$ and $\mathbf{v}_{\alpha\beta mnk}$ are **reference values** and are constant during a nonlinear analysis. This is one of the reasons for the computational efficiency of the element presented here.

4. Comparison with other simple shell elements

For the case of elastic behaviour

$$d\sigma_{ij}/d\varepsilon_{lk} = 2G\left(\delta_{il}\delta_{jk} + (\nu/(1-2\nu))\delta_{lk}\delta_{ij}\right)^{1)} \tag{15}$$

the element was compared to the elements in references /1/ and /3/. Implementation of elastic plastic material laws (in strain space) and the use in contact problems will be presented in another extended paper.

The geometrically nonlinear problem in Fig. 3 shows, that our element has a bending behaviour like the elements of the autors in /1/ and /3/, but ,with its upper and lower nodes and the suitable definition of the transverse strain ε^T_{33}, it also can describe double side contact.

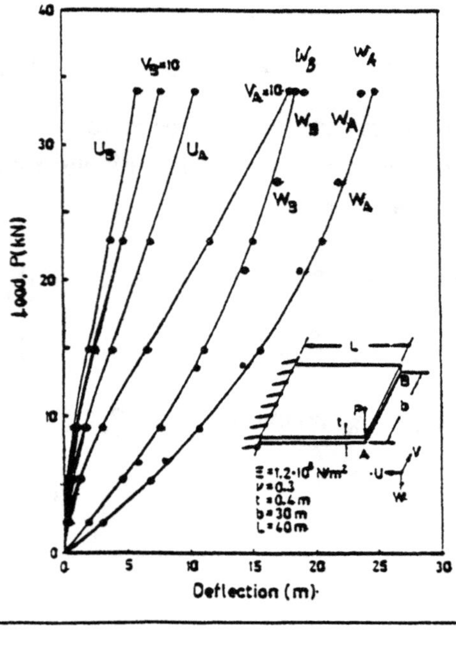

Fig. 3. Geometrically nonlinear problem in /3/.

(● results with new element)

1) σ_{ij} is the 2. Piola Kirchhof stress tensor

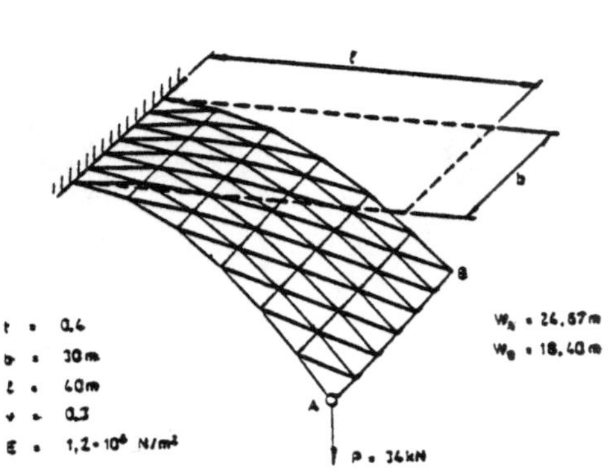

element in /1/

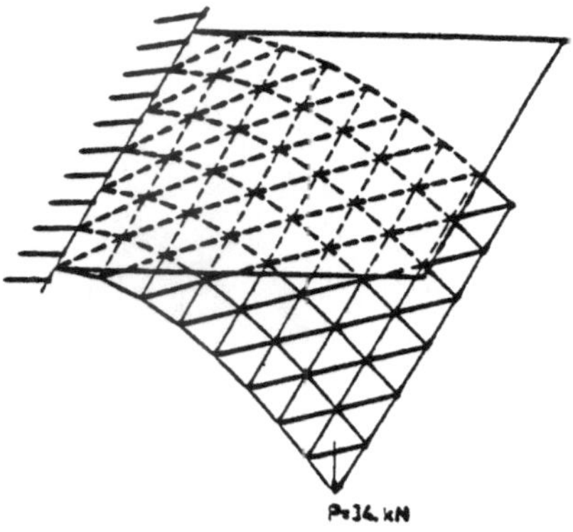

element in /3/

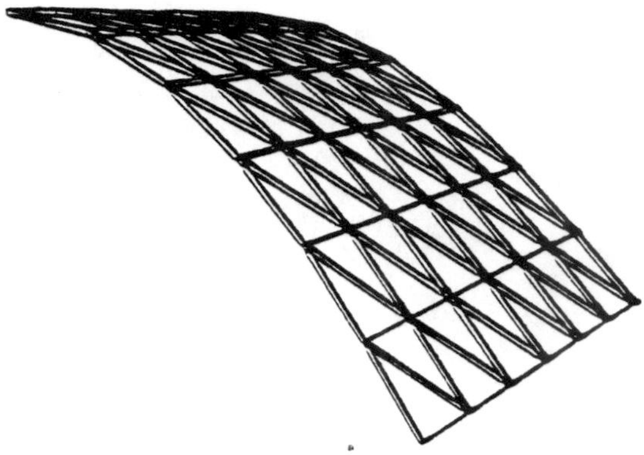

new element

Fig. 4. New element in comparison to /1/ and /3/

5. References

/1/ Schoop H.: A simple nonlinear flat element for large displacement structures. Comput. Struct. 32, 379-385, 1989.

/2/ K.J. Bathe and L.W. Ho.: A simple and effective element for the analysis of general shell structures. Comput. Struct. 13, 673-681, 1981.

/3/ Kuo-Mo Hsiao.: Nonlinear analysis of general shell structures by flat triangular shell element. Comput. Struct. 25, 665-675, 1987.

Determination of Nonassociated Constitutive Equation for Rock Salt from Experiments

N. CRISTESCU* & U. HUNSCHE**

* University of Bucharest, 14 Academiei Str., Bucharest 70109, Romania
** Federal Institute for Geosciences and Natural Resources, Stilleweg 2, W-3000 Hannover 51, Germany

Summary

Starting from precise true triaxial failure tests on cubic rock salt samples a triaxial elastic/viscoplastic constitutive equation has been formulated which comprises: creep (including work hardening and unloading), volume dilatancy and compressibility during deformation, instantaneous failure, or damage with subsequent creep-failure. The main steps to formulate such a constitutive equation are: determination of the instantaneous response of rock salt, the yield function and the viscoplastic potential. All constitutive parameters and functions are determined from the experimental data in three dimensions without any a priori assumption concerning their form; a preliminary uniaxial form of the constitutive equation is not needed. Finally it is discussed under which conditions this constitutive equation can be simplified and when an associated law is possible.

Introduction

The paper continues the efforts done in the last years by the authors for the formulation of a general, triaxial, constitutive equation for rock salt. Many other authors have also shown a great deal of interest in the formulation of such a constitutive equation. For corresponding references see Cristescu & Hunsche [1] or Hunsche [2]. The great interest shown is generated by mining engineering reasons but also by the problem of deep underground storage cavities in rock salt either for oil or natural gas, or for permanent repositories for radioactive or toxic wastes.

Since rock salt deforms significantly by creep, various rheological models have been developed mainly inspired by the already existing models for metals. Many of them are based on the existence of several deformation mechanisms for creep dominating in different areas of effective stress and temperature. Within the framework of classical metal plasticity the mean stress does not play any role and the volumetric deformation of metals is considered to be purely elastic. Therefore, the idea has been to obtain a triaxial constitutive equation also for rock salt via this concept of an "universal creep-curve". This concept is possible as long as the volume

D. Besdo, E. Stein (Eds.)
Finite Inelastic Deformations –
Theory and Applications
IUTAM Symposium Hannover/Germany 1991

change is purely elastic. However, the deformation behaviour of rock salt is complicated by the fact that it exhibits irreversible dilatancy and compressibility due to opening and closing of microcracks in a broad range of stress states. Therefore, an advanced constitutive equation has to be developed which considers dilatancy, damage, and failure, too.

In an earlier paper by the authors [1] a general constitutive equation was given. It is able to describe the time-dependent behaviour of rock salt, and is established on the base of triaxial experimental data. The purpose of the present paper is to improve this version. The improvement consists in considering a greater quantity of experimental data; also the quality of the experimental data was increased, especially what concerns the precise calibration for an improved determination of the volumetric strain. For detailed description of the experiments and their results see Hunsche & Albrecht [3] and Hunsche [2,4 (this volume)]. Finally the analysis used for the formulation of the constitutive equation was made more accurate. As in the earlier version we made no a priori assumption concerning the form of the constitutive functions. We have taken into account such mechanical properties of rock salt as elasticity, creep (including work-hardening and unloading), irreversible compressibility and dilatancy, damage, failure and creep-failure. We have examined if the irreversible volumetric strain is sometimes negligible, if the potentials used in classical plasticity (Mises or Tresca) can be used for rock salt, if the compressible/dilatant boundary can be determined with the same accuracy as for other rocks (see Cristescu [5]), if rock salt would accept an associated constitutive equation or nonassociated laws are to be used. Nonassociated viscoplastic constitutive equations have already been developed for sand [6] by a similar procedure.

It has to be mentioned that in this stage of development the main emphasis was placed on the description of the effects related to dilatancy and failure, not on long-term creep. The experiments on which the constitutive equation is based are quasi-static but no long-term creep tests. At present only transient creep is included but not steady state creep. The description of transient creep after unloading is included.

Constitutive equation

Before starting with the formulation of a constitutive equation first let us make clear what kind of constitutive equation will be the most appropriate for rock salt, i.e. how its main mechanical properties are to be described reasonably well. The guidance ideas are discussed below. Generally we expect from such a constitutive equation to be able to describe besides the creep phenomena, quite present in rock salt even at very small applied stresses, irreversible dilatancy and compressibility, damage and failure, creep-failure, as well as its instantaneous response.

Generally one starts by checking what kind of instantaneous response is exhibited by rock salt. Because the two seismic body waves (longitudinal and transverse) are propagating in rock salt the instantaneous response can be described by

$$\dot{\boldsymbol{\varepsilon}}^{E} = \frac{\dot{\boldsymbol{\sigma}}}{2G} + \left(\frac{1}{3K} - \frac{1}{2G}\right) \dot{\sigma}\, \mathbf{1} \tag{1}$$

where $\dot{\boldsymbol{\varepsilon}}^{E}$ is the elastic part of the rate of strain tensor, $\boldsymbol{\sigma}$ is Cauchy's stress tensor, σ is the mean stress, $\mathbf{1}$ is the Kronecker unit tensor and G and K are the two elastic parameters possibly depending on stress state and on a damage parameter. This dependence is not considered in this paper.

For the non-instantaneous response we choose the irreversible rate of strain in the form (see Cristescu [5])

$$\dot{\boldsymbol{\varepsilon}}^{I} = k \left\langle 1 - \frac{W^{I}(t)}{H(\boldsymbol{\sigma})} \right\rangle \frac{\partial F}{\partial \boldsymbol{\sigma}} \tag{2}$$

Here $F(\boldsymbol{\sigma})$ is a viscoplastic potential, i.e. a function depending in the stress invariants.

$$H(\boldsymbol{\sigma}(t)) = W^{I}(t) \tag{3}$$

is the equation of the stabilization boundary for transient creep or a yield function, with the irreversible stress work

$$W^{I}(t) = \int_{0}^{t} \boldsymbol{\sigma}(s) \cdot \dot{\boldsymbol{\varepsilon}}^{I}(s)\, ds \tag{4}$$

The equation of the stabilization boundary was chosen in the form (3) since it was found in creep tests that stabilization depends on the loading history, thus creep of the rock salt cannot be described accurately by viscoelastic models but by viscoplastic ones (in our model history is involved via (4)). As work-hardening parameter (or internal variable) the irreversible stress work (4) was chosen since it can well describe irreversible dilatancy and irreversible compressibility, if we decompose W^I as

$$W^I(t) = \int_0^t \sigma(s)\ \dot{\varepsilon}_v^I(s)\ ds + \int_0^t \sigma'(s)\cdot\dot{\varepsilon}^{I'}(s)\ ds \tag{5}$$

where prime stands for "deviator" and ε_v^I is the irreversible volumetric strain. Here the first right-hand term is the work W_v^I related to volume change and is positive during compression but negative during dilatancy. Therefore, this term can be used as a scalar damage parameter as well (see [5]) since it is related to the energy of microcrack opening or closing. For the same reason other classical parameters used to describe hardening (as for instance the equivalent strain) would be unable to distinguish between irreversible dilatancy and irreversible compressibility of rocks. The second right hand term W_d^I in (5) is mainly related to shape change.

Let us give an example how creep can be described by (1). Let us assume that up to the initial moment t_0 is $\dot{\varepsilon}^I = 0$ i.e. $H(\sigma^P) = W^{IP}$ for $t < t_0$, where the superscript P marks the "primary" state. If a stress variation is produced just at time t_0 and $H(\sigma(t_0) > H(\sigma^P)$, and for $t > t_0$ the stress is kept constant, then $W^I(t)$ varies according to a law, which is obtained straight from (2) by differentiation (see [5], p. 219):

$$\dot{\varepsilon}^I(t) \sim 1 - \frac{W^I(t)}{H(\sigma(t_0))} = \left(1 - \frac{W^{IP}}{H(\sigma(t_0))}\right)\cdot\left(\exp\left[\frac{k(t_0-t)}{H(\sigma(t_0))}\ \sigma(t_0)\ \cdot\ \frac{\partial F}{\partial\sigma}\bigg|_{\sigma(t_0)}\right]\right) \tag{6}$$

Thus for $t \to \infty$, $W^I(t)$ tends towards $H(\sigma(t_0))$ and in this way one can describe transient creep. How fast the variation of $\dot{\varepsilon}^I$ takes place depends on the magnitude of the variable viscosity coefficient $k\,\frac{\partial F}{\partial\sigma}$, i.e. on the viscoplastic potential F. The derivative $\frac{\partial F}{\partial \sigma}$ is determining also the orientation of $\dot{\varepsilon}^I$, not only its magnitude. For metals is $F \equiv H$ and thus the orientation of $\dot{\varepsilon}^I$ is governed by the yield function, and since this yield function is independent of σ for metals, the volumentric deformation is purely elastic. With geomaterials, generally, we have to assume that F is distinct from H (see [6]) and thus that the constitutive equation is non-associated (to the yield function). That is why in the present paper we start with a nonassociated constitutive equation and we examine later if an associated one can sometimes be applied (i.e. $F \equiv H$) and also if classical plasticity yield surfaces (independent on σ as Mises or Tresca) can be used.

Finally, k in equation (2) may depend on stress and damage. It is one part of the viscosity coefficient (the reciprocal of $k \frac{\partial F}{\partial \sigma}$) and will be determined together with F; in most practical cases is k = const. The bracket $\langle\rangle$ is the positive part of the mentioned argument function, i.e.

$$\langle A \rangle = A^+ = ½ (A + |A|) \tag{7}$$

Thus the constitutive equation which hopefully can be adapted for rock salt will be written in the form

$$\dot{\varepsilon} = \frac{\dot{\sigma}}{2G} + \left(\frac{1}{3K} - \frac{1}{2G}\right) \dot{\sigma}\, \mathbf{1} + k \left\langle 1 - \frac{W^I(t)}{H(\sigma)} \right\rangle \frac{\partial F}{\partial \sigma} \tag{8}$$

From here follows that the variation of the irreversible volumetric rate of strain is described by

$$\dot{\varepsilon}^I_v = k \left\langle 1 - \frac{W^I(t)}{H(\sigma)} \right\rangle \frac{\partial F}{\partial \sigma} \tag{9}$$

Therefore, as long as $H(\sigma) > W^I(t)$, the volume is compressible if $\frac{\partial F}{\partial \sigma} > 0$ and is dilatant if $\frac{\partial F}{\partial \sigma} < 0$. Moreover $\frac{\partial F}{\partial \sigma} = 0$ is the equation of the compressible/dilatant boundary. It follows from here that function F is to be determined so that this volumetric behaviour would be described by (9). Further, when approaching ultimate failure the volume is increasing steeply by dilatancy so that failure is also to be described by peculiar properties of function F. Generally, failure strength depends on the loading rate or on the strain rate [5]. However, in additional tests it was found that strain rates below about $10^{-1}s^{-1}$ influence strength of rock salt only slightly; loading rates below about 300 $MPa \cdot min^{-1}$ have a somewhat larger but not very important influence.

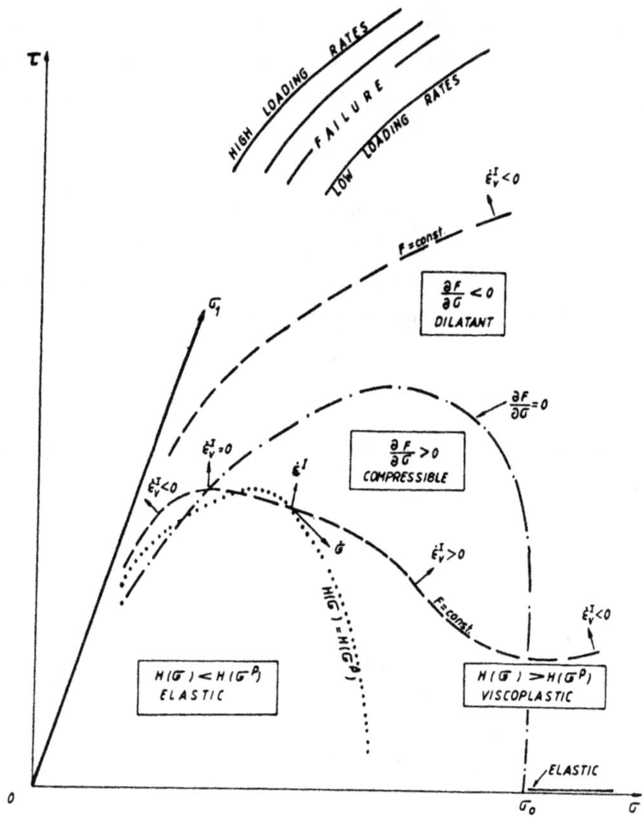

Fig. 1: Schematic of domains and surfaces in the (σ–τ)-plane.
σ_1: direction of uniaxial tests

The schematic in Fig. 1 shows some typical surfaces: H = const, F = const, the compressible/dilatant boundary and various possible failure surfaces. Let us assume that before an excavation the rock is in equilibrium ($\dot{\boldsymbol{\varepsilon}}^I = 0$) and therefore that the initial in situ stress $\boldsymbol{\sigma}^P$ satisfies $H(\boldsymbol{\sigma}^P) = W^{IP}$. If a variation of the stress is produced (due to an excavation, say) from

σ^P to $\sigma(t)$ and if $H(\sigma(t)) < H(\sigma^P)$ an elastic unloading will take place. If however $H(\sigma(t_0)) > H(\sigma^P)$ then a viscoplastic deformation occurs with the three subcases discussed above and mentioned in Fig. 1: dilatant, compressible or constant volume. The small arrows in Fig. 1 indicate various normals to the surfaces F = const., i.e. various possible directions of $\dot{\varepsilon}^I$. Generally, along the lower surface F = const. shown it is $\dot{\varepsilon}^I_v < 0$ for small σ, then $\dot{\varepsilon}^I_v = 0$, then $\dot{\varepsilon}^I_v > 0$ when σ is increasing, then again $\dot{\varepsilon}^I_v = 0$ and finally $\dot{\varepsilon}^I_v < 0$ for very high σ.

Further we will try to determine the constitutive functions H and F for rock salt so that all the previously mentioned mechanical properties would be described by one constitutive equation. Since it is assumed that rock salt is isotropic, both constitutive functions H and F will depend on the stress invariants only. At the present stage of the development of this model these functions depend only on mean stress (first invariant) and octahedral shear stress (second invariant)

$$\tau = \frac{1}{3}\left[(\sigma_1-\sigma_2)^2+(\sigma_2-\sigma_3)^2+(\sigma_3-\sigma_1)^2\right]^{1/2} \tag{10}$$

The third stress invariant will be taken into account in a future variant of the model since it is well known that this invariant is involved in the equation of the failure surface (Hunsche [4]). Therefore, in this publication it can be written: $H(\sigma) = H(\sigma,\tau)$ and $F(\sigma) = F(\sigma,\tau)$.

In conjunction with the formulation of a constitutive equation for rock and also for an accurate definition of the concept of damage, it is important to define precisely the reference configuration, i.e. the state of the rock (including the existing microcracking and pores) before any human intervention. The reason is that all rocks, before being tested in the laboratory, have been subjected to a long history of loadings and unloadings, as well as to other influences; all of them are in some way stored in the memory of the material. For rocks there is no stress-free, strain-free and damage-free configuration (often considered in metal plasticity). Even the rocks tested in the laboratory and which may be considered to be stress-free, contain a significant microcracking distribution and damage from the very beginning. When mining applications are envisaged the reference configuration is the state in situ before excavation, including stress-state, microcracks, temperature etc.. It is only the change of this primary stress-state produced by an excavation that is responsible for the strains around the excavation. The memory of all irreversible deformation to which the rock was subjected before excavation is summarized in the term W^{IP} which can be determined in practice (see Cristescu [5]).

Determination of constitutive functions for rock salt

All constitutive functions and parameters are determined from the 14 true triaxial tests described by Hunsche [4]. In these tests cubical specimens are subjected first to a hydrostatic loading ($\tau = 0$) followed by a deviatoric one (σ = const). During both stages of the test all principal stress and strain components are recorded. From unloading tests the elastic parameters were determined: K = 21.7 GPa, G = 11.8 GPa, E = 30 GPa and $\nu = 0.27$.

The yield function $H(\sigma,\tau)$ is determined from the data by numerical estimation of both integrals from (5). Following a procedure described in detail by Cristescu [5] we obtain

$$H(\sigma,\tau) := \sigma_* \left\{ \exp\left(a_1 \left(\frac{\sigma}{\sigma_*}\right)^2 + a_2 \frac{\sigma}{\sigma_*} + a_3 \right) \right\} \left(\frac{\tau}{\sigma_*}\right)^9 + \left\{ b_1 \left(\frac{\sigma}{\sigma_*}\right)^2 + b_2 \frac{\sigma}{\sigma_*} + b_3 \right\} \left(\frac{\tau}{\sigma_*}\right)^2 + \left\{ c_1 \left(\frac{\sigma}{\sigma_*}\right)^2 + c_2 \frac{\sigma}{\sigma_*} + c_3 \right\} \frac{\tau}{\sigma_*} + \begin{cases} h_1 \sigma^2 & \text{if } \sigma \leq \sigma_A \\ h_2 \sigma + h_3 & \text{if } \sigma \geq \sigma_A \end{cases} \tag{11}$$

$a_1 = 4.5\times10^{-3}$ $\quad b_1 = -4.7\times10^{-6}$MPa $\quad c_1 = 4.46\times10^{-5}$MPa $\quad h_1 = 4.6\times10^{-4}$MPa^{-1}

$a_2 = -0.35$ $\quad b_2 = 3.13\times10^{-4}$MPa $\quad c_2 = -2.65\times10^{-3}$MPa $\quad h_2 = 5.56\times10^{-3}$

$a_3 = -21.9$ $\quad b_3 = -2.648\times10^{-3}$MPa $\quad c_3 = 4.78\times10^{-2}$MPa $\quad h_3 = -1.67\times10^{-2}$MPa

$\sigma_* = 1$ MPa $\quad \sigma_A = 6$ MPa

This new version of the function H differs from the one given in [1] not only by the fact that the new version is based on a larger quantity and quality of the experimental data, but also by other forms of the coefficient functions involved in (11). The former expression of the function H was not suitable for computation of mining engineering design.

The viscoplastic potential is determined following a more involved procedure, similar to that previously used for sand (Cristescu [6]). It is obtained using two main formulae which follow from (1):

$$k \frac{\partial F}{\partial \sigma} = \frac{\dot{\varepsilon}_v^I}{\left\langle 1 - \frac{W^I(t)}{H(\sigma,\tau)} \right\rangle} \tag{12.1}$$

$$k \frac{\partial F}{\partial \tau} = \frac{1}{\left\langle 1 - \frac{W^I(t)}{H(\sigma,\tau)} \right\rangle} \sqrt{\left(\dot{\varepsilon}_1^I - \dot{\varepsilon}_2^I\right)^2 + \left(\dot{\varepsilon}_2^I - \dot{\varepsilon}_3^I\right)^2 + \left(\dot{\varepsilon}_3^I - \dot{\varepsilon}_1^I\right)^2} \tag{12.2}$$

Shortly the successive steps are the following. From hydrostatic tests and using formula (12.1) we first determine

$$k \left. \frac{\partial F}{\partial \sigma} \right|_{\tau=0} = q_1 \left(\frac{\sigma}{\sigma_*}\right)^3 + q_2 \left(\frac{\sigma}{\sigma_*}\right)^2 + q_3 \frac{\sigma}{\sigma_*} =: \phi(\sigma) \tag{13}$$

where $q_1 = 2.26\times10^{-8}$, $q_2 = -2.34\times10^{-6}$ and $q_3 = 6.04\times10^{-5}$, all given in sec^{-1}. Then the compressible/dilatant boundary is determined directly from the tests as

$$X(\sigma,\tau) := -\frac{\tau}{\sigma_*} + f_1 \left(\frac{\sigma}{\sigma_*}\right)^2 + f_2 \frac{\sigma}{\sigma_*} = 0 \tag{14}$$

where $f_1 = -1.68\times10^{-2}$ and $f_2 = 0.86$. Where this boundary intersects the hydrostatic axis, that is for $\sigma = 51.4$ MPa, all microcracks and pores are already closed so that in hydrostatic tests done for $\sigma > 51.4$ MPa, the rock salt will behave purely elastically. It has to be mentioned that the experimental values of this boundary are not very precisely defined. The error of the values for τ can reach $\pm$ 2 MPa in some tests (see [4]). Further, again straight from the data, the equation of the failure surface is determined in terms of stresses as

$$Y(\sigma,\tau) := -\left(\frac{\tau}{\sigma_*}\right)^{\frac{m}{n}} + h \frac{\sigma}{\sigma_*} = 0 \tag{15}$$

where $\frac{m}{n} = 1.1$ and $h = 1.6$ for $0 \leq \sigma \leq 22$ MPa while for $\sigma \geq 22$ MPa is $\frac{m}{n} = 2.2$ and $h = 55$.

Furthermore, we look for a function for $k \frac{\partial F}{\partial \sigma}$, depending on the invariants σ and τ, and possessing the following properties:

$$k \frac{\partial F}{\partial \sigma} \begin{cases} = \phi(\sigma) \quad \text{(hydrostatic loading)} & \text{for } \tau = 0 \\ > 0 \iff X > 0 \text{ (compressible)} & \\ = 0 \iff X = 0 \text{ (compressible/dilatant boundary)} & \text{for } \tau > 0 \\ < 0 \iff X < 0 \text{ (dilatant)} & \\ \to -\infty \iff Y \to 0 \text{ (failure)} & \end{cases} \tag{16}$$

An example of a function possessing all the above mentioned properties is

$$k \frac{\partial F}{\partial \sigma} = \frac{\left[-\dfrac{\tau}{\sigma_*} + f_1 \left(\dfrac{\sigma}{\sigma_*}\right)^2 + f_2 \dfrac{\sigma}{\sigma_*}\right]\left[p_1 \left(\dfrac{\sigma}{\sigma_*}\right)^2 + p_2 \dfrac{\sigma}{\sigma_*}\right]}{-\left(\dfrac{\tau}{\sigma_*}\right)^{\frac{m}{n}} + h \dfrac{\sigma}{\sigma_*}} \tag{17}$$

Function $\phi(\sigma)$ defined by (13) is used to determine $p_1 = -1.346\times10^{-6}\cdot h$ and $p_2 = 7\times10^{-5}\cdot h$. We now integrate (17) with respect to σ to get

$$k\, F(\sigma,\tau) = k \int \frac{\partial F}{\partial \sigma} d\sigma + g(\tau) = F_1(\sigma,\tau) + g(\tau) \tag{18}$$

By differentiating here with respect to τ using formula (12.2) in conjunction with the empirical data we can determine the function $g(\tau)$ as

$$g(\tau) = g_0 \frac{\tau}{\sigma_*} + g_1 \left(\frac{\tau}{\sigma_*}\right)^2 + g_2 \left(\frac{\tau}{\sigma_*}\right)^3 \tag{19}$$

with $g_0 = 1.5\times10^{-3}$ MPa·s^{-1}, $g_1 = 1.12\times10^{-4}$ MPa·s^{-1} and $g_2 = 8.9\times10^{-6}$ MPa·s^{-1}.

Thus the viscoplastic potential function is entirely determined as:

$$k\ F(\sigma,\tau) = \frac{f_1 p_1}{4h^5}\left[-\left(\frac{\tau}{\sigma_*}\right)^{\frac{m}{n}} + h\frac{\sigma}{\sigma_*}\right]^4 + \left[\frac{4}{3}\frac{f_1 p_1}{h^5}\left(\frac{\tau}{\sigma_*}\right)^{\frac{m}{n}} + \frac{f_2 p_1 + f_1 p_2}{3h^4}\right]\cdot$$

$$\cdot\left[-\left(\frac{\tau}{\sigma_*}\right)^{\frac{m}{n}} + h\frac{\sigma}{\sigma_2}\right]^3 + \left[3\frac{f_1 p_1}{h^5}\left(\frac{\tau}{\sigma_*}\right)^{\frac{2m}{n}} + \frac{3}{2}\frac{f_2 p_1 + f_1 p_2}{h^4}\left(\frac{\tau}{\sigma_*}\right)^{\frac{m}{n}} + \frac{-\frac{\tau}{\sigma_*}p_1 + f_2 p_2}{2h^3}\right]\cdot$$

$$\cdot\left[-\left(\frac{\tau}{\sigma_*}\right)^{\frac{m}{n}} + h\frac{\sigma}{\sigma_*}\right]^2 + \left[4\frac{f_1 p_1}{h^5}\left(\frac{\tau}{\sigma_*}\right)^{\frac{3m}{n}} + 3\frac{f_2 p_1 + f_1 p_2}{h^4}\left(\frac{\tau}{\sigma_*}\right)^{\frac{2m}{n}} +\right.$$

$$\left. + 2\frac{-\frac{\tau}{\sigma_*}p_1 + f_2 p_2}{h^3}\left(\frac{\tau}{\sigma_*}\right)^{\frac{m}{n}}\right]\left[-\left(\frac{\tau}{\sigma_*}\right)^{\frac{m}{n}} + h\frac{\sigma}{\sigma_*}\right] + \left[\frac{f_1 p_1}{h^5}\left(\frac{\tau}{\sigma_*}\right)^{\frac{4m}{n}} + \frac{f_2 p_1 + f_1 p_2}{h^4}\cdot\right.$$

$$\left.\cdot\left(\frac{\tau}{\sigma_*}\right)^{\frac{3m}{n}} + \frac{-\frac{\tau}{\sigma_*}p_1 + f_2 p_2}{h^3}\left(\frac{\tau}{\sigma_*}\right)^{\frac{2m}{n}} - \left(\frac{\tau}{\sigma_*}\right)^{\frac{m+n}{n}}\frac{p_2}{h^2}\right]\ln\left[-\left(\frac{\tau}{\sigma_*}\right)^{\frac{m}{n}} + h\frac{\sigma}{\sigma_*}\right] - \frac{\tau\, p_2}{\sigma_* h}\cdot\frac{\sigma}{\sigma_*} +$$

$$+ g_0\frac{\tau}{\sigma_*} + g_1\left(\frac{\tau}{\sigma_*}\right)^2 + g_2\left(\frac{\tau}{\sigma_*}\right)^3 \qquad (20)$$

In Fig. 2 the real shapes of various important surfaces are shown: failure surface (15), H = const. (11), compressible/dilatant boundary (14), F = const. (20). From this figure we can conclude that in the neighborhood of the failure surface and of the hydrostatic axis H = const. and F = const. are distinct and also departing significantly from the classical Mises or Tresca surfaces (parallel to the σ-axis). For high values of σ and not too small or too large values of τ the two kinds of surfaces approach the classical one and one can envisage here to use an associated constitutive equation ($F \equiv H$). It would mean a significant simplification of the constitutive equation in that particular stress domain. From Fig. 2 follows also that the domain of dilatancy is quite broad. However, for high values of σ the dilatancy rate is relatively small (since the surfaces F = const. are nearly parallel to the σ-axis).

The general discussion concerning the simplified variants of the present model as well as mining applications will be considered elsewhere. Examples are shown in [1] using the previous model.

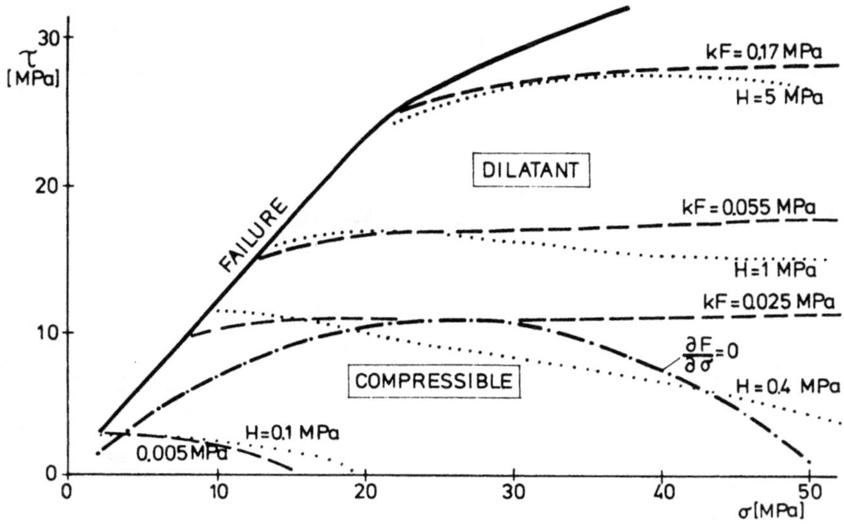

Fig. 2: Domains and surfaces determined from the nonassociated constitutive equation for rock salt

Conclusions

A general triaxial constitutive equation for rock salt has been formulated from true triaxial tests. The cumbersome generalization of a uniaxial model is thus avoided.

The nonassociated constitutive equation is general and comprises: elastic response, transient creep (including work-hardening and unloading), irreversible volumetric compressibility and dilatancy, damage, instantaneous failure and creep failure. All these properties are described by the constitutive functions F and H. The present version of the model can be used for the design of any underground structure, of any complexity, in a quite broad interval of stress variation.

Special emphasis has been layed on the effects related to volume change and failure and not on long-term behaviour. Therefore, at this state of evolution, only transient creep is included but not steady state creep. Improvement into this direction has to be done. Also, the influences of the third invariant of the stress deviator and temperature have to be included in future.

References

1. Cristescu, N. & U. Hunsche, 1991: A constitutive equation for salt. In: Proc. 7. Int. Congr. on Rock Mech., Workshop on Rock Salt Mech., v. 3, Aachen, Sept. 1991. Balkema: Rotterdam. (in press)

2. Hunsche, U., 1991: Volume change and energy dissipation in rock salt during triaxial failure tests. In: Mechanics of Creep Brittle Materials - 2, Proc. of Coll. 'Mechanics of Creep Brittle Materials 2', Leicester (UK), Sept. 1991; A. Cocks & A. Ponter (eds), p. 172-182. Elsevier: Amsterdam.

3. Hunsche, U. & H. Albrecht, 1990: Results of true triaxial strength tests on rock salt. Engineering Fracture Mechanics 35 (4/5): 867-877.

4. Hunsche, U., 1991: True triaxial failure tests on cubic rock salt samples - experimental methods and results. In: Proc. of IUTAM-Symposium 1991, Hannover. Springer: Berlin. (in press)

5. Cristescu, N., 1989: Rock Rheology. Kluwer Academic Publ.: Dordrecht.

6. Cristescu, N., 1991: Nonassociated elastic/viscoplastic constitutive equations for sand. Int. J. Plasticity 7: 41-64.

True Triaxial Failure Tests on Cubic Rock Salt Samples. Experimental Methods and Results

Udo E. Hunsche

Federal Institute for Geosciences and Natural Resources,
Stilleweg 2, W-3000 Hannover 51

Summary

A constitutive equation for natural rock salt should take into consideration stress, strain, and volume change until failure, as well as failure itself, under various conditions. Therefore, equipment for precise true triaxial tests on cubic specimens and a special test procedure for quasi-static failure tests on rock salt have been developed. Deformation, volume change, and irreversible stress work were determined from the empirical data as a function of time. Failure strength was determined as well. First, a conservative equation for failure strength was determined from 12 test series.

The detailed results of 14 compression tests at various mean stresses are discussed. Stress conditions for the minimum volume (i.e. the compressible/dilatant boundary) were determined from the tests. The volume change at failure was between 1.5 and 7 %. The calculated stress work at failure that is related to volume change and the stress work related to the shape change are also shown. The results have been used to derive a nonassociated elasto-viscoplastic constitutive equation for rock salt as shown in another paper.

Introduction

Knowledge of the geomechanical behaviour of rock salt is important for the dimensioning and safety analysis of mines, caverns, and repositories for toxic and radioactive wastes in natural salt formations. Rock salt is also often used as a model material in the materials sciences. Therefore, much experimental and theoretical work has been carried out in this field during the last two decades, mainly on creep and failure behaviour. For corresponding references see Cristescu & Hunsche [1] or Hunsche [2]. None of published equations for failure can be considered as really satisfactory because the relevant deformation and failure mechanisms are not yet fully understood.

Several kinds of tests are carried out to evaluate failure under various conditions: quasi-static tests on cylindrical or cubic samples at many load or deformation rates under triaxial conditions at intermediate or high mean stress σ, and uniaxial or triaxial tension tests, as well as biaxial and uniaxial compression tests at low mean stress. The final aim is to develop reliable physically and experimentally based constitutive equations for rock salt which describe (1)

D. Besdo, E. Stein (Eds.)
Finite Inelastic Deformations –
Theory and Applications
IUTAM Symposium Hannover/Germany 1991

elastic behaviour, (2) failure and related phenomena, such as dilatancy and compressibility due to the opening and closing of microcracks, and (3) creep phenomena.

In this publication the experimental procedures and some results of the true triaxial quasi-static failure tests on cubic samples are described. Emphasis is placed on the phenomena mentioned under (2). They are evaluated in a manner that they could be used for determination of the constitutive equation described by Cristescu & Hunsche [3] in this volume. As shown below, the experiments had been previously evaluated with regard to the effect of mean stress, load geometry und temperature on strength, residual strength and fracture pattern. (For variables and basic formulae see Table 1.) Results of failure tests carried out at low mean stresses σ, i.e. uniaxial and triaxial tension tests and biaxial tests, are described by Hunsche [4].

Equipment and test procedures

To obtain a reliable constitutive equation, experiments have to be performed with high precision. Therefore, a rig for precise true triaxial tests on cubic rock salt specimens, a sophisticated test procedure, and an evaluation method have been developed to provide reliable data for derivation of a constitutive equation for rock salt that will include stress, strain, and volume

Table 1: Variables discussed in this paper. Compression: positive values

Variable	Meaning
$\sigma_1, \sigma_2, \sigma_3$	principal stresses
$\sigma = 1/3(\sigma_1 + \sigma_2 + \sigma_3)$	mean stress, octahedral normal stress
$s_i = \sigma_i - \sigma$	stress deviator
$\tau = [1/3(s_1^2 + s_2^2 + s_3^2)]^{½}$	octahedral shear stress
$m = \dfrac{3\,s_2}{s_1 - s_3}$; $s_1 > s_2 > s_3$	stress path, load geometry Lode parameter
$J_m = \dfrac{m(9-m^2)}{(3+m^2)^{1.5}}$	invarient of load geometry
$\varepsilon_i = (l_o - l_i)/l_o$	principal strains
$\varepsilon v = (l_1 \cdot l_2 \cdot l_3)/V_o$ + correction	relative volume change
$\varepsilon_{eff} = (\varepsilon_1^2 + \varepsilon_2^2 + \varepsilon_3^2)^{½}$	effective deformation
l_1, l_2, l_3	actual edge lengths
l_o	initial edge length
V_o	initial (reference) Volume
W^I_v, W^I_d, W^I_g	different kinds of irreversible stress work (Eqs. 1, 2, & 3)
$\dot{\sigma}, \dot{\tau}$	stress rates
τ_B	failure strength

change until failure, as a function of time. These measurements also require a good team and precise calibration. The rig used for these tests consists of a rigid frame in which three pairs of double-acting pistons are arranged orthogonally to each other about the center of the frame. A load is applied to the cubic specimen via square steel plates (platens) which can be heated. The loads in the three directions are controlled independently by servohydraulic valves. Thus, the three principal stresses or strains can be controlled. The important advantage of these true triaxial tests is that all load geometries m that are possible in nature can be applied to a sample. The acoustic emission rate (AE) is measured in many tests, too. A more detailed description of the apparatus has been published by Hunsche & Albrecht [5]. The most important specifications of the rig and the tests are as follows:

- Maximum force (per axis): 2000 kN
- Sample size (edge length), this study: 57.5 and 53 mm
  maximum: ca. 200 mm
- Platen size in standard tests: 50 mm
- Temperature maximum: 400 °C
- Friction of piston: < 1 % of load
- Independent load or deformation control along the 3 axes is possible.
- Loading rates in standard tests:
  hydrostatic phase: $\dot{\sigma}$ = 7.6 MPa/min
  deviatoric phase, compression tests: $\dot{\tau}$ = 21.4 MPa/min
  other load geometries: 12.3 MPa/min $\leq \dot{\tau} \leq$ 21.4 MPa/min
- Deformation rate at failure, standard tests: $\dot{\varepsilon}_{eff} \approx 0.007\ s^{-1}$
- Digital data acquisition, standard interval: 1 s
- Lubrication with a thin film of paraffin wax (for tests at room temperature) or graphite (elevated temperatures)

All tests in this paper are load controlled, the stress is first applied hydrostatically (hydrostatic phase). When the desired σ level has been reached, the three principal stresses, and thus τ, are changed linearly with time so that σ is held constant (deviatoric phase). The procedure is shown in Fig. 1 schematically for three load geometries m. This is done until the octahedral shear stress τ reaches the failure strength τ_B of the sample and fracturing occurs as shown in Figs. 4 and 5 for two sets of compression tests. τ_B is defined as the local maximum of τ in a test. Fig. 2 shows three samples after failure. The fracture patterns are analysed in Hunsche [6]. It should be noticed that the shear stress does not go to zero at failure; the fractured sample has a considerable residual strength τ_R. In a number of tests it was determined by controlled unloading after failure and subsequent reloading as shown in Fig. 4.

A standard test series is carried out on each rock salt type. It usually consists of 28 tests at various values of σ and m, including four uniaxial and biaxial tests and five tests at elevated temperature between 60 and 260 °C (m = -1). Deformation and forces in the three principal directions are measured during the tests. The equipment has been very precisely calibrated. Especially the calibration for the determination of the volume change εv has been difficult. It takes

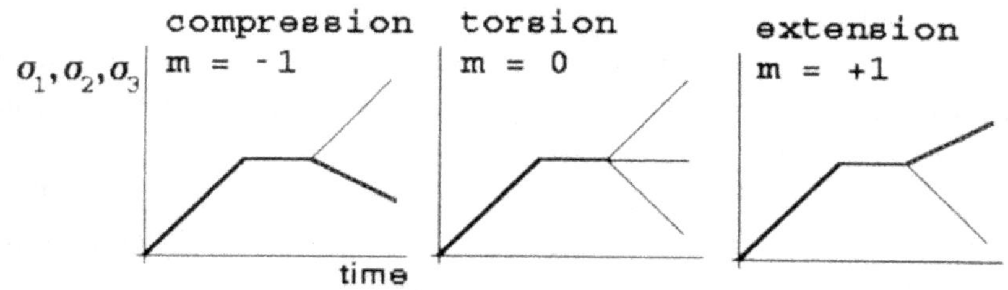

Fig. 1: Scematic of the test procedures for three load geometries m

m = -1 m = 0 m = +1

Fig. 2: Three specimens after failure tests

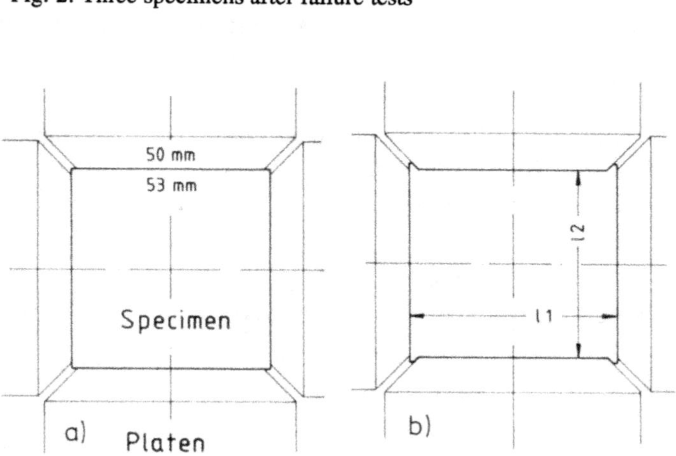

Fig. 3: Cross section of the arrangement of specimen and platens (to scale); a) at beginning; b) at failure (test 3389)

into account the deformation of the platens and pistons, and the small amount of salt which is squeezed out at the edges of the sample as shown in Fig. 3 (see also Fig. 2). A precision of better than 0.3 % has been achieved for the determination of εv. One of the objectives of the tests is to determine experimentally the failure surface in the σ_1, σ_2, σ_3 coordinate system.

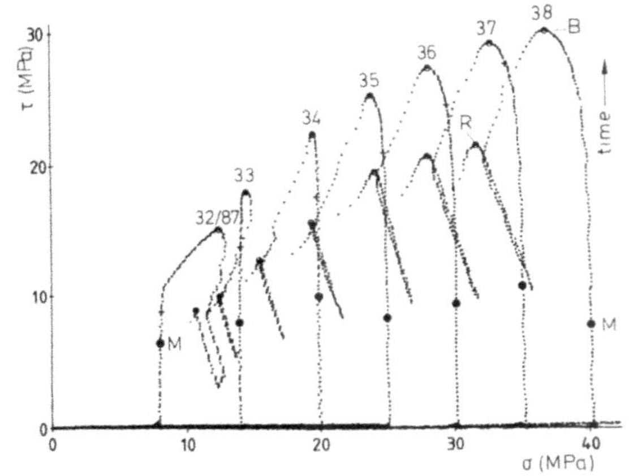

Fig. 4: Seven true triaxial compression tests of series S8; T = 30 °C; spacing of points: 1s
M: minimum volume; B: failure; R: residual strength

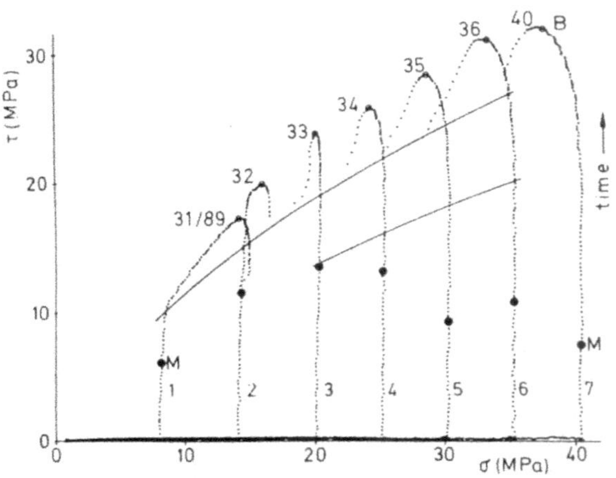

Fig. 5: Seven true triaxial compression tests of series S9; T = 30 °C; spacing of points: 1s
M: minimum volume; B: failure.
The lines represent the conservative failure equation at m = -1 and m = + 1.

<u>Tested samples</u>

Twelve test series totaling 350 tests were carried out on twelve different types of rock salt from a salt dome in northern Germany. Evaluation yielded a conservative failure equation in terms of stress. It is presented in brief in the next section; the equation has been published by Hunsche & Albrecht [5] and Hunsche [6,7]. The results of a more detailed reevaluation of two recent test series (S8 and S9) are presented, too. The samples of these sets consist of rather clean halite with a grain size up to 5 mm, in some cases 20 mm. All specimens were machined on a lathe and geologically described before the tests.

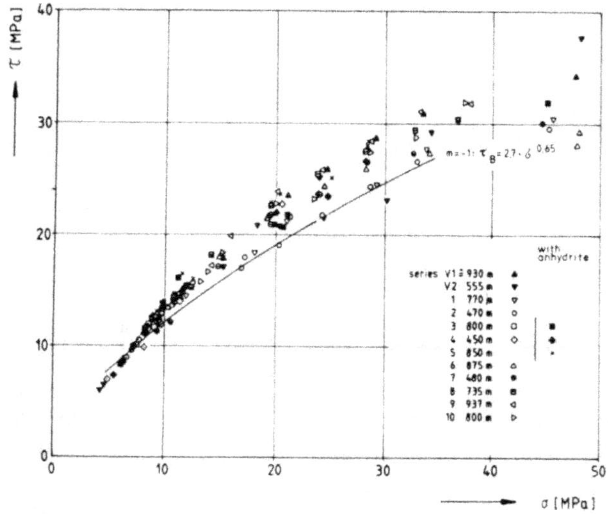

Fig. 6: Failure strength for 12 series; compression tests; T = 30 °C

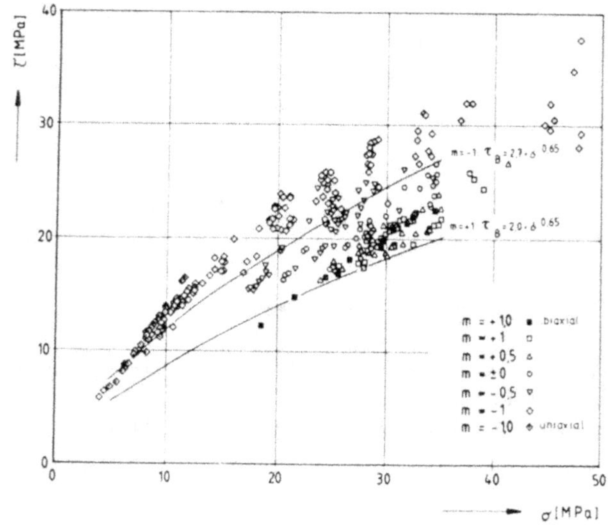

Fig. 7: Failure strength for 12 series, grouped according to the value of m; T = 30 °C. The lines represent the conservative failure equation at m = -1 and m = + 1.

The conservative failure equation

The twelve test series have been used to determine the above-mentioned conservative failure equation. The results of the compression tests at room temperature of the two recent series S8 and S9 are shown as examples in Figs. 4 and 5. The failure strength of all compression tests at room temperature are summarized in Fig. 6. This graph demonstrates that the strength of rock salt varies nonlinearly with increasing mean stress σ. It also shows that there are distinct differences between the different types of salt: The strength τ_B varies up to 4 MPa at the same conditions. Fig. 7 shows all the tests at room temperature grouped according the value of m. It is important to note that the extension tests (m = +1) exhibit the lowest strengths, averaging about 30 % less than the compression tests in each series. This is different from the behaviour of metals but usual in rocks. The effect decreases at high σ, however. Generally, failure strength depends on loading rate or strain rate. However, in additional tests it was found that strain rates below about 10^{-1} s^{-1} influence strength of rock salt only slightly; loading rates below about 300 MPa/min have a somewhat larger but not very important influence.

Figs. 6 and 7 also demonstrate how the conservative failure equation was fitted to the lower strength boundary of all tests with the same load geometry m. The analysis results in the formula shown in Table 2. The dependence of the failure strength on temperature, as well as the result for the residual strength τ_R, are also given. τ_R is not dependent on m and has about the same value as the strength for extension at the same σ. Fig. 8 displays the failure equation in the the $\tau - \sigma$ plane and in the octahedral plane. The upper and lower limits of validity are caused by lack of tests beyond these lines; below the lower limit, tension has to be applied in one or two directions for failure (see Hunsche [4]). However, in general it is a safe assumption that the tensional strength is equal to zero below the lower limit of validity. Thus, this statement can be included in the conservative failure equation.

Table 2: Conservative failure equation for rock salt

$$\tau_B = f(\sigma_0)\ g(m)\ h(T) \qquad : \underline{\textbf{failure strength}}$$

$$f(\sigma) = b\ (\sigma/\sigma_*)^p$$

$$g(m) = 2k/[(1 + k) + (1 - k)J_m]$$

$$h(T) = \begin{cases} 1 & \text{for } 20°C \leq T \leq 100°C \\ 1 - c\ (T-100°C) & \text{for } 100°C \leq T \leq 260°C \end{cases}$$

$b = 2.7$ MPa, $p = 0.65$,

$k = 0.74$, $\sigma_* = 1$ MPa,

$c = 0.002\ K^{-1}$

$$\tau_R = \tau_R(m=+1) \qquad : \underline{\textbf{residual strength}}$$

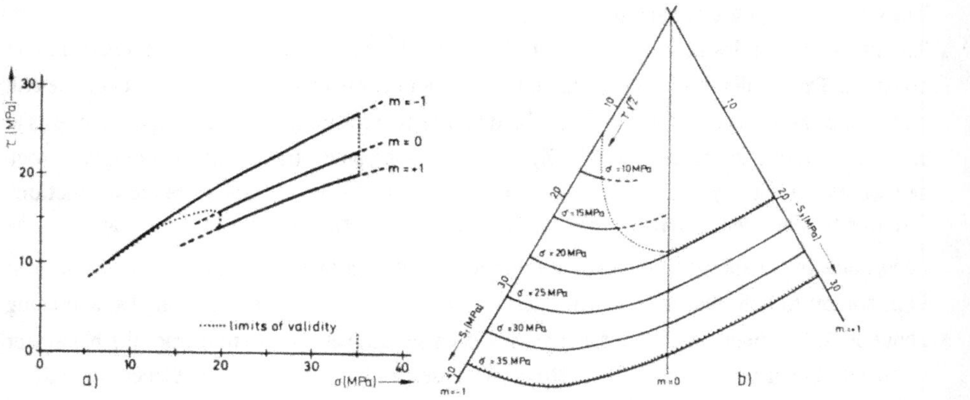

Fig. 8: Plots of the conservative failure equation for rock salt (Table 2) at T ≤ 100 °C: (a) in a graph of τ vs. σ and (b) in the octahedral plane.

Detailed evaluation and results

A constitutive equation for rock salt should be able be predict not only the elastic response, creep, and failure strength, but also volume change and strain caused by microcracks over the whole range as a function of time. If the equation is to do this, the energies involved must be taken into consideration. A good method for deriving such an equation has been described by Cristescu [8] and Cristescu & Hunsche [1,3]. It contains a procedure for obtaining an appropriate elasto-viscoplastic constitutive equation. Data on volume change, irreversible stress work during a test, and failure strength, which will be discussed in this chapter are required to derive this equation.

The total irreversible stress work W^I_g (i.e. energy used for deformation minus the elastic energy) consists of two components:

$$W^I_v(t) = \int_0^t \sigma \, d\varepsilon v - \sigma^2/(2\,K) \tag{1}$$

$$W^I_d(t) = \int_0^t \mathbf{s} \cdot \dot{\mathbf{e}} \, dt - \mathbf{s} \cdot \mathbf{s} \,/(4\,G) \tag{2}$$

$$W^I_g = W^I_v + W^I_d \tag{3}$$

$\sigma = \sigma(t)$

$\mathbf{s} = \boldsymbol{\sigma} - \sigma \mathbf{1}$: tensor of stress deviator

$\mathbf{e} = \boldsymbol{\varepsilon} - \varepsilon \mathbf{1}$: tensor of strain deviator

G = 11.8 GPa (shear modulus)

K = 21.7 GPa (compression modulus)

W^I_v is related to volume change and consequently to the closure or opening of microcracks, which finally merge and thus cause failure. W^I_d is caused by the stress deviator and is related to creep, initiation and propagation of microcracks. W^I_d also supplies the energy consumed by microcrack opening, which is expressed by negative values for W^I_v (which means energy loss by the specimen). Therefore, W^I_g only is related to creep and microcrack initiation and propagation, which cannot be distinguished without model calculations.

The principal stresses σ_i and principal strains ε_i were calculated from the measurements taking into account the changing dimensions of the sample. These values were used to calculate τ and σ. The results of the compression tests at room temperature for series S8 and S9 (Figs. 4 and 5) will be now discussed in more detail. The two series exhibit similar failure behaviour and high failure strength. They were used for the constitutive model by Cristescu & Hunsche [3]. Some results for series S9 are also reported in Hunsche [2].

The three principal strains are the basis for the calculation of the volume change εv. Reference volume V_0 for the calculation of εv is the hydrostatically compacted but unloaded sample. Examples of the strain components and volume change as a function of τ are shown in Fig. 9. Failure stress τ_B and the point M at which the minimum volume εv_M occurs are marked. Considerable compaction of 0.5 to 1 % occurs during the hydrostatic loading phase. This is caused by loosening of the specimen during coring and machining. A small initial decrease in volume can be observed also during the deviatoric phase until M. The volume increases nonlinearly after point M due to microcrack opening and reaches several percent at failure. Fig. 10 shows the volume changes in a test together with acoustic emission rate (AE). It demonstrates clearly that AE is closely related to the volume change and, hence, to the opening of microcracks. In the next step, the irreversible stress work W^I_v, W^I_d, W^I_g are calculated using Equations (1), (2) and (3): The equations are integrated from the time of the beginning of the deviatoric phase. The results for the three kinds of irreversible stress work are presented in Fig. 11. Note that all W^I are nonlinear functions of τ.

A summary of the results of all compression tests of series S8 and S9 in Table 3 shows the following: The initial volume decrease to the minimum at point M is $\varepsilon v_M = 0.1$ to 0.2 % at an octahedral shear stress τ that is half of the failure strength τ_B or less as shown in Figs. 4 and 5. The value of τ_M decreases at high σ and becomes 0 at σ = 50 to 60 MPa. Because the empirical curve for εv is quite flat at minimum volume, the value for τ_M is not well defined. It has to be determined by means of a fitted function. The uncertainty of τ_M reaches about ±2 MPa sometimes. The volume increase at failure εv_B is of great interest; it is plotted in Fig. 12. εv_B is larger at small σ than above about 18 MPa, where it is between 1.6 and 2.2 %. The ongoing analysis of the test series on other rock salt types yield similar results. They are also in good

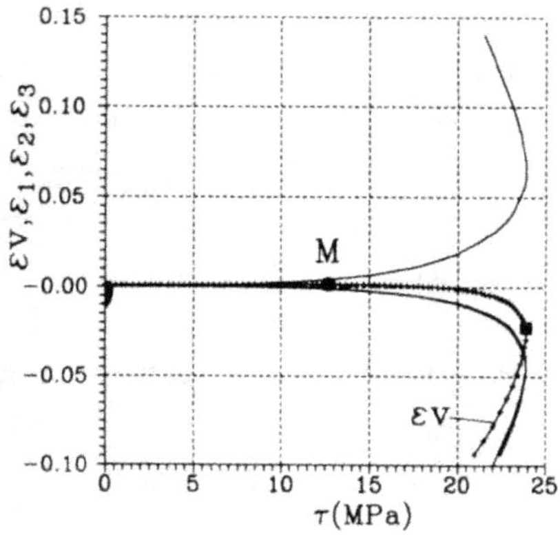

Fig. 9: Strains and volume change in test 3389

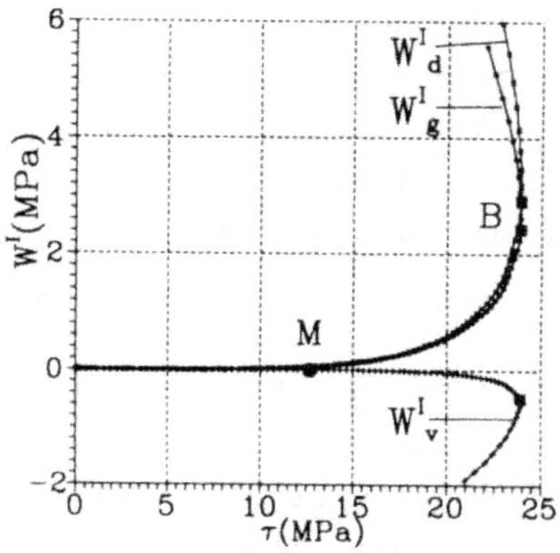

Fig. 11: Irreversible stress work in test 3389

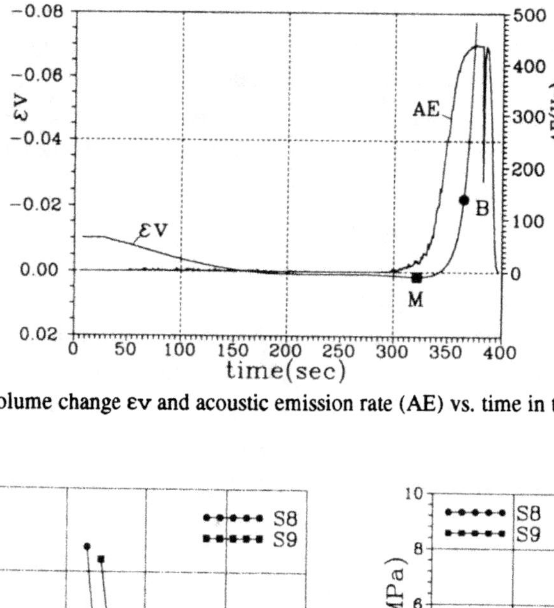

Fig. 10: Volume change εv and acoustic emission rate (AE) vs. time in test 3389

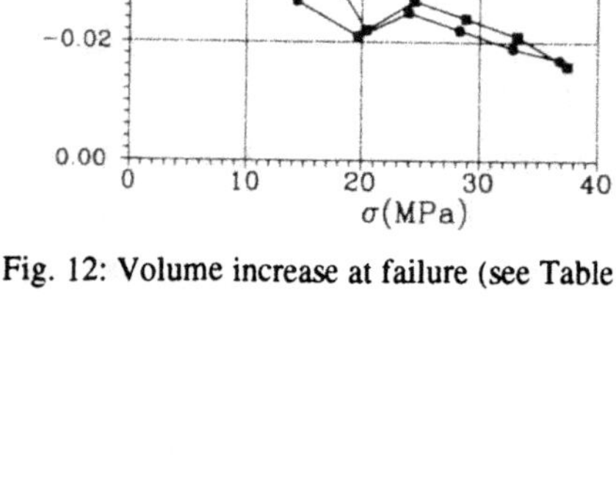

Fig. 12: Volume increase at failure (see Table 3)

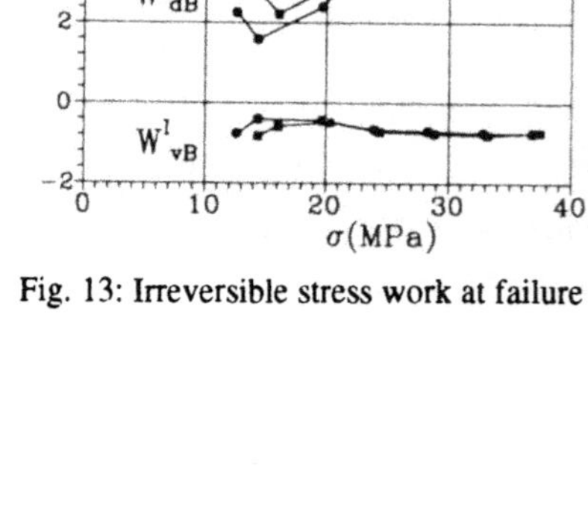

Fig. 13: Irreversible stress work at failure

Table 3: Test results; series S8 und S9; Compression tests; T = 30 °C
M: values at minimum volume; B: values at failure

Test	τ_M (MPa)	σ_M (MPa)	ε_{vM}	τ_B (MPa)	σ_B (MPa)	ε_{vB}	W^I_{vB} (MPa)	W^I_{dB} (MPa)	W^I_{gB} (MPa)	ε_{effB} (%)
Series S8										
3287	6.43	8.00	0.0010	15.31	12.58	-0.066	-0.76	2.26	1.50	10.23
3387	8.00	13.99	0.0010	18.15	14.38	-0.027	-0.40	1.59	1.20	6.06
3487	10.03	19.98	0.0014	22.56	19.69	-0.021	-0.44	2.40	1.96	7.40
3587	8.50	25.06	0.0015	25.48	23.93	-0.025	-0.66	3.98	3.32	10.69
3687	9.62	30.18	0.0017	27.59	28.30	-0.022	-0.70	4.30	4.60	13.20
3787	11.20	35.12	0.0019	29.44	32.82	-0.019	-0.74	6.68	5.94	15.49
3887	8.00	40.19	0.0021	30.41	36.84	-0.017	-0.75	8.84	8.09	19.71
Series S9										
3189	6.08	8.16	0.0015	17.29	14.38	-0.063	-0.82	2.80	1.98	11.53
3289	10.21	14.48	0.0013	19.95	16.08	-0.036	-0.58	2.22	1.64	7.56
3389	12.69	20.46	0.0015	23.89	20.33	-0.022	-0.49	2.92	2.44	8.31
3489	12.11	25.42	0.0015	25.88	24.40	-0.027	-0.72	4.90	4.18	12.70
3589	9.00	30.35	0.0016	28.48	28.82	-0.024	-0.76	6.17	5.41	14.66
3689	9.67	35.32	0.0017	31.39	33.22	-0.021	-0.80	7.06	6.26	15.24
4089	7.06	40.32	0.0020	32.97	37.47	-0.016	-0.73	8.40	7.64	17.94

agreement with the results obtained by Wawersik & Hannum [9] (εv_B = 2.8 to 7 %). The volume increase at failure for different kinds of hard rock is between 1 and 3 % (Cristescu, [8]).

Irreversible stress work at failure is shown in Fig. 13. W^I_{vB} lies between -0.44 and -0.82 MPa (1 MPa = 1 MJ/m³). It can be noticed that it does not depend much on the mean stress σ, which is why W^I_v can be used as a damage parameter. The results for W^I_{vB} are comparable to the energies measured by Cristescu [8] in uniaxial tests. W^I_{gB} is between 1.2 and 8 MPa; Alheid & Rummel [10] obtained about 0.5 MPa at failure in triaxial tests on sandstone. These comparisons indicate that the results shown above are reasonable. The stress work related to volume change at failure $|W^I_{vB}|$ (crack-opening energy) is as much as 1/3 to 1/10 of the total energy W^I_{gB} used for creep and microcrack initiation and propagation. Of course, the part of W^I_{gB} related to creep is a function of test duration and the amount of deformation.

Conclusions

It has been shown how the apparatus for true triaxial tests, the test procedure, and the evaluation method are performed and that they yield comprehensive and reliable results. They were used to determine a conservative failure equation. The volume changes and the irreversible stress work derived from the true triaxial tests on two sets of rock salt samples are reasonable. Therefore, the results are a sound basis for of the constitutive equation. The results shown have been used in the determination of the nonassociated elasto-viscoplastic constitutive equation carried out by Cristescu & Hunsche [3] in this volume. The effects of load path, temperature, load or deformation rate, and rock salt type will be studied in future analyses.

Acknowldgements

The investigations are supported by the Federal Ministry of Research and Technology. I thank C. Caninenberg, G. Notbohm, and R. Baumann for the careful performance of the tests.

References

1. Cristescu, N. & U. Hunsche, 1991: A constitutive equation for salt. In: Proc. of 7. Int. Congr. on Rock Mech., Aachen, Sept. 1991, v. 3: Workshop on Rock Salt Mech. Balkema: Rotterdam. (in press)

2. Hunsche, U., 1991: Volume change and energy dissipation in rock salt during triaxial failure tests. In: Mechanics of Creep Brittle Materials - 2, Proc. of Coll. 'Mechanics of Creep Brittle Materials 2', Leicester (UK), Sept. 1991; A. Cocks & A. Ponter (eds.), 172-182. Elsevier: Amsterdam.

3. Cristescu, N. & U. Hunsche, 1991: Determination of a nonassociated constitutive equation for rock salt from experiments. In: Proc. IUTAM-Symposium 1991, Hannover. Springer: Berlin. (in press)

4. Hunsche, U., 1991: Strength of rock salt at low mean stresses. In: Proc. of 7. Int. Congr. on Rock Mech., Aachen, Sept. 1991, v. 3: Workshop on Rock Salt Mech. Balkema: Rotterdam. (in press)

5. Hunsche, U. & H. Albrecht, 1990: Results of true triaxial strength tests on rock salt. Engineering Fracture Mechanics 35 (4/5), 867-877.

6. Hunsche, U., 1990: On the fracture behavior of rock salt. In: Constitutive Laws of Plastic Deformation and Fracture, Proc. 19th Canadian Fracture Conf., Ottawa (Canada) 1989; A. S. Krausz, J. I. Dickenson, J.-P. A. Immarigeon & W. Wallace (eds.); 155-163. Kluwer Acad. Publ.: Dordrecht.

7. Hunsche, U., 1990: A failure criterion for natural polycrystalline rock salt. In: Advances in Constitutive Laws for Engineering Materials, Proc. Int. Conf. on Constitutive Laws for Eng. Mat. (ICCLEM), Chongquing (China) 1989; Fan Jinghong (ed.), v. II, 1043-1046. Int. Academic Publ.: Beijing.

8. Cristescu, N., 1989: Rock Rheology. Kluwer Acad. Publ.: Dordrecht.

9. Wawersik, W. R. & D. W. Hannum, 1980: Mechanical behavior of New Mexico rock salt in triaxial compression up to 200°C. J. Geophys. Res. 85 (B2), 891-900.

10. Alheid, H.-J. & F. Rummel, 1976: Energieregelung bei Druckversuchen an Gesteinen. Sonderforschungsbereich 77 - Felsmechanik, Univ. Karlsruhe, Jahresber. 1975, Karlsruhe.

Finite Elastic-Plastic Deformations and Shakedown Analysis of Bellows

J. SKRZYPEK, B. SKOCZEŃ

Institute of Mechanics and Machine Design
Cracow University of Technology, Kraków, Poland

Summary

Shakedown and failure modes due to inadaptation of S-shaped bellows within the framework of large deflection theory are investigated. The bellows is subject to constant internal pressure and variable repeated axial force of a given amplitude. Two inadaptation modes: ultimate load mechanism (mixed alternating plasticity and incremental collapse) and formation of plastic hinge (incremental collapse) are discussed. Transition from the first mechanism to the second is explained provided that the unloading starts from the point of maximum load.

Introductory remarks

Piping systems are usually equipped with flexible, thin- walled, rotationally- symmetric shell segments, called bellows. These resilient joints compensate thermal expansion of pipes heated by a medium passing through the system. Combined loadings, the bellows are subjected to, depend strictly on the exploitation conditions. Generally the loading consists of three components i.e. axial force, hydrostatic pressure and a temperature field. Each of these components may vary within a certain range so that the structure is subjected to variable repeated loads. The process of cyclic loading and unloading beyond the elastic limit leads to the energy dissipation and the formation of a sequence of plastic strain fields inside the shell. The question whether the energy dissipation is bounded or not appears of the highest importance for an engineer since it is the question whether the structure will shake down or collapse.

D. Besdo, E. Stein (Eds.)
Finite Inelastic Deformations -
Theory and Applications
IUTAM Symposium Hannover/Germany 1991

The shakedown problems within the framework of geometrically non-linear continuum mechanics were discussed by Weichert [1], [2], [3], Weichert and Gross-Weege [4]. The fatigue-strength of U-shaped bellows under variable repeated axial force in the elastic-plastic range was studied by Hamada and Takezono [5]. The formula for estimation of the fatigue-life of U-shaped bellows, based upon the experimental data, was proposed by the authors. The low-cycle fatigue failure of U-shaped bellows subject to completely reversed loading was also investigated by Hamada and Tanaka [6] on the ground of large deflection approach. Cyclic hardening property of material due to the alternation of external load was also discussed. Another fundamental mode of inadaptation, incremental collapse (called also ratchetting), consists in accumulation of plastic strains within the structure exposed to cyclic loads. After a number of cycles the structure becomes unserviceable due to an incremental failure mechanism. The evaluation method of plastic ratchetting was proposed by Porowski and O'Donnell [7] and incorporated into the design code of the expansion joints by American Society of Mechanical Engineers [8].

B. Skoczeń, J. Skrzypek and J.Bielski [9] analysed adaptation and inadaptation of S-shaped bellows (Fig.1) subject to cyclic axial force imposed upon constant internal pressure. To this end the Reissner type

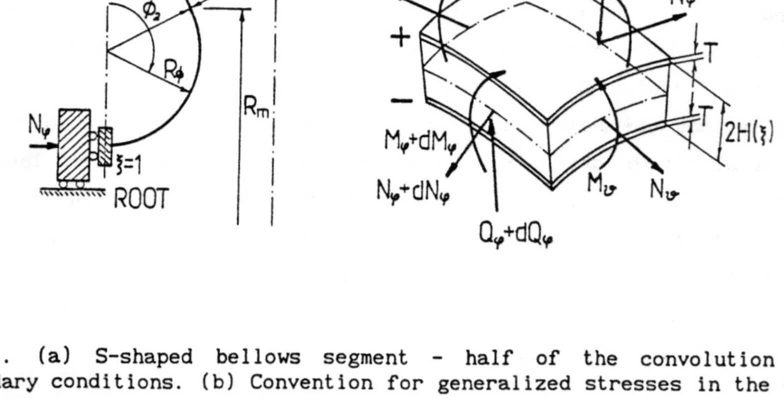

Fig.1. (a) S-shaped bellows segment - half of the convolution and boundary conditions. (b) Convention for generalized stresses in the case of nonlinear axisymmetric bending.

equations were formulated on the basis of the geometrically nonlinear III-rd order theory of small strains but large displacements and rotations. In order to simplify the elastic- plastic analysis the substitutive sandwich section (Fig.1), considered as a discrete approximation of the uniform section of the wall was introduced. The material of working layers was assumed to be perfectly elastic- plastic, incompressible, described by the associated flow rule and the Huber- Mises- Hencky (HMH) yield condition. The core was assumed as perfectly rigid in the direction normal to the middle surface and perfectly flexible with respect to bending. The present paper is direct extension of the above discussed work. The notation and the governing equations as well as the numerical integration method are to be found in the quoted paper.

In the previously mentioned paper the ultimate load mechanism was considered as a termination of the process of cyclic loading of bellows under presumed axial force amplitude. The aim of the present analysis is an extension of this process provided that the unloading from the maximum load point is admissible.

Shakedown

Let us assume a complex, cyclic loading comprising the internal pressure (of a constant magnitude) and the variable repeated axial force. For certain pairs of the pressure and the axial force magnitudes (p_n,N) the structural behaviour, after some plastic deformation in the initial load cycles, becomes eventually elastic. According to the cycle- by- cycle decreasing plastic energy dissipation the permanent deflection stabilizes and the structure shakes down. In other words, if the structure shakes down, the total amount of plastic energy dissipated within the process of elastic- plastic deformation must be finite. Another local criterion imposes a limit upon the functional expressing length of the trajectory in the plastic strains space, namely to the Odqvist parameter:

$$\int_0^t \sqrt{\dot{\varepsilon}^p_{ij}\dot{\varepsilon}^p_{ij}}\, d\tau < C \;. \tag{1}$$

If we now calculate the Odqvist parameter increment at a given point, for each cycle separately, we notice that its magnitude is a monotonic function of the number of cycles and eventually, after a finite number of cycles, approaches zero. If the same phenomenon takes place at each point, the structure shakes down. Stabilization of the permanent deflection as well as the distribution of the Odqvist parameter increments, calculated cycle by cycle at the crest of bellows for positive axial force increments only, are plotted in Fig.2 for the following data:

$$\rho=0.1,\ \lambda=1.0,\ p_n/p_e=0.4,\ \mathbb{N}/\mathbb{N}_e=1.05 \tag{2}$$

where $\rho=R_\phi/R_m$, $\lambda=R_m h/R_\phi^2$, p_e and $\mathbb{N}_e$ are the appropriate elastic carrying capacities.

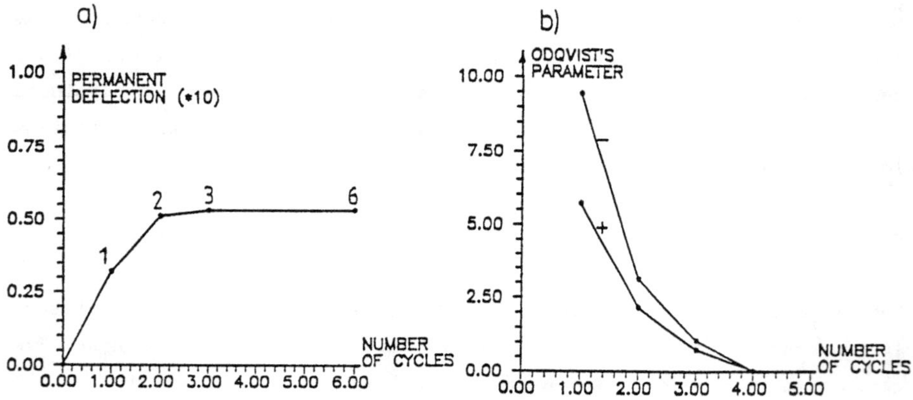

Fig.2. (a) Stabilization of the permanent deflection. (b) Evolution of the Odqvist parameter increments.

Failure mechanisms due to inadaptation

The elastic- plastic structure which doesn't shake down will finally collapse due to an inadaptation process. Generally the mechanism of inadaptation is based upon the alternating plasticity or/and incremental collapse phenomena. If both the phenomena take simultaneously place we face the effect of nonsymmetric alternating plasticity (mixed alternating plasticity and incremental collapse). One can anticipate that the failure mechanism for bellows subject to the assumed loading programm depends strongly on the pressure contribution. We expect that for lower pressures the nonsymmetric alternating plasticity occurs whereas for higher pressures the incremental collapse phenomenon should manifest strongly. In the light of these remarks two basic collapse modes may be distinguished: 'ultimate load mechanism' associated with the nonsymmetric alternating plasticity and 'formation of plastic hinge' associated with domination of the incremental collapse effect. Let us consider these two basic collapse mechanisms in detail.

Ultimate load mechanism

Ultimate load mechanism for cyclic loading may be regarded as analogous to the effect of reaching the maximal load for quasistatically loaded bellows. After a certain number of cycles, the presumed axial force level can not be reached (Fig.3). The process of plastic flow is localized at the crest and the root of the segment under consideration. Plastic strain increments increase at root, cycle by cycle, both in the inner and outer layers (Fig.4a,b). Simultaneously the expansion of plastic zone around the root is observed. The structural softening due to the intensive plastic flow yields a decrease of the carrying capacity with each loading cycle and makes the structure become unserviceable. From the point of

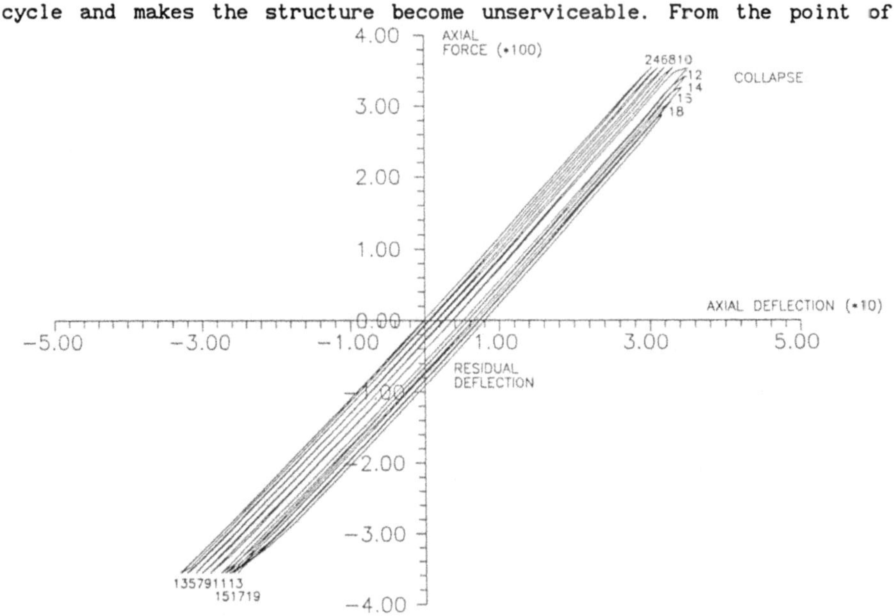

Fig.3. Ultimate load mechanism. Continuation of the process of cyclic loading up to the points of maximal load. Rapid decrease of the maximal carrying capacity.

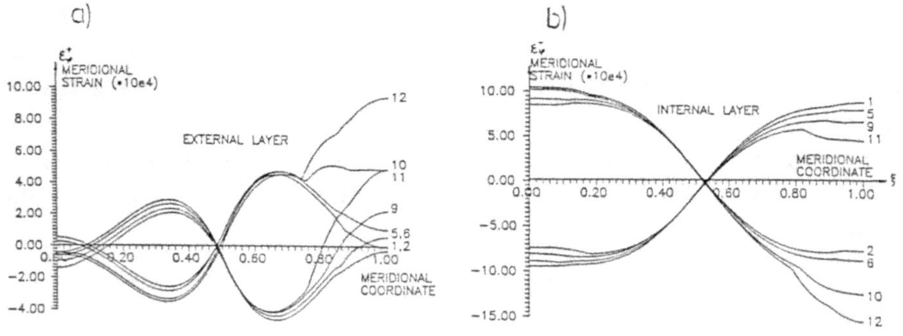

Fig.4. Plastic strain localization in the layers due to the ultimate load mechanism.

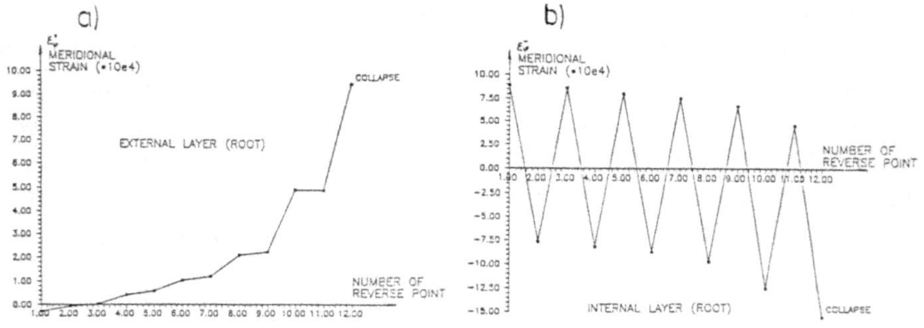

Fig.5. Meridional strains evolution at root. (a) Outer layer (incremental collapse effect). (b) Inner layer (nonsymmetric alternating plasticity).

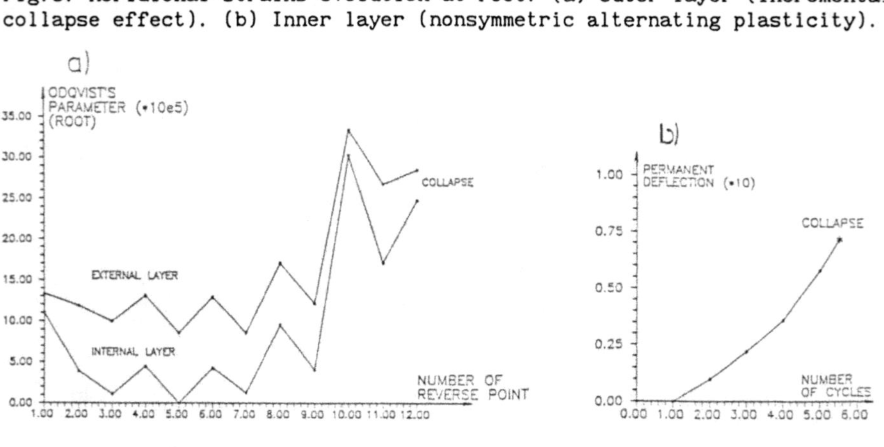

Fig.6. (a) Evolution of the Odqvist parameter increments at root. (b) Nonlinear increase of permanent deflection.

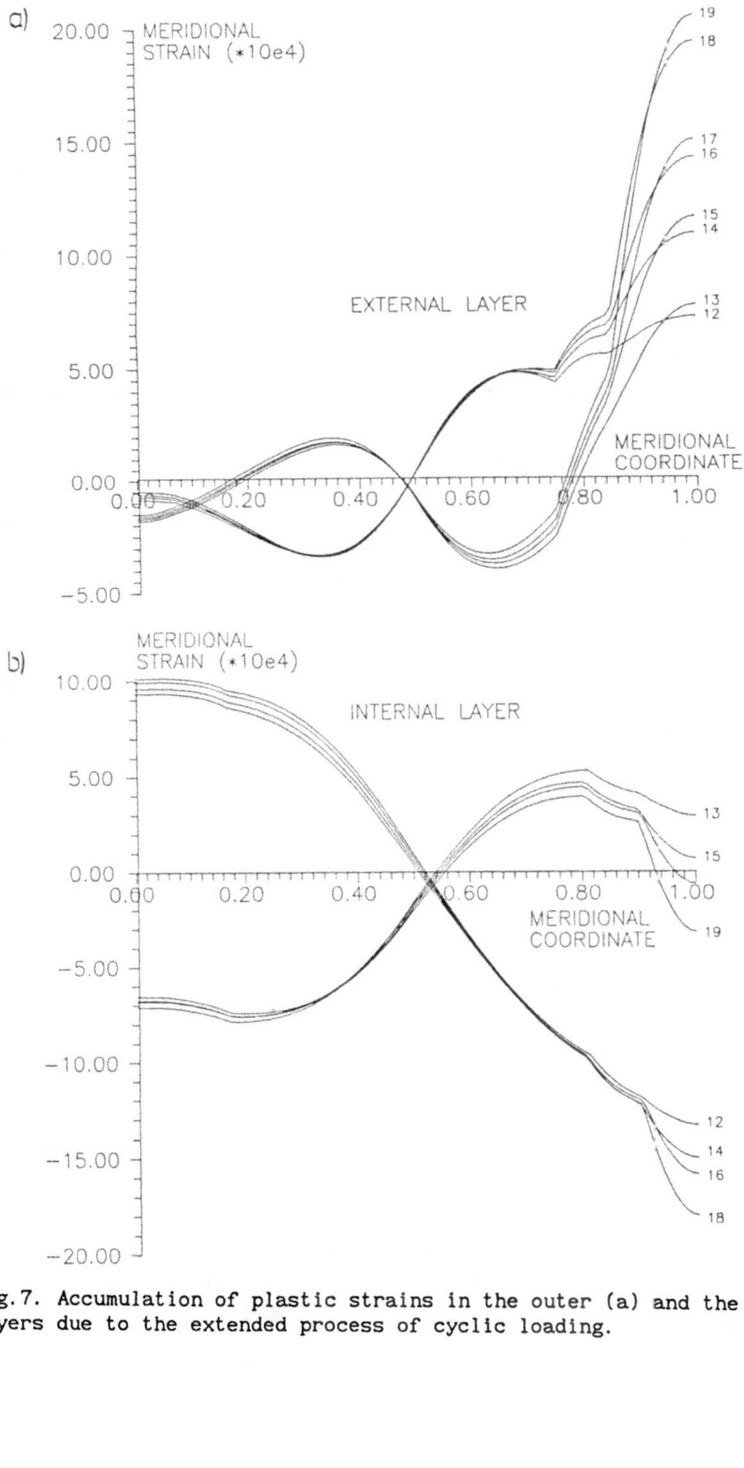

Fig.7. Accumulation of plastic strains in the outer (a) and the inner (b) layers due to the extended process of cyclic loading.

view of the load controlled process this phenomenon must be regarded as collapse. The diagrams 5a,b, plotted at the root for each layer separately , indicate that the effect of nonsymmetric alternating plasticity predominates in the inner layer whereas the incremental collapse phenomenon is observed in the outer one. In other words a mixed type inadaptation mechanism, consisting of alternating plasticity and incremental collapse, occurs. If the process were controlled kinematically it possibly would lead to the formation of plastic hinge at the root. The Odqvist parameter increments, calculated both for the positive and the negative axial force increments, as well as the permanent deflection of the structure are depicted in Fig.6a,b. The reverse point 12 which terminates the load controlled process corresponds to the 'ultimate load effect'. The diagrams for Figs 3-6 were prepared for the following data:

$$\rho=0.1,\ \lambda=1.0,\ p_n=0.0,\ \mathbb{N}/\mathbb{N}_e=1.08 \tag{3}$$

Let us now formulate the following question. What happens if we unload the structure at the point of ultimate load (12) and make an attempt to continue the process of cyclic loading? An answer to this question is illustrated in Figs 3 and 7a,b. Decrease of the carrying capacity with each loading cycle, for the positive axial load increments (points 12, 14, 16, 18), is now even more distinct. Expansion of the plastic zone around the root stops developing. On the other hand, a considerable accumulation of the plastic strain increments in the layers is observed. Such a mechanism, described precisely in the next section, is called 'formation of plastic hinge'.

Formation of plastic hinge

The same mechanism which terminates the processes of quasistatic loading of bellows for higher magnitudes of pressure, appears also for cyclic loading. Analysis of the meridional strain distribution for each cycle leads to the conclusion that the width of plastic zone around the crest of the segment doesn't change. Plastic strain increments accumulate within the zone which yields the strain concentration, various for both layers. Strong concentration of positive strains at the inner layer around the crest eventually results in the jump of slope of normal to the middle surface (Fig.8). This phenomenon is restricted to the comparatively narrow and stable zone of yielding at the crest. The occurence of

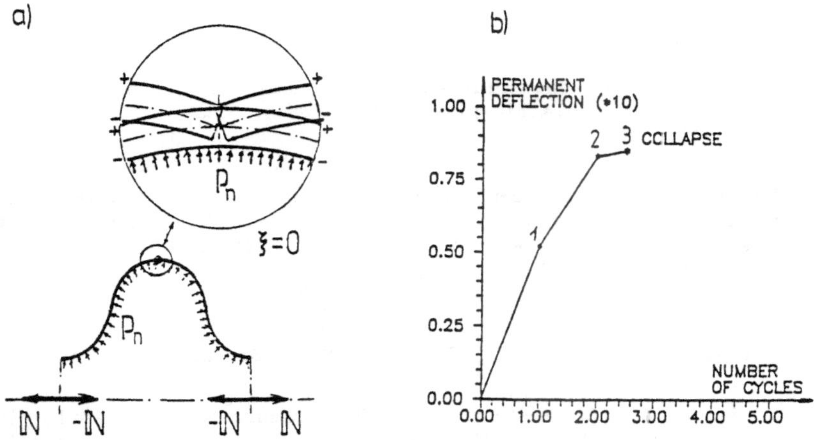

Fig.8. (a) Collapse mechanism due to the formation of plastic hinge. (b) Residual collapse deflection.

the hinge either at the root or at the crest depends, for a given geometry, on the value of internal pressure. In this case failure is associated with predominance of the incremental collapse effect. The following data was applied for the example presented in Fig.8:

$$\rho=0.1, \quad \lambda=1.0, \quad p_n/p_e=0.5, \quad \mathbb{N}/\mathbb{N}_e=1.06 \tag{4}$$

Conclusions

Hence, when both the inadaptation modes of S-shaped bellows are explained, the following conclusion may be formulated. Structural softening of the bellows associated with the ultimate load mechanism may turn out to be deeper, provided that the unloading from the point of maximum load is admissible. Continuation of the process of cyclic loading leads to cycle-by-cycle rapid decrease of the maximum carrying capacity (cyclic softening) as a result of considerable accumulation of plastic strain increments in the layers. Transition from ultimate load mechanism (Figs 3,4,5,6) to formation of plastic hinge (Fig.8) is observed. Eventually, the structure becomes unserviceable due to the incremental collapse phenomenon.

References

1. Weichert, D.: On the influence of the geometrical nonlinearities on the shakedown of elastic-plastic structures. Int. J. Plasticity. 2(2) (1986) 135-148.

2. Weichert, D.: Elastic-plastic structures under variable loads at small strains and moderate rotations. In: Finite Rotations in Structural Mechanics, Ed. W. Pietraszkiewicz, Lecture Notes 19, Springer, Berlin 1986, 343-356.

3. Weichert, D.: Zum Problem geometrischer Nichtlinearitäten in der Plastizitätstheorie. Mitt. Inst. für Mech. **53**, Ruhr-Uni. Bochum 1987.

4. Weichert, D., Gross-Weege, J.: The assessment of elastic-plastic sheets under variable mechanical and thermal loads using a simplified two-surface yield condition. Int. J. Mech. Sci. **30**(10) (1989) 757-767.

5. Hamada, M., Takezono, S.: Strength of U-shaped bellows (5th Report, Fatigue strength under axial loading). Bull. of JSME **10**(40) (1967) 611.

6. Hamada, M., Tanaka, S.: A consideration on the low-cycle fatigue life of bellows (The fatigue life under completely-reversed deflection cycles). Bull. of JSME **17**(103) (1974) 41-49.

7. Porowski, J., O'Donnell, W.J.: Upper bounds for creep ratchetting in convoluted bellows. ASME PVP-PB-029 (1978) 15-45.

8. ASME: Expansion joints for class 1 liquid metal piping, Code Case N-290-1, ASME, New York, April 1983.

9. Skoczeń, B.; Skrzypek, J.; Bielski, J.: Shakedown and inadaptation mechanisms of bellows subject to cyclic axial force and constant pressure. Mech. Struct. Mach. **20**(1) (1992) 119-143.

Termination of Process of Finite Plastic Deformations in Disks under Combined Loadings

M. ŻYCZKOWSKI, K. SZUWALSKI, B. SKOCZEŃ

Cracow University of Technology, Kraków, Poland

Summary

Finite deformations of perfectly elastic-plastic annular disk with rigid circular inclusion, subject to combined: thermal and surface loadings in presence of body forces (rotation of disk) are investigated. The termination of the deformation process is connected with occurance of some inadmissible discontinuities of the stress field. The corresponding surfaces of the decohesive carrying capacity in the loadings space, as well, as the surfaces of elastic carrying capacity are found for various parameters of disks.

Finite Plastic Deformations

For structures made of perfectly elastic-plastic material it is usually assumed, that they can work up to the exhaustion of the limit carrying capacity, connected with onset of certain mechanism of plastic collapse. However, Życzkowski and Szuwalski [1] have proved, that for some structures process of elastic-plastic deformations ends earlier, due to inadmissible discontinuities of displacements. For disks with rigid circular inclusion [2] it is connected with infinite growth of the radial strain at the joint of the disk with the inclusion. To avoid the internal inconsistency of the theory (infinitely large strains within the framework of the small strain theory) we should adopt the finite strain theory [3].

We shall discuss elastic plastic deformations of the axially-symmetric disk with rigid inclusion (Fig. 1).

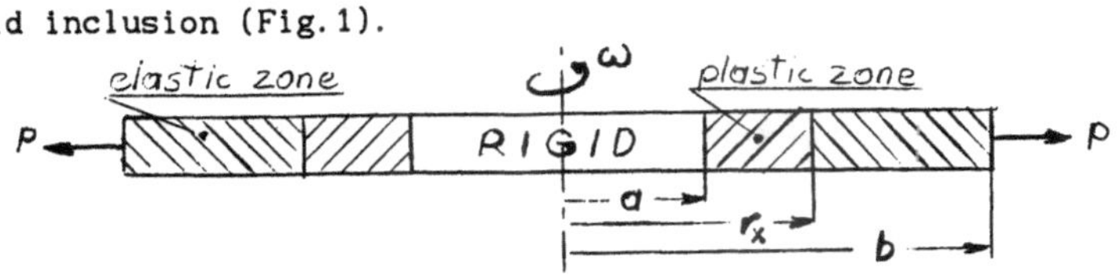

Fig. 1. Disk with rigid inclusion

D. Besdo, E. Stein (Eds.)
Finite Inelastic Deformations –
Theory and Applications
IUTAM Symposium Hannover/Germany 1991

Disk made of perfectly elastic-plastic material, characterized by the elastic modulus E, Poisson's ratio ν, and the yield stress σ_0, rotates with constant angular velocity ω. Moreover, it is subject to uniform traction (compression) p at the outer radius b, and uniform heating (cooling). The influence of temperature on the material constants is neglected.

Resigning of the rigidification theorem, we consider internal equilibrium of already deformed disk. In Lagrangean (material) coordinates, denoted by capital letter R, we obtain:

$$\frac{1}{h}\,\frac{1}{1+\frac{du}{dr}}\,\frac{d}{dR}(h\,\sigma_r) + \frac{\sigma_r - \sigma_\vartheta}{R+u} + \frac{\gamma}{g}\,\omega^2(R+u) = 0 \tag{1}$$

where u stands for the radial displacement, h - for the current thickness (initial thickness H is assumed to be constant) and γ for the specific weight of the material. Both stresses: radial σ_r and circumferrential σ_ϑ in this equation are true stresses.

The same equation written in the Eulerian (spatial) coordinates, denoted by small letter r, takes the form:

$$\frac{1}{h}\,\frac{d}{dr}(h\,\sigma_r) + \frac{\sigma_r - \sigma_\vartheta}{r} + \frac{\gamma}{g}\,\omega^2 r = 0 \tag{2}$$

In present paper we shall use the material coordinates, more frequently used in the solids mechanics. In the plastic zone, that will form after exceeding of the elastic carrying capacity around the inclusion, stresses must satisfy the yield condition. Here we adopt the Huber-Mises-Hencky yield condition in the parametrizied form proposed by Nadai-Sokolovsky ($\sigma_z \equiv 0$):

$$\sigma_r = \frac{2}{\sqrt{3}}\,\sigma_0 \sin\zeta \tag{3}$$

$$\sigma_\vartheta = \frac{2}{\sqrt{3}}\,\sigma_0 \sin\left(\zeta+\frac{\pi}{3}\right) \tag{4}$$

where ζ is a parameter, which distribution within the plastic zone should be found. The problem cannot be solved without the strain analysis. We define them as logarithmic ones:

$$\varepsilon_r = \ln\left(1 + \frac{du}{dR}\right) \tag{5}$$

$$\varepsilon_{\vartheta} = \ln\left(1 + \frac{u}{R}\right) \tag{6}$$

$$\varepsilon_{z} = \ln\frac{h}{H} \tag{7}$$

and eliminating from (5) and (6) the radial displacement u, we come to the compatibility equation

$$\varepsilon_{r} = \varepsilon_{\vartheta} + \ln\left[1 + R\,\frac{d\,\varepsilon_{\vartheta}}{dR}\right] \tag{8}$$

The physical law we adopt in form of the similarity of deviators of true stresses and logarithmic strains (Nadai-Davis theory):

$$\frac{\varepsilon_{r} - \varepsilon_{m}}{\varepsilon_{\vartheta} - \varepsilon_{m}} = \frac{\sigma_{r} - \sigma_{m}}{\sigma_{\vartheta} - \sigma_{m}} \tag{9}$$

where ε_{m} stands for mean logarithmic strain and σ_{m} - mean true stress. The law of the volume change takes form:

$$\varepsilon_{r} + \varepsilon_{\vartheta} + \varepsilon_{z} = \frac{1 - 2\nu}{E}\,(\sigma_{r} + \sigma_{\vartheta}) + 3\alpha T \tag{10}$$

where α is the coefficient of linear thermal expansion, and T - uniform temperature change.

The set of eight equations :(1), (3), (4), (5), (6), (7), (9) and (10), contains eight unknowns: true stresses σ_{r} and σ_{ϑ}, logarithmic strains ε_{r}, ε_{ϑ} and ε_{z}, radial displacement u, current thickness h and parameter ζ. Subsequent elimination of unknowns leads finally to the set of two equations:

$$\frac{d\varepsilon_{\vartheta}}{d\Xi} = \exp\left\{\frac{\sin(\zeta-\frac{\pi}{3})}{\sqrt{3}\,\cos\zeta}\left[3(\varepsilon_{\vartheta}-\tau) - 2s(1-2\nu)\,\cos(\zeta-\frac{\pi}{3})\right]\right\} - 1 \tag{11}$$

$$\frac{d\zeta}{d\Xi} = \frac{\frac{d\varepsilon_{\vartheta}}{d\Xi}\left\{3\cos\zeta-2\sqrt{3}w^{2}\cos^{2}\zeta\exp[2(\Xi+\varepsilon_{\vartheta})]\right\}-2\sqrt{3}\cos^{2}\zeta\left\{\sin(\zeta-\frac{\pi}{3})+w^{2}\exp[2(\Xi+\varepsilon_{\vartheta})]\right\}}{s(1-2\nu)(\sqrt{3}\sin^{2}\zeta+4\sin\zeta\cos^{3}\zeta) + 2\sqrt{3}\cos^{3}\zeta + 3\sin\zeta(\tau-\varepsilon_{\vartheta})} \tag{12}$$

with respect to two unknowns: circumferrential strain ε_{ϑ} and parameter ζ. These equations are here written in dimensionless form, convenient for further numerical integration, where was introduced the dimensionless logarithmic material radius:

$$\Xi = \ln \frac{a}{R} \tag{13}$$

and other dimensionless quantities:

$$s = \frac{\sigma_0}{E}; \qquad q = \frac{p}{\sigma_0}; \qquad \tau = \alpha T; \qquad w = \omega\, a \sqrt{\frac{\sqrt{3}}{2}\,\frac{\gamma}{g\,\sigma_0}} \tag{14}$$

The above described solutions are valid only within the plastic zone and must be combined with elastic ones:

$$\sigma_r^{(e)} = A - \frac{B}{R^2} - \frac{1}{2}E\tau - \frac{(3+\nu)\gamma\omega^2}{8g}R^2 \tag{15}$$

$$\sigma_r^{(e)} = A + \frac{B}{R^2} - \frac{1}{2}E\tau - \frac{(1+3\nu)\gamma\omega^2}{8g}R^2 \tag{16}$$

where A and B are constants that should be determined using boundary conditions. Here we use elastic solutions obtained from the small strain theory, because elastic strains are assumed to be small in comparison with plastic ones.

Decohesive Carrying Capacity

The process of elastic-plastic deformations can be continued only up to the moment, when denominator in (12) tends to zero. This effect can be observed always at the inner radius a, for the parameter ζ_0 calculated from the equation:

$$s(1-2\nu)(\sqrt{3}\sin^2\zeta_0+4\sin\zeta_0\cos^3\zeta_0) + 2\sqrt{3}\cos^3\zeta_0 + 3\tau\sin\zeta_0 = 0 \tag{17}$$

independent from angular velocity, traction and width of the disk. Further increase of external loadings must lead to occurence of inadmissible discontinuities of the stress field, as:

$$\frac{d\sigma_r}{d\Xi} = \frac{2}{\sqrt{3}}\,\sigma_0\,\cos\zeta\,\frac{d\zeta}{d\Xi} \longrightarrow \infty \tag{18}$$

The corresponding set of external loadings determines the decohesive carrying capacity of the disk. To evaluate it we have to integrate numerically equations (11) and (12) at initial conditions:

$$\text{for } \Xi = 0 \qquad\qquad \varepsilon_\vartheta = 0; \qquad \zeta = \zeta_0 \tag{19}$$

at each step of integration we must check up whether the elastic zone was not reached yet. To do it, we must investigate the ratio between components of strain and stress deviators. In the plastic zone it it is larger than $\frac{1+\nu}{E}$ (constant value for the elastic zone). So if inequality:

$$\frac{1}{\cos\zeta}\left\{\sqrt{3}(\varepsilon_\vartheta-\tau) - 2s\left[\sin(\zeta+\frac{\pi}{3}) - \nu\sin\zeta\right]\right\} \geq 0 \tag{20}$$

is satisfied we can make the next step of integration. When it is not fulfilled, we must use further the elastic solutions (15) and (16).
The equations (11) and (12) for spatial (Eulerian) coordinate

$$\xi = \ln\frac{r}{a} \tag{21}$$

take the form

$$\frac{d\varepsilon_\vartheta}{d\xi} = 1 - \exp\left\{\frac{\sin(\zeta-\frac{\pi}{3})}{\sqrt{3}\cos\zeta}\left[2s(1-2\nu)\cos(\zeta-\frac{\pi}{3}) + 3(\tau-\varepsilon_\vartheta)\right]\right\} \tag{22}$$

$$\frac{d\zeta}{d\xi} = \frac{\frac{d\varepsilon_\vartheta}{d\xi}\,2\sqrt{3}\sin\zeta\cos\zeta\cos(\zeta-\frac{\pi}{3}) - 2\sqrt{3}\cos^2\zeta[\sin(\zeta-\frac{\pi}{3})+w^2\exp(2\xi)]}{s(1-2\nu)\sin\zeta(\sqrt{3}\sin\zeta+4\cos^3\zeta) + 2\sqrt{3}\cos^3\zeta + 3\sin\zeta(\tau-\varepsilon_\vartheta)} \tag{23}$$

What is the most important, denominator of the second of these equations is the same as in (12), it will vanish for the parameter ζ_0 described by (17) and all conclusions are the same as above. It is obvious, as we describe the same process of large elastic-plastic deformations, only using now the different coordinates.

Numerical Results

To calculate the decohesive carrying capacity of the disk, first we must find parameter ζ_0 from (17), for given values of temperature τ, and material constants s and ν. Making use of the initial conditions (19) we can now integrate numerically set of two equations (11) and (12), for taken value of angular velocity w. At each step of integration we check up inequality (20), to know whether we are still in the plastic zone and can make the next step. When the elastic-plastic boundary is reached, we evaluate the constants A and B in elastic solutions (15) and (16), and calculating the radial stress (15) at the external radius of the disk we obtain the third loading - traction.

In Fig.2 the part of the surface of elastic carrying capacity is shown. Each cross-section for constant angular velocity has the shape of ellipse.

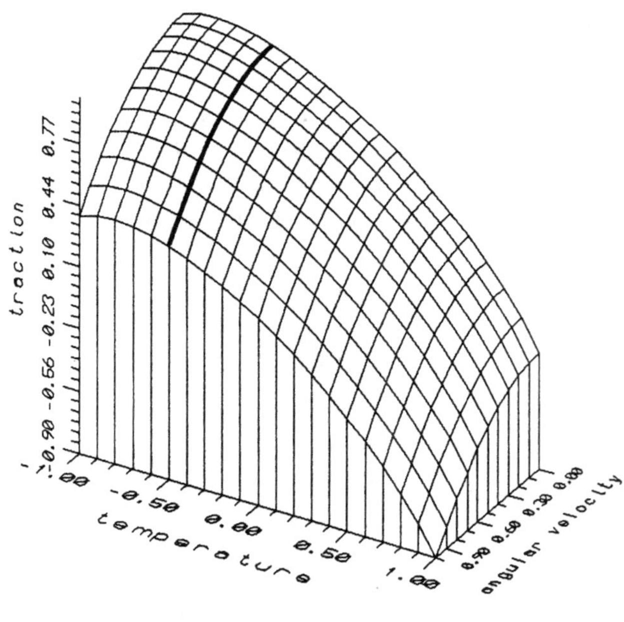

Fig.2 The surface of elastic carrying capacity for β=1.5, s=0.001, ν=0.3

The thick line on this surface is common for the surface of decohesive carrying capacity in Fig.3. It corresponds to instant decohesion, at the moment of onset of the first plastic deformations. Then the width of the plastic zone at the moment of decohesion is equal zero.

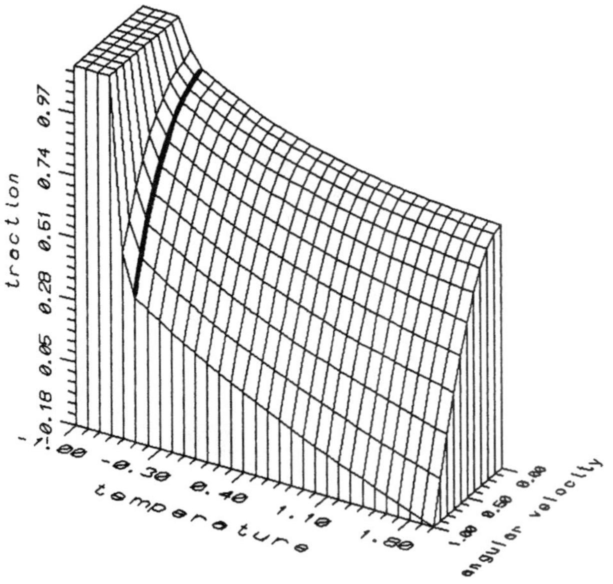

Fig.3 The surface of decohesive carrying capacity for β=1.5, s=0.001. ν=0.3.

The flat surface at the left hand side of this diagram (for negative temperatures - cooling), where the traction is significantly higher than elsewhere, corresponds to full plastification of the disk (limit carrying capacity).

The very important parameter describing the disk is ratio of the outer and inner radii:

$$\beta = \frac{b}{a} \tag{24}$$

Its influence on the decohesive carrying capacity is shown in Fig.4. Rotation of the broad disk (for large β), moves the curves of elastic (e.c.c.) and decohesive (d.c.c.) carrying capacities to negative values of traction.

The similar effect can be observed for increasing values of angular velocity (Fig.5), especially for sufficiently broad disks.

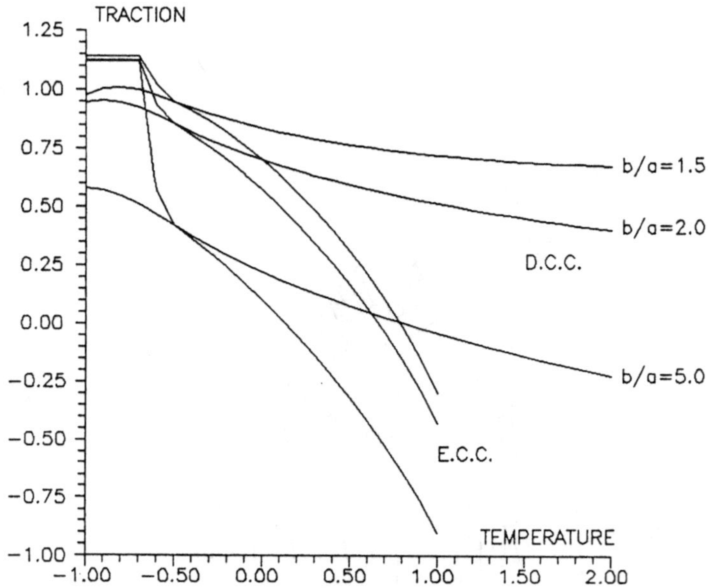

Fig.4 Influence of the width of the disk on elastic and decohesive carrying capacities for s=0.001, ν=0, w=0.2.

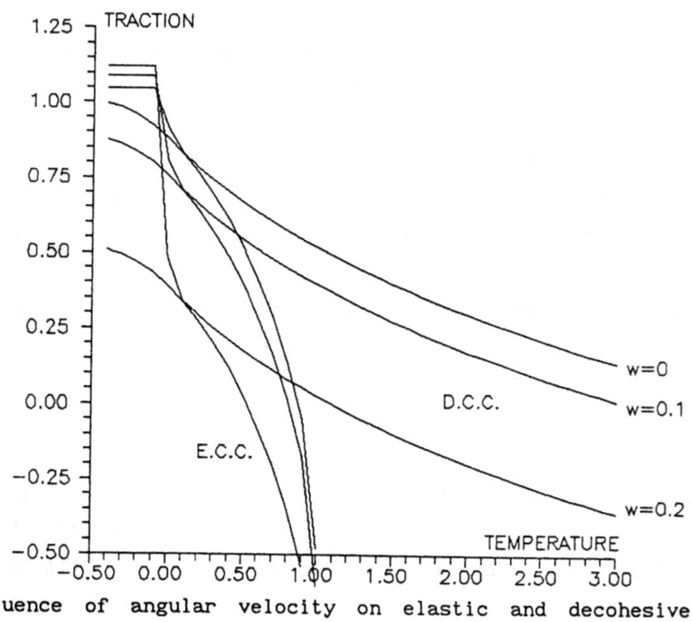

Fig.5 Influence of angular velocity on elastic and decohesive carrying capacities for β=5.0, s=0.001, ν=0.5.

The Poisson's ratio ν also significantly changes especially curves of elastic carrying capacity (Fig.6). The incompressibility of material (ν close to 0.5) allows for higher temperatures and stronger tractions at the outer radius, for given angular velocity.

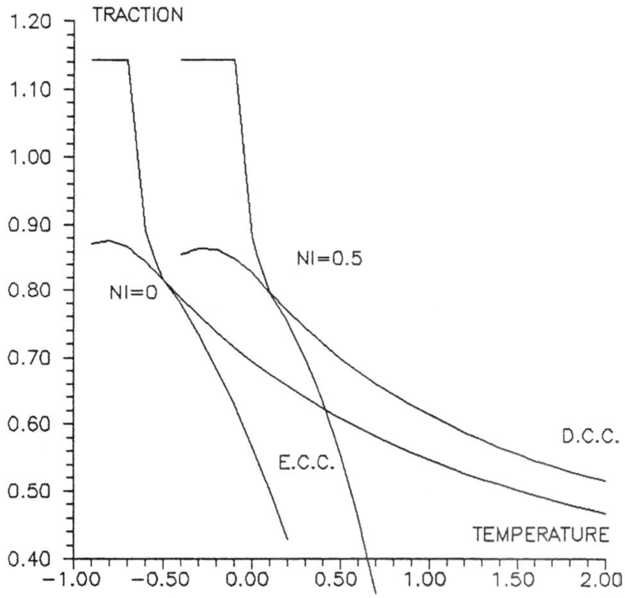

Fig.6 Elastic and decohesive carrying capacity for vorious values of Poisson's ratio ν (for β=1.5, s=0.001, w=0.5).

Thus, the process of finite elastic-plastic deformations not always must lead to exhaustion of limit carrying capacity (mechanism of plastic collapse). As it was proved it sometimes ends earlier because of onset of inadmissible discontinuities of the stress field. This effect may be of special importance in case of thermal loadings [4], [5].

In present paper, for the first time, the influence of body forces, connected with rotation, on the decohesive carrying capacity was investigated. It was also the reason of application of the finite strain theory, as with increase of plastic deformations, the body forces become larger, too.

Let us notice, that curves of decohesive carrying capacity are here concave, while almost all limiting curves in theory of plasticity are convex.

References

1. Szuwalski, K.; Życzkowski, M.: On the phenomenon of decohesion in perfect plasticity, *Int. J. Solids Struct.* 7 (1973), 85-98.
2. Szuwalski, K.: Influence of material characteristics on decohesive carrying capacity of an infinite plate with a rigid circular inclusion. *Bull. Ac. Pol. Sci. Techn.* **34** (1986), 11-12, 655-664.
3. Życzkowski, M.; Szuwalski, K.: On the termination of the process of finite plastic deformations. *J. Mech. Theor. Appliquee,* Numero special (1982), 175-186.
4. Szuwalski, K.; Życzkowski, M.: A concave interaction curve corresponding to decohesive carrying capacity under thermal loadings. *J. Therm. Stresses* 7 (1984), 121-134.
5. Skoczeń, B.; Szuwalski, K.: Elastic-plastic disks with jumplike variable thickness under purely thermal loadings. *J. Therm Stresses*, **11** (1988), 3, 273-285

www.ingramcontent.com/pod-product-compliance
Lightning Source LLC
Chambersburg PA
CBHW071406230926
51846CB00001B/25
* 9 7 8 3 6 4 2 8 4 8 3 5 3 *